Lecture Notes in Computer Science 9225

Commenced Publication in 1973
Founding and Former Series Editors:
Gerhard Goos, Juris Hartmanis, and Jan van Leeuwen

De-Shuang Huang · Vitoantonio Bevilacqua
Prashan Premaratne (Eds.)

Intelligent Computing Theories and Methodologies

11th International Conference, ICIC 2015
Fuzhou, China, August 20–23, 2015
Proceedings, Part I

 Springer

Editors
De-Shuang Huang
Tongji University
Shanghai
China

Prashan Premaratne
University of Wollongong
North Wollongong, NSW
Australia

Vitoantonio Bevilacqua
Polytecnic of Bari
Bari
Italy

ISSN 0302-9743   ISSN 1611-3349   (electronic)
Lecture Notes in Computer Science
ISBN 978-3-319-22179-3   ISBN 978-3-319-22180-9   (eBook)
DOI 10.1007/978-3-319-22180-9

Library of Congress Control Number: 2015945123

LNCS Sublibrary: SL3 – Information Systems and Applications, incl. Internet/Web, and HCI

Printed on acid-free paper

Springer International Publishing AG Switzerland is part of Springer Science+Business Media
(www.springer.com)

Preface

The International Conference on Intelligent Computing (ICIC) was started to provide an annual forum dedicated to the emerging and challenging topics in artificial intelligence, machine learning, pattern recognition, bioinformatics, and computational biology. It aims to bring together researchers and practitioners from both academia and industry to share ideas, problems, and solutions related to the multifaceted aspects of intelligent computing.

ICIC 2015, held in Fuzhou, China, August 20–23, 2015, constituted the 11th International Conference on Intelligent Computing. It built upon the success of ICIC 2014, ICIC 2013, ICIC 2012, ICIC 2011, ICIC 2010, ICIC 2009, ICIC 2008, ICIC 2007, ICIC 2006, and ICIC 2005 that were held in Taiyuan, Nanning, Huangshan, Zhengzhou, Changsha, China, Ulsan, Korea, Shanghai, Qingdao, Kunming, and Hefei, China, respectively.

This year, the conference concentrated mainly on the theories and methodologies as well as the emerging applications of intelligent computing. Its aim was to unify the picture of contemporary intelligent computing techniques as an integral concept that highlights the trends in advanced computational intelligence and bridges theoretical research with applications. Therefore, the theme for this conference was "Advanced Intelligent Computing Theories and Applications." Papers focusing on this theme were solicited, addressing theories, methodologies, and applications in science and technology.

ICIC 2015 received 671 submissions from 25 countries and regions. All papers went through a rigorous peer-review procedure and each paper received at least three review reports. On the basis of the review reports, the Program Committee finally selected 233 high-quality papers for presentation at ICIC 2015, included in three volumes of proceedings published by Springer: two volumes of *Lecture Notes in Computer Science* (LNCS) and one volume of *Lecture Notes in Artificial Intelligence* (LNAI).

This volume of *Lecture Notes in Computer Science* (LNCS) 9225 includes 75 papers.

The organizers of ICIC 2015, including Tongji University and Fujian Normal University, China, made an enormous effort to ensure the success of the conference. We hereby would like to thank the members of the Program Committee and the reviewers for their collective effort in reviewing and soliciting the papers. We would like to thank Alfred Hofmann, executive editor at Springer, for his frank and helpful advice and guidance throughout and for his continuous support in publishing the proceedings. In particular, we would like to thank all the authors for contributing their papers. Without the high-quality submissions from the authors, the success of the conference would not have been possible. Finally, we are especially grateful to the International Neural Network Society and the National Science Foundation of China for their sponsorship.

June 2015

De-Shuang Huang
Vitoantonio Bevilacqua
Prashan Premaratne

ICIC 2015 Organization

General Co-chairs

De-Shuang Huang	China
Changping Wang	China

Program Committee Co-chairs

Kang-Hyun Jo	Korea
Abir Hussain	UK

Organizing Committee Co-chairs

Hui Li	China
Yi Wu	China

Award Committee Chair

Laurent Heutte	France

Publication Co-chairs

Valeriya Gribova	Russia
Zhi-Gang Zeng	China

Special Session Co-chairs

Phalguni Gupta	India
Henry Han	USA

Special Issue Co-chairs

Vitoantonio Bevilacqua	Italy
Mike Gashler	USA

Tutorial Co-chairs

Kyungsook Han	Korea
M. Michael Gromiha	India

International Liaison

Prashan Premaratne   Australia

Publicity Co-chairs

Juan Carlos Figueroa   Colombia
Ling Wang   China
Evi Syukur   Australia
Chun-Hou Zheng   China

Exhibition Chair

Bing Wang   China

Organizing Committee Members

Jianyong Cai   China
Qingxiang Wu   China
Suping Deng   China
Lin Zhu   China

Program Committee Members

Andrea Francesco Abate, Italy
Waqas Haider Khan Bangyal, Pakistan
Shuhui Bi, China
Qiao Cai, USA
Jair Cervantes, Mexico
Chin-Chih Chang, Taiwan, China
Chen Chen, USA
Huanhuan Chen, China
Shih-Hsin Chen, Taiwan, China
Weidong Chen, China
Wen-Sheng Chen, China
Xiyuan Chen, China
Yang Chen, China
Cheng Cheng, China
Ho-Jin Choi, Korea
Angelo Ciaramella, Italy
Salvatore Distefano, Italy
Jianbo Fan, China
Minrui Fei, China
Juan Carlos Figueroa, Colombia

Huijun Gao, China
Shan Gao, China
Dunwei Gong, China
M. Michael Gromiha, India
Zhi-Hong Guan, China
Kayhan Gulez, Turkey
Phalguni Gupta, India
Fei Han, China
Kyungsook Han, Korea
Laurent Heutte, France
Wei-Chiang Hong, Taiwan, China
Yuexian Hou, China
Peter Hung, Ireland
Saiful Islam, India
Li Jia, China
Zhenran Jiang, China
Xin Jin, USA
Joaquín Torres-Sospedra, Spain
Dah-Jing Jwo, Taiwan, China
Vandana Dixit Kaushik, India

Sungshin Kim, Korea
Yoshinori Kuno, Japan
Takashi Kuremoto, Japan
Xiujuan Lei, China
Bo Li, China
Guo-Zheng Li, China
Kang Li, UK
Peihua Li, China
Shi-Hua Li, China
Shuai Li, Hong Kong, China
Shuwei Li, USA
Xiaodi Li, China
Xiaoou Li, Mexico
Chengzhi Liang, China
Bingqiang Liu, China
Ju Liu, China
Shuo Liu, USA
Weixiang Liu, China
Xiwei Liu, China
Yunxia Liu, China
Chu Kiong Loo, Malaya
Geyu Lu, China
Ke Lu, China
Yingqin Luo, USA
Pabitra Mitra, India
Tarik Veli Mumcu, Turkey
Roman Neruda, Czech Republic
Ben Niu, China
Sim-Heng Ong, Singapore
Seiichi Ozawa, Japan
Vincenzo Pacelli, Italy
Shaoning Pang, New Zealand
Francesco Pappalardo, Italy
Surya Prakash, India
Prashan Premaratne, Australia
Yuhua Qian, China
Daowen Qiu, China
Ivan Vladimir Meza Ruiz, Mexico
Seeja.K.R, India
Fanhuai Shi, China

Wilbert Sibanda, South Africa
Jiatao Song, China
Stefano Squartini, Italy
Wu-Chen Su, Taiwan, China
Zhan-Li Sun, China
Antonio E. Uva, Italy
Mohd Helmy Abd Wahab, Malaysia
Bing Wang, China
Jingyan Wang, USA
Ling Wang, China
Shitong Wang, China
Xuesong Wang, China
Yi Wang, USA
Yong Wang, China
Yuanzhe Wang, USA
Yufeng Wang, China
Yunji Wang, USA
Wei Wei, Norway
Zhi Wei, China
Ka-Chun Wong, Canada
Hongjie Wu, China
Qingxiang Wu, China
Yan Wu, China
Junfeng Xia, USA
Shunren Xia, China
Bingji Xu, China
Xin Yin, USA
Xiao-Hua Yu, USA
Shihua Zhang, China
Eng. Primiano Di Nauta, Italy
Hongyong Zhao, China
Jieyi Zhao, USA
Xiaoguang Zhao, China
Xing-Ming Zhao, China
Zhongming Zhao, USA
Bojin Zheng, China
Chun-Hou Zheng, China
Fengfeng Zhou, China
Yong-Quan Zhou, China

Additional Reviewers

Haiqing Li
JinLing Niu
Li Wei
Peipei Xiao
Dong Xianguang
Shasha Tian
Li Liu
Xuan Wang
Sen Xia
Jun Li
Xiao Wang
Anqi Bi
Min Jiang
Wu Yang
Yu Sun
Bo Wang
Wujian Fang
Zehao Chen
Chen Zheng
Hong Zhang
Huihui Wang
Liu Si
Sheng Zou
Xiaoming Liu
Meiyue Song
Bing Jiang
Min-Ru Zhao
Xiaoyong Bian
Chuang Ma
Yujin Wang
Wenlong Hang
Chong He
Rui Wang
Zhu Linhe
Xin Tian
Mou Chen
F.F. Zhang
Gabriela Ramírez
Ahmed Aljaaf
Neng Wan
Abhineet Anand
Antonio Celesti
Aditya Nigam
Aftab Yaseen

Weiyuan Zhang
Shamsul Zulkifli
Alfredo Liverani
Siti Amely Jumaat
Muhammad Amjad
Angelo Ciaramella
Aniza Mohamed Din
En-Shiun Annie Lee
Anoosha Paruchuri
Antonino Staiano
Antony Lam
Alfonso Panarello
Alfredo Pulvirenti
Asdrubal Lopez-Chau
Chun Kit Au
Tao Li
Mohd Ayyub Khan
Azizi Ab Aziz
Azizul Azhar Ramli
Baoyuan Wu
Oscar Belmonte
Biao Li
Jian Gao
Lin Yong
Boyu Zhang
Edwin C. Shi
Caleb Rascon
Yu Zhou
Fuqiang Chen
Fanshu Chen
Chenguang Zhu
Chin-Chih Chang
Chang-Chih Chen
Changhui Lin
Cheng-Hsiung Chiang
Linchuan Chen
Bo Chen
Chen Chen
Chengbin Peng
Cheng Cheng
Jian Chen
Qiaoling Chen
Zengqiang Chen
Ching-Hua Shih

Chua King Lee
Quistina Cuci
Yanfeng Chen
Meirong Chen
Cristian Rodriguez Rivero
Aiguo Chen
Aquil Mirza Mohammed
Sam Kwong
Yang Liu
Yizhang Jiang
Bingbo Cui
Yu Wu
Chuan Wang
Wenbin Chen
Cheng Zhang
Dhiya Al-Jumeily
Hang Dai
Yinglong Dai
Danish Jasnaik
Kui Liu
Dapeng Li
Dan Yang
Davide Nardone
Dawen Xu
Dongbo Bu
Liya Ding
Shifei Ding
Donal O'Regan
Dong Li
Li Kuang
Zaynab Ahmed
David Shultis
Zhihua Du
Chang-Chih Chen
Erkan İMal
Ekram Khan
Eric Wei
Gang Wang
Fadzilah Siraj
Shaojing Fan
Mohammad Farhad
  Bulbul
Mohamad Farhan
    Mohamad Mohsin

Fengfeng Zhou
Feng Jiqiang
Liangbing Feng
Farid García-Lamont
Chien-Yuan Lai
Filipe de O. Saraiva
Francesco Longo
Fabio Narducci
Francesca Nardone
Francesco Camastra
Nhat Linh Bui
Hashim Abdellah
Hashim Moham
Gao Wang
Jungwon Yu
Ge Dingfei
Geethan Mendiz
Na Geng
Fangda Guo
Gülsüm Gezer
Guanghua Sun
Guanghui Wang
Rosalba Giugno
Giovanni Merlino
Xiaoqiang Zhang
Guangchun Cheng
Guanglan Zhang
Tiantai Guo
Weili Guo
Yanhui Guo
Xiaoqing Gu
Hafizul Fahri Hanafi
Haifeng Wang
Mohamad Hairol Jabbar
H.K. Lam
Khalid Isa
Hang Su
Guangjie Han
Hao Chu
Ben Ma
Hao Men
Yonggang Chen
Haza Nuzly Abdull
  Hamed
Mohd Helmy Abd Wahab
  Pang
Hei Man Herbert

Hironobu Fujiyoshi
Guo-Sheng Hao
Huajuan Huang
Hiram Calvo
Hongjie Wu
Hongjun Su
Hitesh Kumar Sharma
Haiguang Li
Hongkai Chen
Qiang Huang
Tengfei Zhang
Zheng Huai
Joe Huang
Jin Huang
Wan Hussain Wan Ishak
Xu Huang
Ying Hu
Ho Yin Sze-To
Haitao Zhu
Ibrahim Venkat
Jooyoung Lee
Josef Moudřík
Li Xu
Jakub Smid
Jianhung Chen
Le Li
Jianbo Lu
Jair Cervantes
Junfeng Xia
Jinhai Li
Hongmei Jiang
Jiaan Zeng
Jian Lu
Jian Wang
Jian Zhang
Jianhua Zhang
Jie Wu
Jim Jing-Yan Wang
Jing Sun
Jingbin Wang
Wu Qi
Jose Sergio Ruiz Castilla
Joaquín Torres
Gustavo Eduardo Juarez
Jun Chen
Junjiang Lin
Junlin Chang

Juntao Liu
Justin Liu
Jianzhong Guo
Jiayin Zhou
Abd Kadir Mahamad
K. Steinhofel
Ka-Chun Wong
Ke Li
Kazuhiro Fukui
Sungshin Kim
Klara Peskova
Kunikazu Kobayashi
Konstantinos Tsirigos
Seeja K.R.
Kwong Sak Leung
Kamlesh Tiwari
Li Kuang
K.V. Arya
Zhenjiang Lan
Ke Liao
Liang-Tsung Huang
Le Yang
Erchao Li
Haitao Li
Wei Li
Meng Lei
Guoqi Luo
Huihui Li
Jing Liang
Liangliang Zhang
Liang Liang
Bo Li
Bing Li
Dingshi LI
Lijiao Liu
Leida Li
Lvzhou Li
Min Li
Kui Lin
Ping Li
Liqi Yi
Lijun Quan
Bingwen Liu
Haizhou Liu
Jing Liu
Li Liu
Ying Liu

Zhaoqi Liu
Zhe Liu
Yang Li
Lin Zhu
Jungang Lou
Lenka Kovářová
Shaoke Lou
Li Qingfeng
Qinghua Li
Lu Huang
Liu Liangxu
Lili Ayu Wulandhari
Xudong Lu
Yiping Liu
Yutong Li
Junming Zhang
Mohammed Khalaf
Maria Musumeci
Shingo Mabu
Yasushi Mae
Manzoor Lone
Liang Ma
Manabu Hashimoto
Md. Abdul Mannan
Qi Ma
Martin Pilat
Asad Khan
Maurizio Fiasché
Max Talanov
Mohd Razali MD Tomari
Mengxing Cheng
Meng Xu
Tianyu Cao
Minfeng Wang
Muhammad Fahad
Michele Fiorentino
Michele Scarpiniti
Zhenmin Zhang
Ming Liu
Miguel Mora-Gonzalez
Aul Montoliu
Yuanbin Mo
Marzio Pennisi
Binh P. Nguyen
Mingqiang Zhang
Muhammad Rashid
Shenglin Mu

Musheer Ahmad
Monika Verma
Naeem Radi
Aditya Nigam
Aditya Nigam
Nagarajan Raju
Mohd Najib Mohd Salleh
Yinan Guo
Zhu Nanli
Mohammad Naved
  Qureshi
Su Rina
Sanders Liu
Chang Liu
Patricio Nebot
Nistor Grozavu
Zhixuan Wei
Nobuyuki Nezu
Nooraini Yusoff
Nureize Arbaiy
Kazunori Onoguchi
R.B. Pachori
Yulei Pang
Xian Pan
Binbin Pan
Peng Chen
Klara Peskova
Petra Vidnerová
Qi Liu
Prashan Premaratne
Puneet Gupta
Prabhat Verma
Peng Zhang
Haoqian Huang
Qiang Fu
Qiao Cai
Haiyan Qiao
Qingnan Zhou
Rabiah Ahmad
Rabiah Abdul Kadir
R. Rakkiyappan
Ramakrishnan
Chandrasekaran
Yunpeng Wang
Hao Zheng
Radhakrishnan Delhibabu
Rohit Katiyar

Wei Cui
Rozaida Ghazali
Raghuraj Singh
Rey-Sern Lin
S.M. Zakariya
Sabooh Ajaz
Alexander Tchitchigin
Kuo-Feng Huang
Shao-Lun Lee
Wei-Chiang Hong
Toshikazu Samura
Sandhya Pundhir
Jin-Xing Liu
Shahzad Alam
Mohd Shamrie Sainin
Shanye Yin
Shasha Tian
Hao Shen
Chong Shen
Jingsong Shi
Nobutaka Shimada
Lu Xingjia
Shun Chen
Silvio Barra
Simone Scardapane
Salvador Juarez Lopez
Shenshen Liang
Sheng Liu
Somnath Dey
Rui Song
Seongpyo Cheon
Sergio Trilles-Oliver
Subir Kumar Nandy
Hung-Chi Su
Sun Jie
QiYan Sun
Shiying Sun
Sushil Kumar
Yu Su
Hu Zhang
Yan Qi
Hotaka Takizawa
Tanggis Bohnuud
Yang Tang
Lirong Tan
Yao Tuozhong
Tian Tian

Tianyi Wang
Toshiaki Kondo
Tofik Ali
Tomáš Ken
Peng Xia
Gurkan Tuna
Tutut Herawan
Zhongneng Xu
Danny Wu
Umarani Jayaraman
Zhichen Gong
Vibha Patel
Vikash Yadav
Esau Villatoro-Tello
Prashan Premaratne
Vishnu Priya Kanakaveti
Vivek Srivastava
Victor Manuel
    Landassuri-Moreno
Hang Su
Yi Wang
Chao Wang
Cheng Wang
Jiahai Wang
Jing Wang
Junxiao Wang
Linshan Wang
Waqas Haider Bangyal
Herdawatie Abdul Kadir
Yi Wang
Wen Zhang
Widodo Budiharto
Wenjun Deng
Wen Wei
Wenbin Chen
Wenzhe Jiao
Yufeng Wang
Haifeng Wang
Deng Weilin
Wei Jiang
Weimin Huang
Wufeng Tian
Wu-Chen Su
Daiyong Wu
Yang Wu

Zhonghua Wu
Jun Zhang
Xiao Wang
Wei Xiong
Weixiang Liu
Wenxi Zhang
Wenye Li
Wenlin Zhang
Li-Xin He
Ming Tan
Yi-Gang Zhang
Xiangliang Zhang
Nan Xiang
Xiaobo Zhang
Xiaohu Wang
Xiaolei Wang
Xiaomo Liu
Lei Wang
Yan Cui
Xiaoyong Zhang
Xiaozhen Xue
Jianming Xie
Xin Gao
Xinhua Xiao
Xin Lu
Xinyu Zhang
Liguang Xu
Hao Wu
Xun Li
Jin Xu
Xin Xu
Yuan Xu
Xiaoyin Xu
Yuan Xu
Shiping Chen
Xiaoyan Sun
Xiaopin Zhong
Atsushi Yamashita
Yanen Guo
Xiaozhan Yang
Zhanlei Yang
Yaqiang Yao
Bei Ye
Yan Fu
Yehu Shen

Yu-Yen Ou
Yingyou Wen
Ying Yang
Yingtao Zhao
Yanjun Zhao
Yong Zhang
Yoshinori Kobayashi
Yesu Feng
Yuan Lin
Lin Yuan
Yugandhar Kumar
Yujia Li
Yupeng Li
Yuting Yang
Yuyan Han
Yong Wang
Yingying Wei
Yingxin Guo
Xiangjuan Yao
Guodong Zhao
Huan Zhang
Bo Zhang
Gongjie Zhang
Yu Zhang
Zhao Yan
Wenrui Zhao
Zhao Yan
Tian Zheng
Zhengxing Hang
Xibei Yang
Zhenxin Zhan
Zhipeng Cai
Lingyun Zhu
Yanbang Zhang
Zheng-Ling Yang
Juan Li
Zongxiao He
Zhengyu Ouyang
Will Zhu
Xuebing Zhang
Zhile Yang
Yi Zhang

Contents – Part I

Contents – Part II

Parallel Collaborative Filtering Recommendation Model Based on Two-Phase Similarity

Hongyi Su[✉], Xianfei Lin, Caiqun Wang, Bo Yan, and Hong Zheng

Key Lab of Intelligent Information Technology,
Beijing Institute of Technology, Beijing, China
{henrysu,xflin,caiqunwang,yanbo,hongzheng}@bit.edu.cn

Abstract. Problems such as cold startup, accuracy, and scalability are faced by traditional collaborative filtering recommendation algorithm if the system is expanded continuously. To resolve these issues, we propose a parallel collaborative filtering recommendation model on the basis of two-phase similarity (PCF-TPS) and weighted distance similarity measure (WDSM). In accordance with WDSM, the users' similarity is calculated and their similarity matrix is obtained. At the same time, the items' similarity is counted and its similarity matrix is got in line with Tanimoto Coefficient Similarity. For the users' similarity matrix, their preferences are endowed with weights and in this way their new preferences matrix is received. In addition, the nearest neighbor item is found and a more accurate recommendation to the target user is given on the basis of the items' similarity matrix and users' new preferences matrix. Besides, in regard to the parallel computing framework, the parallel implementation of the model is completed. All these experiments are done on MovieLens dataset. The results show that PCF-TPS solves the problem of cold startup and increases the accuracy concerning CF. Compared with PCF-EV, PCF-TPS's parallel realization can be improved to nearly 125 times on the whole. That is to say, it will be more meaningful to complex model using GPU than a small model. What's more, PCF-EV's distributed implementation is much more efficient than PCF-EV's.

Keywords: Recommend mechanism · Collaborative filtering · Two-phase similarity · GPU

1 Introduction

The collaborative filtering [1] proposed in 1992 by Goldberg [2] has become one of the most widely used methods to users' recommend items. It is used in Tapestry and can recommend some articles for users according to their comments after reading some articles. However, with the constant enlargement of the systematic scale, the traditional collaborative filtering recommendation algorithm (CF) always suffers some problems including cold startup, sparsity and scalability.

To overcome these problems, many optimization methods are proposed. In this paper, a new model based on previous researches is introduced to emphasize three key

© Springer International Publishing Switzerland 2015
D.-S. Huang et al. (Eds.): ICIC 2015, Part I, LNCS 9225, pp. 1–10, 2015.
DOI: 10.1007/978-3-319-22180-9_1

issues that the traditional algorithms face. The innovations of this paper are embodied in the following aspects:

1. A parallel collaborative filtering recommendation model based on two-phase similarity (PCF-TPS) is proposed to make the recommendations become more efficient and it is oriented to the whole group of users.
2. Based on the programming mode of CUDA, the parallel implementation of PCF-TPS is realized.
3. On the MovieLens data set, the contrast tests of the recommendation algorithm in terms of PCF-TPS, CF [2] and PCF-EV [3] are made, and the qualitative improvement of accuracy, recall and efficiency are verified. The experimental results show that the new approach is obviously valid.

2   Related Works and Background

In this section, we will review the background of the traditional collaborative filtering recommendation algorithm and make an introduction of the CUDA basics.

2.1   Collaborative Filtering Recommendation Algorithm

As the main recommendation technology, the fundamental assumption of CF is that if user X and user Y rate n items similarly, or have similar behaviors (e.g., buying, watching, and listening), then they tend to rate or act on other items similarly [4]. Generally, there are three types of CF, which are CF based on item, CF based on user and CF based on model. CF's process include three parts: fetching data, searching nearest neighborhood and making recommendation. In the following, CF based on user is taken as an example in details:

Fetching data. In a typical CF scenario, there is a list of m users $\{u_1, u_2, \ldots, u_m\}$ and a list of n items $\{i_1, i_2, \ldots, i_n\}$, and each user, u_i, has a list of items, I_{ui}. The ratings can be explicit indications, for example, on a 1 5 scale [5]. The list of the items concerning users' like or dislike can be converted to $m \times n$ user-item ratings matrix B. In matrix B, as in (1), p_{ij} represents the preference of user i to item j as below.

$$B_{m \times n} = \begin{bmatrix} p_{11} & \cdots & p_{1n} \\ p_{21} & \cdots & p_{2n} \\ \vdots & & \vdots \\ p_{m1} & \cdots & p_{mn} \end{bmatrix} \tag{1}$$

Searching Nearest Neighbor. For a user-based CF algorithm, firstly we mainly calculate the similarity between user u and user v who have both rated the same items, and we give the nearest neighbor list of the target user by sorting the similarity results. Tanimoto Coefficient Similarity is one of the most popular measures to compute similarity and can be shown in formula (2):

$$sim_{(u,v)} = \frac{|I_u \cap I_v|}{|I_u \cup I_v|} \tag{2}$$

In this formula, I_u represents the set of the items that the preference user u has provided, I_v means the set of the items which user v has rated, $sim_{(u,v)}$ stands for the value of the similarity between user u and user v, and it is the ratio of the size of the intersection to the size of the union of their preferred items. Then, we can sort the sim and receive the target user's nearest neighbor.

Making Recommendation. As is shown in formula (3), $R_{(u,j)}$ is used to estimate the preference value that user u may give to item j. It is estimated based on nearest neighborhoods' preference.

$$R_{(u,j)} = \frac{1}{|I_u|} \sum_{j' \in I_u} sim_{(j,j')} p_{(u,j')} \tag{3}$$

Specifically, I_u represents the set of the items which user u has given the preference. $p_{(u,j')}$ means preference from user u to item j' and $sim_{(j,j')}$ stands for the similarity between item j and item j'. Besides, the best recommendation (the most similar to item j') can be chosen according to the value of the $R_{(u,j)}$.

2.2 Programming Model of CUDA

The programmable graphics processing unit (GPU) is able to transfer the calculation from CPU to GPU thanks to the development in science and technology.

To make a better use of the programmability, CUDA (Compute Unified Device Architecture), which is a common parallel computing architecture and parallel programming model, improves the programmability of GPU. The CUDA programming model include one or more sequential threads running on the host CPU, and one or more parallel kernels suitable for execution on GPU. A grid executed by a kernel contains a limited number of threads, and for the Tesla C2050 GPU, this number is 1024.

3 Parallel Implementation of PCF-TPS

In the following subsection, we will describe three parts of the parallel implementation of PCF-TPS in CUDA:

1. Get the New Preference Matrix in CUDA
2. Nearest Neighbor Searching in CUDA
3. Get the expected preference vector in CUDA
4. Make recommendations in CUDA.

The detailed steps are as follows:

3.1 Get the New Preference Matrix in CUDA

In the process of searching for the nearest neighbor, it is necessary to calculate their similarity. This article uses Weighted Distance Similarity Measure [6] containing two factors which are similarity and weight to do it. The process of the similarity calculating between user u and user v is presented in the formula (4) and Fig. 1:

$$sim_{(u,v)} = \sum\nolimits_{i=1}^{n} w_i * \left(\overline{a_i \oplus a'_i}\right) \tag{4}$$

We suppose the attributes of user u can be expressed as the vector $A_u = (a_1, a_2, \ldots, a_n)$ and allocate each attribute with a weight to get the corresponding vector of weight $W_u = (w_1, w_2, \ldots, w_n)$, . Then the formula (4) can be used to obtain similarity between user u and user v. In this way, we get the users' weighted similarity matrix and by combining it with the users' preference, we can get the users' new preference vector.

3.2 Nearest Neighbor Searching in CUDA

To get the target user's nearest neighborhood, we can make each thread on GPU responsible for the calculation of similarity between two users, such as user u and user v. The corresponding pseudo-code is presented as in Fig. 2.

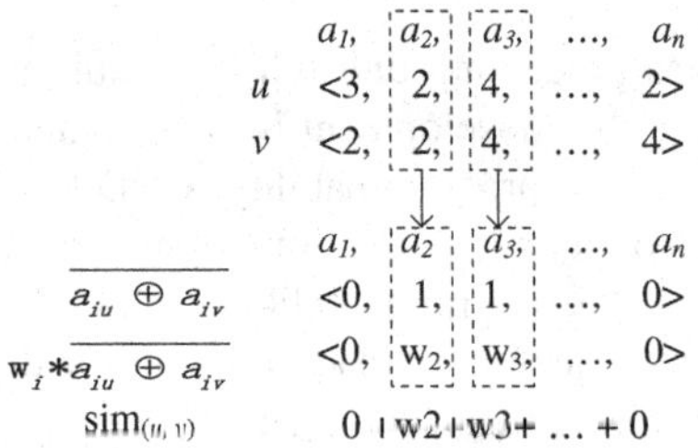

Fig. 1. Weighted distance similarity measure

```
Input: two-dimensional array pref[m][n], m is the count of the user and
n is the count of item and userID i;
Output: linear array neighborhood[m]
Require: for i(0<i<m), get neighborhood[i] on a separate GPU thread
Method getNeighborhood() //Execute on each thread
    Int threadID = getThreadID(); // GPU provides this method
    Int j=0;
    While(j<n) do
  /*Both the user threadID[j] and i give the preference to item j */
        If(pref[threadID][j]>0&&pref[i][j]>0)
            Neighborhood [threadID] = 1;
```

Fig. 2. The pseudo-code of nearest neighbor searching in CUDA

3.3 Get the Candidate Item Sets and Expected Preference Vector in CUDA

According to the nearest neighborhood's preference vector, the candidate recommendation items can be got. Its appropriate pseudo-code is shown in [3].

3.4 Make Recommendations in CUDA

For the output in step 3.3, the target user's expected preference vector can be received by formula (3) and we use top-N algorithm to get the top N elements and the items will be recommended to the target user.

Each thread on GPU is responsible to recommend items to a user. The thread reads the user's preference vector and expected preference vector. It is proceeded as follows: if the element's value isn't 0, the corresponding element's value in expected preference vector will be modified to 0. Then we can get the top N element by these vectors. Its appropriate pseudo-code is shown in [3].

4 Distributed Implementation of PCF-TPS

Due to some advantages of Map-Reduce in the big data processing, we can apply it and Hadoop to the further optimization on PCF-TPS successfully. Description of the concrete process applied to the item is shown in quote [6].

To be specific, the distributed implementation of PCF-EV includes five steps: transform file format and deploy the input file, generate user preference vector, get item co-occurrence matrix, get the user expected preference matrix and make recommendation [6]. All the files generated in the process are placed on the HDFS server steadily.

5 Experiment and Results

In this section, we will evaluate the PCF-TPS in comparison with the traditional collaborative filtering recommendation algorithm, PCF-EV [3] and DCF-EV [7].

5.1 Datasets

In our experiments, three datasets of MovieLens (http://www.movielens.umn.edu) are used. The first set in MovieLens is ML-100 K which has 100,000 ratings including 943 persons and 1682 movies; the second set is ML-1 M, which has 1,000,209 ratings including 6040 users and 3952 movies; the last set is ML-10 M, which consists of 10 million ratings and 100,000 tag applications of 10,000 movies from 72,000 users. The users' profile includes their age, sex, and profession and the movie in the item profile includes 19 types. Besides, density of the user-item matrix is 6.3 % in ML-100 K and 4.1 % in ML-1 M. These datasets are widely used by researchers and developers in collaborative filtering domain.

Our experiments are carried out respectively on three datasets mentioned above and more related details are as follows.

5.2   The Comparison of the Recommendation Based on PCF-TPS and CF

The experiment' results in quote [7] shows clearly that the performance of Tanimoto Coefficient is the best choice for CF. The comparison results of the recommendation algorithm based on PCF-TPS and CF is shown in Table 1:

From Table 1, we know that the precision and recall ratio of CF-TPS have qualitative leaps in comparison with CF, and in this way it can obtain a better performance in the whole range.

The experiment' results in quote [3] show that due to the new user's similarity matrix which can join in the similarity calculation, the problem of cold startup is resolved. This is PCF-TPS's Efficiency Evaluation Comparison and PCF-EV's Parallel Implementation.

5.3   The Efficiency Evaluation Comparison of the PCF-TPS and PCF-EV's Parallel Implementation

The system used in the text is a PC with an Intel Core 6 Exon 2620 CPU. The GPU is a NVIDIA Tesla k20 with 2496 stream processors and CUDA is 5.5 version. We write programs in Java and use Windows 7 Pro (32 bit) and Eclipse as system and development environment respectively. The experiment allocation is the same as quote [3].

To verify the efficiency of PCF-TPS's parallel implementation, we make a comparison of CPU between GPU and PCF-TPS. The results of the two algorithms' each step are shown in Tables 2, 3, 4 and their performance is shown in Table 5. Besides, the comparison result of PCF-TPS and PCF-EV [3] is shown as Fig. 3.

The results show that compared with the serial implementation on the high-end dual-core CPU, the overall parallel implementation on the low and middle-end GPU

Table 1. The precision and recall ratio of CF and PCF-TPS

Algorithm	Precision		Recall	
	100 k	1 m	100 k	1 m
CF	0.001 169	0.001 094	0.003 770	0.003 890
DCF-TP	0.062 147	0.063 875	0.437 550	0.323 601

Table 2. The speedup for the step of nearest neighbor and candidate

Users(m)	Run Time		
	100 k	1 m	10 m
CPU(sec)	146 435	12 664 585	438 402 920
GPU(sec)	7 187	138 379	20 535 774
Speedup	20.37	91.52	21.35

Table 3. The speedup for the step of getting the new preference matrix in CUDA

Users(m)	Run Time		
	100 k	1 m	10 m
CPU(sec)	30	11 108 330	16 862 431 200
GPU(sec)	710	91 850	168 133 402
Speedup	23.67	120.93	100.29

Table 4. The speedup for the step of getting the expected preference matrix and making recommendations

Users(m)	Run Time		
	100 k	1 m	10 m
CPU(sec)	176 455	15 222 790	1 148 477 400
GPU(sec)	6 029	88 463	3 468 463
Speedup	29.27	172.08	331.12

Table 5. The speedup for the whole algorithm

Users(m)	Run Time		
	100 k	1 m	10 m
CPU(sec)	323 780	29 443 734	1 845 0867 800
GPU(sec)	16 084	234 911	193 041 094
Speedup	20.13	125.34	95.58

can reach nearly 125 times speedup. Specifically, the local optimization speedup can reach nearly 329 times.

The Fig. 3 shows that for a complex calculation using GPU, its speedup can be increased many times obviously. This is due to the fact that with increase of the operand, it is more efficient to use GPU.

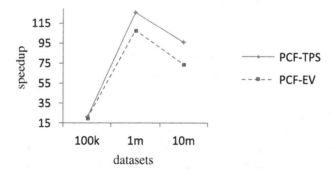

Fig. 3. The efficiency evaluation comparison of the PCF-TPS and PCF-EV's parallel implementation

Table 6. Speedup of tasks on HOD cluster

Node Number	Speedup		
	100 k	1 m	10 m
1	1	1	1
2	1.144 470	1.158 427	1.581 044
3	1.218 750	1.234 731	2.101 324

5.4 The Comparison Efficiency Evaluation of the PCF-TPS and Distributed Implementation of PCF-EV

In order to verify the speed performance of the recommendation algorithm based on PCF-TPS, the test is done on single machine and a small HOD (Hadoop On Demand) cluster respectively [6]. The experiment allocation is the same as quote [7]. To test the performance, we use three datasets orderly in the experiment. The detailed result is shown in Table 6 and the comparison result is shown in Fig. 4:

Table 6 shows that with increase of the computing nodes, the run time of all recommendation tasks are shortened and a higher linear speedup is achieved.

Figure 4 shows that with increase of the algorithm's complexity, distributed algorithm using Mapreduce is optimized as well.

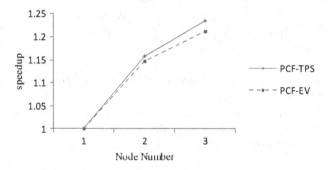

Fig. 4. The efficiency evaluation comparison of the PCF-TPS and PCF-EV's distributed implementation on datasets of 1 m

In a word, the parallel collaborative filtering recommendation model based on two-phase similarity is of practical significance.

6 Conclusion

The paper firstly analyzes disadvantages of the existed similarity measures, and then we propose a parallel collaborative filtering recommendation model based on two-phase similarity to overcome these shortages. The new similarity measures take two users' proportion of the same value into account. Besides, to demonstrate its efficiency, several comparison experiments are performed on three widely used datasets. The experimental results show PCF-TPS's better performance and higher efficiency compared with other recommendation algorithm. In the future, we will further investigate different techniques to improve our system's performance. For example, we can use timestamp to alleviate the problem of data sparse and combine CPU and Hadoop in close association.

References

1. Breese, J.S., Heckerman, D., Kadie, C.: Empirical analysis of predictive algorithms for collaborative filtering. In: Proceedings of the Fourteenth conference on Uncertainty in artificial intelligence, pp. 43–52. Morgan Kaufmann Publishers Inc.(1998)
2. Goldberg, D., Nichols, D., Oki, B.M., et al.: Using collaborative filtering to weave an information tapestry. Commun. ACM **35**(12), 61–70 (1992)
3. Su, H., Wang, C., Zhu, Y. et al.: Parallel collaborative filtering recommendation model based on expand-vector. In: 2014 International Conference on Multisensor Fusion and Information Integration for Intelligent Systems (MFI), pp. 1–6. IEEE (2014)
4. Goldberg, K., Roeder, T., Gupta, D., et al.: Eigentaste: a constant time collaborative filtering algorithm. Inf. Retrieval **4**(2), 133–151 (2001)
5. Miller, B.N., Konstan, J.A., PocketLens, R.J.: Toward a personal recommender system. ACM Trans. Inf. Syst. (TOIS) **22**(3), 437–476 (2004)
6. Wang, C.Q., Su, H.Y., Zhu, Y. et al.: Distributed collaborative filtering recommendation model based on two-phase similarity. Future Communication, Information and Computer Science (2015)
7. Zhu, Y., Su, H.Y., Wang, C.Q., et al.: Distributed collaborative filtering recommendation model based on expand-vector. Adv. Mater. Res. **989**, 2188–2191 (2014)
8. Zheng, N., Li, Q., Liao, S., et al.: Which photo groups should I choose? a comparative study of recommendation algorithms in Flickr. J. Inf. Sci. **36**(6), 733–750 (2010)
9. Brynjolfsson, E., Hu, Y., Smith, M.D.: Consumer surplus in the digital economy: estimating the value of increased product variety at online booksellers. Manage. Sci. **49**(11), 1580–1596 (2003)
10. Baluja, S., Seth, R., Sivakumar, D. et al.: Video suggestion and discovery for youtube: taking random walks through the view graph. In: Proceedings of the 17th International Conference on World Wide Web, pp. 895–904. ACM (2008)
11. Zhang, X., Li, Y.: Use of collaborative recommendations for web search: an exploratory user study. J. Inf. Sci. **34**(2), 145–161 (2008)
12. Cong, L., Changyong, L.: Li21 M. A collaborative filtering recommendation algorithm based on domain nearest neighbor. Journal of Computer Research and Development (2008–09)
13. Dongyan, J., Fuzhi, Z.: A collaborative filtering recommendation algorithm based on double neighbor choosing strategy. J. Comput. Res. Dev. **5**, 020 (2013)
14. Malucelli, F., Cremonesi, P., Rostami, B.: An application of bicriterion shortest paths to collaborative filtering. In: Federated Conference on Computer Science and Information Systems (FedCSIS), pp. 423–429. IEEE (2012)
15. He, B., Fang, W., Luo, Q. et al.: Mars: a mapreduce framework on graphics processors. In: Proceedings of the 17th International Conference on Parallel Architectures and Compilation Techniques, pp. 260–269. ACM (2008)

16. Dean, J., Ghemawat, S.: Mapreduce: simplified data processing on large clusters. Commun. ACM **51**(1), 107–113 (2008)
17. Borthakur, D.: HDFS architecture guide. Hadoop Apache Project, 53 (2008)
18. Krishnan, S., Tatineni, M., Baru, C.: myHadoop-Hadoop-on-Demand on Traditional HPC Resources. San Diego Supercomputer Center Technical report TR-2011–2, University of California, San Diego (2011)
19. Guo, G., Zhang, J., Thalmann, D.: Merging trust in collaborative filtering to alleviate data sparsity and cold start. Knowl.-Based Syst. **57**, 57–68 (2014)
20. Choi, K., Suh, Y.: A new similarity function for selecting neighbors for each target item in collaborative filtering. Knowl.-Based Syst. **37**, 146–153 (2013)
21. Wei, S., Ye, N., Zhang, S. et al.: Collaborative filtering recommendation algorithm based on item clustering and global similarity. In: 2012 Fifth International Conference on Business Intelligence and Financial Engineering (BIFE), pp. 69–72. IEEE (2012)
22. Zhao, Z.D, Shang, M.S.: User-based collaborative-filtering recommendation algorithms on hadoop. In: WKDD 2010 Third International Conference on Knowledge Discovery and Data Mining, pp. 478–481. IEEE (2010)
23. Ali, M., Johnson, C.C., Tang, A.K.: Parallel collaborative filtering for streaming data. University of Texas Austin, Technical report (2011)
24. Adamopoulos, P., Tuzhilin, A.: Recommendation opportunities: Improving item prediction using weighted percentile methods in collaborative filtering systems. In: Proceedings of the 7th ACM Conference on Recommender systems, pp. 351–354. ACM (2013)

An Improved Ant Colony Algorithm to Solve Vehicle Routing Problem with Time Windows

Yi Yunfei[1,2,3(✉)], Lin Xiaodong[1], Sheng Kang[2], and Cai Yongle[2]

[1] College of Computer and Information Engineering,
Hechi University, Yizhou 546300, China
gxyiyf@163.com
[2] Computer School, Wuhan University, Wuhan 430079, China
[3] Guangxi Key Laboratory of Hybrid Computation and IC Design Analysis,
Nanning 530006, China

Abstract. This paper presents an improved ant colony optimization algorithm (ACO algorithm) based on Ito differential equations, the proposed algorithm integrates the versatility of Ito thought with the accuracy of ACO algorithm in solving the vehicle routing problem (VRP), and it executes simultaneous move and wave process, and employs exercise ability to unify move and wave intensity. Move and wave operator rely on attractors and random perturbations to set the motion direction. In the experiment part, this improved algorithm is implemented for solving vehicle routing problem with soft time windows (VRPSTW), and tested by Solomon Benchmark standard test dataset, the result shows that the proposed algorithm is effective and feasible.

Keywords: Ant colony optimization algorithm · Vehicle routing problem · Move operator · Wave operator

1 Introduction

Vehicle Routing Problem is firstly proposed by G. Dantzig and J. Ramser in 1959, ever since extensive attention are given on this subject by large number of experts and scholars, and the issue has become one of the hot topic in combination optimization. As a result, the problem is well expanded in accordance with various facts in life by experts, and lots of branches have been developed, such as Split and Simultaneous Pickup and Delivery Vehicle Routing Problem with Time Windows Constraint [1] (SVRPS-PDTW), Optimal Vehicle Routing With Real-Time Traffic Information [2], Capacitated Vehicle Routing Problem [3] (CVRP), Heterogeneous Fleet Vehicle Routing Problem [4] (HFVRP), etc. In summary, there exists a bunch of algorithms aimed to solve these problems, which can be classified into three categories, thus precise algorithm, heuristics algorithm and meta-heuristics algorithm. In recent years, meta-heuristic algorithm has

This work was supported by the Open Foundation of Guangxi Key Laboratory of Hybrid Computation and IC Design Analysis (No.HCIC201411),the National Undergraduate Training Programs for Innovation and Entrepreneurship(No.201410605055),and the Education Scientific Research Foundation of Guangxi Province (No. KY2015YB254).

D.-S. Huang et al. (Eds.): ICIC 2015, Part I, LNCS 9225, pp. 11–22, 2015.
DOI: 10.1007/978-3-319-22180-9_2

become the most commonly used algorithm for solving VRP problems, these high-precision algorithms can effectively solve lots of large-scale problems, such as particle swarm optimization (PSO), genetic algorithm (GA), ant colony optimization algorithm (ACO), simulated annealing algorithm (SA) and so on. However, as a general optimizer in practical applications, meta-heuristic optimization algorithm inevitably confront the contradiction of common versus proficient, exploration versus exploitation and efficiency versus accuracy. Based on these facts, a hybrid IT-AC algorithm is proposed by combining the ITO algorithm, which is first introduced by Professor Dong Wenyong and Professor Li Yuanxiang in Wuhan University [2], with ACO algorithm which has been deeply developed in solving VRP problems. By the means of combining the generality of ITO algorithm with the high accuracy of the ACO algorithm in solving VRP, the proposed optimization algorithm is applied in solving VRPSTW [5].

ITO algorithm (Evolutionary Algorithm inspired by Ito stochastic process, in order to reflect the origin of the algorithm, referred to as ITO) by Dong and Li [6] is based on random process of Ito process, it analyzes the movement of particles from the microscopic point of view, and mimic the mechanism of kinetics equation that particles colliding and interacting with others to design algorithm and solve problems, which reflects the features of group search in bionics. At present, time series modeling [7], function optimization [2], combinatorial optimization [8], multi-objective optimization [9] and other problems can obtain comparatively good solutions by utilizing ITO algorithm. ACO algorithm, for its natural advantages, ever since the time it was proposed in the field of combinatorial optimization, especially in the study of vehicle routing problem, has achieved much success. In this paper, therefore, based on ITO algorithm framework, the selection strategy in ACO algorithm is utilized to design ITO-ACO algorithm.

In the experiment part, Solomon Benchmark standard test dataset[1] are used as experimental data. The dataset includes a total of 56 data subsets, the dataset nodes can be divided based on the distribution of geographical position into three categories: class R, class C and class RC, while class R data nodes are randomly distributed, nodes in class C are in clustered distribution, and nodes in RC class are between class R and class C, which means some nodes are in randomly distribution, the others are in clustered distribution. Additionally, test dataset can be further classified into six different categories according to the level of the time schedule, named R1, R2, C1, C2, RC1, RC2.

2   Vehicle Routing Problem Description

Vehicle Routing Problem (VRP) is a well-known NP-hard problem, solving the problem effectively is of great importance in real life. Generally, VRP contains a warehouse, N clients and NV vehicles, each vehicle is required to be of same specifications that the cargo capacity is Q. The desired scheduling scheme is that the warehouse is designed as the starting point and finishing point in all of the suitable vehicle routes to ensure that all customers are visited only once by one vehicle, and the total transportation costs (e.g., path length, transit time, needed vehicles, etc.) are minimized. Certain constraints should be met when designing the route program as following:

---

[1] Source: http://web.cba.neu.edu/∼msolomon/problems.htm.

$$x_{kij} = \begin{cases} 1, & \text{vehicle } k \text{ move from } i \text{ to } j \\ 0, & \text{otherwise} \end{cases} \quad (i,j = 0, 1, \ldots, N; k = 1, 2, \ldots, NV) \quad (1)$$

$$y_{ki} = \begin{cases} 1, & \text{vehicle } k \text{ finish task } i \\ 0, & \text{otherwise} \end{cases} \quad (i,j = 0, 1, \ldots, N; k = 1, \ldots, NV) \quad (2)$$

$$\sum_{i=0}^{N} q_i y_{ki} \leq Q \, (k = 1, 2, \ldots, N) \quad (3)$$

$$\sum_{k=0}^{NV} y_{ki} = 1 \, (i = 1, 2, \ldots, N) \quad (4)$$

$$\sum_{j=0}^{N} x_{kij} = y_{ki} \, (i = 1, 2, \ldots, N; \; k = 1, 2, \ldots, NV) \quad (5)$$

$$\sum_{i=0}^{N} x_{kij} = y_{ki} \, (j = 0, 1, \ldots, N; k = 1, 2, \ldots, N) \quad (6)$$

Among them, the warehouse is numbered 0, customer demand point are numbered 1,2,3, ..., N, $\sum_{i=0}^{N} d_{i,i+1}/speed + \sum_{i=1}^{N} \delta_i + \sum_{i}^{N} wt_i \leq T$ denotes the demand for customer $\sum_{i=0}^{N} d_{i,i+1}/speed + \sum_{i=1}^{N} \delta_i + \sum_{i}^{N} wt_i \leq T$, Eq. (3) indicates capacity constraints of vehicle; formula (4) represents each client point must be visited once, and only by one vehicle; formulas (5) and (6) represents the points $\sum_{i=0}^{N} d_{i,i+1}/speed + \sum_{i=1}^{N} \delta_i + \sum_{i}^{N} wt_i \leq T$ in the path of vehicle $\sum_{i=0}^{N} d_{i,i+1}/speed + \sum_{i=1}^{N} \delta_i + \sum_{i}^{N} wt_i \leq T$ are to be visited by vehicle $\sum_{i=0}^{N} d_{i,i+1}/speed + \sum_{i=1}^{N} \delta_i + \sum_{i}^{N} wt_i \leq T$.

3 Vehicle Routing Problem with Time Windows

When solving VRPSTW problem, it usually have to be considered that the vehicle service time $\sum_{i=0}^{N} d_{i,i+1}/speed + \sum_{i=1}^{N} \delta_i + \sum_{i}^{N} wt_i \leq T$ for each customer, and working hours (driving time + service time + wait time) of each vehicle cannot exceed the maximum limit (i.e. the time of going off work). Suppose a scheduling scheme $\sum_{i=0}^{N} d_{i,i+1}/speed + \sum_{i=1}^{N} \delta_i + \sum_{i}^{N} wt_i \leq T$, where $\sum_{i=0}^{N} d_{i,i+1}/speed + \sum_{i=1}^{N} \delta_i + \sum_{i}^{N} wt_i \leq T$ (7), which represents that a vehicle will fulfill all of the customer's distribution tasks.

Now the working time cannot exceed the maximum operating time $\sum_{i=0}^{N} d_{i,i+1}/speed + \sum_{i=1}^{N} \delta_i + \sum_{i}^{N} wt_i \leq T$ can be formulated as following:

$$\sum_{i=0}^{N} d_{i,i+1}/speed + \sum_{i=1}^{N} \delta_i + \sum_{i}^{N} wt_i \leq T \tag{7}$$

Where $wt_i = \begin{cases} \infty & \text{if } CurTime > li \quad (\text{arrive late}) \\ ei - CurTime & \text{if } CurTime < ei \quad (\text{arrive early}) \\ 0 & \text{else } (\text{comform to the time window}) \end{cases}$ denotes the

path   length   between   clients $wt_i =$

$\begin{cases} \infty & \text{if } CurTime > li \quad (\text{arrive late}) \\ ei - CurTime & \text{if } CurTime < ei \quad (\text{arrive early}) \\ 0 & \text{else } (\text{comform to the time window}) \end{cases}$ and

$wt_i = \begin{cases} \infty & \text{if } CurTime > li \quad (\text{arrive late}) \\ ei - CurTime & \text{if } CurTime < ei \quad (\text{arrive early}), \\ 0 & \text{else } (\text{comform to the time window}) \end{cases}$ speed stands for the

velocity, in addition, despite of the common limits of VRP problem, there exists other constraints, such as: customer demand service time window [ei, li], if the demand is not met, there will be some punishment. The wait time function for serving customer

$wt_i = \begin{cases} \infty & \text{if } CurTime > li \quad (\text{arrive late}) \\ ei - CurTime & \text{if } CurTime < ei \quad (\text{arrive early}) \\ 0 & \text{else } (\text{comform to the time window}) \end{cases}$ is designed as follows.

$$wt_i = \begin{cases} \infty & \text{if } CurTime > li \quad (\text{arrive late}) \\ ei - CurTime & \text{if } CurTime < ei \quad (\text{arrive early}) \\ 0 & \text{else } (\text{comform to the time window}) \end{cases} \tag{8}$$

Where CurTime represents the time that a vehicle reaches the customer point $f_{\min} =$

$NV \times VecCost + PC \times \sum_{i=0}^{N}\sum_{j=0}^{N}\sum_{k=1}^{NV} d_{ijk}x_{ijk} + Cw \times \sum_{i=1}^{N} wt_i + Cs \times \sum_{i=1}^{N} \delta_i$ General VRP

problems strive to find the shortest path, but there exists many limitations, for example, when in severe traffic congestion, the shortest path usually lead to long time operation to vehicle, cause high transportation costs. Instead, we set the minimum total cost as the goal, and the object function is designed as follows:

$$f_{\min} = NV \times VecCost + PC \times \sum_{i=0}^{N}\sum_{j=0}^{N}\sum_{k=1}^{NV} d_{ijk}x_{ijk} + Cw \times \sum_{i=1}^{N} wt_i + Cs \times \sum_{i=1}^{N} \delta_i \tag{9}$$

Where NV, VecCost, PC, Cw, Cs are the number of vehicles, vehicle costs, unit path cost, unit waiting time penalties and unit time service fee respectively.

4   The Design of Algorithm

When initially designed, ITO algorithm has been carefully designed by considering all elements of swarm intelligence algorithms, in order to be able to give a unified swarm intelligence algorithm model. Conventional swarm intelligence algorithm uses the exploration and exploitation process, namely firstly employ particles to explore the solution space, then update fitness value, utilize new fitness value for exploitation, and update the fitness value again. Such a process-oriented approach is neither consistent with the actual situation of mining industry, nor with the Brownian motion process on which ITO algorithm based, So starting from the particle movement and ITO algorithm itself, there comes out the thought of executing move process and wave process simultaneously, that is, particle movement is not conducted following mechanical steps (thus, moves are executed around the most attractive element, and waves are around the random attractors, i.e., two separate processes), instead, it's continuous motion under joint effects, that is, it is the movement under common effects of the most attractive element and random elements. This process is not only consistent with the facts of movement, it can also reduce costs effectively by lower the calculation times on the fitness value. Therefore, a solution based on ITO algorithm model is given in this paper. Additionally, we take into account of the natural advantages of ACO algorithm when applied in solving VRP problems and our in-depth study of merging it with VRP problems, some mechanisms of ACO algorithm are also borrowed, such as route update strategies, to improve our ITO algorithm which is not yet mature, in order to integrate the universality of ITO algorithm with the accuracy of ACO algorithm for solving VRP problems, there comes out the algorithm which named "improved ant optimization algorithm with soft time windows for solving vehicle routing problem" (referred to as IAO).

The proposed algorithm mainly includes three operators, i.e., radius operator which maintains the personality of particles, environmental temperature operator to exert macro-control, move and wave operators which optimize the routines by continuous learning from the feedbacks. Firstly, each particle utilizes own radius to maintain its personality, the size is related to the merits of its solution, and the personality of groups is employed to ensure the diversity of particles; Secondly, ambient temperature exert macro control on particle motion capability, that is, environmental temperature decreases with the increasing of iteration times to ensure that the algorithm converges gradually; then particle move and wave operators are designed based on particle radius and ambient temperature, they utilize respectively the optimal particle and random particle as attractor to implement move and wave movement, get new solution, and search for more optimal solutions in the motion. When designing move and wave operators (includes strength and process), we consider the intensity of move and wave as the reflection of particle exercise capability, therefore, these two operators, to some extent, are of one concept, so we no longer differentiate the move intensity and wave intensity when two operators are unified in this paper, it is indicated by exercise ability instead, which simplifies ITO algorithm design a lot. The move process and wave process are all attracted by attractors, while the attractor of move process is global optimal solution and that of wave process is randomly chosen solution. Finally, we use the path construction rules of ACO algorithm, thus the influence of moving and waving have unified effects on changing pheromone concentration, then scheduling scheme are constructed under the influence of the pheromone concentration. Afterwards, execute the process of move and wave simultaneously, so in the proposed it is further promoted for the integration of exploration and exploitation process.

Based on the above analysis, we conclude the design ideas as follows:

(1) Move and wave intensity are only the reflection of particle exercise capacity, in this paper the two intensity is replaced with unified exercise intensity, thus no longer treat them differently.
(2) Move refers to the process of attractor particles attracting current particles, so what need to do is increase the pheromone concentration of attractors.
(3) Wave means generating random perturbations under the influence of environment, so we just need to randomly select several paths from the environment to increase the pheromone concentration.
(4) Pheromone evaporates on all paths to ensure that pheromone concentration will not be too high in the environment.

The operator and pheromone concentration on each path are designed as follows:
Particle radius:

$$r_{n_i} = \frac{n - n_i}{n - 1} \tag{10}$$

Environment temperature:

$$T = \exp\left(-\frac{1}{T}\right) \tag{11}$$

Motion capability:

$$f(r, T) = \gamma_{\min} + f_1(r) \times f_2(T) \times (\gamma_{\max} - \gamma_{\min}) \tag{12}$$

All explanation and design ideas of the above parameters see reference [8], and will not be described here.

All paths execute the evaporation process in line with the volatilization factor $\tau(i,j) = (1-\rho) \times \tau(i,j)$ (13), the formula is:

$$\tau(i,j) = (1-\rho) \times \tau(i,j) \tag{13}$$

Where (i,j) represents all paths.

The formula of increasing the pheromone concentration on optimal paths is as following:

$$\tau(i,j) = \tau(i,j) + \gamma \tag{14}$$

if $(i,j) \in$ all paths in optimal solutions. Where $\tau(i,j) = \tau(i,j) + \gamma$ represents the motion capability.

The increase of pheromone concentration on randomly selected path is according to the following formula:

$$\tau(i,j) = \tau(i,j) + \gamma \tag{15}$$

$$\tau(i,j) = \rho \times \tau(i,j) + \begin{cases} \gamma \ if \ e(i,j) \in \sigma' \\ \gamma \ if \ e(i,j) \in \sigma \ and \ rand() < p \end{cases}.$$ Where the rand() is a

function that generates numbers from 0 to 1, p is the probability of selecting random path, which is to control the wave intensity.

In summary, the pheromone updating formula is:

$$\tau(i,j) = \rho \times \tau(i,j) + \begin{cases} \gamma \ if \ e(i,j) \in \sigma' \\ \gamma \ if \ e(i,j) \in \sigma \ and \ rand() < p \end{cases} \tag{16}$$

Where $p^k(i,j) = \begin{cases} \dfrac{[\tau(i,j)]^\alpha [\eta(i,j)]^\beta}{\sum\limits_{l \notin tabu_k} [\tau(i,l)]^\alpha [\eta(i,l)]^\beta}, i \in tabu_k \cap j \notin tabu_k \\ \\ 0, else \end{cases}$ denotes the paths

that haven't been visited by certain particle and the optimal solution particle neither.

The solution construction method is as follows:

Learn the path generating method from ant colony algorithm, that is, according to the pheromone concentration and distance to calculate the probability of selecting each candidate edge, then utilize the roulette approach to select a candidate edge as the next path, iteratively run this method until the scheduling scheme is constructed. The probability is calculated as following:

$$p^k(i,j) = \begin{cases} \dfrac{[\tau(i,j)]^\alpha [\eta(i,j)]^\beta}{\sum\limits_{l \notin tabu_k} [\tau(i,l)]^\alpha [\eta(i,l)]^\beta}, i \in tabu_k \cap j \notin tabu_k \\ \\ 0, else \end{cases} \tag{17}$$

Where $\eta(i,j)$ is the reciprocal of the distance, i.e., $1/d_{ij}$, denotes as empirical knowledge called visibility. $tabu_k$ is a taboo table for storing customer demand points have been searched by vehicle k. α is a factor for controlling the importance measurement of edge weight in the probabilistic choice, and β is for controlling the effect of visibility (i.e., edge length factor).

IAO algorithm flow

Initialization: parameters, particle radius, ambient temperature and pheromone concentration

WHILE (termination condition is not satisfied)

step1. According to the pheromone concentration and path distance, follow the path selection rule, generate the scheduling scheme of each particle, and finally according to the probability to decide whether to accept this update or not.

step2. Update the fitness value of each particle, and according to the size of fitness value to sort all the scheduling schemes generated by current iteration in descending order

step3. Select the optimal solution, and update the global optimal solution

step4. Update each particle radius, ambient temperature and pheromone concentration, and calculate out each particle's exercise ability

step5. Increase the concentration of pheromone on the optimal path, lead the particles move towards the optimal solution; increase the pheromone concentration on random path to implement random wave process in the environment.

5 Experimental Results and Analysis

In this section, the VRPSTW problem is solved, and the solution is compared with the results in [10, 11] respectively, according to the comparison and analysis, effectiveness of the proposed algorithm is proved.

The experiment environment: Eclipse Helios Service Release 2. The related cost is calculated according to Solomon method: rental fee per vehicle is 500, wait fee per unit is 10, travel cost per path unit is 40, service cost per unit is 5, the speed is 1. Algorithm related parameters are set as follows: population size $M = 50$, maximum evolution generation GEN = 500, initial temperature of annealed table $T = 1000$, annealing table length TLength = 2, annealing rate $\rho = 0.99$, edge weight importance factor $\alpha = 5$, the importance factor of the distance between customer points $\beta = 3$, the probability of selecting a random path $p = 0.3$. The results are showed in Tables 1 and 2; Figs. 1, 2, 3 are illustrative diagram of the distance, the optimum value and the average value respectively, which are depicted based on the experimental data in Tables 1 and 2.

From Table 1, we compare the proposed algorithm with the experimental results in [10], mainly are the number of vehicles, path length, and total cost of the proposed algorithm is listed as well. As can be seen from Table 1, the proposed algorithm for C class problems can always get the optimal results or better results, while for R and RC class problems there exists certain bias to the optimal solution, mainly due to the design of the proposed algorithm is a macro-design based on our thoughts of the algorithm, while the specific design details such as initialization method, local search methods haven't yet been in detailed design, in addition, all of the running time of the proposed algorithm is limited in 50 s, while in [10] the algorithm running time are typically between 2−7 h. As a result, the effectiveness of the proposed IAO algorithm can be proved according to the results of Table 1.

From Table 2, comparisons have been drawn on the calculation of fitness value between the proposed algorithm and the algorithm in [11]. Literature [11] uses the "first-expired-first-serviced" algorithm (FEFS) to work out initial solution, then simulated annealing algorithm (SA) is employed to improve the initial solution, and

Table 1. Comparison between the algorithm in [10] and IAO

Pro	Best known		Algorithm in [10]		The proposed algorithm (IAO)			
	VN	Distance cost	VN	Distance cost	VN	Distance cost	Waiting cost	Total cost
C101	10	828.94	10	828.94	**10**	**828.94**	**0**	**83 157.47**
C106	10	828.94	10	828.94	10	835.42	49.64	83 913.36
C201	3	591.56	3	591.56	**3**	**591.56**	**0**	**70 162.26**
R103	14	1237.05	14	1287.0	17	1522.21	1149.64	85 884.80
			15	1264.2				
R108	10	960.26	10	971.91	12	1123.87	467.96	60 634.52
R203	4	935.04	3	1041.0	5	1320.55	1436.07	74 682.60
			5	995.8				
			6	978.5				
R204	3	789.72	3	1130.1	**3**	**950.53**	**780.77**	**52 328.90**
			4	927.7				
			5	831.8				
			6	826.2				
RC101	15	1636.92	15	1690.6	18	1871.01	731.33	96 153.80
			16	1678.9				
RC102	13	1470.26	15	1493.2	16	1817.48	618.87	91 888.09
RC105	16	1590.25	15	1611.5	18	1875.87	826.65	97 301.12
			16	1589.4				
RC108	10	1142.66	11	1156.5	12	1335.28	227.90	66 690.29

Table 2. Comparison between the algorithm in [11] and IAO

Pro	TS		Algorithm in [11] (SA)		The proposed algorithm (IAO)	
	Optimal	Average	Optimal	Average	Optimal	Average
R101	128 005.12	129 862.66	121 122.50	123 229.51	104 808.41	110 506.79
R102	132 038.25	137 941.96	121 516.61	127 056.92	100 104.02	102 570.48
R201	120 492.04	125 031.48	110 738.33	114 076.51	88 834.95	94 294.84
R202	110 849.52	120 857.59	101 296.37	111 122.22	89 127.64	91 597.10
C101	153 800.02	160 337.92	136 719.63	143 741.30	83 157.47	88834.47
C102	164 917.74	166 576.88	146 608.96	154 561.76	91 751.71	104 462.04
C201	117 154.81	120 497.33	102 681.24	111 295.40	70 162.26	91 561.28
C202	129 560.34	133 189.78	120 051.88	124 762.01	70 966.79	99 670.66
RC101	119 678.16	121 419.28	108 689.43	113 520.80	96 835.02	102 032.03
RC102	128 343.59	133 229.97	112 415.05	120 523.93	88 693.09	95 680.04
RC201	155 422.90	156 767.73	150 829.97	154 549.38	99 663.79	103 191.14
RC202	132 486.31	139 866.27	119 488.16	126 091.85	92 590.15	99 073.89

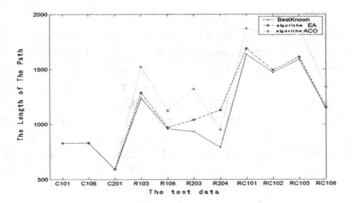

Fig. 1. Distance comparison

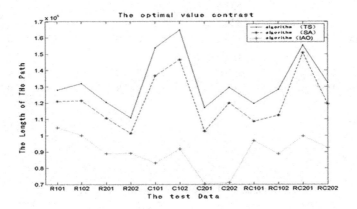

Fig. 2. Optimal value comparison

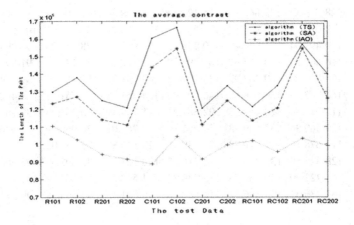

Fig. 3. Average value comparison

ultimately the satisfactory solution is obtained, this corresponds to the results with SA algorithm in Table 2; TS refers to the tabu search algorithm used for comparison with the algorithm proposed in [11]. The bold ones indicate the best solutions among the three algorithms listed. Each data set is executed for 20 times respectively, "Optimal" and "Avérage" mean the optimal solution and average solution respectively among the 20 iterations. In the experiment, first two cases of the Solomon's six categories are selected, it can be seen from the experimental results that the optimal solution and the average value of the proposed algorithm are significantly better than the two algorithms in [11], but in [11] the execution time of the algorithm are about 3 s, while it is comparatively longer in the proposed algorithm, which are between 22.5-48.5 s, in fact, it's also acceptable. Despite of the time, the obvious better results can also prove the effectiveness of IAO algorithm proposed in this paper.

6 Conclusion

In this paper, Vehicle Routing Problem is firstly analyzed, then the Ito stochastic differential thought is integrated with ACO algorithm, and the intensity of move operator and wave operator are unified to be reflected by exercise ability. In addition, a new pheromone updating rule is given to be more in line with Brownian motion and ant optimization rule. From the comparison results between the proposed algorithm and that in [10, 11], it is demonstrated that the proposed IAO algorithm can effectively solve VRPSTW problem. Surely, there remains certain settings to be further researched, for example, the parameters of the proposed algorithm can be better adjusted, thereby enhancing the efficiency and accuracy of the algorithm. Additionally, when setting the parameters of the algorithm, it is mostly based on certain assumptions and analysis on large amounts of data, rather than on scientific setting rule, these will be further researched in the future.

References

1. Kim, S., Lewis, M.E., White III, C.C.: Optimal vehicle routing with real-time traffic information. IEEE Trans. Intell. Transp. Syst. **6**(2), 178–188 (2005)
2. Dong, W., et al.: BBOB-benchmarking: a new evolutionary algorithms inspired by ITO process for noiseless function tested. J. Comput. Inf. Syst. pp. 2195–2203 (2011)
3. Korenevskiy, N., Al-Kasasbeh, R.T., Ionescouc, F., Alshamasin, M., Alkasasbeh, E., Smith, A.P.: Fuzzy determination of the human's level of psycho-emotional. In: Toi, V.V., Toan, N. B., Dang Khoa, T.Q., Lien Phuong, T.H. (eds.) 4th International Conference on Biomedical Engineering in Vietnam. IFMBE Proceedings, vol. 40, pp. 213–216. Springer, Heidelberg (2013)
4. Hongbin, L.: Study on multi-depots and multi-vehicles vehicle scheduling problem based on improved particle swarm optimization. Comput. Eng. Appl. **50**(7), pp. 251−253 (2014) (in Chinese)
5. Qiuyun, W., Wenbao, J.: Solving algorithm of vehicle routing problem with soft time windows. J. Beijing Inf. Sci. Technol. Univ. **28**(4), pp. 57−59 (2013) (in Chinese)

6. Wen, M., Cordeau, J.-F., Laporte, G., et al.: The Dynamic Multi-PeriodVehicle Routing Problem. DTU Management Engineering, Denmark (2009)
7. Dong, W.: Time series modeling based on ITO algorithm. In: Proceedings of the International Conference on Natural Computation, pp. 398–402 (2007)
8. Wenyong, D., Wensheng, Z., Ruiguo, Y.: Convergence and runtime analysis of ITO algorithm for one class of combinatorial optimization. chinese. J. Comput. **34**(4), pp. 636–646 (2011) (in Chinese)
9. Dong, W.: The multi-objective ITO algorithms. In: Proceedings of the International Conference on Intelligence Computation and Application, pp. 21–23 (2007)
10. Ghannadpour, S.F., Noori, S., Tavakkoli-Moghaddam, R.: Multiobjective dynamic vehicle routing problem with fuzzy travel times and customers' satisfaction in supply chain management. IEEE Trans. Eng. Manage. **60**(4), 777–790 (2013)
11. MA, H., Jin, P., Yang, S.: Heuristic methods for time-dependent vehicle routing problem. J. Syst. Eng. **27**(2), pp. 256–262 (2012) (in Chinese)

Evolutionary Nonnegative Matrix Factorization for Data Compression

Liyun Gong[1($\boxtimes$)], Tingting Mu[2], and John Y. Goulermas[2]

[1] School of Computer Science and School of Engineering,
The University of Lincoln, Lincoln LN6 7TS, UK
lgong@lincoln.ac.uk
[2] Electrical Engineering, Electronics and Computer Science,
The University of Liverpool, Brownlow Hill, Liverpool L69 3GJ, UK
{t.mu,j.y.goulermas}@liverpool.ac.uk

Abstract. This paper aims at improving non-negative matrix factorization (NMF) to facilitate data compression. An evolutionary updating strategy is proposed to solve the NMF problem iteratively based on three sets of updating rules including multiplicative, firefly and survival of the fittest rules. For data compression application, the quality of the factorized matrices can be evaluated by measurements such as sparsity, orthogonality and factorization error to assess compression quality in terms of storage space consumption, redundancy in data matrix and data approximation accuracy. Thus, the fitness score function that drives the evolving procedure is designed as a composite score that takes into account all these measurements. A hybrid initialization scheme is performed to improve the rate of convergence, allowing multiple initial candidates generated by different types of NMF initialization approaches. Effectiveness of the proposed method is demonstrated using Yale and ORL image datasets.

Keywords: Non-negative matrix factorization · Data compression · Evolutionary computation

1 Introduction

Non-negative matrix factorization (NMF) is an algorithm based on decomposition by parts the input data matrix, which can reduce the dimensionality of the datasets while keep the most information about the datasets. It is suitable for redundancy reduction in image data, known as image compression, to optimize storage space and increase transmission rate. Different from compression methods such as principal component analysis (PCA) and independent component analysis (ICA), NMF introduces non-negative constraints which offer more clear interpretation. Recent advances on NMF are focused on formulating more sophisticated objective function for NMF to better serve a dimensionality reduction, clustering or classification task, by incorporating extra concerns into the original reconstruction error such as preservation of local data geometry [12] and enhancement of class separation [5].

In traditional NMF, the initial values of the factorization variables of NMF are usually set as random values. However, this is not the most effective setup. Many algorithms have

© Springer International Publishing Switzerland 2015
D.-S. Huang et al. (Eds.): ICIC 2015, Part I, LNCS 9225, pp. 23–33, 2015.
DOI: 10.1007/978-3-319-22180-9_3

been proposed to obtain the initial values of the factorization variables in a more sophisticated way, in order to improve the rate of convergence. For example, spherical k-means clustering is used to initialize one factorization matrix, and then nonnegative least square is used to derive the other factorization matrix [8]. Principal component analysis (PCA) can also be used to perform such initialization, for which non-negativity of the factorization variables can be enforced by either converting all the negative elements of the PCA output to zero [11] or keeping the absolute values [10]. Another way is to utilize fuzzy c-means clustering (FCM), where the FCM cluster centroids can be used as one factorization matrix while the cluster membership degrees to derive the other factorization matrix [7, 11]. Performances of six initialization methods of random, centroid, singular value decomposition (SVD) centroid, random acol, random C and co-occurrence based ones is compared in [3]. When applying NMF to data compression, an appropriate initialization method has the potential to enhance both convergence rate and compression performance. On the other hand, sensitivity of the NMF performance to different initialization methods makes it challenging for the user to choose an appropriate one for a given task.

Instead of choosing one particular initialization scheme, we propose an evolutionary NMF updating procedure, which learns from multiple seed candidates initialized in the solution space, and effectively updates the candidate set along multiple directions in order to obtain better quality of factorization matrices to facilitate data compression. The proposed method is general, and there is no limitation on the used number and types of the initialization methods. The major contributions of the proposed design include to improve the most commonly used multiplicative NMF update [4] so that it can serve better the data compression task, and to take advantage of the hybrid of different NMF initialization setups to produce NMF approximations that suits better the data compression purpose and save the users' effort on initialization determination. Effectiveness of the proposed method will be demonstrated thoroughly through benchmark testing and comparison with existing approaches using Yale and ORL image datasets.

The rest of the paper is organized as follows: Sect. 2 explains the pro- posed method. Performance evaluation and comparative analysis are conducted in Sect. 3 by experimenting with the proposed and state-of-the-art methods. Finally, a conclusion is drawn in Sect. 4.

2 Proposed Method

Given a $d \times n$ non-negative matrix $\mathbf{X} = [x_{ij}]$ with each element $x_{ij} \geq 0$, its columns represent images to be analyzed. NMF seeks two non-negative matrices, a $d \times k$ one $\mathbf{W} = [w_{ij}]$ and an $n \times k$ one $\mathbf{H} = [h_{ij}]$, so that the following factorization error is minimized:

$$\min_{\substack{w_{ij} \geq 0, \\ h_{ij} \geq 0}} \left\| \mathbf{X} - \mathbf{WH}^T \right\|_F^2 \tag{1}$$

where $\| . \|_F$ denotes the Frobenius norm. Each column of $\mathbf{W}$ is known as the basic vector, while each column of $\mathbf{H}$ as the encoding coefficient vector. For the image compression analysis, each column of $\mathbf{W}$ indicates one basic image and $k \leq \min(d, n)$ is often assumed as the number of the basic images determined by the user or the specific purpose.

We propose an evolutionary strategy to improve the iterative updating procedure of NMF, named as ENMF. It aims at producing higher-quality basis and encoding coefficient matrices $\mathbf{W}$ and $\mathbf{H}$ to suit the data compression purpose. The algorithm starts from multiple pairs of initialization matrices for the basis and encoding coefficient matrices, which form an initial candidate set denoted as $S_0 = \left\{ \left(\mathbf{W}_0^i, \mathbf{H}_0^i \right) \right\}_{i=1}^m$ where $\left\{ \mathbf{W}_0^i \right\}_{i=1}^m$ and $\left\{ \mathbf{H}_0^i \right\}_{i=1}^m$ are referred as the seed matrices. The algorithm then evolves, creating an updated candidate set at each iteration, denoted as $S_t = \left\{ \left(\mathbf{W}_t^i, \mathbf{H}_t^i \right) \right\}_{i=1}^{m_t}$ for the tth iteration with m_t denoting the new candidate number. In the end, the optimal basis matrix and its corresponding encoding coefficient matrix are selected from the finally evolved candidate set based on a score function formulated to suit data compression.

2.1 Seed Matrix Generation

To take advantage of the state-of-the-art NMF initialization strategies and to achieve local improvement of the optimal solution, multiple NMF initialization approaches are utilized to construct the initial candidate set, which contains various seed matrices of the basis and encoding coefficient ones: (1) The clustering-based initialization (CI) approach is first conducted via performing k-means clustering [2]. The resulting binary cluster membership matrix is used as $\mathbf{H}_0^1$, and the resulting clustering centroid matrix as $\mathbf{W}_0^1$. (2) A similar CI approach is conducted again but via FCM clustering [1]. The obtained cluster membership and centroid matrices are used as $\mathbf{H}_0^2$ and $\mathbf{W}_0^2$, respectively. (3) The random initialization (RI) [4] and random acol initialization (RAI) [3] are used to generate the two candidates of $\left(\mathbf{W}_0^3, \mathbf{H}_0^3 \right)$ and $\left(\mathbf{W}_0^4, \mathbf{H}_0^4 \right)$. It is worth to note that the proposed NMF updating algorithm is a general method. The users can freely include any type and any number of initial candidates to suit their needs apart from the above ones.

2.2 Evolving Strategy

In each iteration, three new subsets of candidates $S_{t+1}^{(M)}, S_{t+1}^{(F)}$ and $S_{t+1}^{(S)}$ are generated from the previous set S_t, according to three types of evolving rules proposed, including the multiplicative, firefly and the survival of the fittest rules. The three subsets together constitute the updated set $S_{t+1} = S_{t+1}^{(M)} \cup S_{t+1}^{(F)} \cup S_{t+1}^{(S)}$ at the $(t+1)$th iteration, from which the best candidate is selected as the final output in the last iteration. In the following, we explain the three rules in detail.

Multiplicative Rule: This rule is constructed to take advantage of the classical multiplicative update rules for NMF approximation [4]. It generates the new candidate subset by

$$S_1^{(M)} = \Phi_1(S_0, \mathbf{X}) \tag{2}$$

For the first iteration and

$$S_{t+1}^{(M)} = \Phi_1\left(S_t^{(M)}, \mathbf{X}\right) \tag{3}$$

for the $(t+1)$th iteration $(t \geq 1)$. The operation $S' = \Phi_1(S, \mathbf{X})$ takes one set of matrix pairs $S = \{\mathbf{W}_i, \mathbf{H}_i\}_{i=1}^{m}$ and one $d \times n$ matrix $\mathbf{X}$ as the input, outputs a set of matrix pairs denoted as $S' = \left\{(\mathbf{W}_i', \mathbf{H}_i')\right\}_{i=1}^{m}$ and is formulated as

$$\mathbf{H}_i' = \mathbf{H}_i \circ (\mathbf{X}^T \mathbf{W}_i) \emptyset (\mathbf{H}_i \mathbf{W}_i^T \mathbf{W}_i) \tag{4}$$

$$\mathbf{W}_i' = \mathbf{W}_i \circ (\mathbf{X} \mathbf{H}_i) \emptyset (\mathbf{W}_i \mathbf{H}_i^T \mathbf{H}_i) \tag{5}$$

where $\circ$ and $\emptyset$ denote the Hadamard product and division, respectively. This rule updates the candidates separately from the other rules in order to enable the inclusion of multiple NMF solutions obtained by the multiplicative update rules to the final evolved candidate set. These solutions are driven by the same factorization error minimization but initialized through different ways.

Firefly Rule: This rule encourages the generation of new candidate matrix pairs that may contain higher quality of basic matrix than those obtained by the previous multiplicative rule, in order to facilitate the data compression task more effectively.

In the first iteration, the firefly rule operates on the candidate subset of $S_1^{(M)}$ of the multiplicative rule, and further creates another candidate subset by

$$S_1^{(F)} = \Phi_2\left(S_1^{(M)}, \mathbf{W}_1^*\right), \tag{6}$$

The operation $S' = \Phi_2(S, \mathbf{A})$ takes a set $S = \{(\mathbf{W}i, \mathbf{H}i)\}_{i=1}^{m}$ and an $d \times k$ matrix A as input, while outputs a new set $S' = \{(\mathbf{W}_i', \mathbf{H}_i')\}_{i=1}^{m}$. The corresponding relationship between its input and output is defined by

$$\mathbf{W}_i' = \mathbf{W}_i + \beta(\mathbf{A} - \mathbf{W}_i) \tag{7}$$

$$\mathbf{H}_i' = \mathbf{H}_i, \tag{8}$$

where $0 < \beta \leq 1$ is set by the same user. It is obvious that, given $0 < \beta \leq 1$, $\mathbf{W}_i'$ is always non-negative when $\mathbf{W}_i$ and $\mathbf{A}$ are both non-negative. These guarantee that the generated matrix pairs $(\mathbf{W}_i', \mathbf{H}_i')$ are eligible to be used as NMF candidates. The matrix

$\mathbf{W}_1^*$ used in Eq. (6) is selected through searching within the combined set of $S_1^{(M)} \cup S_0$ based on a predefined score function $O(\cdot)$ for assessing the compression quality, which will be explained in Sect. 2.3.

Equation (7) drives $\{\mathbf{W}i\}_{i=1}^m$ generated by the multiplicative rule to move towards a pre-selected optimal basis matrix $\mathbf{W}_1^*$. This design is motivated by a recent evolutionary optimization algorithm inspired by the flashing behavior of firefly algorithm [9]. It assumes that attractiveness between fireflies is proportional to their brightness, thus, given any two fireflies, one will move towards the other that glows brighter. Following Eq. (7), each candidate in $S_1^{(M)}$ is viewed as a firefly. The quality of the basic matrix for each candidate, evaluated by the score function $O(\cdot)$, represents the brightness degree of the firefly. The evolving rule is constructed by letting all the fireflies move towards the brightest one in each iteration. This procedure offers an opportunity to evolve higher quality of basic matrices to better serve the data compression task.

From the second iteration, the firefly rule starts to create new candidate subset $S_{t+1}^{(F)}$ by operating on its previously generated subset $S_t^{(F)}$. It first modifies $S_t^{(F)}$ by the multiplicative rule Φ_1, and then updates the resulting set based on the firefly operation Φ_2. This gives the following new candidate subset for the $(t+1)$th iteration $(t \geq 0)$:

$$S_{t+1}^{(F)} = \Phi_2\left(\Phi_1\left(S_t^{(F)}, \mathbf{X}\right), \mathbf{W}_t^*\right), \tag{9}$$

where $\mathbf{W}_t^*$ is selected according to the score function $O(\cdot)$, through searching within not only the whole previous candidate set but also its update via multiplicative rule $S_t \cup \Phi_1(S_t, \mathbf{X})$ to maintain its quality. Instead of directly updating $S_t^{(F)}$ with Φ_2, Eq. (9) uses the multiplicative rule to smoothen out the given candidates, which may potentially reduce the factorization error. The mixed application of Φ_1 and Φ_2 attempts to evolve matrix pairs offering good quality of basic matrix while alternatively ensuring the joint quality of the basis and encoding coefficient matrices.

Survival of the Fittest Rule: This rule ensures the candidates containing the best basic matrix are always included in the evolved set. At the first iteration, the candidate subset $S_1^{(S)}$ is generated by

$$S_1^{(S)} = \Phi_3\left(S_1^{(M)}, \mathbf{W}_1^*\right), \tag{10}$$

After that, it modifies its previously generated subset $S_t^{(S)}$ by

$$S_{t+1}^{(S)} = \Phi_3\left(\Phi_1\left(S_t^{(S)}, \mathbf{X}\right), \mathbf{W}_t^*\right). \tag{11}$$

Here, the operation $S' = \Phi_3(S, \mathbf{A})$ creates m matrix pairs $S' = \{(\mathbf{W}_i', \mathbf{H}_i')\}_{i=1}^m$ from the input set $S = \{(\mathbf{W}_i, \mathbf{H}_i)\}_{i=1}^m$, and is formulated as $\mathbf{W}_i' = A$ and $\mathbf{H}_i' = \mathbf{H}_i$. It combines the best basic matrix $\mathbf{W}_t^*$ selected in each iteration with various encoding coefficient matrices. The use of $\mathbf{W}'_i = \mathbf{A}$ can be viewed as a special case of Eq. (7) with the fixed parameter $\beta = 1$, equivalent to forcing all the weaker fireflies to eliminate themselves but let the brightest one survive. Thus, this rule is named as the survival of the fittest.

2.3   Score Function

Since the primary goal of this work is to improve NMF so that it can serve better the data compression task, it is important to design an appropriate score function to assess compression quality. Usually, in addition to factorization error, data compression performance is also indicated by sparsity and orthogonality of the resulting basis matrix. Given the data matrix $\mathbf{X}$ and the basis matrix $\mathbf{W}$, the following measurements are usually computed [11]:

$$error(\mathbf{W}) = \frac{\| \mathbf{X} - \mathbf{WH}_W^T \|_F}{\mathbf{X}_F} \tag{12}$$

$$sparsity(\mathbf{W}) = \frac{\sum_{i=1}^{d} \sum_{j=1}^{k} |w_{ij}|}{k} \tag{13}$$

$$orthogonality(\mathbf{W}) = \sqrt{\frac{d \sum_{i,j=1,i\neq j}^{k} w_i^T w_j}{k(k-1)/2}}, \tag{14}$$

where w_{ij} denotes the ij-th element of the basis matrix $\mathbf{W}$ while the vector w_i denotes its ith column, and the encoding matrix $\mathbf{H}_W^T$ is computed from W by applying non-negative least square analysis [4]. Based on the formulations of all the three measurements, the lower value they possess, the better compression quality they indicate. To take into account all these three measurements, the following composite score function is proposed, given as

$$O(\mathbf{W}) = \begin{cases} \frac{sparsity(W) + orthogonality(\mathbf{W})}{2}, & \text{if } error(\mathbf{W}) \leq \alpha, \\ \frac{sparsity(W) + orthogonality(\mathbf{W})}{2\beta}, & \text{otherwise.} \end{cases} \tag{15}$$

The parameter $0 < \alpha < 1$ is defined by the user, providing a threshold of the minimum allowed factorization error rate. This proposed score examines the averaged sparsity and orthogonality performance within a pre-defined factorization error range. When the allowed error threshold α is exceeded, the other two quantities are heavily penalized by $1/\beta$, where β can be set as a very small positive value to impose high penalty, e.g., 10^{-6}. The overall data flow of the proposed ENMF is shown in Fig. 1.

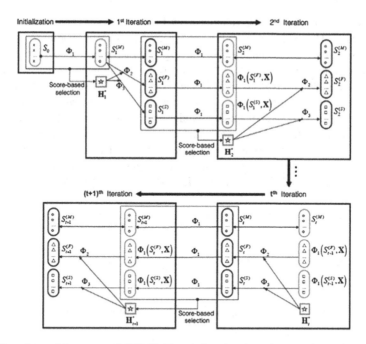

Fig. 1. Data flow of the proposed ENMF. The circle, triangle and rectangle symbols represent candidates derived during the generation of the $S_t^{(M)}$, $S_t^{(F)}$ and $S_t^{(S)}$ subset, respectively.

3   Experimental Results and Analysis

Two face image datasets of Yale and ORL are used to examine the performance. Yale contains grayscale images representing 15 subjects with 11 images per subject including center-light, w/glasses, happy, left-light, w/no glasses, normal, right-light, sad, sleepy, surprised and wink. ORL contains 400 images representing 40 distinct subjects, with 10 images per subject taken at different times varying the lighting, facial expressions (open/closed eyes, smiling/not smiling) and facial details (glasses/no glasses). Each image is scaled by its maximum pixel value so that the resulting data matrix $\mathbf{X}$ possesses elements between 0 and 1.

The images are compressed by $k = 25$, which was empirically observed to be sufficient for representing image data in our study. The proposed ENMF is compared with the most commonly used multiplicative NMF approach based on four different initialization methods including RI, RAI, CI by k-means (CI1) and CI by FCM (CI2). The error threshold α is set as 0.3 and 0.15 for Yale and ORL, respectively, which was chosen by comparing the output sample images with the original ones given different threshold values. The maximum iteration number for ENMF is set as 500 for Yale and 1000 for ORL to allow good convergence, for which we have observed that the score measure converges approximately after 200 interactions for Yale while after 500 iterations for ORL (see Figs. 2 and 3). Each algorithm is run five times for each dataset and the averaged performance is reported. In addition to the three measurements for data compression, we also report the RAND index [6], which evaluates the quality of the encoding matrix **H** to see how well it preserves the ground truth partition of the data.

In Figs. 2 and 3, we compare the convergence of the proposed ENMF with the competing methods, in terms of the three quality measurements of sparsity, orthogonality and factorization error, as well as the proposed composite score function. The x-axis of each plot represents the iteration number while the y-axis represents the values of the relevant score function as indicated in the figures. It can be seen that ENMF offers the fastest convergence and the best data compression performance including much better sparsity and orthogonality and equally low factorization error. We also provide examples of the learned basis images by different methods for Yale and ORL in Figs. 4 and 5, respectively. It can be seen that ENMF achieves higher

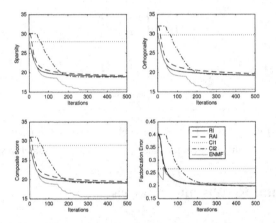

Fig. 2. Convergence and compression performance comparison for different methods with Yale in terms of different measurements

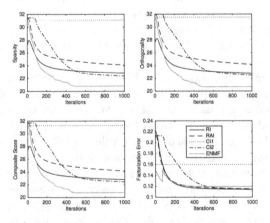

Fig. 3. Convergence and compression performance comparison for different methods with ORL in terms of different measurements.

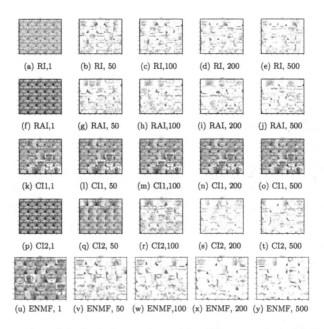

(a) RI,1 (b) RI, 50 (c) RI,100 (d) RI, 200 (e) RI, 500

(f) RAI,1 (g) RAI, 50 (h) RAI,100 (i) RAI, 200 (j) RAI, 500

(k) CI1,1 (l) CI1, 50 (m) CI1,100 (n) CI1, 200 (o) CI1, 500

(p) CI2,1 (q) CI2, 50 (r) CI2,100 (s) CI2, 200 (t) CI2, 500

(u) ENMF, 1 (v) ENMF, 50 (w) ENMF,100 (x) ENMF, 200 (y) ENMF, 500

Fig. 4. Demonstration of the basis images learned by different methods for Yale. The rows represent different methods including multiplicative NMF with RI, RAI, CI1 and CI2 initializations and ENMF from top to bottom. The columns represent the compared iteration numbers of 1, 50, 100, 200 and 500 from left to right.

sparsity than the others after 50 iterations for Yale and after 100 iterations for ORL. In general, the basis matrix computed by CI1 possesses pretty low sparsity. Also, observing in detail each column of the learned basis matrix that is corresponding to one patch of the 5×5 face patches in each subfigure, ENMF offers more distinct face patches that indicate good orthogonality between the columns of the basis matrix.

Table 1 reports the averaged performance of the proposed ENMF and the competing methods by repeating the experiments five times for each method. Sparsity, orthogonality and factorization error evaluate the quality of the basis matrix $\mathbf{W}$ that is important for data compression, while the RAND index measures the possible information loss of the encoding matrix $\mathbf{H}$ in terms of cluster structure preservation. It can be seen that the proposed method possesses significantly better sparsity and orthogonality, meanwhile comparative factorization error and RAND index as compared to the others. The competitive performance of the proposed method demonstrated in Figs. 2, 3 and Table 1 benefits from its three-rule-driven update procedure. It inherits good quality of candidates derived by the classical multiplicative rule, while introduces more diversified offsprings generated from the strongest parents (the brightest fireflies) to avoid local optimum but maintain competent searching direction.

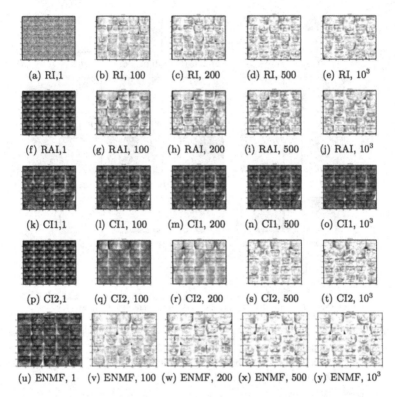

Fig. 5. Demonstration of the basis images learned by different methods for ORL. The rows represent different methods including multiplicative NMF with RI, RAI, CI1 and CI2 initializations and ENMF from top to bottom. The columns represent the compared iteration numbers of 1, 100, 200, 500 and 1000 from left to right.

Table 1. Performance comparison for different methods.

Dattatasets	Sparsity	Orthogonality	Factorization Error	RAND
Yale	RI: 18.8 RAI:19.3 CI1:28.1 CI2:18.6 ENMF: **15.3**	RI:19.2 RAI:19.8 CI1:29.8 CI2:19.0 ENMF: **15.4**	RI:0.2 RAI:0.2 CI1:0.3 CI2:0.2 ENMF:**0.2**	RI:91.9 RAI:91.8 CI1:93.1 CI2:91.6 ENMF: 90.8
ORL	RI: 22.9 RAI:24.1 CI1:31.0 CI2:22.5 ENMF: **20.6**	RI:22.9 RAI:24.2 CI1:31.5 CI2:22.6 ENMF:**20.4**	RI:0.1 RAI:0.1 CI1:0.2 CI2:0.1 ENMF: **0.1**	RI:95.0 RAI:95.1 CI1:95.1 CI2:94.9 ENMF: 94.1

4   Conclusions

We have proposed a novel evolutionary updating strategy to improve NMF for data compression. The hybrid initialization takes advantages of multiple initialized seed candidates in the solution space and saves the users' effort on selection of an appropriate initialization scheme. Three sets of updating rules including multiplicative, firefly and survival of the fittest rules have been proposed driven by a composite score function for assessing data compression quality in terms of sparsity, orthogonality and factorization error. This enables effective searching of an optimal candidate along multiple directions directly controlled by the performance of the targeted data compression task. Experiments have been carried out using two image datasets of Yale and ORL and the results have shown that the proposed ENMF outperforms the most popular multiplicative NMF methods for data compression in terms of both convergence and compression performance.

References

1. Bezdek, J.C., Ehrlich, R., Full, W.: FCM: the fuzzy c-means clustering algorithm. Comput. Geosoci **10**(2–3), 191–203 (1984)
2. Forgy, E.W.: Cluster analysis of multivariate data: efficiency versus interpretability of classifications. Biometrics **21**, 768–769 (1965)
3. Langville, A.N., Meyer, C.D., Albright, R.: Initializations for the nonnegative matrix factorization. In: Proceeding of the Twelfth ACM SIGKDD International Conference on Knowledge Discovery and Data Mining (2006)
4. Lawson, C.L., Hanson, R.J.: Solving Least Squares Problems. Prentice Hall, Englewood Cliffs (1974)
5. Nikitidis, S., Tefas, A., Pitas, I.: Projected gradients for subclass discriminant nonnegative subspace learning. IEEE Transactions on Cybernetics. In press (2014). doi:10.1109/TCYB.2014.2317174
6. Rand, W.M.: Objective criteria for the evaluation of clustering methods. J. Am. Stat. Assoc. Theor. Methods Sect. **66**, 846–850 (1971)
7. Rezaei, M., Boostani, R., Rezaei, M.: An efficient initialization method for non- negative matrix factorization. J. Appl. Sci. **11**, 354–359 (2011)
8. Wild, S.: Seeding Non-negative Matrix Factorizations with The Spherical K-means Clustering. Master's thesis, Master of Science thesis, University of Colorado (2003)
9. Yang, X.S.: Firefly Algorithm, Stochastic Test Functions and Design Optimization
10. Zhao, L., Zhuang, G., Xu, X.: Facial expression recognition based on PCA and NMF. In: Proceedings of the 7th World Congress on Intelligent Control and Automation, pp. 6826–6829 (2008)
11. Zheng, Z., Yang, X., Zhu, Y.: Initialization Enhancer for Non-negative Matrix Factorization. Eng. Appl. Artif. Intell. **20**, 101–110 (2007)
12. Zhi, R., Flierl, M., Ruan, Q., Kleijn, W.B.: Graph-preserving sparse nonnegative matrix factorization with application to facial expression recognition. IEEE Trans. Cybern. **41**, 38–52 (2011)

Learning-Based Evolutionary Optimization
for Optimal Power Flow

Qun Niu[(⊠)], Wenjun Peng, and Letian Zhang

Shanghai Key Laboratory of Power Station Automation Technology,
School of Mechatronic Engineering and Automation, Shanghai University,
Shanghai 200072, China
comelycc@hotmail.com

Abstract. This paper proposes a learning-based evolutionary optimization
(LBEO) for solving optimal power flow (OPF) problem. The LBEO is a simple
and effective algorithm, which simplifies the structure of teaching-learning-based
optimization (TLBO) and enhances the convergence speed. The performance of
this method is implemented on IEEE 30-bus test system with the minimized fuel
cost objective function, and the results show that LBEO is practicable for OPF
problem compared with other methods in the literature.

Keywords: Optimal power flow · Learning-based evolutionary optimization ·
Teaching-learning-based optimization

1 Introduction

Optimal power flow (OPF) problem has received much attention over past two decades
and has established its position as one of the main tools for optimal operation processes
in the modern power system. The OPF is a large-scale, highly constrained nonlinear
optimization problem which uses optimal power flow to optimize resource allocation,
reduce the cost of electricity transmission and improve the service of power system.
The objective of the OPF problem is to minimize a given objective function such as
total fuel cost of all generator units, total fuel cost with value point effects, voltage
profile improvement, voltage stability enhancement, piecewise quadratic cost curve
function, through adjustments of control variables while satisfying various equality and
inequality constraints at the same time.

The optimal power flow was first proposed by Dommel and Tinney in 1968 [1].
After that, much effect has been focused on a wide variety of conventional optimization
techniques to solve the OPF problem such as linear programming, Lagrangian relaxation, interior point method and quadratic programming. However, the conventional
optimization techniques have some defects. For example, these methods are usually
based on differentiation and easy to fall into local minimum or even result into
divergence. To overcome these problems, evolutionary algorithms are used to solve
OPF. Genetic algorithm (GA) was used to solve OPF problem [2]. In [3], particle
swarm optimization (PSO) has been successfully implemented for solution of OPF
problem. Cai [4] has adopted the differential evolution algorithm (DE) to solve a OPF
problem with a constrained objective function. Despite the searching capability has

© Springer International Publishing Switzerland 2015
D.-S. Huang et al. (Eds.): ICIC 2015, Part I, LNCS 9225, pp. 34–45, 2015.
DOI: 10.1007/978-3-319-22180-9_4

been improved, stability and scalability of these intelligent approaches for various OPF problems are still the major issues to tackle for satisfactory application. Therefore, it is necessary to develop effective and simplified methods to solve OPF problems.

In this paper, a novel yet simple learning-based evolutionary algorithm, namely LBEO is proposed based on teaching-learning-based optimization (TLBO) to solve OPF problems. In the proposed LBEO, the teacher phase is removed to enhance the efficiency of the searching capability. Similar to the basic TLBO, LBEO retains the advantages of simple in concept, easy to implement and having no adjustable parameters. In comparison with basic TLBO, the structure is further simplified, and the function evaluation number in each generation is significantly reduced. To evaluate the performance of LBEO, it is tested on three large scale benchmark functions, and compared with the basic TLBO and five commonly used methods, including GA-TB [5], PSO-IW [5], SaDE [5], GL-25 [5], HIS [5] and TLBO [5]. Then, LBEO is employed to solve the OPF problem in a standard IEEE 30-bus test system with the objective of the minimum total fuel cost. The performance and potential capabilities of this approach are presented and the results are presented in a comparison to above-mentioned methods in this paper.

2   Optimal Power Flow Formulation

As mentioned earlier, the OPF is a nonlinear optimization problem to minimize a certain objective subject to several equality and inequality constraints. OPF is a highly nonlinear, large scale static optimization problem due to large number of variables and constraints. The general OPF problem can be mathematically formulated using the following standard form [6]:

$$\text{Minimize } F(x, u) \tag{1}$$

$$\text{Subject to } g(x, u) = 0 \tag{2}$$

$$\text{and } h(x, u) \leq 0 \tag{3}$$

In the above equation, F is the objective function of total generator fuel cost to be minimized and it is usually a scalar.

x is the vector of dependent variables (state variables) consisting of:

1. Generator active power output at slack bus P_{G_1}.
2. Voltage magnitude at PQ bus V_L.
3. Reactive power output of all generator units Q_G.
4. Transmission line loading (or line flow) S_l.

Hence, x can be expressed as:

$$x^T = [P_{G_1}, V_{L_1} \cdots V_{L_{NL}}, Q_{G_1} \cdots Q_{G_{NG}}, S_{l_1} \cdots S_{l_{nl}}] \tag{4}$$

where NL, NG, and nl define the number of load buses, generators and transmission lines, respectively.

In a similar way, u is the vector of independent variables (control variables) consisting of:

1. Generators active power output P_G at PV buses except at the slack bus P_{G_1}.
2. Voltage magnitude at PV bus V_G.
3. Tap setting of the tap regulating transformers T.
4. Shunt VAR compensation Q_C.

Hence, u can be expressed as:

$$u^T = [P_{G_2} \cdots P_{G_{NG}}, V_{G_1} \cdots V_{G_{NG}}, Q_{C_1} \cdots Q_{C_{NC}}, T_1 \cdots T_{NT}] \tag{5}$$

where NT and N_C define the number of the regulating transformers and VAR compensators, respectively.

2.1 Objective Constraints

The OPF problems have several equality and inequality constraints, which are listed as follows.

2.1.1 Equality Constraints

The equality constraints of the OPF represent the typical load flow equations as follows:

(a) Active power balance in the network

$$P_{G_i} - P_{D_i} - V_i \sum_{j=1}^{NB} V_j[G_{ij} \cos(\delta_i - \delta_j) + B_{ij} \sin(\delta_i - \delta_j)] = 0 \tag{6}$$

(b) Reactive power balance in the network

$$Q_{G_i} - Q_{D_i} - V_i \sum_{j=1}^{NB} V_j[G_{ij} \sin(\delta_i - \delta_j) + B_{ij} \cos(\delta_i - \delta_j)] = 0 \tag{7}$$

where V_i and V_j are the voltage magnitudes of ith and jth bus respectively, P_{G_i} and Q_{G_i} are the active and reactive power output of ith generator unit, P_{D_i} and Q_{D_i} are the loading demand of active and reactive power of ith bus, G_{ij} and B_{ij} are the conductance and susceptance of voltages between ith and jth bus, δ_i and δ_j are the voltage phase angles of ith and jth bus and NB is the total number of buses.

2.1.2 Inequality Constraints

The inequality constraints of the OPF represent the limits on physical devices in the power system as well as the limits to satisfy system security:

(c) Generator constraints

For all generator units include the slack bus: generator unit bus voltage magnitudes, active power outputs and reactive power outputs are restricted by their lower and upper limits as follows:

$$V_{G_i}^{\min} \le V_{G_i} \le V_{G_i}^{\max}, \ i = 1, \ldots, NG \tag{8}$$

$$P_{G_i}^{\min} \le P_{G_i} \le P_{G_i}^{\max}, \ i = 1, \ldots, NG \tag{9}$$

$$Q_{G_i}^{\min} \le Q_{G_i} \le Q_{G_i}^{\max}, \ i = 1, \ldots, NG \tag{10}$$

(d) Transformer constraints

Transformer tap settings are restricted by their lower and upper limits as follows:

$$T_i^{\min} \le T_i \le T_i^{\max}, \ i = 1, \ldots, NT \tag{11}$$

(e) Shunt *VAR* compensator constraints

Shunt *VAR* compensations are restricted by their limits as follows:

$$Q_{C_i}^{\min} \le Q_{C_i} \le Q_{C_i}^{\max}, \ i = 1, \ldots, NC \tag{12}$$

(f) Security constraints

Security constraints include the constraints of voltage magnitudes at load buses and transmission line loading as follows:

$$V_{L_i}^{\min} \le V_{L_i} \le V_{L_i}^{\max}, \ i = 1, \ldots, NL \tag{13}$$

$$S_{l_i} \le S_{l_i}^{\max}, \ i = 1, \ldots, nl \tag{14}$$

2.2 Handing of Constraints

There are a great number of ways to handle constraints in evolutionary algorithms. In this paper, the constraints are incorporated into fitness function by means of penalty function method. A specific penalty factor multiplied with the square of the violated value of variable is added into the objective function so that any infeasible solution obtained is declined.

To handle the inequality constraints of state variables including load bus voltage magnitudes and output variables with active power generation output at slack bus, reactive power generation output, and line loading, the extended objective function is mathematically formulated as:

$$F_{aug} = \sum_{i=1}^{NG} F_i(P_{G_i}) + \lambda_P(P_{G_1} - P_{G_1}^{lim})^2 + \lambda_V \sum_{i=1}^{NL} (V_{L_i} - V_{L_i}^{lim})^2$$
$$+ \lambda_Q \sum_{i=1}^{NG} (Q_{G_i} - Q_{G_i}^{lim})^2 + \lambda_S \sum_{i=1}^{nl} (S_{l_i} - S_{l_i}^{max})^2 \tag{15}$$

where $\lambda_P, \lambda_V, \lambda_Q$ and λ_S are penalty factors and x^{lim} is the limit value of the dependent variable x. If x is higher than the upper limit, x^{lim} takes the value of this one, likewise if x is lower than the lower limit x^{lim} takes the value of this limit as follows:

$$x^{lim} = \begin{cases} x; & x^{min} \leq x \leq x^{max} \\ x^{max}; & x > x^{max} \\ x^{min}; & x < x^{min} \end{cases} \tag{16}$$

3   Learning-Based Evolutionary Optimization (LBEO)

In 2011, Rao and his colleagues [7] proposed a novel evolutionary algorithm for global optimization, namely TLBO owing to a special kind of teaching and learning processes, which can create new off-spring from parent chromosomes instead of classical crossover and mutation. TLBO has been applied to various optimization problems, such as reserve constrained dynamic economic dispatch problem, parameter optimization of modern machining process, optimal distribution generation location and size.

In order to improve the searching efficiency and capacity of basic TLBO, a simplified LBEO based on mutual learning and greed evolutionary selection is proposed in this paper. In LBEO, the teacher phase is removed, while the learner phase remains unchanged. The basic LBEO is presented in detail in the next section.

3.1   LBEO

The procedures of LBEO algorithm are presented as follows.

3.1.1   Initialization
The initial population is randomly selected based on uniform probability distribution for all variables to cover the entire search. Each individual X_i is a solution that contains all control variables. The initial population is given by the following equation:

$$X_i^0 = X_{i,\min} + rand(\,) \cdot (X_{i,\max} - X_{i,\min}), \ i = 1, \ldots, NP \tag{17}$$

where $rand(\,)$ represents a uniformly distributed random number within the range [0,1]. This produces NP individuals of X_i^0 randomly.

3.1.2   Evaluation
Run power flow using Newton-Raphson method and evaluate the fitness value of each individual (in this paper, the objective function is to minimize the total cost function of all generator power outputs).

3.1.3   Learning and Recombination
For each control vector $X_{i,G}$, randomly select two different individuals $X_{j,G}$ and $X_{k,G}$, a new mutated recombinant individual is produced by

$$V_{i,G} = \begin{cases} X_{i,G} + rand(\,) * (X_{j,G} - X_{k,G}), & if \ F(X_{j,G}) \leq F(X_{k,G}) \\ X_{i,G} + rand(\,) * (X_{k,G} - X_{j,G}), & if \ F(X_{j,G}) > F(X_{k,G}) \end{cases} \tag{18}$$

where $i, j, k \in \{1, 2, \ldots, NP\}$ are randomly chosen and must be different from each other.

3.1.4   Estimation and Selection
In order to ensure the robustness and convergence of the algorithm, the fitness of the offspring is in competition with its parent. The parent is replaced by its offspring if the fitness of the offspring is better than that of its parent. On the other hand, the parent is retained in the next generation if the fitness of the offspring is worse than that of its parent. LBEO actually involves the survival of the fittest principle in its selection process. The selection process can be expressed as follows:

$$X_{i,G+1} = \begin{cases} V_{i,G} & if \ F(V_{i,G}) \leq F(X_{i,G}) \\ X_{i,G} & if \ F(V_{i,G}) > F(X_{i,G}) \end{cases} \tag{19}$$

Hence the population either gets better in terms of the fitness function or remains constant but never deteriorates. The flow chart for the proposed ELBO algorithm is given in Fig. 1.

4   Results and Discussions

The proposed LBEO algorithm was written in MATLAB 2008a computing environment and applied on an Intel® Core™ i3-2100 CPU @ 3.10 GHz personal computer with 4.00 GB-RAM. To make a fair comparison on the performance of LBEO with other algorithm, the same total number of function evaluations (FES) is selected in this paper.

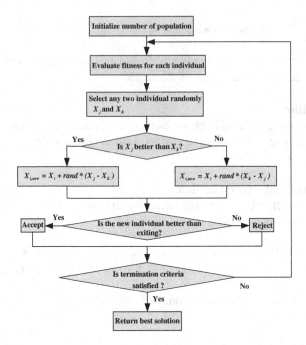

Fig. 1. The flow chart of LBEO

4.1 Benchmark Test

To ensure the proposed LBEO achieve relative better results on problems with different characteristics, a comprehensive evaluation of the population size was carried out on three commonly used benchmark functions with 100 dimensions, including Shifted Sphere, Shifted Griewank and Shifted Salomon. The proposed LBEO was compared with six recent popular algorithms, which are GA-TB [5], PSO-IW [5], SaDE [5], GL-25 [5], HIS [5] and TLBO [5], respectively.

The detail of 3 commonly used benchmark functions are given in Table 1. Range denotes the search ranges of the variables, F is the objective value of the global optimum, and all these functions have an optimal value $F_{min} = 0$. For each run, the maximal FES was set to 600,000, which is same as [5].

Table 1. Formulation of benchmark functions

No.	Function	Formulation	Type	Dim	Range	Min F
1	Shifted Sphere	$F_1(x) = \sum\limits_{i=1}^{Dim} x_i^2$	Unimodal	100	$[-100,100]$	0
2	Shifted Griewank	$F_2 = \sum\limits_{i=1}^{Dim} \frac{x_i^2}{4000} - \prod\limits_{i=1}^{Dim} \cos(\frac{x_i}{\sqrt{i}}) + 1$	Multimodal	100	$[-600,600]$	0
3	Shifted Salomon	$F_3 = 1 - \cos\left(2\pi\sqrt{\sum\limits_{i=1}^{Dim} x_i^2}\right) + 0.1\sqrt{\sum\limits_{i=1}^{Dim} x_i^2}$	Multimodal	100	$[-100,100]$	0

Table 2. Comparison of seven methods on average results over 50 runs for 100 Dim

No.	Function	GA-TB [5]	PSO-IW [5]	SaDE [5]	GL-25 [5]	HIS [5]	TLBO [5]	LBEO
1	Shifted Sphere	3.27e−07	1.36e+02	4.46e−28	2.23e−09	7.40e+ 03	4.50e−26	**1.9e−28**
2	Shifted Griewank	1.32e−01	5.21e+00	2.03e−02	8.76e−02	7.07e+01	6.17e−02	**5.5e−03**
3	Shifted Salomon	8.08e+00	2.38e+00	1.32e+00	1.54e+00	1.13e+01	1.74e+00	**4.35e−01**

The mean results over 50 independent runs are summarized in Table 2, and the best results are marked in boldface. It is clear that the proposed algorithm outperformed than GA-TB, PSO-IW, GL-25 and HIS. SaDE and TLBO also can get a set of good solutions, however still worse than LBEO. Obviously, the LBEO was demonstrated the best performance in all listed algorithms.

4.2 IEEE-30 Bus Test System

The proposed LBEO algorithm has been implemented to solve the standard IEEE 30-bus test system was shown in Fig. 2.

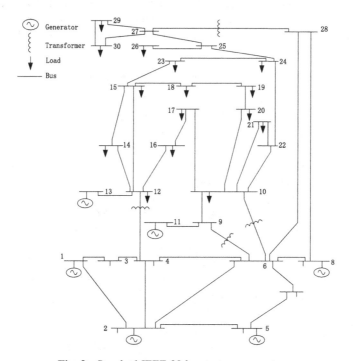

Fig. 2. Standard IEEE 30-bus test power system.

The IEEE 30-bus test system has six generators at the buses 1, 2, 5, 8, 11 and 13, four transformers with off-nominal tap ratio at lines 6–9, 6–10, 4–12 and 28–27. In addition, buses 10, 12, 15, 17, 20, 21, 23, 24, and 29 have been chosen as shunt VAR compensation buses. The total system demand is 283.4 p.u. at 100 MVA base. In this paper, 50 test runs were performed for solving the OPF problem using the LBEO approach.

The simulation results are presented as follows.

4.2.1   Minimization of Fuel Cost

The minimization of the total fuel cost is considered in this paper, which is the most commonly used objective for OPF. The minimization of total fuel cost F is defined as quadratic cost function of generator unit power output which is expressed as follows:

$$F = \sum_{i=1}^{NG} F_i(P_{G_i}) = \sum_{i=1}^{NG} (a_i + b_i P_{G_i} + c_i P_{G_i}^2) \tag{20}$$

Where F_i and P_{G_i} are the fuel cost and power output of the ith generator unit, respectively. a_i, b_i and c_i are the basic, the linear and the quadratic cost coefficients of the ith generator unit, respectively. The values of these coefficients are given in [8, 9].

A comparison between the results of fuel cost obtained by the proposed LBEO approach and some other techniques are reported in the literature as shown in Table 3. It can be seen that the minimum fuel cost is 699.07071 $/h, with an average cost of 699.07078 $/h and a maximum cost of 699.07085 $/h obtained by LBEO which are all minimum in comparison to reported results in the literature. The results showed that the proposed LBEO has strong robustness for OPF problem. The optimal of control

Table 3. Comparison of the simulation results for this case

Algorithm	Min	Average	Max
LBEO	**799.07071**	**799.07078**	**799.07085**
TLBO [10]	799.0715	NA	NA
BBO [11]	799.1116	799.1985	799.2042
DE [12]	799.2891	NA	NA
LTLBO [13]	799.4369	799.7186	800.2578
SA [14]	799.4500	NA	NA
EEA [15]	800.0831	800.1730	800.2123
DE-PS [16]	800.1475	800.7962	801.2417
EADDE [17]	800.2041	800.2412	800.2784
PSO [18]	800.41	NA	NA
ABC [19]	800.66	800.8715	801.8674
NPSO [20]	800.6815	800.9024	801.37
Fuzzy-GA [21]	801.0554	801.627	802.1158
MDE [22]	802.376	802.382	802.404
IEP [23]	802.4650	802.5210	802.5810
EP [24]	802.6200	803.5100	805.6100

Table 4. Optimal settings of control variables

Control variables	Min	Max	Initial case	This paper	Control variables	Min	Max	Initial case	This paper
P_1	50	200	99.2230	176.9005	T_{12}	0.9	1.1	1.069	0.9000
P_2	20	80	80	48.6975	T_{15}	0.9	1.1	1.032	0.9862
P_5	15	50	50	21.3043	T_{36}	0.9	1.1	1.068	0.9657
P_8	10	35	20	21.0806	QC_{10}	0	5	0	5.0000
P_{11}	10	30	20	11.8840	QC_{12}	0	5	0	5.0000
P_{13}	12	40	20	12.0000	QC_{15}	0	5	0	5.0000
V_1	0.95	1.1	1.05	1.1000	QC_{17}	0	5	0	5.0000
V_2	0.95	1.1	1.04	1.0878	QC_{20}	0	5	0	5.0000
V_5	0.95	1.1	1.01	1.0616	QC_{21}	0	5	0	5.0000
V_8	0.95	1.1	1.01	1.0694	QC_{23}	0	5	0	3.8428
V_{11}	0.95	1.1	1.05	1.1000	QC_{24}	0	5	0	5.0000
V_{13}	0.95	1.1	1.05	1.1000	QC_{29}	0	5	0	2.7394
T_{11}	0.9	1.1	1.078	1.0447	Cost	–	–	901.9516	**799.07071**

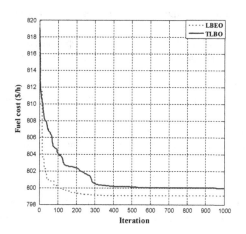

Fig. 3. Fuel cost characteristics of LEBO and TLBO

variables obtained by the proposed LBEO are given in Table 4. The total fuel cost obtained is reduced to 799.07071 $/h compared to the initial fuel cost of 901.99 $/h which gives a reduction equals to 11.41 %. The convergence curve of the total fuel cost over iteration is shown in Fig. 3. The figure shows that the proposed LBEO has excellent convergence characteristics.

5 Conclusion

In this paper, a novel nature-inspired optimization approach which is learning-based evolutionary optimization (LBEO) has been proposed to solve OPF problem, and implemented on IEEE 30-bus test system. The comparison of results obtained from proposed LBEO and other mentioned methods in the literature confirm the superiority of the proposed LBEO algorithm in terms of solution quality for the OPF problems. In the future, we would like to applied the proposed LBEO to more complex OPF problems, such as IEEE 57-bus and IEEE 118-bus test systems.

Acknowledgements. This work is supported by the National Natural Science Foundation of China (61273040), and Shanghai Rising-Star Program (12QA1401100).

References

1. Dommel, H.W., Tinney, T.F.: Optimal power flow solutions. IEEE Trans. Power Appar. Syst. **87**(5), 1866–1876 (1968)
2. Paranjothi, S.R., Anburaja, K.: Optimal power flow using refined genetic algorithm. Electric Power Compon. Syst. **30**, 1055–1063 (2002)
3. Liang, R.H., Tsai, S.R., Chen, Y.T., Wan-Tsun, T.: Optimal power flow by a fuzzybased hybrid particle swarm optimization approach. Electr. Power Syst. Res. **81**(7), 1466–1474 (2011)
4. Cai, H.R., Chung, C.Y., Wong, K.P.: Application of differential evolution algorithm for transient stability constrained optimal power flow. IEEE Trans. Power Syst. **23**(2), 719–728 (2008)
5. Niu, Q., Zhang, H.Y., Li, K.: An improved TLBO with elite strategy for parameters identification of PEM fuel cell and solar cell models. Int. J. Hydrogen Energy **39**(8), 3837–3854 (2014)
6. Frank, S., Steponavice, I., Rebennack, S.: Optimal power flow: a bibliographic survey I. formulations and deterministic methods. Energy Syst. **3**(3), 221–258 (2012)
7. Rao, R.V., Savsani, V.J., Vakharia, D.P.: Teaching-learning-based optimization: a novel method for constrained mechanical design optimization problems. Comput. Aided Des. **43**(3), 303–315 (2011)
8. Alsac, O., Stott, B.: Optimal load flow with steady-state security. IEEE Trans. Power Appar. Syst. PAS **93**(3), 745–751 (1974)
9. Lee, K.Y., Park, Y.M., Ortiz, J.L.: A united approach to optimal real and reactive power dispatch. IEEE Trans. Power Appar. Syst. **104**(5), 1147–1153 (1985)
10. Bouchekara, H.R.E.H., Abido, M.A., Boucherma, M.: Optimal power flow using teaching-learning-based optimization technique. Electr. Power Syst. Res. **114**, 49–59 (2014)
11. Bhattacharya, A., Chattopadhyay, P.K.: Application of biogeography-based optimisation to solve different optimal power flow problems. IET Gener. Transm. Distrib. **5**, 70–80 (2011)
12. Abou El Ela, A.A., Abido, M.A., Spea, S.R.: Optimal power flow using differential evolution algorithm. Electr. Power Syst. Res. **80**, 878–885 (2010)
13. Ghasemi, M., Ghavidel, S., Gitizadeh, M., Akbari, E.: An improved teaching-learning-based optimization algorithm using LÉVy mutation strategy for non-smooth optimal power flow. Electr. Power Energy Syst. **65**, 375–384 (2015)

14. Roa-Sepulveda, C.A., Pavez-Lazo, B.J.: A solution to the optimal power flow using simulated annealing. Electr. Power Energy Syst. **25**, 47–57 (2003)
15. Reddy, S.S., Bijwe, P.R., Abhyankar, A.R.: Faster evolutionary algorithm based optimal power flow using incremental variables. Electr. Power Energy Syst. **54**, 198–210 (2014)
16. Gitizadeh, M., Ghavidel, S., Aghaei, J.: Using SVC to economically improve transient stability in long transmission lines. IETE J. Res. **60**, 319–327 (2014)
17. Vaisakh, K., Srinivas, L.R.: Evolving ant direction differential evolution for opf with non-smooth cost functions. Eng. Appl. Artif. Intell. **24**, 426–436 (2011)
18. Abido, M.: Optimal Power Flow Using Particle Swarm Optimization. Int. J. Electr. Power Energy Syst. **24**, 563–571 (2002)
19. Adaryani, M.R., Karami, A.: Artificial bee colony algorithm for solving multi-objective optimal power flow problem. Int. J. Electr. Power Energy Syst. **53**, 219–230 (2013)
20. Selvakumar, A.I., Thanushkodi, K.: A new particle swarm optimization solution to nonconvex economic dispatch problems. IEEE Trans. Power Syst. **22**, 42–51 (2007)
21. Hsiao, Y.-T., Chen, C.-H., Chien, C.-C.: Optimal capacitor placement in distribution systems using a combination fuzzy-GA method. Int. J. Electr. Power Energy Syst. **26**, 501–508 (2004)
22. Sayah, S., Zehar, K.: Modified differential evolution algorithm for optimal power flow with non-smooth cost functions. Energy Convers. Manage. **49**, 3036–3042 (2008)
23. Ongsakul, W., Tantimaporn, T.: Optimal powers flow by improved evolutionary programming. Electr. Power Comput. Syst. **34**, 79–95 (2006)
24. Yuryevich, J., Wong, K.P.: Evolutionary programming based optimal power flow algorithm. IEEE Trans. Power Syst. **14**(4), 1245–1250 (1999)

Blind Nonparametric Determined
and Underdetermined Signal Extraction
Algorithm for Dependent Source Mixtures

Fasong Wang[1(⊠)], Rui Li[2], Zhongyong Wang[1],
and Xiangchuan Gao[1]

[1] School of Information Engineering, Zhengzhou University,
Zhengzhou 450001, China
fasongwang@126.com
[2] School of Sciences, Henan University of Technology, Zhengzhou, China

Abstract. Blind extraction or separation statistically independent source signals from linear mixtures have been well studied in the last two decades by searching for local extrema of certain objective functions, such as nonGaussianity (NG) measure. Blind source extraction (BSE) algorithm from underdetermined linear mixtures of the statistically dependent source signals is derived using nonparametric NG measure in this paper. After showing that maximization of the NG measure can also separate or extract the statistically weak dependent source signals, the nonparametric NG measure is defined by statistical distances between different source signals distributions based on the cumulative density function (CDF) instead of traditional probability density function (PDF), which can be estimated by the quantiles and order statistics using the L^2 norm efficiently. The nonparametric NG measure can be optimized by a deflation procedure to extract or separate the dependent source signals. Simulation results for synthesis and real world data show that the proposed nonparametric extraction algorithm can extract the dependent signals and yield ideal performance.

Keywords: Blind source separation (BSS) · Nongaussian measure · Independent component analysis (ICA) · Dependent component analysis (DCA) · Blind source extraction (BSE)

1 Introduction

Blind source separation (BSS) as an active topic in signal processing community aims to separate linearly or nonlinearly mixtures in both noise-free and noisy environments mixed latent source signals. It has become an important topic of research and development in many areas [1–3]. Various algorithms have been proposed in the last two decades to separate or extract source signals from their mixtures [1–7]. The propose of BSS is to separate the potential source signals from the mixtures obtained by the sensors without a priori about the source signals and the mixing process. This is realized by a variety of criteria, including the minimization of mutual information (MI), maximization of nonGaussianity (NG) and maximization of likelihood (ML) [8, 9]. A key factor in BSS is the assumption about the statistical properties of source signals

© Springer International Publishing Switzerland 2015
D.-S. Huang et al. (Eds.): ICIC 2015, Part I, LNCS 9225, pp. 46–57, 2015.
DOI: 10.1007/978-3-319-22180-9_5

such as statistical independence among different ones. When the source signals are mutual independent, the BSS problem can be solved by using the so called independent component analysis (ICA) method which has attracted considerable attention in the signal processing fields and several efficient algorithms have been proposed [1–3].

Despite the success of using standard ICA for BSS in many applications, original ICA algorithms are in the sense that all sources are assumed to be statistically independent random variables, but the basic assumptions of ICA may not hold for some real-world situations. Several extented ICA models have been considered based on the basic ICA framework. This type of model can be called dependent component analysis (DCA) model. Multidimensional independent component analysis (MICA) model [11, 12] as the first DCA model of BSS problem, instead of assuming all the source signals to be mutually statistically independent, all the source signals are divided into several groups and the size of the groups can be different, where the signals between different groups are statistical independent and signals within the same group have dependencies, this DCA model can also be called independent subspace analysis (ISA). DCA related algorithm concern mostly the estimation of the entropy or of the MI. Various BSS algorithms have been developed in response to different DCA models [13–24].

Another extension of the original BSS task is the blind source extraction (BSE) problem. Unlike common BSS algorithms which consist of separating all the source signals simultaneously by means of the maximization of an independence measure between the output estimated signals, however, in some situations, it may be more appropriate to extract only a single source of interest based on a certain fundamental signal property, which is the task of BSE [2, 10]. One of the main advantages of BSE compared to traditional BSS is that it decreases computational cost since the degrees of freedom are reduced, the possibility to relax the need for preprocessing or postprocessing. Furthermore, this procedure has a great potential when the number of sensors and sources are not equal, even when unknown or underdetermined.

In addition, as compared with overdetermined or determined BSS problem, the underdetermined BSS one, where the number of available recorded mixtures is less than the underlying source signals, is more difficult to treat and attracts much attention in recent years. In this case, even if the mixing matrix is known or has been estimated, it is impossible to estimate the source signals directly. Therefore, in order to realize the source extraction from the mixed signals, some a *priori* knowledge about the whole system must to be exploited, such as independence or sparsity [25, 26].

The simultaneous assumption of the two extensions of BSS, that is, DCA combined with BSE in underdetermined situation seems to be a more realistic model than any of the two models alone. For example, at the biomedical signal processing, only small number of sources should be extracted which may be weak correlated in spatial. Cardoso showed that strong relationship existed among MI, correlation and NG of source estimates [27]. Inspired from this conclusion, we get that one can not resort minimization the MI, but on the other hand, according to maximization the NG, the dependent sources can be separated or extracted.

Here, we exploit some weaker conditions for extraction or separation source signals assuming that they have statistically dependent properties in the underdetermined situation. Based on the generalization of the central limit theorem (CLT) to special dependent variables, we will try to track the DCA model by maximization NG measure. The proposed NG measure is defined in terms of cumulative distribution function (CDF) instead of the widespread probability density function (PDF) using the nonparametric estimation method, afterwards, the NG distance between the given CDF and the standard normal CDF is proposed which can be estimated by the order statistics using the L^2 norm efficiently. The NG distance based cost function will be optimized resorting to a deflation procedure by gradient iterative algorithm, whose local maximization performs the extraction of one dependent component.

2 Theory Fundamentals

The problem of linear instantaneous BSS problem can be formulated as Eq. (1) (see [1–3] for overview):

$$\mathbf{x}(t) = \mathbf{A}\mathbf{s}(t) + \mathbf{n}(t) \tag{1}$$

where $\mathbf{s}(t) = (s_1(t), s_2(t), \cdots, s_N(t))^{\mathrm{T}}$ is an unknown source vector which contains N source signals. The M observed mixtures $\mathbf{x}(t) = (x_1(t), x_2(t), \cdots, x_M(t))^{\mathrm{T}}$ are sometimes called as sensor outputs. Matrix $\mathbf{A} = [a_{ij}] \in \mathbb{R}^{M \times N}$ is an unknown full column rank mixing matrix. $\mathbf{n}(t) = (n_1(t), n_2(t), \cdots, n_N(t))^{\mathrm{T}}$ is a vector of additive noise.

The task of BSS contains estimation of the mixing matrix $\mathbf{A}$ or its pseudoinverse separating (unmixing) matrix $\mathbf{W} = \mathbf{A}^{\dagger}$ in order to estimate the original source signals $\mathbf{s}(t)$, given only a finite number of observation data $\{\mathbf{x}(t), t = 1, \cdots, T\}$.

In order to simplify the problem, most of the algorithms of BSS problem contain a spatial decorrelation procedure $\mathbf{V}$ over noiseless $\mathbf{x}(t)$ to obtain the decorrelated signals $\mathbf{z}(t) = (z_1(t), z_2(t), \cdots, z_M(t))^{\mathrm{T}}$, that is,

$$\mathbf{z}(t) = \mathbf{V}\mathbf{x}(t) \tag{2}$$

So, the global mixture can be expressed as,

$$\mathbf{z}(t) = \mathbf{U}\mathbf{s}(t) \tag{3}$$

where $\mathbf{U} = \mathbf{V}\mathbf{A}$ is an unknown orthogonal matrix.

After some standard BSS methods (such as ICA) are used to the preprocessed decorrelated data $\mathbf{z}(t)$, one can obtain an unitary linear transformation $\mathbf{B}$ and the estimation of the source signals $\mathbf{y}(t)$:

$$\mathbf{y}(t) = \mathbf{B}\mathbf{z}(t) = \mathbf{B}\mathbf{V}\mathbf{x}(t) = \mathbf{B}\mathbf{V}\mathbf{A}\mathbf{s}(t) \tag{4}$$

Denote $\mathbf{W} = \mathbf{B}\mathbf{V}$, then one can get

$$y(t) = \mathbf{WA}s(t) \tag{5}$$

Recall that two indeterminacies cannot be resolved in BSS without some priori knowledges: scaling and permutation ambiguities. Thus, if the estimate of the mixing matrix $\hat{\mathbf{A}}$ satisfies

$$\mathbf{P} = \mathbf{WA} = \hat{\mathbf{A}}\mathbf{A} = \mathbf{GD} \tag{6}$$

where $\mathbf{P}$ is a global transformation which combines the mixing and separating system, $\mathbf{G}$ is permutation matrix and $\mathbf{D}$ is some nonsingular scaling diagonal matrix, then $(\hat{\mathbf{A}}, \hat{s})$ and $(\mathbf{A}, s)$ are said to be related by a waveform-preserving relation.

The propose of BSE is to design an extracting vector $\mathbf{w}$ to extract an expected source signal from the mixtures $\mathbf{x}(t)$,

$$y(t) = \mathbf{w}^T\mathbf{x}(t) = \mathbf{w}^T\mathbf{A}s(t) \tag{7}$$

where $y(t)$ is an estimated of a source signal with scalar ambiguity.

In applications, the *priori* information of the expected source signals can be utilized to design proper extracted algorithms, so, any of the source signals could come out as the first one with particular property, such as absolute normalized kurtosis value [2], temporal structure [28–30], sparseness [31], morphological structure [32] and so on. In this paper, as shown in Sect. 4, the *priori* information of the desired source signal to be extracted is the maximum NG measure.

3   Generalized CLT and Nonparametric NG Measure

The Gaussian distribution has the maximum Shannon differential entropy (maximum uncertainty) over all the continuous distributions defined on the real space with the same variance. This fact makes the Gaussianity measure a very useful tool for the characterization of data. In recent years, a connection between NG and ICA has been suggested. It can also be explained by the CLT theory. Since CLT is not valid for any set of dependent variables, we must be aware that one may not always recover the original dependent source signals using maximum NG criteria. Caiafa et al. give a very special condition on sources, for which the linear combinations of dependent signals are not more Gaussian than the components and therefore the maximum NG criteria fails, but fortunately this is not the case in most of real world scenarios [33]. Moreover, the independence of source signals is not required when we solve blind deconvolution problem [34]. Additionally, based on minimum entropy, the dependent source signals can be recovered [35].

Conclusion 1. The maximum NG method can be described as exploring a linear transformation of the mixed signals in the unit-variance signals space, so that the transformed signals (source signal estimates) have maximum NG distributions.

According to this conclusion, if we choose a robust and efficient NG measure, the source signals can be extracted or separated properly.

A natural measure of NG based on the L^2 distance of an estimated PDF to the Gaussian PDF is introduced in [19, 33]. The NG measure is defined as Eq. (8):

$$d(y, g) = \left(\int [p_g(y) - p_y(y)]^2 dy \right)^{1/2} \tag{8}$$

where the integral is defined in Lebesgue sense and is taken on all the range of variable y, and $p_g(y)$ is the Gaussian PDF with the same variance of variable y whose PDF is $p_y(y)$. In this paper, we will build the NG measure using the concept of CDF instead of traditional PDF. Let us call F_y and F_g are the CDFs of random variable y to be analyzed and its equivalent Gaussian one, respectively, then, the NG measure based on CDF can defined as Eq. (9):

$$d(F_{y_i}, F_g) = \left(\int_{-\infty}^{\infty} [F_{y_i}(x) - F_g(x)]^2 dx \right)^{1/2} \tag{9}$$

The definition of NG measure $d(F_{y_i}, F_g)$ possesses the following property that the distance measure should have:

$$\begin{cases} d(F_{y_i}, F_g) = 0, & c = 2 \\ d(F_{y_i}, F_g) > 0, & c \neq 2 \end{cases} \tag{10}$$

where c is the shape parameter in the generalized Gaussian distribution (GGD).

The PDF of GGD can be described as: $p(y) = \frac{c}{2\gamma \, \Gamma(1/c)} \exp \left[-\left(|y - \mu_y|/\gamma \right)^c \right]$, where $\Gamma(z) = \int_0^\infty e^{-t} t^{z-1} dt$ is Gamma function and $\gamma = \sqrt{\sigma^2 \Gamma(1/c)/\Gamma(3/c)}$ is the scale parameter. By changing the values $c(c > 0)$, a family of distributions with different sharpness will be given. From the relationship between the CDF based NG measure $d(F_{y_i}, F_g)$ and the shape parameter or Gaussian parameter c, we can conclude that the distance measure $d(F_{y_i}, F_g)$ can be used as the measure of NG since it offers a global minimum when the Gaussian parameter c various from 0^+ to infinite and it reaches its global minimum when $c = 2$. In other words, the measure of NG obtains its global minimum when the analyzed distribution is Gaussian.

As the definition of $d(F_{y_i}, F_g)$, we must estimate the CDF F_{y_i}. So, the next question is how to efficient get the estimation of CDF $\hat{F}_{y_i}$. As we know, it would need high computational cost through nonparametric histograms. Alternatively, an equivalent measure can be established in terms of inverse CDF, which is defined as:

$$Q_{y_i} = F_{y_i}^{-1}, \quad Q_g = F_g^{-1} \tag{11}$$

The relationship between CDF F and its inverse Q can be generalized to the NG distance, which is formulated as:

$$\begin{cases} D(Q_{y_i}, Q_g) = 0, & c = 2 \\ D(Q_{y_i}, Q_g) > 0, & c \neq 2 \end{cases} \tag{12}$$

Since the relationship between Q and F, they also present the monotone properties in the proper intervals of c, consequently, the distance $d(\cdot)$ and its correspondent $D(\cdot)$ preserve the same properties, as a result, one can conclude that

$$D(Q_{y_i}, Q_g) = \left(\int_0^1 [Q_{y_i}(x) - Q_g(x)]^2 dx \right)^{1/2} \tag{13}$$

is also a proper NG measure. In order to estimate the NG measure $D(Q_{y_i}, Q_g)$ from the discrete samples, we must estimate Q_{y_i} firstly. The estimation of Q_{y_i} can be performed robustly in a simple practical way by using the order statistics (OS) from a large set of discrete time samples. Then the quantiles of the CDF can be constructed using OS, which is a consistent estimator of the distribution [34]:

$$\hat{Q}_{y_i}\left(\frac{k}{T}\right) = y_{i(k)} \Leftrightarrow \hat{F}\left(y_{i(k)}\right) = \frac{k}{T} \tag{14}$$

As a result, the estimation of NG measure using the OS can be expressed as:

$$\hat{D}(Q_{y_i}, Q_g) = \frac{1}{T} \left(\sum_{k=1}^{T} \left[y_{i(k)} - Q_g\left(\frac{k}{T}\right) \right]^2 \right)^{1/2} \tag{15}$$

where $Q_g(k/T)$ is the k/T quantile of the equivalent Gaussian distribution.

4 Nonparametric NG Algorithm for Dependent Source Signals

Conclusion 2. The nonparametric NG measure $\hat{D}(Q_{y_i}, Q_g)$ will reach a local maximum at any output channel for each component if $\mathbf{b}_i$ is forced to be unitary [36].

To extract a different source signal at each output channel, a multistage deflation procedure must be applied to the separation system. The NG measure $\hat{D}(Q_{y_i}, Q_g)$ is maximized at each output channel successively under the constriction that the vector $\mathbf{b}_i$ has to be orthonormal to the previously obtained vectors, the separation matrix is composed of all the vectors $\mathbf{b}_i$. Taking into account Eq. (4) in vector form:

$$y_i(t) = \mathbf{b}_i^T \mathbf{z}(t) \tag{16}$$

where $\mathbf{b}_i^T$ is the i-row of the separation matrix $\mathbf{B}$, the goal is to update $\mathbf{b}_i$ at each stage by optimizing a cost function $J(\mathbf{b}_i)$. We take the objective function $J(\mathbf{b}_i)$ as:

$$J(\mathbf{b}_i) = D(Q_{y_i}, Q_g) \tag{17}$$

$J(\mathbf{b}_i)$ will be optimized by the stochastic gradient rule of the constrained optimization method with the constrains [36]:

$$\begin{cases} \mathbf{b}_i(k+1) = \mathbf{b}_i(k) + \mu \nabla J|_{\mathbf{b}_i(k)} \\ s.t. \ \mathbf{b}_i \text{ is orthonormal to } \{\mathbf{b}_1, \cdots, \mathbf{b}_{i-1}\} \end{cases} \tag{18}$$

Denote $Y_i(k) = y_{i(k)} - Q_g(k/T)$, then the gradient of $J(\mathbf{b}_i)$ in Eq. (15) is:

$$\nabla J|_{\mathbf{b}_i(k)} = \frac{1}{2T} \left(\sum_{t=1}^{T} [Y_i(t)]^2 \right)^{-1/2} \frac{d\left(\sum_{t=1}^{T} [Y_i(t)]^2 \right)}{d\mathbf{b}_i} \Bigg|_{\mathbf{b}_i(k)}$$

$$= \frac{1}{T} \left(\sum_{t=1}^{T} [Y_i(t)]^2 \right)^{-1/2} \left(\sum_{t=1}^{T} Y_i(t) \right) \mathbf{z} \frac{dy_{i(t)}}{dy_i} \Bigg|_{\mathbf{b}_i(k)} \tag{19}$$

where $\frac{dy_{i(t)}}{dy_i}\Big|_{\mathbf{b}_i(k)} = \mathbf{e}_t = [0,0,\cdots,0,1,0,\cdots,0]^T$ and $e_t(l) = \begin{cases} 1 & \text{if } y_i(l) = y_{i(t)} \\ 0 & \text{else} \end{cases} \Big|_{t=1,\cdots,T}$.

After the ith source signal is extracted, $\mathbf{b}_i$ must be normalized and projected over the subspace orthonormal $\mathbf{C}_{i-1}$ to the vectors obtained at every previous stage. Let us quote that the $\mathbf{C}_{i-1}$ expression is

$$\mathbf{C}_{i-1} = \mathbf{I} - (\mathbf{B}_{i-1}\mathbf{B}_{i-1}^T)^{-1}\mathbf{B}_{i-1}^T \tag{20}$$

where $\mathbf{B}_{i-1} = (\mathbf{b}_1, \cdots, \mathbf{b}_{i-1})$.

5 Computer Simulations

In order to show the performance and the validity of the proposed algorithm, simulations using Matlab are given below. The simulation results presented in this section are divided into four Examples. The statistical performance, or accuracy, was measured by the index signal-to-interference ratio (SIR) as,

$$\text{SIR}(s_i, y_j) = 10 \log \frac{\sum_{t=1}^{T} (s_i(t))^2}{\sum_{t=1}^{T} (|s_i(t)| - |y_j(t)|)^2}, \quad i,j = 1,\cdots,N$$

where $y_i(t)$ is the estimation of $s_i(t)$.

5.1 Simulations on Determined BSS Case

In this simulation, we use $N = 4$ source signals which are extracted from the real world photo. It should be noted that the source signals are extracted from different pixel columns of real world images and then stack these columns one by one to get a one dimensional source signals. By selecting different intervals between the columns of the image, we can control the level of dependence between the source signals. We choose the columns of the photo which are relatively far away, therefore they are mutually weak correlated. The source' correlation coefficients are shown in Table 1.

Table 1. The correlation coefficients between different source signals

	Source 1	Source 2	Source 3	Source 4
Source 1	1.0000	0.1396	−0.0305	−0.2182
Source 2	0.1396	1.0000	−0.0987	−0.0074
Source 3	−0.0305	−0.0987	1.0000	0.1967
Source 4	−0.2182	−0.0074	0.1967	1.0000

The input mixed signals of the algorithm are generated by mixing the four source signals with a 4×4 random mixing matrix in which the elements are distributed with $N(0, 1)$. After convergence of the proposed algorithm, the average results of the performance criteria evaluated by SIR over 10 experiments are shown in Table 2.

Table 2. Average SIR for different source signals using the proposed algorithm over 10 experiments

Performance index	Source signals			
	s_1	s_2	s_3	s_4
SIR	24.07	21.95	14.98	27.86

5.2 Simulations on Underdetermined BSE Case

In this example, we use 4 source signals which are extracted from the same real world photo as example 1. The 3*4 mixing matrix are generated by the randn function of Matlab. The source' correlation coefficients are shown in Table 3.

Table 3. The correlation coefficients between different source signals

	Source 1	Source 2	Source 3	Source 4
Source 1	1.0000	0.1923	0.0473	−0.1803
Source 2	0.1923	1.0000	0.1298	−0.1571
Source 3	0.0473	0.1298	1.0000	−0.1184
Source 4	−0.1803	−0.1571	−0.1184	1.0000

After convergence of the proposed algorithm, the average results of the performance criteria evaluated by SIR over 10 experiments are shown in Table 4. The correlation coefficients of source signals and their corresponding extracted signals using the proposed algorithm are shown in Table 5.

Table 4. Average SIR for different source signals using the proposed algorithm over 10 experiments

Performance index	Source signals		
	s_2	s_3	s_4
SIR	16.43	19.90	21.83

Table 5. The correlation coefficients of source signals and their corresponding extracted signals using the proposed algorithm

Extracted signals using the Proposed Algorithms	Source signals			
	s_1	s_2	s_3	s_4
y_1	0.1765	0.1232	0.0411	*-0.9967*
y_2	-0.2711	*-0.9154*	-0.0798	0.2098
y_3	-0.1476	0.2224	*0.9392*	-0.2493

5.3   Simulations on Effect of Strong Correlations

In order to verify the performance of the proposed algorithm for strong correlations between source signals, we choose four face images extracted from face databases of [37] as the source signals. The source' correlation coefficients are shown in Table 6. The input mixed signals of the algorithm are generated by mixing the four source signals with a 4×4 random mixing matrix in which the elements are distributed with $N(0, 1)$. After convergence of the proposed algorithm, the average results of the performance criteria evaluated by SIR over 10 experiments are shown in Table 7. The definition of SIR for images can be found in [38].

Table 6. The correlation coefficients between four different face images

	Source 1	Source 2	Source 3	Source 4
Source 1	1.0000	0.9106	0.8421	0.8388
Source 2	0.9106	1.0000	0.8379	0.8363
Source 3	0.8421	0.8379	1.0000	0.9279
Source 4	0.8388	0.8363	0.9279	1.0000

Table 7. Average SIR for different source signals using the proposed algorithm over 10 experiments

Performance index	Source signals			
	s_1	s_2	s_3	s_4
SIR	62.22	54.48	52.87	54.96

5.4   Simulations on Comparison with Other Algorithms

For comparison, we display the this simulation, at the same convergent conditions, the proposed algorithm was compared along two sets of criteria, statistical and computational with other popular BSS algorithms such as FastICA (the nonlinearity function to be chosen as y^3), COMBI, SOBI and JADEop algorithm [39]. The computational load was measured as CPU time needed for convergence (using Matlab R2010b and run in the CPU 3.0 GHz Pentium4 computer). The four source signals and mixed signals are the same as Sect. 5.1.

The statistical performance, or accuracy, was measured using an alternative popular BSS performance index called cross-talking error index E defined as [2]. The separation results of the 4 different sources are shown in Table 8 for various BSS algorithms (averaged over 100 Monte Carlo simulations).

From Table 8 we conclude that the proposed algorithm can make an ideal separation results for statistically dependent source signals, the BCA algorithm can also get ideal separation results for these source signals, the other four popular BSS algorithms can not all work well in this condition, but when the boundaries of the source signals are not satisfied properly, the proposed algorithm can also works well, that is to say our proposed method has a further wide field of applications. Then we will discuss the computational load for convergence, as a nonparametric method, the proposed algorithm requires more computation than all the other algorithms, but it has a better convergent performance, moreover the nonparametric estimation method has the robust properties, so from the convergent performance and computation load two aspects, the proposed algorithm can work in the DCA situation.

Table 8. The separation results of various BSS algorithms

Different Algorithms	FastICA	COMBI	*SOBI*	JADEop	BCA	Proposed
Performance index E	0.1999	0.1913	0.3972	0.3218	0.1671	0.1677
Computation time (s)	0.22	0.32	0.09	0.08	0.1374	0.96

6   Conclusions

Most of state-of-the-art algorithms for solving BSS or BSE problem rely on independence or at least second order statistical assumption of the source signals. In this paper, we developed a nonparametric BSE algorithm for statistical dependent source signals using the nonparametric NG measure. We show that maximization of the NG measure can not only separate the statistically independent but also dependent source signals, even in the underdetermined BSE situation. The NG measure is defined by statistical distances between distributions based on the CDF instead of traditional PDF which can be realized by the order statistics efficiently. Simulation results on both synthetic and real world data show that the proposed nonparametric algorithm is able to extract the dependent source signals and yield ideal performance. The next purpose of this study is utilizing the proposed method to extract ERP signals in the brain-computer interfaces (BCI) applications.

Acknowledgments. This research is financially supported by the National Natural Science Foundation of China (No. 61401401, 61172086, 61402421, U1204607), the China Postdoctoral Science Foundation (No. 2014M561998) and the young teachers special Research Foundation Project of Zhengzhou University (No. 1411318029).

References

1. Comon, P., Jutten, C.: Handbook of Blind Source Separation: Independent Component Analysis and Applications. Elsevier, Oxford (2010)
2. Cichocki, A., Amari, S.: Adaptive Blind Signal and Image Processing: Learning Algorithms and Applications. Wiley, New York (2003)
3. Hyvarinen, A.: Independent component analysis: recent advances. Philos. Trans. R. Soc. A **371**, 20110534 (2013)
4. Cardoso, J.: Blind signal separation: statistical principles. Proc. IEEE **86**(10), 2009–2025 (1998)
5. Särelä, J., Valpola, H.: Denoising source separation. J. Mach. Learn. Res. **6**, 233–272 (2005)
6. Leong, W., Mandic, D.: Noisy component extraction (NoiCE). IEEE Trans. Circuits Syst. I. **57**(3), 664–671 (2010)
7. Deville, Y., Hosseini, S.: Recurrent networks for separating extractable-target nonlinear mixtures. Part I: Non-blind Configurations. Signal Process. **89**(4), 378–393 (2009)
8. Bell, A.J., Sejnowski, T.J.: An information-maximisation approach to blind separation and blind deconvolution. Neural Comput. **7**(6), 1129–1159 (1995)
9. Amari, S., Cichocki, A., Yang, H.: A new learning algorithm for blind signal separation. In: Advances in Neural Information Processing Systems, pp. 757–763. MIT Press, Cambridge (1996)
10. Bloemendal, B., Laar, J., Sommen, P.: A single stage approach to blind source extraction based on second order statistics. Signal Process. **93**(2), 432–444 (2013)
11. Cardoso, J.F.: Multidimensional independent component analysis. In: ICASSP 1998, Seattle, WA, USA, pp. 1941–1944. IEEE (1998)
12. Lahat, D., Cardoso, J.F., Messer, H.: Second-order multidimensional ica: performance analysis. IEEE Trans. Signal Process. **60**(9), 4598–4610 (2012)
13. Gutch, H.W., Theis, F.J.: Uniqueness of linear factorizations into independent subspaces. J. Multivar. Anal. **112**, 48–62 (2012)
14. Kawanabe, M., Muller, K.R.: Estimating functions for blind separation when sources have variance dependencies. J. Mach. Learn. Res. **6**, 453–482 (2005)
15. Hyvarinen, A., Hoyer, P.O., Inki, M.: Topographic independent component analysis. Neural Comput. **13**(7), 1527–1558 (2001)
16. Bach, F.R., Jordan, M.I.: Kernel independent component analysis. J. Mach. Learn. Res. **3**, 1–48 (2002)
17. Zhang, K., Chan, L.W.: An adaptive method for subband decomposition ICA. Neural Comput. **18**(1), 191–223 (2006)
18. Wang, F.S., Li, H., Li, R.: Novel nongaussianity measure based bss algorithm for dependent signals. In: Dong, G., Lin, X., Wang, W., Yang, Y., Yu, J.X. (eds.) APWeb/WAIM 2007. LNCS, vol. 4505, pp. 837–844. Springer, Heidelberg (2007)
19. Caiafa, C.: On the conditions for valid objective functions in blind separation of independent and dependent sources. EURASIP J. Adv. Signal Process. **2012**, 255 (2012)
20. Aghabozorgi, M.R., Doost-Hoseini, A.M.: Blind separation of jointly stationary correlated sources. Signal Process. **84**(2), 317–325 (2004)

21. Abrard, F., Deville, Y.: A time-frequency blind signal separation method applicable to underdetermined mixtures of dependent sources. Signal Process. **85**(7), 1389–1403 (2005)
22. Kopriva, I., Jeric, I., Brkljacic, L.: Nonlinear mixture-wise expansion approach to underdetermined blind separation of nonnegative dependent sources. J. Chemom. **27**(1), 189–197 (2013)
23. Cruces, S.: Bounded component analysis of linear mixtures: a criterion for minimum convex perimeter. IEEE Trans. Signal Process. **58**(4), 2141–2154 (2010)
24. Erdogan, A.T.: A class of bounded component analysis algorithms for the separation of both independent and dependent sources. IEEE Trans. Signal Process. **61**(22), 5730–5743 (2013)
25. Li, Y., Amari, S.I., Cichocki, A.: Underdetermined blind source separation based on sparse representation. IEEE Trans. Signal Process. **54**(2), 423–437 (2006)
26. Almeida, A., Luciani, X., Stegeman, A., Comon, P.: CONFAC decomposition approach to blind identification of underdetermined mixtures based on generating function derivatives. IEEE Trans. Signal Process. **60**(11), 5698–5713 (2012)
27. Cardoso, J.F.: Dependence, correlation and gaussianity in independent component analysis. J. Mach. Learn. Res. **4**, 1177–1203 (2003)
28. Cichocki, A., Thawonmas, R.: On-line algorithm for blind signal extraction of arbitrarily distributed, but temporally correlated sources using second order statistics. Neural Process. Lett. **12**(1), 91–98 (2000)
29. Barros, A.K., Cichocki, A.: Extraction of specific signals with temporal structure. Neural Comput. **13**(9), 1995–2003 (2001)
30. Anderson, M., Adali, T., Li, X.L.: Joint blind source separation with multivariate gaussian model: algorithms and performance analysis. IEEE Trans. Signal Process. **60**(4), 1672–1683 (2012)
31. Zibulevsky, M., Zeevi, Y.Y.: Extraction of a source from multichannel data using sparse decomposition. Neurocomputing **49**(1–4), 163–173 (2002)
32. Zhang, Z.L.: Morphologically constrained ICA for extracting weak temporally correlated signals. Neurocomputing **71**, 1669–1679 (2008)
33. Caiafa, C.F., Proto, A.N.: Separation of statistically dependent sources using an l^2 distance non-gaussianity measure. Signal Process. **86**(11), 3404–3420 (2006)
34. Hyvarinen, A.: Fast and robust fixed-point algorithms for independent component analysis. IEEE Trans. Neural Networks **10**(3), 626–634 (1999)
35. Friedman, J.H.: Exploratory projection pursuit. J. Am. Stat. Assoc. **82**(397), 249–266 (1987)
36. Blanco, Y., Zazo, S.: New gaussianity measures based on order statistics: application to ica. Neurocomputing. **51**, 303–320 (2003)
37. Psychology Department of University of Stirling. http://pics.psych.stir.ac.uk/
38. Vincent, E., Araki, S., Theis, F.J., Nolte, G., Bofill, P., et al.: The signal separation evaluation campaign (2007-2010): achievements and remaining challenges. Sig. Process. **92**, 1928–1936 (2012)
39. Cichocki, A., Amari, S., Siwek, K.: ICALAB toolboxes. http://www.bsp.brain.riken.jp/ICALAB

The Chaotic Measurement Matrix
for Compressed Sensing

Shihong Yao[1], Tao Wang[1], Weiming Shen[1], Shaoming Pan[1],
and Yanwen Chong[2(✉)]

[1] State Key Laboratory for Information Engineering in Surveying, Mapping and
Remote Sensing, Wuhan University, 129 Luoyu Road, Wuhan 430079, China
yao_shi_hong@whu.edu.cn
[2] Lmars, Wuhan University, Wuhan 430079, China
ywchong@whu.edu.cn

Abstract. How to construct a measurement matrix with good performance and
easy hardware implementation is the core research problem in compressed
sensing. In this paper, we present a simple and efficient measurement matrix
named Incoherence Rotated Chaotic (IRC) matrix. We take advantage of the
well pseudorandom of chaotic sequence, introduce the concept of the incoherence factor and rotation, and adopt QR decomposition to obtain the IRC measurement matrix which is suited for sparse reconstruction. Simulation results
demonstrate IRC matrix has a better performance than Gaussian random matrix,
Bernoulli random matrix and other state-of-the-art measurement matrices. Thus
it can efficiently work on both natural image and remote sensing image.

Keywords: Measurement matrix · Compressed sensing · Chaotic sequence ·
Restricted Isometry Property (RIP)

1 Introduction

Compressed sensing (CS) technique brings great convenience for data storage, transmission and processing and has already attracted broad attention in the fields of remote
sensing and medical imaging and become a research issue in recent years. In CS theory,
there are three core issues that are signal sparse representation, measurement matrix
design and reconstruction algorithm construction [1]. The designing level of measurement matrix directly influences signal reconstruction error. In order to evaluate the
performance of measurement matrices, Restricted Isometry Property (RIP) [2] is proposed by Candes et al. in 2007, which described three mathematics criterions that the
measurement matrix should satisfy. In practice, it is difficult to prove whether the
measurement matrices satisfy RIP, thus we adopt an equivalent condition of RIP
criterion, which is the incoherence between the measurement matrix and the compressible signal, to guide the design of measurement matrix.

Random measurement matrix [3, 4] has high computational complexity and difficult hardware implementation and needs large storage space. Deterministic measurement matrix [5–7] can make up the shortfall on hardware implementation of random

© Springer International Publishing Switzerland 2015
D.-S. Huang et al. (Eds.): ICIC 2015, Part I, LNCS 9225, pp. 58–64, 2015.
DOI: 10.1007/978-3-319-22180-9_6

measurement matrix. On this account the contradiction between determinacy and randomness come into being.

Chaos theory reveals the unity of determinacy and randomness exactly. The sequences generated by chaotic system have outstanding pseudo-randomness and are easy to generate and reproduce. Therefore, chaotic sequence [8, 9] can be used to construct measurement matrix in CS. So this paper makes use of strong pseudo-randomness of chaotic system, introduces the concept of the incoherence factors and rotation based on the discrete chaotic sequence, and adopts QR decomposition to obtain a measurement matrix suited for sparse reconstruction. This measurement matrix is named Incoherence Rotated Chaotic (IRC) matrix. With the IRC matrix, we can reconstruct the sparse signal with a high accuracy.

2 Background of CS Theory

CS is an efficient signal acquisition framework for signals that are sparse or compressible in an appropriate domain. Suppose that a signal X is sparse in a certain set of orthogonal basis or a tight frame domain ψ (such as wavelet domain), where $X \in R^N$. In other words, there exists a sparse vector α such that $X = \psi_\alpha$, where the vector α is equivalent or approximate sparse representation of X in the domain ψ. We use a measurement basis Φ which is incoherent with transformation basis ψ to measure signal X, where $\Phi \in R^{M \times N}$ and $M < <N$, hence we obtain the observed vector $Y \in R^{M \times 1}$. Finally we can make use of optimization algorithm to reconstruct the original signal X from the observed sets accurately or with a high probability.

It is difficult to directly construct a measurement matrix Φ to make Φv obey RIP criterion. v is strictly s-sparse N- dimensional vector. We can only assure the sensing matrix Φv to obey RIP as much as possible. Only when matrix Φ is adequately incoherent with sparse coefficient v, L1 norm optimization method can be used to solve α.

$$\hat{\alpha} = \min\|\alpha\|_1 \text{ s.t } Y = A\alpha \tag{1}$$

Where $A = \Phi\Psi$. The problem of L1 norm minimization is equivalent to the problem of signal reconstruction. At present, OMP algorithm is an effective in signal reconstruction and chosen as the reconstruction algorithm in this paper.

3 Construction of Incoherence Rotated Chaotic Matrix

Chaotic is a kind of seeming random or irregular movement, which appears in a deterministic nonlinear system. Chaotic system has both determinacy and randomness. Therefore adopting a chaotic system to construct measurement matrix in CS is feasible.

Logistic chaotic system is the most popular used nonlinear dynamics chaotic systems [10]. In this paper, we select Logistic mapping system to create chaotic measurement matrix. The Logistic map function is expressed as:

$$x_{n+1} = \mu x_n (1 - x_n) \tag{2}$$

where $\mu \in (0,4), x_n \in (0,1)$. In order to enhance the randomness of chaotic factors, we remove the first 1000 elements, and then perform downsampling on the chaotic factors with sampling interval d. Hence the chaotic factors are $v_i = 1 - 2x_{1000+id}, (i = 1, 2, \ldots)$.

The IRC matrix Φ_k of the kth compression ratio can be denoted as

$$\Phi = \sqrt{\frac{1}{2\lambda_k}} \begin{bmatrix} v_{1+kn} & v_{2+kn} & \cdots & v_{n-1+kn} & v_{n+kn} \\ \eta v_{n+kn} & v_{1+kn} & v_{2+kn} & \cdots & v_{n-1+kn} \\ \eta v_{n-1+kn} & \eta^2 v_{n+kn} & v_{1+kn} & \cdots & v_{n-2+kn} \\ \eta v_{n-2+kn} & \eta^2 v_{n-1+kn} & \eta^3 v_{n+kn} & \cdots & v_{n-3+kn} \\ \vdots & \vdots & \vdots & \ddots & \vdots \\ \eta v_{n-\lambda_k+2+kn} & \eta^2 v_{n-\lambda_k+3+kn} & \eta^3 v_{n-\lambda_k+4+kn} & \cdots & v_{n-\lambda_k+1+kn} \end{bmatrix} \tag{3}$$

where λ_k $(k = 1, 2, \ldots M)$ is observed frequency of the kth compression ratio. As shown in formula (3), when the elements move to the beginning of a row from the end, they are multiply by the fixed incoherent coefficient η, where $\eta \in [0.6, 1]$ for each rotation. To construct a measurement matrix suited for variety compression ratios, we only need to determine the first row elements on different compression ratios and adopt the same construction method with formula (3). Then we have $\Phi_k = R_k^T Q_k^T$. The elements of the main diagonal in R_k are reserved and off-diagonal elements are set to zero, thus a new diagonal matrix R_k' is generated. The IRC matrix can denoted as $\Phi_k' = R_k'^T Q_k^T$. The IRC matrix Φ_k' of variety compression ratios can be obtained by the above steps.

4 Experimental Results

According to the construction procedures of IRC matrix presented in Sect. 3, we choose sampling distance $d = 10$, then generate the IRC matrix by formula (3). In this section, we perform simulations using variety images of size 256×256 by Intel(R), Core(TM), dual-core i3 processor, 3.07 GHz, RAM 2.99 G, Matlab 2009a. The image sparse algorithm is the wavelet transform and the image reconstruction algorithm is the OMP reconstruction algorithm [11].

Figure 1 shows the varied original standard test images including Remote sensing image, natural scenery, fingerprints, as well classical Lena, Barbara, Peppers, Tiffany, Baboon.

Figure 2 The reconstruction results are different on different compression ratios for one image. Figure 2 shows that the peak signal-to-noise ratios (PSNRs) between original test images and reconstruction images on different compression ratios using Gaussian random measurement matrix and IRC matrix. The X-axis shows the serial number of images and the Y-axis shows the PSNRs between original test images and

Fig. 1. The original standard test images

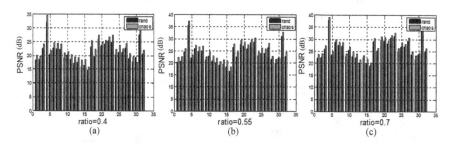

Fig. 2. The reconstructed image PSNRs on different compression ratios using Gaussian random measurement matrix and IRC matrix. (compression ratios are 0.4, 0.55, 0.7 respectively)

reconstruction images. 'Chaos' represent the PSNRs using IRC matrix and 'rand' represent the PSNRs using Gaussian random measurement matrix.

As shown in Fig. 2, on different compression ratios, we can see that the PSNRs of the reconstructed images, except the 30th image named testpat, using the IRC matrix are better than Gaussian random measurement and the PSNRs are improved by. 1.5 to 2.5 dB.

To further test the reconstruction quality of IRC matrix, we take structure similarity measure (SSIM) that takes into account some of the spatial relationships between regions as another important measure of image reconstruction performance [12]. Table 1 shows that the mean values of SSIM for all the test images on different compression ratios using Gaussian random measurement matrix and IRC matrix and it indicates that the SSIM under IRC matrix are higher than the SSIM under Gaussian random measurement matrix. From the view of the results of two measurable indicators, the

Table 1. The mean values of SSIM for all the test images on different compression ratios using Gaussian random measurement matrix and IRC matrix. (compression ratios are 0.4, 0.55, 0.7 respectively)

M/N The measurement matrices	0.4	0.55	0.7
SSIM of Gaussian random measurement matrix	0.6277	0.7654	0.8410
SSIM of IRC matrix	0.7584	0.8358	0.8880

reconstruction performance of IRC is significantly better than that of Gaussian random measurement matrix.

Figure 3 shows when the compression ratio is 0.4 or 0.8, the reconstruction result comparisons of classical images, including Barbara、Lena、Peppers. The original images are in the first column, the obvious distinction between reconstruction images using Gaussian random measurement matrix are in the second column and fourth column and the obvious distinction between reconstruction images using IRC matrix are in the third column and fifth column. We can see that the visual quality of reconstructed images using IRC matrix is much better than Gaussian random measurement matrix and the noise points of reconstructed images using IRC matrix are less than Gaussian random measurement matrix.

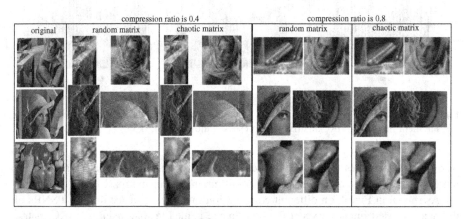

Fig. 3. The reconstruction result comparisons of classical images

The comparison of the reconstructed image PSNR using different measurement matrix under different compression ratios are shown in Table 2. The experimental processes refer to the reference [13]. We select the discrete wavelet transform as the sparse algorithm and the OMP algorithm as the image reconstruction algorithm. The incoherence coefficient η is set to 0.85. The experiment of each measurement matrix under different compression ratios to reconstruct the 256 × 256 Lena.bmp 2-D image has been conducted. As shown in Table 2, the reconstructed image PSNRs of these eight measurement matrices increase with compression ratio increase and the PSNR of

Table 2. Each measurement matrix under different compression ratios of reconstructed image PSNR values

M/N	Chao(dB)	Gaus(dB)	Four(dB)	Bern(dB)	Ploy(dB)	Tope(dB)	Rota(dB)	Kron(dB)
0.4	**26.3461**	24.2485	24.0759	23.4901	17.9578	18.3825	18.9907	22.7438
0.55	**29.2168**	27.5921	27.3359	26.9858	20.4652	20.5625	20.9537	24.6491
0.7	**31.5148**	29.9208	28.7732	29.3375	23.5479	23.6343	23.5233	27.7758
0.8	**33.1058**	31.3538	30.1493	30.5902	24.7167	25.0692	25.2591	29.0746

Chao, Chaotic noncorrelation matrix; Gaus, Gaussian random matrix; Four, Fourier random matrix; Bern, Bernoulli matrix;Poly, polynomial matrix; Tope, Toeplitz matrix; Rota, rotation matrix; Kron, based on orthogonal basis Kronecker product of matrix.

Table 3. Each measurement matrix reconstruction effect and time consumed.

The measurement matrices	Chao	Gaus	Four	Bern	Ploy	Tope	Rota	Kron
PSNR(dB)	30.1309	28.4385	27.9497	28.0258	22.0542	21.3743	22.4629	26.2274
The image reconstruction time consumed (s)	7.0509	13.7541	13.9357	11.0960	13.0644	7.8477	7.2542	7.5738

the reconstructed image using the IRC matrix is higher than the other measurement matrices. In the same simulation environment and simulation process, Table 3 shows that IRC matrix performs better than the other measurement matrices in the image reconstruction effect of PSNR under the fixed compression ratio of 0.6, and IRC matrix has less time consumed in reconstructing the image.

All the experiment results show that the reconstruction performance of the IRC matrix is better than the other state-of-the-art measurement matrices. The construction procedures of IRC matrix only need a small amount of calculations and storage space thus it has an easy embedded hardware implementation.

5 Conclusion

With the development and promotion of compressed sensing theory in practical application, data acquisition has significantly overcome the limitations of Nyquist sampling theory. However, the existing random measurement matrices and deterministic measurement matrices have their own deficiency respectively. In this paper, we take advantage of the well pseudorandom of chaotic sequence, introduce the concept of the incoherence factor and rotation, and adopt QR decomposition to obtain the IRC measurement matrix which is suited for sparse reconstruction. Simulated results demonstrate IRC matrix has a better performance than Gaussian random measurement matrix and the other state-of-the-art measurement matrices. The reconstructed image PSNRs of IRC matrix are improved at 1.5 dB to 2.5 dB at least and IRC matrix can efficiently work on both natural image and remote sensing image.

Acknowledgments. This paper was supported by the Foundation Research Funds for the central Universities (204201kf0242, 204201kf0263), National Natural Science Foundation of China (41271398), Shanghai Aerospace Science and Technology Innovation Fund Projects (SAST201425).

References

1. Candès, E.J.: Compressive sampling. In: Proceedings of the International Congress of Mathematicians, vol. 3, pp. 1433 -1452, (2006)
2. Candes, E.J., Tao, T.: Near-optimal signal recovery from random projections: universal encoding strategies? IEEE Trans. Inf. Theor. **52**(12), 5406–5425 (2006)
3. Sun, R., Zhao, H., Xu, H.: The application of improved Hadamard measurement matrix in compressed sensing. In: Proceedings of International Conference on Systems and Informatics (ICSAI), pp. 1994–1997 (2012)
4. Zhang, G., Jiao, S., Xu, X., et al.: Compressed sensing and reconstruction with bernoulli matrices. In: Proceedings of International Conference on Information and Automation (ICIA), pp. 455–460 (2010)
5. Yin, W., Morgan, S., Yang, J., et al.: Practical compressive sensing with Toeplitz and circulant matrices. In: Proceedings of International Society for Optics and Photonics on Visual Communications and Image Processing, pp. 77440K–77440K (2010)
6. DeVore, Ronald A.: Deterministieeon structions of compressed sensing matrices. J. Complex. **23**(4–6), 918–925 (2007)
7. Bajwa, W., Haupt, J., Raz, G., Wright, S.J., Nowak, R.D.: Toeplitz-structured compressed sensing matrices. In: Proceedings of the IEEE Workshop on Statistical Signal Processing, Washington D. C., USA, pp. 294–298 (2007)
8. Yu, L., Barbot, J.P., Zheng, G., et al.: Compressive sensing with chaotic sequence. Signal Process. Lett. **17**(8), 731–734 (2010)
9. Linh-Trung, N., Van Phong, D., Hussain, Z.M., et al.: Compressed sensing using chaos filters. In: Proceedings of IEEE Conference on Telecommunication Networks and Applications (ATNAC), pp. 219–223 (2008)
10. Huang, W., Ye, X.: Topological complexity, return times and weak disjointness. Ergodic Theor. Dyn. Syst. **24**(3), 825–846 (2004)
11. Donoho, D.L., Tsaig, Y., Drori, I., et al.: Sparse solution of underdetermined linear equations by stagewise orthogonal matching pursuit. Technique report TR-2006-2, Standford University, Department of Statistics (2006)
12. Wang, H., Maldonado, D., Silwal, S.: A nonparametric-test-based structural similarity measure for digital images. Comput. Stat. Data Anal. **55**(11), 2925–2936 (2011)
13. Zhang, B., Tong, X., Wang, W., et al.: The research of Kronecker product-based measurement matrix of compressive sensing. EURASIP J. Wirel. Commun. Networking **2013**(1), 1–5 (2013)

Leveraging Semantic Labeling for Question Matching to Facilitate Question-Answer Archive Reuse

Tianyong Hao[1,2(✉)], Xinying Qiu[2], and Shengyi Jiang[2]

[1] Key Lab of Language Engineering and Computing of Guangdong Province,
Guangdong University of Foreign Studies, Guangzhou, China
[2] Cisco School of Informatics, Guangdong University of Foreign Studies,
Guangzhou, China
{haoty, qiuxinying, jiangshengyi}@gdufs.edu.cn

Abstract. A new question representation method is proposed for automated question matching over accumulated question-answer data archive. The representation defines four kinds of question words as question-type words, user-centered words, shareable-pattern words, and irrelevant words for question analysis. These question words are further annotated by a semantic labeling ontology to enhance the semantic representation for the purpose of word ambiguity reduction. We tested the matching precision on 5,000 questions with respect to various generators and the result demonstrated the stability of the method. We further compared the method with Cosine similarity and WordNet-based semantic similarity as baselines on a standard TREC dataset containing 5,536 questions. The results presented that our method improved MRR by 8.6 % and accuracy by 9.6 % on average, indicating its effectiveness.

Keywords: Question representation · Question matching · Semantic labeling

1 Introduction

Though just emerging a couple of years, User-Interactive Question Answering (UIQA) systems (also known as Knowledge Sharing Communities or Community Question Answering systems) are becoming popular information services to meet the ever-increasing information need of users [1, 2]. For example, Baidu Zhidao[1] has millions of active users per day and help users answered 3.58 billon questions as to Feb 2015. As a result, a large amount of questions and choose answers either by users or by systems are accumulated.

The resulting of question and answer archives can help answer newly coming similar questions, particularly, frequently asked questions (FAQ) [3–5]. Therefore, they are potentially useful data resources. However, current automated question-answering techniques fail to satisfy users' information demand to some extent. Especially in UIQA systems, users still need to post similar questions and wait for answers for a time period, regardless of the accumulated large amount of questions and answers.

---

[1] http://zhidao.baidu.com/.

© Springer International Publishing Switzerland 2015
D.-S. Huang et al. (Eds.): ICIC 2015, Part I, LNCS 9225, pp. 65–75, 2015.
DOI: 10.1007/978-3-319-22180-9_7

Consequently, similar or even exact same questions are answered repeatedly, which is a waste of the accumulated data resources. Moreover, in industry, commercial companies may need to pay huge cost to manually answer a large number of customers' online questions every day. Therefore, there is a high demand to develop cutting-edge QA techniques to instantly fulfill users' information needs by reusing the existing data archives to reduce training and operating cost for the online custom services [6].

For reusing the accumulated question and answer (Q&A) archives, a new question representation method is proposed. The representation is designed to analyze question components by their functionalities so as to reuse them across question sets. The representation defines question-type words, user-centered words, shareable-pattern words, and irrelevant words, existing or potentially existing within a question context. To evaluate the method, we firstly tested the question matching performance based on a dataset containing 5,000 questions. Experiment results showed that the precisions of pattern matching were stabilized between 0.71 and 0.93 depending on different question generators. We then tested the performance based on a public available Text REtrieval Conference (TREC) dataset containing 5,536 questions from University of Illinois at Urbana-Champaign (UIUC). Compared with baselines Cosine similarity and WordNet-based semantic similarity methods, our method achieved best performance with MRR increased by 9.8 % and 8.6 % while accuracy increased by 9.9 % and 9.6 % on average. The comparison presented the effectiveness of our method.

The rest of this paper is organized as follows: Sect. 2 introduces related work. Section 3 presents the question representation method and Sect. 4 shows the question matching strategy. The detailed experiment results are demonstrated in Sect. 5 and Sect. 6 summarizes the paper.

2   Related Work

In an automated QA system, a Frequently Asked Question (FAQ) retrieval method obtains the existing questions/answers (Q&A) pairs from the frequently-asked question files [7] rather than generating desired answers directly. There are a lot of researches focusing on FAQ retrieval in the last two decades [8]. FAQ Finder and Auto-FAQ represents two types of FAQ answering systems. FAQ Finder [9, 10] was designed to improve navigation through existing external FAQ collections. The system has an index – FAQ text files organized into questions, section headings, and keywords etc. FAQ-Finder adopted two major approaches, i.e., concept expansion using hypernyms defined in WordNet [2] and TF-IDF weighted score in the retrieval process. Auto-FAQ [11] relied on a shallow, surface-level analysis for similar question retrieval, rather than performing indexing of an external FAQ collection. The system uses language understanding with keyword comparison enhanced by limited language processing. It maintains its own FAQ set. In FAQ-Finder, certain question types may not be detected correctly, e.g., "what" and "how" are incorrectly identified as interrogative words regardless the interrogative phrases "for what" and "how large".

---

[2] http://wordnet.princeton.edu/.

To solve the above problems, Tomuro [12] combined lexicon and semantic features to automatically extract the interrogative words from a corpus of questions. Typically, Jeon et al. [13] applied the IBM machine translation model on a collection of similar question pairs to learn word translation probabilities, denoting semantic similarities between words. Based on the model, Xue et al. [14] compared the pure IBM model with a query likelihood language model and proposed a retrieval model that combines a translation-based language model with a query likelihood approach. Wang et al. [15] presented a retrieval framework - syntactic tree matching to tackle the similar question matching problem. The method demonstrated better question matching but still has room to be improved, particularly, on the aspect long questions. Cao et al. [16] proposed a new approach named as "category enhanced retrieval model" to exploiting category information of questions for improving the performance of question retrieval. The approach has advantage in application to existing question retrieval models, e.g., Vector Space Mode.

In the recent 2 years (2014–2015), there are several related works. Zhang et al. [17] addressed the issue that questions and answers are heterogeneous and there are a particularly large number of low quality answers. A supervised question-answer topic modeling approach was proposed based on the assumption that questions and answers share some common latent topics and are generated following the topics, which are used to determine an answer quality signal. Zhang et al. [18] analyzed the limitations of syntactic based question match approaches (e.g., [19]) and the traditional Bag-of-Word methods (e.g., [20]) on capturing the semantic similarity of CQA questions due to the complexity of the CQA questions and the variety of the users' expressions. They introduced a topic modeling approach to capturing the word semantic similarity between query and question. Afterwards, an unsupervised machine-learning approach to finding similar questions on Q&A archives is proposed. Experimental results show that the approach outperformed four baseline methods including the syntactic tree matching method. Toba et al. [21] addressed the problem that most existing works use the same model on quality prediction for all answers, despite the fact that the answers are intrinsically different. They proposed a hybrid hierarchy-of-classifiers framework to model the QA pairs, particularly, question type analysis to guide the selection of the right answer quality model. The method was tested on 50 thousand QA pairs from Yahoo! Answer and results proved the effectiveness of the method. The recent works addressed that question type is beneficial for applying right model to determine high quality answers. This motivated us to work on question representation integrating the question type analysis.

3   The Question Representation Method

The question representation method is to enhance question understanding capability according to the definitions of the functionalities of question components, for supporting answering FAQs through reusing accumulated data. To that end, we firstly define a new question representation method consisting of four kinds of words:

question-type word, user-centered word, shareable-pattern word, and irrelevant word, existing or potentially existing within a question context. The descriptions of the four kinds of word are presented in Table 1.

Table 1. The descriptions of the four word kinds defined in the question representation method.

Word kind	Description
Question-type words	A *Question-type Word* is a word that represents the target of the question.
User-centered words	A *User-centered Word is* a word that conveys the essence of user intention in the question.
Shareable-pattern words	A *Shareable-pattern Word is* a word inside the structure/pattern of the question and the structure/pattern is shareable with other questions in the same question collection.
Irrelevant words	An *Irrelevant Word* is a word not belongs to question-type words, user-centered words, or shareable-pattern words.

To represent question target, the question-type (QT) words are a commonly used question type word list, e.g., "What", "Who", and "When", or nouns/noun phrases. The QT words are further extended to hierarchically semantic categories (e.g., "*[Quantity]/[Speed]*") through a well-defined QT category by UIUC [22] in this paper. The user-centered words mainly contain named-entities (e.g., organization name, individual name, time, location, etc.), nouns, and noun phrases. The designing initiative of "irrelevant words" is different from "stop words" that are commonly used in Information Retrieval area. The stop words are assumed to be always "unimportant" in a given collection of documents, while some stop words could be "important" according to their usage in a certain question context. This indicates that the label of "irrelevant word" for a same word is dynamic across questions and inside the same question. The irrelevant words do not share words with other kinds of question words but are also usefully in the question matching and filtering. Please note a word can be assigned with different question word kinds in the same question in the framework.

Semantic labeling is an essential part of the representation for representing the word kinds semantically. We redesigned and extended an ontology named SL ontology based on the ontology reported in [23] to generate multi-level semantic labels as the semantic annotations for the three word kinds. For convenience use in real user-interactive question answering systems, the SL ontology is designed into two levels of concepts hierarchy with defined relations (e.g., IS_A) for representing the relationship between those concepts. A semantic label thus can be generated in the format of [*Concept A*] \ [*Concept B*], where *Concept A* and *Concept B* have the relationship of *SubClassOf* (*Concept A, Concept B*). The ontology, containing 7 first level concepts and 63 second level concepts in total, is flexible to be modified or extended.

To label question words semantically, as for a given word or phrase, the ontology is applied to retrieve its super concepts (hyponyms). For extending the usage of the SL ontology, we leverage WordNet to retrieve the super concept collections for a given word (noun/verb) and generate semantic labels by mapping the concepts to WordNet. For example, the semantic label "*[location]\[city]*" is mapped to "*[physical_entity] \[city]*" in WordNet.

Question context is considered to determine the best semantic label in case that multiple semantic labels are generated for the same word. Derived from Naive Bayesian classification, a calculation method is proposed for best semantic label determination. For a given word w_i, set l_k as a semantic label in label set L, which refers to all candidate semantic labels. Assuming the word w_i co-occurs with another word w_j, the suitable label l^* for w_i is determined by select the maximum product value between the probability of word w_i to be assigned with semantic label l_k and the probability of word w_j occurred in the same question when w_i is assigned with label l_k.

By defining the four question components and the semantic labeling of the components, a formal definition of the question representation is given as follows:

Definition: A question q_i is represented as $< QTW_i, UCW_i, SPW_i, IRW_i >$ and their corresponding semantic labels, where $QTW_i = w_i^0, \ldots, w_i^k$ is a collection of j question-type words, UCW_i is a collection of k user-center words, SPW_i is a collection of m shareable-pattern words, and IRW_i is a collection of n irrelevant words.

With the defined word kinds and their semantic labels, a question therefore can be represented by integrating each word kind and semantic labels together. For example, as for a question "*What is the distance between Paris and London?*", the question-target words are "What" and "distance"; user-centered words are "*distance*", "*Paris*", and "*London*". Shareable-pattern words are "*be*", "*distance*", "*between*", and "*and*". Finally, the irrelevant words "*the*" are identified. According to semantic labeling, the words "*distance*", "*Paris*", and "*London*" are annotated with labels "*Numeric/Distance*", "*Physical_entity\City*", and "*Physical_entity\City*", respectively. In the example, the word "*distance*" exists among question-target words, user-centered words, and shareable-pattern words.

4   Comprehensive Question Matching Strategy

To answer newly coming questions through reusing accumulated questions/answers, an essential step is the matching between new questions and existing FAQs. We propose a comprehensive strategy based on the above question representation by matching every question components. The question-type words (QTW), user-centered words (UCW), and shareable-pattern words (SPW) are separately calculated and merged as a final similarity score. For each part of similarity, a different strategy is developed according to their characteristics for improving QA performance.

There are some widely used similarity calculation methods including Cosine, Dice, and Jaccard coefficients. Among them, we select Dice coefficient as the base of our word kind matching based on two considerations: the convenience of semantic similarity/synonymy integration and the calculation efficiency. Dice coefficient ($2 \times |X \cap Y|)/(|X| + |Y|)$ returns the ratio of the number of words that can be matched over the total of words.

(1) For the question-type similarity, semantic similarity is applied since question targets can be classified into semantic categories through semantic labeling, as

described in the previous section. The similarity is calculated by comparing the similarity of these semantic labels. The calculation formula is shown as Eq. (2).

$$Simi_{QT}(q_i, faq_j) = \frac{2 \times \left| q_i(QTW \rightarrow L^*) \cap faq_j(QTW \rightarrow L^*) \right|}{\left| q_i(QTW \rightarrow L^*) \right| + \left| faq_j(QTW \rightarrow L^*) \right|} \qquad (2)$$

$Simi_{QT}$ denotes the similarity score of QT word aspect between the user question q_i and the existing FAQ question faq_j. $q_i(QTW \rightarrow L^*)$ and $faq_j(QTW \rightarrow L^*)$ represent the semantic labels of QT words in q_i and faq_j through semantic labeling, respectively.

(2) For user-centered word matching, we believe that semantic similarity strategy may affect the answer retrieval negatively. For example, "Sydney" and "New York" may have high semantic similarity as they both means "location" or "city" but the answer for a question asking about "the history of Sydney" fails to answer another question about "the history of New York". Therefore, we use synonymy words with exact same semantic meaning (semantic equivalence) as the strategy for the user-centered (UC) word matching. The similarity is calculated using Eq. (3).

$$Simi_{UC}(q_i, faq_j) = \frac{2 \times \sum SMatch(q_i(UCW), faq_j(UCW))}{\left| q_i(UCW) \right| + \left| faq_j(UCW) \right|} \qquad (3)$$

The matching $Simi_{UC}$ denotes the user-centered word kind matching between the given question q_i and the existing FAQ question faq_j. $SMatch$ denotes synonymy-based word matching of two words w_m and w_n. The matching relies on synonymy extension of word w_m by adding its synonymy word collection. We use WordNet for the synonymy extension in the paper. Once a shared word between two extended word lists is identified, the $SMatch$ matching immediately stop and set the value to be 1 for further calculation.

(3) Differently, the similarity calculation of shareable-pattern words (SPW) is integrated into semantic pattern matching as we already have semantic pattern framework, generation method, and annotated FAQ data. The semantic pattern and its related techniques are to accumulate the questions and answers in a more structured, and more semantic way. It is designed to generalize a class of questions with the same sentence structure and relevant semantics. Due to the complexity of semantic pattern matching, we simplified the matching as Eq. (4) so as to focus on SPW matching, where $SPW \rightarrow pattern^*$ represents the linking from shareable-pattern words to the corresponding semantic patterns.

$$Simi_{SP}(q_i, faq_j) = \frac{2 \times \left| q_i(SPW \rightarrow pattern^*) \cap faq_j(SPW \rightarrow pattern^*) \right|}{\left| q_i(SPW \rightarrow pattern^*) \right| + \left| faq_j(SPW \rightarrow pattern^*) \right|} \qquad (4)$$

Through integrating the previous part-by-part matching, we calculate the overall matching score $SimiSC(q_i, faq_j)$ of the two questions q_i and faq_j through balancing the similarity for each part, as shown in formula (5). Two coefficients α and β ($\alpha \geq 0, \beta \geq 0, \alpha + \beta \leq 1$) are set as weights for question-type word part $Simi_{QT}$ and user-centered word part $Simi_{UC}$, respectively. The two parameter values are described in next section.

$$MatchSC(q_i, faq_j) = \alpha \times Simi_{QT} + \beta \times Simi_{UC} + (1 - \alpha - \beta) \times Simi_{SP} \qquad (5)$$

The matching strategy including three matching calculations is based on the consideration that different word/pattern functionalities may have different importance for use in the same question. Particularly, the strategy enables us to review the question matching on different matching aspects by controlling the parameters. For example, only question-type word similarity is considered to view questions with the same target when α is set to 1, while only shareable-pattern word similarity is investigated when $\alpha + \beta$ is set to 0.

5   Evaluation

To evaluate the performance of the proposed question matching approach, two experiments have been conducted: the first to evaluate the stability of question matching, and the second to evaluate the effectiveness by comparing with baselines methods. The evaluation metrics are commonly used precision, recall, MRR, and accuracy in Information Retrieval area.

Due to the difficulty of achieving a fully agreed similar question dataset as reference standard, we developed a program to modify original questions automatically by changing or adding words (controlled by a generator λ) and used the modified questions as testing dataset. The words were randomly selected from a lexicon containing 93,696 words with Part-of-Speech tags. Those revised questions with one word modified ($\lambda = 1$) were collected as test dataset C_1. Those revised questions with two words modified ($\lambda = 2$) were collected as test dataset C_2, and so on. Obviously, the generator λ affected the question matching as the various numbers of words were modified. The experiment datasets were pre-annotated with semantic labels using the question representation method, as shown in Fig. 1. The parameters were set as $\alpha = 0.2$ and $\beta = 0.4$ according to our empirical studies.

We further conducted the evaluation of question matching stability by increasing the number of testing questions. 5,000 questions were randomly selected and used to generate testing datasets C_1, C_2, C_3, and C_4. For each question in a testing dataset, our system analyzed its question word kinds and matched it with the 5,000 questions dataset to obtain best matched question for calculating correctness. The matric Precision was used (recall and F1 were not used since the experiment focused on the stability of matching precision). The results on the testing datasets are shown in the Fig. 2.

Fig. 1. Examples of question dataset annotated with semantic labels.

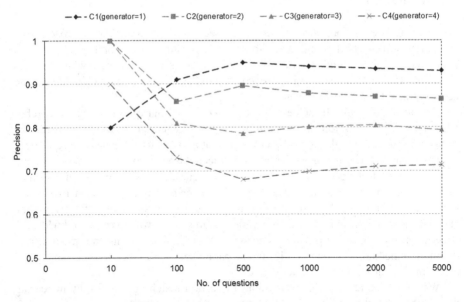

Fig. 2. The evaluation of question matching stability on testing datasets C_1, C_2, C_3, and C_4 by increasing the number of questions.

From the experiment results, the curves for all the testing datasets tend to be more flat with the increasing question number. Particularly, the precision performance is relatively stable when the question number is equal to or larger than 1,000. Eventually, we reviewed stable precisions for question matching at: 0.93 on C_1, 0.879 on C_2, 0.802 on C_3, and 0.71 on C_4. It presented that our approach achieved 0.93 precision at best and 0.71 at least, indicating that our approach combining question-type words (with semantic extension), shareable pattern words, and user-centered words (with semantic extension) is more stable in question matching compared with the unreliable word-based matching.

We further compared our approach with two baseline methods. Cosine similarity [24] and WordNet-based semantic similarity measure [25] are two commonly used text similarity methods. The comparison was based on a public available dataset - TREC 10 and 11 questions from UIUC [22, 26]. It contains 5,366 questions (though 5,500 questions were claimed on the website) and each question is already annotated with question target. The evaluation measures are normally used Mean Reciprocal Rank (MRR) and Accuracy@N.

We first applied same strategy to generate four new testing datasets C_1, C_2, C_3, and C_4 on the TREC dataset using different generators (≤ 4). Each question in the testing datasets was then matched with all the original questions C_0 as reference standard to determine the matching correctness and the rank of the correct match for calculating MRR and Accuracy. The results are shown in Table 2 by running on the four testing datasets.

From the MRR comparison results, WordNet-based approach achieved 0.852, 0.772, and 0.673 on dataset C_1, C_2, and C_3, respectively, better than Cosine similarity. However, the gap was shorted along with the increasing of the generator (testing datasets). When the generator equaled to 4 (C_4), the Cosine similarity obtained better MRR as 0.594 than WordNet-based approach as 0.568, indicating that the variety of question words increasingly affected the WordNet-based similarity calculation. Our approach achieved the best MRR performance as 0.906, 0.840, 0.770, and 0.691 on the the datasets among the approaches. Compared with WordNet-based approach, 8.6 % MRR was increased on average. Particularly, our approach was more resistant to the increasing generator than WordNet-based approach by reviewing the performance

Table 2. The MRR and accuracy comparison on the 5,536 TREC questions using Cosine similarity and WordNet-based similarity, and our question matching approach.

Approach	Testing dataset	MRR	Accuracy				
			@1	@2	@3	@4	@5
Cosine-based	C_1	0.811	0.771	0.804	0.833	0.841	0.843
	C_2	0.740	0.688	0.747	0.769	0.779	0.787
	C_3	0.670	0.611	0.662	0.685	0.706	0.717
	C_4	0.594	0.545	0.573	0.618	0.629	0.635
WordNet-based	C_1	0.852	0.840	0.861	0.874	0.882	0.898
	C_2	0.772	0.738	0.747	0.762	0.775	0.783
	C_3	0.673	0.649	0.661	0.688	0.693	0.717
	C_4	0.568	0.527	0.552	0.579	0.590	0.619
Our approach	C_1	0.906	0.868	0.918	0.931	0.936	0.943
	C_2	0.840	0.788	0.847	0.869	0.879	0.887
	C_3	0.770	0.707	0.769	0.793	0.810	0.819
	C_4	0.691	0.619	0.679	0.707	0.726	0.738

dropping velocity. From the accuracy comparison, our approach also achieved best performances at @1, @2, @3, @4, and @5, exceeding Cosine similarity and WordNet-based similarity approaches by 9.9 % and 9.6 % on average. The comparison demonstrated the effectiveness of our approach.

6 Summary

This paper presents a new question representation method for understanding question by leveraging semantic labeling to automatically answering similar questions. The approach represents a user question according to the automatic processing of four kinds: question-target words, user-centered words, shareable-pattern words, and irrelevant words. A question matching method was further presented based on the representation. Experiments on 5,000 questions presented the question matching performance reached 0.93 at most. The further experiment on a standard 5,536 TREC question set demonstrated the effectiveness of the approach through the comparison with Cosine similarity and WordNet-based semantic similarity as baseline methods.

Acknowledgements. This work was supported by National Natural Science Foundation of China (grant No. 61403088 and No.61305094).

References

1. Liu, W.Y., Hao, T.Y., Chen, W., Feng, M.: A web-based platform for user-interactive question-answering. World Wide Web **12**(2), 107–124 (2009)
2. Hao, T.Y., Xu, F.F., Lei, J.S., Liu, W.Y., Li, Q.: Toward automatic answers in user-interactive question answering systems. Int. J. Softw. Sci. Comput. Intell. **3**(4), 52–66 (2011)
3. Liu, Y., Bian, J., Agichtein, E.: Predicting information seeker satisfaction in community question answering. In: Proceedings of the 31st Annual International ACM SIGIR Conference on Research and Development in Information Retrieval, pp.483–490 (2008)
4. Seo, J.W., Croft, W.B., Smith, D.A.: Online community search using conversational structures. Inf. Retrieval **14**(6), 547–571 (2011)
5. Agichtein, E., et al.: Finding high-quality content in social media. In: Proceedings of the 2008 International Conference on Web Search and Data Mining (2008)
6. Hao, T.Y., Agichtein, E.: Finding similar questions in collaborative question answering archives: toward bootstrapping-based equivalent pattern learning. Inf. Retrieval **15**(3–4), 332–353 (2012)
7. Voorhees, E.: The TREC-8 Question Answering Track Report, NIST Special Publication of the Eighth Text REtrieval Conference TREC 8, National Institute of Standards and Technology, pp. 743–751 (1999)
8. Wu, C.H., Yeh, J.F., Chen, M.J.: Domain-specific FAQ retrieval using independent aspects. ACM Transactions Asian Language Information Processing **4**(1), 1–17 (2005)
9. Hammond, K., Bruke, R., Martin, C., Lytinen, S.: FAQ-Finder: a case based approach to knowledge navigation. In: Working Notes of the AAAI Spring Symposium on Information Gathering from Heterogeneous Distributed Environments, AAAI (1995)

10. Burke, R.D., Hammond, K., Kulyukin, V., Lytinen, S.L., Tomuro, N., Schoenberg, S.: Question answering from frequently-asked-question files: experiences with the FAQ finder system, Technique report TR-97-05, University of Chicago, Chicago (1997)
11. Whitehead, S.D.: Auto-FAQ: an experiment in cyberspace leveraging. J. Comput. Netw. ISDN Syst. **28**, 137–146 (1995)
12. Tomuro, N.: Question terminology and representation for question type classification. In: Proceedings of the 2nd International Workshop on Computational Terminology (COMPUTERM02), Taipei (2002)
13. Jeon, J., Croft, W.B., Lee, J.H.: Finding similar questions in large question and answer archives. In Proceedings of the 14th ACM International Conference on Information and Knowledge Management, pp. 84–90 (2005)
14. Xue, X., Jeon, J., Croft, W.B.: Retrieval models for question and answer archives. In: Proceedings of SIGIR, pp. 475–482 (2008)
15. Wang, K., Ming, Z., Chua, T.-S.: A syntactic tree matching approach to finding similar questions in community-based qa services. In: Proceedings of SIGIR, pp. 187–194 (2009)
16. Cao, X., Cong, G., Cui, B., Jensen, C.S.: A generalized framework of exploring category information for question retrieval in community question answer archives. In: Proceedings of the 19th International Conference on World Wide Web, pp. 201–210 (2010)
17. Zhang, K., Wu, W., Wu, H., Li, Z., Zhou, M.: Question retrieval with high quality answers in community question answering. In Proceedings of the 23rd ACM International Conference on Conference on Information and Knowledge Management, pp. 371–380 (2014)
18. Zhang, W.N., Liu, T., Yang, Y., Cao, L., Zhang, Y., Ji, R.: A topic clustering approach to finding similar questions from large question and answer archives. PLoS ONE **9**(3), e71511 (2014)
19. Cui, H., Sun, R., Li, K., Kan, M.Y., Chua, T.S.: Question answering passage retrieval using dependency relations. In: Proceedings of the 28th Annual International ACM SIGIR Conference on Research and Development in Information Retrieval, pp. 400–407 (2005)
20. Gao, Y., Wang, M., Zha, Z.J., Shen, J., Li, X., Wu, X.: Visual-textual joint relevance learning for tag-based social image search. IEEE Trans. Image Process. **22**(1), 363–376 (2013)
21. Toba, H., Ming, Z.Y., Adriani, M., Chua, T.S.: Discovering high quality answers in community question answering archives using a hierarchy of classifiers. Inf. Sci. **261**, 101–115 (2014)
22. Li, X., Roth, D.: Learning question classifiers. In: Proceedings of the 19th International Conference on Computational Linguistics, vol. 1, pp. 1–7 (2002)
23. Hao, T.Y., Ni, X.L., Quan, X.J., Liu, W.Y.: Automatic construction of semantic dictionary for question categorization. J. Syst. Cybern. Inform. **7**(6), 86–90 (2009)
24. Singhal, A.: Modern information retrieval: a brief overview. Bull. IEEE Comput. Soc. Tech. Comm. Data Eng. **24**(4), 35–43 (2001)
25. Source code of WordNet-based semantic similarity measurement. http://www.codeproject.com/Articles/11835/WordNet-based-semantic-similarity-measurement. Accessed 2015
26. Experimental Data for Question Classification. http://cogcomp.cs.illinois.edu/Data/QA/QC/. Accessed 2015

How to Detect Communities in Large Networks

Yasong Jiang[1]([✉]), Yuan Huang[2], Peng Li[2], Shengxiang Gao[2],
Yan Zhang[1], and Yonghong Yan[1]

[1] The Key Laboratory of Speech Acoustics and Content Understanding Institute
of Acoustics, Chinese Academy of Sciences, Beijing, China
{jiangyasong, zhangyan, yanyonghong}@hccl.ioa.ac.cn
[2] National Computer Network Emergency Response Technical
Team/Coordination Center of China, Beijing, China
{huangyuan, gao.shengxiang}@cert.org.cn,
12685581@qq.com

Abstract. Community detection is a very popular research topic in network science nowadays. Various categories of community detection algorithms have been proposed, such as graph partitioning, hierarchical clustering, partitional clustering. Due to the high computational complexity of those algorithms, it is impossible to apply those algorithms to large networks. In order to solve the problem, Blondel introduced a new greedy approach named lovian to apply to large networks. But the remained problem lies in that the community detection result is not unstable due to the random choice of seed nodes. In this paper, we present a new modularity optimization method, LPR, for community detection, which chooses the node in order of the PageRank value rather than randomly. The experiments are executed by using medium-sized networks and large networks respectively for community detection. Comparing with lovian algorithm, the LPR method achieves better performance and higher computational efficiency, indicating the order of choosing seed nodes greatly influences the efficiency of community detection. In addition, we can get the importance values of nodes which not only is part of our algorithm, but also can be used to detect the community kernel in the network independently.

Keywords: Pagerank · Community detection · Modularity · Large network

1 Introduction

We know that real networks are not random and they usually exhibit inhomogeneity, indicating the coexistence of order and organization. The most famous character of the networks is the community structure. Community structure embodies the famous saying that "the birds of a feather flock together". In society, individuals with similar interests are more likely to become friends [6, 16]. In the Web, web pages with related topics are often hyperlinked together [5]. In the protein interaction network, communities are composed of proteins with the same specific function for chemical reactions [3, 17]. In metabolic networks, communities may correspond to functional modules such as cycles and pathways [9]. In food webs, compartments can be viewed as communities [12, 19]. Hence, the community becomes the entry point of researches of

© Springer International Publishing Switzerland 2015
D.-S. Huang et al. (Eds.): ICIC 2015, Part I, LNCS 9225, pp. 76–84, 2015.
DOI: 10.1007/978-3-319-22180-9_8

networks structure and functionality. Community detection is a fundamental research issue and attracts much interest over the last decade.

Community detection is to recognize the inherent structure of networks, i.e., dividing a network into several communities which have high density of edges within communities and low density between them. Nowadays, the most often used method is the modularity optimization-based community detection approach. Precise formulations of this optimization problem are known to be computationally intractable.

Several algorithms have therefore been proposed to find reasonably good partitions efficiently. The first algorithm devised to maximize modularity was a greedy method proposed by Newman [4]. It is an agglomerative hierarchical clustering method, where groups of vertices are successively joined to form larger communities such that modularity increases after the merging. A different greedy approach has been introduced by Blondel [1], for the general case of weighted graphs, which is the best algorithm that can be used in large networks. We take the two methods for comparison on different kinds of sizes of networks.

Besides, we make use of the PageRank algorithm [10] which is applied widely in community kernel detection to evaluate the importance of nodes. PageRank is a link analysis algorithm and it assigns a numerical weighting to each element of a hyperlinked set of documents, such as the World Wide Web, with the purpose of "measuring" its relative importance within the set. As the structure of the World Wide Web is very similar to the structure of network, in which we can regard the nodes as the web pages and the edges as the hyperlink.

Our new algorithm, which we refer to as a smart local moving algorithm, takes advantage of both local moving heuristic and PageRank algorithm. Furthermore, the experimental result verifies the superior performance in modularity and computational efficiency in compare with the lovian algorithm.

The remainder of this paper is organized as follows. In Sect. 2, we present our new algorithm. We analyze the result of community kernel detection and compare the performance of the lovian algorithm with the fastgreedy algorithm in Sect. 3. We first consider small and medium-sized networks, and then focus on large networks. We summarize the conclusions of our research in Sect. 4.

2 Algorithm

Before we introduce our algorithm, we should make a detailed introduction on two crucial concepts, namely PageRank and modularity.

2.1 PageRank

PageRank [10] is a probability distribution used to represent the likelihood that a person would randomly visit a particular webpage. The idea is to imagine a random web surfer visiting a page and randomly clicking links to visit other pages then randomly going to a new page and repeating the process. The original PageRank of page A can be expressed as:

$$PR(A) = \frac{1-d}{n} + d \sum_{a \in W_b} \frac{PR(a)}{L(a)} \qquad (1)$$

Where n is the total number of pages in the system and d is the dampening factor that has been tried and tested in numerous studies happens to be about 0.85. W_b is the set of pages connected to page A, $PR(A)$ is the *PageRank(A)* and $L(a)$ is the number of outbound links on page A.

But the difference from the original PageRank in web page is that the network we study here is undirected. So, we have to change the formula to:

$$PR(i) = \frac{1-d}{N} + d \sum_{j \neq i} PR(j) \frac{w_{ij}}{\sum_{k \in adj[i]} w_{ik}} \qquad (2)$$

Where N denotes the number of nodes, $adj[i]$ denotes the set of neighbors of i, and W_{ij} denotes the weight of edge ij.

2.2   Modularity

The modularity function [4] of Newman and Girvan is based on the idea that a random graph is not expected to have a cluster structure, thus the possible existence of clusters is revealed by the comparison between the actual density of edges in a subgraph and the density one would expect in the subgraph if the vertices of the graph were attached regardless of community structure. So it is written as:

$$Q = \frac{1}{2m} \sum_{ij} (A_{ij} - \frac{k_i k_j}{2m}) \delta(c_i, c_j) \qquad (3)$$

Where c_i denotes the community to which node i has been assigned; A_{ij} denotes whether there is an edge between nodes i and j $(A_{ij} = 1)$ or not $(A_{ij} = 0)$; $k_i = \sum_j A_{ij}$ denotes the degree of node i, and m $= 1/2 \sum_{ij} A_{ij}$ denotes the total number of edges in the network. The function $\delta(c_i, c_j)$ indicates whether nodes i and j belong to the same community, which equals 1 if $c_i = c_j$ and 0 otherwise.

The gain in modularity $\varDelta Q$ obtained by moving an isolated node i into a community C can be easily computed by

$$\varDelta Q = [\frac{\sum_{in} + k_{i,in}}{2m} - (\frac{\sum_{tot} + k_i}{2m})^2] - [\frac{\sum_{in}}{2m} - (\frac{\sum_{tot}}{2m})^2 - (\frac{k_i}{2m})^2] \qquad (4)$$

Where $\sum_{in}$ is the sum of the weights of the links inside C, $\sum_{tot}$ is the sum of the weights of the links incident to nodes in C, and $k_{i,in}$ is the sum of the weights of the links from i to nodes in C.

2.3   Algorithm Description

The main steps of the algorithm named LPR(Local PageRank) are shown in Algorithm 1:

Input: $G = (V, E)$: The edges of the network
Output: c: Final assignment of nodes to communities;

// assign the initial communities to nodes
$c \leftarrow Initial(G)$

//compute the PageRank of each node

$pageRankValue \leftarrow PageRank(G)$
//sort the nodes in reversed order by pageRankValue
$nodesInOrder \leftarrow Sort(pageRankValue, G)$

// Run the local moving heuristic.
$c \leftarrow LocalMove(G, nodesInOrder)$

// Construct a new network.
$G_{new} \leftarrow GetNewNetwork(G, c)$

if $Modularity(G) < Modularity(G_{new})$ **then**
   | $c_{new} \leftarrow LR(G_{new})$
end

Algorithm 1. The main steps of the LPR algorithm

Our algorithm consists of three phases that are repeated iteratively. Assume that we start with a weighted network of N nodes. Initially, we assign a different community to each node of the network, so each node in a network is assigned to its own singleton community.

Firstly, we calculate the PageRank value in which assigning initial PageRank value of each node with one, and recalculating each value according to the Eq. 6 until each PageRank value does not change any more. The PageRank value is not only used as the measurement to evaluate the importance of each node, but also used to determine the order in which nodes are chosen in the second phase.

In the second phase, we take out the node i in the reversed order according to the PageRank value calculated in first phase. Then, considering the neighbours j of node i, we evaluate the gain of modularity that would take place by removing i from its current community and by placing it in the community of j. If the maximal gain is positive, the node i is then placed in the community for which this gain, otherwise, node i stays in its original community. This process is applied repeatedly and sequentially for all nodes until a local maxima of the modularity is attained and the phase is then complete.

In the last phase of the algorithm, we rebuild a new network whose nodes are now the communities found during the second phase. To do so, the weights of the links between the new nodes are given by the sum of the weight of the links between nodes in the corresponding two communities. Links between nodes of the same community lead to self-loops for this community in the new network. After the last phase is completed, it is then possible to reapply the first two phases of the algorithm to the resulting weighted network and to iterate.

3 Experiments and Results

In this section, we study the performance of our LPR algorithm in contrast to the lovian algorithm and the fastgreedy algorithm. To quantitatively evaluate our algorithms, we take the modularity Q as the measurement and compare the computational time. Empirically, higher values of the Q function have been shown to correlate well with better graph clusterings [18]. In addition, we apply the LPR algorithm to the karate club network.

3.1 Data Set

We have selected ten small and medium-sized networks and three large networks commonly used in community detection, originating from a number of different domains. Although the real system is more complex, most directed networks can be transformed to undirected networks. Therefore all networks considered are undirected, shown in Tables 1 and 2.

Table 1. Number of nodes and edges of ten small and medium-sized networks.

Network	Nodes	Edges
Karate club [21]	34	78
Les Miserables [11]	77	254
Football [7]	115	613
Jazz [8]	198	2,742
Ego-Facebook [15]	4,039	88,234
Ca-GrQc [13]	5,242	14,496
PGP [2]	10,680	24316
Ca-AstroPh [2]	18,772	198,110
Condmat2003 [16]	27,519	116,181
Email-enron [14]	36,692	183,831

Table 2. Number of nodes and edges of three large networks

Network	Nodes	Edges
com-DBLP [20]	317,080	1,049,866
com-amazon [20]	334,863	925,872
com-LiveJournal [20]	3,997,962	34,681,189

3.2 Result Analyses

Result of Community Kernel Detection. In this section, we adopt the PageRank algorithm independently which is the first step in our algorithm to detect the community kernel in the karate club network. We show the original karate club network in Fig. 1

and the PageRank values in Fig. 2. We could get some information from Fig. 2 that the PageRank values of node 0 and node 33 are the greatest and obviously much higher than other's which is consistent with the fact that at some point, a conflict between the club president (indicated by node 33) and the instructor (indicated by node 0) led to the fission of the club into two separate groups, supporting the instructor and the president respectively.

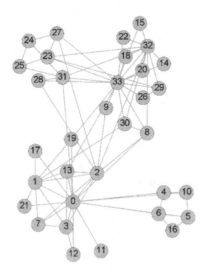

Fig. 1. The original network of karate club

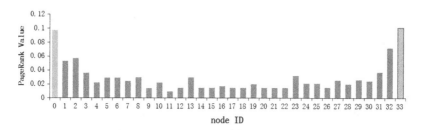

Fig. 2. The PageRank values of karate club

Quantitative Performance. We can get modularity results when applying all algorithms to each network, shown in the Tables 3 and 4. It indicates that the modularity of our algorithm is always higher than the others two in all network data source. Especially comparing with the fastgreedy algorithm, our algorithm is obvious higher, but our algorithm gets slightly higher modularity when compared to lovian algorithm. However, as we can see from the Table 5, our algorithm has an advantage in computational time when applying in large networks, and especially in the com-LiveJournal network, our algorithm gets a 25 % reduction in comparison with the lovian algorithm.

Table 3. Results for 10 small and medium-sized networks.

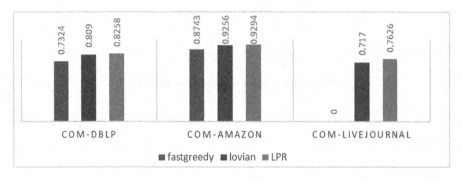

Table 4. Results for 3 large networks. For the com-LiveJournal network, the result of fastgreedy is not available because the computational complexity of those algorithm is too high to get a result in reasonabletime.

Table 5. The time each algorithm takes in three large networks.

Network	LPR	fastgreedy	lovian
com-DBLP	20.33(s)	4896.62(s)	20.85(s)
com-amazon	10.66(s)	1545.24(s)	11.73(s)
com-LiveJournal	4870(s)	–	6520(s)

Application Case Study. From our experiment, we know that the modularity of the karate club network is 0.4198, which is the best compared to the other algorithms'. We show the original karate club network in Fig. 1 and the division results in Fig. 3. The division result of our algorithm is the same as the famous lovian algorithm's that splits the network into four parts.

Fig. 3. The application of our algorithm to the karate club network

4 Conclusion

In this paper, we have introduced the LPR algorithm for modularity-based community detection. Our algorithm is intended primarily for community detection in large networks, and is combined with the PageRank algorithm to evaluate the importance of the nodes. Compared with five other algorithms, our algorithm gets a better result in the modularity and performs well in the division result of karate club network.

In future work, we would like to investigate the effect of other methods of community detection using the seed set expansion. Also, it is very interesting to detect community using the distributed computation method when dealing with super large networks.

Acknowledgement. This work is partially supported by the National Natural Science Foundation of China (Nos. 11161140319, 91120001, 61271426), the Strategic Priority Research Program of the Chinese Academy of Sciences (Grant Nos. XDA06030100, XDA06030500), the National 863 Program (No. 2012AA012503) and the CAS Priority Deployment Project (No. KGZD-EW-103-2).

References

1. Blondel, V.D., Guillaume, J.-L., Lambiotte, R., Lefebvre, E.: Fast unfolding of communities in large networks. J. Stat. Mech. Theor. Exp. **2008**, 10008 (2008). http://iopscience.iop.org/11742-15468/12008/10010/P10008
2. Boguñá, M., Pastor-Satorras, R., Díaz-Guilera, A., Arenas, A.: Models of social networks based on social distance attachment. Phys. Rev. E **70**, 056122 (2004). http://journals.aps.org/pre/abstract/056110.051103/PhysRevE.056170.056122
3. Chen, J., Yuan, B.: Detecting functional modules in the yeast protein–protein interaction network. Bioinformatics **22**, 2283–2290 (2006). http://bioinformatics.oxfordjournals.org/content/2222/2218/2283.short

4. Clauset, A., Newman, M.E.J., Moore, C.: Finding community structure in very large networks. Phys. Rev. E **70**, 066111 (2004). http://journals.aps.org/pre/abstract/066110. 061103/PhysRevE.066170.066111

5. Flake, G.W., Lawrence, S., Giles, C.L., Coetzee, F.M.: Self-organization and identification of web communities. Computer **35**, 66–70 (2002). http://ieeexplore.ieee.org/xpls/abs_all. jsp?arnumber=989932

6. Freeman, L.: The Development Of Social Network Analysis: A Study in the Sociology of Science. Empirical Press, Vancouver (2004). http://www.researchgate.net/profile/Linton_ Freeman/publication/239228599_The_Development_of_Social_Network_Analysis/links/ 54415c650cf2e6f0c0f616a8.pdf

7. Girvan, M., Newman, M.E.J.: Community structure in social and biological networks. Proc. Nat. Acad. Sci. **99**(12), 7821–7826 (2002). http://www.pnas.org/content/7899/7812/7821. short

8. Gleiser, P.M., Danon, L.: Community structure in jazz. Advances in complex systems **6**(4), 565–573 (2003). http://www.worldscientific.com/doi/abs/510.1142/S0219525903001067

9. Guimera, R., Amaral, L.A.N.: Functional cartography of complex metabolic networks. Nature **433**, 895–900 (2005). http://www.nature.com/articles/nature03288

10. Haveliwala, T.: Efficient computation of PageRank (1999)

11. Knuth, D.E.: The Stanford GraphBase: a platform for combinatorial computing. Addison-Wesley, Reading (1993). http://tex.loria.fr/sgb/abstract.pdf

12. Krawczyk, M.J.: Differential equations as a tool for community identification. Phys. Rev. E **77**, 065701 (2008). http://journals.aps.org/pre/abstract/065710.061103/PhysRevE.065777. 065701

13. Leskovec, J., Kleinberg, J., Faloutsos, C.: Graph evolution: densification and shrinking diameters. ACM Trans. Knowl. Disc. Data (TKDD) **1**(1), 2 (2007). http://dl.acm.org/ citation.cfm?id=1217301

14. Leskovec, J., Lang, K.J., Dasgupta, A., Mahoney, M.W.: Community structure in large networks: natural cluster sizes and the absence of large well-defined clusters. Internet Math. **6**(1), 29–123 (2009). http://www.tandfonline.com/doi/abs/110.1080/15427951.15422009. 10129177

15. Leskovec, J., McAuley, J.J.: Learning to discover social circles in ego networks. Advances in neural information processing systems **25**, 539–547 (2012). http://papers.nips.cc/paper/ 4532-learning-to-discover-social-circles-in-ego-networks

16. Newman, M.E.J.: The structure of scientific collaboration networks. Proc. Natl. Acad. Sci. **98**(?), 404–409 (2001). http://www.pnas.org/content/498/402/404.short

17. Rives, A.W., Galitski, T.: Modular organization of cellular networks. Proc. Nat. Acad. Sci. **100**(3), 1128–1133 (2003). http://www.pnas.org/content/1100/1123/1128.short

18. White, S., Smyth, P.: A Spectral Clustering Approach To Finding Communities in Graph. SDM. SIAM **5**, 76–84 (2005). http://epubs.siam.org/doi/abs/10.1137/1131.9781611972757. 9781611972725

19. Williams, R.J., Martinez, N.D.: Simple rules yield complex food webs. Nature **404**, 180–183 (2000). http://www.nature.com/articles/35004572

20. Yang, J., Leskovec, J.: Defining and evaluating network communities based on ground-truth. In: Proceedings of the ACM SIGKDD Workshop on Mining Data Semantics, p. 3. ACM (2012). http://dl.acm.org/citation.cfm?id=2350193

21. Zachary, W.W.: An information flow model for conflict and fission in small groups. J. Anthropol. Res. **33**(4), 452–473 (1977)

Self-adaptive Percolation Behavior Water Cycle Algorithm

Shilei Qiao[1], Yongquan Zhou[1,2(✉)], Rui Wang[1], and Yuxiang Zhou[1]

[1] College of Information Science and Engineering,
Guangxi University for Nationalities, Nanning 530006, China
yongquanzhou@126.com
[2] Guangxi High School Key Laboratory of Complex System
and Computational Intelligence, Nanning 530006, China

Abstract. Water cycle algorithm is a new meta-heuristic optimization algorithm based on the observation of water cycle and how rivers and streams flow downhill towards the sea in the real world. In this paper, a new self-adaptive water cycle algorithm with percolation behavior is proposed. The percolation behavior is introduced to accelerate the convergence speed of proposed algorithm. At the same time, a self-adaptive rainfall process can generate the new stream, more and more new position can be explored, consequently, increasing the diversity of population. Eight typical benchmark functions are tested, the simulation results show that the proposed algorithm is feasible and effective than basic water cycle algorithm, and demonstrate that this proposed algorithm has superior approximation capabilities in high-dimensional space.

Keywords: Water cycle algorithm · Percolation behavior · Rainfall · Self-adaptive water cycle algorithm · Function optimization

1 Introduction

Nowadays, due to the evolutionary algorithm can solve some problem that the traditional optimization algorithm cannot do easy, more and more modern meta-heuristic algorithms inspired by nature or social phenomenon are emerging and they become increasingly popular. For example, genetic algorithm (GA) [1] or particle swarm optimization (PSO) [2], and the physical annealing which is generally known as simulated annealing (SA) [3], glowworm swarm optimization (GSO) [4], harmony search (HS) [5], bacterial foraging optimization algorithm (BFOA) [6], and invasive weed optimization (IWO) [7], and so on [8, 9].

The water cycle algorithm (WCA) is proposed by Eskandar etc. (2012) [10]. Similar to other meta-heuristic algorithms, the proposed method begins with an initial population so called the raindrops. First, we assume that we have rain or precipitation. The best individual (best raindrop) is chosen as a sea. Then, a number of good raindrops are chosen as a river and the rest of the raindrops are considered as streams which flow to rivers and sea. In addition, rivers flow to the sea which is the most downhill location. This algorithm gradually aroused people's close attention, and which is increasingly applied to different area. Ail Sadollah etc. (2014) [11] proposed an

© Springer International Publishing Switzerland 2015
D.-S. Huang et al. (Eds.): ICIC 2015, Part I, LNCS 9225, pp. 85–96, 2015.
DOI: 10.1007/978-3-319-22180-9_9

improved WCA to solve weight optimization of truss structures problem. Ardeshir Bahreininejad etc. (2014) [12] apply WCA to solve multi-objective optimization problems. Zhang, Liu etc. (2014) [13] apply WCA to solve practical engineering optimization problems. Although the basic WCA have remarkable property compared against several traditional optimization methods, the phenomenon of slow convergence rate and low accuracy is still existed. Therefore, in this article, we present an improved self-adaptive percolation behavior water cycle algorithm (SPWCA); the purpose is to improve the convergence rate and precision of water cycle algorithm. At the end of this paper, we tested eight typical benchmark functions [14], the simulation results showed that the proposed algorithm is feasible and effective, which is more robust, also demonstrated the superior approximation capabilities in high-dimensional space.

The rest of this paper is organized as follows. In the Sect. 2, the basic WCA were described. In Sect. 3, a new percolation behavior and self-adaptive process of rainfall is introduced, at the same time, we gave the process of SPWCA. The numerical simulation experiment and comparison of improved algorithm are presented in Sect. 4. Finally, we concluded this paper in Sect. 5.

2   The Background and Concept of Basic WCA Algorithms

The idea of the WCA is inspired from nature and base on the observation of water cycle and how rivers and streams flow downhill towards the sea in the real world. First, we assume that we have rain or precipitation. The best individual (best raindrop) is chosen as a sea. Then, a number of good raindrops are chosen as a river and the rest of the raindrops are considered as streams which flow to the rivers and sea.

As in nature, the streams are created from the raindrops and join each other to form new rivers. All rivers and streams end up in sea (best optimal point). Figure 1 shows the schematic view of stream's flow towards a specific river. As shown in Fig. 1, star and circle represent river and stream, respectively.

In Fig. 1, a stream flows to the river along the connecting line between them using a randomly chosen distance given as follow:

$$X \in (0, C \times d) \qquad C > 1 \tag{1}$$

where C is a value between 1 and 2 (near to 2). The best value for C may be chosen as 2. The current distance between stream and river is represented as d. The value of X in Eq. (1) corresponds to a distributed random number (uniformly or may be any appropriate distribution) between 0 and ($C \times d$).

The value of C being greater than one enables streams to flow in different directions towards the rivers. This concept may also be used in flowing rivers to the sea. Therefore, the new position for streams and rivers may be given as:

$$X_{Stream}^{i+1} = X_{Stream}^i + rand \times C \times (X_{River}^i - X_{Stream}^i) \tag{2}$$

$$X_{River}^{i+1} = X_{River}^i + rand \times C \times (X_{Sea}^i - X_{River}^i) \tag{3}$$

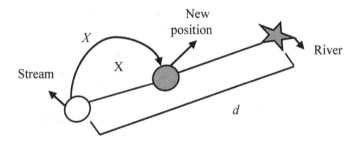

Fig. 1. Schematic view of stream's flow to a specific river

where rand is a uniformly distributed random number between 0 and 1. If the solution given by a stream is better than its connecting river, the positions of river and stream are exchanged (i.e. stream becomes river and river becomes stream). Such exchange can similarly happen for rivers and sea.

Rainfall process is one of the most important factors that can prevent the algorithm from rapid convergence (immature convergence). In the WCA, this assumption is proposed in order to avoid getting trapped in local optima. In the raining process, the new raindrops form streams in the different locations. For specifying the new locations of the newly formed streams, the following equation is used:

$$X_{Stream}^{new} = LB + rand \times (UB - LB) \tag{4}$$

Where LB and UB are lower and upper bounds defined by the given problem, respectively.

The schematic view of the WCA is illustrated in Fig. 2 where circle, stars, and the diamond correspond to streams, rivers, and sea, respectively. From Fig. 2, the white (empty) shapes refer to the new positions found by streams and rivers. Here, Fig. 2 is an extension of Fig. 1.

3   Percolation Behavior and Self-adaptive Process of Rainfall

3.1   Percolation Behavior

The WCA algorithm uses the whole update and evaluation strategy on solutions. For solving multidimensional function optimization problems, this strategy may deteriorate the convergence speed and the quality of solution due to interference phenomena among dimensions, to overcome this shortage, we uses the percolation behavior that dimension by dimension to update the position of the stream. The percolation behavior of dimension by dimension can be expressed in the following:

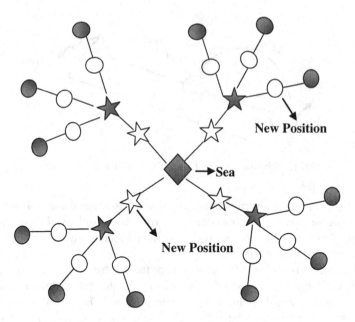

Fig. 2. Schematic view of WCA processes

for v = 1:d
    temp(i,:) = stream(i,:);
      select a stream randomly stream(k,:)
    temp(i,v) = temp(i,v) + rand * (stream(k,v) − stream(i,v));
    Generate new solutions by change the dimension
      if fitness(temp(i,v)) < fitness(stream(i,v))
            stream(i,:) = temp(i,:);
        Accept the new solutions
      endif
endfor
stream = temp;

Pseudocode1: percolation behavior

So the improved update strategy may be given as:

$$X_{Stream}^{i+1} = \begin{cases} X_{Flow}^{i}; & \text{if } fitness(X_{Flow}^{i}) < fitness(X_{Stream}^{i}) \\ X_{Percolation}^{i}; & \text{else by the percolation operator} \end{cases} \tag{5}$$

where $X_{Flow}^{i} = X_{Stream}^{i} + rand \times C \times (X_{River}^{i} - X_{Stream}^{i})$.

3.2   Self-adaptive Process of Rainfall

In order to prevent the algorithm trapped in local optimal value, we proposed a new strategy of rainfall to replace the old. If the position of new streams didn't update, we will think the stream trapped in local optimal, and we recorded. At the end each of iteration, we will check which stream up to the maximum limit. At the same time, we will update these streams by the process of rainfall. In addition, the process of rainfall is based on the simplex method [15, 16] (Fig. 3):

The rainfall process based on simplex method is described as below:

Step 1:   Find the best point Xg (sea) and the second best point Xb in the population the bad point (the stream will be replaced) Xs, their fitness values are recorded as $f(Xg)$, $f(Xb)$, $f(Xs)$ respectively.

Step 2:   Calculate the middle position of Xg and Xb:

$$Xc = \frac{Xg + Xb}{2} \tag{6}$$

Step 3:   Perform reflection operations and get the reflection point Xr, the reflection coefficient α usually take 1:

$$Xr = Xc + \alpha(Xc - Xs) \tag{7}$$

Step 4:   If $f(Xr) < f(Xg)$ it prove that reflected in the right direction, and we will expand the point, the expand coefficient usually take 2:

$$Xe = Xc + \gamma(Xr - Xc) \tag{8}$$

If $f(Xe) < f(Xg)$, we will replace Xs into Xe, else replace Xs into Xr.

Step 5:   If $f(Xr) > f(Xs)$ it proved that the reflection operation is wrong, we perform the condense operations, and get the condense point Xt, the condense coefficient β usually take 0.5:

$$Xt = Xc + \beta(Xs - Xc) \tag{9}$$

If $f(Xt) < f(Xs)$, we will replace Xs into Xt, else replace Xs into Xr.

Step 6:   If $f(Xg) < f(Xr) < f(Xs)$ we perform the shrink operations, and get the condense point Xw, the shrink coefficient take β:

$$Xw = Xc - \beta(Xs - Xc) \tag{10}$$

If $f(Xw) < f(Xs)$, we will replace Xs into Xw, else replace Xs into Xr.

Through the above process, we can get a better raindrop, and these raindrops return to the ground to update the streams and rivers. Experiments prove that these methods above greatly increased the diversity of population.

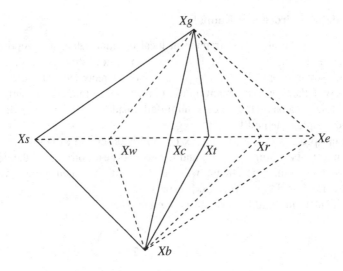

Fig. 3. Schematic view of simplex method

3.3 The Basic Process of SPWCA

Based on the above content, the steps of SPWCA are summarized as follows:

Step 1:   Choose the initial parameters of the SPWCA the number of rivers and maximum running times etc.

Step 2:   Generate random initial population and form the initial streams rivers, and sea.

Step 3:   The streams flow to the rivers. If the position of streams has not been updated $limit(i) = limit(i) + 1$, turn to the step4 otherwise turn to the step 5.

Step 4:   Implement percolation operator update the position of stream.

Step 5:   Exchange positions of river with a stream which gives the best solution.

Step 6:   The river flow to the sea which is the most downhill place.

Step 7:   Similar to Step 5 if a river finds better solution than the sea, the position of river is exchanged with the sea.

Step 8:   Check the matrix of *limit* find all elements that satisfy the rainfall condition $(limit(i) > 5)$.
Implement the process of rainfall.

Step 9:   Running times add 1 if the number of running is less than the pre-specified number of times, turn to Step 3.

Step 10:   End.

4   Numerical Simulation Experiments

In this section, we applied 8 standard test functions to evaluate the optimal performance of the SPWCA algorithm. The mean and standard deviation results of 25 independent runs for each algorithm have been summarized in Tables 2, 3.

Table 1. Benchmark functions based in our experimental study

Benchmark Test Functions	D	Range	Optimal	Iterations
$f_{01} = \sum_{i=1}^{D} x_i^2$	1000	[-100,100]	0	600
$f_{02} = \sum_{i=1}^{D} \|x_i\| + \prod_{i=1}^{D} \|x_i\|$	30	[-10,10]	0	1200
$f_{03} = \sum_{i=1}^{D} \left(\sum_{j=1}^{i} x_j \right)^2$	30	[-100,100]	0	800
$f_{04} = \max\{\|x_i\|, 1 \le i \le D\}$	30	[-100,100]	0	1200
$f_{05} = \sum_{i=1}^{D} [x_i^2 - 10\cos 2\pi x_i + 10]$	30	[-5.12,5.12]	0	100
$f_{06} = -20\exp\left(-0.2\sqrt{\dfrac{1}{D}\sum_{i=1}^{D} x_i^2}\right) - \exp\left(\dfrac{1}{D}\sum_{i=1}^{D}\cos 2\pi x_i\right) + 20 + e$	30	[-32,32]	0	100
$f_{07} = \dfrac{1}{4000}\sum_{i=1}^{D} x_i^2 - \prod_{i=1}^{D}\cos\left(\dfrac{x_i}{\sqrt{i}}\right) + 1$	30	[-600,600]	0	50
$f_{08} = \dfrac{\pi}{D}\left\{ 10\sin^2(\pi y_1) + \sum_{i=1}^{D-1}(y_i - 1)^2\left[1 + \sin^2(\pi y_{y+1})\right] + (y_D - 1)^2 \right\}$ $+ \sum_{i=1}^{D} u(x_i, 10, 100, 4)$ $y_i = 1 + \dfrac{x_i + 1}{4}$ $u(x_i, a, k, m) = \begin{cases} k(x_i - a)^m & x_i > a \\ 0 & -a \le x_i \le a \\ k(-x_i - z)^m & x_i < a \end{cases}$	30	[-50,50]	0	1000

The 8 standards benchmark functions have been widely used in the literature. The space dimension, scope of the variables, maximum number of iterations and the optimal value of 8 functions are in Table 1. Table 1 has shown the details of these functions.

4.1 Experimental Setup

The entire algorithm was programmed in MATLAB R2012a, numerical experiment was set up on AMD Athlon™Π*4 640 processor and 2 GB memory.

4.2 Parameters Setting

The proposed SPWCA algorithm compared with swarm intelligence algorithms DE [17], CS [18], GSO [4], ABC [21], HGAPSO [19], FSFLA [20], WCA [10], respectively using the mean and standard deviation to compare their optimal performance. The setting values of algorithm control parameters of the mentioned algorithms as follows.

DE parameters setting: $F = 0.5$ and $CR = 0.9$ in accordance with the suggestions given in [17], the population size is 50; CS parameters setting: $\beta = 1.5$, $\rho_0 = 1.5$ have been used as recommended in [18], the population size is 50; GSO parameters setting: $G_0 = 100$, $\alpha = 20$, K_0 is set to NP and is decreased linearly to 1 have been used as recommended in [4], the population size is 50; ABC parameters setting: $Limit = 5D$ has been used as recommended in [21], the population size is 50; HGAPSO parameters setting: $\omega = 0.6$, $c_1 = c_2 = 2$, $Kp = 0.5$, $Cp = 0.7$ has been used as recommended in [12] the population size is 50; FSFLA parameters setting: $m = 10$, $n = 5$, $Ne = 15$ has been used as recommended in [20] the population size is 50; WCA parameters setting: $Nsr = 8$ (the number of the river), $C = 2$ has been used as recommended in [10] the population size is 50; SPWCA parameters setting: $Nsr = 8$, $C = 2$ the population size is 50.

In order to show the performance of our proposed SPWCA, we compare it with the DE, CS, GSO, ABC, HGAPSO, FSFLA and WCA. In the experiment, the mean results of 25 independent runs for f_{01} - f_{04} are summarized in Table 2. Functions f_{05} - f_{08} summarized in Table 3.

From the Table 2, we can conclude that SPWCA provides the most of the better results than WCA and other algorithms. For f_{01}, f_{02}, f_{03} and f_{04} the mean and standard deviation of SPWCA are much higher than WCA. Figures 4, 5 shows the graphical analysis results of ANOVA tests. As can be seen in Fig. 5, when solving function f_{03}, most of the algorithms can obtain the stable optimal value except HGAPSO and FSFLA. Figures 6, 7 show the fitness function curve evolution of each algorithm for the functions of f_{01} and f_{03}. From the two figures, we can conclude that SPWCA has the faster convergence rate and the higher optimizing precision. These show that SPWCA has a certain degree of improvement than WCA in solving multidimensional functions.

As can be seen in Table 3, for f_{05} and f_{07} SPWCA all achieved the optimal value and the standard deviations are all 0. For f_{06}, the mean and standard deviation of SPWCA are 18 orders of magnitude higher than WCA and 19 orders of magnitude higher than DE. For f_{08}, the mean and standard deviation of SPWCA are 30 orders of

Table 2. Experiment results of benchmark functions f_{01} - f_{04}

Functions		f_{01}	f_{02}	f_{03}	f_{04}
DE	Mean	2.0453E + 04	0.3373	4.7215	2.0452E-05
	Std	1.8407E + 03	1.0385	21.1152	3.2507E-05
CS	Mean	1.8071E + 03	1.8707E-13	3.2552E-06	2.0452E-05
	Std	267.5824	2.4805E-13	3.7924E-16	3.2507E-05
GSO	Mean	79.7846	1.6049E-09	2.6473E-07	7.1236E-07
	Std	26.4423	2.5760E-09	6.2457E-07	5.5270E-07
ABC	Mean	635.2784	4.0778E-17	0.1470	2.7321E-12
	Std	1.1039E + 03	1.3006E-17	0.2246	5.6446E-12
HGAPSO	Mean	1.8421E + 05	5.3668E-11	1.3576E + 02	5.1106
	Std	2.5281E + 04	2.1425E-10	2.2672E + 02	6.8890
FSFLA	Mean	1.7972E + 04	2.9704	2.3978E + 02	16.7551
	Std	2.5645E + 03	1.0946	1.7099E + 02	4.5902
WCA	Mean	1.3867E + 03	1.1329E-05	0.9035	0.0167
	Std	333.4621	7.1920E-06	1.0129	0.0131
SPWCA	Mean	0	0	0	0
	Std	0	0	0	0

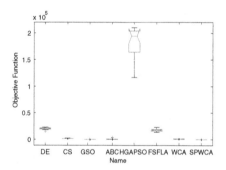

Fig. 4. ANOVA tests for f_{01} **Fig. 5.** ANOVA tests for f_{03}

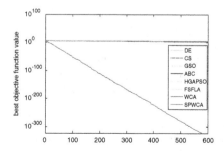
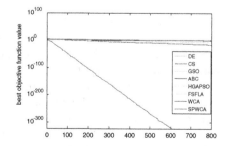

Fig. 6. Fitness function curve evolution for f_{01} **Fig. 7.** Fitness function curve evolution for f_{03}

Table 3. Experiment results of benchmark functions f_{05} - f_{08}

Functions		f_{07}	f_{08}	f_{09}	f_{10}
DE	Mean	1.5747E + 02	12.9680	35.8180	7.7655E-10
	Std	18.2073	1.4750	15.4353	2.8625E-09
CS	Mean	4.8305	9.1661	3.4776	7.3152E-06
	Std	1.1212	1.1434	1.5388	1.0132E-05
GSO	Mean	0.7483	0.3168	0.2650	4.1341E-12
	Std	0.7830	0.4519	0.1553	1.000E-11
ABC	Mean	0.2833	3.3083E-06	0.3019	1.4745E-16
	Std	0.4385	5.4202E-06	0.1083	9.8500E-17
HGAPSO	Mean	14.6217	0.2539	0.9929	7.2701E + 03
	Std	9.6010	0.7933	1.0021	2.3052E + 04
FSFLA	Mean	8.3609	8.1650	36.6720	2.1370E + 03
	Std	2.8872	1.4557	19.9110	4.4197E + 03
WCA	Mean	35.4568	4.7059	14.6001	0.4782
	Std	14.0251	2.1225	10.9864	1.3394
SPWCA	Mean	0	8.8818E-16	0	2.8507E-31
	Std	0	0	0	9.4702E-31

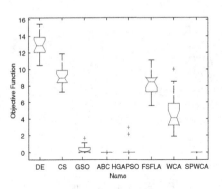

Fig. 8. ANOVA tests for f_{06} **Fig. 9.** ANOVA tests for f_{08}

magnitude higher than WCA. Figures 8 and 9 show the graphical analysis results of the ANOVA test. Figure 8 shows the ABC and SPWCA can obtain the stable optimal values. Figure 9 shows when solving function f_{08}, most of the algorithms can obtain the stable optimal value except HGAPSO and FSFLA. Figures 10 and 11 show the fitness function curve evolution. For the two Figures, we can conclude that SPWCA has the faster convergence rate and the higher optimizing precision.

In previous sections, 8 standard test functions are applied to evaluate the optimal performances of the SPWCA in the case of low dimension. In order to evaluate the performances of SPWCA comprehensively, we also do some high-dimensional tests in f_{01}, f_{02}, f_{04}, f_{07}. Table 4 is the experimental result performs 25 times independently

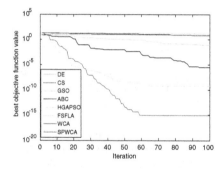 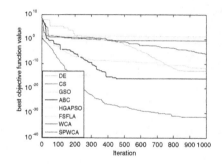

Fig. 10. Fitness function curve evolution for f_{06} **Fig. 11.** Fitness function curve evolution for f_{08}

Table 4. High-dimensional functions test results

Functions	Dimensions	Means	Std	Iterations
f_{01}	2000	0	0	1000
f_{02}	500	0	0	1500
f_{04}	1500	0	0	1500
f_{07}	500	0	0	150

under the high-dimensional situation. As can be seen in Table 4, SPWCA not only has superior approximation ability in low-dimensional space, but also has excellent global search ability in high-dimensional situation.

5   Discussion and Conclusion

In this paper, we present an improved water cycle algorithm with percolation behavior and self-adaptive process of rainfall. Firstly, we define the percolation behavior when the stream cannot flow to the river (the position of stream cannot update), it is more graphic to simulate the process of water cycle in the real world. Moreover, it can make the algorithm effectively jump out of the local optimum. Secondly, the self-adaptive process of rainfall can generate the new stream, more and more new position can be explored, consequently, increasing the diversity of population. By eight typical standard test functions simulation, the result show that SPWCA algorithm generally has strong global searching ability and local optimization ability, and can effectively avoid the deficiencies that conventional algorithms are easily fall into local optimum. Besides, it also has excellent global search ability in high-dimensional situation. Therefore, it is very practical and effective to solve complex functions optimization problems

Acknowledgments. This work is supported by National Science Foundation of China under Grants No.61165015; 61463007.

References

1. Cocke, T., Moscicki, Z., Agarwal, R.: Optimization of hydrofoils using a genetic algorithm. J. Aircraft **51**, 78–89 (2014)
2. Kennedy, J., Eberhart, R.: Particle swarm optimization. In: Proceedings of the IEEE International Conference on Neural Networks 4, pp. 1942–1948 (1995)
3. Kirkpatrick Jr., S., Gelatt, C.D., Vecchi, M.P.: Optimization by simulated annealing. Science **220**, 650–671 (1983)
4. Yongquan, Z., Jiakun, L., Guangwei, Z.: Leader glowworm swarm optimization algorithm for solving nonlinear equations systems. Przeglad Elektrotechniczny **88**, 101–106 (2012)
5. Alatas, B.: Chaotic harmony search algorithms. Appl. Math. Comput. **216**, 2687–2699 (2010)
6. Passino, K.M.: Biomimicry of bacterial foraging for distributed optimization and control. IEEE Control Syst. Mag. **22**, 52–68 (2002)
7. Mehrabian, A.R., Lucas, C.: A novel numerical optimization algorithm inspired from weed colonization. Ecol. Inform. **1**, 355–366 (2006)
8. Yang, X.-S.: A new metaheuristic Bat-inspired algorithm. Studies Comput. Intell. **284**, 65–74 (2010)
9. Oftadeh, R., Mahjoob, M.J., Shariatpanahi, M.: A novel meta-heuristic optimization algorithm inspired by group hunting of animals: hunting search. Comput. Math. Appl. **60**(7), 2087–2099 (2010)
10. Eskandar, H., Sadollah, A., Bahreininejad, A., et al.: Water cycle algorithm-a novel metaheuristic optimization method for solving constrained engineering optimization problems. Comput. Struct. **1**, 151–166 (2012)
11. Eskandar, H., Sadollah, A., Bahreininejad, A.: Weight optimization of truss structures using water cycle algorithm. Int. J. Optim. Civil Eng. **3**, 115–129 (2013)
12. Sadollah, A., Eskandar, H.: Water cycle algorithm for solving multi-objective optimization problems. Appl. Soft Comput. **27**, 279–298 (2014)
13. Chun, Z., Wei, L.G., Chun, L.L.: Optimizations of space truss structures using WCA algorithm, vol. 16, pp. 35–38 (2014)
14. A. Hedar, function http://www-optima.amp.i.kyoto-u.ac.jp/member/student/hedar/Hedar files/TestGofiles/
15. Kitahara, T., Mizuno, S.: A bound for the number of different basic solutions generated by the simplex method **137**, 579–586 (2013)
16. Kheirfam, B., Verdegay, J.L.: The dual simplex method and sensitivity analysis for fuzzy linear programming with symmetric trapezoidal numbers. Fuzzy Optim. Decis. Making **12**, 171–189 (2013)
17. Das, S., Suganthan, P.N.: Differential evolution: A survey of the state-of-the-art. IEEE Trans. Evol. Comput. **15**, 4–31 (2011)
18. Yang, X.-S., Deb, S.: Cuckoo search: recent advances and applications. Neural Comput. Appl. **24**, 169–174 (2014)
19. Xing, Y., Wang, Y.: Assembly sequence planning based on a hybrid particle swarm optimization and genetic algorithm. Int. J. Prod. Res. **50**, 7303–7315 (2012)
20. Wang. L., Gong. Y.: A fast shuffled frog leaping algorithm. In: 2013 Ninth International Conference on Natural Computation (ICNC). IEEE, pp. 369–373 (2013)
21. Karaboga, D., Bastuk, B.: A powerful and efficient algorithm for numerical function optimization: artificial bee colony (ABC) algorithm. J. Global Optim. **39**, 459–477 (2007)

Oscillatory Behavior in An Inertial Six-Neuron Network Model with Delays

Chunhua Feng[1] and Zhenkun Huang[2(✉)]

[1] College of Mathematical Science, Guangxi Normal University,
Guilin 541004, Guangxi, People's Republic of China
[2] School of Science, Jimei University,
Xiamen 361021, Fujian, People's Republic of China
hzk974226@jmu.edu.cn

Abstract. This paper discusses the existence of oscillatory solutions in an inertial six neurons BAM neural network model with delays. By means of Chafee's criterion of limit cycle, some sufficient conditions to ensure the existence of oscillatory solutions for this delayed system are provided. Computer simulations verify the correctness of the results.

Keywords: Inertial BAM neural network model · Equilibrium · Delay · Oscillatory solution

1 Introduction

In the last few decades, the dynamics properties of delayed bidirectional associative memory (BAM) models have been widely studied by many researchers, and various interesting results have been reported [1–15]. In [8], Zhang et al. have considered the following symmetric BAM neural network model with one delay:

$$
\begin{cases}
x_1'(t) = -ax_1(t) + f(x_4(t-\tau)) + g(x_5(t-\tau)) + g(x_6(t-\tau)) \\
x_2'(t) = -ax_2(t) + f(x_5(t-\tau)) + g(x_6(t-\tau)) + g(x_4(t-\tau)) \\
x_3'(t) = -ax_3(t) + f(x_6(t-\tau)) + g(x_4(t-\tau)) + g(x_5(t-\tau)) \\
x_4'(t) = -ax_4(t) + f(x_1(t-\tau)) + g(x_2(t-\tau)) + g(x_3(t-\tau)) \\
x_5'(t) = -ax_5(t) + f(x_2(t-\tau)) + g(x_3(t-\tau)) + g(x_1(t-\tau)) \\
x_6'(t) = -ax_6(t) + f(x_3(t-\tau)) + g(x_1(t-\tau)) + g(x_2(t-\tau))
\end{cases}
\tag{1}
$$

For this symmetric BAM neural network model, the existence of multiple periodic solutions is established by using the symmetric Hopf bifurcation theory. Recently, Xu et al. have discussed the stability and bifurcation for the following six-neuron BAM neural network model [12]:

$$
\begin{cases}
x_1'(t) = -\mu_1 x_1(t) + c_{11}f_{11}(y_1(t-\tau_4)) + c_{12}f_{12}(y_2(t-\tau_4)) + c_{13}f_{13}(y_3(t-\tau_4)) \\
x_2'(t) = -\mu_2 x_2(t) + c_{21}f_{21}(y_1(t-\tau_5)) + c_{22}f_{22}(y_2(t-\tau_5)) + c_{23}f_{23}(y_3(t-\tau_5)) \\
x_3'(t) = -\mu_3 x_3(t) + c_{31}f_{31}(y_1(t-\tau_6)) + c_{32}f_{32}(y_2(t-\tau_6)) + c_{33}f_{33}(y_3(t-\tau_6)) \\
y_1'(t) = -\mu_4 y_1(t) + c_{41}f_{41}(x_1(t-\tau_1)) + c_{42}f_{42}(x_2(t-\tau_2)) + c_{43}f_{43}(x_3(t-\tau_3)) \\
y_2'(t) = -\mu_5 y_2(t) + c_{51}f_{51}(x_1(t-\tau_1)) + c_{52}f_{52}(x_2(t-\tau_2)) + c_{53}f_{53}(x_3(t-\tau_3)) \\
y_3'(t) = -\mu_6 y_3(t) + c_{61}f_{61}(x_1(t-\tau_1)) + c_{62}f_{62}(x_2(t-\tau_2)) + c_{63}f_{63}(x_3(t-\tau_3))
\end{cases}
\tag{2}
$$

© Springer International Publishing Switzerland 2015
D.-S. Huang et al. (Eds.): ICIC 2015, Part I, LNCS 9225, pp. 97–105, 2015.
DOI: 10.1007/978-3-319-22180-9_10

Assume that $\tau_1 + \tau_4 = \tau_2 + \tau_5 = \tau_3 + \tau_6 = \tau$, and make the change of variables as $u_1(t) = x_1(t - \tau_1), u_2(t) = x_2(t - \tau_2), u_3(t) = x_3(t - \tau_3),\ u_4(t) = y_1(t),\ u_5(t) = y_2(t),$ $u_6(t) = y_3(t)$. System (2) has been changed to have only one time delay model. By analyzing the associated characteristic transcendental equation, the linear stability of the model is considered and Hopf bifurcation is demonstrated. In [13], Ge and Xu have investigated the stability switches and fold-Hopf bifurcations in an inertial four-neuron network model with coupling delay as follows:

$$\begin{cases} v_1''(t) = -v_1'(t) - \mu_1 v_1(t) + a_1 f(v_3(t - \tau)) + a_2 f(v_4(t - \tau)) \\ v_2''(t) = -v_2'(t) - \mu_2 v_2(t) + b_1 f(v_3(t - \tau)) + b_2 f(v_4(t - \tau)) \\ v_3''(t) = -v_3'(t) - \mu_3 v_3(t) + c_1 f(v_1(t - \tau)) + c_2 f(v_2(t - \tau)) \\ v_4''(t) = -v_4'(t) - \mu_4 v_4(t) + d_1 f(v_1(t - \tau)) + d_2 f(v_2(t - \tau)) \end{cases} \tag{3}$$

By employing the extended perturbation-incremental scheme, Local stability for the trivial equilibrium is analyzed. Fold-Hopf bifurcations are completely analyzed in the parameter space of the coupling delay and the connection weights.

Motivated by the above models, in this paper we consider the following inertial six-neuron BAM network model with different time delays:

$$\begin{cases} v_1''(t) = -\alpha_1 v_1'(t) - \mu_1 v_1(t) + \tilde{c}_{11} f_{11}(v_4(t - \tilde{\tau}_4)) + \tilde{c}_{12} f_{12}(v_5(t - \tilde{\tau}_5)) + \tilde{c}_{13} f_{13}(v_6(t - \tilde{\tau}_6)) \\ v_2''(t) = -\alpha_2 v_2'(t) - \mu_2 v_2(t) + \tilde{c}_{21} f_{21}(v_4(t - \tilde{\tau}_4)) + \tilde{c}_{22} f_{22}(v_5(t - \tilde{\tau}_5)) + \tilde{c}_{23} f_{23}(v_6(t - \tilde{\tau}_6)) \\ v_3''(t) = -\alpha_3 v_3'(t) - \mu_3 v_3(t) + \tilde{c}_{31} f_{31}(v_4(t - \tilde{\tau}_4)) + \tilde{c}_{32} f_{32}(v_5(t - \tilde{\tau}_5)) + \tilde{c}_{33} f_{33}(v_6(t - \tilde{\tau}_6)) \\ v_4''(t) = -\alpha_4 v_4'(t) - \mu_4 v_4(t) + \tilde{c}_{41} f_{41}(v_1(t - \tilde{\tau}_1)) + \tilde{c}_{42} f_{42}(v_2(t - \tilde{\tau}_2)) + \tilde{c}_{43} f_{43}(v_3(t - \tilde{\tau}_3)) \\ v_5''(t) = -\alpha_5 v_5'(t) - \mu_5 v_5(t) + \tilde{c}_{51} f_{51}(v_1(t - \tilde{\tau}_1)) + \tilde{c}_{52} f_{52}(v_2(t - \tilde{\tau}_2)) + \tilde{c}_{53} f_{53}(v_3(t - \tilde{\tau}_3)) \\ v_6''(t) = -\alpha_6 v_6'(t) - \mu_6 v_6(t) + \tilde{c}_{61} f_{61}(v_1(t - \tilde{\tau}_1)) + \tilde{c}_{62} f_{62}(v_2(t - \tilde{\tau}_2)) + \tilde{c}_{63} f_{63}(v_3(t - \tilde{\tau}_3)) \end{cases}$$
$$\tag{4}$$

By using the Chafee's limit cycle theorem [16], some sufficient conditions are provided to guarantee the existence of oscillation solutions of system (4). It is well known that bifurcation can induce periodic solution. However, if the time delays $\tilde{\tau}_1, \tilde{\tau}_2, \ldots, \tilde{\tau}_6$ in system (4) are different each other, Xu's approach in [12] will be hard to deal with the bifurcation of system (4) because there is a complex associated characteristic transcendental equation of (4).

2   Preliminaries

For notational convenience, let $v_1(t) = x_1(t),\ v_1'(t) = x_2(t),\ v_2(t) = x_3(t),\ v_2'(t) = x_4(t),\ v_3(t) = x_5(t),\ v_3'(t) = x_6(t),\ v_4(t) = x_7(t),\ v_4'(t) = x_8(t),\ v_5(t) = x_9(t),\ v_5'(t) = x_{10}(t),\ v_6(t) = x_{11}(t),\ v_6'(t) = x_{12}(t)$. Then system (4) is equivalent to the following system:

$$
\left\{
\begin{aligned}
x_1'(t) &= x_2(t),\\
x_2'(t) &= -a_1 x_1(t) - b_2 x_2(t) + \tilde{c}_{11} f_{11}(x_7(t - \tau_7)) + \tilde{c}_{12} f_{12}(x_9(t - \tau_9)) + \tilde{c}_{13} f_{13}(x_{11}(t - \tau_{11}))\\
x_3'(t) &= x_4(t)\\
x_4'(t) &= -a_3 x_3(t) - b_4 x_4(t) + \tilde{c}_{21} f_{21}(x_7(t - \tau_7)) + \tilde{c}_{22} f_{22}(x_9(t - \tau_9)) + \tilde{c}_{23} f_{23}(x_{11}(t - \tau_{11}))\\
x_5'(t) &= x_6(t)\\
x_6'(t) &= -a_5 x_5(t) - b_6 x_6(t) + \tilde{c}_{31} f_{31}(x_7(t - \tau_7)) + \tilde{c}_{32} f_{32}(x_9(t - \tau_9)) + \tilde{c}_{33} f_{33}(x_{11}(t - \tau_{11}))\\
x_7'(t) &= x_8(t)\\
x_8'(t) &= -a_7 x_7(t) - b_8 x_8(t) + \tilde{c}_{41} f_{41}(x_1(t - \tau_1)) + \tilde{c}_{42} f_{42}(x_3(t - \tau_3)) + \tilde{c}_{43} f_{43}(x_5(t - \tau_5))\\
x_9'(t) &= x_{10}(t)\\
x_{10}'(t) &= -a_9 x_9(t) - b_{10} x_{10}(t) + \tilde{c}_{51} f_{51}(x_1(t - \tau_1)) + \tilde{c}_{52} f_{52}(x_3(t - \tau_3)) + \tilde{c}_{53} f_{53}(x_5(t - \tau_5))\\
x_{11}'(t) &= x_{12}(t)\\
x_{12}'(t) &= -a_{11} x_{11}(t) - b_{12} x_{12}(t) + \tilde{c}_{61} f_{61}(x_1(t - \tau_1)) + \tilde{c}_{62} f_{62}(x_3(t - \tau_3)) + \tilde{c}_{63} f_{63}(x_5(t - \tau_5))
\end{aligned}
\right.
\tag{5}
$$

where $a_1 = \mu_1$, $a_3 = \mu_2$, $a_5 = \mu_3$, $a_7 = \mu_4$, $a_9 = \mu_5$, $a_{11} = \mu_6$; $b_2 = \alpha_1$, $b_4 = \alpha_2$, $b_6 = \alpha_3$, $b_8 = \alpha_4$, $b_{10} = \alpha_5$, $b_{12} = \alpha_6$; $\tau_1 = \tilde{\tau}_1$, $\tau_3 = \tilde{\tau}_2$, $\tau_5 = \tilde{\tau}_3$, $\tau_7 = \tilde{\tau}_4$, $\tau_9 = \tilde{\tau}_5$, $\tau_{11} = \tilde{\tau}_6$. Assume that (C1) For $i = 1, 2, \ldots, 6$, $j = 1, 2, 3$, constants α_i, $\mu_i > 0$, $f_{ij}(u)$ are continuous bounded monotone increasing (or decreasing) activation functions. Let

$$
|f_{ij}(u)| \le L \text{ for all } u \in R, \quad f_{ij}(0) = 0, \text{ and } \lim_{u \to 0} \frac{f_{ij}(u)}{u} = \beta_{ij} > 0 \ (\text{or} < 0)
\tag{6}
$$

The general activation functions like $\tanh(u)$, $\arctan(u)$, $\frac{1}{2}(|u + 1| - |u - 1|)$ satisfy condition (6). Since $f_{ij}(0) = 0$, using condition (6), the linearized system of (5) is the follows:

$$
\left\{
\begin{aligned}
x_1'(t) &= x_2(t),\\
x_2'(t) &= -a_1 x_1(t) - b_2 x_2(t) + c_{11} x_7(t - \tau_7) + c_{12} x_9(t - \tau_9) + c_{13} x_{11}(t - \tau_{11})\\
x_3'(t) &= x_4(t)\\
x_4'(t) &= -a_3 x_3(t) - b_4 x_4(t) + c_{21} x_7(t - \tau_7) + c_{22} x_9(t - \tau_9) + c_{23} x_{11}(t - \tau_{11})\\
x_5'(t) &= x_6(t)\\
x_6'(t) &= -a_5 x_5(t) - b_6 x_6(t) + c_{31} x_7(t - \tau_7) + c_{32} x_9(t - \tau_9) + c_{33} x_{11}(t - \tau_{11})\\
x_7'(t) &= x_8(t)\\
x_8'(t) &= -a_7 x_7(t) - b_8 x_8(t) + c_{41} x_1(t - \tau_1) + c_{42} x_3(t - \tau_3) + c_{43} x_5(t - \tau_5)\\
x_9'(t) &= x_{10}(t)\\
x_{10}'(t) &= -a_9 x_9(t) - b_{10} x_{10}(t) + c_{51} x_1(t - \tau_1) + c_{52} x_3(t - \tau_3) + c_{53} x_5(t - \tau_5)\\
x_{11}'(t) &= x_{12}(t)\\
x_{12}'(t) &= -a_{11} x_{11}(t) - b_{12} x_{12}(t) + c_{61} x_1(t - \tau_1) + c_{62} x_3(t - \tau_3) + c_{63} x_5(t - \tau_5)
\end{aligned}
\right.
\tag{7}
$$

where $c_{ij} = \tilde{c}_{ij} \beta_{ij}$ ($i = 1, 2, \ldots, 6$, $j = 1, 2, 3$). The matrix form of system (6) is the following:

$$X'(t) = -AX(t) + CX(t - \tau) \tag{8}$$

where $\quad X(t) = (x_1(t), x_2(t), \ldots, x_{12}(t))^T, \quad X(t - \tau) = (x_1(t - \tau_1), 0, x_3(t - \tau_3), \ldots, x_{11}(t - \tau_{11}), 0)^T.$

$$A = \begin{bmatrix} 0 & 1 & 0 & 0 & 0 & 0 & 0 & 0 & 0 & 0 & 0 & 0 \\ a_1 & b_2 & 0 & 0 & 0 & 0 & 0 & 0 & 0 & 0 & 0 & 0 \\ 0 & 0 & 0 & 1 & 0 & 0 & 0 & 0 & 0 & 0 & 0 & 0 \\ 0 & 0 & a_3 & b_4 & 0 & 0 & 0 & 0 & 0 & 0 & 0 & 0 \\ 0 & 0 & 0 & 0 & 0 & 1 & 0 & 0 & 0 & 0 & 0 & 0 \\ 0 & 0 & 0 & 0 & a_5 & b_6 & 0 & 0 & 0 & 0 & 0 & 0 \\ 0 & 0 & 0 & 0 & 0 & 0 & 0 & 1 & 0 & 0 & 0 & 0 \\ 0 & 0 & 0 & 0 & 0 & 0 & a_7 & b_8 & 0 & 0 & 0 & 0 \\ 0 & 0 & 0 & 0 & 0 & 0 & 0 & 0 & 0 & 1 & 0 & 0 \\ 0 & 0 & 0 & 0 & 0 & 0 & 0 & 0 & a_9 & b_{10} & 0 & 0 \\ 0 & 0 & 0 & 0 & 0 & 0 & 0 & 0 & 0 & 0 & 0 & 1 \\ 0 & 0 & 0 & 0 & 0 & 0 & 0 & 0 & 0 & 0 & a_{11} & b_{12} \end{bmatrix}, \text{ and}$$

$$C = \begin{bmatrix} 0 & 0 & 0 & 0 & 0 & 0 & 0 & 0 & 0 & 0 & 0 & 0 \\ 0 & 0 & 0 & 0 & 0 & 0 & c_{11} & 0 & c_{12} & 0 & c_{13} & 0 \\ 0 & 0 & 0 & 0 & 0 & 0 & 0 & 0 & 0 & 0 & 0 & 0 \\ 0 & 0 & 0 & 0 & 0 & 0 & c_{21} & 0 & c_{22} & 0 & c_{23} & 0 \\ 0 & 0 & 0 & 0 & 0 & 0 & 0 & 0 & 0 & 0 & 0 & 0 \\ 0 & 0 & 0 & 0 & 0 & 0 & c_{31} & 0 & c_{32} & 0 & c_{33} & 0 \\ 0 & 0 & 0 & 0 & 0 & 0 & 0 & 0 & 0 & 0 & 0 & 0 \\ c_{41} & 0 & c_{42} & 0 & c_{43} & 0 & 0 & 0 & 0 & 0 & 0 & 0 \\ 0 & 0 & 0 & 0 & 0 & 0 & 0 & 0 & 0 & 0 & 0 & 0 \\ c_{51} & 0 & c_{52} & 0 & c_{53} & 0 & 0 & 0 & 0 & 0 & 0 & 0 \\ 0 & 0 & 0 & 0 & 0 & 0 & 0 & 0 & 0 & 0 & 0 & 0 \\ c_{61} & 0 & c_{62} & 0 & c_{63} & 0 & 0 & 0 & 0 & 0 & 0 & 0 \end{bmatrix}$$

Lemma 1. If the determinant of matrix $K = A + C$ for all given values is not equal to zero, then system (5) has a unique equilibrium point.

Proof. The linearization of system (5) around $x = 0$ is (7). Noting that $f_{ij}(u)$ are monotone increasing (or decreasing) bounded continuous activation functions. Hence, if system (7) has a unique equilibrium point which implies that system (5) also has a unique equilibrium point. An equilibrium point $X^* = [x_1^*, x_2^*, \ldots, x_{12}^*]^T$ is the solution of the following algebraic equation

$$AX^* + CX^* = 0.$$

Assume that X^* and Y^* are equilibrium points of system (9), then we have

$$(A + C)(X^* - Y^*) = 0. \tag{9}$$

Since the determinant of matrix $K = A + C$ for all values is not equal to zero, based on basic algebraic knowledge, this means that matrix K is a nonsingular matrix, implying that $X^* - Y^* = 0$, and hence $X^* = Y^*$. Therefore, system (7) has a unique equilibrium point implying that system (5) also has a unique equilibrium point. Obviously, this equilibrium point is exactly the zero point.

Lemma 2. Each solution of system (4) is bounded.

Proof. Since $|f_{ij}(u)| \leq L$ for all $u \in R$, from (4) we get

$$v_i''(t) + \alpha_i v_i'(t) + \mu_i v_i(t) \leq N_i \qquad (i = 1, 2, \ldots, 6) \qquad (10)$$

where N_i $(i = 1, 2, \ldots, 6)$ are constants. Noting that (10) is a linear nonhomogeneous equation. It is easy to know that all solutions of (10) are bounded based on the theory of linear nonhomogeneous differential equation since $\alpha_i, \mu_i (i = 1, 2, \ldots, 6)$ are positive constants.

3   Existence of Oscillatory Solutions

It is known that if the unique equilibrium point of linearized system (7) is unstable, then this implies the instability of the unique equilibrium point of system (5).

Theorem 1. Suppose that system (7) has a unique equilibrium point. Let $\rho_1, \rho_2, \ldots,$ and ρ_{12} and $\beta_1, \beta_2, \ldots,$ and β_{12} be the eigenvalues of the matrices A and B respectively. If for some complex number β_i, $i \in \{1, 2, \ldots, 12\}$, there is at least one real ρ_j such that $\rho_j - \mathrm{Re}\, \beta_j < 0$, $j \in \{1, 2, \ldots, 12\}$, then the unique equilibrium point of system (7) is unstable, implying that system (4) generates an oscillatory solution.

Proof. We first consider the case that $\tau_1 = \tau_3 = \ldots = \tau_{11} = \tau_*$ in system (7) as the following:

$$
\begin{cases}
x_1'(t) = x_2(t), \\
x_2'(t) = -a_1 x_1(t) - b_2 x_2(t) + c_{11} x_7(t - \tau_*) + c_{12} x_9(t - \tau_*) + c_{13} x_{11}(t - \tau_*) \\
x_3'(t) = x_4(t) \\
x_4'(t) = -a_3 x_3(t) - b_4 x_4(t) + c_{21} x_7(t - \tau_*) + c_{22} x_9(t - \tau_*) + c_{23} x_{11}(t - \tau_*) \\
x_5'(t) = x_6(t) \\
x_6'(t) = -a_5 x_5(t) - b_6 x_6(t) + c_{31} x_7(t - \tau_*) + c_{32} x_9(t - \tau_*) + c_{33} x_{11}(t - \tau_*) \\
x_7'(t) = x_8(t) \\
x_8'(t) = -a_7 x_7(t) - b_8 x_8(t) + c_{41} x_1(t - \tau_*) + c_{42} x_3(t - \tau_*) + c_{43} x_5(t - \tau_*) \\
x_9'(t) = x_{10}(t) \\
x_{10}'(t) = -a_9 x_9(t) - b_{10} x_{10}(t) + c_{51} x_1(t - \tau_*) + c_{52} x_3(t - \tau_*) + c_{53} x_5(t - \tau_*) \\
x_{11}'(t) = x_{12}(t) \\
x_{12}'(t) = -a_{11} x_{11}(t) - b_{12} x_{12}(t) + c_{61} x_1(t - \tau_*) + c_{62} x_3(t - \tau_*) + c_{63} x_5(t - \tau_*)
\end{cases}
\qquad (11)
$$

The matrix form as follows

$$X'(t) = -AX(t) + CX(t - \tau_*) \tag{12}$$

The characteristic equation associated with (12) given by

$$\prod_{i=1}^{12} \left(\lambda + \rho_i - \beta_i e^{-\lambda \tau_*} \right) = 0 \tag{13}$$

Since there exists some ρ_j such that $\rho_j - \operatorname{Re} \beta_j < 0$, so we consider the equation

$$\lambda = -\rho_j + \beta_j e^{-\lambda \tau_*} \tag{14}$$

Since β_j is a complex number, this implies that λ is a complex number. Let $\lambda = \sigma + i\omega$, then from (14) we have the following:

$$\begin{cases} \sigma = -\rho_j + (\operatorname{Re} \beta_j) e^{-\sigma \tau_*} \cos \omega \tau_* + (\operatorname{Im} \beta_j) e^{-\sigma \tau_*} \sin \omega \tau_* \\ \omega = (\operatorname{Im} \beta_j) e^{-\sigma \tau_*} \cos \omega \tau_* - (\operatorname{Re} \beta_j) e^{-\sigma \tau_*} \sin \omega \tau_* \end{cases} \tag{15}$$

Consider the function

$$F(\sigma) = \sigma + \rho_j - (\operatorname{Re} \beta_j) e^{-\sigma \tau_*} \cos \omega \tau_* - (\operatorname{Im} \beta_j) e^{-\sigma \tau_*} \sin \omega \tau_* \tag{16}$$

Obviously, $F(\sigma)$ is a continuous function of σ, $F(+\infty) = +\infty$, and

$$F(0) = \rho_j - (\operatorname{Re} \beta_j) \cos \omega \tau_* - (\operatorname{Im} \beta_j) \sin \omega \tau_* \tag{17}$$

When $\tau_* \to 0$ we have $\cos \omega \tau_* \to 1$ and $\sin \omega \tau_* \to 0$. thus, under the condition $\rho_j - \operatorname{Re} \alpha_j < 0$ we get $F(0) < 0$ when $\tau_* > 0$ is sufficiently small. This implies that there exists a $\sigma > 0$ such that $F(\sigma) = 0$ by the continuity of function $F(\sigma)$. Therefore, λ has a positive real part. This suggests that the trivial solution of system (11) is unstable when $\tau_* > 0$ is sufficiently small. Because increasing time delay cannot change the instability of the equilibrium point. So for any time delays in system (7), thus system (4) the trivial solution is unstable. According to Chafee's criterion, system (4) generates an oscillatory solution.

Theorem 2. Suppose that system (7) has a unique equilibrium point. Assume that at least one of the following characteristic equations has a positive root:

$$\lambda^2 + \alpha_i \lambda + \mu_i - (\tilde{c}_{i1} \beta_{i1} + \tilde{c}_{i2} \beta_{i2} + \tilde{c}_{i3} \beta_{i3}) e^{-\tau_*} = 0, \ (i = 1, 2, \ldots, 6). \tag{18}$$

where $\tau_* = \min\{\tau_1, \tau_3, \ldots, \tau_{11}\}$. Then the unique equilibrium point of system (7) is unstable, implying that system (5) has an oscillatory solution.

Proof. Based on the basic theory of differential equation, if at least one of the characteristic Eq. (18) has a positive root, then the trivial solution is unstable. Since system (7) has only one equilibrium point and all solutions are bounded, then system (7) thus system (4) generates an oscillatory solution according to the Chafee's criterion.

Theorem 3. Suppose that system (7) has a unique equilibrium point. If all eigenvalues of matrix K (= A + C) are complex numbers or there is at least one positive real eigenvalue, then the unique equilibrium point of system (7) is unstable, implying that system (5) has an oscillatory solution.

Proof. We consider without time delays case of system (7). Based on the differential equation theory, all eigenvalues of matrix K (= A + C) are complex numbers or there is at least one positive real eigenvalue of K, then the trivial solution of without time delays case of system (7) is unstable. This instability will maintain when system (7) has time delay. Since system (7) has a unique equilibrium point, it will generate an oscillatory solution, suggesting (5) has an oscillatory solution.

4 Computer Simulations

In order to analyze the effect of activation functions for the oscillatory behavior, in system (5), we fix time delays as $(6, 6, 6, 6, 6, 6)$, first choose the activation functions $f_{ij}(u) = \tanh(u)$, thus $\beta_{ij} = 1 (i = 1, 2, \ldots, 6, j = 1, 2, 3)$. The parameters are as follows: $a_1 = 0.05$, $a_3 = 0.04$, $a_5 = 0.02$, $a_7 = 0.04$, $a_9 = 0.03$, $a_{11} = 0.04$; $b_2 = 0.02$, $b_4 = 0.03$, $b_6 = 0.04$, $b_8 = 0.05$, $b_{10} = 0.05$, $b_{12} = 0.04$; $\tilde{c}_{11} = -0.45$, $\tilde{c}_{12} = -0.26$, $\tilde{c}_{13} = -0.45$, $\tilde{c}_{21} = -0.28$, $\tilde{c}_{22} = -0.38$, $\tilde{c}_{23} = -0.45$, $\tilde{c}_{31} = -0.5$, $\tilde{c}_{32} = -0.6$, $\tilde{c}_{33} = -0.24$, $\tilde{c}_{41} = -0.28$, $\tilde{c}_{42} = -0.48$, $\tilde{c}_{43} = 0.12$, $\tilde{c}_{51} = -0.25$, $\tilde{c}_{52} = -0.24$, $\tilde{c}_{53} = -0.05$, $\tilde{c}_{61} = 0.28$, $\tilde{c}_{62} = 0.18$, $\tilde{c}_{63} = -0.05$. Thus the eigenvalues of matrix

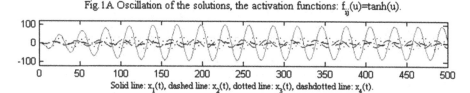

Fig. 1A Oscillation of the solutions, the activation functions: $f_{ij}(u){=}\tanh(u)$.

Solid line: $x_1(t)$, dashed line: $x_2(t)$, dotted line: $x_3(t)$, dashdotted line: $x_4(t)$.

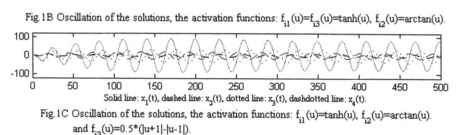

Fig. 1B Oscillation of the solutions, the activation functions: $f_{11}(u){=}f_{13}(u){=}\tanh(u)$, $f_{12}(u){=}\arctan(u)$.

Solid line: $x_1(t)$, dashed line: $x_2(t)$, dotted line: $x_3(t)$, dashdotted line: $x_4(t)$.

Fig. 1C Oscillation of the solutions, the activation functions: $f_{11}(u){=}\tanh(u)$, $f_{12}(u){=}\arctan(u)$. and $f_{13}(u){=}0.5*(|u{+}1|{-}|u{-}1|)$.

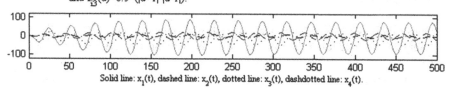

Solid line: $x_1(t)$, dashed line: $x_2(t)$, dotted line: $x_3(t)$, dashdotted line: $x_4(t)$.

Fig. 1. Oscillation results.

K(= A+C) are: -0.6868, $0.0186 \pm 0.6518i$, $-0.14354 \pm 0.1830i$, $0.2066 \pm 0.3842i$, 0.7243, $0.4232 \pm 0.3857i$, $0.4715 \pm 0.1820i$. Since there is a positive real root $0.7243 > 0$ of matrix K, from Theorem 3, system generates an oscillatory solution (see Fig. 1A). Then we select the activation function $f_{i1}(u) = f_{i3}(u) = \tanh(u)$, $f_{i2}(u) = \arctan(u)$, and $f_{i1}(u) = \tanh(u)$, $f_{i2}(u) = \arctan(u)$, $f_{i3}(u) = \frac{1}{2}(|u + 1| - |u - 1|)$ $(i = 1, 2, \ldots, 6)$ respectively, the time delay and the other parameters are the same as Fig. 1A. We see the oscillatory solutions are still appeared (Figs. 1B and C). In Fig. 2, we analyze the effect of the time delays, the activation functions are fixed to tanh (u), and the other parameters are the same as Fig. 1. When the time delays are (6, 6, 6, 6, 6, 6) and (10, 10, 10, 10, 10, 10) respectively, system generates an oscillatory solution. However, when we select time delays are (20, 10, 10, 10, 110, 120), the oscillation is still maintained, implying that the time delays can be selected different values. We pointed out that the bifurcating approach is hard to discuss the existence of bifurcation periodic solutions according to [12, 13] in this different time delays.

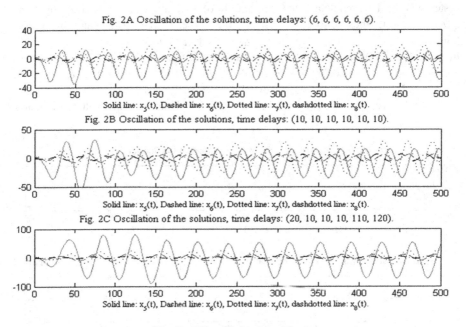

Fig. 2A Oscillation of the solutions, time delays: (6, 6, 6, 6, 6, 6).

Solid line: $x_5(t)$, Dashed line: $x_6(t)$, Dotted line: $x_7(t)$, dashdotted line: $x_8(t)$.

Fig. 2B Oscillation of the solutions, time delays: (10, 10, 10, 10, 10, 10).

Solid line: $x_5(t)$, Dashed line: $x_6(t)$, Dotted line: $x_7(t)$, dashdotted line: $x_8(t)$.

Fig. 2C Oscillation of the solutions, time delays: (20, 10, 10, 10, 110, 120).

Solid line: $x_5(t)$, Dashed line: $x_6(t)$, Dotted line: $x_7(t)$, dashdotted line: $x_8(t)$.

Fig. 2. Effect of the time delays

5 Conclusion

This paper discusses a time delay inertial six-neuron network model. By studying the instability of the unique equilibrium point of the system, two criteria to ensure the existence of periodic oscillatory solutions are derived based on the Chafee's criterion. Computer simulations are provided to demonstrate the correctness of the result.

Acknowledgement. This research work was supported by NNSF of China (11361010), and Scientific Research Foundation of the Education Department of Guangxi Zhuang Autonomous Region (N0. KY2015ZD103)

References

1. Zheng, B., Zhang, Y., Zhang, C.: Global existence of periodic solutions on a simplified BAM neural network model with delay. Chaos, Solitons Fractals **37**, 1397–1408 (2008)
2. Cao, J., Wang, L.: Periodic oscillatory solution of bidirectional associative memory networks with delays. Phys. Rev. E **61**, 1825–1828 (2000)
3. Yu, W., Cao, J.: Stability and Hopf bifurcation on a four-neuron BAM neural network with delays. Phys. Lett. A **351**, 64–78 (2006)
4. Xu, C., Zhang, Q.: Existence and global exponential stability of anti-periodic solutions for BAM neural networks with inertial term and delay. Neurocomputing **153**, 108–116 (2015)
5. Wang, F., Sun, D., Wu, H.: Global exponential stability and periodic solutions of high-order bidirectional associative memory (BAM) neural networks with time delays and impulses. Neurocomputing **155**, 261–276 (2015)
6. Liu, M., Xu, X., Zhang, C.: Stability and global Hopf bifurcation for neutral BAM neural network. Neurocomputing **145**, 122–130 (2014)
7. Zhang, Z., Quan, Z.: Global exponential stability via inequality technique for inertial BAM neural networks with time delays. Neurocomputing **151**, 1316–1326 (2015)
8. Zhang, C., Zheng, B., Wang, L.: Multiple Hopf bifurcations of symmetric BAM neural network model with delay. Appl. Math. Lett. **22**, 616–622 (2009)
9. Wang, C.: Almost periodic solutions of impulsive BAM neural networks with variable delays on time scales. Commu. Nonl. Sci. Numer. Simul. **19**, 2828–2842 (2014)
10. Berezansky, L., Braverman, E., Idels, L.: New global exponential stability criteria for nonlinear delay differential systems with applications to BAM neural networks. Applied Math. Comput. **243**, 899–910 (2014)
11. Li, Y., Wang, C.: Existence and global exponential stability of equilibrium for discrete-time fuzzy BAM neural networks with variable delays and impulses. Fuzzy Sets Syst. **217**, 62–79 (2013)
12. Xu, C., Tang, X., Liao, M.: Stability and bifurcation analysis of a six-neuron BAM neural network model with discrete delays. Neurocomputing **74**, 689–707 (2011)
13. Ge, J., Xu, J.: Stability switches and fold-Hopf bifurcations in an inertial four-neuron network model with coupling delay. Neurocomputing **110**, 70–79 (2013)
14. Wang, D., Huang, L., Cai, Z.: On the periodic dynamics of a general Cohen-Grossberg BAM neural networks via differential inclusions. Neurocomputing **118**, 203–214 (2013)
15. Zhao, Z., Jian, J., Wang, B.: Global attracting sets for neutral-type BAM neural networks with time-varying and infinite distributed delays. Nonlinear Anal. HS **15**, 63–73 (2015)
16. Chafee, N.: A bifurcation problem for a functional differential equation of finitely retarded type. J. Math. Anal. Appl. **35**, 312–348 (1971)

An Online Supervised Learning Algorithm Based on Nonlinear Spike Train Kernels

Xianghong Lin$^{(\boxtimes)}$, Ning Zhang, and Xiangwen Wang

School of Computer Science and Engineering,
Northwest Normal University, Lanzhou 730070, China
linxh@nwnu.edu.cn

Abstract. The online learning algorithm is shown to be more appropriate and effective for the processing of spatiotemporal information, but very little researches have been achieved in developing online learning approaches for spikingneural networks. This paper presents an online supervised learning algorithm based on nonlinear spike train kernels to process the spatiotemporal information, which is more biological interpretability. The main idea adopts online learning algorithm and selects a suitable kernel function. At first, the Laplacian kernel function is selected, however, in some ways, the spike trains expressed by the simple kernel function are linear in the postsynaptic neuron. Then this paper uses nonlinear functions to transform the spike train model and presents the detail experimental analysis. The proposed learning algorithm is evaluated by the learning of spike trains, and the experimental results show that the online nonlinear spike train kernels own a super-duper learning effect.

Keywords: Spiking neural networks · Supervised learning · Spike train kernels · Online learning

1 Introduction

Artificial neural networks (ANNs) have got great progress and successfully applied in many fields [1]. In recent years, the focus on ANNs is gradually turning to the spiking neural networks (SNNs) which are more biological plasticity, especially the learning methods and theoretical researches of the SNNs [2–4]. According to the learning rule, supervised learning methods based on temporal coding of SNNs mainly contain the supervised learning algorithm based on the gradient descent rule and the supervised learning algorithm based on the Hebb rule. The supervised learning algorithms based on the gradient descent rule are mainly divided into single spike learning and multi-spike learning (spike trains learning). The most typical single spike learning is the SpikeProp [5] and its various modified forms such as increasing the momentum item [6, 7], QuickProp and RProp [8] etc. However these methods operate pure single spike, which requires all of neurons in the input layer, hidden layer and output layer only fire one spike. Booij and Nguyen [9] continued to improve the SpikeProp, their algorithm was not restricted to the number of firing spikes for the neurons in the input layer and hidden layer, which achieved the spike trains learning in the input layer and hidden layer. Similarly, Ghosh-Dastidar et al. [10] derived a gradient descent spike

© Springer International Publishing Switzerland 2015
D.-S. Huang et al. (Eds.): ICIC 2015, Part I, LNCS 9225, pp. 106–115, 2015.
DOI: 10.1007/978-3-319-22180-9_11

trains learning rule for the synaptic weights of the output layer and hidden layer using Chain rules, named the Multi-SpikeProp algorithm. But the neurons in the output layer can only fire one spike. Recently Xu et al. [11] proposed a new supervised multiple-spike learning algorithm based on the gradient descent rule for SNNs. The algorithm does not restrict the number of firing spikes for neurons in all layers of the network, which realizes the spike trains spatial-temporal mode learning for multilayer feedforward spiking neural networks.

However, the learning efficiency and the accuracy of supervised learning algorithms based on gradient descent rule are not high. These algorithms adopt pure mathematical methods, completely ignore the concept for neurons running and lack of biological plasticity. Then in order to simulate real biological neurons well, researchers have proposed Hebb supervised learning algorithms which own more biological plasticity. Hebb [12] firstly put forward a synaptic plasticity hypothesis: "If two neurons are excited at the same time, the synapses between them can be strengthened ". The hypothesis emphasizes the importance of the synergistic activity and synaptic strengthening between the pre/postsynaptic neurons. In fact, spike trains can not only cause the continuous change of synapses, but also satisfy Spike Timing-Dependent Plasticity (STDP) [13] mechanism. Ponulak and Kasiński [14] represented a supervised learning method, named Remote Supervised Method (ReSuMe), which the basic idea of the method came from Widrow-Hoff rules. The method is composed of the Widrow-Hoff rule expression which uses the STDP rules to deduce the actually learning method. And the adjustment of the synaptic weights relies on the combination of the STDP and anti-STDP. The results of the study show that the ReSuMe owns a very good learning performance and widely applicable areas. Florian [15] proposed Chrontron learning methods, which included two supervised learning algorithms: the E-Learning with higher learning ability and I-Learning with more biological plasticity.

Recently, with the continuous deepening of researches, the supervised learning algorithm based on kernel function gradually has formed a new system, SPAN and PSD are the representative. Mohemmed and Schliebs [16, 17] proposed SPAN (Spike Pattern Association Neuron) algorithm based on kernel function. Its main features are to use kernel function convert spike trains to the convolution signals, and then the transformed input spike trains, desired spike trains and actual output spike trains apply Widrow-Hoff rules to regulate the synaptic weights. Inspired by SPAN algorithm, Yu et al. [18] had given a different explanation which put the traditional Widrow-Hoff rules to SNNs and put forward the PSD (Precise-Spike-Driven) supervised learning algorithm. Adjustment of the synaptic weights is driven by the error between the desired output spike trains and the actual output spike trains, positive error will lead to enhanced long-term potentiation, negative error will lead to long-term depression.

In this paper, we propose an online supervised learning algorithm based on the nonlinear spike train kernels (STK). Online learning is different from the offline learning or the batch learning, which is a supervised learning method with dynamic learning performance [19]. In the real world, we obtain the data with time and space characteristics, and the spatiotemporal data is generally represented as a continuous spike train current [20]. We require the learning algorithm to process real-time learning for neural networks, and synaptic weights are dynamic change with the spike input process. Therefore, online learning algorithm will be more appropriate and effective for the processing of this kind of real-time tasks [21]. To implement the learning of the complex spatiotemporal model on the spike trains, the key problem is to construct a suitable kernel function to express the spike trains. In the STK learning algorithm, the Laplacian kernel function is used to the input spike trains which are transformed into analog convolution signals. The spike trains expressed by these kernel functions acting on postsynaptic neuron are linear, therefore, this paper adopt the nonlinear model with more biological interpretability.

2 Online Supervised Learning Rules for Spiking Neuron

In this section, we present a new online learning algorithm based on STK. The learning methods of traditional SNNs mostly adopt the way of offline learning, its synaptic weights will be adjusted after a large number of spike trains submit training. The offline learning can make neural networks form a network structure before the systems work, but the parameters and temporal of the actual systems are often unpredictable and varying, which causes the offline learning not to be spatiotemporal pattern learning. Nevertheless, the online learning rule of the spike neural networks is to train each spike train after submitting, and only learns one spike train at one moment, then in real time adjusts the network parameter values according to the training results. We main process spatiotemporal data in the real world, it demands the real-time network learning, synaptic weights process dynamic change with the input of the spike trains. Hence, online learning algorithm will be more appropriate and effective for this kind of real-time tasks.

2.1 Learning Rule Based on Linear Spike Train Kernels

In the learning process of spike trains, after the presynaptic membrane released neurotransmitter into the synaptic cleft, neurotransmitter diffuses into the postsynaptic membrane, which causes the change of postsynaptic membrane for some ion permeability and produces electric potential difference. Then it makes some charged ions in and out of the postsynaptic membrane, which forms a postsynaptic potential. Therefore we must simulate the process of spike trains occurrence in the synapse, and convert spike trains into a functional form model.

For a spike train, the spike train $s = \{t^f: f = 1, 2 \dots F\}$ stands for the orderly sequence of neurons firing spike time, the spike train model can be shown as follows:

$$s(t) = \sum_{f=1}^{F} \delta\left(t - t^f\right) \tag{1}$$

where F is the number of spiking for a spike train, $\delta(t)$ is the Dirac delta function, if $t = 0$, $\delta(t) = 1$, else $\delta(t) = 0$, t^f is the fth spiking firing time. So we represent the formalized spike trains as follows:

$$v_i(t) = \sum_f^F k(t - t^f) \tag{2}$$

where $k(t)$ is the kernel function which includes the Gaussian, Laplacian and α-function and so on, we choose the Laplacian kernel in our paper. Firstly, we employ the Widrow-Hoff rules for the ith synapse in the input neurons O, as the follow function:

$$\Delta\omega_{oi} = \rho x_i(s_d - s_o) \tag{3}$$

where ρ is a positive constant on behalf of the learning rate, x_i is an input spike train of the input neurons O, s_d and s_o refer to the desired and actual output neurons spike trains. In order to reduce the complexity of the calculation and improve learning efficiency, in this paper, we process convolution kernels computation just for the input spike trains. Then we can get the learning rule based on the linear STK like that:

$$\begin{aligned}\Delta\omega_i(t) &= \rho v_i(t)(s_d(t) - s_o(t)) \\ &= \rho \sum_f k(t - t^f)(s_d(t) - s_o(t))\end{aligned} \tag{4}$$

2.2 Learning Rule Based on Nonlinear Spike Train Kernels

However, the above description is flawed from the biological perspective. One of the basic defects is that it ignores the capability limitation of the dendrite receptor. When two spikes arrive to the postsynaptic membrane within a short period of time or at the same time, the first spike makes the dendrite receptor receive stimulation and lead to increase dendritic conductance. Because dendrite receptor is being influenced by the neurotransmitters which are released by the previous spike, the second spike will only induce a little influence. Therefore, the spike trains expressed by the Eq. (2) are linear to the postsynaptic neuron from another aspect. We use a nonlinear function $f(x)$ to transform the spike train model. Here the nonlinear functions which we choose are $f_1(x) = \tanh(x/\sigma)$ and $f_2(x) = (1 - e^{-x/\sigma})$. We apply it to the nonlinear STK, the equation can be expressed as:

$$v_i^{\dagger}(t) = f\left(\sum_f^F k(t - t^f)\right) \tag{5}$$

where the $k(t)$ is still the kernel function like the Eq. (2), and we adopt the Laplacian kernel, $k(t) = e^{-|t|/\tau}H(t)$, $H(t)$ is the Heaviside function, if $t < 0$, $H(t) = 0$, else $H(t) = 1$.

In Fig. 1, we give the different STK simulation processes and the STK synaptic weights adjustment process. In Fig. 1(a), the nonlinear STK1 and the nonlinear STK2 adopt the nonlinear functions $f_1(x)$ and $f_2(x)$ respectively. The synaptic weights adjustment depends on three aspects: When the desired spikes are the same with the output neuron spikes, we will get a zero error; there is not an output spike in times of the desired spike, we can get a positive error; a negative error is received when an output spike is not supposed by the desired spike. The adaptation process of the STK synaptic weights is shown in Fig. 1(b).

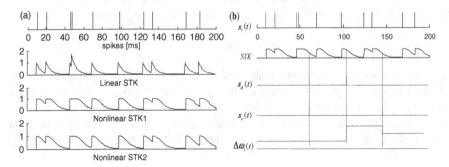

Fig. 1. The STK simulation and weight adaptation process. (a) The kernel function transformation form of the input spike trains linear, the nonlinear STK1 and the nonlinear STK2, and the y-axis is the same scale. (b) The weight adaptation process. $S_i(t)$ is the input spike train, $S_d(t)$ and $S_o(t)$ are the desired and actual output neurons spike trains respectively.

Hence we obtain the nonlinear STK function, and then we use it to the renewal of the synaptic weight. We get a new online learning rule based on nonlinear STK, which can solve spatiotemporal data problems and possess more biological interpretability like that:

$$\Delta\omega_i(t) = \rho v_i^\dagger(t)(s_d(t) - s_o(t))$$
$$= \rho f\left(\sum_f k(t - t^f)\right)(s_d(t) - s_o(t)) \qquad (6)$$

2.3   The Measure Criterion of the Spike Train Learning

In the spike trains learning, the measure criterion of learning result is the close extent between the desired output spike trains and the actually output spike trains at the end of the learning, which is the distance between the desired and the actually output spike trains. In this paper we adopt the method C based on the correlation measurement to describe the distance between the desired output spike trains and the actually output spike trains. According to the Cauchy-Schwarz inequality $|v_d \cdot v_o|^2 \leq |v_d \cdot v_d| \cdot |v_o \cdot v_o|$, we get that $0 \leq |v_d \cdot v_o|^2 / |v_d \cdot v_d| \cdot |v_o \cdot v_o| \leq 1$. Where v_d and v_o are the vectors which represent a convolution of the desired output spike trains and the actually output spike trains with a Gaussian low-pass filter respectively. Then we make the expression of measurement C to define as:

$$C = \frac{v_d \cdot v_o}{\sqrt{|v_d \cdot v_d|} \cdot \sqrt{|v_o \cdot v_o|}} \tag{7}$$

The numerator is the inner product of two vectors in the equation, and the denominator is the product of two the Euclidean norms vectors. When two spike trains are completely same, the value of C is 1, and it gradually tends to zero with the decrease of their correlation.

3   Experimental Results

In this section, we are by the learning problems of the spike trains to verify the performance of the linear STK and the two nonlinear STKs. There are 400 synaptic input neurons in our simulation, the length of the input and desired output spike trains are defined as 200 ms. The input and desired output spike trains are randomly generated by the frequency of 30 Hz Poisson process respectively. Neural model adopt the simple model SRM, the time delay constant of postsynaptic neuron is 5 ms, and the delay constant of the refractory period function selects 50 ms. The firing threshold of the Neurons is 1, the length of absolute refractory period after the spike fires is 1 ms. At first the max fired time of the every neuron is 200 ms, the time step is 0.1 ms. All of the learning rates set 0.05. We use the Laplacian kernel as the kernel function in our simulation, and its parameter τ uses 5. The value of σ is 0.2 in the nonlinear function. In each simulation, we process 100 times iterations, and each experimental result is the average result which runs 100 times experiments.

First of all, as shown in Fig. 2, we analyze the learning process of spike neuron. Figure 2(a) − (c) shows the learning process of spike neuron, Fig. 2(a) represents the learning process of nonlinear STK1, which adopt the nonlinear function $f_1(x)$; Fig. 2(b) represents the learning process of nonlinear STK2, which adopt the nonlinear function $f_2(x)$; Fig. 2(c) represents the learning process of linear STK. In Fig. 2, ∇ represents the desired output spike train, Δ represents the output spike train before the learning of spike neuron, • represents the actual output spike train of some learning cycles in the learning process. From the learning process we can see that Fig. 2(a) and Fig. 2(b) neuron takes about 3 steps from the learning of initial random output spike train to the desired output spike train, while Fig. 2(c) takes about 12 steps. Figure 2(d) represents the change of learning precision curve in the learning process, as shown in Fig. 2(d), the learning precision of nonlinear STK1 which just takes about 2 learning cycles reaches 1, the nonlinear STK2 takes about 3 learning cycles, while linear STK takes about 12 learning cycles.

Figure 3 analyses the learning performance of the algorithm in different length of spike train. In the experiment, the length of input and desired output spike train increases gradually from 100 to 1000 by the interval of 100 ms, other settings remain unchanged. The nonlinear STK1 and the nonlinear STK2 adopt the nonlinear functions $f_1(x)$ and $f_2(x)$ respectively. As we can see from Fig. 3(a), the highest learning accuracy

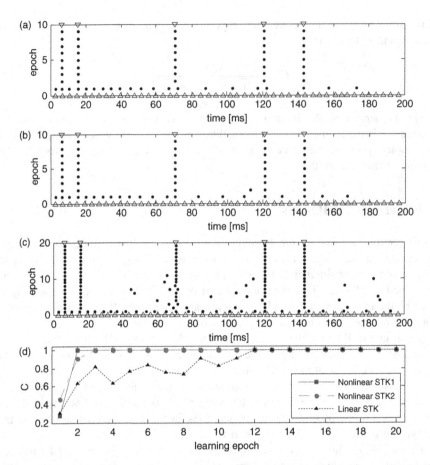

Fig. 2. The learning process of spike train. (a) The learning process of the nonlinear STK1 spike neuron, which adopts the nonlinear function $f_1(x)$. (b) The learning process of the nonlinear STK2 spike neuron, which adopts the nonlinear function $f_2(x)$. (c) The learning process of the linear STK spike neuron. (d) The change curve of the learning precision.

of the two algorithms are decreasing along with the length of the spike train increased, the learning accuracy of the two nonlinear STKs is almost at the same, and the learning precision of the two nonlinear STKs online learning is higher than the linear STK online learning apparently. When the length of spike train reaches 1000 ms, both of the two nonlinear STKs arrive 0.85, and the linear STK is 0.8, then the curve almost remains unchanged. Therefore, online learning is more suitable for dealing with large-scale data problems. Figure 3(b) is the minimum iterations to the highest accuracy. From Fig. 3(b), we know that the iterations of the two nonlinear STKs are smaller than the linear STK and are faster get the highest learning accuracy. For example, when the length of spike train is 400 ms, the minimum iterations of the linear STK, the nonlinear STK1 and the nonlinear STK2 are 35, 26 and 24 respectively.

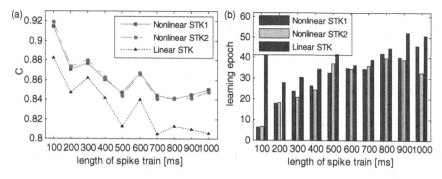

Fig. 3. The learning effects of spike train in different length. (a) The learning accuracy of spike train in different length. (b) The minimum iterations to the highest accuracy.

In Fig. 4, it analyzes the learning performance of input and desired output spike train when they are in different spike frequency. Setting the input and desired output spike trains are generated by the process of Poisson where they are in different firing frequencies. In the experiment, the firing frequency of input and desired output spike train increases gradually from 10 Hz to 100 Hz in 10 Hz intervals, also, the firing frequency between input spike train and the desired output spike train is equal, other settings remain unchanged. The nonlinear STK1 and the nonlinear STK2 adopt the nonlinear functions $f_1(x)$ and $f_2(x)$ respectively. As we can see from Fig. 4(a), the learning accuracy of the three algorithms are first increased and then tend to be gradual declined along with the increase of the firing frequency of the spike train, and the learning accuracy of the two nonlinear STK online learning is always higher than the linear STK online learning. When the firing rate of the spike train reaches 20 Hz, both of the nonlinear STK1 and STK2 arrive 0.88, the linear STK arrives the highest learning precision 0.85, but it is still below the nonlinear STK, and then tends to decline gradually. Figure 4(b) is the minimum iterations to the highest accuracy in different firing rate. From Fig. 4(b), we can get that the nonlinear STKs are faster to achieve the highest accuracy in different firing rate. Such as, when the firing rate of the

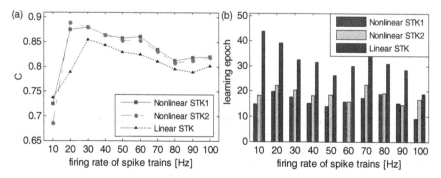

Fig. 4. The learning effects of spike train in different firing rate. (a) The learning accuracy in different firing rate. (b) The minimum iterations to the highest accuracy.

spike train reaches 70 Hz, the nonlinear STK1 achieves the highest accuracy after 17 iterations; the nonlinear STK2 is through 22 iterations; while the linear STK is 35 iterations.

4 Conclusions

This paper comes up with an online supervised learning algorithm based on nonlinear STK. To achieve the learning of complex spatiotemporal pattern on spike trains, the key issue is that we need to construct a proper kernel function to represent the spike train. In our paper, we choose the Laplacian kernel function. However, in some ways, the spike train which is expressed by the simple kernel function has the linear effect on the postsynaptic neurons, therefore, we propose nonlinear STK which has more biological interpretability.

As distinguished from offline learning and batch learning, the learning speed of the online learning is superior to offline learning, because online learning trains one by one according to the order of the spike train, rather than each of iteration would train the entire spike train like the offline learning. So the online learning is more suitable for dealing with problems of large-scale data. This makes it possible that the online learning of spike neural networks can applied to real life. Based on the analysis and summary of advantages and disadvantages of the current spike neural networks online learning algorithms, the main problems we have to solve in the future are as follows: (1) In this paper, the network structure of nonlinear STK only have one layer of network, that is, only the input layer and the output layer, so it can be considered to extend to a multi-layer network. (2) We are only giving a learning of simple spike train, and then we should apply it to solve practical problems later. (3) The research of scholars on the online learning of spike neural networks is very few, so we could consider online learning to be applied to the rules of gradient descent learning.

Acknowledgement. The work is supported by the National Natural Science Foundation of China under Grants No. 61165002 and No. 61363059.

References

1. Yegnanarayana, B.: Artificial Neural Networks. PHI Learning Pvt Ltd., New Delhi (2009)
2. Ahmed, F.Y.H., Yusob, B., Hamed, H.N.A.: Computing with spiking neuron networks a review. Int. J. Adv. Soft Comput. Its Appl.s 6(1), 1–21 (2014)
3. Yu, Q., Goh, S.K., Tang, H., Tan, K.C.: Application of precise-spike-driven rule in spiking neural networks for optical character recognition. In: Handa, H., Ishibuchi, K.-C., Ong, H., Tan, K.-C. (eds.) Proceedings of the 18th Asia Pacific Symposium on Intelligent and Evolutionary Systems, vol. 2, pp. 65–75. Springer International Publishing, NewYork (2015)
4. Carlson, K.D., Nageswaran, J.M., Dutt, N., Krichmar, J.L.: An efficient automated parameter tuning framework for spiking neural networks. Front. Neuroscience 8, 10 (2014)

5. Bohte, S.M., Kok, J.N., La Poutre, H.: Error-backpropagation in temporally encoded networks of spiking neurons. Neurocomputing **48**(1), 17–37 (2002)
6. Xin, J., Embrechts, M.J.: Supervised learning with spiking neural networks. In: Proceedings of International Joint Conference on Neural Networks, IJCNN 2001, vol. 3, pp. 1772–1777. IEEE (2001)
7. Schrauwen, B., Van Campenhout, J.: Extending spikeprop. In: Proceedings of 2004 IEEE International Joint Conference on Neural Networks, vol. 1. IEEE (2004)
8. McKennoch, S., Liu, D., Bushnell, L.G.: Fast modifications of the spikeprop algorithm. In: International Joint Conference on Neural Networks, IJCNN 2006, pp. 3970–3977. IEEE (2006)
9. Booij, O., Nguyen, T.H.: A gradient descent rule for spiking neurons emitting multiple spikes. Inf. Process. Lett. **95**(6), 552–558 (2005)
10. Ghosh-Dastidar, S., Adeli, H.: A new supervised learning algorithm for multiple spiking neural networks with application in epilepsy and seizure detection. Neural Netw. **22**(10), 1419–1431 (2009)
11. Xu, Y., Zeng, X., Han, L., Yang, J.: A supervised multi-spike learning algorithm based on gradient descent for spiking neural networks. Neural Networks **43**, 99–113 (2013)
12. Hebb, D.O.: The Organization of Behavior: a Neuropsychological Theory. Psychology Press, New York (2005)
13. Caporale, N., Dan, Y.: Spike timing-dependent plasticity: a Hebbian learning rule. Annu. Rev. Neurosci. **31**, 25–46 (2008)
14. Ponulak, F., Kasinski, A.: Supervised learning in spiking neural networks with ReSuMe: sequence learning, classification, and spike shifting. Neural Comput. **22**(2), 467–510 (2010)
15. Florian, R.V.: The chronotron: a neuron that learns to fire temporally precise spike patterns. PLoS ONE **7**(8), e40233 (2012)
16. Mohemmed, A., Schliebs, S., Matsuda, S., Kasabov, N.: Span: spike pattern association neuron for learning spatio-temporal spike patterns. Int. J. Neural Syst. **22**(04), 1250012 (2012)
17. Mohemmed, A., Schliebs, S., Matsuda, S., Kasabov, N.: Training spiking neural networks to associate spatio-temporal input–output spike patterns. Neurocomputing **107**, 3–10 (2013)
18. Yu, Q., Tang, H., Tan, K.C., Li, H.: Precise-spike-driven synaptic plasticity: learning hetero-association of spatiotemporal spike patterns. PLoS ONE **8**(11), e78318 (2013)
19. Wang, J., Belatreche, A., Maguire, L., McGinnity, M.: Online versus offline learning for spiking neural networks: a review and new strategies. In: 2010 IEEE 9th International Conference Cybernetic Intelligent Systems (CIS), pp. 1–6. IEEE (2010)
20. Kasabov, N.K.: Neucube: A spiking neural network architecture for mapping, learning and understanding of spatio-temporal brain data. Neural Netw. **52**, 62–76 (2014)
21. Oniz, Y., Kaynak, O.: Variable-structure-systems based approach for online learning of spiking neural networks and its experimental evaluation. J. Franklin Inst. **351**(6), 3269–3285 (2014)

Forecasting Weather Signals Using a Polychronous Spiking Neural Network

David Reid[1], Hissam Tawfik[1(✉)], Abir Jaafar Hussain[2],
and Haya Al-Askar[3]

[1] Mathematics and Computer Science Department,
Liverpool Hope University, Liverpool, UK
{reidd, tawfikh}@hope.ac.uk
[2] Liverpool John Moores University, Byrom Street,
Liverpool L3 3AF, UK
a.hussain@ljmu.ac.uk
[3] Department of Computer Science, Salman Bin Abdulaziz University,
Al-Kharj PO Box 151, 11942, KSA

Abstract. Due to its inherently complex and chaotic nature predicting various weather phenomena over non trivial periods of time is extremely difficult. In this paper, we consider the ability of an emerging class of temporally encoded neural network to address the challenge of weather forecasting. The Polychronous Spiking Neural Network (PSNN) uses axonal delay to encode temporal information into the network in order to make predictions about weather signals. The performance of this network is benchmarked against the Multi-Layer Perceptron network as well as Linear Predictor. The results indicate that the inherent characteristics of the Polychronous Spiking Network make it well suited to the processing and prediction of complex weather signals.

Keywords: Spiking neural network · Axonal delay · Natural event forecasting · Weather time series

1 Introduction

The prediction of specific weather time series data is an important prerequisite in general forecasting of natural weather phenomena that agriculture and other industries depend upon [1]. The importance of being able to accurately predict weather patterns cannot be overstated; beyond predicting catastrophic events such as storms or unusually severe spells of cold or heat, weather has a dramatic effect on a population's day to day health and behaviour and has a significant impact on the economy.

Trying to model the behaviour of an extremely dynamic system such as the weather in the UK or Europe is incredibly challenging. Discovering which the significant parameters to on which to focus and understanding what are the underlying ramifications of changes in specific weather data patterns is extremely difficult.

Physical based numerical models are traditionally used to forecast the weather; this requires very powerful supercomputers in order to recreate and model the specific mechanics of some aspect of the weather. In contrast some efforts have been made to

© Springer International Publishing Switzerland 2015
D.-S. Huang et al. (Eds.): ICIC 2015, Part I, LNCS 9225, pp. 116–123, 2015.
DOI: 10.1007/978-3-319-22180-9_12

only use patterns of data in order to make a prediction. Most of these tend to be linear in nature; for example the Autoregressive (AR) model. This is based on the simple principle that time series data are highly correlated; therefore, previous observations of the data series are used to predict future observations.

For non-stationary time series data, the autoregressive integrated moving average (ARIMA) was developed by Box and Jenkins [2]. This model is an extended model of ARMA, consisting of autoregressive (AR), integrated (I) and moving average (MA) parts. In ARIMA, the time series must be integrated before the forecasts are created so that the predictions are expressed in values matching the input data. The ARIMA model assumes that the data can be stationary after differencing [3]. Both of these models have traditionally been considered the basis for time series analysis and are widely used in time series predictions [4].

Despite the wide applications and easy implementation of these traditional fore-casting models, their ability to understand complex time series is very limited [5]. They assume that the relations between data in time series are linear and are generally used under stationary conditions. In practice however relations in most time series are complex and nonlinear in nature. By their nature traditional linear prediction methods are therefore particularly limited at capturing non-linear signals. The inherent limita-tions of the existing nonphysical based models have led to an effort to find more robust and powerful non-linear prediction methods.

New nonlinear flexible predictor methods extend the power of time series analysis. Often these models are based on Artificial Neural Networks (ANNs) as these are known as nonlinear flexible models that are able to model complex relationships and therefore may be of use in complex time series analysis [6]. Therefore, ANNs have the potential to overcome the inherent limitation of traditional models based on statistical methods. By using such adaptive learning models it may be that it is possible to increase the probability of producing correct predictions by using less "brittle" systems that can classify and generalize data more effectively. A good predictor model will have the ability to discover the correct internal presentation and capture the more subtle patterns in the time series data.

However, in order to use time series data ANNs need to be modified in such a way that the input data presented to the neural network are drawn from a number of sequential previous inputs, rather than a single input vector. The rest of the network is then used to essentially adapt weights in order to minimise the forecasting error. A number of previous studies have proved that ANNs have the potential to produce acceptable results when trying to predict time series based events.

This paper utilises a special type of spiking neural networks called a Polychronous Spiking Neural Network (PSNN) for the prediction of three noisy weather signals representing the maximum temperature, the minimum temperature and the rainfall in the Oak Park weather Station in Ireland in the period between September 2013 up to September 2014.

The reminder of this paper is organized as follows. Section 2 will discuss spiking neural networks, Sect. 3 will shows the simulation results while Sect. 4 will discuss our findings and potential future work.

2   Spiking Neural Networks

2.1   Background

In recent years overwhelming experimental evidence indicates that many biological neural systems use the timing of single action potentials (or "spikes") to encode information [7–9]. It is argued that these new generation of neural networks are potentially much more powerful, and, in reality, a superset, or the "traditional" or rate encoded neural networks hither to used [10].

The "leaky integrate and fire" (LIF) model of Spiking Neural Networks (SNNs) has so far dominated this area of neural network research. However, while this model is biologically appropriate and relatively efficient in terms of computationally cost it suffers from only being able to simulate a relatively small number of spiking behaviours; it cannot simulate a number of spiking (single), bursting (multiple spike burst) or phasic (where a neuron may fire only a single spike at the onset of the input and remain quiescent afterwards) behaviours. Moreover, there is no explicit mechanism for modelling spike latencies or delays. A review of these spiking types is given by Izhikevich [11].

This paper postulates that it is logical to use a neuron model that has the capability for significant temporal control in order to accurately model the temporal nature of the physical time series. For this reason the neural model posited by Izhikevich was favoured over the more traditional LIF model.

2.2   Izhikevich Model

The Izhikevich model uses two variables, a variable representing voltage potential (v) and another representing membrane recovery (activation of potassium currents and inactivation of sodium currents) (u). Voltage is computed by integrating the following two differential equations (usually using Euler's method):

$$\frac{dv}{dt} = 0.04v^2 + 5v + 140 - u + W$$
$$\frac{du}{dt} = a(bv - u)$$

(1)

Where W is the weighted inputs (thalamic input) and a and b are control parameters for the model.

Like LIF when the voltage v exceeds a threshold value (usually 30) both v and u are reset as follows:

$$v \rightarrow c$$
$$u \rightarrow u + d$$

(2)

So in summary a and b are parameters effecting the recovery variable. c is the value for v which occurs after a spike and d is a constant value added to u after a spike.

2.3   Learning and Delays

The learning rule most commonly used for spiking neural networks is derived from traditional Hebbian learning. This is Spike Time Dependent Plasticity (STDP) [12, 13]. STDP adjusts the connection strengths based on the relative timing of a particular neuron's output and input spikes. Synapses increase their efficacy if a pre-synaptic spike arrives just before the postsynaptic neuron is activated but are weakened if they occur afterwards. Repeated pre-synaptic spike arrival a few milliseconds before post-synaptic action potential leads to Long-Term Potentiation (LTP) of the synapses, whereas repeated spike arrival after postsynaptic spikes leads to Long-Term Depression (LTD). The STDP ability to rapidly react to the relative timing of spikes suggests the possibility of temporal coding schemes on a millisecond time scale. This effect has significant biological justification [14, 15].

The weight change Δw_j of a synapse from a pre-synaptic neuron j depends on the relative timing between pre-synaptic spike arrivals and postsynaptic spikes from neuron i. If we label the pre-synaptic spike arrival times at j as t_j^f where f is pre-synaptic spike count at time t, and similarly t_i^n is the postsynaptic spike count at time t then the total weight change Δw_j can be calculated as:

$$\Delta w_j = \sum_{j=1}^{N} \sum_{n=1}^{N} W\left(t_i^j - t_j^f\right) \tag{3}$$

where $W(....)$ denotes a specific STDP function. This function is often:

$$W(x) = A_+ \exp\left(\frac{-x}{\tau_+}\right) for\, x > 0$$
$$W(x) = A_- \exp\left(\frac{x}{\tau_-}\right) for\, x < 0 \tag{4}$$

where parameters A_+ and A_- are parameters dependant on the value of the current synaptic weight and τ_+ and τ_- are time constants/boundaries normally in the order of 10 ms. However it should be noted that other STDP calculations can be used; there is nothing sacrosanct about this particular function [13].

In biological systems axonal conduction delays can be as small as 0.1 ms and as large as 44 ms and yet the reaction delay between any individual pair of linked neurons often has sub millisecond precision. Therefore some timing must be taking place. When both the order and precise timing of firing is taken account an enormous number of spike patterns/trains can be encoded onto even a fairly modest number of neurons. Moreover this 'chain reaction' or cascade of spiking activity means that neurons in such a system are not synchronous but are time locked to each other. Neuronal chains of firing patterns spontaneously emerge in such a network. This is referred to as a gamma rhythm [16, 17]. Moreover groups of neurons contributing to the gamma rhythm are both very precise in their firing pattern activity while also being remarkably tolerant to noise.

This gives the new type of network the capability to very rapidly recognize complex patterns, often with a single neuron's spiking output being a flagged as a result of a pattern of other afferent neuron's reaction to stimulation.

It is argued in this paper that this type of polychromic delay calculation linked with STDP learning are important constructs to use in the effort to make sense of complex natural time series based phenomena.

3 Experiments and Results

The simulation results for the prediction of the physical time series using the proposed PSNN will be presented. Three weather signals are utilized for these experiments representing the maximum and the minimum temperatures and the rainfall from the Oak Weather station in Ireland from the period between September 2013 and September 2014 (the maximum temperature signal is shown in Fig. 1).

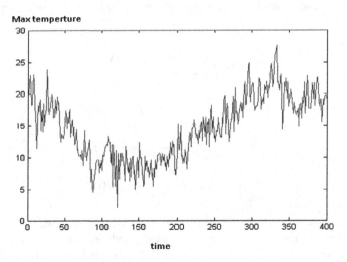

Fig. 1. Daily maximum temperature reading in °C from September 2013 to September 2014 from Oak weather station, Ireland.

In order to encode the daily weather data into the PSNN a very simple metric was used: individual neurons where created that corresponded to a specific weather data values (maximum temperature in the above case). These where then activated modifying the thalamic input to those neurons. In other words the seen data was used to train the network by altering the thalamic input parameters of the individual neurons (corresponding to the values in Y axis in the above graph). As these activations are inherently temporally encoded chains of firing events correspond to the firing events along the X axis. In this way neural network learns the seen part of the signal. As these firing chains are strengthened or weakened a path emerges (by STDP) that corresponds to learning the weather input signal. In practice very many such activations paths are generated as individual neurons firing causes other neurons to fire, and so on. The best performing chain of activations (determined by Euclidean distance) are then used to make a prediction about the weather in the future; the activation patterns of future events are extrapolated by the chain of firing events of the dominant paths that have been learnt in the past.

The performance of our proposed DSMIA neural networks is evaluated using the following predictors:

- A Linear Predictor Coefficients (LPC) model (ARMA based)
- A traditional MLP network

The prediction performance of our networks was evaluated using two statistical metrics which are used to provide accurate tracking of the signals as shown in Table 1. These include the Normalised Mean Square of the Error (NMSE and the Signal to Noise Ratio (SNR).

Normalised Mean Squared Error shows overall deviations between predicted and measured values. NMSE is a useful measure because if a system has a very low NMSE, then it indicates that it is correctly identifying patterns. As it can be shown in Table 2,

Table 1. Performance metrics an their calculations

Metrics	Calculations
NMSE	$$NMSE = \frac{1}{\sigma^2 * N} \sum_{i=1}^{N} (y_i - \hat{y}_i)^2$$ $$\sigma^2 = \frac{1}{N-1} \sum_{i=1}^{N} (y_i - \bar{y}_i)$$
SNR	$$SNR = 10 * log_{10}(sigma)$$ $$sigma = \frac{m^2 * n}{SSE}$$ $$SSE = \sum_{i=1}^{n} (y_i - \hat{y}_i)$$ $$m = max(y)$$

n is the total number of data patterns
y and $\hat{y}$ represent the actual and predicted output value

Table 2. Performance of the PSN for the prediction of the weather signals

Measure	Network	Max temperature	Minimum temperature	Rainfall
SNR	LPC	21.2300	16.1141	18.5367
	MLP	27.1889	24.0815	25.6091
	PSNN	33.5339	26.9713	26.0224
NMSE	LPC	0.2318	0.3844	1.0529
	MLP	0.0331	0.0255	0.0115
	PSNN	0.0104	0.0047	0.0091

the proposed PSNN generated comparable results in terms of the NMSE when compared to the Linear and the MLP network for nonstationary time series prediction.

The Signal to Noise Ratio compares the level of a desired signal to the level of background noise; in this case it is the ratio of useful information about a portfolio compared to false or irrelevant data. The predictions show consistent results. Again, the PSNN has the best SNR for the weather signals. The results also indicate that the proposed PSNN network generates significantly better results than the Linear and the MLP network with an average improvement of 10.2156 dB and 3.2160 dB, respectively as shown in Table 2.

4   Conclusion and Future Works

The Polychronous Spiking Neural Network is proposed for the prediction of weather signals. The simulation results showed significant improvement achieved by the proposed network when compared to the linear and the MLP predictors with an average improvement of 10.2156 dB and 3.2160 dB using the SNR measure, respectively. For the NMSE, the proposed technique showed an average improvement of 0.5483 over the linear predictor and 0.0161 over the MLP network.

However despite improving the SNR and NMSE the training and classification effort of the PSNN compared to the traditional neural networks examined was greater by several orders of magnitude. The PSNN classification stage on a i5 based laptop took several hours; this was because vast quantities of firing paths where generated as the STDP learning algorithm was deployed.

However, this work has both demonstrated the applicability of a particular type of PSNN to weather data forecasting and its potential to perform more effectively than traditional neural network such as the MLP network in non-stationary environments.

Future work will focus on the exploration of improved ways to map the data onto the PSNN, and the adaptation of the classification and grading of candidate solutions for parallel architectures so that different parts of the problem can be solved by decomposition of the search space. Moreover the algorithm itself is an ideal candidate for parallelization; it is hoped that this may significantly improve both training and classification times for the system.

References

1. Mathur, P.S., Kumar, A., Chandra, M.: A feature based neural network model for weather forecasting. In: World Academy of Science, Engineering and Technology (2007)
2. Box, G.E.P., Jenkins, G.W., Reinsel, G.C.: Time Series Analysis - Forecasting and Control, 3rd edn. Prentice-Hall, Englewood Cliffs (1994)
3. Sfetsos, A.: A comparison of various forecasting techniques applied to mean hourly wind speed time series. Renewable Energy 21(1), 23–35 (2000)
4. Ho, S., Xie, M., Goh, T.N.: A comparative study of neural network and Box-Jenkins ARIMA modeling in time series prediction. Comput. Ind. Eng. 42(2–4), 371–375 (2002)

5. Faraway, J., Chatfield, C.: Time series forecasting with neural networks: a comparative study using the airline data. J. R. Stat. Soc. **47**(2), 231–250 (1998)
6. Zhang, G.P., Patuwo, B.E., Hu, M.Y.: A simulation study of artificial neural networks for nonlinear time-series forecasting. Comput. Oper. Res. **28**, 381–396 (2001)
7. Hodgkin, A., Huxley, A.: A quantitative description of membrane current and its application to conduction and excitation in nerve. J. Physiol. **117**, 500–544 (1952)
8. Abeles, M.: Corticonics: Neural circuits of the cerebral cortex. Cambridge University Press, New York (1991)
9. Maass, W.: Networks of spiking neurons the third generation of neural network models. Neural Netw. **10**(9), 1659–1671 (1997). Elsevier Publishing
10. Maass, W., Bishop, C.M.: Pulsed Neural Networks. MIT Press, Cambridge (1998). ISBN 0-262-13350-4
11. Izhikevich, E.M.: Simple model of spiking neurons. IEEE Trans. Neural Netw. **14**(6), 1569–1572 (2003)
12. Legenstein, R., Naeger, C., Maass, W.: What can a neuron learn with spike-timing-dependent-plasticty. J. Neural Comput. **17**(11), 2337–2382 (2005)
13. Nessler, B., Pfeiffer, M., Maass, W.: STDP enables spiking neurons to detect hidden causes of their inputs. In: Proceedings of NIPS 2009: Advances in Neural Information Processing Systems, vol. 22, pp. 1357–1365. MIT Press (2010)
14. Levy, W.B., Steward, O.: Temporal contiguity requirements for long-term associative potentiation/depression in the hippocampus. Neuroscience **8**(4), 791–797 (1983)
15. Markram, H., Lübke, J., Frotscher, M., Sakmann, B.: Regulation of synaptic efficacy by coincidence of postsynaptic APs and EPSPs. Science **275**(5297), 213–215 (1997)
16. Izhikevich, E.M.: Polychronization: computation with spikes. J. Neural Comput. **18**(2), 245–282 (2006)
17. Arthur, J.V., Boahen, K.A.: Synchrony in Silicon The Gamma Rhythm. IEEE Trans. Neural Netw. **18**(6), 1815–1825 (2007)

A Water Wave Optimization Algorithm with Variable Population Size and Comprehensive Learning

Bei Zhang, Min-Xia Zhang, Jie-Feng Zhang, and Yu-Jun Zheng$^{(\boxtimes)}$

College of Computer Science and Technology,
Zhejiang University of Technology, Hangzhou 310023, China
{zhangbei-zjut, zhangjiefeng_zjut}@outlook.com,
zmx@zjut.edu.cn, yujun.zheng@computer.org

Abstract. Water wave optimization (WWO) is a new nature-inspired meta-heuristic by mimicking shallow water wave motions including propagation, refraction, and breaking. In this paper we present a variation of WWO, named VC-WWO, which adopts a variable population size to accelerate the search process, and develops a comprehensive learning mechanism in the refraction operator to make stationary waves learn from more exemplars to increase the solution diversity, and thus provides a much better tradeoff between exploration and exploitation. Experimental results show that the overall performance of VC-WWO is better than the original WWO and other comparative algorithms on the CEC 2015 single-objective optimization test problems, which validates the effectiveness of the two new strategies proposed in the paper.

Keywords: Water wave optimization (WWO) · Variable population size · Comprehensive learning · Global optimization

1 Introduction

Evolutionary algorithms (EAs) are stochastic methods taking inspiration from natural/biological evolution for solving complex, typically non-convex optimization problems. With the increase of the complexity and scale of real-world engineering problems, traditional EAs such as genetic algorithms (GAs) [8] and particle swarm optimization (PSO) [6] often suffer the problems of low convergence speed, premature convergence, and low solution quality. Consequently, a variety of novel EAs has been proposed and has aroused much interest in the community in recent years [3].

Initially proposed by Zheng [18], water wave optimization (WWO) is a relatively new EA borrowing ideas from shallow water wave theory [7] for global optimization. As most other EAs, WWO maintains a population of solutions which are called "waves", and iteratively evolves the solutions to find the optimum by mimicking the phenomena of water waves including propagation, refraction and breaking. The key idea of WWO is that low energy (fitness) waves have large wavelengths and thus explore in larger areas with short life cycles, while high energy (fitness) waves have small wavelengths and thus intensify local search in smaller areas with long life cycles.

© Springer International Publishing Switzerland 2015
D.-S. Huang et al. (Eds.): ICIC 2015, Part I, LNCS 9225, pp. 124–136, 2015.
DOI: 10.1007/978-3-319-22180-9_13

The results of experiments on the CEC 2014 benchmark problems [11] show that the overall performance of WWO is better than a set of state-of-the-art metaheuristic optimization methods. WWO has also shown its applicability and effectiveness on a real-world high-speed train scheduling problem in China [18].

In this paper we propose a new variation of WWO, named VC-WWO, which uses a variable population size strategy to further accelerate the convergence speed, and designs a comprehensive learning mechanism for the refraction operator to enable waves lost their energy to learn from not only the best wave but also other better exemplars to diverse the population and avoid search stagnation. We then conduct numerical experiments on the CEC 2015 single-objective optimization benchmark suite [12], and the results demonstrate that VC-WWO achieves a much better performance than the basic WWO and three other comparative EAs.

The remainder of the paper is organized as follows: Sect. 2 introduces the basic WWO algorithm, Sect. 3 describes the VC-WWO algorithm in detail, Sect. 4 presents the numerical experiments, and Sect. 5 concludes.

2 The Basic WWO Algorithm

Based on shallow water wave models, Zheng [18] developed the WWO algorithm, where the search space is analogous to the seabed area, each solution is analogous to a "wave" with a height h and a wavelength λ, and the solution fitness is measured by its seabed depth: the shorter the distance to the still water level, the higher the fitness is.

When solving an optimization problem, WWO firstly initializes a set of waves, whose heights are all set to a constant h_{max} and wavelengths are all set to 0.5; during the search process, WWO employs three operators, i.e. propagation, refraction and breaking, to evolve the population to the global optimum.

The *propagation* operator propagates each wave once at each generation, and creates a new wave X' by adding a different offset at each dimension d to the original wave X as:

$$X'_d = X_d + rand(-1, 1) \cdot \lambda L_d \tag{1}$$

where *rand* generates a random number uniformly distributed within the specified range, and L_d is the length of the dth dimension of the search space. After forming the new wave X', its fitness is evaluated and compared with the original wave X: If $f(X')$ is higher than $f(X)$, X' replaces X in the population, and the height of X' is reset to h_{max}; Otherwise X remains but its height is reduced by one.

A natural phenomenon is that waves in deep water have long wavelengths and low wave heights, while ones in shallow water have short wavelengths and high wave heights, as illustrated in Fig. 1 [18]. When a wave moving from deep water (low fitness location) to shallow water (high fitness location), its wavelength decreases. WWO imitates the phenomenon by updating the wavelength of each wave X after each generation as:

$$\lambda = \lambda \cdot \alpha^{-(f(X)-f_{\min}+\varepsilon)/(f_{\max}-f_{\min}+\varepsilon)} \tag{2}$$

where $f_{\max}$ and $f_{\min}$ are the maximum and minimum fitness values in current generation respectively, α is the wavelength reduction coefficient, and ε is a small constant to avoid zero-division-error.

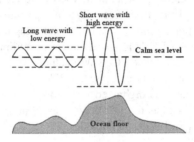

Fig. 1. Different wave shapes in deep and shallow water.

The *refraction* operator only performs on stationary waves whose heights decrease to zero. It makes these waves move to new positions by learning from the best wave in the current population. That is, at each dimension d, a new wave X' is created based on the position of both the original wave X and the best wave X_{best} as:

$$X'_d = norm\left(\frac{X_{\text{best},d} + X_d}{2}, \frac{X_{\text{best},d} - X_d}{2}\right) \tag{3}$$

where *norm* generates a Gaussian random number with specified mean and standard deviation. After refraction, the wave height of X' is reset to $h_{\max}$, and its wavelength is updated as:

$$\lambda' = \lambda\frac{f(X)}{f(X')} \tag{4}$$

The *breaking* operator only breaks a new wave X, whose fitness is higher than the current best wave, into a train of solitary waves. Each offspring wave X' is generated by randomly selecting k dimensions (where k is a random number between 1 and a parameter $k_{\max}$) and shifting each dimension d of the original position as:

$$X'_d = X_d + norm(0, 1) \cdot \beta L_d \tag{5}$$

where β is the breaking coefficient. If the fittest solitary wave X' is better than X, X' takes the place of X in the population.

Algorithm 1 presents the basic procedure of the WWO algorithm.

Algorithm 1. The WWO algorithm.

---
1 Randomly initialize a population P of n solutions (waves);
2 **while** stop criterion is not satisfied **do**
3   **for each** wave $X \in P$ **do**
4   Propagate X to a new X' according to Eq. (1);
5   **if** $f(X') > f(X)$ **then**
6   **if** $f(X') > f(X_{best})$ **then**
7   Break X' into new waves according to Eq. (5);
8   Update X_{best} with X';
9   Replace X with X';
10   **else**
11       $X.h = X.h - 1$;
12   **if** $X.h = 0$ **then**
13   Refract X to a new X' according to Eq. (3) and (4);
14   Update the wavelengths according to Eq. (2);
15 **return** the best solution found so far.

---

3 The VC-WWO Algorithm

The proposed VC-WWO algorithm aims to improve the basic WWO with a variable population size strategy and a comprehensive learning mechanism.

3.1 A Variable Population Size Strategy

Most EAs maintain a fixed population size at each generation, which is easy to implement but may limit the adaptability of the algorithms. Thereby, Smith [15] and Arabas [1] first introduced variable population size to GAs and demonstrated its capability of refining the solutions and reducing the parameter sensitivity. Recently, some researchers [2, 4, 10, 13] also used a dynamic population size in PSO, DE [16], and other hybrid EAs, where the population size is adjusted according to the evaluation of search status or population quality. However, such strategies also increase the complexity and computational burden of the algorithms.

In general, a large population size prefers to explore the whole solution space in early search stages, while a small population size can be used to facilitate local exploitation in later stages. Therefore, we design a simple linear model which decreases the population size n from a maximum size n_{max} to a minimum size n_{min} as [19]:

$$n = n_{max} - (n_{max} - n_{min}) \cdot \frac{t}{t_{max}} \qquad (6)$$

where t is the current generation number (or the number of fitness evaluations, NFEs), and t_{max} is the maximum generation number (or NFEs).

Assume the current population size is n, according to Eq. (6) the size will be updated to n' for the next generation, we can sort the individuals in decreasing order of fitness values and weed out the worst $(n-n')$ individuals from the population. In most cases t_{max} is much larger than n, and thus only the worst individual needs to be removed. Consequently, the proposed linear decreasing population size strategy is very simple and easy to implement. However, it can benefit the algorithm in two aspects:

- More individuals explore larger areas in early stages and fewer ones exploit smaller areas in later stages, which is expected to balance the global and local search better than the fixed population size and thus promote the convergence process;
- With the decrease of population, the required NFEs reduces by generations. Thereby, under the limit of the same computational cost (NFEs), the algorithm can evolve for more generations, which is expected to increase the solution accuracy.

3.2 A Comprehensive Learning Mechanism

As shown in Eq. (3), the key idea of the refraction operator in WWO is to make a wave that fails to find better positions for a certain period learn from the best wave found so far. Such a learning mechanism is similar to the global best model in PSO [9], except that PSO makes all particles fly to the global best while WWO only enables those zero-height waves to move towards the best wave. Similarly, such waves can be strongly attracted by the current best wave which may be trapped in a local optimum, and thus the basic WWO often takes the risk of premature convergence.

In order to tackle this limitation, we design a comprehensive learning mechanism that makes those zero-height waves learn from more other exemplars in the current population rather than only the best one. That is, when performing refraction on a wave X, we select a different exemplar, denoted as X_{ex}, at each dimension d, and update the dimension of X as:

$$X'_d - X_d \mid rand(0, 1) \cdot (X_{ex,d} - X_d) \tag{7}$$

where X_{ex} is probabilistically selected according to its fitness. In VC-WWO, we set the selection probability of each other wave X of being the exemplar as:

$$p(X) = \frac{f(X) - f_{min} + \varepsilon}{f_{max} - f_{min} + \varepsilon} \tag{8}$$

Thus high-fitness individuals have high selection probabilities but low-fitness ones still have chances to be exemplars, and stationary waves can learn from different exemplars at different dimensions so that the solution diversity can be greatly improved and the risk of premature convergence can be decreased.

3.3   The Framework of VC-WWO

Algorithm 2 presents the framework of the VC-WWO algorithm.

Algorithm 2. The VC-WWO algorithm.

1 Randomly initialize a population P of n solutions (waves);
2 **while** stop criterion is not satisfied **do**
3   **for each** wave $X \in P$ **do**
4   Propagate X to a new X' according to Eq. (1);
5   **if** $f(X') > f(X)$ **then**
6   **if** $f(X') > f(X_{best})$ **then**
7   Break X' into new waves according to Eq. (5);
8   Update X_{best} with X';
9   Replace X with X';
10   **else** set $X.h = X.h - 1$;
11   **if** $X.h = 0$ **then**
12   **for** $d = 1$ to D **do**
13   Select an exemplar wave X_{ex} with probability according to Eq. (8);
14   Refract X at the dth dimension according to Eq. (7);
15   Set n according to Eq. (6) and Update the wavelengths according to Eq. (2);
16 **return** the best solution found so far.

---

4   Numerical Experiments

4.1   Experimental Settings

We conduct numerical experiments on the CEC 2015 single-objective optimization test suite [12], which consists of 15 benchmark functions, denoted by $f_1 - f_{15}$, as summarized in Table 1 (where f^* denotes the exact optimal function value). The functions are all scalable high-dimensional problems, covering four types including unimodal, simple multimodal, hybrid, and composition. In this paper, we use 30-D functions.

To test the effectiveness of our new strategies, we compare the proposed VC-WWO with the basic WWO [18] and the following three popular metaheuristics:

- The particle swarm optimization (PSO) algorithm [8].
- The biogeography-based optimization (BBO) algorithm [14].
- The harmony search (HS) algorithm [5].

For both WWO and VC-WWO, we set $h_{max} = 12$, $\alpha = 1.0026$, $k_{max} = 12$, and β linearly decreases from 0.25 to 0.001. Empirically, we set $n = 10$ for WWO and $n_{max} = 50$ and $n_{min} = 3$ for VC-WWO. The control parameters of the other three algorithms are set as in the literature. Besides, it should be noted that we just simply use a fixed parameter setting for each algorithm on all the test problems instead of

Table 1. A summary of the CEC 2015 benchmark functions.

Type	ID	Function	f^*
Unimodal	f_1	Rotated High Conditioned Elliptic Function	100
	f_2	Rotated Bent Cigar Function	200
Multimodal	f_3	Shifted and Rotated Ackley's Function	300
	f_4	Shifted and Rotated Rastrigin's Function	400
	f_5	Shifted and Rotated Schwefel's Function	500
Hybrid	f_6	Hybrid Function 1 (N = 3)	600
	f_7	Hybrid Function 2 (N = 4)	700
	f_8	Hybrid Function 3 (N = 5)	800
Composition	f_9	Composition Function 1 (N = 3)	900
	f_{10}	Composition Function 2 (N = 3)	1000
	f_{11}	Composition Function 3 (N = 5)	1100
	f_{12}	Composition Function 4 (N = 5)	1200
	f_{13}	Composition Function 5 (N = 5)	1300
	f_{14}	Composition Function 6 (N = 7)	1400
	f_{15}	Composition Function 7 (N = 10)	1500

fine-tuning the parameter values. To ensure a fair comparison, all the algorithms have a maximum NFEs of 300,000 on each test problem. The experimental environment is a computer of Intel Core i5-2430M processor and 4 GB memory.

4.2   Comparative Experiment

We run the comparative algorithms for 50 times (with different random seeds) on each test problem, and record the maximum (max), minimum (min), median and standard deviation (std) values of function values among the 50 runs. Table 2 presents the experimental results of the algorithms on the unimodal and simple multimodal benchmark functions, and Table 3 gives the results on the hybrid and composition functions. On each problem, the best median values among the five algorithms are marked with boldface (the bold values are better than those seemingly same values not in boldface because the digits after the second decimal place are omitted).

From the experimental results we can see that:

- On the unimodal group of 2 functions, VC-WWO obtains the best median values on both the two functions.
- On the simple multimodal group of 3 functions, VC-WWO obtains the best median value on f_3, HS does so on f_4, and BBO does so on f_5.
- On the hybrid group of 3 functions, VC-WWO obtains the best median values on f_7, while WWO does so on f_6 and f_8.

Table 2. Experimental results on the unimodal and simple multimodal benchmark functions.

ID	Metric	PSO	BBO	HS	WWO	VC-WWO
f_1	max	2.59E + 08	1.57E + 07	1.55E + 07	3.07E + 06	1.56E + 06
	min	2.46E + 06	1.80E + 06	9.96E + 05	2.45E + 05	2.40E + 05
	median	2.87E + 07	5.99E + 06	4.62E + 06	1.13E + 06	**7.70E + 05**
	std	4.71E + 07	3.20E + 06	3.22E + 06	7.45E + 05	3.09E + 05
	h	1^+	1^+	1^+	1^+	
f_2	max	1.76E + 10	1.86E + 06	5.17E + 05	6.19E + 03	1.66E + 03
	min	1.52E + 04	2.50E + 05	1.25E + 03	2.01E + 02	2.02E + 02
	median	4.61E + 09	7.85E + 05	5.81E + 04	8.04E + 02	**4.58E + 02**
	std	3.76E + 09	3.87E + 05	9.96E + 04	1.87E + 03	3.55E + 02
	h	1^+	1^+	1^+	1^+	
f_3	max	3.21E + 02	3.20E + 02	3.21E + 02	3.20E + 02	3.20E + 02
	min	3.20E + 02	3.20E + 02	3.21E + 02	3.20E + 02	3.20E + 02
	median	3.21E + 02	3.20E + 02	3.21E + 02	3.20E + 02	**3.20E + 02**
	std	1.51E-01	2.62E-02	7.06E-02	1.03E-05	3.91E-06
	h	1^+	1^+	1^+	0	
f_4	max	5.92E + 02	4.81E + 02	4.96E + 02	5.55E + 02	4.78E + 02
	min	4.72E + 02	4.21E + 02	4.17E + 02	4.56E + 02	4.26E + 02
	median	5.26E + 02	4.47E + 02	**4.28E + 02**	4.96E + 02	4.45E + 02
	std	2.65E + 01	1.28E + 01	1.17E + 01	2.32E + 01	1.23E + 01
	h	1^+	0	1^-	1^+	
f_5	max	5.29E + 03	3.60E + 03	6.43E + 03	4.70E + 03	6.62E + 03
	min	2.69E + 03	1.66E + 03	1.29E + 03	1.54E + 03	2.40E + 03
	median	4.03E + 03	**2.44E + 03**	3.96E + 03	3.20E + 03	3.31E + 03
	std	5.79E + 02	4.63E + 02	1.19E + 03	6.16E + 02	6.55E + 02
	h	1^+	1^-	1^+	0	

- On the composition group of 7 functions, except that on f_{13} HS obtains the best median value, VC-WWO obtains the best median values on all the remaining 6 functions, and WWO obtains the same value with VC-WWO on f_{15}.

As a result, among the 15 benchmark functions, VC-WWO obtains the best median values on 10 problems, WWO and HS respectively do so on 3 problems, and BBO does so on 1 problem. In particular, VC-WWO exhibits promising performance advantage over the other algorithms on the most complex composition group, while WWO shows the best performance on the hybrid group, which demonstrates their effectiveness for solving different types of functional optimization problems.

We have also performed nonparametric Wilcoxon rank sum tests between the results of VC-WWO and each of the other four algorithms on each problem, and given the resulting h values in Tables 2 and 3, where an h value of 1^+ implies the performance of VC-WWO is significantly better than the corresponding algorithm (with a 95 %

Table 3. Experimental results on the hybrid and composition benchmark functions.

ID	Metric	PSO	BBO	HS	WWO	VC-WWO
f_6	max	3.39E + 07	1.42E + 07	4.90E + 06	3.57E + 05	3.45E + 05
	min	1.80E + 05	2.53E + 05	3.40E + 05	8.89E + 03	1.19E + 04
	median	1.02E + 06	1.63E + 06	1.26E + 06	**6.76E + 04**	9.87E + 04
	std	7.56E + 06	2.73E + 06	1.14E + 06	8.32E + 04	7.47E + 04
	h	1^+	1^+	1^+	1^-	
f_7	max	7.56E + 02	7.16E + 02	7.64E + 02	7.19E + 02	7.16E + 02
	min	7.09E + 02	7.07E + 02	7.08E + 02	7.07E + 02	7.07E + 02
	median	7.20E + 02	7.13E + 02	7.12E + 02	7.12E + 02	**7.11E + 02**
	std	1.04E + 01	1.54E + 00	7.54E + 00	2.54E + 00	1.74E + 00
	h	1^+	1^+	0	1^+	
f_8	max	5.81E + 06	2.71E + 06	2.44E + 06	1.73E + 05	8.31E + 05
	min	8.67E + 03	6.35E + 04	4.76E + 04	2.28E + 03	3.09E + 03
	median	3.43E + 05	8.04E + 05	3.23E + 05	**5.04E + 04**	1.54E + 05
	std	8.88E + 05	6.89E + 05	4.19E + 05	3.52E + 04	1.80E + 05
	h	1^+	1^+	1^+	1^-	
f_9	max	1.27E + 03	1.01E + 03	1.12E + 03	1.00E + 03	1.00E + 03
	min	1.00E + 03	1.00E + 03	1.00E + 03	1.00E + 03	1.00E + 03
	median	1.03E + 03	1.00E + 03	1.00E + 03	1.00E + 03	**1.00E + 03**
	std	4.63E + 01	4.88E-01	1.67E + 01	3.56E-01	2.00E-01
	h	1^+	1^+	1^+	1^+	
f_{10}	max	5.12E + 07	4.85E + 03	2.43E + 03	2.33E + 03	2.02E + 03
	min	2.77E + 03	1.79E + 03	1.23E + 03	1.22E + 03	1.22E + 03
	median	1.15E + 06	2.88E + 03	2.09E + 03	1.43E + 03	**1.40E + 03**
	std	1.36E + 07	5.30E + 02	2.95E + 02	3.60E + 02	1.71E + 02
	h	1^+	1^+	1^+	0	
f_{11}	max	2.40E + 03	2.00E + 03	1.89E + 03	2.19E + 03	1.90E + 03
	min	1.41E + 03	1.74E + 03	1.61E + 03	1.41E + 03	1.40E + 03
	median	2.12E + 03	1.86E + 03	1.75E + 03	1.75E + 03	**1.42E + 03**
	std	1.79E + 02	6.33E + 01	6.06E + 01	2.67E + 02	2.00E + 02
	h	1^+	1^+	1^+	0	
f_{12}	max	1.40E + 03	1.31E + 03	1.31E + 03	1.31E + 03	1.30E + 03
	min	1.31E + 03	1.31E + 03	1.31E + 03	1.30E + 03	1.30E + 03
	median	1.40E + 03	1.31E + 03	1.31E + 03	1.31E + 03	**1.30E + 03**
	std	3.40E + 01	1.35E + 00	6.51E-01	1.67E + 00	4.62E-01
	h	1^+	1^+	1^+	1^+	
f_{13}	max	1.30E + 03	1.30E + 03	1.30E + 03	1.30E + 03	1.30E + 03
	min	1.30E + 03	1.30E + 03	1.30E + 03	1.30E + 03	1.30E + 03
	median	1.30E + 03	1.30E + 03	**1.30E + 03**	1.30E + 03	1.30E + 03
	std	4.11E-02	3.67E-03	1.91E-03	9.81E-03	7.03E-03
	h	0	1^+	1^-	0	

(*Continued*)

Table 3. (*Continued*)

ID	Metric	PSO	BBO	HS	WWO	VC-WWO
f_{14}	max	6.34E + 04	3.72E + 04	3.61E + 04	3.91E + 04	3.72E + 04
	min	3.47E + 04	3.29E + 04	3.33E + 04	3.27E + 04	3.26E + 04
	median	3.92E + 04	3.50E + 04	3.51E + 04	3.47E + 04	**3.42E + 04**
	std	5.99E + 03	9.12E + 02	5.55E + 02	1.54E + 03	1.32E + 03
	h	1^+	0	1^+	0	
f_{15}	max	2.10E + 03	1.60E + 03	1.60E + 03	1.60E + 03	1.60E + 03
	min	1.61E + 03	1.60E + 03	1.60E + 03	1.60E + 03	1.60E + 03
	median	1.62E + 03	1.60E + 03	1.60E + 03	**1.60E + 03**	**1.60E + 03**
	std	7.95E + 01	1.67E-01	5.65E-02	7.11E-13	1.04E-12
	h	1^+	1^+	1^+	0	

confidence level), 1^- vice versa, and 0 denotes there is no significant difference. As we can see, in comparison with WWO, VC-WWO shows significantly better performance on 6 functions, worse performance on 2 functions (f_6 and f_8), and there is no significant difference on the remaining 7 functions. This validates the effectiveness of the combination of the two new strategies used in VC-WWO.

By comparing with the other three algorithms, we can see that VC-WWO performs significantly better than PSO on 14 functions, better than BBO on 12 functions and worse than BBO on 1 function (f_5), better than HS on 12 functions and worse than HS on 2 functions (f_4 and f_{13}), and there is no significant difference in other cases.

Moreover, we present the convergence curves of the five algorithms on the 15 test problems in Fig. 2(a)–(o) respectively. As we can see, the two WWO versions converge much faster than the others on a majority of the problems, which indicates their performance advantages on the suite. On the other hand, by comparing the VC-WWO with the basic WWO, we can find out that, on most functions, WWO converges faster than VC-WWO at the early stages, but its curves fall behind the curves of VC-WWO in later stages where VC-WWO continues to refine the solutions and finally reaches better results than WWO. This is mainly because of the following two reasons:

- VC-WWO maintains a large population size and thus converges slower than WWO in early stages, but with the decrease of population size, VC-WWO accelerates its search process better than WWO;
- VC-WWO makes stationary individuals learn from different exemplars at different dimensions so that the algorithm has capability of jumping out of the local optima where WWO may be trapped.

In summary, the test results show that the overall performance of VC-WWO is better than other comparative algorithms (including the basic WWO) on the benchmark suite, which demonstrates that the combination of the two strategies can provide VC-WWO with a much better tradeoff between exploration and exploitation, and thus suppress premature convergence and improve solution quality.

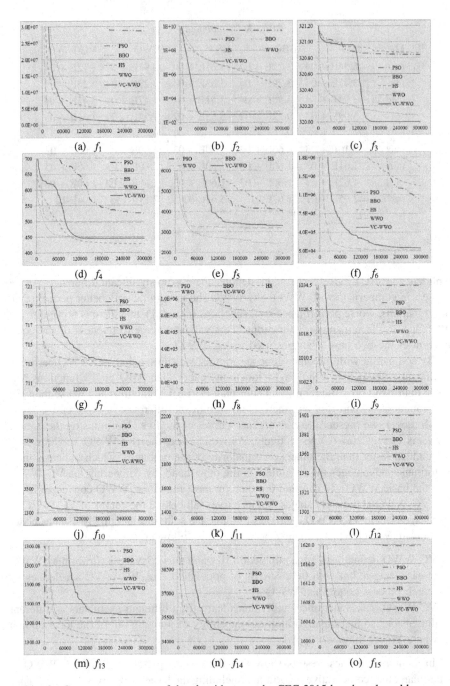

Fig. 2. Convergence curves of the algorithms on the CEC 2015 benchmark problems.

5 Conclusion

Borrowing ideas from shallow water wave theory, WWO is a new metaheuristic optimization method that has a simple algorithmic framework and great search capability for high dimensional problems. This paper proposes VC-WWO, a variation of WWO which utilizes a variable population size to speed up the convergence and designs a comprehensive learning mechanism to diversify the population and avoid search stagnation. Numerical experiments on the CEC 2015 benchmark suite show that the overall performance of VC-WWO is better than WWO and other three popular algorithms.

According to the numerical results on the CEC 2015 test functions, VC-WWO and WWO both have advantages and disadvantages for coping with different types of optimization problems. As there is no free lunch for optimization [17], we do not aim to replace the basic WWO with the new VC-WWO, but to provide VC-WWO as a novel alternative for global optimization, especially for those unknown problems. Our future work includes adapting WWO to constrained and/or multiobjective optimization problems.

Acknowledgements. The work is supported by grants from National Natural Science Foundation (No. 61473263) and Zhejiang Provincial Natural Science Foundation (No. LY14F030011) of China.

References

1. Arabas, J., Michalewicz, Z., Mulawka, J.: GAVaPS-a genetic algorithm with varying population size. In: Proceedings of the First IEEE Conference on Evolutionary Computation, pp. 73–78. IEEE Press, New York (1994)
2. Brest, J., Maucec, M.S.: Population size reduction for the differential evolution algorithm. Appl. Intell. **29**(3), 228–247 (2008)
3. Boussaid, I., Lepagnot, J., Siarry, P.: A survey on optimization metaheuristics. Inf. Sci. **237**(1), 82–117 (2013)
4. Chen, D.B., Zhao, C.X.: Particle swarm optimization with adaptive population size and its application. Appl. Softw. Comput. **9**(1), 39–48 (2009)
5. Geem, Z.W., Kim, J.H., Loganathan, G.V.: A new heuristic optimization algorithm: harmony search. Simulation **76**(2), 60–68 (2001)
6. Holland, J.H.: Adaptation in natural and artificial systems: an introductory analysis with applications to biology, control and artificial intelligence. University of Michigan Press, Michigan (1975)
7. Huang, H.: Dynamics of surface waves in coastal waters: wave-current-bottom interactions. Springer, Berlin-Heidelberg (2009)
8. Kennedy, J., Eberhart, R.: Particle swarm optimization. In: Proceeding of the IEEE International Conference on Neural Networks, vol. 4, pp. 1942–1948. IEEE Press, New York (1995)
9. Kennedy, J., Mendes, R.: Neighborhood topologies in fully informed and best-of-neighborhood particle swarms. IEEE Trans. Syst. Man Cybern. Part C **36**(4), 515–519 (2006)

10. Koumousis, V.K., Katsaras, C.P.: A saw-tooth genetic algorithm combining the effects of variable population size and reinitialization to enhance performance. IEEE Trans. Evol. Comput. **10**(1), 19–28 (2006)
11. Liang, J.J., Qu, B.Y., Suganthan, P.N.: Problem definitions and evaluation criteria for the CEC 2014 special session and competition on single objective real-parameter numerical optimization. Technical report 201311, Computational Intelligence Laboratory, Zhengzhou University, Zhengzhou, China (2014)
12. Liang, J. J., Qu, B. Y., Suganthan, P. N., Chen, Q.: Problem definitions and evaluation criteria for the CEC 2015 competition on learning-based real-parameter single objective optimization. Technical report 201411A, Computational Intelligence Laboratory, Zhengzhou University, Zhengzhou, China (2014)
13. Shi, X.H., Wan, L.M., Lee, H.P., Yang, X.W., Wang, L.M., Liang, Y.C.: An improved genetic algorithm with variable population-size and a PSO-GA based hybrid evolutionary algorithm. In: 2003 International Conference on Machine Learning and Cybernetics, vol. 3, pp. 1735–1740 (2003)
14. Simon, D.: Biogeography-based optimization. IEEE Trans. Evol. Comput. **12**(6), 702–713 (2008)
15. Smith, R.E.: Adaptively resizing populations: an algorithm and analysis. In: Proceedings of the 5th International Conference on Genetic Algorithms (ICGA 1993), pp. 653–653 (1993)
16. Storn, R., Price, K.: Differential evolution-a simple and efficient heuristic for global optimization over continuous spaces. J. Global Optim. **11**(4), 341–359 (1997)
17. Wolpert, D.H., Macready, W.G.: No free lunch theorems for optimization. IEEE Trans. Evol. Comput. **1**(1), 67–82 (1997)
18. Zheng, Y.J.: Water wave optimization: a new nature-inspired metaheuristic. Comput. Oper. Res. **55**, 1–11 (2015)
19. Zheng, Y.J., Zhang, B.: A simplified water wave optimization algorithm. In: Proceedings of the 2015 IEEE Congress on Evolutionary Computation, pp. 807–813. IEEE Press, New York (2015)

Water Wave Optimization for the Traveling Salesman Problem

Xiao-Bei Wu[1](✉), Jie Liao[2](✉), and Zhi-Cheng Wang[1](✉)

[1] College of Electronics and Information Engineering,
Tongji University, Shanghai 201804, China
xb.wu@ymail.com, zhichengwang@tongji.edu.cn
[2] AVIC Jiangxi Hongdu Aviation Industry Group Co., Ltd.,
Nangchang 330024, Jiangxi, China
liao-jie@hongdu.com.cn

Abstract. Water wave optimization (WWO) is a novel evolutionary algorithm borrowing ideas from shallow water wave models for global optimization problems. This paper presents a first study on WWO for a combinatorial optimization problem — the traveling salesman problem (TSP). We adapt the operators in the original WWO so as to effectively exploring in a discrete solution space. The results of simulation experiments on a set of test instances from TSPLIB show that the proposed WWO algorithm is not only applicable and efficient for TSP, but also has significant performance advantage in comparison with two other methods, genetic algorithm (GA) and biogeography-based optimization (BBO).

Keywords: Water wave optimization (WWO) · Traveling salesman problem (TSP) · Combinatorial optimization · Propagation

1 Introduction

Nature-inspired computing has been fascinating computer scientists for a long time, giving rise to a variety of evolutionary algorithms including genetic algorithms (GA) [1], particle swarm optimization (PSO) [2], differential evolution (DE) [3], ant colony optimization (ACO) [4], biogeography-based optimization (BBO) [5], bat algorithm (BA) [6], etc. Very recently, inspired by wave motions controlled by the wave-current bottom interactions [7], Zheng proposed a novel optimization algorithm called water wave optimization (WWO) [8], which derives effective mechanisms from water wave phenomena such as propagation, refraction, and breaking for solving high-dimensional global optimization problems. Numerical tests and real-world applications have shown that WWO is very competitive to other state-of-the-art heuristic algorithms.

WWO is originally proposed for global optimization in continuous search spaces. However, as other global optimization methods such as PSO and DE that have been adapted to solve many combinatorial optimization problems [9, 10], we believe that WWO also has a great potential in combinatorial optimization.

© Springer International Publishing Switzerland 2015
D.-S. Huang et al. (Eds.): ICIC 2015, Part I, LNCS 9225, pp. 137–146, 2015.
DOI: 10.1007/978-3-319-22180-9_14

In this paper, we develop a new WWO algorithm for a well-known combinatorial optimization problem, the traveling salesman problem (TSP) [11]. TSP is a classical *NP*-hard problem in which a salesman tries to find the shortest closed tour to visit a set of n cities (vertices) under the condition that each city is visited exactly once. As we can see, a solution to the TSP is a permutation of the set of n cities problem, and thus the studies on TSP also provide a foundation for a wide range of permutation-based optimization problems in areas such as scheduling and routing [12].

The methods for solving TSP so far can be divided into two parts: the exact methods and the heuristic methods. The exact methods are mainly based on enumeration search [13] and branch-and-bound [14]. They guarantee to achieve the actual optimal (shortest) tour, but their computational time increases exponentially with the problem size n, which makes them impossible for solving large-size problem instances in a reasonable time. On the contrary, heuristic methods typically cannot guarantee the optimal solution, and their aim is to obtain a satisfactory near-optimal solution in a reasonable time. In recent years, evolutionary algorithms have been a mainstream for solving large-size TSP [4, 15–18].

In this paper we present a first study on WWO for the TSP. We redefine the propagation, refraction, and breaking operators for effectively exploring in the solution space of TSP, and test the performance on a set of problem instances. The experimental results show that the proposed algorithm has significant performance advantage in comparison with two promising algorithm in the literature.

The rest of our paper is organized as follows. Section 2 gives a brief introduction to WWO. Section 3 describes how to adapt the operators of the original WWO for the TSP. Section 4 presents the simulation experiments, and Sect. 5 concludes.

2   Water Wave Optimization

WWO [8] is a novel global optimization algorithm based on the shallow water wave theory, which uses numerical techniques to study the evolution of wave heights, periods, and propagation directions under various conditions such as wind forcing, nonlinear wave interactions, and frictional dissipation [7, 19]. In WWO, the solution space is analogous to the seabed area, and the fitness of a point in the space is measured inversely by its seabed depth: the shorter the distance to the still water level, the higher the fitness. It should be noted that by analogy the 3-D space of the seabed is generalized to a high-dimensional space.

As most evolutionary algorithms, WWO maintains a population of solutions, each of which is analogous to a "wave" with a height h in the integer domain and a wavelength λ in the real domain (initialized as a constant $h_{\max}$ and 0.5 respectively). A good wave (high quality solution) located in shallow water has a large height (high energy) and a short wavelength, while a bad wave (poor quality solution) located in shallow water has a small height (low energy) and a long wavelength, as illustrated in Fig. 1. The search process of the WWO algorithm in the solution space is modeled as the propagation, refraction, and breaking of the waves.

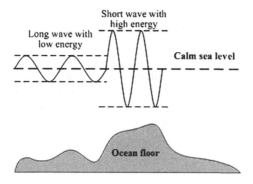

Fig. 1. Illustration of the WWO algorithm model

2.1 Propagation

In propagation, each wave x propagates exactly once at each generation, which creates a new wave x' by shifting each dimension d of the original wave x as follows:

$$x'(d) = x(d) + rand(-1, 1) \cdot \lambda \cdot L(d) \qquad (1)$$

Where $rand(-1,1)$ is a random number uniformly distributed in the range [-1,1], and $L(d)$ is the length of the dth dimension of the search space ($1 \leq d \leq n$). If the new position is outside the feasible range, it will be reset to a random position in the range. After propagation, if the fitness of the offspring wave x' is higher than the original x, then x is replaced by x' whose height is reinitialized to h_{max}. Otherwise, the height of the original x is decreased by one

In this way, high fitness waves tend to perform more local search in small areas and low fitness waves tend to perform more global search in wide areas. After each generation, each wave x updates its wavelength as follows:

$$\lambda = \lambda \cdot \alpha^{-(f(x)-f_{min}+\varepsilon)/(f_{max}-f_{min}+\varepsilon)} \qquad (2)$$

Where f_{max} and f_{min} are the maximum and minimum fitness values in the current population respectively, α is a parameter called the wavelength reduction coefficient, and ε is a small positive number to avoid division-by-zero.

2.2 Refraction

WWO uses the wave height property h to control the life span of the waves: If a wave propagates from deep water (low fitness location) to shallow water (high fitness location), its height increases; otherwise its height decreases, which mimics energy dissipation. When the height of a wave x decreases to zero, WWO uses the refraction operator to replace x with a new generated wave x'. Each dimension of x' is set as a random number centered halfway between the original position and the corresponding position of the best known solution x^*:

$$x'(d) = norm(\frac{x^*(d) + x(d)}{2}, \frac{|x^*(d) - x(d)|}{2}) \tag{3}$$

Where $norm(\mu,\sigma)$ is a Gaussian random with mean μ and standard deviation σ. The purpose of the operator is to remove those waves that fail to find better solutions for a given period so as to avoid search stagnation, and generates new waves to improve the diversity.

After each refraction operation, the height of new x' is also set to h_{max}, and its λ is updated as:

$$\lambda' = \lambda\frac{f(x)}{f(x')} \tag{4}$$

2.3   Breaking

The third operator in WWO is breaking, which simulates the breaking of a steepest wave into a train of solitary waves so as to enhance local search around a promising point. Whenever a new wave x is found to be better than the best known solution x^* (i.e., x becomes a new current best), WWO applies the breaking operation to conduct local search around x several times. In details, the breaking operator first randomly chooses k dimensions of x, and then generates a solitary wave x' at each chosen dimension d as follows:

$$x'(d) = x(d) + norm(0, 1) \cdot \beta \cdot L(d) \tag{5}$$

Where β is a parameter called the breaking coefficient, k is a random number between 1 and a predefined number k_{max}. Empirically, it is suggest to set k_{max} as min $(12, n/2)$, where n is the dimension of the problem. If none of the solitary waves is better than x, x is remained; else it is replaced by the best one among the solitary waves.

Algorithm 1 shows the algorithm framework of the original WWO.

Algorithm 1. The original WWO algorithm.

---
1 Randomly initialize a population P of waves (solutions) to the problem;
2 **while** stop criterion is not satisfied **do**
3   **for each** $x \in P$ **do**
4   Propagate x to a new x' based on Eq. (1);
5   **if** $f(x')<f(x)$ **then**
6   **if** $f(x')<f(x^*)$ **then**
7   Break x' based on Eq. (5);
8   Update x^* with x';
9   Replace x with x';
10   **else**
11         $x.h = x.h$ 1;
12   **if** $x.h==0$ **then**
13   Refract x to a new x' based on Eq. (3) and Eq. (4);
14   Update the wavelengths based on Eq. (2);
15   **return** x^*.

---

3 WWO Algorithm for TSP

In this paper, we adapt WWO to solve the TSP. As we know, each solution to TSP is a path or sequence of cities, and we model each solution as a wave, whose fitness is measured inversely by the length of a path: the shorter the path, the higher the solution fitness, the smaller wavelength λ is.

As the original WWO, we first initialize a population of solutions to the problem, and then use propagation, refraction and breaking operators to continually evolve the population until the termination condition is satisfied. But TSP is a combinatorial optimization problem, and thus the algorithm described in Sect. 2 cannot be applied directly. Therefore, we redefine the three operators to manipulate TSP solutions.

3.1 Propagation for TSP

In the propagation operator of WWO, we regard the wavelength λ of each wave x as the probability of the mutation on the solution: A poor solution has a large λ, and thus has a large probability of being mutated; on the contrary, a quality solution has a small probability of being mutated.

When applying to the TSP, we check at each dimension d whether to mutate x ($1 \leq d \leq n$). That is, for each dimension d we generate a random real number r between 0 and 1 and compare it with λ: If $r < \lambda$, we reverse $x[d, d + l]$, the subsequence from the dth dimension to the order of the $(d + l)$th dimensions, where l is a random integer between 1 and $(n\text{-}d)$.

Using a TSP instance with 6 cities for example, suppose $d = 2$ and $l = 3$, Fig. 2 illustrates the reversion on a solution when $r < \lambda$ is satisfied.

Equation (2) performs well in WWO for global optimization. But for TSP we design a new equation for updating wavelength of each solution x in the population:

$$\lambda = \lambda_{\max} \cdot \frac{1 + f(x)}{\sum_{i=1}^{NP} f(x_i)} \tag{6}$$

Where NP is the population size and $\lambda_{\max}$ is the maximum value of λ. Note here $f(x)$ denotes the objective function value (i.e., the tour length) of solution x, instead of the fitness value as in Eq. (2).

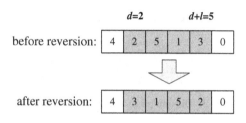

Fig. 2. Illustration of a reversion step in the propagation operator.

3.2   Refraction for TSP

The main role of the refraction operator of WWO is to make poor solution absorb some desirable features from the known best solution. In the TSP, a subsequence of a solution records the information about the relations between the solution components (i.e., cities), which can be regarded as more important features of the solution than the absolute positions of the cities. Therefore, in our WWO for TSP, a refraction operation moves a randomly selected subsequence $x^*[a, a + l]$ from the known best solution x^* to the corresponding part of the refracted solution x, where a is a random integer in $[1, n]$ and l is a random integer in $[0, n-a]$.

To satisfy the permutation constraint on TSP solutions, when moving the dth component form x^* to x, we find the position of $x^*[d]$ in x, denoted as d_1, and swap the dth and d_1th components of x.

Figure 3 illustrates such a refraction operation, where the subsequence $[1, 2, 4]$ of x^* moves to x:

(1) The first component of x is 4, which is as the same as x^*, so no operation is needed.
(2) The second component of x is 2, and we find the position of 1 in x, the result of which is 4. Therefore we swap the 2nd and 4th components of x.
(3) The third component of x is 5, and we find the position of 2 in x, the result of which is 4. Therefore we swap the 3rd and 4th components of x.

After refraction, x becomes $[4, 1, 2, 5, 3, 0]$, which is expected to improve its fitness by learning the sub-sequence $[1, 2, 4]$ from x^*.

Similarly, we change the wavelength update Eq. (4) of the original WW as below:

$$\lambda' = \lambda + \frac{f(x') - f(x)}{f(x^*)} \tag{7}$$

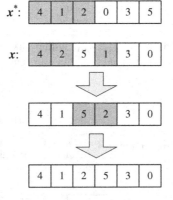

Fig. 3. Illustration of a refraction operator.

3.3 Breaking for TSP

Regarding the breaking operation on every newly found best solution x, we directly generate k_{max} solitary waves, each of which is obtained by swapping two randomly chosen components, as shown in Fig. 4.

Algorithm 2 shows the algorithm framework of the proposed WWO for TSP.

Algorithm 2. The proposed WWO algorithm for TSP.

---
1 Randomly initialize a population P of waves (solutions) to the problem;
2 **while** stop criterion is not satisfied **do**
3   **for each** $x \in P$ **do**
4   **for** $i=1$ to n **do**
5   **if** $rand()<x.\lambda$ **then**
6   let $l=randi(1,n-d)$;
7   reverse the subsequence $x[d, d+l]$ of x;
8   **if** $f(x') < f(x)$ **then**
9   **if** $f(x') < f(x^*)$ **then**
10   Break x' by tentatively swapping two random components x';
11   Update x^* with x';
12   Replace x with x';
13   **else**
14     $x.h = x.h-1$;
15   **if** $x.h == 0$ **then**
16   Refract x by moving a subsequence from x^* to x;
17   Update the wavelengths based on Eq. (6);
18 **return** x^*.

---

4 Computational Experiment

To verify the performance of the proposed WWO algorithm for the TSP, we conduct simulation experiments on a set of TSP instance chosen from TSPLIB [20]. The set of seven instances are bays29, eil51, eil76, kroA100, ch130, ch150, and kroA200, which cover a wide range of problem features and relatively small, middle, and large size instances. We compare our WWO algorithm with two other evolutionary algorithms for TSP in the literature:

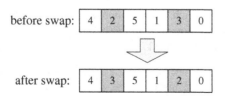

Fig. 4. Illustration of a swap step in the breaking operator.

Table 1. The results of comparative experiments on the 7 instances from TSPLIB.

Instance	Best known value	Results	GA	BBO	WWO
bays29	2020	min	2026	**2020**	**2020**
		max	2075	2427	2072
		average	2040.667	2153.533	**2033.600**
		std	13.223	101.299	12.381
		per run time	48 s	41 s	25 s
eil51	426	min	437	459	**427**
		max	502	554	454
		average	458.300	500.900	**442.100**
		std	12.332	22.248	6.261
		per run time	69 s	66 s	67 s
eil76	538	min	564	642	**557**
		max	646	789	587
		average	587.500	705.400	**571.700**
		std	20.158	36.310	8.243
		per run time	91 s	96 s	102 s
kroA100	21282	min	24089	27741	**21668**
		max	28613	39330	24144
		average	25495.000	34111.100	**22607.500**
		std	1030.587	2384.727	595.003
		per run time	114 s	130 s	114 s
ch130	6110	min	7129	10067	6338
		max	8742	13306	6915
		average	7910.467	11534.600	**6618.400**
		std	397.354	795.200	168.212
		per run time	142 s	178 s	187 s
ch150	6528	min	8235	12406	**7014**
		max	10697	16248	7630
		average	9192.600	14393.870	**7311.967**
		std	560.116	987.429	148.875
		per run time	164 s	189 s	126 s
kroA200	29368	min	45767	89225	**31064**
		max	63052	109993	33847
		average	51923.330	97735.970	**32777.930**
		std	4314.07	5330.973	646.186
		per run time	217 s	252 s	218 s

- A GA method [16] which is an improvement of the classic GA by using greedy-crossover, swap mutation, and Eshelman's selection method [21].
- A relatively new BBO method [17].

The parameters of WWO are set as: $h_{max} = 12$, $\lambda_0 = 1$, $\lambda_{max} = 1$, and the population size $NP = 0.5n$ where n is the instance size, i.e., the number of cities of the instance. For GA, the crossover probability is set to 1, the mutation probability is 0.015, and the population size is 100. For BBO, the maximum immigration rate and maximum emigration rate are both set to 1, the initial mutation probability is 0.001, and the population size is 50. The details about the parameter settings of the other two methods are discussed in [16, 17] respectively.

For WWO, we set the maximum iteration number of the algorithm to 400000 for bays29 and eil51, 300000 for eil76, 200000 for kroA100 and ch130, and 100000 for ch150 and kroA200. For the sake of fairness, for GA and BBO we set the maximum number of objective function evaluations as the same as WWO on each test instance. The experiments are conducted on a computer of Intel(R) Pentium(R) CPU and 2 GB memory.

Each algorithm is run 30 times on each test instance, and their experimental results over the 30 runs are given in Table 1, where the best min/average values among the three algorithms are shown in bold. As we can see from the results, on the simplest instance bays29, both BBO and WWO achieve the optimal value 2020. Except this, the min, max, and average values obtained by WWO are always better than GA and BBO on all the seven instances. The larger the instance size, the more performance advantage WWO has. Moreover, the standard deviations of WWO are also always the smallest among the three algorithms, which show that WWO is much more stable than GA and BBO. On the other hand, except on eil76 the running time of WWO is a little more than GA and BBO, on most instances the running time BBO is the largest, and WWO and GA consumes similar time. In summary, our WWO exhibits a great performance advantage over GA and BBO in the experiments, which demonstrates the effectiveness of the WWO algorithm framework and our adapted operators for TSP.

5 Conclusion

WWO is a relatively new metaheuristic optimization algorithm which has a great potential for a wide range of optimization problems. In this paper, we adapt WWO for effectively solving the TSP by designing new propagation, refraction, and breaking operators according to characteristics of the TSP solution space. The results of experiments show that WWO significantly outperforms GA and BBO on a set of instances from TSPLIB. In our further work, we will further improve the WWO performance on TSP and its variants, and adapt WWO for more other combinatorial optimization problems.

References

1. Holland, J.H.: Adaptation in natural and artificial systems: an introductory analysis with applications to biology, control and artificial intelligence. U Michigan Press (1975)
2. Kennedy, J., Eberhart, R.: Particle swarm optimization. In: IEEE International Conference on Neural Networks, vol. 4, pp. 1942–1948. IEEE Press, New York (1995)
3. Storn, R., Price, K.: Differential evolution-a simple and efficient heuristic for global optimization over continuous spaces. J. Global Optim. 11(4), 341–359 (1997)
4. Dorigo, M., Gambardella, L.M.: Ant colony system: a cooperative learning approach to the traveling salesman problem. IEEE Trans. Evol. Comput. 1(1), 53–66 (1997)

5. Simon, D.: Biogeography-based optimization. IEEE Trans. Evol. Comput. **12**(6), 702–713 (2008)
6. Yang, X.S.: Hossein Gandomi, A.: Bat algorithm: a novel approach for global engineering optimization. Eng. Comput. **29**(5), 464–483 (2012)
7. Mei, C.C., Liu, P.L.: Surface waves and coastal dynamics. Annu. Rev. Fluid Mech. **25**(1), 215–240 (1993)
8. Zheng, Y.J.: Water wave optimization: a new nature-inspired metaheuristic. Comput. Oper. Res. **55**, 1–11 (2015)
9. Banks, A., Vincent, J., Anyakoha, C.: A review of particle swarm optimization. Part II: hybridisation, combinatorial, multicriteria and constrained optimization, and indicative applications. Nat. Comput. **7**(1), 109–124 (2008)
10. Prado, R.S., Silva, R.C., Guimarães, F.G., Neto, O.M.: Using differential evolution for combinatorial optimization: a general approach. In: 2010 IEEE International Conference on Systems Man and Cybernetics, pp. 11–18. IEEE Press, New York (2010)
11. Lawler, E.L., Lenstra, J.K., Shmoys, D.B.: The Traveling Salesman Problem: A Guided Tour of Combinatorial Optimization. Wiley and Sons, New York (1985)
12. Wang, H.F., Wu, K.Y.: Hybrid genetic algorithm for optimization problems with permutation property. Comput. Oper. Res. **31**(14), 2453–2471 (2004)
13. Smith, T.H.C., Thompson, G.L.: A LIFO implicit enumeration search algorithm for the symmetric traveling salesman problem using Held and Karp's 1-tree relaxation. Ann. Discret. Math. **1**, 479–493 (1977)
14. Laporte, G.: The traveling salesman problem: an overview of exact and approximate algorithms. Eur. J. Oper. Res. **59**(2), 231–247 (1992)
15. Potvin, J.Y.: Genetic algorithms for the traveling salesman problem. Ann. Oper. Res. **63**(3), 337–370 (1996)
16. Wu, J.: Genetic Algorithm for the Traveling Salesman Problem. PhD Thesis, Oklahoma State University, Stillwater, Oklahoma (2000)
17. Shi, X.H., Liang, Y.C., Lee, H.P., et al.: Particle swarm optimization-based algorithms for TSP and generalized TSP. Inf. Process. Lett. **103**(5), 169–176 (2007)
18. Song, Y., Liu, M., Wang, Z.: Biogeography-based optimization for the traveling salesman problems. In: Third International Joint Conference on Computational Science and Optimization, vol. 1, pp. 295–299. IEEE, New York (2010)
19. Huang, H.: Dynamics of Surface Waves in Coastal Waters: Wave-current-bottom Interactions. Springer, Berlin-Heidelberg (2009)
20. Reinelt, G.: TSPLIB—A traveling salesman problem library. ORSA Journal on Computing **3**(4), 376–384 (1991)
21. Eshelman, L.J.: The CHC adaptive search algorithm: how to have safe search when engaging in nontraditional genetic recombination. In: Rawlins, G.J.E. (ed.) Foundations of Genetic Algorithms, pp. 265–283. Morgan Kauffman, Los Altos (1991)

Improved Dendritic Cell Algorithm with False Positives and False Negatives Adjustable

Song Yuan[1,2(✉)] and Xin Xu[1,2]

[1] College of Computer Science and Technology,
Wuhan University of Science and Technology, Wuhan, China
[2] Hubei Province Key Laboratory of Intelligent Information Processing
and Real-time Industrial System, Wuhan, China
yuansong_2002@163.com

Abstract. In order to overcome the blindness of the evaluation on contexts in the classical Dendritic Cell Algorithm (DCA), how weight matrixes influence detection results is analyzed, and two kinds of DCA which can adjust false positives and false negatives are proposed. The first one is the improved voting DCA, the Tendency Factor (TF) is involved in the Dendritic Cell (DC) state transition to assess contexts fairly, and through the fine adjustment of TF false positives and false negatives of detection results are controlled; the other one is the scoring DCA, in the DC state transition phase the evaluation of contexts is ignored, instead, the antigen is directly given a score, then according to the distribution of average scores of antigens the anomaly threshold value can be adjusted to control false positives and false negatives. Experiments show that the two algorithms can both effectively realize results controlled, comparatively the scoring DCA is more intuitive.

Keywords: Dendritic cell algorithm · Anomaly detection · Tendency factor · Artificial immune · Data fusion

1 Introduction

The Dendritic Cell Algorithm (DCA) [1] is an artificial immune algorithm inspired by the function of Dendritic Cells (DCs) [2, 3] in natural immune system. This algorithm is derived from the abstract model of the antigen presenting behavior of DCs, and has been applied to solve all kinds of problems, especially in the field of anomaly detections [4, 5]. There are a large number of biological elements in the biological mechanism of DCs, among which the danger signal fusion and the internal signal generation are both very complicated processes, so far there are still many unknown fields [6]. The DCA is a highly random algorithm based on such a complicated biological mechanism, involving many interactional components and parameters, so it is difficult to be analyzed and controlled. At present studies on the DCA mainly focus on improving the detection accuracy and simplifying parameters [7, 8], but there are few studies on analyzing the parameter sensitivity and controlling the result accuracy flexibly. In order to control false positives and false negatives of detection results effectively, in this article, the influences of the weight matrixes on detection results are

© Springer International Publishing Switzerland 2015
D.-S. Huang et al. (Eds.): ICIC 2015, Part I, LNCS 9225, pp. 147–158, 2015.
DOI: 10.1007/978-3-319-22180-9_15

analyzed, the concept of the Tendency Factor (TF) is put forward contraposing the evaluation criterion on the cellular environment by DCs, the TF is integrated into the state transition mechanism of DCs, and an improved voting DCA is designed; as a natural extension, changing the evaluation methods on the cellular environment by DCs from voting to marking, a scoring DCA is designed. The design ideas and regulatory mechanisms of the two algorithms are detailed, the validity and the controllability are verified through experiments.

2   The Classical DCA

In the immune process a DC exists in one of three different states of maturity, i.e. immature, semi-mature or mature state according to the different living environment. Each DC is abstracted as a signal processor. The DC fuses input signals to generate output signals which decide the state of the DC. According to the state of the DC the abnormal degree of the antigen is evaluated. Input signals include Pathogen Associated Molecular Patterns (PAMP), Danger Signals (DS), Safe Signals (SS) and Inflammatory Cytokines (IC). Output signals include co-stimulatory molecules (csm), semi-mature DC cytokines (semi) and mature DC cytokines (mat).

The DCA flow is as follows:

Step 1. Each DC gathers environmental signals (PAMP, DS, SS, IC) and antigens;
Step 2. The DC processes input signals to generate three output signals (csm, semi, mat) accumulated respectively;
Step 3. If $\sum$csm is less than the set migration threshold (MT), continue step 1 and step 2, otherwise proceed to next step;
Step 4. If $\sum$semi is greater than $\sum$mat, the DC turns into semi-mature state, the antigen environment value will be labeled 0, otherwise, the DC turns into mature state, the antigen environment value will be labeled 1.

The antigen environment of semi-mature state means that the antigen is collected under normal conditions, and the antigen environment of mature state means that there is some potential anomaly. The Mature Context Antigen Value (MCAV) of each antigen is calculated, i.e. the proportion of the times that the antigen is presented as a mature environmental antigen to the total presenting times, which reflects the abnormal degree of the antigen in order to detect whether there is any anomaly. More details please refer to [9, 10].

All antigens sampled by the semi-mature DCs are labeled as normal, all antigens sampled by the mature DCs are labeled as anomaly, it's like that each DC as a judge gives a 'normal' vote or an 'anomaly' vote to the antigens sampled by it according to its own state, and finally judgments by multiple DCs will be integrated to decide the abnormal degree of the antigen. Therefore for each antigen two data will be written down: total votes and 'anomaly' votes, when the total votes reach a threshold 'N', calculate the MCAV of the antigen, i.e. the proportion of the 'anomaly' votes to the total votes, then compare the MCAV with the anomaly threshold for a final evaluation. In this article the classical DCA is called voting DCA, the voting criterion is to compare the accumulated semi and mat concentration, whether fair depends on the mechanism of signal fusion process.

3   The Fusion Process of Signals

3.1   The Signal Process Formula and Weight Matrixes of DCA

The fusion process of signals is very complicated, in order to avoid the modeling of real biological signal conversion mechanism, at present the weighted summation formula may be applied most [11]. The influence of IC is ignored in this article, the most simple signal conversion process formula such as (1) is used in order to analyze the fair degree of the cell environment conversion more clearly.

$$O_j = \sum_{i=0}^{2} (W_{ij} \times S_i), (j = 0, 1, 2) \tag{1}$$

O_j are the output signals (O_0-csm, O_1-semi, O_2-mat), S_i are the input signals (S_0-PAMP, S_1-DS, S_2-SS). W_{ij} is the transforming weight from S_i to O_j. The weights are derived from biological immune experiments by the immunologists, and the values can be user adjusted according to the specific application scene, but they must correspond with the interaction between the input signals and the output signals. PAMP influences csm and mat; DS influences csm and mat; SS influences csm, semi and mat; a high level of SS will increase semi, and reduce mat which will be increased by the receipt of either PAMP or DS. The most common weight matrixes are shown in Tables 1, 2 and 3:

The difference of the three weight matrixes lies in the weights of SS to semi and mat.

Table 1. Weight matrix 1

W_{ij}	PAMP ($i = 0$)	DS ($i = 1$)	SS ($i = 2$)
csm ($j = 0$)	2	1	2
semi ($j = 1$)	0	0	**1**
mat ($j = 2$)	2	1	**−1.5**

Table 2. Weight matrix 2

W_{ij}	PAMP ($i = 0$)	DS ($i = 1$)	SS ($i = 2$)
csm ($j = 0$)	2	1	2
semi ($j = 1$)	0	0	**2**
mat ($j = 2$)	2	1	**−2**

3.2   Effects of Weight Matrixes on Detection Results

The standard UCI Wisconsin Breast Cancer data set mentioned in [9] is used in this article, containing 700 items, 240 items of them are labeled as class 1 (normal), the

Table 3. Weight matrix 3

W_{ij}	PAMP ($i = 0$)	DS ($i = 1$)	SS ($i = 2$)
csm ($j = 0$)	2	1	2
semi ($j = 1$)	0	0	**3**
mat ($j = 2$)	2	1	**−3**

other 460 items are labeled as class 2 (anomaly), some of the attributes of the items are abstracted as input signals. The three weight matrixes are applied to the signal conversion formula (1), data sets of different sequence are used for experiments:

Order 1: use all of class 1 (240 items) followed by all of class 2 (460 items), let the algorithm undergo one change in contexts.

Order 2: use 230 data items of class 2, all 240 items of class 1 followed by the remaining 230 items of class 2, let the algorithm undergo two changes in contexts.

Order 3: use 115 data items of class 2, 120 items of class 1, 115 data items of class 2, 120 items of class 1 followed by the remaining 230 items of class 2, let the algorithm undergo four changes in contexts.

The DCA is a highly random algorithm, in order to make the experimental results comparable, all the parameters are the same, including MT which is set a fixed value. But because of the randomness of DCs in sampling antigens, the results of one experiment are still not the same every time. The average accuracy, the average rate of false positives and the average rate of false negatives of each experiment for 20 times are shown in Table 4.

Table 4. Effects of weight matrixes on detection results

Data sets of different sequence	The three weight matrixes	Accuracy	False positives	False negatives
Order 1: one change in contexts	Weight matrix 1	99.542 %	0.235 %	0.223 %
	Weight matrix 2	99.329 %	0.100 %	0.571 %
	Weight matrix 3	98.750 %	0.035 %	1.215 %
Order 2: two changes in contexts	Weight matrix 1	99.389 %	0.345 %	0.266 %
	Weight matrix 2	99.000 %	0.243 %	0.757 %
	Weight matrix 3	97.759 %	0.049 %	2.192 %
Order 3: four changes in contexts	Weight matrix 1	98.570 %	0.479 %	0.951 %
	Weight matrix 2	97.700 %	0.150 %	2.150 %
	Weight matrix 3	95.614 %	0.057 %	4.329 %

The experiment results in Table 4 indicate the effects of the three weight matrixes on detection results: the effect of weight matrix 1 is best which has the highest accuracy and the lowest false negative; weight matrix 3 has the worst effect, although the false positive is the lowest, but the false negative is the highest; the effect of weight matrix 2 is middling. The different classification results are just because of the different weights of SS to semi and mat.

Using the weight matrix 2 as an example, the input signals of 700 items and the weights in weight matrix 2 are put into formula (1) to get 700 groups of results containing 3 kinds of output signals (csm, semi, mat), the average value of each kind of output signals for 240 normal antigens and the average value of each kind of output signals for 460 abnormal antigens are calculated respectively, and finally the average value of each kind of output signals for the above two sets is calculated, which is called the 'neutral data' in this article, as shown in Table 5:

Table 5. Average output values of normal and abnormal antigens

	csm	semi	mat
Average output value of 240 normal antigens	9.53	6.69	−3.85
Average output value of 460 abnormal antigens	8.6	0.59	7.42
Output value of neutral data	9.065	3.640	1.785

It can be seen that semi > mat in the neutral data which indicates that DCs tend to semi, that is to say, the evaluations on neutral data by DCs as judges are biased towards safety, which will result in the high false negatives. Subtract semi from mat in the neutral data: $1.785 - 3.640 = -1.855$, this value shows the tendency degree of DCs to semi, so the concept of Tendency Factor is put forward:

Definition 1: Input signals and the weight matrix are put into the signal conversion formula to generate output signals (csm, semi, mat), the average values of each kind of output signals for normal antigens and abnormal antigens are calculated respectively, and then the average values of each kind of output signals for the above two sets are calculated respectively to generate the neutral data of csm, semi and mat, the result of subtracting semi neutral data from mat neutral data is called Tendency Factor (TF), reflecting the tendency degree of DCs to safety or danger.

According to the above concept of TF, put weight matrix 3 into the signal conversion formula to get TF = −5.5, which shows that the evaluations on environment by DCs are more biased towards safety, and result in higher false negatives. Put weight matrix 1 into the signal conversion formula to get TF = +0.87, which shows that the evaluations on environment by DCs are biased towards danger, and result in lower false negatives, and the value is nearest to zero, indicating that the evaluation on environment is more fair.

It can be concluded that the different weight matrix will produce the different TF which reflects the tendency degree of DCs to safety or danger, random weights will lead to blind judgment on the environment, so the state conversion rule in the classical DCA is to be improved.

4 The Improved Voting DCA

In the classical DCA, when a DC samples enough antigens and its csm reaches MT, there are some limitations in deciding the DC conversion state through comparing $\sum$mat and $\sum$semi. Because using the different weight matrix and the different signal conversion formula in the fusion process will result in the different tendency degree on cell environment evaluation. In order to be fair in the cell environment evaluation, and have greater flexibility to control false positives and false negatives of detection results, the TF will be involved to improve the DC state conversion criterion, the idea is as follows:

(1) Preprocess the data set, calculate the TF according to the weight matrix and signal conversion formula;
(2) Initialize each DC and add a variable n = 0, recording the number of antigens that the DC samples, n plus 1 whenever the DC samples another antigen;
(3) When the accumulated csm reaches MT, if $\sum$mat-$\sum$semi > n*TF, the DC turns into mature state, otherwise the DC turns into semi-mature state.

In the classical DCA, the DC state transition criterion is whether the difference between $\sum$mat and $\sum$semi is greater than zero; however, the improved DC state transition criterion is whether the difference between $\sum$mat and $\sum$semi is greater than n*TF, n refers to the number of antigens that the DC samples. Just think, if the DC samples one antigen, it will cause 1*TF, if the DC samples n antigens, it will result in n*TF, so using n*TF as the standard to measure the difference between $\sum$mat and $\sum$semi will make the evaluation on environment more fair.

With the improved DC state transition criterion, the above three kinds of experiments are redone, of which the experiment 2 and experiment 3 have got obvious improvement effects, as shown in Table 6.

Table 6. Detection results of the improved DC state transition criterion

Data sets of different sequence	The three weight matrixes	Accuracy	False positives	False negatives
Order 1: one change in contexts	Weight matrix 1	99.500 %	0.170 %	0.330 %
	Weight matrix 2	99.642 %	0.121 %	0.237 %
	Weight matrix 3	99.500 %	0.286 %	0.215 %
Order 2: two changes in contexts	Weight matrix 1	99.342 %	0.265 %	0.393 %
	Weight matrix 2	99.305 %	0.379 %	0.316 %
	Weight matrix 3	99.191 %	0.386 %	0.423 %
Order 3: four changes in contexts	Weight matrix 1	98.350 %	0.300 %	1.350 %
	Weight matrix 2	98.312 %	0.365 %	1.323 %
	Weight matrix 3	98.513 %	0.337 %	1.151 %

The TF in experiment 1 is +0.87, this value is most close to zero, indicating that the evaluation on environment by DCs are more fair, so the improvement effect is not obvious, and even a bit inferior to the previous experimental results, which shows that the TF is not certain to lead to optimum solution, but can be used as reference to seek the optimization. Further experiments demonstrate this point, the adjusting ranges of TF for the optimum solution are shown in Table 7.

Table 7. Adjustment ranges of TF for optimal solutions

	TF	Adjustment range of TF
Experiment 1: weight matrix 1	+0.87	$-0.435 \sim 0.609$
Experiment 2: weight matrix 2	-1.855	$-1.855 \sim -3.71$
Experiment 3: weight matrix 3	-5.5	$-5.5 \sim -7.15$

Further, (n*TF) can be regarded as the critical value for safety or danger, setting a value slightly greater than (n*TF) as the evaluation standard will be biased towards safety, but can reduce false positives; and setting a value slightly less than (n*TF) as the evaluation standard will be biased towards danger, but can reduce false negatives. So the value of TF can be adjusted according to the specific application requirements to control false positives and false negatives of detection results.

5 The Scoring DCA

5.1 Algorithm Ideas

In the improved voting DCA, the TF is aimed at the fairness of evaluation criterion, through adjusting the DC state transition criterion to control false positives and false negatives of detection results, but the adjusting ranges (shown in Table 7) are derived from a lot of tentative experiments, which are not intuitive, if the adjustment can be done on the basis of the final abnormal degree of antigens, more intuitive effects can be obtained. So instead of the improved voting DCA a scoring DCA is proposed, in which the focus is no longer on the DC state transition criterion. The concrete ideas are as follows: each DC samples signals and antigens in the antigen-signal pool, when its $\sum csm$ reaches MT, the difference between $\sum mat$ and $\sum semi$ will be given to each antigen collected by the DC as a score. Two data will be written down for each antigen: the times of scoring and the accumulated score, when the times of scoring reach a threshold N, calculate the average score of the antigen, i.e., the accumulated score/N as MCAV of the antigen to reflect how abnormal the antigen is. That is to say, in the scoring DCA, each DC no longer votes for the antigen, but gives it a score, there isn't the DC state transition criterion, instead, a score line, i.e., the anomaly threshold will be determined according to the score distribution to regulate false positives and false negatives of detection results.

5.2   Pseudo Code of the Scoring DCA

```
Input: antigen and signal feature vector
Output: average score of each antigen
Initialize an antigen-signal pool of length m;
Put the front m items of the data set into the pool;
While (all of antigens have not been detected)
{
  Initialize each DC;
  While ( Σcsm < MT )
  {
    The DC samples from the antigen-signal pool randomly,
    Accumulate three output values: Σcsm, Σsemi, Σmat;
  }
  For each antigen that the DC samples, (scoring times)
  plus 1, (accumulated score) plus (Σmat-Σsemi);
  Deal with the antigen whose scoring times reach N,
  (average score) = (accumulated score)/N, and remove
  this antigen from the antigen-signal pool, then add a
  new antigen;
}
```

5.3   Experiments and Result Analysis

Set m = 20, N = 10, MT = 39, respectively using three kinds of weight matrixes to perform experiments on order 3 data set (115 data items of class 2 + 120 items of class 1 + 115 data items of class 2 + 120 items of class 1 + 230 items of class 2), and the results are shown in Figs. 1, 2 and 3, the abscissa is the ID of 700 antigens, the ordinate is the average score of each antigen.

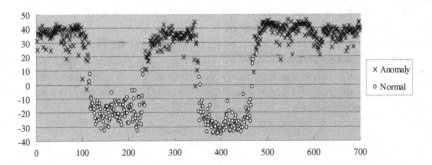

Fig. 1. The effect of the scoring DCA using weight matrix 1

It can be seen that the average scores of abnormal antigens are mainly distributed in the upside 10∼45 and the centers move down successively and slightly; the average scores of normal antigens are mainly distributed in the bottom and the centers move

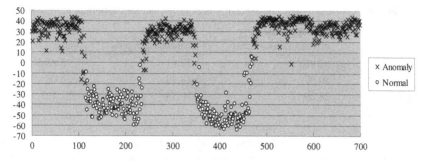

Fig. 2. The effect of the scoring DCA using weight matrix 2

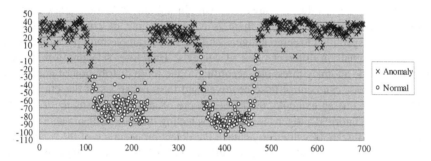

Fig. 3. The effect of the scoring DCA using weight matrix 3

down successively and significantly. This is because that in the dangerous environment, SS is 0 in general, PAMP and DS influence the output signals mainly, in the three weight matrixes the weights from PAMP and DS to output signals are the same, so the abnormal antigen scores are influenced slightly; in the safe environment, the main factor affecting the output signals is SS, the differences in the three weight matrixes are exactly the weighs from SS to semi and mat, and the positive influence on semi are increased successively, as well as the negative influence on mat, as a result, the centers of average scores of normal antigens move down significantly.

The average score of the same kind of antigens will be calculated as shown in the formula (2):

$$average = (mat - semi) * \frac{MT}{csm} \tag{2}$$

Put the weight matrix 2 into the signal conversion formula (1) to get the average output values of normal and abnormal antigens shown in Table 5, according to the formula (2) the average value of abnormal antigens can be obtained, $(7.42-0.59)$ $*39/8.6 = 30.97$; the average value of normal antigens is $(-3.85-6.69)*39/9.53 =$ -43.13.

In addition, respectively using weight matrix 1 and weight matrix 3 to calculate the average values are summarized in Table 8:

Table 8. Estimated values of average scores of normal and abnormal antigens

	Average score of abnormal antigens	Average score of normal antigens	Intermediate value
Weight matrix 1	32.98	−22.61	5.186
Weight matrix 2	30.97	−43.13	−6.08
Weight matrix 3	28.29	−70.55	−21.127

The values in Table 8 are calculated according to the average output values of the antigens, on this basis the score of actual antigens will fluctuate as shown in Figs. 1, 2 and 3.

No matter choosing any weight matrix, the normal antigens and abnormal antigens can be divided, the confusions always occur during the transition phases, this is because during a transition phase there is a small degree of confusion regarding temporally and spatially clustered antigens and DCs may sample multiple antigens and many sets of signals in different types of contexts. Therefore, selecting a reasonable score line, i.e. the anomaly threshold, according to the score distribution of normal and abnormal antigens, can control false positives and false negatives of detection results flexibly. Figure 3 as an example: the highest score of normal antigens is −2.737, the lowest score of abnormal antigens is −42.533, selecting the intermediate value −21.127 as the anomaly threshold value can achieve the detection accuracy of 99 %, the false positive rate of 0.571 %, and the false negative rate of 0.429 %. Adjusting the anomaly threshold value up to −2.737 can realize zero false positive rate, and adjusting the anomaly threshold down to −42.533 can realize zero false negative rate. Of course, this is the experimental results of the scoring DCA for one time, a large number of random factors in the algorithm will lead to different experimental results, the adjustable ranges of the anomaly threshold value are not the same every time, but the intermediate values are not very different.

Calculate the intermediate value according to the average output value of antigens as follows:

$$median = \frac{1}{2}\sum_{j}(mat_j - semi_j)^* \frac{MT}{csm_j} \quad (j = 0, 1; 0 \text{ - normal, 1- anomaly})$$

The intermediate value can be used as the reference value of the anomaly threshold, and then according to the specific application the anomaly threshold value can be adjusted in order to control false positives and false negatives of detection results.

The intermediate value can also be expressed as a linear function of the TF, as shown in formula (3):

$$median = TF^* \frac{MT}{csm_0} + \frac{1}{2}(mat_1 - semi_1)^*(\frac{MT}{csm_1} - \frac{MT}{csm_0}) \quad (3)$$

Using the weight matrix 2 as an example, put the data in Table 5 into the formula (3) as follows:

$$median = -1.855^* \frac{39}{9.53} + \frac{1}{2}(7.42 - 0.59)^*(\frac{39}{8.6} - \frac{39}{9.53}) = -6.08$$

Because the influences of the three weight matrixes on abnormal antigens are not very different, $(mat_1 - semi_1)$ will be treated as a constant, formula (3) can be transformed into:

$$median = n^*TF + t \quad (n = 4.09, t \approx 1.55)$$

Put the TF values of the three weight matrixes into the above formula to get three intermediate values are: 5.108, −6.037 and −20.945, which are consistent with the intermediate values shown in Table 8.

6   Conclusions

The influences of weight matrixes on detection results in the DCA are presented, the fusion processing of signals, the DC state transformation, the comprehensive evaluation on antigens are analyzed, two result-controllable DCA are designed: the improved voting DCA in which the DC state transition criterion is used as a control mechanism; the scoring DCA in which through reasonably selecting the score line false positives and false negatives can be controlled. But this is only the preliminary study on the DCA which is a highly random algorithm involving many interactive compositions and parameters, the influences of the related parameters on detection results, the essence of the DCA, and how to adjust the dynamic parameters to realize algorithm optimization according to detection results will be further studied.

Acknowledgement. This work was supported by the Natural Science Foundation of Hubei Provincial of China (2014CFB247), and the National Natural Science Foundation of China (No. 61440016).

References

1. Greensmith, J., Aickelin, U., Twycross, J.: Articulation and clarification of the dendritic cell algorithm. In: Bersini, H., Carneiro, J. (eds.) ICARIS 2006. LNCS, vol. 4163, pp. 404–417. Springer, Heidelberg (2006)
2. Oates, R., Kendall, G., Garibaldi, J.: Frequency analysis for dendritic cell population tuning. Evol. Intel. 1(2), 145–157 (2009)
3. Greensmith, J., Aickelin, U.: Artificial Dendritic Cells: Multi-faceted Perspectives. In: Bargiela, A., Pedrycz, W. (eds.) Human-Centric Information Processing. SCI, vol. 182, pp. 375–395. Springer, Heidelberg (2009)
4. Greensmith, J., Aickelin, U., Tedesco, G.: Information fusion for anomaly detection with the dendritic cell algorithm. Inf. Fusion 11(1), 21–34 (2010)

5. Twycross, J.: Integrated innate and adaptive artificial immune systems applied to process anomaly detection. Ph.D. thesis. University of Nottingham (2007)
6. Ni, J.C., Li, Z.S., Sun, J.R., Zhou, L.P.: Research on differentiation model and application of dendritic cells in artificial immune system. Acta Electronica Sin. **36**(11), 2210–2215 (2008)
7. Yang, C.X., Wu, G.F., Hu, M.: Improved dendritic cells algorithm. Comput. Eng. **35**(23), 194–200 (2009)
8. Greensmith, J., Aickelin, U.: The deterministic dendritic cell algorithm. In: Bentley, P.J., Lee, D., Jung, S. (eds.) ICARIS 2008. LNCS, vol. 5132, pp. 291–302. Springer, Heidelberg (2008)
9. Greensmith, J.: The dendritic cell algorithm. Ph.D. thesis. University of Nottingham (2007)
10. Greensmith, J., Aickelin, U., Cayzer, S.: Detecting danger: the dendritic cell algorithm. In: Schuster, A. (ed.) Robust Intelligent Systems, pp. 89–112. Springer, Heidelberg (2008)
11. Chen, Y.B., Feng, C., Zhang, Q., Tang, C.J.: Principles and application of dendritic cell algorithm. Comput. Eng. **36**(8), 173–176 (2010)

Topographic Modulations of Neural Oscillations in Spiking Networks

Jinli Xie[✉], Jianyu Zhao, and Qinjun Zhao

School of Electrical Engineering, University of Jinan, Jinan 250022, China
cse_xiejl@ujn.edu.cn

Abstract. We present a computational model evoked by electrosensory system which is able to display oscillatory activity, and focus on the coherence of the spectral power of the ELL neurons with the topographic modulations for different spatial scale regimes. Numerical simulations reveal that the spatial scale is a very important determinant of neural oscillations in gamma band. The spectral power is enhanced by decreasing feedback spatial spread. This enhancement can also occur if the feedforward is global. However, when the feedforward is topographic, the oscillations saturate to a steady state. In brief, the topographic feedback alone enables the system to modulate gamma activity with the spatial scale, while the introduction of topography in feedforward brings little effect on oscillations. What our results further indicate is that the topographic feedback can induce and enhance oscillations even when the external stimulus is local.

Keywords: Oscillation · Topographic · Feedback · Feedforward

1 Introduction

Electrosensory systems are organized in a feedforward manner, where the information of stimulus is relayed from primary sensory neurons to higher brain centers, and these in turn feed back to primary areas [1–3]. There has been much interest recently in the gain modulation in networks developed from sensory systems [4–9]. It is known theoretically and numerically that oscillations can be induced and enhanced in networks with global delayed feedback [10–13]. However, these studies focused on the neural dynamics with connections which are spatially uniform, and thus lack a notion of the topography of the connections. The connections of neural network can act globally, or locally, i.e., be organized topographically [14,15]. There is debate as to the dynamic role played by topographic connections which are widely existed in electrosensory pathways, which will be our focus here.

To this end our work considers how spatially organized connections affect neural firing activity using a model based on the electrosensory system of electric fish [1,5], and focuses closely on gamma oscillations. The respective domains of neurons are modeled by a neighborhood of cells projects near itself via feedforward or feedback connections, and the topography of network pathways are studied over a wide range of spatial scales. Numerical results provide support to the effects of the spatial scale of connections on the spectral coherence. Accordingly, the topographic modulations of oscillatory activity caused by feedback and feedforward are revealed.

D.-S. Huang et al. (Eds.): ICIC 2015, Part I, LNCS 9225, pp. 159–166, 2015.
DOI: 10.1007/978-3-319-22180-9_16

2 Model and Methods

A simplified scheme of the neural network that we consider in this study in shown in Fig. 1. This network is made up of two neural population, the electrosensory lateral-line lobe (ELL), and the nucleus praeminentialis (nP). The ELL projects to nP topographically. The nP is involved strictly in the feedback regulation of electrosensory input in a topographic manner.

The simulations were carried out with two neural populations of leaky integrate-and-fire neurons with local coupling subjected to spatial correlated forcing, as in [5], except that we used the spike output of a population of LIF neurons as local inhibitory feedback to the excitatory neurons, rather than the convolution of a linear summation of action potentials of the excitatory neurons.

The LIF neuron evolves according to:

$$\frac{dE_i}{dt} = -E_i + \mu + \eta_i(t) + I(i,t) + I_g \tag{1}$$

$$\frac{dP_j}{dt} = -P_j + \mu + \eta_j(t) + I_f \tag{2}$$

where E and P denote the membrane potential of principal neurons in ELL and nP, respectively. μ is the sensory input which is constant and the same for all cells, $\eta_i(t)(\eta_j(t))$ is the internal noise modeled by Gaussian white noise of intensity D that represents synaptic and channel noise in vivo, and $I(i, t)$ is a time-varying input to ELL follows:

$$\langle I(x,t_1)I(y,t_2)\rangle = D_E d(t_1 - t_2)C(x - y) \tag{3}$$

with

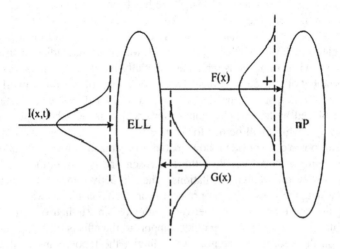

Fig. 1. Organization of feedforward and feedback pathways

$$C(z) = \exp(-z^2/2\sigma_i^2)/\sqrt{2\pi} \tag{4}$$

where D_E is the noise variance, σ_i is the standard deviation of the spatial extent of the external stimulus. The external stimulus consist of Gaussian low-pass filtered (0–150 Hz, eight-order Butterworth filter) noise process of unit variance. Every time the potential reaches the fixed firing threshold V_{th}, the neuron fires, and the voltage is reset to a value V_r where it remains for the absolute refractory time τ_R. The output spike train of the LIF model is obtained by collecting the instants of threshold crossing and modeling the associated spikes as δ-functions:

$$x(t) = \sum_l \delta(t - t_l) \tag{5}$$

where t_l are the successive firing times. The last terms in Eqs. (1) and (2) represents the inhibitory feedback and excitatory feedforward respectively:

$$I_g = \frac{g_{in}}{M} \int_{\tau_{in}}^{\infty} dt' h(t' - \tau_{in}) \sum_{k=1}^{M} G(i - k)x_k(t - t') \tag{6}$$

$$I_f = \frac{g_{ex}}{N} \int_{\tau_{ex}}^{\infty} dt' h(t' - \tau_{ex}) \sum_{k=1}^{N} F(j - k)x_k(t - t') \tag{7}$$

where M and N are the population size of principal neurons in ELL and nP. The excitatory (inhibitory) synapses of the afferents to the nP (ELL) have the synaptic efficacy g_{ex} (g_{in}) and the response time delay τ_{ex} (τ_{in}). In addition, $h(t) = \alpha^2(t)\exp(-\alpha t)$ is a standard α-function. The topography of the feedforward is denoted by normalized connectivity kernel:

$$F(x) = \exp(-x^2/2\sigma_f)/\sqrt{2\pi}\sigma_f \tag{8}$$

where σ_f is the standard deviation of Gaussian distribution. The feedback kernel $G(x)$ is chosen similarly as Eq. (8), but the standard deviation is σ_b.

As sketched in Fig. 1, the external stimulus, the feedforward and feedback connections are topographic, which will be studied over a wide range of spatial scales in this paper. For simplicity, there are no local horizontal connections within the population.

The oscillatory activity of feedback spiking neural network is measured by calculating the spike train power spectrum by numerical simulations:

$$S_c(\omega) = \left\langle \tilde{x}_i \tilde{x}_j^* \right\rangle, i \neq j \tag{9}$$

$$\tilde{x}_i(\omega) = \frac{1}{\sqrt{L}} \int_0^L e^{-i\omega t} x_i(t) dt \tag{10}$$

where $\tilde{x}_i$ is the Fourier transform of the spike train of the ith neuron, $\tilde{x}_j^*$ is the complex conjugate of $\tilde{x}_j$. Since the statistics of spike trains are the same for all excitatory neurons, it suffices to show the firing rate and power spectrum for one neuron. The brackets represent an average over multiple realizations of the internal noise, which change from trial to trial.

Since the peak height of the cross spectrum varies, we then measure the oscillation by the coherence:

$$\beta = \frac{h_p}{\Delta\omega} \cdot \omega_p \tag{11}$$

where h_p and ω_p are the height and central frequency of the first spectral peak, respectively. $\Delta\omega = \omega_R - \omega_L$ is the half-width, i.e., $\omega_{R,L}$ are the closest frequency values to the right and left of ω_p, respectively, where $S_c(\omega_{R,L}) = S_c(\omega_p)/2$.

We choose the following values for the time constants: $\mu = 0.8$, $D = 0.08$, $D_E = 0.12$, $g_{ex} = 1$, $g_{in} = -1$, $\tau_R = 3\,ms$, $\tau_{in} = 6\,ms$, $\alpha_{in} = 18\,ms^{-1}$, $\tau_{ex} = 1\,ms$, $\alpha_{ex} = 3\,ms^{-1}$, here time was measured in units of the membrane time constant, which is assumed to be 6 ms. For parameters related to the membrane potential we considered dimensionless units by setting $V_{th} = 1$, $V_r = 0$. For simulations, we used a ELL population of $M = 80$ neurons and a nP population of $N = 20$. We simulate 10^2 realizations with a time duration of 10 s and a time step of 10^{-4} using a simple Euler procedure to obtain the spike trains.

3 Results

Since larger σ leads to increased spatial extent of the connections, smaller σ represents less neighborhoods pulled together, namely, topographic feedback. Thus, we use $1/\sigma_f$ and $1/\sigma_f$ as the ratio of the characteristic length scales or feedforward and feedback. Figure 2 provides the sketch of the spatial extent of the connections for different values of σ. Note that, when $1/\sigma < 0.1$, the connection is global. While the connection is local (topographic) once $1/\sigma$ exceeds 0.1.

3.1 Power Spectrum

The response measured in temporal frequency range further illustrates the effects of the spatial scales of the network. The power spectrum of one randomly chosen ELL neuron is shown in Fig. 3. The spatial parameters are $\sigma_b = 5$, $\sigma_i = 5$ for upper panel, $\sigma_f = 5$, $\sigma_i = 5$ for lower panel.

The power spectrum in the ELL for a large, medium and low value of the spatial scales σ_f and σ_b was shown in Fig. 3. We observe a clear power peak at frequency in the gamma range for $\sigma_f = 100$ (Fig. 3(a)). The height of the gamma peak h_p exhibits an increase for $\sigma_f = 5$ (Fig. 3(b)), but increases only slightly from $\sigma_f = 5$ to $\sigma_f = 1$ (Fig. 3(c)). Besides, the small value $\sigma_b = 1$ yields a peak at the same frequency (Fig. 3(f)), and the same is true for $\sigma_b = 5$ with lower power (Fig. 3(e)). In contrast, a large value $\sigma_b = 100$ makes the peak vanish.

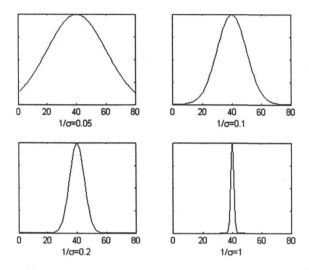

Fig. 2. Spatial extent of connections

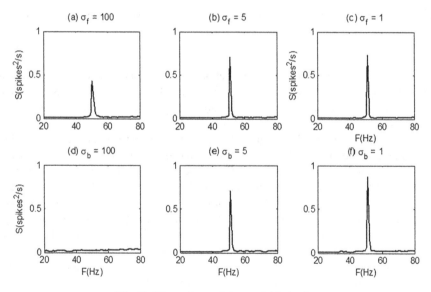

Fig. 3. Illustrations of the model behavior

3.2 Spectral Coherence

To better understand the influence of the topographic connections on oscillatory activities, the magnitude of the power spectrum is computed numerically from Eq. (11). The spectral coherence with the topographic modulation, for different spatial scale regimes, is depicted in Fig. 4. When $1/\sigma_b$ is small, the oscillation is weak. This can

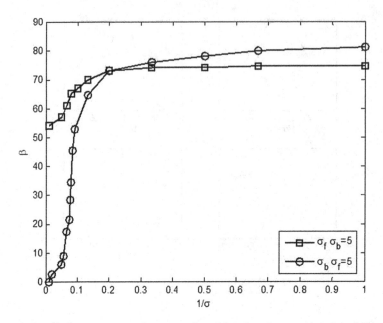

Fig. 4. Relationship between spectral power and spatial scales of connections in spiking neurons networks.

change if $1/\sigma_b$ increases. Significant oscillations arise and saturate when $1/\sigma_b$ approaches 1. Besides, for $1/\sigma_f$, the behaviors are qualitatively the same. So we see that the spatiotemporal structure of the network sets up temporal resonances that depend on the spatial scales.

In addition to these results, we note that the variation range of the numerical value of coherence is different for the two cases. When the feedforward is global and the feedback is topographic ($1/\sigma_f < 0.2$, $\sigma_b = 5$), the spectral coherence β is relatively higher than that with topographic feedforward and global feedback ($1/\sigma_b < 0.2$, $\sigma_f = 5$) over the same range. Furthermore, when the feedforward and feedback connections are both topographic, β is increased in power as $1/\sigma_b$ increases, and this continues until saturation. However, the parameter $1/\sigma_f$ slightly changes the value of β. Thus, the attenuation or enhancement of the oscillations is determined mainly by the spatial scale of feedback and less by that of feedforward, which is show good accordance with experiment. Even the feedforward is global ($1/\sigma_f = 0.01$), the network shows epochs of oscillatory activity. Here, the emergence of a network oscillation is associated with the topographic feedback. Although β increases with increasing $1/\sigma_f$, the topographic feedforward contributes little to the raise of the spectral power, especially when σ_f is small in comparison with the feedback spread ($1/\sigma_f > 0.2$, $\sigma_b = 5$). Consequently, topographic feedforward do not seem to account for the network coherence during cortical gamma oscillations.

4 Conclusions

Despite the fact that oscillatory activity of groups of neurons has been the subject of intense research efforts in many studies, the functional role of spatial characters in neural oscillations is still a topic of debate. In this work, we present a computational model evoked by electrosensory system which is able to display oscillatory activity, and focus on the coherence of the spectral power of the ELL neurons with the topographic modulation for different spatial scale regimes. There is a parameter that specifies the spatial spread of the connections. We have shown that the power of the gamma oscillation is suppressed as the spatial extent increases, i.e., the topographic connections play an important role in modulating the oscillatory behavior of the system.

Supposing the feedforward and feedback interactions were collapsed to a single spatial interaction, recent research findings revealed the effects of topographic feedback on the oscillatory activity [5]. What our results further indicate is that the respective contributions of spatial spread of feedforward and feedback to the oscillatory activity. We point out that the vanishing of oscillations occurs when the feedback is global, while the feedforward and input is local, which show good accordance with former studies [10–12]. Moreover, the spectral power increases rapidly with decreasing feedback length scale with medium input correlation length. Therefore, the topographic feedback can induce and enhance oscillations even when the external stimulus is local. Nevertheless, when the feedback length scale exceeds the feedforward range of connectivity, the decrease in feedforward length scale results in a saturation of spectral power, providing evidence of no topographic modulation of feedforward.

Acknowledgements. This work was supported by the National Natural Science Foundation of China under Grant No. 61203375, and the Doctoral Foundation of University of Jinan under Grant No. XBS1240.

References

1. Maler, L.: Neural strategies for optimal processing of sensory signals. Prog. Brain Res. **165**, 135–154 (2007)
2. Chacron, M.J., Bastian, J.: Population coding by electrosensory neurons. J. Neurophysiol. **99**, 1825–1835 (2008)
3. Marsat, G., Longtin, A., Maler, L.: Cellular and circuit properties supporting different sensory coding strategies in electric fish and other systems. Curr. Opin. Neurobiol. **22**(4), 686–692 (2012)
4. Battaglia, D., Brunel, N., Hansel, D.: Temporal decorrelation of collective oscillations in neural networks with local inhibition and long-range excitation. Phys. Rev. Lett. **99**(23), 238106 (2007)
5. Hutt, A., Sutherland, C., Longtin, A.: Driving neural oscillations with correlated spatial input and topographic feedback. Phys. Rev. E **78**, 021911 (2008)
6. Rothma, J.S., Cathala, L., Steuber, V., Silver, R.A.: Synaptic depression enables neuronal gain control. Nature **457**(7232), 1015–1018 (2009)

7. Ly, C., Doiron, B.: Divisive gain modulation with dynamic stimuli in integrate-and-fire neurons. PLoS Comput. Biol. **5**(4), e1000365 (2009)

8. Serrano, E., Nowotny, T., Levi, R., Smith, B.H., Huerta, R.: Gain control network conditionsin early sensory coding. PLoS Comput. Biol. **9**, 7 (2013)

9. Mejias, J.F., Payeur, A., Selin, E., Maler, L., Longtin, A.: Subtractive, divisive and non-monotonic gain control in feedforward nets linearized by noise and delays. Front. Comput. Neurosci. **25**, 8–19 (2014)

10. Doiron, B., Linder, B., Longtin, A., Maler, L., Bastian, J.: Oscillatory activity in electrosensory neurons increases with the spatial correlation of the stochastic input stimulus. Phys. Rev. Lett. **93**(4), 048101 (2004)

11. Lindner, B., Doiron, B., Longtin, A.: Theory of oscillatory firing induced by spatially correlated noise and delayed inhibitory feedback. Phys. Rev. E **72**(6), 061919 (2005)

12. Marinazzo, D., Kappen, H.J., Gielen, S.C.A.M.: Input-driven oscillations in networks with excitatory and inhibitory neurons with dynamic synapses. Neural Comput. **19**, 1739–1765 (2007)

13. Xie, J.L., Wang, Z.J., Longtin, A.: Correlated firing and oscillationsin spiking networks with global delayed inhibition. Neurocomput. **83**, 146–157 (2012)

14. Hansel, D., Sompolinsky, H.: Synchrony and computation in a chaotic neural network. Phys. Rev. Lett. **68**, 718–721 (1992)

15. Roxin, A., Brunel, N., Hansel, D.: Role of delays in shaping spatiotemporal dynamics of neuronal activity in large networks. Phys. Rev. Lett. **94**, 238103 (2005)

Non-negative Approximation with Thresholding for Cortical Visual Representation

Jiqian Liu$^{(\boxtimes)}$, Chunli Song, and Chengbin Zeng

School of Information Engineering, Guizhou Institute of Technology,
1st, Caiguan Road, Yunyan District, Guiyang 550003,
People's Republic of China
{liujiqian, songchunli, zengchengbin}@git.edu.cn

Abstract. This paper presents a neurally plausible algorithm for the representation of visual inputs by cortical neurons. It has been demonstrated in previous theoretical studies that the main goal of the encoding of the input from lateral geniculate nucleus (LGN) by simple cell is to minimize the representation error. Based on the existing methods, we propose a non-negative approximation algorithm using thresholding. We validate the algorithm via simulation of several known response properties of simple cells, including the sharp and contrast invariant orientation tuning and surround suppression, and as cross orientation suppression.

Keywords: Sparse representation · Predictive coding · Thresholding · Non-negative approximation

1 Introduction

A large number of theoretical models for cortical visual representation have been developed during the last decades. The existing methods, including sparse coding [1] and predictive coding [2, 3], are very successful in accounting for the response properties of both neuronal populations and single neurons in the primary visual cortex (V1). Ever since Field [4] pointed out that retinal stimuli might be encoded in a sparse manner by cortical neurons, sparse representation of sensory stimuli in cortex has been widely validated [5–9]. Predictive coding was demonstrated to be very successful in simulating many V1 classical receptive field (CRF) and non-classical receptive field (nCRF) properties [10–12]. Recently, it was reported that some CRF and nCRF response properties can also be simulated by sparse coding [13–15]. Despite differences in the coding strategies, both of these two kinds of models consider representation

---

Jiqian Liu−This work was supported by the Science Research Foundation for High-level Talents of Guizhou Institute of Technology XJGC20130902, and in part by the Science and Technology Foundation of Guizhou Province J[2014]2081 and by the Innovation Team of Guizhou Provincial Eduction Department under Grant No [2014] 34.

D.-S. Huang et al. (Eds.): ICIC 2015, Part I, LNCS 9225, pp. 167–176, 2015.
DOI: 10.1007/978-3-319-22180-9_17

errors as an optimization objective. It indicates that the main goal of visual coding is to approximate the input.

Sparse approximation using iterative thresholding, which are known as iterative thresholding algorithms (ITA), have been presented [16–18]. In these methods, the input is approximated iteratively while the representation is made sparser by a thresholding mechanism at each iteration step. In this paper we present a similar algorithm to ITA for non-negative visual inputs. Differently from non-negative spare coding [19] and the modified predictive coding by Spratling [3] which have been developed on the basis of non-negative matrix factorization (NMF) [20], the presented algorithm minimize Euclidean distance other than divergence. To validate this algorithm, we simulate several well-known response properties of single neurons, including cross orientation suppression within CRF, and surround suppression from nCRF, and also the sharp and contrast invariant orientation tuning.

2 Error-Driven Micro-Adjustment Approximation

The issue to be addressed here is to find the best approximation of an input signal from a given basis set. The optimization problem can be expressed as

$$E = \min_y (x - Wy)^2 \tag{1}$$

where x is the input vector, W is a matrix whose ith column is the basis vector w_i, and y is a vector of coefficients. Here, Wy is an approximation of x in the column space of W, denoted as $\hat{x}$. $e = x - \hat{x}$ is the residual error. Solving Eq. (1) by gradient descent gives

$$\Delta y = \mu We, \tag{2}$$

where μ is the step length which is usually chosen to be small enough to decrease the optimization objective in Eq. (1).

According to Eq. (2), the modulation of the coefficients is driven by the residual error in a micro-adjustment manner. We call a model adopting such a rule an Error-driven macro-adjustment (EmA) model. The iterative form of EmA is given by

$$\begin{cases} e(n+1) = x - Wy(n); \\ y(n+1) = y(n) + \mu W^T e(n+1) \end{cases} \tag{3}$$

where W^T is the transpose of the matrix W. In the neural network implementation of the model, the response of the ith output neuron is the ith component of y, denoted by y_i, and the connection weight to the neuron is w_i. It was suggested that negative feedback connections [21, 22] or error detection neurons [3] could be responsible for the calculation of the residual error. Combining Eq. (3), we obtain

$$y(n+1) = y(n) + \mu \left(W^T x - W^T Wy(n) \right) \tag{4}$$

Equation (4) is the lateral inhibition version of EmA in which feedforward excitation $W^T x$ is suppressed by the lateral inhibition $W^T W y(n)$. Here, $W^T W$ is the lateral inhibitory coefficient matrix. As proven in the appendix, $\|e(n)\|$ is monotone decreasing given that

$$\mu < \frac{1}{\|W^T W\|} \tag{5}$$

where $\|W^T W\|$ is the norm of the matrix $W^T W$. When $\|W^T W\| < 1$, the inequality (5) is true for $\mu = 1$. Thus Eq. (4) can be rewritten as

$$y(n+1) = W^T x - \left(W^T W - I\right) y(n) \tag{6}$$

where I is the identity matrix.

Equations (4) and (6) have been adopted in many previous studies. In NMF using Euclidean distance, the parameter μ of Eq. (4) is set according to each output neuron. For the ith output neuron, the step length $\mu_i = y_i(n)/W^T W y(n)$ [20]. The locally competitive algorithm (LCA) uses a rule similar to Eq. (6) to update internal states which determine the spike rates of the output neurons [23, 24]. Inspired by their study, we adopt integration and spiking mechanisms in the presented work.

3   Non-negative Approximation with Thresholding

In visual systems, natural stimuli are first preprocessed in the retina and LGN. Then the positive and negative parts of the result are transferred to the visual cortex through ON and OFF channels separately. That is, the input signal from sub-cortical is $\chi = [x^+ x^-]$ instead of x, where x^+ and x^- are vectors containing the absolute values of the positive and negative components of x respectively. The connection weights should also be modified by this non-negative data processing. But still, we denote the connection weight matrix as W.

To guarantee the non-negativity of the output, we apply the hard thresholding mechanism as follows:

$$S_\theta(y) = \begin{cases} y, y > \theta \\ 0, y < \theta \end{cases} \tag{7}$$

where $\theta > 0$ is the threshold. Equation (7) could be implemented in neural systems by spiking: neurons integrating feedforward excitation and lateral inhibition will fire spikes when the integration exceed a threshold, otherwise will not be activated. Combining the above results, the main steps of our presented non-negative approximation with thresholding (NAT) method is given as follows:

1. **Preprocessing** Filter the input with difference of Gaussian (DoG). Operate on the result with an input saturation nonlinearity and non-negative data processing.

2. **Integration** Calculate the integration of feedforward excitation and lateral inhibition:

$$v(n+1) = \mu\big(W^T\chi - W^T W y(n)\big), \tag{8}$$

3. **Spiking** Modulate spike responses of neurons whose integration exceeds the threshold θ:

$$y(n+1) = S_\theta(y(n) + v(n+1)). \tag{9}$$

Please note that the total inhibition item in Eq. (8) is equal to $W^T W y(n)$, which includes lateral inhibition from other spiking neurons and self-inhibition. The integration at each iteration step is small which is scaled down by the parameter μ. This simulates the spiking activity in neural systems: every time a neuron fires a spike, it will slightly modify the activities of neighbor neurons.

4   Experimental Results

We validate the presented NAT algorithm by simulating several well-known response properties of single neurons, including cross orientation suppression within CRF, and surround suppression from nCRF, and the sharp and contrast invariant orientation tuning. The visual stimuli and Gabor-like connection weights are generated using the same code published by Spratling [10]. A family of 32 Gabor functions covering 8 orientations and 4 phases is used as the connection weights $\{w_i\}$, which are normalized to unit length. The DoG filtered visual stimulus x is subject to a saturating nonlinearity:

$$x' = \tanh(2\pi x) \tag{10}$$

Both the visual stimuli $\{x'\}$ and the Gabor functions are then processed to be non-negative to generate χ and W.

4.1   Sparsity of Outputs

We first measure the sparsity of outputs. The results are shown in Fig. 1, in which the sparsity is defined as the ratio of the number of inactive neurons to the total number of neurons, the representation error is measured by the root square error and the inhibition level is calculated as the average of the total inhibition received by each neuron. We see that the population response becomes sparser when cortical inhibition gradually increases with the iteration times. It indicates that the activities of some output neurons are suppressed to be inactive by the presence of cortical inhibition.

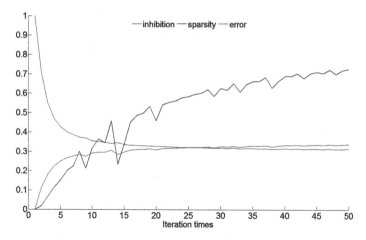

Fig. 1. Trend of the sparsity, representation error and inhibition level.

4.2 Orientation Tuning

The simulation result of orientation tuning (OT) is shown in Fig. 2, from which we see that the selected output neuron exhibits sharp orientation tuning and the tuning curve calculated only with the LGN inputs is be much broader than that with lateral inhibition (LI). It indicates that the feedforward excitation is sensitive to a wide range of stimuli and exhibits less orientation selectivity than the spike response. This is consistent with the facts that the sharp orientation selectivity of the spike responses can be reduced by blocking cortical inhibition [25, 26].

It has been suggested that the orientation selectivity of simple cells originates purely from the excitatory convergence of LGN afferents [27]. According to this theory, the tuning width of orientation selectivity should be widened with increasing contrast of stimuli, which on the contrary was not observed in experiments [28]. This phenomenon is known as the contrast invariance of orientation selectivity. The simulation result given by NAT is shown in Fig. 3. The tuning with remains unchanged with increasing stimulus contrast.

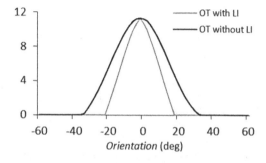

Fig. 2. Lateral inhibition (LI) sharpens orientation tuning (OT).

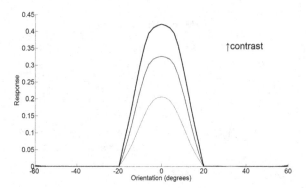

Fig. 3. Illustration of the contrast invariance of orientation tuning.

4.3   Cross Orientation Suppression

Cross orientation suppression is known as a phenomenon in which simple cell responses to the preferred orientation are suppressed by a superimposed orthogonal stimulus [29, 30]. It was once thought that cross orientation suppression should be responsible for contrast invariance of orientation selectivity [31, 32]. But intracellular recording observations rule out this hypothesis. Finn et al. [33] proposed a purely feedforward model in an attempt to attribute cross orientation suppression to the contrast saturation and rectification in LGN which, however, failed to predict the suppression at low contrasts [34]. We suggest that the non-negative data processing of LGN input provides an explanation for this disagreement. Simulating results of cross orientation suppression by NAT at different contrasts are shown in Fig. 4. From top to

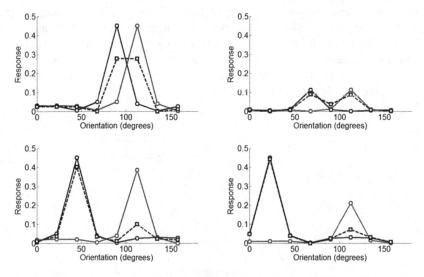

Fig. 4. Illustration of cross orientation suppression using stimuli with different orientation and contrast.

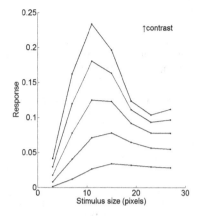

Fig. 5. Contrast dependent size tuning of surround suppression.

bottom and left to right, the contrasts of the two components of the superimposed stimulus are: 1, 1; 0.1, 0.1; 1, 0.5; 1, 0.2. As illustrated in Fig. 4, the suppression effect is still observable at low contrasts. This demonstrates our suggestion that non-negative data processing besides non-linear saturation of input is capable to predict an occurrence of the suppression effect at low contrasts.

4.4 Surround Suppression

Another phenomenon relating to cortical inhibition is surround suppression where the presence of stimuli in the non-classical receptive field of cortical neurons can suppress their spiking responses. We simulate the contrast dependent size tuning of surround suppression [35]. Only the Gabor function with the same orientation as the center stimulus is used for the simplicity of computation. This is reasonable because surround suppression is narrowly tuned to the preferred orientation. As shown in Fig. 5, the optimal stimulus sizes at high contrasts are smaller than those at low contrasts. The optimal size of the tuning curve is the one that elicits the peak response. The optimal sizes at low contrasts are larger than those at high contrasts.

5 Conclusion

In this paper, we apply the NAT algorithm, which is similar to the iterative thresholding algorithms, to simulate several well known response properties of cortical neurons. NAT modulates outputs in a macro-adjustment manner. That is, the spike responses of output neurons are modified slightly by the feedforword error. Each step of the presented algorithm is neutrally plausible. Experimental results validate our method.

Appendix

According to the second equation in (3), we obtain $y(n) = y(n - 1) + \mu W^T e(n)$. Substitution of this result into the first equation in (3) yields

$$e(n + 1) = x - Wy(n - 1) - \mu WW^T e(n)$$
$$= e(n) - \mu WW^T e(n).$$

Then we have

$$\|e(n + 1)\|^2$$
$$= (e^T(n) - \mu e^T(n)WW^T) \bullet (e(n) - \mu WW^T e(n))$$
$$= \|e(n)\|^2 - \mu(2z^T z - \mu z^T W^T Wz)$$

where $z = W^T e(n)$. Obviously, $\|e(n)\|$ is monotone decreasing when

$$2z^T z - \mu z^T W^T Wz > 0.$$

Considering that

$$\|z^T W^T Wz\| \leq \|z\| \cdot \|W^T Wz\| \leq \|W^T W\| \cdot \|z\|^2,$$

That is,

$$z^T W^T Wz \leq \|W^T W\|^2 z^T z.$$

Thus, $\|e(n)\|$ is monotone decreasing given that

$$\mu \|W^T W\|^2 \leq 2.$$

Then we can set $\mu < \frac{1}{\|W^T W\|^2}$.

References

1. Olshausen, B.A., Field, D.J.: Emergence of simple-cell receptive field properties by learning a sparse code for natural images. Nature **381**, 607–609 (1996)
2. Rao, R.P., Ballard, D.H.: Predictive coding in the visual cortex: a functional interpretation of some extra-classical receptive-field effects. Nat. Neurosci. **2**, 79–87 (1999)
3. Spratling, M.W.: Predictive coding as a model of biased competition in visual attention. Vision. Res. **48**, 1391–1408 (2008)
4. Field, D.: What is the goal of sensory coding? Neural Comput. **6**, 559–601 (1994)
5. Vinje, W.E., Gallant, J.L.: Sparse coding and decorrelation in primary visual cortex during natural vision. Science **287**, 1273 (2000)

6. Perez-Orive, J., Mazor, O., Turner, G.C., Cassenaer, S., Wilson, R.L., Laurent, G.: Oscillations and sparsening of odor representations in the mushroom body. Science **297**, 359 (2002)
7. Lewicki, M.S.: Efficient coding of natural sounds. Nat. Neurosci. **5**, 356–363 (2002)
8. Laurent, G.: Olfactory network dynamics and the coding of multidimensional signals. Nat. Rev. Neurosci. **3**, 884–895 (2002)
9. Smith, E.C., Lewicki, M.S.: Efficient Auditory Coding. Nature **439**, 978–982 (2006)
10. Spratling, M.W.: Predictive coding as a model of response properties in cortical area V1. J. Neurosci. **30**, 3531–3543 (2010)
11. Spratling, M.W.: A single functional model accounts for the distinct properties of suppression in cortical area V1. Vision. Res. **51**, 563–576 (2011)
12. Spratling, M.W.: Predictive coding accounts for V1 response properties recorded using reverse correlation. Biol. Cybern. **2**, 1–13 (2012)
13. Hunt, J.J., Dayan, P., Goodhill, G.J.: Sparse coding can predict primary visual cortex receptive field changes induced by abnormal visual input. PLoS Comput. Biol. **9**, e1003005 (2013)
14. Giorno, A.D., Zhu, M., Rozell, C.J.: A sparse coding model of V1 produces surround suppression effects in response to natural scenes. BMC Neurosci. **14**, P335 (2013)
15. Zhu, M., Rozell, C.J.: Visual nonclassical receptive field effects emerge from sparse coding in a dynamical system. PLoS Comput. Biol. **9**, e1003191 (2013)
16. Herrity, K.K., Gilbert, A.C., Tropp, J.A.: Sparse approximation via iterative thresholding. In: IEEE International Conference on Acoustics, Speech and Signal Processing (ICASSP 2006) (2006)
17. Fornasier, M., Rauhut, H.: Iterative thresholding algorithms. Appl. Comput. Harmonic Anal. **25**, 187–208 (2008)
18. Blumensath, T., Davies, M.E.: Iterative thresholding for sparse approximations. J. Fourier Anal. Appl. **14**, 629–654 (2008)
19. Hoyer, P.O.: Modeling receptive fields with non-negative sparse coding. Neurocomput. **52**, 547–552 (2003)
20. Lee, D.D., Seung, H.S.: Algorithms for non-negative matrix factorization. Adv. Neural Inf. Process. Syst. **13**, 556–562 (2001)
21. Harpur, G.F., Prager, R.W.: Development of low entropy coding in a recurrent network. Netw. Comput. Neural Syst. **7**, 277–284 (1996)
22. Harpur, G.F.: Low entropy coding with unsupervised neural networks. Ph. D. dissertation, Queen's College, University of Cambridge (1997)
23. Rozell, C.J., Johnson, D.H., Baraniuk, R.G., Olshausen, B.A.: Sparse coding via thresholding and local competition in neural circuits. Neural Comput. **20**, 2526–2563 (2008)
24. Charles, S., Garrigues, P., Rozell, C.J.: A common network architecture efficiently implements a variety of sparsity-based inference problems. Neural Comput. **1**, 1–23 (2012)
25. Sillito, A.: Inhibitory mechanisms influencing complex cell orientation selectivity and their modification at high resting discharge levels. J. Physiol. **289**, 33–53 (1979)
26. Sillito, A., Kemp, J., Patel, H.: Inhibitory interactions contributing to the ocular dominance of monocularly dominated cells in the normal cat striate cortex. Exp. Brain Res. **41**, 1–10 (1980)
27. Hubel, D.H., Wiesel, T.N.: Receptive fields, binocular interaction and functional architecture in the cat's visual cortex. J. Physiol. **160**, 106–154 (1962)
28. Sclar, G., Freeman, R.D.: Orientation selectivity in the cat's striate cortex is invariant with stimulus contrast. Exp. Brain Res. **46**, 457–461 (1982)

29. Morrone, M.C., Burr, D.C., Maffei, L.: Functional implications of cross-orientation inhibition of cortical visual cells. Proc. R. Soc. Lond. Series B Biol. Sci. I. Neurophysiol. Evid. **216**, 335–354 (1982)
30. DeAngelis, G.C., Robson, J.G., Ohzawa, I., Freeman, R.D.: Organization of suppression in receptive fields of neurons in cat visual cortex. J. Neurophysiol. **68**, 144–163 (1992)
31. Sompolinsky, H., Shapley, R.: New perspectives on the mechanisms for orientation selectivity. Curr. Opin. Neurobiol. **7**, 514–522 (1997)
32. Lauritzen, T.Z., Miller, K.D.: Different roles for simple-cell and complex-cell inhibition in V1. J. Neurosci. **23**, 10201 (2003)
33. Finn, I.M., Priebe, N.J., Ferster, D.: The emergence of contrast-invariant orientation tuning in simple cells of cat visual cortex. Neuron **54**, 137–152 (2007)
34. MacEvoy, S.P., Tucker, T.R., Fitzpatrick, D.: A precise form of divisive suppression supports population coding in the primary visual cortex. Nat. Neurosci. **12**, 637–645 (2009)
35. Sceniak, M.P., Ringach, D.L., Hawken, M.J., Shapley, R.: Contrast's effect on spatial summation by macaque V1 neurons. Nat. Neurosci. **2**, 733–739 (1999)

Targeting the Minimum Vertex Set Problem with an Enhanced Genetic Algorithm Improved with Local Search Strategies

Vincenzo Cutello[1] and Francesco Pappalardo[2(✉)]

[1] Department of Mathematics and Computer Science,
University of Catania, Catania, Italy
cutello@dmi.unict.it
[2] Department of Drug Sciences, University of Catania, Catania, Italy
francesco.pappalardo@unict.it

Abstract. The minimum feedback vertex set in a directed graph is a *NP*-hard problem i.e., it is very unlikely that a polynomial algorithm can be found to solve any instances of it. Solutions of the minimum feedback vertex set can find several real world applications. For this reason, it is useful to investigate heuristics that might give near-optimal solutions. Here we present an enhanced genetic algorithm with an ad hoc local search improvement strategy that finds good solutions for any given instance. To prove the effectiveness of the algorithm, we provide an implementation tested against a large variety of test cases. The results we obtain are compared to the results obtained by greedy and randomized algorithms for finding approximate solutions to the problem.

Keywords: Artificial intelligence · Genetic algorithms · Graphs

1 Introduction

The Feedback Vertex Set (FVS) problem i.e., the problem of generating a maximal acyclic subgraph from a given graph, is NP-complete [8, 12]. Independently, a general proof of the NP-completeness of all the feedback set problems for planar graphs, was given in [26]. The FVS problem has many applications: deadlock prevention [25], program verification [21], Bayesian inference [1, 2] and Computer Aided Design (CAD) applications [15], and, more recently, in genome sequence assembly [19] and in large-scale biological network [22].

The FVS problem can be defined as follows. Let G(V, E) be a directed graph, where V denotes the set of vertices, and E the set of arcs, n = |V|, m = |E|. A Feedback Vertex Set or cutset is a subset V' of V which contains at least one vertex from each cycle in G. The problem is to find a cutset of minimal cardinality. (Note that in view of the above definition of a cutset, if G has no cycle then the only cutset is the empty set.) If we define a weight function w : V → R + , the FVS solution is a feedback vertex set of minimal weight.

The FVS problem is APX-hard i.e., it is very difficult to find good approximate solutions (see [13, 16] for the definition of APX; and [11] for the proof of APX-hardness

© Springer International Publishing Switzerland 2015
D.-S. Huang et al. (Eds.): ICIC 2015, Part I, LNCS 9225, pp. 177–188, 2015.
DOI: 10.1007/978-3-319-22180-9_18

of FVS). An algorithm to find approximate solutions in the weighted case in [5]. The results rely on the theoretical property saying that the integrality gap in case of unweighted feedback vertex set problem can be $O(\log \tau* \log \log \tau*)$, where $\tau*$ is the weight of an optimal fractional set [20]. Further improvements of the algorithm produce a polynomial approximation scheme which can be applied to the weighted case. The algorithm finds a FVS with weight $O(\min\{\tau* \log \tau* \log \log \tau*, \tau* \log |V| \log \log |V|\})$ in time $O(|E| \cdot |V|2)$.

In [3] the authors present an algorithm for deciding whether a given mixed graph on n vertices contains a feedback vertex set of size at most k, in time $O(47.5 k \cdot k! \cdot n4)$. One successful example is represented by the greedy randomized adaptive search procedure (GRASP) [6, 18]. GRASP is a multistart method having two phases, a construction phase and a local search phase. During the construction phase, a feasible solution is constructed iteratively, one element at time. Each element of the solution is randomly selected from a restricted candidate list (RCL) that contains elements that are well-ranked according to some greedy function. The local search phase tries to improve the constructed solution and produces a solution that is locally optimal with respect to the specified neighbourhood structure. This twostage process is applied repeatedly, and the best solution found is kept as an approximation of the optimal.

To our knowledge, no evolutionary approaches have been applied to FVS problem, and it appears that the best approximation algorithms are based on greedy techniques [7]. Here we present a genetic algorithm (GA) which makes use of a local search improvement strategy. The paper is organized as follows. In Sect. 2, we present and study general properties that hold for the FVS problem. Section 3 presents some greedy algorithms that we will compare with our GA. In Sect. 4, we describe in depth our algorithm. Section 5 presents computational results and, in Sect. 6, we will give conclusions and final remarks.

2 Properties of the FVS Problem

In the following, we prove some properties of the FVS problem that we will use in the paper. Let $G = (V,E)$ be a directed graph. We recall that for each vertex $v \in V$, the *in_degree(v)* is the number of arcs of E entering v, and the *out_degree(v)* is the number of arcs leaving v. It can be easily proven that if there is no vertex with *in_degree* equal to zero, and if there is no vertex with out-degree equal to zero, then the graph has a cycle (we are excluding arcs from one vertex to itself). Thus, in and out degrees of the vertices are a good starting point in search of a greedy algorithm. A Strongly Connected Component (SCC) of the graph G, is a maximal sub- graph $G' = (V',E')$ such that for any two vertices $u,v \in V'$ we have a path in G' from u to v, and a path from v to u. We denote with scc(G) the number of strongly connected components of a graph G. In particular, if the graph is acyclic each single vertex is a strongly connected component of the graph i.e., $scc(G) = |V|$. If the graph has a cutset of cardinality 1, there exists a vertex v whose removal will make the graph acyclic, that is to say, it will produce a graph G' with $|V| - 1$ strongly connected components. In what follows, given a graph $G = (V,E)$ and a subset of vertices $S \subset V$, we will denote by $G_{V\backslash S}$ the graph obtained from G by eliminating the vertices in S and all the arcs that have those vertices as one of the endpoints. The following holds.

Lemma 1. *Let $G = (V, E)$ be a directed graph which contains at least one cycle. Let v be a vertex on a cycle of G then $scc(G_{V \setminus \{v\}}) \geq scc(G)$.*

Proof. Since v is in a cycle, then v belongs to a strongly connected component of G with at least two vertices, and its removal clearly produces a graph $G_{V \setminus \{v\}}$ such that scc $(G_{V \setminus \{v\}}) \geq scc(G)$.

We now have:

Theorem 1. *Let $G = (V, E)$ be a directed graph. If $C \subset V$ is a cutset of minimal cardinality, then $\forall S, S \subset V, scc(G_{V \setminus C}) \geq scc(G_{V \setminus S})$.*

Proof. Let C be a cutset of minimal cardinality and let S be any subset of V. Since C is a cutset, then $G_{V \setminus C}$ is acyclic. Therefore, $scc(G_{V \setminus C}) = |V \setminus C|$. If $G_{V \setminus S}$ is acyclic, S is a cutset. For the minimality of the cardinality of C, $|C| \leq |S|$ and, in turn, $scc(G_{V \setminus C}) \geq scc$ $(G_{V \setminus S})$.

Let us suppose more that $G_{V \setminus S}$ contains some cycles. Let $T \supset S$ be a minimal cutset which properly contains S. $G_{V \setminus T}$ can be obtained by $G_{V \setminus S}$, by eliminating, one at a time, the vertices in $T \setminus S$. Applying lemma 1 and for the minimality of C, we have $scc(G_{V \setminus C}) \geq scc(G_{V \setminus T}) \geq scc(G_{V \setminus S})$.

The converse of theorem 1 is not true. Consider for instance the graph sketched in Fig. 1, with $C = \{1\}$. $\forall S \subseteq V$ we have that $scc(G_{V \setminus C}) \geq scc(G_{V \setminus S})$ but C is not a cutset for G.

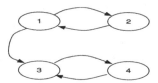

Fig. 1. A graph with $C = \{1\}$

2.1 FVS on a Special Class of Graphs

In the following, we introduce tournaments graphs and a special class of tournaments: Recursive Triangle Tournaments, RTT, for which we are able to know what is the optimum cutset. A tournament is a directed graph obtained by imposing a direction for each edge in an undirected complete graph. For instance, in Fig. 2 we show a tournament on 4 vertices.

The name "tournament" originates from such a graph's interpretation as the outcome of some sports competition in which every player encounters every other player exactly once, and in which no draws occur. We could, for instance, interpret a directed edge from x to y as a game won by x over y.

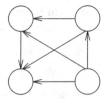

Fig. 2. A tournament graph with four vertices

A player who wins all games would naturally be the tournament's winner. However, as the above example shows, there might not be such a player. A tournament for which there is no winner is called a 1-paradoxical tournament. More generally, a tournament $T = (V, E)$ is called k-paradoxical if for every k- subset V' of V there is a $v_0 \in V \backslash V'$ such that $v_0 \to v$ for all $v \in V'$. By means of probabilistic methods Erdös showed that if $|V|$ is sufficiently large, then almost every tournament on V is k-paradoxical, for $k > 1$. Tournaments enjoy many properties, among which the fact the any tournament on a finite number n of vertices contains very likely a Hamiltonian path, i.e., directed path on all n vertices.

The FVS Problem remains NP-hard even for tournaments [23]. The proof is based on a linear reduction to the Vertex Cover Problem. Therefore, any approximation result on the Vertex Cover Problem gives us a corresponding result on the FVS Problem on tournaments. In particular, since Vertex Cover can be approximated with a constant factor of 2 [10, 25], we have a constant factor approximation for FVS on Tournaments, namely 2.5 as proven in [4]. Therefore, the FVS on Tournaments is not APX-hard.

Recursive Triangular Tournaments (RTT). We will now introduce a special class of tournaments which we will call Recursive Triangular Tournaments (RTT for short). A RTT is a Tournament with 3^k vertices, with $k \geq 1$, and the latter is the degree of the RTT. If we denote with A^k the adjacency matrix of a RTT with degree k and with 1^h (respectively 0^h) the $h \times h$ square matrix whose entries are all 1's (respectively 0's), we have:

$$A^1 = \begin{bmatrix} 0 & 1 & 0 \\ 0 & 0 & 1 \\ 1 & 0 & 0 \end{bmatrix}$$

$$A^2 = \begin{bmatrix} A^1 & 1^1 & 0^1 \\ 0^1 & A^1 & 1^1 \\ 1^1 & 0^1 & A^1 \end{bmatrix}$$

and, in general,

$$A^k = \begin{bmatrix} A^{k-1} & 1^{k-1} & 0^{k-1} \\ 0^{k-1} & A^{k-1} & 1^{k-1} \\ 1^{k-1} & 0^{k-1} & A^{k-1} \end{bmatrix}$$

In Fig. 3 we can see a RTT of degree 2 where directed arcs from a triangle to another, represent the fact that there is a directed arc from any vertex of the first triangle to every vertex of the second triangle.

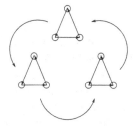

Fig. 3. RTT of degree k = 2

The following holds.

Theorem 2. *The optimal cutset cardinality of a RTT of degree k, is 3 k − 2 k.*

Proof. From the definition of the RTT of degree k, it is clear that in order to obtain an acyclic tournament we need to eliminate entirely one of the three subtournaments of degree k − 1, for otherwise we would still have a cycle. If we denote by T (k − 1) the cutset cardinality of a RTT of degree k − 1, we have:

$$T(k) = \begin{cases} 1 & \text{if } k = 1 \\ 2T(k-1) + 3^{k-1} & \text{if } k > 1 \end{cases}$$

The above recurrence equation can be solved exactly as follows:
We divide by 2^k, and obtain:

$$\frac{T(k)}{2^k} = \frac{2T(k-1)}{2^k} + \frac{3^{k-1}}{2^k};$$

$$\frac{T(k)}{2^k} = \frac{T(k-1)}{2^{k-1}} + \frac{3^{k-1}}{2^k};$$

We put $S(k) = \frac{T(k)}{2^k}$. So,

$$S(k) - S(k-1) = \frac{1}{2}\left(\frac{3}{2}\right)^{k-1};$$

$$S(k-1) - S(k-2) = \frac{1}{2}\left(\frac{3}{2}\right)^{k-2};$$

$$\cdots$$

$$S(2) - S(1) = \frac{1}{2}\left(\frac{3}{2}\right);$$

$$S(1) = \frac{1}{2};$$

Summing up:

$$S(k) = \frac{1}{2}\left[1 + \ldots + \left(\frac{3}{2}\right)^{k-2} + \left(\frac{3}{2}\right)^{k-1}\right];$$

$$S(k) = \frac{1}{2}\sum_{h=0}^{k-1}\left(\frac{3}{2}\right)^{h};$$

$$S(k) = \left(\frac{3}{2}\right)^{k} - 1.$$

Finally, since $S(k) = T(k)/2^k$, we obtain $T(k) = S(k) \cdot 2^k$, from which it follows:

$$T(k) = 3^k - 2^k.$$

3 Greedy Algorithms

The exact solution of the minimum FVS is, as well known, exponential so of no interest. Greedy algorithms can be defined for FVS problems, but their convergence to the minimum is not guaranteed. In the following we recall some of them which we will use for comparison with the proposed GA. First of all, let we consider the general greedy strategy [5, 14], denoted G*: (*i*) start from the given graph; (*ii*) delete at every step the vertex whose removal maximizes the number of strongly connected components; (*iii*) stop when the obtained graph is acyclic. The algorithm G* will find a cutset which is not always minimal. We show this in the following example. Let consider the graph shown in Fig. 4. It has a minimal cutset of cardinality 4, namely vertices 2, 3, 4 and 5. G* will first remove vertex 1, maximizing the number of SCC's, and then it will remove other four vertices from the external triangles. So, it will produce a cutset of cardinality 5.

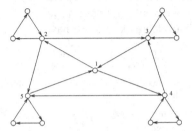

Fig. 4. Bad scenario for G*

We now recall three different greedy algorithms, which are well used in literature [2]: (*i*) The first greedy algorithm, G1, removes at every step the vertex v such that in degree(v) · out degree(v) is maximum; (*ii*) the second greedy algorithm, G2, removes at every step the vertex v such that in degree(v) + out degree(v) is maximum; (*iii*) the third greedy algorithm, G3, removes at every step the vertex v such that max{in degree(v), out degree(v)} is maximum. Using the G* strategy one can define the greedy algorithms G1*, G2* and G3* that act like the algorithms defined above, but that compute the in_degree() and out_degree() of a vertex "within" the strongly connected component to which the vertex belongs.

4 The Proposed Algorithm

Our genetic algorithm is based on the following key ideas: (*i*) a partition of the genetic population into males (graphs with cycles) and females (acyclic graphs, possible solutions): reproduction will be allowed only between individuals of different sex; (*ii*) a mechanism to assure that the male-female subsets remain balanced in size; (*iii*) a virus mechanism instead of the usual mutation operator; (*iv*) a strategy of local search improvement applied to some female individuals. Given a directed graph G = (V,E) such that $|V| = n$, we can represent V as the set$\{0,1,...,n-1\}$. Therefore, any subset $V' \subseteq V$ can be seen as a binary array of length n in such a way that $V'[i] = 1$ if and only if $i \in V'$. Each chromosome in the population is a binary array of fixed length i.e., a subset of the set of vertices V and, in turn, it uniquely identifies a subgraph of G. The corresponding cutset is the set of all elements which have value zero in the array. The population is partitioned into two subsets: cyclic (males) and acyclic (females) graphs. Sexual distinction allows to: (i) have a rich genetic population; (ii) have in the population, throughout evolution, always candidate solutions (female individuals). The selection operator is tournament selection (see [9]) and the selected individuals mate with probability $p = 1$. Reproduction uses uniform crossover (however this does not involve the virus part as we will describe later). We use two different fitness functions (one for the males and one for the females). If $I \subset V$ is an element of the population, putting $C = V \setminus I$, the algorithm will try to maximize:

$$f(I) = |I| \text{ iff } G_{V \setminus C} \text{ is acyclic}$$
$$f(I) = scc(G_{V \setminus C}) \text{ iff } G_{V \setminus C} \text{ contains a cycle}$$

where $scc(G_C)$ is the number of strongly connected components of the subgraph G_C. In the latter we use the results of theorems proposed above.

For males, i.e., graphs with cycles, maximizing $scc(C)$ will tend to produce females, i.e. acyclic graphs. For females, maximizing $f(C)$ will give us the mini- mum feedback vertex set. Elitism is used on two specific elements of the population: (i) best fitness male element; (ii) best fitness female element (also a candidate solution).

4.1   Virus Description

Chromosomes contain some extra genes, specifically $2 + \lceil \log |V| \rceil$. These genes represent the genetic patrimony of the virus. As a consequence, the total length of a chromosome is $|V| + 2 + \lceil \log |V| \rceil$. We have (i) the extra $\lceil \log |V| \rceil$ bits are the virus modifying agents; and (ii) viruses will hit an individual if the remaining two extra bits, called control bits, have both value 1. Thus, chromosomes can be partitioned into three groups: (i) healthy (control bits are both 0); (ii) disease carrier (control bits have different values); (iii) sick (control bits are both 1).

However, all individuals, including the healthy ones, carry in their genes the virus genes. Two disease carrier chosen for reproduction, will produce a sick offspring with probability 1/4. If a virus hits an individual, it will modify the individual genes using a logical XOR (exclusive OR) with its genetic patrimony as mask. To do that, we act as following: consider a bitstring of length $|V|$. Consider an integer k such that $k \cdot \lceil \log(V) \rceil < |V| < (k+1) \cdot \lceil \log(V) \rceil$. We now fill up the bitstring with k- repeated values of length $\lceil \log(V) \rceil$ plus the first $|V| - k \cdot \lceil \log(V) \rceil$ bits of the virus. We use the constructed bitstring against the chromosome using the XOR logical operator. For instance, if the chromosome is [1 0 1 1 0 0 1 1] and the virus is [0 1 1], the bitstring will be [0 1 1 0 1 1 0 1]. The new obtained individual will be: [1 1 0 0 1 1 1 0]. During the first generation, the virus bits are uniformly random generated. Control bits are instead set to 1 with a given probability p_{v1}. In all tests, we choose $p_{v1} = 0.1$. Virus reproduction is slightly different than uniform crossover. Basically, when the virus bits differ at one position, the offspring are given a random value (in the uniform crossover, such a case would imply that one offspring will get the value 1 and the other the value 0). Control bits have also their specific reproduction procedure. If c_1, c_2 are the offspring control bits and c_{11}, c_{12} are the first parent control bits, while c_{21}, c_{22} are the second parent control bits, we choose randomly $c_1 \in \{c_{11}, c_{12}\}$ and $c_2 \in \{c_{21}, c_{22}\}$. Then, with probability p_{v2}, we set first child control bit to 1 if control bits are both 0. In all tests, we choose $p_{v2} = 0.1$.

The above defined virus has already produced quite satisfactory result, as showed in [17] where a comparison with a general mutation operator has been made.

4.2   Local Search Improvement

During evolution, some elements of the female population will be the starting point of a local search improvement strategy. In details, a female individual is chosen with probability ρ, a parameter which can be set by the user. Randomly, a locus with value zero will be chosen. Such a bit is set to 1 if the individual will remain "female" (i.e. the subgraph will remain acyclic). Otherwise, it will not be changed. Yet the individual has missed one chance for improvement. It is worthwhile to recall that the procedure LOIS is not computationally expensive. In particular, once the constant number of allowed chances is fixed, the overall cost will be $O(|V| + |E|)$.

4.3   Genesis

The genetic algorithm allows reproduction only between individuals of different sex. Initially, the population is created randomly with 50 % males and 50 % females. The ratio between the two sub-populations varies dynamically, since nothing can be said about the "sex" of an offspring. In order to keep sexual distinction we force both populations to be over a given threshold value (10 % of total population size). For this we introduce a genesis procedure which acts when the populations undergo the threshold.

The genesis procedure adds new females female-genesis and males male-genesis individuals since initial values are reached. female-genesis constructs a female adding new randomly chosen nodes until no cycles are present. male-genesis generates new male individuals choosing a node with probability $p = 0.5$. If such a created subgraph is acyclic, then it continues to add randomly chosen nodes until a cycle appears.

4.4   Computing Strongly Connected Components as a Pre-Processing Step

A simple, yet fruitful idea is to compute first the strongly connected components of the input graph, and then apply the algorithm for the FVS to those components. As we can see later, t his pre-processing step improves the results obtained for both our GA and GRASP.

5   Computational Results

The algorithm has been tested on a large number of graphs, with different vertex set cardinalities and arc densities. We compared the obtained results with the ones obtained by GRASP, and some greedy approaches. Tests were performed on various test suites. For all tests we performed, for our GA, 30 runs; for all of there, we reported best obtained result along with its percentage, and average objective values with standard deviations. We used an IBM PC-compatible, Pentium class. It is worthwhile to mention that our GA returns results in few minutes, while GRASP takes a lot of time, especially during tests on high density edges graphs. In Table 1 we propose a first test suite. For sake of clarity, the names given to the graphs, follow the rule *graph-number_of_nodes-density.txt*. The density parameter indicates the probability that a given arc (i, j), with $i \in \{1,...,|V|\}, j \in \{1,...,|V|\}$, belongs to E. So, for instance, the graph "graph-200-25" is a graph with two hundred vertices. Edges are randomly generated. For each ordered pair of vertices, the probability that there exists an arc is 0.25. For each graph, the shown numbers for the edges are, obviously, approximately close to the expected number, given the density and the number of vertices. As one can easily check, our approach is always better than GRASP (or at least not worse).

Finally we test the proposed algorithm on general tournaments and RTT. We introduce here a new strategy, G4, which removes the vertex with minimal |in degree (k) − out degree(k)| value. Such a greedy technique seems to be quite natural when dealing with tournaments. In the Table 2, the name of the tournament $T - k$ gives us the

Table 1. The first test suite, and cardinality of cutsets produced by GRASP and GA

graph − test	#nodes	#arcs	GRASP	GA	best %	$\bar{x}$	σ
graph-600-1	600	3581	212	207	63.333	207.666	1.043
graph-600-5	600	17290	466	459	46.666	460.066	1.152
graph-600-25	600	78813	570	562	56.666	562.880	1.107
graph-600-50	600	134858	584	573	53.333	574.933	1.181
graph-600-75	600	168508	587	568	50	568.833	1.035
graph-700-1	700	4868	291	286	60	286.833	1.127
graph-700-5	700	23841	565	559	56.666	560	1.238
graph-700-25	700	107033	667	660	60	660.733	1.062
graph-700-50	700	183622	685	673	56.666	673.733	0.963
graph-700-75	700	229390	688	668	53.333	668.700	0.900
graph-800-1	800	6346	366	358	40	359.033	1.048
graph-800-5	800	31084	662	652	56.666	653.666	0.869
graph-800-25	800	139687	766	761	56.666	761.833	1.067
graph-800-50	800	239977	780	772	56.666	772.733	1.030
graph-800-75	800	299853	789	767	56.666	768	1.238
graph-900-1	900	8063	442	435	50	436.100	1.247
graph-900-5	900	39599	766	753	43.333	754.066	1.062
graph-900-25	900	177441	869	861	70	861.566	0.955
graph-900-50	900	304196	885	873	53.333	873.966	1.196
graph-900-75	900	379413	889	870	56.666	870.933	1.152
graph-1000-1	1000	9943	529	523	60	523.800	1.107
graph-1000-5	1000	48691	955	849	30	850.200	0.979
graph-1000-25	1000	218826	985	960	50	961.133	1.284
graph-1000-50	1000	374281	981	972	43.333	973.06	1.123
graph-1000-75	1000	468181	984	966	33.333	967.233	1.085

Table 2. Cutsets cardinality for tournament graphs

GRAPH	GA	best %	$\bar{x}$	σ	GRASP	BEST GREEDY
T-10	5	100	5	0	5	5
T-20	12	100	12	0	12	13
T-30	20	83.333	20.233	0.558	20	23
T-40	30	73.333	60.433	0.760	31	32
T-50	39	76.666	39.300	0.585	40	41
T-60	48	70	48.400	0.663	50	50
T-70	58	70	58.433	0.715	60	61
T-80	67	70	67.533	0.845	69	70
T-90	77	73.333	77.733	0.596	79	80
T-100	87	76.666	87.300	0.585	90	90
T-200	186	76.666	186.266	0.512	190	190
T-250	236	70	236.433	0.715	240	239
T-300	286	70	286.433	0.715	288	287

Table 3. utsets cardinality for RTT graphs

GRAPH	GA	GRASP	BEST GREEDY
RTT-3	19	19	19
RTT-4	65	65	65
RTT-5	211	211	214
RTT-6	665	713	669
RTT-7	2059	2250	2062

number of vertices, the shown results are the "best" obtained results. We underline that in 98 % of the experiments, G4 obtained the best results. GRASP was very slow. Moreover, the greedy approaches, quite fast, give us results which are comparable to the ones obtained by GRASP. As one can see, our GA gives the best results and maintains a good computational efficiency. When dealing with RTT's, our GA always finds the best cutset. GRASP obtains good results for RTT's up to degree 5. Greedy finds the optimum up to degree 5. The results of these experiments are shown in Table 3.

6   Conclusions and Final Remarks

In this paper we propose a genetic algorithm with local search improvement strategies for the minimum feedback vertex set. The algorithm, to our knowledge, is the first example of genetic algorithm which successfully attacks such a hard combinatorial problem. To verify the good performances of our GA, we compared it with several greedy algorithms and with a good greedy-randomized algorithm called GRASP. Our GA obtains always results which are no worse, in many cases better, than all the other algorithms and, it has a much better computational performance. Our approach makes a heavy use of strongly connected components. It evolves tentative solutions which maximize strongly connected components. In case of acyclic graphs, this means cutsets of minimal cardinality. In case of graphs with cycles, it means that the number of vertices to eliminate, to reach acyclicity, is small. The greedy algorithm G* which at every step chooses the vertex whose removal maximizes the number of strongly connected components, is also presented and partially studied. Numerical experiments show that is the best greedy algorithm among the presented ones. We believe that such a greedy algorithm is worth a deeper investigation and that, by means of ad hoc heuristics, could be extended to a very competitive algorithm.

References

1. Becker, A., Bar-Yehuda, R.: Randomized algorithms for the loop cutset problem. J. Artif. Intell. Res. **12**, 219–234 (2000)
2. Becker, A., Geiger, D.: Optimization of pearl's method of conditioning and greedy like approximation algorithms for the vertex feedback set problem. Artif. Intell. **83**, 167–188 (1996)
3. Bonsma, P., Lokshtanov, D.: Feedback vertex set in mixed graphs. Lect. Notes Comput. Sci. **6844**, 122–133 (2011)
4. Cai, M.C., Deng, X., Zang, W.: An approximation algorithm for feedback vertex sets in tournaments. SIAM J. Comput. **30**(6), 1993–2007 (2001)
5. Even, G., Naor, S., Schieber, B., Sudan, M.: Approximating minimum feedback sets and multicuts in directed graphs. Algorithmica **20**, 151–174 (1998)
6. Feo, T.A., Resende, M.G.C.: Greedy randomized adaptive search procedures. J. Global Opt. **6**, 109–133 (1995)
7. Festa, P., Pardalos, P.M., Resende, M.G.C.: Feedback set problems. In: Du, D.Z., Pardalos, P.M. (eds.) Handbook of Combinatorial Optimization, Supplement, vol. A, pp. 209–259. Kluwer Academic Publishers, Dordrecht (1999)
8. Garey, M.R., Johnson, D.S.: Computers and Intractability - A guide to the theory of NP-completeness. W. H. Freeman and Company, San Francisco (1979)
9. Goldberg, D.E.: A comparative analysis of selection schemes used in genetic algorithms. Morgan Kaufmann Publishers, gregory rawlins (edn) (1991)
10. Hockbaum, D.S.: Approximating covering and packing problems: set cover, vertex cover, indipendent set, and related problems. In: Hockbaum, D.S. (ed.) Approximation Algorithms for NP-Hard Problems, pp. 94–143. PWS Publishing Company, Boston (1997)
11. Kann, V.: On the Approximability of NP-complete Optimization Problems. Ph.D. thesis, Department of Numerical Analysis and Computing Science, Royal Institute of Technology, Stockholm (1992)
12. Karp, R.M.: Reducibility among combinatorial problems. In: Miller, R.E., Thatcher, J.W. (eds.) Complexity of Computer Computations, pp. 85–103. Plenum Press (1972)
13. Khanna, S., Motwani, R., Sudan, M., Vazirani, U.: On syntactic versus computational views of approximability. SIAM J. Comp. **28**, 164–191 (1999)
14. Lin, H.M., Jou, J.Y.: Computing minimum feedback vertex sets by contraction operations and its applications on cad. In: International Conference on Computer Design, (ICCD 1999). pp. 364–369 (10–13 October 1999)
15. Lin, H.M., Jou, J.Y.: On computing the minimum feedback vertex set of a directed graph by contraction operations. IEEE Trans. Comput. Aided Des. Integr. Circuits Syst. **19**(3), 295–307 (2000)

16. Papadimitriou, C.H., Yannakakis, M.: Optimization, approximation, and complexity classes. J. Comput. System Sci **43**, 425–440 (1991)
17. Pappalardo, F.: Using Viruses to Improve GAs. In: Wang, L., Chen, K., S. Ong, Y, (eds.) ICNC 2005. LNCS, vol. 3612, pp. 161–170. Springer, Heidelberg (2005)
18. Pardalos, P.M., Qian, T., Resende, M.G.C.: A greedy randomized adaptive search procedure for feedback vertex set. J. Comb. Opt. **2**, 399–412 (1999)
19. Pop, M., Kosack, D., Salzberg, S.: Hierarchical scaffolding with bambus. Genome Res. **14** (1), 149–159 (2004)
20. Seymour, P.: Packing directed circuits fractionally. Combinatorica **15**, 281–288 (1995)
21. Shamir, A.: A linear time algorithm for finding cutsets in reduced graphs. J. Comput. **8**, 645–655 (1979)
22. Soranzo, N., Ramezani, F., Iacono, G., Altafini, C.: Decompositions of large-scale biological systems based on dynamical properties. Bioinform. **28**(1), 76–83 (2012)
23. Speckenmeyer, E.: On feedback problems in digraphs. Graph-Theoretic Concepts in Computer Science. Lecture Notes in Computer Science, vol. 411, pp. 218–231. Springer-Verlag, Berlin (1989)
24. Taoka, S., Watanabe, T.: Performance comparison of approximation algorithms for the minimum weight vertex cover problem. In: SCAS 2012 - 2012 IEEE International Symposium on Circuits and Systems, vol. 6272111, pp. 632–635 (2012)
25. Wang, C., Lloyd, E., Soffa, M.: Feedbackvertexsetsandcyclicallyreduciblegraphs. J. ACM **32**, 296–313 (1985)
26. Yannakakis, M.: Node and edge-delition np-complete problems. In: Proceedings of the 10-th Annual ACM Symposium on Theory of Computing. pp. 253–264 (1978)

Lie Detection from Speech Analysis Based on K–SVD Deep Belief Network Model

Yan Zhou[1,2(⊠)], Heming Zhao[2], and Xinyu Pan[2]

[1] College of Electronic Information Engineering, Suzhou Vocational University,
Suzhou 215104, Jiangsu, China
zhyan@jssvc.edu.cn
[2] College of Electronic and Information Engineering, Soochow University,
Suzhou 215006, Jiangsu, China

Abstract. Considering the task of lie detection relates some nonlinear characteristics, such as psychological acoustics and auditory perception, which are difficult to be extracted and have high computational complexity. So this paper proposes a deep belief network based on the K-singular value decomposition (K-SVD) algorithm. This method combined the multi-dimensional data linear decomposition ability of sparse algorithm and the deep nonlinear network structure of deep belief network. It is aim to extract the significant time dynamic deep lie structure characteristics. Based on these deep characteristics, the lie database of Arizona University at United States was used to test. The experimental results show that, compared with the K-SVD sparse characteristics and basic acoustic characteristics, the deep characteristics proposed in this paper has better recognition rate. Furthermore, this paper provides a new exploration for psychology calculation.

Keywords: Correct Detection Rate · Deep belief network · Feature extraction · K-singular value decomposition · Speech lie detection

1 Introduction

Speech lie detection is a typical research of psychophysiological calculation problem. It has been widely used in the fields such as military, justice, forensic etc. Speech signal detection is simple to operate. It is convenient to conduct the normal measurement to test the physical parameters under some kind of psychological activity. Recently, the research team at Purdue University study speech lie detector by the speech modulation model, and the work was supported by the company of Lockheed Martin with military background [1–4]. The researchers at Oklahoma State University of American established the speech polygraph system through the analysis of speech tension, and it can get the efficiency of 76 % [5]. In 2012, researchers in the York University published a paper in "Applied Ergonomics", which showed that speech lie detector is a complex topic, only depend on the characteristic parameters such as resonant and with the traditional classification method cannot effectively implement the speech lie detection [6–8]. But, the research of speech lie detection is just at the primary and exploration

© Springer International Publishing Switzerland 2015
D.-S. Huang et al. (Eds.): ICIC 2015, Part I, LNCS 9225, pp. 189–196, 2015.
DOI: 10.1007/978-3-319-22180-9_19

stage. The problems such as the quantitative calculation of psychological acoustic features, the analysis of data fusion, the constructing of time dynamic feature model etc., all of these problems needed to be solved [9–11]. In recent years, the Deep Belief Network (DBN) model has been paid more attention due to its nonlinear network structure [12–19]. However, K-SVD sparse coding (SC) model is an effective method to express high-dimensional data. It can get the sparse representation of signal based on a complete dictionary and effectively reduce the dimension of speech signal. In addition, it also has good robustness for the speech signal with noise.

Considering the advantages of DBN and K-SVD algorithm above, the speech lie detection model which is the combination of DBN and K-SVD sparse coding algorithm is proposed in this paper. The model introduces the speech signal sparse representation into the deep learning framework and builds multiple hidden layers learning model. Thus the inner structure of information in the lie speech signal can be extracted. The speech lie detection method proposed in this paper was tested by the scene database samples, and here, the practicality and effectiveness was also discussed. The method performed in this paper provides a new way of thinking for the research of speech lie detection.

2 Deep Belief Network Model

Deep belief network model is referenced the expression of hierarchy, that is, integrating the underlying characteristics to form the more abstract expression. So the internal structure of the high dimensional characteristics can be found and extracted. DBN is a kind of multilayer neural network which is composed of multiple Restricted Boltzmann Machines (RBM). The DBN structure is shown in Fig. 1, in which, V is the show layer and the H_j are hidden layers.

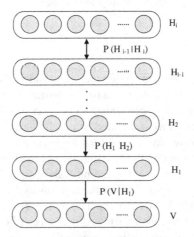

Fig. 1. DBN network structure composed of RBM

DBN training process mainly includes two steps: firstly, training the network initial parameters by using unsupervised way. Secondly, fining tune the overall network parameters by using the supervision way. The calculation progress is as this:

$$P_\theta(v, h) = \exp(-E(v, h; \theta)) \Big/ \sum_{h,v} \exp(-E(v, h; \theta))$$

$$= \prod_{ij} e^{W_{ij}v_i h_j} \prod_i e^{b_i v_i} \prod_j e^{a_j h_j} \Big/ \sum_{h,v} \exp(-E(v, h; \theta)) \tag{1}$$

The weights of the RBM network weights are updated as this:

$$\Delta w_{ij} = \varepsilon \frac{\partial \log P_\theta(v, h)}{\partial \theta} - \lambda \left\| w_{i,j} \right\|_F^2$$

$$= \varepsilon \left(\frac{\sum_h e^{-E(v_i, h_j)} \left(-\frac{\partial E(v_i, h_j)}{\partial \theta} \right)}{\sum_h e^{-E(v_i, h_j)}} - \frac{\sum_{v,h} e^{-E(v_i, h_j)} \left(-\frac{\partial E(v_i, h_j)}{\partial \theta} \right)}{\sum_{v,h} e^{-E(v_i, h_j)}} \right) - \lambda \left\| w_{i,j} \right\|_F^2 \tag{2}$$

$$= \varepsilon E_{P_{data}}[v_i h_j] - \varepsilon E_{P_\theta}[v_i h_j] - \lambda \left\| w_{i,j} \right\|_F^2$$

In which, w_{ij} is the connection weights between the show layer nodes and hidden layer nodes of RBM, $E(v_i, h_j)$ is the energy function of RBM, $E_{P_{data}}(v_i h_j)$ is the expectation for the free energy function of the input data, and the $E_{P_\theta}(v_i h_j)$ is the expectation for the free energy function of the sample data produced by the models. Generally, these values are obtained by the method of Contrast Divergence (CD). ε is the vector parameter, the coefficient of λ can be estimated in the training.

When the training of RBM is completed, the activation probability vector of the hidden layer nodes is used as the input values for the next RBM layer. The multilayer RBM are trained by this way, and the initial parameter values of each RBM can be calculated. Based on this, an initial deep belief networks can be constructed by connecting the trained RBM network and adding a SVM classifier at the top of the network. However, the labeled data is used to train by the mean of monitor, and the conjugate gradient descent method is introduced to fine-tune the network parameters, finally, the global optimal network can be achieved. The structure of the input speech data is unsupervised learned by the deep learning model, so that the initial value is much closer to the global optimal and obtains better effect.

3 Sparse Representation of the Speech Signal

In order to reflect the process of dynamic change of the basic acoustic parameters, here, the spectrogram signal is analyzed. Assumed that the length of lie speech signal sample is t, Add the window to the speech signal and length of subsection speech is t_k, and the corresponding spectrogram is Y_k. The spectrogram dimension should be reduced by the PCA procession. However, in order to reduce the amount of calculation, the spectrogram should be divided into the overlap of small pieces y_{ij}^k whose size is $p \times p$. The Discrete Cosine Transform (DCT) complete dictionary is chosen as the initial dictionary and the dictionary redundancy is set as 4. Firstly, a complete dictionary D can be adaptively trained from the sub-spectrogram. Then, the y_{ij}^k can be decomposed as the corresponding sparse coefficients by using the K-SVD algorithm. For the multiple estimate value of the overlap spectrogram, the average value is calculated. Finally, the sparse decomposition coefficients of each sub- spectrogram are put into the DBN model for the deep learning to obtain deep features. The goal function of K-SVD sparse decomposition is as following:

$$\{\tilde{s}_{ij}, \tilde{D}, \tilde{Y}\} = \underset{s_{ij}, D, Y}{\arg\min}[\left\|Y - \tilde{Y}\right\|_F^2 + \sum_{k,i,j} \mu_i \left\|s_{ij}^k\right\|_0 + \sum_{k,i,j} \left\|Ds_{ij}^k - F_{ij}Y_k\right\|_2^2] \quad (3)$$

In which, Y is the original spectrogram, $\tilde{Y}$ is the reconstruction value of the original spectrogram, $\tilde{D}$ is the estimated dictionary of D, s_{ij}^k is the K-SVD sparse vector of the k th sub-spectrogram with the position of (i,j), $F_{ij}Y_k$ is the sub-spectrogram with the position of (i,j) and the size of $p \times p$ which is extracted from the k th sub- spectrogram, $\sum_i \mu_i \left\|s_{ij}\right\|_0$ is the sparse penalty term.

4   The Lying Speech Feature Extraction Model

The lie speech feature extraction model based on the combination of K-SVD sparse representation and DBN is proposed in this paper. It is mainly includes the sparse representation of process for the temporal characteristics and the deep representation of the sparse features. Feature extraction model includes two main phases:

(1) The complete dictionary training and the sparse feature extraction. Firstly, adding the Hamming window and reducing the dimension, then, performing the sparse decomposition for the lie speech signal use K-SVD method. With the alternating iterative calculation, the sparse matrix of the training speech and the complete dictionary D can be achieved.
(2) The DBN characteristics learning based on the sparse coefficient matrix which is obtained by K-SVD sparse decomposition is conducted in this step. Here, the deep lie feature can be extracted.

The lying speech feature extraction model is shown as following (Fig. 2)

5   Simulation and Conclusion Analysis

5.1   The Experimental Data

As the experimental data, the performance lie speech database established by Arizona University at USA was introduced in this paper. Speech samples are chosen from this database which is marked and the length is 60s. In This paper, speech samples of 10 testers are randomly selected from the database of PLSD. In which, 5 male and 5

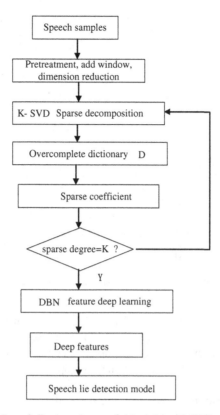

Fig. 2. Speech lie detection model based on K-SVD and DBN

female, 10 sections from each tester, 5 sections are used for learning and 5 used for test samples. In the experiment, the parameters are set as following: the sampling frequency is 8 KHZ, frame length is 256 points, Hamming window length is set as 8000 points, the speech signal with 60s is sequence fragmented as the length of 1s, the time domain segments signal is converted to 128×128 pixels of spectrogram, part of the spectrogram after the segmentation in this experiment are shown in Fig. 3.

Limited space, this article only gives part of the lying speech samples in PLSD. In order to reduce the amount of calculation, using the window 8×8 pixels to cut out as sub spectrogram, each spectrogram randomly be segmented 100 times, thus, a 64×6000 training matrix X is constructed, and performing PCA to reduce the dimension.

5.2 Experiment Result and Analysis

In order to illustrate the computability of K-SVD deep learning characteristics, here, the deep lying characteristics, sparse characteristics and the basic acoustic characteristics were compared the effectiveness for lying detection. The experiment was design as following, the experiment data was from PLSD lying database. There are five layers in the DBN training network and the number of hidden layer nodes are respectively

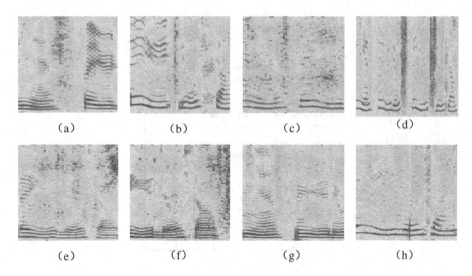

(a)   (b)   (c)   (d)

(e)   (f)   (g)   (h)

Fig. 3. Part of the experiment samples

1600,900,400,225, 49 from low to high. The K-SVD algorithm is used to calculate the complete dictionary for the lying speech signals. In the sparse decomposition processing, DCT dictionary is used as the initial over complete dictionary, the redundancy is 4.

The speech samples were decomposed on the complete dictionary to obtain a sparse coefficient characteristic matrix A, the five layer DBN network is applied in order to obtain deep learning features S. The learning parameters of RBM in DBN model is $\varepsilon = 0.005$, the number of iterations for weights adjustment is set to 100 times, Furthermore, the number of iterations for the adjustment of DBN weight parameters is set to 80 times. In addition, the speech signal formant, short-time energy and 12 order MFCC feature parameters is extracted as the basic acoustic parameters. Here, the DBN model is used to identify the characteristics. The Correct Detection Rate (CDR) is designed to measure the performance of the model. Table 1 shows the speech lie detection results respectively based on the basic acoustic parameters, sparse

Table 1. Comparison of CDR with different characteristic parameters (%)

Samples	Deep feature	Sparse feature	Basic feature
N = 1000	60.12	50.46	37.63
N = 2000	61.82	51.25	37.87
N = 3000	63.25	52.13	38.09
N = 4000	65.34	52.30	38.24
N = 5000	65.89	52.97	38.96
N = 6000	66.02	53.05	39.59
N = 7000	66.86	53.24	40.27
N = 8000	67.23	54.32	41.48
N = 9000	68.67	55.43	42.56
N = 10000	69.83	56.42	45.78

characteristic parameters and the deep characteristic parameters, here, the training samples come from the same tester. The table shows that the detection rate of the model is all increased with the increase of the training sample, and the highest correct lie speech detection rate is 69.83 %. This means that adopting deep feature achieved better recognition performance than that of using sparse characteristics and basic acoustic characteristics. Due to the deep learning method is a kind of learning method based on unsupervised feature learning and hierarchy feature structure. Through the deep non-linear network structure learning, it can search for the hidden inner lies information structure from the speech which can improve the accuracy of detection. However, the independent characteristics of speech signals can be obtained by the K-SVD method. At the same time, the sparse distribution coefficients can increase the distance between all kinds of characteristics. Therefore, the features extracted the sparse coding algorithm can yield good classification performance. In a word, it is superior to the basic acoustic characteristic parameters in the lying speech recognition.

6   Conclusion

Considering the characteristics of lying speech structure and the advantages of DBN model, the speech lie detection method based on the combination of K-SVD sparse representation and DBN model is proposed in this paper which realized the speech feature extraction and lie speech testing. In order to validate the effectiveness of the proposed algorithm, the performance scene database was introduced. The basic acoustic parameters, the sparse features as well as the deep characteristics were extracted separately. Then, the effectiveness of the three features to the identification model was compared. Experimental results show that, the deep lying characteristics obtained by the proposed algorithm can successfully express the different lying state. Further, this reflects the deep learning algorithm has the ability of extracting the inner structure information in the lie speech. In addition, in order to validate the advantages of DBN model for lying speech detection, the deep features of the speech samples and the DBN recognition model were used to test the PLSD database. Here, the SBN recognition model and RBF identification model were used for comparison, and the CDR is used to judge the recognition rate of the model. The results confirmed that the lie speech detection method proposed can achieve better lie feature extracting and lying state identification.

Acknowledgment. This work was supported by the funding of the National Natural Science Foundations of China (Grant No. 61372146, No. 61373098), Innovative team foundation of Suzhou vocational university and the Innovative Plan Project for Graduate students of Jiangsu province (No. CXZZ13_0812).

References

1. Hu, B.: The frontier science problems and key technologies of the psychophysiological calculation. In: The 431th Academic Seminar of Xiangshan Science Conference, Beijing (2012) (in Chinese)

2. Shikler, T., Robinson, P.: Classification of complex information: inference of co-occurring affective states from their expressions in speech. J. IEEE Trans. Pattern Anal. Mach. Intell. **32**, 1284–1297 (2010)

3. Enos, F.: Detecting deception in speech, Ph.D. thesis, Columbia University (2010)

4. Liu, D., Shi, G., Zhou, S.: A method of signal sparse decomposition on the redundant dictionary. J. Xian Electron. Sci. Technol. Univ. (Nat. Sci. Ed.) **35**, 228–232 (2008). (in Chinese)

5. Candès, E., Wakin, M.: An introduction to compressive sampling. J. IEEE Signal Process. Mag. **25**, 21–30 (2008)

6. Jin, J., Gu, Y., Mei, S.: Compressed sampling technique and application. J. Electron. Inf. **32**, 470–475 (2010). (in Chinese)

7. Candès, E., Romberg, J., Tao, T.: Robust uncertainty principles: exact signal reconstruction from highly incomplete frequency information. J IEEE Trans. Inf. Theory. **52**, 489–509 (2006)

8. Kirchhübel, C., Howard, D.: Detecting suspicious behaviour using speech: Acoustic correlates of deceptive speech An exploratory investigation. J. Applied Ergonomics. **43**, 561–569 (2012)

9. Zhiliang, W., Zheng, S., Wang, X.: Research status and development trend of psychological cognitive computing. J. Pattern Recogn. Artif. Intell. **24**, 215–223 (2011). (in Chinese)

10. Gopalan, P., Wenndt, S.: Speech analysis using modulation-based features for detecting deception. In: The 15th International Conference on Digital Signal Processing. pp. 619–622 (2007)

11. Michal, A., Elad, M., Alfred, B.: K-SVD: an algorithm for designing over-complete dictionaries for sparse representation. J. IEEE Trans. Signal Process. **54**, 4311–4322 (2006)

12. Anton, N.: Computational deception and noncooperation. J. IEEE Intell. Syst. **27**, 60–75 (2012)

13. Anolli, L., Ciceri, R.: The Voice of deception: vocal strategies of naïve and able liars. J. Nonverbal Behav. **21**, 259–284 (1997)

14. Christin, K., David, M.: Detecting suspicious behavior using speech: acoustic correlates of deceptive speech - an exploratory investigation. J. Appl. Ergon. 1–9 (2012)

15. Patton, M.W.: Decision support for rapid assessment for truth and deception using automated assessment technologies and kiosk-based embordied conversational agents. Ph.D. thesis, The University of Arizona (2009)

16. Lee, H., Largman, Y., Pham, P.: Unsupervised feature learning for audio classification using convolutional deep belief networks. In: Neural Information Processing Systems, pp. 1–9. MIT Press, Vancouver (2009)

17. Dong, Y., Deng, L.: Deep learning and its applications to signal and information processing. J. IEEE Signal Process. **28**, 145–154 (2011)

18. Ma, Y., Bao, C., Xia, B.: Speaker segmentation based on the distinctiveness deep belief network. J. Tsinghua Univ. (Nat. Sci. Ed.) **53**, 804–807 (2013). (in Chinese)

19. Sun, Z., Xue, L., Xu, Y.: The marginal Fisher analysis feature extraction algorithm based on deep learning. J. Electron. Inf. **35**, 805–811 (2013). (in Chinese)

Automatic Seizure Detection in EEG Based on Sparse Representation and Wavelet Transform

Shanshan Chen[1,2], Qingfang Meng[1,2(✉)], Yuehui Chen[1,2], and Dong Wang[1,2]

[1] School of Information Science and Engineering,
University of Jinan, Jinan 250022, China
[2] Shandong Provincial Key Laboratory of Network Based Intelligent Computing,
Jinan 250022, China
ise_mengqf@ujn.edu.cn

Abstract. Sparse representation has been widely applied to pattern classification in recent years. In the framework of sparse representation based classification (SRC), the test sample is represented as a sparse linear combination of the training samples. Due to the epileptic EEG signals are non-stationary and transitory, wavelet transform as a time-frequency analysis method is widely used to analyze EEG signals. In this work, a novel EEG signal classification method based on sparse representation and wavelet transform was proposed to detect the epileptic EEG from EEG recordings. The frequency subbands decomposed by wavelet transform provided more information than the entire EEG. The experimental results showed that the proposed method could classify the ictal EEG and interictal EEG with accuracy of 98 %.

Keywords: Epileptic EEG · Sparse representation based classification (SRC) · Wavelet decomposition · Classification

1 Introduction

Epilepsy is a chronic neurological disorder that affects about 50 million people worldwide [1]. Brain activities during seizure may present transient signs and symptoms with respect to the abnormal, excessive or synchronous neuronal firing. The electroencephalogram (EEG) contains important physiological and pathological information about the conditions and functions of the brain. Usually, two categories of epileptic EEG signals can be observed: ictal EEG (during an epileptic seizure) and interictal EEG (between seizures). However, visual inspection of long-term EEG recordings is tedious and time-consuming. So the automatic seizure detection in EEG is significant for diagnosing epilepsy in clinical practice.

Many techniques developed for epileptic activity detection in the past few years. In the seizure detection schemes, feature extraction and selection determine the performance of the algorithm. Many feature extraction methods such as frequency domain based, time-frequency domain based methods have been proposed and improved [2–7].

© Springer International Publishing Switzerland 2015
D.-S. Huang et al. (Eds.): ICIC 2015, Part I, LNCS 9225, pp. 197–205, 2015.
DOI: 10.1007/978-3-319-22180-9_20

Since wavelet transform (WT) has the capability to analyze EEG signals at different scales and captures the subtle changes in nonstationary signals, it has been used as a powerful signal processing tool. Considering the EEG signals are nonlinear, the non-linear dynamical methods have been widely used for seizure detection and diagnosis [8–15]. Besides the feature extraction methods, the designs of classifiers also have an important effect on the epileptic EEG classification. Various effective classifiers [16–19] have been used to improve the performance of epileptic EEG classification automatically.

Sparse representation comes from compressed sensing, potentially using lower sampling rates than the Shannon-Nyquist bound [21]. Unlike the traditional representation method of signals, sparse representation selects the most compact subset with the least number of base elements to express signals. In 2009, Wright exploited sparse representation to perform classification in face recognition and proposed the framework of the sparse representation based classification (SRC) [21]. SRC algorithm has the successful application in image processing and pattern recognition filed, such as speech signal recovery [20], face recognition [21, 22], image super-resolution [23], EEG signals analysis [24–26] etc.

In this paper, we propose a novel EEG signal classification method for epilepsy diagnosis based on sparse representation and wavelet transform. Since EEG signals are nonstationary, wavelet analysis is used to decompose the signals into multi-scales frequency subbands. The subbands are sparse in wavelet domain. So the classification method based on sparse representation could be potentially used for seizure detection.

2 Automatic Classification of Epileptic EEG Based on SRC and WT

2.1 Signal Preprocessing

Majumdar et al. [27] proposed that differentiation can enhance certain features of EEG signals. Based on the differential operator, the preprocessing procedure follows this definition:

$$y(t) - \exp((1/w)|D'y_0(t)|) \tag{1}$$

y_0 is original time series with respect to time t. D' is the first order derivative with respect to t and w is a normalization constant ($w = 100000$).

For clinical epileptic EEG, the ictal EEGs generally contain many spike and sharp waves, while the interictal EEGs are considered as normal signals. Differential operator is able to enhance the spike, and the ictal EEG signals get enhanced considerably relative to the interictal EEG signals.

2.2 Sparse Representation Based Classification (SRC)

Sparse representation has aroused an increasing amount of interest in pattern classification recently. Based on sparse representation, the test sample is categorized as either "ictal" or "interictal" under the constraint of sparse regularized. Given the training dataset with 2 object classes $A = [A_1, A_2] \in \Re^{m \times (n_1 + n_2)}$, A_1 represents the interictal EEG and A_2 represents the ictal EEG. $A_i = [s_{i1}, s_{i2}, \cdots, s_{in}]$, $s_{i,j}$ is the jth sample vector of the ith class. For a test sample $y \in \Re^m$, it could be well approximated by the training samples associated with the same class i:

$$y = \sum_{j=1}^{n_i} \alpha_{ij} s_{ij} = A_i a_i \tag{2}$$

If the test sample occurs during an epileptic seizure, the coefficient vector should be $a = [0, \cdots, 0, \alpha_{21}, \alpha_{22}, \cdots, \alpha_{2n_i}]^T$. The sparsest solution of $y = Aa$ is defined as the following l_0-optimization problem:

$$\hat{a} = \arg \min \|a\|_0 \ subject \ to \ Aa = y \tag{3}$$

However, the l_0-minimization solution is NP-hard problem. It is generally known that if just a few coefficients in vector a are nonzero entries, the sparsest solution can be formulated as the following l_1-optimization problem:

$$\hat{a} = \arg \min \|a\|_1 \ subject \ to \ Aa = y \tag{4}$$

Since EEG signals are usually contaminated with noise, the test sample may not be represented sparsely by all the training samples. The constraint is modified with an error tolerance ε to solve the stable l_1-minimization problem:

$$\hat{a} = \arg \min \|a\|_1 \ subject \ to \ \|Aa - y\|_2 \leq \varepsilon \tag{5}$$

The sparse coefficient solution can be regarded as a convex optimization problem with linear matrix inequalities constraints. For classification problem, a new vector δ_i is defined, whose nonzero entries are associated with the ith class in $\hat{a}$. Then, the test sample can be reconstructed by the coefficient of the same classes as:

$$\tilde{y} = A\delta_i(\hat{a}) \tag{6}$$

If test sample y occurs during seizure, the reconstructed signal $\tilde{y}$ by $\delta_2(\hat{a})$ is closer to y. To identify a new test sample, the minimum redundancy is defined as:

$$\min_i r_i(y) = \|y - A\delta_i(\hat{a})\| \ for \ i = 1, 2, \cdots, k \tag{7}$$

2.3   Wavelet Transform (WT)

Wavelet transform (WT) is a widespread time-frequency analysis method. It achieves a complete representation of the localized and transient phenomena occurring at different temporal scales. In long time windows, wavelet transform gives a finer low frequency resolution. In short time windows, it gives high frequency information. Those characteristics make the wavelet transform suitable for the decomposition of non-stationary epileptic EEG signals in different scales. EEG signals were decomposed into frequency subbands using discrete wavelet transform (DWT).

The subbands of original series under level j can be reconstructed by the approximation coefficient c_j and the detail coefficients d_k, k is time translation factor, h_{k-2n} and g_{k-2n} are the complex conjugate. The decomposition formula is defined as:

$$c_{jk} = \sum_{n \in z} c_{j+1n} h_{k-2n} + \sum_{n \in z} d_{j+1n} g_{k-2n} \ (k \in z) \tag{8}$$

2.4 Seizure Detection Process in Clinical EEG

In the framework of sparse representation based classification (SRC), a test EEG sample is sparsely represented on the training samples by solving l_1-minimization problem. A novel automatic identification of epileptic EEG signals is presented based on SRC and WT. The main procedure can be summarized as follows:

(1) Data preprocessing: using differential operator to enhance ictal activities of epileptic EEG signals.
(2) Select one sample from the test dataset as a test vector y and the training samples form A.
(3) Applying wavelet transform to decompose the training samples and test samples at multi-scales respectively.
(4) Normalize the columns of A to have unit l_2-norm.
(5) Solve the sparse representation problems in Eq. (5).
(6) Reconstruct the test vector y by the coefficient vector associated with the ith class as Eq. (6).
(7) Compute residue $r_i(y)$ of each class.
(8) Identify test vector y to the class with the least residue.
(9) Iterate Step 2-8 until there are no new test samples.

The flow chart is showed in Fig. 1.

3 Experiment Results and Analysis

The data used in this research come from Department of Epileptology, Bonn University, Germany. The complete dataset consists of five sets of data denoted sets A-E, each containing 100 single-channel EEG segments of 23.6 s duration. The classification of interictal (Set D) and ictal (Set E) EEGs is more difficult to solve but closer to clinical diagnosis than others. In this work, each of the segments is separated into equal four samples, each sample containing 1024 data points. For the classification problem of ictal and interictal EEGs, half of the data per class (200 samples from each class) are chosen as the training samples, while 50 samples from each class are used to test the performance of this method.

Generally, the individual EEG frequency subbands can provide more information than the entire EEG. In this work, wavelet transform (WT) is designed to decompose the epileptic EEG signals. Compared with Fourier transform, the main advantage of WT is

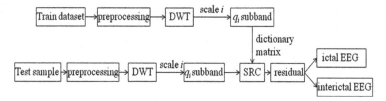

Fig. 1. Flow chart of EEG classification algorithm

that it has a varying window size changing with the frequency, thus leading to an optimal time-frequency resolution in all frequency ranges. So the combination with wavelet transform is superior to Fourier transform. The wavelet "db4" is chosen such that those subbands reconstructed by wavelet coefficients could have excellent performance for EEG classification. The number of decomposition levels is chosen to be 5. Figure 2 shows that the decomposition of a interital EEG sample (a) and a ictal EEG sample (b).

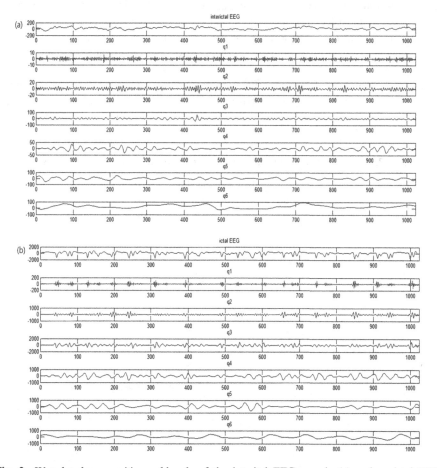

Fig. 2. Wavelet decomposition subbands of the interital EEG sample (a) and an ictal EEG sample (b)

Table 1. The classification accuracy of wavelet subbands in the scheme of SRC with different parameter ε

ε	0.05	0.015	0.005	0.0015	0.001	0.0005	0.0001
q1	0.59	0.72	0.67	0.87	0.76	0.87	–
q2	0.79	0.85	0.88	0.92	0.92	0.92	–
q3	0.84	0.90	0.92	0.94	0.94	0.94	–
q4	0.81	0.85	0.83	0.91	0.88	0.91	0.85
q5	0.86	0.86	0.86	0.88	0.88	0.88	0.86

Afterwards, the subbands component of test dataset is sparsely coded by the training samples. The parameter ε represents the tolerance of reconstruction error. The l_1-minimization solution of sparse representation is implemented by the CVX toolbox. The classification results of wavelet subbands are shown in Table 1.

From the classification accuracy of ictal EEG and interictal EEG, the lower frequency subbands, such as q2, q3, q4 components have better performance. To improve the recognition accuracy, the combined subbands frequency characteristics of epileptic EEG are used to test the performance of the SRC. The classification results of combined wavelet subbands component in ictal EEG and interictal EEG are shown in Table 2.

The reconstruction of an ictal EEG sample combined with the frequency characteristic in q3, q4 and q5 components based on SRC (ε = 0.0015) is presented in Fig. 3. We can see that the first one of figures is subband signal of the test sample, the second

Table 2. The classification results of combined subbands based on SRC and WT

	Sensitivity	Specificity	Accuracy
q2 + q3 + q4	0.94	0.98	0.96
q3 + q4 + q5	1.00	0.96	0.98

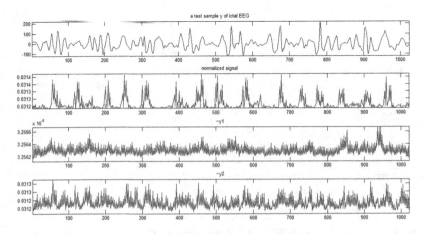

Fig. 3. The reconstruction of an ictal EEG sample

Table 3. The classification results of combined subbands based on SRC

Methods	Accuracy (%)
Wavelet coefficients + neural network [28]	94.25
Wavelet analysis + SVM [17]	97.25
This method	98.00

one is normalized signal with unit l2-norm, the third one is the reconstruction of the ictal EEG signal using the interictal training samples, and the last one is the reconstruction of the ictal EEG signal using the ictal training samples. The reconstruction with the ictal training samples is more similar to the ictal EEG. Compared with other advanced technologies, the classification accuracy of different methods on the epileptic EEG database is showed in Table 3.

4 Conclusions

The sparse representation has been widely applied to pattern classification in recent years. In the scheme of sparse representation based classification (SRC), choice and calculation of EEG features are avoided, and the new test EEG sample is sparsely represented on the training dataset.

In this work, we propose a seizure detection method using SRC to classify the epileptic ictal EEG and interictal EEG. To make full use of subband frequency characteristics, wavelet transform is used for analysis of epileptic EEG. Wavelet analysis is a common time-frequency analysis method which is a powerful tool to deal with nonstationary signals. The EEG signals can be decomposed into many frequency subband features. These features are sparse in wavelet domain, which could be sparsely encoded by SRC to characterize the behavior of EEG signals. Then, the paper combines subbands component of wavelet transform based on SRC, which have better performance in the classification of epileptic EEG signals.

Acknowledgement. This work was supported by the National Natural Science Foundation of China (Grant No. 61201428, 61302090), the Natural Science Foundation of Shandong Province, China (Grant No. ZR2010FQ020, ZR2013FL002), the Shandong Distinguished Middle-aged and Young Scientist Encourage and Reward Foundation, China (Grant No. BS2009SW003, BS2014DX015), the Graduate Innovation Foundation of University of Jinan (Grant No. YCX13011).

References

1. Iasemidis, L.D.: Epileptic seizure prediction and control. IEEE Trans. Biomed. Eng. **50**(5), 549–558 (2003)
2. Khan, Y.U., Gotman, J.: Wavelet based automatic seizure detection in intracerebral electroencephalogram. Clin. Neurophysiol. **114**(5), 898–908 (2003)

3. Saab, M.E., Gotman, J.: A system to detect the onset of epileptic seizure in scalp EEG. Clin. Europhysiol. **116**(2), 427–442 (2005)
4. Subasi, A.: EEG signal classification using wavelet feature extraction and a mixture of expert model. Expert Syst. Appl. **32**(4), 1084–1093 (2007)
5. Adeli, H., Zhou, Z., Dadmehr, N.: Analysis of EEG records in an epileptic patient using wavelet transform. J. Neurosci. Methods **123**, 69–87 (2003)
6. Adeli, H., Ghosh, D.S., Dadmehr, N.: A wavelet-chaos methodology for analysis of EEGs and EEG subbands to detect seizure and epilepsy. IEEE Trans. Biomed. Eng. **54**(2), 205–211 (2007)
7. Altunaya, S., Telatarb, Z., Erogulc, O.: Epileptic EEG detection using the linear prediction error energy. Expert Syst. Appl. **37**(8), 5661–5665 (2010)
8. Wang, X.Y., Meng, J., Qiu, T.S.: Research on chaotic behavior of epilepsy electroencephalogram of ehildren based on independent component analysis algorithm. J. Biomed. Engin. **24**, 835–841 (2007)
9. Swiderski, B., Osowski, S., Rysz, A.: Lyapunov exponent of EEG signal for epileptic seizure characterization. Chaos. **5**, 82–87 (1995)
10. Hasan, O.: Automatic detection of epileptic seizures in EEG using discrete wavelet transform and approximate entropy. Expert Syst. Appl. **36**, 2027–2036 (2009)
11. Achary, U.R., Molinari, F., Sree, S.V., Chattopadhyay, S., Ng, K.H., Suri, J.S.: Automated diagnosis of epileptic eeg using entropies. Biomed. Signal Process. Control **7**, 401–408 (2012)
12. Meng, Q., Chen, S., Zhou, W., Yang, X.: Seizure detection in clinical EEG based on entropies and EMD. In: Guo, C., Hou, Z.-G., Zeng, Z. (eds.) ISNN 2013, Part II. LNCS, vol. 7952, pp. 323–330. Springer, Heidelberg (2013)
13. Thomasson, N., Hoeppner, T.J., Webber, C.L., Zbilut, J.P.: Recurrence quantification in epileptic EEGs. Phys. Lett. A **279**, 94–101 (2001)
14. Acharya, U.R., Sree, S.V., Chattopadhyay, S., Yu, W.W., Alvin, P.C.: Application of recurrence quantification analysis for the automatic EEG signals. Int. J. Neural Syst. **21**, 199–211 (2011)
15. Acharya, U.R., Sree, S.V., Suri, J.S.: Automatic detection of epileptic EEG signals using higher order cumulant features. Int. J. Neural Syst. **21**(5), 403–414 (2011)
16. Guo, L., Rivero, D., Seoane, J., Pazos, A.: Classification of EEG signals using relative wavelet energy and artificial neural networks. In: The First ACM/SIGEVO Summit on Genetic and Evolutionary Computation, pp. 177–183 (2009)
17. Zhao, J.L., Zhou, W.D., Lin, K., Cai, D.M.: EEG signal classification based on SVM and wavelet analysis. Comput. Appl. Softw. **28**, 114–116 (2011)
18. Clodoaldo, A., Lima, M., André, L.V., Coelho, S.C.: Automatic EEG signal classification for epilepsy diagnosis with relevance vector machines. Expert Syst. Appl. **36**, 10054–10059 (2009)
19. Yuan, Q., Zhou, W.D., Li, S.F., Cai, D.M.: Epileptic EEG classification based on extreme learning machine and nonlinear features. Epilepsy Res. **96**, 29–38 (2011)
20. Candes, E., Romberg, J., Tao, T.: Stable signal recovery from incomplete and inaccurate measurements. Comm. Pure Appl. Math. **59**(8), 1207–1223 (2006)
21. Wright, J., Yang, A.Y., Ganesh, A., Sastry, S.S., Ma, Y.: Robust face recognition via sparse representation. IEEE Trans. Pattern Anal. Mach. Intell. **31**(2), 210–227 (2009)
22. Yang, M., Zhang, L., Feng, X.C., Zhang, D.: Fisher discrimination dictionary learning for sparse representation. In: Proceedings of the 13th IEEE International Conference on Computer Vision (ICCV), pp. 543–550 (2011)

23. Yang, J., Wright, J., Huang, T., Ma, Y.: Image super-resolution as sparse representation of raw patches. In: Proceedings of the IEEE International Conference on Computer Vision and Pattern Recognition. (2008)

24. Ren, Y.F., Wu, Y., Ge, Y.B.: A cotraining algorithm for EEG classification with biomimetic pattern recognition and sparse representation. Neurocomputing. **137**, 212–222 (2014)

25. Wu, M., Wei, Z.H., Tang, L.M., Sun, Y.B., Xiao, L.: The reconstruction study of EEG signal based on sparse approximation and compressive sensing. Chinese J. Medi. Instrumentation. **34**(4), 241–245 (2010)

26. Faust, O., Acharya, U.R., Adeli, H., Adeli, A.: Wavelet-based EEG processing for computer-aided seizure detection and epilepsy diagnosis. SEIZURE: european. J. Epilepsy (2015). doi:10.1016/j.seizure.2015.01.012

27. Majumdar, K.K., Vardhan, P.: Automatic seizure detection in ECoG by differential operator and windowed variance. IEEE Trans. Neural Syst. Rehabil. Eng. **19**(4), 356–365 (2011)

28. Ubeyli, E.D.: Combined neural network model employing wavelet coefficients for EEG signals classification. Digit. Sig. Proc. **19**(2), 297–308 (2009)

Design of Serial Communication Module Based on Solar-blind UV Communication System

Chunpei Li[✉], Xiangdong Luo[✉], and Huajian Wang

Jiangsu Key Laboratory of ASIC Design,
Nantong University, Nantong 226019, China
pursuinging@sina.com, luoxd@ntu.edu.cn

Abstract. UV communication is a wireless optical communication technology which bases on the scattering and absorption of UV light in atmospheric. It combines the advantages of traditional optical communications and wireless communications, including non-line-of-sight (NLOS), anti-interference, low wiretapping and electromagnetic silent. With a wide range of serial communications for remote monitoring and control, the demand for serial communications is increasing in engineering applications. This paper presents the application of serial port in non line-of-sight UV communications to meet the requirements of low-rate, short-range communication. A serial communication interface based on Altera Company's Cyclone II series FPGA chip (EP2C35F672C6) has been designed. The Universal Asynchronous Receiver Transmitter (UART) module is generated by the Verilog HDL programming language in the integrated software development environment (Quartus II 9.0). Simulation and debugging results indicate that it shows good functions and satisfies protocol requirements.

Keywords: UV communication · Serial communication · FPGA

1 Introduction

Solar-blind UV communication, a new communication technology with good confidentiality and strong tolerance to environmental factors, can be applied to a variety of short range and anti-interference environment. It has characteristics of flexibility and high reliability [1], which can communicate in different UV communications platforms. Solar-blind UV communication is similar to the traditional free-space optical communication, including the kind way of line of sight (LOS) communication. While the UV light also has the unique characteristic of a non-line-of-sight communication (NLOS) [2]. When ultraviolet light transmits in NLOS, it is effected by the factors of signal attenuation, pulse spreading and multipath propagation effect, resulting in a lower transmission rate than other light in circumstances of LOS communication [3]. With the development of UV communication technology, transmitting high-quality signal in the limit of transmission rate becomes one of the important issues. In this system, a method of asynchronous serial communication is presented, which has advantages of less transmission line, high reliability, low speed and low-cost [4]. Therefore, it can be well used for data exchange in UV communication.

D.-S. Huang et al. (Eds.): ICIC 2015, Part I, LNCS 9225, pp. 206–215, 2015.
DOI: 10.1007/978-3-319-22180-9_21

Universal Asynchronous Receiver Transmitter (UART) is a serial transmission interface that was used extensively, which allows full-duplex communication in a serial link. It is widely used in data exchange between computers and peripherals, such as mouses, keyboards, modulator-demodulators and printers. Meanwhile, in the design of today's embedded microprocessor chips, UART interface has become the standard configuration [2], including the interface standards and bus standards of RS232, RS449, RS423, RS422, RS485 and so on. UART can convert the computer's internal parallel data into serial data stream output. It also can convert serial data from peripherals into parallel data for the use of computer's internal devices. At the same time, according to the serial communication protocol, the transmitter should add start bit, parity and stop bits before the message out from port. The design of UART is based on FPGA, which is in line with RS232 standard. The standard RS232 chip MAX232 is adopted in this work. In order to exchange the serial data between UV communication systems and PC, the TOP-DOWN design method is used in this solar-blind UV communication system. Separated modules are designed for our system, keeping the signals be connected each other. The solar-blind UV communication system diagram is shown in Fig. 1. In transmitting terminal, the message is transferred to the FPGA modulation by UART serial port, and then the signal drives the LED emitting UV photons to the atmosphere through the driving circuit. In atmospheric channel, the UV light reaches the receiving terminal and is converted into electrical signal by the ultraviolet photo multiplier tube (PMT) detector. In receiving terminal, the data are transferred to photo electron conversion and FPGA demodulation. Finally, the data are transferred to PC through UART serial port and the communication is finished [5].

The physical photos of the system show in Fig. 2. The hardware circuit structures are still in a progress, so the simulation of the software parts will be presented in this paper.

2   The Description of Serial Port Communication

2.1   Serial Communication Protocol

Characters transmitted by serial communication are generally referred to data frames. In order to obtain the correct data frames in the process of transmitting and receiving, synchronous communication in the data communication is needed. Synchronous

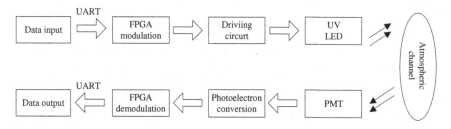

Fig. 1. Structural diagram of a UV system to achieve serial communication

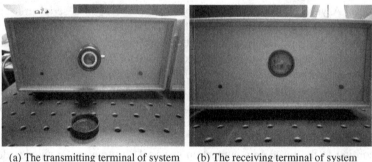

(a) The transmitting terminal of system   (b) The receiving terminal of system

Fig. 2. The physical photos of the system

communication is bilateral coordination between transmitting and receiving operation. Both parts of communication must make a set of regulations and abide to achieve synchronization in the aspects of the data transfer mode, synchronous control mode, error handling, response mode, and data formats and other issues, such provisions was named communication protocol. Serial communication protocol includes asynchronous and synchronization communication protocol. And the asynchronous communication is the most commonly used in data transmission of computers. Receiving and transmitting parts do not use a common reference clock, but requires formatting every message before transmitting. The data should add "start bit" and "stop bit" when sending in characters, thus the transmission efficiency is lower than synchronous communication, while the requirement of transmitting and receiving clock synchronization is reduced [6].

2.2   The Communication Principle of UART

In asynchronous serial communication, the message is transmitted byte by byte. The transmitter and receiver must communicate in the same format of byte frame and baud rate. The format of byte frame in order is start bit, data bits, parity bit and stop bit. The start bit means the beginning of byte frames. It is active-low, which is used to indicate that the receiver starts to send data. Data bits are transmitted from low to high bit. Parity bit is a simple way of error detection, which is used for determining the errors in the process of receiving and transmitting data. Stop bit means the termination of data frames. It is active-high, which is used for indicating that the data frame transmission is completed in receiver [7]. This design uses a baud rate of 9600 bps.

UART frame format includes line idle state (high level), start bit (low level), data bits (5–8 bits), parity bit (optional), and stop bit (1 bit or 1.5 bits or 2 bits), as shown in Fig. 3. In general, the UART has internal configuration registers, in which users can more easily set data bits, as well as the type of parity and stop bits [8]. The hardware is implemented with less resources, and it can be more flexibly embedded into FPGA/CPLD's environments.

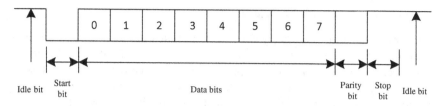

Fig. 3. The frame format of UART

2.3 RS232 Interface

RS232 standard is a communication protocol that developed by the US EIA (Electronic Industries Association) and BELL companies. The standard specifies a 25-pins connector: DB25, and stipulates the contents of each pin and the level of various signals of the connector. With the continuous improvement of equipment, the DB25 is replaced by DB9 interface. Now RS232 interface is also called DB9.

Some devices don't use transmit-control-signals under the condition of connecting to RS-232C interface in the PC. To complete transmission of data, just need three wires: pin 2 is used to receive data, pin 3 is used to send data, pin 5 is used to connect ground. This paper adopts the DB9 interface with 9 pins, and the definition of connector port is shown in Fig. 4.

3 The Design of Communication Modules

Here, the serial communication based on solar-blind UV communication system consists of four modules: baud rate generator module, receiver module, optical modem module and transmission module. UART baud rate generator generates clock frequency control module, receiver module switches the data from the serial input to parallel output, optical modem module will process the data with light pulse modulation data, transmission module transform the data from the parallel input to the serial output. In order to assure stability and reliability of solar-blind UV communication system, frequency division from the system clock (50 MHz) is needed. Here, a global clock BPS_CLK is created. In order to achieve larger message like image data and video data in this system, the method of modulation and demodulation is a challenging point.

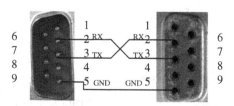

Fig. 4. The definition of RS-232 port

3.1 Baud Rate Generator Module Design

The baud rate, the number of binary data bits transmitted per second, is an important indicator of serial communications, which shows the speed of data transmission. In asynchronous communication, it mainly produces a synchronous baud rate clock signal between the receiving and transmitting sides of UART. The baud rate of 9600 bps is used here. Baud rate generator is actually a frequency divider [9]. Here, the frequency clock produced by the baud rate generator that is using 16 times the baud rate clock. Thus, the frequency coefficient of the baud rate generator is: NUM = 15. It is very important to determine where to sample the data in serial communication. The sample point is designed in the midpoint of the data information, so as to improve the sampling accuracy of data. The realization code is following:

```
reg [3:0]Count_Mid;
reg mid_data_r;
parameter NUM=15;
always @ ( posedge BPS_CLK or negedge RSTn)
   begin
      if ( !RSTn )
         Count_Mid <= 4'd0;
      else if (Count_Mid < NUM && CONTROL_EN)
         Count_Mid <= Count_Mid + 1'b1;
      else
         Count_Mid <= 4'd0;
   end
always @ ( posedge BPS_CLK or negedge RSTn )
      if( !RSTn )
         mid_data_r <= 1'b0;

      else if( Count_Mid == NUM/2 && CONTROL_EN)
         mid_data_r <= 1'b1;
      else if( Count_Mid == 4'd8 )
         mid_data_r <= 1'b0;
      else
         mid_data_r <= mid_data_r;
assign mid_data = mid_data_r;
```

3.2 The Design of Receiver Modules

When receiving module detects a RS-232C bus transmission lines with the low level of start bit, the receiving process is triggered. The UART receiving module interface signals are shown in Fig. 5. Firstly, the receiver must capture the start bit, to determine the real start bit with high pulse of mid_data (the sample signal generated by baud rate generator) and low level of RX_in. The following 8 bits is useful data. They were sent to the receive register Data_out [1..0]. Then the data are outputted. When the data receiving is finished, the signal of RX_Done symbols receiving is ended.

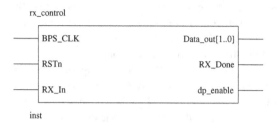

Fig. 5. The interface signals of UART receiving module

In Fig. 5, the signal of BPS_CLK is the global clock of the modules in serial communication. RSTn is a reset signal, and RX_in means the input data. Data_out [1..0] is the register to receive and store data, and RX_Done is a flag of completing data reception. When RX_Done equals to 1, the data receiving is done. dp_enable is an enable signal which means the start of modulation module. After transmitting module completes, the dp_enable is set to 1, receiving module start working again. The flow diagram of receiving data is shown in Fig. 6.

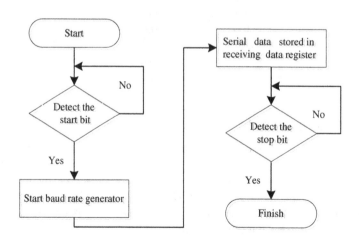

Fig. 6. The flow diagram of receiving data

3.3 Light Modulation and Demodulation Modules

Modulation and demodulation is a key technology in ultraviolet light communication, un-modulated light waves do not carry any information. Modulation will make the measured parameters in light waves varies with the signals. UV communication mainly carries information to the emitted pulses through the base-band modulation. Here, the modulation method of digital pulse interval modulation (DPIM) is adopted, which is based on a group unit of binary information. The modulation schemes of OOK, PPM and DPIM are suitable for optical wireless communication systems. In aspects of the bandwidth, power efficiency, transmission capacity and error probability, DPIM offers higher transmission capacity compared to OOK/PPM, which can be employed to improve either the bandwidth efficiency or power efficiency of the system [10]. In addition, it has higher average power efficiency than other optical modulation method. Therefore, DPIM is used in this ultraviolet light communication system, meeting the requirements of bandwidth of UV communication.

In the modulation schemes of DPIM, each group of data source has the same number of bits, represented by M, the time slot corresponding to the binary bits of each group called a DPIM symbol, each DPIM frame consists of a string of high and low levels, the number of low level is determined by the decimal value of the binary, while the number of slots contained in each symbol is not fixed [11]. When the modulation order is equal to 3, the corresponding relationship between the transmission symbol of DPIM and bit information, is given in Table 1.

Table 1. The relationship of DPIM and bit information

Source bits	DPIM symbol
000	10
001	100
010	1000
011	10000
100	100000
101	1000000
110	10000000
111	100000000

3.4   The Design of Transmission Modules

Transmitting module is used to read the data from the demodulation module. Driving by baud clock, the start bit, parity bit and stop bits should be added in data according to the information of configuration register. According to the serial protocol, the packaged data will be sent to the serial output port TX_Out. When the shift serial data transmission is completed, TX_Done as the finished flag signal turns into high level. The interface signals of UART transmitting module are shown in Fig. 7.

In Fig. 8, BPS_CLK is the global clock and RSTn is reset signal, as same as in the receiver module. The signal of dedpim_enable controls the start of transmitting module, and the module detects a start signal when the signal of dedpim_enable changes from logic 0 to logic 1. As sending data register, TX_in [7..0] mainly receives the parallel data in demodulation data register ded_out [7..0]. The TX_Out is the output of serial data. When TX_Done is equal to 1 means the completion of sending data, the data receiver module can be loaded other data again. Figure 8 is the flow of sending data.

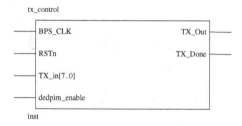

Fig. 7. The interface signals of UART transmitting module

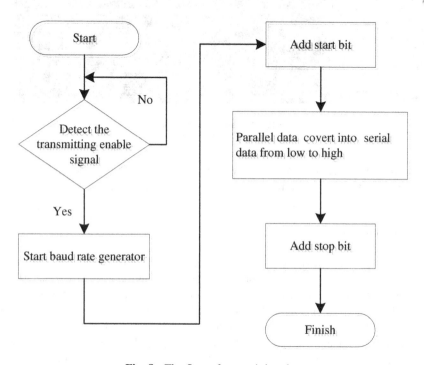

Fig. 8. The flow of transmitting data

4 Simulation of Modules

The RTL of the top file shows in Fig. 9. The RTL viewer is generated by the software Quartus II 9.0. It includes the baud rate generator, receiver, modem and transmitter modules.

After the design has been completed, we carry on the simulation to the function. During simulation, the input sequence is in turn: 10100000, 10010011, 10101010 and so on. The received data is stored into the register Data_out with two bits so as to make convenient for modulation and demodulation. Then the register TX_Out outputs final results. From the waveform diagram in Fig. 10, the data are sent and received

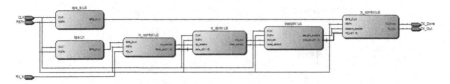

Fig. 9. The RTL of the top file

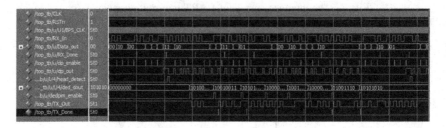

Fig. 10. Simulation result

successfully in this design. At the same time, data transmission is stable. The data output satisfies the serial communication protocol requirements.

5 Conclusion

The working principle and the concrete realization method of each module in serial transmission of ultraviolet communication system are introduced here. The design language of Verilog HDL is used to implement the modules of UART and optical modem. The softwares of Quartus II and Modelsim are adopted. Design and simulation on Altera's Cyclone series FPGA chip EP2C35F672C6 are completed. The FPGA has great flexibility and high integration. The design of serial communication has great significance in UV communication which has special advantages relative to other optical communication. It has a bright prospect in terms of military and traffic short-range communications.

Acknowledgment. This work was supported by the applied basic science project of Ministry of Transport of China (Grant no. 2013319813100), the National Natural Science Foundation of China (Grant no. 61474067).

References

1. Ding, H., Chen, G., Xu, Z., Sadler, B.M.: Channel modeling and performance of non-line-of-sight ultraviolet scattering communications. IEEE Inst. Eng. Technol. **6**(5), 514–524 (2012)

2. Shaw, G.A., Nischan, M.: Short-range NLOS ultraviolet communication test bed and measurements. In: Proceedings of the SPIE 2001, vol. 4396, pp. 31–40 (2001)
3. Gary, A.S., Melissa, N., Mrinal, I., et al.: NLOS UV communication for distributed sensor systems. In: Proceedings of the SPIE 2000, vol. 4126, pp. 83–96 (2000)
4. Liu, T., Lou, X.: The FPGA Digital Electronic System Design and Development Examples Navigation, pp. 100–230. Posts & Telecom Press, Beijing (2005)
5. Zhao, T., Zhang, X., Liu, L., et al.: Fast frame synchronization for wireless ultraviolet image unidirectional communication. In: 25th IET Irish Signals & Systems Conference 2014 and 2014 China-Ireland International Conference on Information and Communications Technologies (ISSC 2014/CIICT 2014), pp. 322–327 (2013)
6. Ali, L., Sidek, R., Aris, I.: Design of a Micro-UART for SOC application. Comput. Electr. Eng. **30**(04), 257–268 (2004)
7. Bhadra, D., Vij, V.S., Stevens, K.S.: A low power UART design based on asynchronous techniques. In: 2013 IEEE 56th International Midwest Symposium on IEEE Circuits and Systems (MWSCAS), pp. 21–24 (2013)
8. Tang, S.H.W.: Design and test of the multiple serial ports extension system based on CPLD. Chin. J. Electron Devices Instrum. **29**(3), 981–984 (2006)
9. Fang, Y., Chen, X.: Design and simulation of UART serial communication module based on VHDL. In: 2011 3rd International Workshop on Intelligent Systems and Applications, pp. 2311–2316, May 2011
10. Mahdiraji, G.A., Zahedi, E.: Comparison of selected digital modulation schemes (OOK, PPM and DPIM) for wireless optical communications. In: 4th Student Conference on Research and Development, SCOReD 2006, pp. 5–10. IEEE (2006)
11. Jin, B., Zhang, S., Zhang, X., et al.: Digital pulse interval modulation for ultra-wideband transmission with energy detection. In: 2010 6th International Conference on Wireless Communications Networking and Mobile Computing (WiCOM), pp. 1–4. IEEE (2010)

Target Tracking via Incorporating Multi-modal Features

Huan Zhang[1(⊠)] and Xiankai Chen[2]

[1] China Security and Surveillance Technology, Inc., Room 8E, Building 97,
Taoyuan Village, Xili Town, Nanshan District, Shenzhen, China
zhanghuanhuan-1990@163.com
[2] Room 202, Qilinge Building, Lianxinyuan Garden, Luohu District,
Shenzhen, China
xiankaichen@gmail.com

Abstract. The challenge of visual tracking is to develop a robust target's appearance model, the core of which involves an appropriate selection and an effective assembly of a cluster of features. In this paper, we propose a novel model to adaptively choose reasonable combination of feature sets to represent the target by employing the multi-kernel ridge regression. This model will update the weights distributions of different kernel groups for feature sets automatically and the regularization parameter's value of the kernel regression objective function as well. For traditional multi-kernel based algorithm would cost too much time on training model, we develop a very simple and efficient algorithm by adapting feature sets to circulant structure so as to make use of the Fast Fourier Transform (FFT). Thus our algorithm can provide more robust tracking while maintaining real-time effects. To the best of our knowledge, this is the first time the multiple kernel learning algorithms is applied to real-time visual tracking. We evaluate the proposed algorithm on the popular benchmark including 50 image sequences and compare it with 9 state-of-art methods. Implemented in Matlab, the experiment results show that the proposed tracker runs at 45.4 frames per second on an i3 machine and outperforms the state-of-the-art trackers on the benchmark with respect to accuracy. Particularly, the average precision of our algorithm achieves 76.7 % under OPE curve at 20px.

Keywords: Multi-kernel ridge regression · Visual target tracking · Fast fourier transform · Circulant structure

1 Introduction

Visual tracking is a comprehensive research field, gathering together all the advanced research fruits from mathematics, machine learning, computer vision and et al. there has been accumulating a lot of work in visual tracking, such as the well-known tracker TLD [9], Struck [5], MIL [1], CT [13] and et al. Currently, the main ideology in visual tracking is tracking-by-detection, namely learning a classifier online which could discriminate target from its background, thus achieving the detection of the target. In

© Springer International Publishing Switzerland 2015
D.-S. Huang et al. (Eds.): ICIC 2015, Part I, LNCS 9225, pp. 216–226, 2015.
DOI: 10.1007/978-3-319-22180-9_22

the framework of tracking-by-detection, the most crucial thing is the representation of the target's appearance model.

MIL [1] proposes a method to incorporate multiple instances to learn a classifier to diminish the drift problem. CT [13] adopts a sparse matrix to represent the features of the targets for the appearance model. Recently, there are some novel methods about feature representation. MOSSE [2] builds an appearance model of the target by adopting an adaptive correlation filter from a few templates in the image. CSK [7] develops the kernelized correlation filters having the same properties as the correlation filters like MOSSE to represent the target's model. CSK [7] and MOSSE [2] both only extract the raw pixels as feature, which is not robust enough to represent the target. KCF [6] enhances the representation of the target by combining multi-channel features, like Variant HOG [3], with the single kernel while preserving real-time performance.

MOSSE, CSK, KCF, they all share the idea of minimizing the Sum of Squared Error by using the ridge regression function. It is simpler and can achieve the similar performance to complex classifiers like SVM, Boosting.

In this paper, we propose an algorithm that furthers the above research by naturally exploiting multi-kernel learning for the representation of the object's appearance model in the real-time tracking while taking advantage of the current technology, called multi-kernel visual tracking (MKVT). Our model applies different weighted kernel combinations to different feature subsets through multi-kernel ridge regression.

The proposed algorithm could accelerate the speed of the multiple kernel learning by applying the Fourier transform in calculation, because of the kernel matrixes' circulant property. Moreover, the proposed algorithm could avoid the tremendous storage of the feature sets. In addition, the proposed algorithm could acquire the regularization parameter's value of the kernel regression objective function synchronously while learning the weights distributions of different kernel groups for feature sets iteratively, thus providing more accurate tracking effect.

2 Appearance Model

For the visual tracking task, a good representation of the target is of significance. And feature extraction and selection are the most key steps to building a reasonable tracking model. In this section, all the useful information covering contour, color, and texture of the tracked object are extracted to help build the tracking model. Moreover, we expect to adopt a good selection of different kernels with different weights combining with the extracted features to establish a robust appearance model.

2.1 Construct Sample Set

Assume the proposed region with the target centered in certain frame of the video is given, and then we extract features, such as contour, color and texture from this region. Suppose that there are R kinds of features, which are defined by $\mathbf{x} = \{\mathbf{x}_1, \mathbf{x}_2, \ldots, \mathbf{x}_R\}$. Note that all the features have the same dimensions N. According to [6], given a base sample $\mathbf{x}_r, r = 1, 2, \ldots, R$, we can construct a sample set $\mathbf{X}_r = C(\mathbf{x}_r)$. Call $\mathbf{X}_r$ as a circulant matrix, of which the first row is denoted by $\mathbf{x}_r$ and the rest of rows are generated based on $\mathbf{x}_r$. The details are referred to [6]. The fabulous property of the circulant matrix is that it could be diagonalized by the Discrete Fourier Transform (DFT), which could be expressed as

$$\mathbf{X_r} = \mathbf{F}^H diag(\tilde{\mathbf{x}}_r)\mathbf{F} \tag{1}$$

where $\tilde{\mathbf{x}}_r = \mathbf{F}\mathbf{x}_r$, $\mathbf{F}$ is the Discrete Fourier transform (DFT) matrix and $\mathbf{F}^H$ represents the Hermitian transpose DFT matrix. They are a little different from what we usually use in the actual DFT transform. There is a difference in coefficients by a factor, which could be referred to [6].

Consider for every $\mathbf{X}_r, r = 1, 2, \ldots, R$, we use it to generate all kinds of kernel matrix with different kernel function or various kernel parameters [10], thus we can generate a set of kernel matrixes $\{\mathbf{K}_1, \mathbf{K}_2, \cdots, \mathbf{K}_M\}$. Besides, if these kernel matrixes are generated by Additive kernels, Gaussian kernels, polynomial kernels, or linear kernels, they are circulant matrixes as well. It suggests that the circulant kernel matrixes can be used to combine all kinds of feature sets in an efficient way, which is discussed in the Sect. 3.

2.2   Construct Sample Set

We build the model of the target by incorporating the multi-kernel structure into the ridge regression. In this way, we could find an optimal model that can indicate the implicit way different features co-work and depict the precise contributions that different features play in the process of predicting the target. Define the decision function as follows

$$f(\mathbf{x}) = \sum_m f_m(\mathbf{x})$$
$$f_m(\mathbf{x}) = \sum_n \alpha_n d_m k_m(\mathbf{x}_n, \mathbf{x}), \alpha_n \in \mathbf{R}$$

where $f_m(\mathbf{x})$ denotes the hypothesis on the reproduce kernels Hilbert space (RKHS) $\mathcal{H}_m$ with kernel $d_m k_m(\mathbf{x}_n, \mathbf{x})$ where $\mathbf{d} = \{d_1, d_2, \ldots, d_M\}$, which denotes kernels weight. $\alpha = \{\alpha_1, \alpha_2, \ldots, \alpha_N\}$ indicates the Lagrangian multiplier. Then the optimal appearance model can be resolved under the following regression optimal problem:

$$\min_{\{f_m\}, \mathbf{d}} \sum_i (f(\mathbf{x}_i) - y_i)^2 + \lambda \sum_m \frac{\|f_m\|^2}{d_m} \tag{2}$$
$$s.t. \|\mathbf{d}\|_p^2 \leq 1, \ 1 < p < \infty$$

where λ is the regular parameter. $\mathbf{d}$ is restricted by the p norm, if p is equal to 1, then it will yield the sparse solution of $\mathbf{d}$, if $\mathbf{d}$ is larger than 1, then yield the dense solution.

Table 1. The procedure of Algorithm 1

Algorithm 1:the update of the appearance model

1. Initialize parameters $\tilde{\alpha}$, $\mathbf{d}$

   if $t = 0$, $\mathbf{d} = \left\{ \frac{1}{M+1}, \frac{1}{M+1}, \cdots, \frac{1}{M+1} \right\}$ otherwise $\mathbf{d} = \mathbf{d}_{t-1}$

2. **while** $k < MaxInteration$ and $\|\Delta \mathbf{d}\|_2 < T$

3. Fix $\mathbf{d}$,update $\tilde{\alpha}$ with formulation (6)

4. Fix $\tilde{\alpha}$,update the weight vector $\mathbf{d}$

   a. Calculate the normal vector $\mathbf{w}$ with equation (8).

   b. Calculate the residual error $\Delta \mathbf{d}$ with equation (9)

   c. Update kernel weight $\mathbf{d} = \mathbf{d} + \Delta \mathbf{d}$

5. **end while.**

6. Output $\tilde{\alpha}$ and $\mathbf{d}$

By employing the Lagrangian multiplier technology, the above primal problem indicated in Eq. (2) can be easily derived to the dual formulization as follows

$$\min_{\mathbf{d},\alpha} \left((\sum_m d_m \mathbf{K}_m)\alpha - y \right)^T \left((\sum_m d_m \mathbf{K}_m)\alpha - y \right) + \lambda \alpha^T (\sum_m d_m \mathbf{K}_m)\alpha$$

$$s.t. \ \|\mathbf{d}\|_p^2 \leq 1, \ 1 < p < \infty \tag{3}$$

where $\mathbf{K}_m$ indicates the kernel matrix with respect to $\mathcal{H}_m$.

We can figure out the solution to the multi-kernel ridge regression by solving the minimizer of the function (3). The framework we would apply in acquiring the solution is referred to [10], namely it can be handled by wrapped-based iterate algorithm. At each iteration we would fix $\mathbf{d}$ to update at first, and then fix α to update $\mathbf{d}$. The procedure will repeat until optimal solution meets the stopping criterion. So consider the fixed kernel weight $\mathbf{d}$, the solution of α is then given as follows:

$$\alpha = (\sum_m d_m \mathbf{K}_m + \lambda \mathbf{I})^{-1} y \tag{4}$$

where $\sum_m d_m \mathbf{K}_m$ is the combined kernel matrix, the solution of $\mathbf{d}$ will be interpreted with details in Sect. 3.

However, it is difficult to directly apply MKL into real time visual tracking, since it involves the computation of multiple kernel matrixes, the excavation of optimal kernel weights and the high frequency of solving the single kernel problem. In the next section we will show how to employ circulant matrix to accelerate the training process.

3 Multi-kernel Tracker

3.1 Fast Model Update

As discussed in above section, the bottle neck of MKL is that it involves a great deal of kernel matrixes. So if the kernel matrix can be constructed by the special sample set to possess some special properties like MOSSE, KCF, thus it is possible to speed up the training algorithm. Naturally Inspired by [6, 10], we will show how to achieve the goal.

Let's look at the solving of $\mathbf{d}$ and $\tilde{\alpha}$. The optimal solution $\mathbf{d}$ and $\tilde{\alpha}$ can be solved alternatively. So by fixing the kernel weight $\mathbf{d}$, we can apply the circulant matrix's property to Eq. (4), making it diagonalizable, obtaining

$$\tilde{\alpha} = \frac{\tilde{\mathbf{y}}}{\sum_{m=1} d_m \tilde{\mathbf{k}}_m^{xx} + \lambda} \tag{5}$$

Where $\tilde{\mathbf{k}}_m^{xx}$ is the Fourier transform of the first row of kernel matrix $\mathbf{K}_m$. That means, we can use set $\left\{\tilde{\mathbf{k}}_1^{xx}, \tilde{\mathbf{k}}_2^{xx}, \cdots, \tilde{\mathbf{k}}_M^{xx}\right\}$ to compute $\tilde{\alpha}$ and $\mathbf{d}$ instead of kernel matrixes set $\{\mathbf{K}_1, \mathbf{K}_2, \cdots, \mathbf{K}_M\}$. And $\mathbf{y}$, the regression target values of the corresponding samples, distributes in a Gaussian function with the first element being the center of the Gaussian function. $\tilde{\mathbf{y}}$ is also the Fourier transform of the first row of vector $\mathbf{y}$. Here, λ the regularization parameter can be incorporated into the form $\sum_m d_m \tilde{\mathbf{k}}_m^{xx}$ and d_0 could be added into vector $\mathbf{d}$ naturally, thus the Eq. (5) can be written as follow.

$$\tilde{\alpha} = \frac{\tilde{\mathbf{y}}}{\sum_{m=0} d_m \tilde{\mathbf{k}}_m^{xx}} \tag{6}$$

where $\tilde{\mathbf{k}}_0^{xx} = (1, 1, 1, \ldots, 1)^T$. That means the regularization parameter λ can be learned by solving d_0 instead of being given an empirical value. And actually we can use set $\left\{\tilde{\mathbf{k}}_0^{xx}, \tilde{\mathbf{k}}_1^{xx}, \tilde{\mathbf{k}}_2^{xx}, \quad , \tilde{\mathbf{k}}_M^{xx}\right\}$.

After figuring out a better solution of $\tilde{\alpha}$, fix it and update the kernel weight $\mathbf{d}$. Firstly we need to calculate the normal vector $\mathbf{w} = (w_0, w_1, w_2, \ldots, w_M)^T$ of $\mathbf{d}$, the formulation shows as follows

$$w_m = d_m^2 \alpha^T \mathbf{K}_m \alpha, \forall m = 0, 1, 2, \ldots M \tag{7}$$

In order to accelerate the computation, the normal vector can be calculated under the frequency domain, as $\mathbf{K}_m$ is the circulant matrix, so the normal vector can be written as

$$w_m = d_m^2 \alpha^T \mathbf{F}^H diag(\tilde{\mathbf{k}}_m^{xx}) \tilde{\alpha}$$
$$= d_m^2 (\mathbf{F}^* \alpha)^T (\tilde{\mathbf{k}}_m^{xx} \odot \tilde{\alpha})$$

Algorithm 2: *the tracking process of the proposed tracker*

1.Input the video

2.Learn an initial model of the target in the first frame, denoted by $Model(\tilde{\alpha}_1, \mathbf{d}_1)$

3.**for** $t = 2$ to N **do**

4.   Detect the new target with $Model(\tilde{\alpha}_{t-1}, \mathbf{d}_{t-1})$:

      a. Compute set $\left\{ \tilde{\mathbf{k}}_1^{xo}, \tilde{\mathbf{k}}_2^{xo}, \cdots, \tilde{\mathbf{k}}_M^{xo} \right\}$

      b. Calculate the detection values in (11)

5.   Locate the target and update the appearance model $Model(\tilde{\alpha}_t, \mathbf{d}_t)$ with equation (12)

6. **end for**

Taking the complex-conjugate of both sides, and because α is a real vector and w_m is real number, equation will be derived as follow

$$
\begin{aligned}
w_m &= d_m^2 (\mathbf{F}\alpha)^T (\tilde{\mathbf{k}}_m^{xx} \odot \tilde{\alpha})^* \\
&= d_m^2 \tilde{\alpha}^T \left(\tilde{\mathbf{k}}_m^{xx} \odot \tilde{\alpha} \right)^*
\end{aligned}
\tag{8}
$$

Then the kernel weight $\mathbf{d}$ is calculated by

$$
d_m = \frac{\|w_m\|^{\frac{1}{p}}}{\left(\sum_{m'=0}^{M} \|w_{m'}\|^{\frac{p}{p+1}} \right)^{\frac{1}{p}}}, \forall m = 0, 1, 2, \dots M
\tag{9}
$$

$\tilde{\alpha}$ and $\mathbf{d}$ will be obtained until the stopping criteria is satisfied. Then the appearance model will be updated with $\tilde{\alpha}$ and $\mathbf{d}$. We have to pay attention to this that each type of features in feature sets we extract has the same dimension as others, so that the kernel matrixes with different features have the same sizes, otherwise they could not be applied with the algorithm to speed up their computing.

3.2   Tracking Algorithm

The tracking process of the proposed tracker is given by Table 2. In the prediction stage, the detection values of the patches could be calculated by

$$
\mathbf{F}(\mathbf{o}) = \sum_m d_m \mathbf{K}_m^{xo} \alpha
\tag{10}
$$

where $\mathbf{K}_m^{xo}$ represents the kernel matrix generated by one kernel between last frame's training samples of one feature and the current frame's patches of the same feature. The patches are the cyclic shifts of the base candidate patch centered at the location of the target detected in the previous frame. Here it is easy to observe that $\mathbf{K}_m^{\text{xo}}$ is also circulant. So apply the circulant matrix's property to Eq. (10), obtaining

$$\tilde{\mathbf{F}}(\mathbf{o}) = \sum_{m=1} d_m \tilde{\mathbf{k}}_m^{\text{xo}} \odot \tilde{\alpha} \qquad (11)$$

Thus we can just compute set $\left\{\tilde{\mathbf{k}}_1^{\text{xo}}, \tilde{\mathbf{k}}_2^{\text{xo}}, \cdots, \tilde{\mathbf{k}}_M^{\text{xo}}\right\}$ for the detection rather than $\{\mathbf{K}_1, \mathbf{K}_2, \cdots, \mathbf{K}_M\}$. Moreover, we only need to save one sample of each feature set of last frame due to the circulant nature of the feature sets for computing set $\left\{\tilde{\mathbf{k}}_1^{\text{xo}}, \tilde{\mathbf{k}}_2^{\text{xo}}, \cdots, \tilde{\mathbf{k}}_M^{\text{xo}}\right\}$. That means we do not need to keep the entire feature sets in storage as traditional method does when using MKL. In this way, we successfully apply MKL into real time visual tracking. After using (11), the patch with the maximal value will be selected as the new position of the target, where the appearance model has to be updated. In Sect. 3.1, we have interpreted the update procedure of appearance model. And in tracking process, the parameter $\tilde{\alpha}_t$, $\mathbf{d}_t$ is learned in the following way.

$$\tilde{\alpha}_t = \eta \tilde{\alpha}_t' + (1 - \eta) \tilde{\alpha}_{t-1} \qquad (12)$$

where η denotes the learning rate of $\tilde{\alpha}$. $\tilde{\alpha}_t$ represents the output parameter $\tilde{\alpha}$ of $Model(\tilde{\alpha}_t, \mathbf{d}_t)$. $\tilde{\alpha}_{t-1}$ represents the parameter $\tilde{\alpha}$ obtained at time $t-1$ of $Model(\tilde{\alpha}_{t-1}, \mathbf{d}_{t-1})$. $\tilde{\alpha}_t'$ represents the parameter $\tilde{\alpha}$ updated with algorithm 1 at time t. Then the tracking proceeds to detect the target and update the tracker to detect.

3.3   Complexity

The computation of the method we put forward lies mainly in the calculation of kernels in the frequency domain. As discussed in [10], the computation consumption of the single kernel is at the level of $N \log N$. For the proposed algorithm in this paper, if Q kernels are obtained and computed for every single base sample, then we can first transform the samples from spatial domain to frequency domain to compute the dot-product with all Q kernels. In this way, the complexity of the proposed algorithm is $O(N \log N + NQ)$. Since actually the dimensions of the features are much larger than the number of the kernels involved, which is usually a few dozen, our algorithm is equally effective.

4   Experiments

We evaluate the proposed algorithm on the popular benchmark including 50 image sequences [11]. The datasets contain all kinds of situation, for instance, partial occlusion, illumination variance, non-rigid deformation, scale variation, motion blur and background cluster, making them more impartial to evaluate various tracking algorithms. Then we will evaluate the proposed tracking algorithm (MKVT) using the Location Error Threshold vs Precision curve as the performance criteria, which is widely used recently [11]. Based on the results running on the benchmark we compare MKVT with 9 state-of-art methods including HOG-KCF [6] (kernel on Gaussian with HOG), RAW-KCF [6] (kernel on Gaussian with raw pixels), MIL [1], Struck [5], TLD [9], OAB [4], SCM [14], ORIA [12], ASLA [8] in terms of efficiency and accuracy. Implemented in MATLAB, our tracker runs on i3-2100 3.09 GHz machine with 4 GB RAM.

4.1 Setting

For our algorithm, the adaptation rate η is set to 0.02 and spatial bandwidth is equal to 0.1. The *MaxInteration* and T of Table 1 is separately set to 10 and 0.001. Variant HOG [3], Hue, Gray raw pixel features are extracted as base sample. In order to keep the same dimensions for all features, we set the cell size as 4 for Hue, Gray raw pixel features. More specifically, the average intensity value is calculated on the cell. For each base sample, the circulant kernels are composed of the kernels: (1). The polynomial kernel function with the parameter degree's range among 2, 3, 4, 5 to be selected and the coefficient set to 1; (2). The Gaussian kernel function with the parameter sigma's range among 0.1, 0.2, 0.4, 0.6, 0.8 and 1 to be selected. Then there are 30 kernels generated in total (Table 3).

For HOG-KCF [6] and RAW-KCF [6], we use the same parameter settings as its paper [6] suggests. And the MIL [1], Struck [5], TLD [9], OAB [4], SCM [14], ORIA [12], ASLA [8] settings are the same as [11].

4.2 Discussion

Table 1 shows the experimental results of various trackers over all data sets on the Benchmark in terms of the precision of the algorithm and tracking FPS. The trackers

Table 3. The table shows the average precision of fifty standard sequence (AUC area OPE curve), the frames per second (FPS) and the features MKVT and KCF (Feature) employed. Note that MKVT has increased by 2.7 % in the average precision, in spite of its intergration of multiple kernels; the proposed method still achieves 45.4 FPS.

Method	Feature	Average precision (20 px)	FPS
MKVT	Multi	**0.767**	**45.4**
KCF [6]	HOG	0.740	138.7
KCF [6]	Raw	0.560	**150.7**
Struck [5]	–	0.656	20.2
SCM [14]	–	0.649	0.51
TLD [9]	–	0.608	28.1
ASLA [8]	–	0.532	8.5
OAB [4]	–	0.504	22.4
MIL [1]	–	0.475	38.1
ORIA [12]	–	0.457	9.0

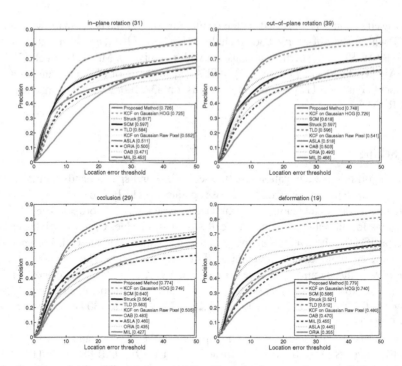

Fig. 1. Precision plot for different characteristic videos: in-plane rotation, out-of-plane rotation, occlusion, non-rigid deformation. The proposed method performs favorably against other trackers.

we compare with include the recently popular tracking algorithms. With respect to time, as discussed in Sect. 3.3, due to the aid of the fast Fourier transform algorithm, the proposed method avoids the computation of the kernel matrixes for traditional multi-kernel learning, thus greatly reducing the complexity of the algorithm. Although MKL is integrated into the algorithm, the tracking speed of the proposed algorithm still achieves 45.4 fps. With regard to precision, due to the integration of a variety of features and different kernel parameters, and the inheritance of the characteristics of strong generalization ability of non-sparse MKL [10], this method can be more effective in combining a variety of features so as to track the target. Over all the test datasets, the average precision of MKVT increases by 2.7 % than KCF with HOG.

As for non-rigid deformation, as is shown in Fig. 1, MKVT presents a superior tracking performance. MKVT has adopted color which is a good anti-deformation character, combining with the highly robust HOG contour feature, with the ability to adaptively learn the kernel weights of the two types of features, thus leading to the superior tracking precision of MKVT. MKVT demonstrates a higher precision by 3.9 % than HOG-KCF, and 29.9 % than RAW-KCF and such a similar experimental result also appears in other three graphs of Fig. 1, thus indirectly confirming the fact that a good representation of the target is the key factor in tracking. The stronger the ability of the features shows in representing the target, the easier the target will be recognized and detected.

As for in-plane or out-of-plane rotation, as shown Fig. 1, because of color feature's anti-rotation character, MKVT exhibits a superior tracking performance than HOG-KCF. Especially for the sequence of out-of-plane rotation, the precision of MKVT is higher than HOG-KCF by 1.9 %, and 15.1 % higher than Struck.

In terms of the occlusion, as shown in Fig. 1, Owing to the anti-occlusion nature of color and MKL's feature fusion framework, compared with HOG-KCF, MKVT has increased the tracking performance by 2.5 % and even 13.4 % than SCM which ranks third.

5 Conclusions

In the paper, we have proposed a real time and robust algorithm which employs multiple kernels for visual tracking. A multi-kernel based visual tracking framework is developed in which all kinds of features are incorporated, and then a more robust appearance model for the target is acquired. We apply the fast Fourier Transform Algorithm to the model update and prediction, yielding a fast tracking algorithm which runs at 45.4 fps with MATLAB implementation. The experiments show that the proposed algorithm outperforms the state-of-the-art trackers on challenging sequence with respect to accuracy and efficiency. In the future, we plan to generalize the proposed framework to other loss function and apply it to other problems, like event detection and object segmentation.

References

1. Babenko, B., Yang, M.H., Belongie, S.: Robust object tracking with online multiple instance learning. IEEE Trans. Pattern Anal. Mach. Intell. **33**(8), 1619–1632 (2011)
2. Bolme, D.S., Beveridge, J.R., Draper, B.A.: Visual object tracking using adaptive correlation filters. In: CVPR, pp. 2544 − 2550 (2010)
3. Felzenszwalb, P., Girshick, R., McAllester, D., et al.: Object detection with discriminatively trained part-based models. IEEE Trans. Pattern Anal. Mach. Intell. **32**, 1627–1645 (2010)
4. Grabner, H., Grabner, M., Bischof, H.: Real-time tracking via on-line boosting. In: BMVC (2006)
5. Hare, S., Saffari, A., Torr, P.H.S.: Struck: structured output tracking with kernels. In: ICCV (2011)
6. Henriques, J.F., Caseiro, R., Martins, P., Batista, J.: High-speed tracking with kernelized correlation filters. IEEE Trans. Pattern Anal. Mach. Intell. **37**, 583–596 (2014)
7. Henriques, J.F., Caseiro, R., Martins, P., et al.: Exploiting the circulant structure of tracking-by-detection with kernels (2011)
8. Jia, X., Lu, H., Yang, M.-H.: Visual tracking via adaptive structural local sparse appearance model. In: CVPR, (2012)
9. Kalal, Z., Mikolajczyk, K., Matas, J.: Tracking-learning-detection. IEEE Trans. Pattern Anal. Mach. Intell. **34**(7), 1409–1422 (2012)
10. Kloft, M., Brefeld, U., Belongie, S., et al.: Lp-norm multiple kernel learning. J. Mach. Learn. Res. **12**, 953–997 (2011)
11. Wu, Y., Lim, J., Yang, M.H.: Online object tracking: a benchmark. In: CVPR (2013)

12. Wu, Y., Shen, B., Ling, H.: Online robust image alignment via iterative convex optimization. In: CVPR (2012)
13. Zhang, K., Zhang, L., Yang, M.-H.: Real-time compressive tracking. In: Fitzgibbon, A., Lazebnik, S., Perona, P., Sato, Y., Schmid, C. (eds.) ECCV 2012, Part III. LNCS, vol. 7574, pp. 864–877. Springer, Heidelberg (2012)
14. Zhong, W., Lu, H., Yang, M.-H.: Robust object tracking via sparsity-based collaborative model. In: CVPR (2012)

Feature Selection Based on Data Clustering

Hongzhi Liu$^{(\boxtimes)}$, Zhonghai Wu, and Xing Zhang

School of Software and Microelectronics, Peking University,
Beijing 102600, People's Republic of China
{liuhz,wuzh,zhx}@pku.edu.cn

Abstract. Feature selection is an important step for data mining and machine learning. It can be used to reduce the requirement of data measurement and storage, and defy the curse of dimensionality to improve the prediction performance. In this paper, we propose a feature selection method via mutual information estimation. It avoids the calculation of high-dimensional mutual information by transforming the high-dimensional feature space into one dimension through a novel supervised clustering method. Experimental results on ten benchmark data sets show that: (1) the performances of kNN, naive Bayes classifier, and C4.5 using much less features selected by the proposed method are similar or even better than those on the original data sets with the whole feature set; (2) different from most of state-of-the-art methods which require to setting the number of features to select in prior, the proposed method can automatically determine the proper size of selected feature subsets.

Keywords: Feature selection · Supervised clustering · Mutual information

1 Introduction

Feature selection is an important step for data mining and machine learning. It can be used to facilitate the visualization and understanding of data, reduce the requirement of data measurement and storage, reduce the time of training and utilization, and defy the curse of dimensionality to improve the prediction performance [1]. Feature selection has been used in various applications, including information retrieval, video analysis, biomedical data analysis, and so on [2].

The goal of feature selection is to remove the irrelevant and redundant features, i.e. selecting the minimal subset of features that contains all the information in the original feature set about the target variable. Feature selection is an NP-hard problem. The number of feature subsets increases exponentially as the size of feature set increases.

The methods of feature selection can be grouped into three main categories: wrapper methods, embedded methods and filter methods [3]. Compared with wrapper methods and embedded methods, which are specific to a chosen predictor, filter methods provide a generic selection of features, not tuned for/by a given predictor.

Different filter methods have been proposed. Among these, methods based on mutual information constitute a broad family [4]. Mutual information is a nonparametric, nonlinear measure of relevance between two variables. It does not rely on any predictors, but provide a bound on the error rate using any predictor for the given distribution.

© Springer International Publishing Switzerland 2015
D.-S. Huang et al. (Eds.): ICIC 2015, Part I, LNCS 9225, pp. 227–236, 2015.
DOI: 10.1007/978-3-319-22180-9_23

The basic idea of feature selection based on mutual information is to select the feature subset with a fixed or minimal size that maximizes the mutual information between selected features and the class variable. The performance of these methods depends on the accuracy of estimating the high-dimensional (joint) mutual information between the candidate feature subsets and the class, which is still a challenging problem. Histograms and continuous kernels are two popular estimators of mutual information. However, estimation using histograms becomes impractical when the dimensionality is high. The complexity of using histograms grows exponentially with the number of features. In additional, the sparse data distribution, which is common in high-dimensional data space, may greatly degrade the reliability of histograms. On the other hand, estimation using a high-dimensional kernel, such as the Parzen window method [5], often demands a large set of training samples, which may be unrealistic for many applications.

To avoid directly calculating of high-dimensional mutual information, many researchers propose to use low-dimensional approximation methods [6–8], under the assumption that there are only lower-order dependencies between features, which may be not true for some real data sets.

In this paper, we propose a feature selection method via mutual information estimation. It avoids direct calculation of high-dimensional mutual information by transforming high-dimensional feature spaces into one dimension through a novel data clustering technique.

2   Background

2.1   Mutual Information

In accordance with Shannon's information theory, the mutual information between two variables X and Y is defined as the amount of information shared by X and Y, i.e.

$$I(X; Y) = H(X) - H(X|Y), \tag{1}$$

where $H(X)$ is the entropy of X which quantifies the uncertainty in the distribution X, $H(X|Y)$ is the conditional entropy of X given Y, i.e. the uncertainty of X after Y is given. $H(X)$ is defined as:

$$H(X) = - \sum_{x \in X} p(x) \log p(x), \tag{2}$$

where $p(x)$ is the marginal probability distribution function of X. $H(X|Y)$ is defined as:

$$H(X|Y) = - \sum_{y \in Y} p(y) \sum_{x \in X} p(x|y) \log p(x|y). \tag{3}$$

From (1), (2) and (3), we can derive that:

$$I(X; Y) = \sum_{x \in X} \sum_{y \in Y} p(x, y) \log \frac{p(x, y)}{p(x)p(y)}. \tag{4}$$

2.2 Relationship Between Bayes Error and Mutual Information

Bayes error is the ultimate criterion (golden standard) for any procedure related to discrimination. Mutual information provides bounds for the Bayes error [9], which is lower-bounded by Fano's inequality [10] and upper-bounded by Hellman-Raviv inequality [11]:

$$\frac{H(Y) - I(X; Y) - 1}{\log(|Y|)} \leq e_{Bayes} \leq \frac{1}{2}(H(Y) - I(X; Y)),$$

where $|Y|$ denotes the size of the set Y, i.e. the number of distinct values in set Y, and $H(Y)$ denotes the entropy of variable Y.

Let Y be the class variable of data sets. The values of $\log(|Y|)$ and $H(Y)$ are fixed for a given data set. To minimize the Bayes error, we only need to select a subset of features F_s that maximizes $I(F_s; Y)$.

2.3 Data Clustering Based on Density Peaks

Clustering is a commonly used data analysis technique. It is aimed at grouping a set of objects so that objects in the same group (called a cluster) are more similar to each other than to those in other groups (clusters).

Recently, Rodriguez and Laio [12] proposed a novel data clustering method based on density peaks (CDP). It assumes that cluster centers are surrounded by neighbors with lower local density and they are at a relative large distance from any data points with a higher local density.

The local density ρ_i of data point i is defined as:

$$\rho_i = \sum_j \exp\left(-\frac{d_{ij}^2}{2d_c^2}\right), \tag{5}$$

where d_{ij} denotes the distance between data points i and j, d_c is the radius of Gaussian kernel. The minimum distance δ_i between data point i and any other point with higher local density is defined as:

$$\delta_i = \min_{j:\rho_j > \rho_i} (d_{ij}). \tag{6}$$

For the data point with the highest local density, we take $\delta_i = \max_{j \neq i}(\delta_j)$. The cluster centers are recognized as points with ρ_i and δ_i are relative large, i.e. $\rho_i > \rho_t$ and $\delta_i > \delta_t$, where ρ_t and δ_t are two thresholds.

One main characteristics of algorithm CDP compared with other clustering methods is that it provides a visualization technique, called decision graph which is the graph of function $\delta(\rho)$, to assist us to determine the number of clusters and the cluster centers.

3 Feature Selection Based on Data Clustering

3.1 Estimation of Mutual Information Using Data Clustering

To estimate the mutual information between a feature set F_s and the class variable C, we first transform the high dimensional feature space F_s into one dimension by data clustering.

We proposed a supervised data clustering method SCDP-MI (Supervised Clustering based on Density Peaks and Mutual Information). It can automatically determine the optimal number of clusters without manual intervention. First, it calculates the local density ρ_i and the associated δ_i for each data point. Then it projects the data points in the decision graph onto the diagonal line and identifies the middle points between adjacent points as candidate threshold points. Finally, it searches the threshold point that maximizes the mutual information between cluster label variable and the class label variable.

Algorithm: SCDP-MI
1. Calculate the distance d_{ij} between each pair of data points.
2. Calculate the local density ρ_i for each data point.
3. Calculate δ_i for each data point.
4. Normalize the values of ρ_i and δ_i into range [0,1], i.e. $\rho_i' = \dfrac{\rho_i - \rho_{min}}{\rho_{max} - \rho_{min}}$, $\delta_i' = \dfrac{\delta_i - \delta_{min}}{\delta_{max} - \delta_{min}}$

5. Project the data points in decision graph onto the diagonal line, i.e. $\tau_i = \min(\rho_i', \delta_i')$.
6. Sort the unique values of τ_i and calculate the middle points between adjacent values.
7. Search the threshold τ_t and set the points with $\tau_i > \tau_t$ as cluster centers to maximize the mutual information between cluster labels and the class variable.

The mutual information between features and the class variable is a byproduct of the proposed clustering method SCDP-MI.

3.2 Towards Optimal Feature Selection via SCDP-MI

The proposed feature selection algorithm FSDC (Feature Selection based on Data Clustering) consists of two main steps. First, the selected feature set S is initialized to the empty set and the candidate feature set F is initialized to the whole feature set. Then, it repeatedly searches the feature f_k from F that maximizes the increment of mutual information between the selected feature set S and the class variable C by adding f_k into S, until it cannot find any feature that increases the mutual information between the selected feature set S and the class variable C. The mutual information between a candidate feature set and the class variable is estimated via the SCDP-MI algorithm. In each iteration, we remove the found feature f_k from F and add it into S.

FSDC is one of forward filter methods based on mutual information. Two key components of this kind of feature selection algorithms are: the calculation of the joint mutual information $I(C; S + f_i)$ and the criterion used to stop the searching. Different from previous methods, FSDC uses a novel algorithm based on data clustering to estimate the joint mutual information. Instead of using the number of selected features as stop criterion, FSDC directly uses the joint mutual information as stop criterion, i.e. stop searching when it cannot find any feature to increase the mutual information between the selected feature set and the class variable.

Algorithm: FSDC
1. Set F to the whole feature set and S to the empty set;
2. Repeat until we cannot find any feature $f_k \in F$ such that $I(C; S+f_k) > I(C; S)$
   2.1 Calculate $I(C; S +f_i)$ for all $f_i \in F$ via SCDP-MI
   2.2 Choose $f_k \in F$ that maximizes $I(C; S+f_k)$
   2.3 Put f_k into S and delete it from F.
3. Output the selected feature set S.

4   Experiments

The performance of the proposed feature selection algorithm FSDC is evaluated and compared with several state-of-the-art feature selection methods.

4.1   Experimental Setup

Ten public benchmark data sets from UCI machine learning repository [13] are used as experimental data. Table 1 gives a summary of the data sets.

To evaluate the performance of feature selection algorithms, we feed the data set before and after feature selection into three well-known classifiers: k-Nearest Neighbor

Table 1. Description of experimental data sets

ID	Data set	# of samples	# of features	# of classes
1	Breast	569	30	2
2	Congress	435	16	2
3	CTG	2126	21	3
4	ImgSeg	2310	19	7
5	Ionosphere	351	34	2
6	Krvskp	3196	36	2
7	PageBlock	5472	10	5
8	PenBased	10992	16	10
9	Texture	5500	40	11
10	Waveform	5000	40	3

(kNN), Naive Bayes classifier (NB) and C4.5, and record their classification accuracies. KNN is a type of instance-based lazy learning: an object is classified by a majority vote of its k nearest neighbors. It is one of the simplest and most effective machine learning methods. A naive Bayes classifier is a simple probabilistic classifier based on the Bayesian theorem with strong (naive) independence assumptions. C4.5 is a state-of-the-art method for inducing decision trees using the concept of information entropy [14].

10-fold cross-validation is adopted to better evaluate the performance of classifiers and feature selection algorithms. Each data set is randomly divided into 10 equally sized subparts. Nine of the ten subparts are used as training data and the remaining one is used as testing data. The cross-validation process is repeated 10 times with each of the 10 subparts used exactly once as testing data. All the experimental results are recorded as the average of the ten runs. All feature selection algorithms and classifiers run on the same data by 10-fold cross-validation to ensure the fairness of comparisons.

4.2   Experimental Design

Using the ten benchmark data sets, we evaluate the performance of feature selection algorithms by the classification accuracy of different classifiers. Five state-of-the-art feature selection methods: JMI [15], DISR [16], MRMR [8], CMI [17], and RELIEF [18], are used as comparison methods. The JMI (Joint Mutual Information) algorithm focuses on increasing complementary information between features. The DISR (Double Input Symmetrical Relevance) algorithm is a modification of JMI algorithm by introducing a normalization term. The MRMR (Minimum-Redundancy Maximum-Relevance) algorithm tries to minimize redundancy among features while maximizing relevance with the class variable. The CMI (Conditional Mutual Information) algorithm uses conditional mutual information to score features. The RELIEF algorithm sets the weights of feature relevance based on the nearest neighbors.

The classification accuracies on the data sets containing only the selected features are expected to be similar or higher as compared to those on the original data sets containing the whole feature set. To evaluate the effects of the feature selection algorithm FSDC, we compare the accuracies of classifiers on both the original data sets and the data sets with only selected features.

4.3   Experimental Results

Table 2 shows the classification accuracies of kNN (k = 3), naive Bayes classifier, and C4.5 with and without feature selection on the ten data sets. The second column shows the numbers of features in the original data sets and the numbers of selected features by FSDC.

The number of selected features by FSDC varies for different data sets. The sizes of the selected feature subsets are much smaller compared with the sizes of the original feature sets.

Table 2. Accuracies of kNN, Naïve Bayes (NB) and C4.5 with and without FSDC(%)

Data set	# of features	Acc. of kNN		Acc. of NB		Acc. of C4.5	
	Orig./Final	Orig.	Final	Orig.	Final	Orig.	Final
Breast	30/4.3	96.3	95.1	**92.8**	**94.6**	**93.9**	**94.4**
Congress	16/3.3	**91.5**	**94.0**	91.7	94.9	96.3	94.9
CTG	21/3.7	90.3	88.1	**81.9**	**83.6**	92.9	90.7
ImgSeg	16/4.7	**90.3**	**90.8**	71.5	73.6	91.7	91.1
Ionosphere	34/3.2	**85.5**	**90.0**	81.7	89.5	**90.3**	**90.9**
Krvskp	36/5.9	94.8	89.7	**84.1**	**92.8**	99.3	93.6
PageBlock	10/3.0	96.8	96.0	**89.5**	**93.0**	97.0	96.5
PenBased	16/9.1	99.4	95.8	85.8	80.1	96.4	93.5
Texture	40/7.0	98.5	94.5	**77.4**	**78.9**	92.7	90.0
Waveform	40/3.2	73.5	70.5	78.8	70.7	74.6	72.3
Avg. ± Std.	25.9 ± 11.3/4.7 ± 2.0	91.7 ± 7.7	90.5 ± 7.6	83.5 ± 6.7	**85.2 ± 9.9**	92.5 ± 6.9	90.8 ± 6.8

The accuracies of kNN, naive Bayes classifier, and C4.5 on the data sets with much less selected features by FSDC are similar to those on the original data sets with the whole feature set. In several cases, the performances of classifiers are better on the data sets with only the selected features than on the original data sets, especially for the naive Bayes classifier. On eight of the ten data sets, naive Bayes classifier performs better with only the selected features by FSDC than with all the original features.

Figures 1, 2, and 3 show the performances of kNN, naive Bayes classifier, and C4.5 with different feature selection methods. The results show that: (1) adding more features does not guarantee the increasing of accuracies; (2) the size of optimal feature subset is different for different data sets; (3) the sizes of selected feature subsets by FSDC are relative smaller than those by CMI.

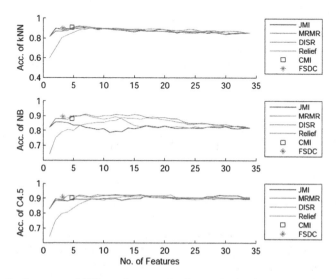

Fig. 1. Performance of different feature selection methods on the ionosphere data set

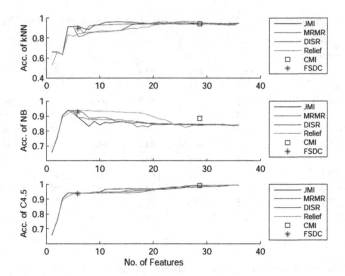

Fig. 2. Performance of different feature selection methods on the Krvskp data set

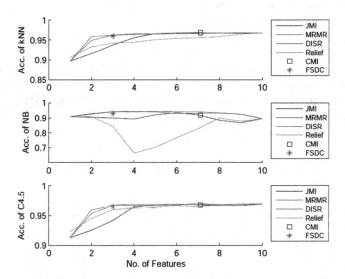

Fig. 3. Performance of different feature selection methods on the PageBlock data set

5 Discussions

Curse of dimensionality is a common phenomenon in machine learning, i.e. adding more features may degrade the performance of learning methods. Table 2, Figs. 1, 2, and 3 show some of these cases. Compared with kNN and C4.5, the naive Bayes classifier is more likely to be affected by the dimensionality. Naive Bayes classifiers assume that all features are conditional independent given the class. The more features we use, the more unreliable of this assumption.

Feature selection is a technique for avoiding the effects of the curse of dimensionality. It can increase the accuracies of classifiers while reducing the number of required features. However, the sizes of the optimal feature subsets are different for different data sets. This makes it difficult to determine the optimal parameter settings for feature selection algorithms that require setting the size of selected feature set in prior.

The supervised clustering algorithm SCDP-MI used in the proposed feature selection method can be seen as a discretization method. It is a supervised multiple variable discretization method. The discretization consists of two phases: first, it is initializes by an unsupervised clustering process; then, it finely tunes the parameters using the class information. This idea is similar to that used in deep autoencoder networks [19], which first initializes the parameters of artificial neural networks by unsupervised learning, then finely tune the parameters using supervised labels information.

6 Conclusions

In this paper, we propose a feature selection method via mutual information estimation. It estimates the high-dimensional mutual information by a supervised clustering algorithm, which transforms the feature set from high dimensionality into one dimension. Experimental results show that: (1) curse of dimensionality is a common phenomenon in machine learning, especially for the naive Bayes classifier; (2) the performances of kNN, naive Bayes classifier, and C4.5 using much less features selected by the proposed method are similar or even better than those on the original data sets with the whole feature set; (3) different from most of state-of-the-art methods which require to setting the number of features to select in prior, FSDC can automatically determine the proper size of selected feature subsets; (4) compared with a recently proposed method CMI [17], which also does not required to setting the number of features to select in prior, the numbers of selected features by FSDC are much less.

Acknowledgment. This work was supported in part by National Natural Science Fund of China (61232005) and National Key Technology R&D Program of China (2012BAH06B01). Liu is partially sponsored by CCF-Tencent Open Research Fund.

References

1. Guyon, I., Elisseeff, A.: An introduction to variable and feature selection. J. Mach. Learn. Res. **3**, 1157–1182 (2003)
2. Guyon, I., Gunn, S., Nikravesh, M., Zadeh, L.A. (eds.): Feature Extraction: Foundations and Applications. Studies in Fuzziness and Soft Computing. Springer, Heidelberg (2006)

3. Liu, H., Motoda, H.: Feature Selection for Knowledge Discovery and Data Mining. Kluwer Academic Publishers, Norwell (1998)
4. Vergara, J.R., Estevez, P.A.: A review of feature selection methods based on mutual information. Neural Comput. Appl. **24**(1), 175–186 (2014)
5. Kwak, N., Choi, C.-H.: Input feature selection by mutual information based on parzen window. IEEE Trans. Pattern Anal. Mach. Intell. **24**(12), 1667–1671 (2002)
6. Balagani, K.S., Phoha, V.V.: On the feature selection criterion based on an approximation of multidimensional mutual information. IEEE Trans. Pattern Anal. Mach. Intell. **32**(7), 1342–1343 (2010)
7. Battiti, R.: Using mutual information for selecting features in supervised neural net learning. IEEE Trans. Neural Netw. **5**(4), 537–550 (1994)
8. Peng, H., Long, F., Ding, C.: Feature selection based on mutual information criteria of max-dependency, max-relevance, and min-redundancy. IEEE Trans. Pattern Anal. Mach. Intell. **27**(8), 1226–1238 (2005)
9. Brown, G.: A new perspective for information theoretic feature selection. In: AISTATS 2009, pp. 49–56 (2009)
10. Fano, R.M.: Transmission of Information: A Statistical Theory of Communications. MIT Press, Cambridge (1961)
11. Hellman, M., Raviv, J.: Probability of error, equivocation, and the chernoff bound. IEEE Trans. Inf. Theory **16**(4), 368–372 (1970)
12. Rodriguez, A., Laio, A.: Clustering by fast search and find of density peaks. Science **344**(6191), 1492–1496 (2014)
13. Asuncion, A., Newman, D.: UCI machine learning repository (2007)
14. Quinlan, J.R.: C4.5: Programs for Machine Learning. Morgan Kaufmann Series in Machine Learning. Morgan Kaufmann, San Francisco (1992)
15. Yang, H.H., Moody, J.E.: Data visualization and feature selection: new algorithms for nongaussian data. In: NIPS 1999, pp. 687–702 (1999)
16. Meyer, P., Schretter, C., Bontempi, G.: Information-theoretic feature selection in microarray data using variable complementarity. IEEE J. Sel. Top. Sig. Process. **2**(3), 261–274 (2008)
17. Brown, G., Pocock, A., Zhao, M.-J., Lujan, M.: Conditional likelihood maximisation: a unifying framework for information theoretic feature selection. J. Mach. Learn. Res. **13**, 27–66 (2012)
18. Kira, K., Rendell, L.A.: The feature selection problem: traditional methods and a new algorithm. In: AAAI 1992, pp. 129–134 (1992)
19. Hinton, G.E., Salakhutdinov, R.R.: Reducing the dimensionality of data with neural networks. Science **313**(5786), 504–507 (2006)

A Comparison of Local Invariant Feature Description and Its Application

Kaili Shi, Qingwei Gao$^{(\boxtimes)}$, Yixiang Lu,
Weiguo Zhang, and Dong Sun

College of Electrical Engineering and Automation,
Anhui University, Hefei, Anhui China
qingweigao@ahu.edu.cn

Abstract. The description of image region draws a lot of attention in the field of computer vision. Recently, many descriptors were proposed for image region description and achieved high achievements. These descriptors are widely used in many fields, such as object recognition, image mosaic, video tracking. In this paper, we first systematically analyze six typical descriptors: SIFT, DAISY, MROGH, MRRID, LIOP and HRI-CSLTP descriptors. Then we conduct experiments in several different situations to evaluate the performance of these descriptors. From the experimental results, we get to make a conclusion and analysis about the advantages and disadvantages of these descriptors. Finally, we make an application of these descriptors in image matching field.

Keywords: Local invariant feature description · Image matching · Comparison

1 Introduction

Image matching is an important research direction in the field of computer vision and a key step from image processing to image analysis. Local invariant features are fundamental and important work in image matching, object recognition, and so on. Local invariant feature refers to the local image f eature detection or description which is invariant to many situations, such as geometric transformation, photometric transformation, convolution transformation, changes of views, etc.

Local invariant feature can not only obtain reliable matching in the big change of the observation condition, shade and noise interference, but also effectively describe the image within the capacity for image retrieval [16],scene or objection recognition [1]. The research of local invariant features includes three fundamental problems: (1) the detection of local invariant features; (2) the description of local invariant features; (3) the application of local invariant features. Many methods have been proposed for detecting local invariant features including Harris-affine [15], Hessian-affine [4], MSER (Maximally Stable Extremal Regions) [3] and intensity and edge-based detectors [5]. Once the local invariant features are detected, the next work is how to describe local invariant features.

We use descriptors to describe the local invariant features. Local image descriptor is an important research field in computer vision, it plays an important role in image matching [17], objection recognition, 3D reconstruction [2] and classification [18] as

© Springer International Publishing Switzerland 2015
D.-S. Huang et al. (Eds.): ICIC 2015, Part I, LNCS 9225, pp. 237–246, 2015.
DOI: 10.1007/978-3-319-22180-9_24

well as video tracking, etc. This paper mainly focuses on the analysis of six typical descriptors: SIFT, DAISY, MROGH, MRRID, LIOP and HRI-CSLTP. Then we give experiments in different situations to evaluate the performance of these descriptors. After experiments we give discussions of each descriptor's advantages and limitations, and make conclusions of them. At last, we give an application of these descriptors in image matching field.

2   Descriptors

2.1   SIFT Descriptor

SIFT (Scale-invariant Feature Transform) [1] which is very famous in image feature description field was proposed by David G. Lowe. Lowe suggests dividing the neighborhood of the keypoints into $d \times d$ (Lowe suggests $d = 4$) sub areas, each seed point has eight directions. So the descriptor is representing as a total of $4 \times 4 \times 8 = 128$ dimensional vector. It contains six steps: (1) Determine the image area that needed for generating the descriptor. Each keypoint contains the following information: the directions, scale and location. (2) Rotate coordinate axis to the direction of the keypoints to ensure the rotation invariance. (3) Distribute the sample points within the neighborhood to the corresponding sub area. Assign the gradient values of the sub area to eight directions, and calculate the weight. (4) Calculate the gradient in eight directions of each seed point using interpolation algorithm. A feature vector will be gotten after counting the 128 gradient information of each keypoint. (5) In order to remove the influence of illumination change of the feature vector, normalize processing is needed. After normalization, give the descriptor vector a threshold. Setting threshold (generally take 0.2) truncated larger gradient value. And then, normalize the processing again, improve the diagnostic of features. (6) Sort the feature description vector according to the scale of the feature points. At this point, SIFT features description vector is generated.

2.2   DAISY Descriptor

DAISY [6] which is highly efficient in computing densely was proposed by Engin Tola et al. This algorithm is the first time for assesses dense depth maps out of wide-baseline image pairs. It maintains the robustness of SIFT and GLOH, and its design is based on quick and efficient computing at every single pixel in the image. It uses gradient convolutions in definite directions with Gaussian filters to replace weighted sums of the gradient.

The presentation of DAISY algorithm [6] as follows: (1) the introduction of vector $h_\Sigma(u, v)$. The vector $h_\Sigma(u, v)$ is defined as the values that convolutions by a Gaussian kernel of standard deviation Σ for location (u, v) in the orientation maps.

$$h_\Sigma(u, v) = \left[G_1^\Sigma(u, v), \cdots, G_H^\Sigma(u, v) \right]^{\mathrm{T}} \tag{1}$$

where G_1^Σ, G_2^Σ and G_H^Σ represent the $\Sigma-$ convolved orientation maps in different directions. Then normalize $h_\Sigma(u, v)$ to unit norm, and denote it as $\tilde{h}_\Sigma(u, v)$; (2) the introduction of DAISY descriptor. The DAISY descriptor $D(u_0, v_0)$ at location (u_0, v_0) is made of the concatenation of normalized h vectors:

$$
\begin{aligned}
D(u_0, v_0) = \Big[& \tilde{h}_{\Sigma_1}^{\mathrm{T}}(u_0, v_0), \\
& \tilde{h}_{\Sigma_1}^{\mathrm{T}}(1_1(u_0, v_0, R_1)), \cdots, \tilde{h}_{\Sigma_1}^{\mathrm{T}}(1_T(u_0, v_0, R_1)), \\
& \tilde{h}_{\Sigma_2}^{\mathrm{T}}(1_1(u_0, v_0, R_2)), \cdots, \tilde{h}_{\Sigma_2}^{\mathrm{T}}(1_T(u_0, v_0, R_2)), \\
& \cdots \\
& \tilde{h}_{\Sigma_Q}^{\mathrm{T}}(1_1(u_0, v_0, R_Q)), \cdots, \tilde{h}_{\Sigma_Q}^{\mathrm{T}}(1_T(u_0, v_0, R_Q)) \Big]^{\mathrm{T}}
\end{aligned}
\tag{2}
$$

where $1_j(u, v, R)$ denotes the location with distance R from (u, v) in direction when the direction is quantized into the T values, and Q denotes the number of the circular layer.

2.3 MROGH Descriptor and MRRID Descriptor

MROGH (Multi-Support Region Order-Based Gradient Histogram) [7] and MRRID (Multi-Support Region Rotation and Intensity Monotonic Invariant Descriptor) [8] was proposed by Bin Fan et al. The difference between MROGH and MRRID is that MROGH is utilized gradient-based feature but the MRRID is utilized intensity-based feature.

MROGH descriptor [7] is based on gradient feature. To point X_i, computing the gradient magnitude $m(X_i)$ and orientation $\theta(X_i)$ according to formula (3).

$$
m(X_i) = \sqrt{D_x(X_i)^2 + D_y(X_i)^2}, \quad \theta(X_i) = \tan^{-1}(D_y(X_i)/D_x(X_i))
\tag{3}
$$

where

$$
D_x(X_i) = I(X_i^1) - I(X_i^5), \quad D_y(X_i) = I(X_i^3) - I(X_i^7)
\tag{4}
$$

where $X_i^j, j = 1, 3, 5, 7$ are X_i's neighboring points in the local $x - y$ coordinate system and $I(X_i^j)$ represents the intensity at X_i^j. Splitting $\theta(X_i)$ which is in the range of $[0, 2\pi)$ into d equal bins as $\mathrm{dir}_i = (2\pi/d) \times (i - 1)$, $i = 1, 2, \cdots, d$, then compute $F_G(X_i) = (f_1^G, f_2^G, \cdots, f_d^G)$ which denote the gradient of X_i, where

$$
f_j^G = \begin{cases} m(X_i) \dfrac{(2\pi/d - \alpha(\theta(X_i), \mathrm{dir}_j))}{2\pi/d}, & \text{if } \alpha(\theta(X_i), \mathrm{dir}_j) < 2\pi/d \\ 0, & \text{otherwise} \end{cases}
\tag{5}
$$

where $\alpha(\theta(X_i), \mathrm{dir}_j)$ denotes the angle between $\theta(X_i)$ and dir_j. $F(R_i) = \sum_{X \in R_i} F_G(X)$ is the accumulated vector of partition R_i.

MRRID descriptor [8] is based on intensity orders. For each sample point X_i, suppose that $X_i^j, j = 1, 2, \cdots, 2m$ are its $2m$ neighboring points. By comparing the intensities of opposite sample points, we get a m dimensional binary vector: $(\text{sign}(I(X_i^{m+1}) - I(X_i^1)), \text{sign}(I(X_i^{m+2}) - I(X_i^2)), \cdots, \text{sign}(I(X_i^{m+m}) - I(X_i^m)))$. Then a local feature $F_I(X_i) = (f_1^l, f_2^l, \cdots, f_{2m}^l)$ of X_i can be obtained by mapping the m dimensional binary vector into a $2m$ dimensional vector, where

$$
f_j^l = \begin{cases} 1, & \text{if } \sum_{k=1}^{m} \text{sign}(I(X_i^{k+m}) - I(X_i^k)) \times 2^{k-1} = (j-1) \\ 0, & \text{otherwise} \end{cases} \tag{6}
$$

$$
\text{sign}(x) = \begin{cases} 1, & x > 0 \\ 0, & \text{otherwise} \end{cases}
$$

then accumulated vector $F(R_i) = \sum_{X \in R_i} F_I(X)$.

At last, the vector $D(R) = (F(R_1), F(R_2), \cdots, F(R_k))$ of MROGH and MRRID are developed. All the vectors computed from the N support regions are concatenated together to form the final descriptor $\{D_1 D_2 \cdots D_N\}$.

2.4   LIOP Descriptor

LIOP [9] was proposed by Zhenhua Wang et al. in 2011, it considers all sample points' intensity in order to describe the local intensity relationships. Since the encoding scheme of permutation-based was proposed for decreasing the dimension, LIOP is more appropriate in local descriptor constructing. On accumulating points of LIOP in every ordinal bin, the descriptor which is based on the local intensity relationships is constructed totally, which ensure that the rotation and monotonic intensity changes invariance. The definition of the LIOP on the point x is as follows:

$$
LIOP(x) = \phi(\gamma(P(x))) = V_{N!}^{Ind(\gamma(P(x)))} = \left(0, \cdots, 0, \underset{(Ind(\gamma(P(x))))}{1}, 0, \cdots, 0\right) \tag{7}
$$

where $P(x) = (I(x_1), I(x_2), \cdots, I(x_N)) \in P^N$ and $I(x_i)$ represents the intensity of the i-th neighboring sample point x_i. The main processing and the calculation of $\gamma(P(x))$ are shown in Fig. 1.

2.5   HRI-CSLTP Descriptor

Before introducing HRI-CSLTP, we would like to introduce CS-LBP (center-symmetric local binary pattern) [10], a kind of algorithm put forward by Heikkila et al. CS-LBP combining the good properties of the SIFT descriptor and the LBP [11] texture operator. It simplifies several steps of the algorithm since it compares center-symmetric pairs of pixels instead of comparing each pixel with the center pixel. Instead of a binary code, CS-LTP (Center-Symmetric Local Ternary Patterns) [12] is a

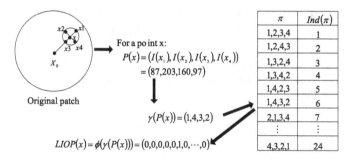

Fig. 1. The construction of the LIOP descriptor.

ternary code which is based on the local gradient information. CS-LTP codes are a variant to the CS-LBP codes, and have some superior performance than CS-LBP codes.

HRI-CSLTP [12] is a combination of two different order-based methods for feature description: the HRI (Histogram of Relative Intensities) and the CS-LTP. The two methods are complementary. It is more robust to Gaussian noise since it uses intensity information together with ordering information.

3 Experiments

3.1 Dataset and Evaluation Criterion

We used standard Oxford dataset which we have downloaded from their website to test these descriptors. It contains six different transformations of images as follows: image blur, viewpoint changes, scale changes, image rotations, illumination changes, and JPEG compression.

We followed the evaluation procedure proposed by Mikolajczyk and Schmid [13] which is based on the number of correct and false matches between two images. The number of correct matches and ground truth correspondences is determined by the overlap error [14]. If the overlap error <0.5 then the match is correct. The results are presented with recall versus 1-precision curves:

$$recall = \frac{\#correct\ matches}{\#correspondences} \ , \ 1 - precision = \frac{\#false\ matches}{\#all\ matches} \qquad (8)$$

where $\#correspondences$ is the ground truth number of matches.

3.2 Performance Evaluation Under Different Situations

In our experiments(Pentium Dual-Core CPU 2.99 GHz), the average time for constructing SIFT, DAISY, MROGH, MRRID, LIOP and HRI-CSLTP descriptors under six different transformations of images is shown in Table 1. We have evaluated the performance of the descriptors using Harris-Affine (haraff) [15] detectors. Then we compared

SIFT, DAISY, MROGH, MRRID, LIOP and HRI-CSLTP experimental results under various image transformations such as image blur, viewpoint changes, rotation and scale changes, illumination changes and JPEG compression. There were six different images in each set of images, so we can have five images matching in each set. Now we just show two images matching under Harris-Affine in each set of images. We show the experimental results in Fig. 2.

Table 1. The time consumption of SIFT, DAISY, MROGH, MRRID, LIOP and HRI-CSLTP under bike, boat, graf, leuven, ubc and wall.

Time Consumption(s)	bike	boat	graf	leuven	ubc	wall
SIFT	2.1459	11.1208	7.6542	3.6458	6.2833	7.6417
DAISY	3.3912	8.6073	9.6259	2.9962	4.4898	6.1282
LIOP	2.2354	6.1962	9.3946	2.0393	3.2945	4.7839
MROGH	4.2126	13.2605	12.2667	4.2297	6.3757	9.4665
MRRID	9.1571	38.5887	26.8483	9.2828	14.2743	22.1747
HRI-CSLTP	4.8728	10.8382	10.4436	3.7425	5.7368	6.8363

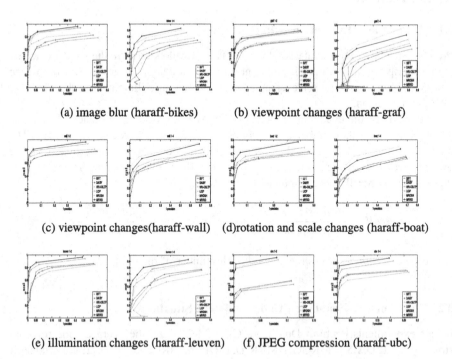

(a) image blur (haraff-bikes)   (b) viewpoint changes (haraff-graf)

(c) viewpoint changes(haraff-wall)   (d)rotation and scale changes (haraff-boat)

(e) illumination changes (haraff-leuven)   (f) JPEG compression (haraff-ubc)

Fig. 2. Experimental results under various image transformations.

3.3 Experiment Summarization

From the experiments above, we can get the advantages and limitations of these algorithms in each situation. We rate the experiment results in four grades, i.e., Best, Better, Good, and Common. SIFT, DAISY, MROGH, MRRID, LIOP and HRI-CSLTP respectively perform Common, Common, Best, Best, Better, Good under image blur, viewpoint changes and rotation and scale changes. SIFT, DAISY, MROGH, MRRID, LIOP and HRI-CSLTP respectively perform Good, Good, Best, Best, Better, Common under illumination changes. SIFT, DAISY, MROGH, MRRID, LIOP and HRI-CSLTP respectively perform Common, Common, Best, Best, Better, Common under JPEG compression. The average time for constructing SIFT, DAISY, MROGH, MRRID, LIOP and HRI-CSLTP respectively perform Best, Best, Common, Common, Best, Better.

4 The Application of These Descriptors

We make an image match application of SIFT, DAISY, MROGH, MRRID, LIOP and HRI-CSLTP using haraff detectors under various image transformations. The application under blur invariance is conducted on 'bikes' set to distinguish the merits and the limitations of each descriptor. For sake of clarity, we just show the first 10 matching lines in Fig. 3. As the same reason, the application under affine invariance that is conducted on 'graf' set is shown in Fig. 4. Finally, we calculate the image matching correct rate in matching application of SIFT, DAISY, MROGH, MRRID, LIOP and HRI-CSLTP under bike, boat, graf, leuven, ubc and wall. The result is shown in Table 2.

Table 2. The image matching correct rate in matching application of SIFT, DAISY, MROGH, MRRID, LIOP and HRI-CSLTP under bike, boat, graf, leuven, ubc and wall.

Matching correct rate(%)	bike	boat	graf	leuven	ubc	wall
SIFT	87.7907	90.1316	73.5577	85.9459	99.1641	94.8718
DAISY	90.0552	89.4389	82.3232	89.6341	98.9605	95.4839
LIOP	99.3103	97.6077	94.6667	97.4359	100	97.7099
MROGH	98.8764	97.6563	94.8187	96.9697	99.9022	99.6855
MRRID	98.9011	99.1266	94.4099	99.3939	100	99.6622
HRI-CSLTP	96.4286	96.6245	87.9121	90.1786	99.6599	98.6301

Fig. 3. The first 10 matching keypoints between the first and the fourth image of the 'bikes' set under blur invariance.

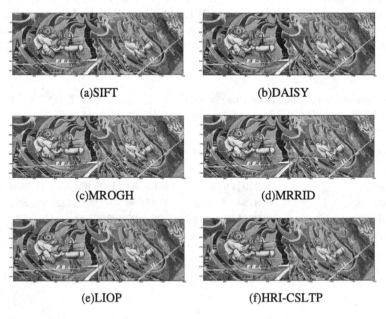

Fig. 4. The first 10 matching keypoints between the first and the fourth image of the 'graf' set under affine invariance.

5 Conclusions

This paper systematically analyzed the performance of SIFT, DAISY, MROGH, MRRID, LIOP and HRI-CSLTP under six different transformations of images. We also researched the descriptors' generated time, the matching correct rate in different situations. Through the above research, we can compare the merits and the limitations of each descriptor.

The experimental results show that MROGH and MRRID performed best on six different transformations of images and matching correct rate, but their descriptors' generated time is common. SIFT and DAISY performed the best on descriptors' generated time, but their matching correct rate is common and the performance under six image transformations is not satisfactory. LIOP is worse than MROGH and MRRID, but usually better than HRI-CSLTP, SIFT and DAISY under six image transformations, and the descriptors' generated time and the matching correct rate of LIOP are satisfactory. The performance of HRI-CSLTP is basically similar to SIFT and DAISY under six image transformations, and its descriptors' generated time and the matching correct rate are good.

After comparing the merits and the limitations of each descriptor, we can choose the best descriptor according to the different requirements. For example, in image matching, we want to get a high matching correct rate, so we choose MROGH or MRRID descriptor. With the development, many new algorithms will be proposed. They will have better performance. We can image that there will be some new algorithms which have better practical values in image matching and object recognition.

Acknowledgements. This work is supported by the National Natural Science Foundations of China (NSFC) (61370110, 61402004 & 61402003).

References

1. Lowe, D.G.: Distinctive image features from scale-invariant keypoints. Int. J. Comput. Vis. **60**(2), 91–110 (2004)
2. Furukawa, Y., Ponce, J.: Accurate, dense, and robust multiview stereopsis. Pattern Anal. Mach. Intell. **32**(8), 1362–1376 (2010)
3. Matas, J., Chum, O., Urban, M., et al.: Robust wide-baseline stereo from maximally stable extremal regions. Image Vis. Comput. **22**(10), 761–767 (2004)
4. Mikolajczyk, K., Schmid, C.: An affine invariant interest point detector. In: Heyden, A., Sparr, G., Nielsen, M., Johansen, P. (eds.) ECCV 2002, Part I. LNCS, vol. 2350, pp. 128–142. Springer, Heidelberg (2002)
5. Tuytelaars, T., Van, G.L.: Matching widely separated views based on affine invariant regions. Int. J. Comput. Vis. **59**(1), 61–85 (2004)
6. Tola, E., Lepetit, V., Fua, P.: Daisy: an efficient dense descriptor applied to wide-baseline stereo. Pattern Anal. Mach. Intell. **32**(5), 815–830 (2010)
7. Fan, B., Wu, F., Hu, Z.: Aggregating gradient distributions into intensity orders: a novel local image descriptor. In: Computer Vision and Pattern Recognition, pp. 2377–2384 (2011)

8. Fan, B., Wu, F., Hu, Z.: Rotationally invariant descriptors using intensity order pooling. Pattern Anal. Mach. Intell. **34**(10), 2031–2045 (2012)
9. Wang, Z., Fan, B., Wu, F.: Local intensity order pattern for feature description[C]. In: Computer Vision (ICCV), pp. 603–610 (2011)
10. Heikkilä, M., Pietikäinen, M., Schmid, C.: Description of interest regions with local binary patterns. Pattern Recogn. **42**(3), 425–436 (2009)
11. Ojala, T., Pietikainen, M., Maenpaa, T.: Multiresolution gray-scale and rotation invariant texture classification with local binary patterns. Pattern Anal. Mach. Intell. **24**(7), 971–987 (2002)
12. Gupta, R., Patil, H., Mittal, A.: Robust order-based methods for feature description. In: Computer Vision and Pattern Recognition, pp. 334–341 (2002)
13. Mikolajczyk, K., Schmid, C.: A performance evaluation of local descriptors. Pattern Anal. Mach. Intell. **27**(10), 1615–1630 (2005)
14. Goswami, B., Chan, C. H., Kittler, J., et al.: Local ordinal contrast pattern histograms for spatiotemporal, lip-based speaker authentication. In: Biometrics: Theory Applications and Systems, pp. 1–6 (2010)
15. Mikolajczyk, K., Schmid, C.: Scale & affine invariant interest point detectors. Int. J. Comput. Vis. **60**(1), 63–86 (2004)
16. Nister, D., Stewenius, H.: Scalable recognition with a vocabulary tree. In: Computer Vision and Pattern Recognition, pp. 2161–2168 (2006)
17. Brown, M., Lowe, D.G.: Automatic panoramic image stitching using invariant features. Int. J. Comput. Vis. **74**(1), 59–73 (2007)
18. Zhang, J., Marszałek, M., Lazebnik, S., et al.: Local features and kernels for classification of texture and object categories: a comprehensive study. Int. J. Comput. Vis. **73**(2), 213–238 (2007)

An Efficient Indexing Scheme Based on K-Plet Representation for Fingerprint Database

Chaochao Bai[1], Tong Zhao[2(⊠)], Weiqiang Wang[1], and Min Wu[3]

[1] School of Computer and Control,
University of Chinese Academy of Sciences, Beijing, China
baichaochao12@mails.ucas.ac.cn, wqwang@ucas.ac.cn
[2] School of Mathematical Sciences,
University of Chinese Academy of Sciences, Beijing, China
zhaotong@ucas.ac.cn
[3] Eastern Golden Finger Technology Co. Ltd, Beijing, China
wumin@egafis.com

Abstract. Fingerprints are now widely employed in the security fields. A typical police fingerprint database may contain millions of template fingerprints. Consequently, fingerprint indexing plays an essential role to improve the performance of matching such a huge database. In this paper, the efficient index tree based on k-plet local patterns of minutiae for fingerprint database is proposed. The proposed algorithm is of robustness since the k-plet is translation-invariant and rotation-invariant, moreover, the multipath indexing strategy is introduced at the stage of indexing. As well, it is quite fast and effective due to look-up operation instead of complex computation. The performance testing was conducted in the datasets of FVC2002 DB1, NIST DB4 and NIST DB14, which concluded that the proposed algorithm is advantageous for fingerprint indexing since it achieves a high correct index performance with a fairly low penetration rate.

Keywords: Fingerprint indexing · Index tree · K-plet local pattern

1 Introduction

Fingerprints, consisting of the ridges and furrows on human fingers, have been one of the most describable biometric traits which are largely employed for authentication and identification tasks in the fields of civil and forensic systems. A number of features could be captured from a fingerprint to identify a person through a fingerprint recognition system [1]. Generally, the fingerprint recognition system operates in either verification mode or identification mode. For the latter one, identification of an unknown template operating in a 1:N matching process over a huge database usually consumes significant time in order to provide a reliable conclusion. Although state-of-the-art fingerprint matching algorithms are fast and accurate, the size of the database can be over one hundred million (e.g., the police database of a country) and it poses much more challenges for the recognition accuracy and efficiency.

D.-S. Huang et al. (Eds.): ICIC 2015, Part I, LNCS 9225, pp. 247–257, 2015.
DOI: 10.1007/978-3-319-22180-9_25

To address the above challenges, fingerprint classification and fingerprint indexing are the most common solutions. In the terms of fingerprint classification, Henry distinguished fingerprints into 5 classes including left loop, right loop, whorl, arch, and tented arch [2]. However, the number of classes is small and fingerprints are unevenly distributed among them. On contrary, fingerprint indexing is a more efficient approach where fingerprints are specified with feature vectors. These feature vectors are generated through a similarity-preserving transformation and similar fingerprints are mapped into close points (vectors) in the multidimensional space. The indexing is performed by matching the query fingerprint against the template fingerprints in the database whose vectors are close to the query one. Subsequently, the top N most similar template fingerprints are returned. Since it is only necessary to match N candidates instead of every fingerprint of the database, the fingerprint indexing narrows the number of the large database effectively.

Fingerprint indexing is challenging and promising so that there have been lots of approaches to improve the related techniques, in which the involved features can be generally classified into minutiae and global features. In global feature methods, singular points or orientation fields represent the global information of ridges. In [3] and [4], the algorithms adopt singular points as features to index fingerprints. However, these techniques mostly require pre-alignment of fingerprints and it is difficult to extract reliable location of singular point from poor fingerprint images. In addition, feature vectors for fingerprints indexing are received by orientation fields [5, 6]. However, these features are still global characters and their discrimination is not as good as minutiae.

In minutia feature methods, the intrinsic idea is to establish a structure of minutiae that is of enough discrimination and robustness in presence of rotation and translation variations. Depending on these minutia features, there mainly exist two indexing techniques, namely hash-based indexing and Approximate Nearest Neighbor (ANN) indexing. The algorithm [7, 8] based on minutiae derives triangle or quadrangle geometric features and adopts simple hashing technique for searching. Unfortunately, these geometric features are more sensitive to noise and distortion. On the other, Minutiae Cylinder Codes (MCC) characterize a minutia neighborhood through encoding the neighborhood of each minutia into a fixed-length bit vector and the algorithm indexes by means of a typical kind of ANN indexing technique, Locality Sensitive Hashing (LSH) [9]. The representation of a minutia's neighborhood in MCC seems quite complicated for the reason that it has a very high-dimensional feature vector. Furthermore, Locality Sensitive Hashing (LSH) is an approximate searching method which is not quite applicable for the cases requiring high accuracy rate.

In this paper, to our knowledge, the k-plet local pattern of minutiae [10] as feature is the first time to be imported in the field of fingerprint indexing. The k-plet representation is invariant under translation and rotation since it is based on its own local coordinate system. In addition, an efficient tree based indexing scheme designed for the k-plet representation is proposed to accelerate retrieval. In order to make this indexing scheme more robust, the multipath indexing strategy is employed which means that brother bins located in a certain range of the hit bin are all browsed at the stage of indexing. Furthermore, this indexing scheme is quite fast since it just need to look up index trees avoiding the complex computation. The k-plet local pattern is enrolled in k 3-level index trees, then a query only need to orderly index k trees and count the matched votes.

The rest of this paper is organized as follow: Sect. 2 introduces the k-plet local pattern representation while Sect. 3 describes the efficient indexing approach. In Sect. 4, experiments for testing the proposed algorithm are conducted on public datasets. Finally, Sect. 5 draws some conclusions.

2 K-Plet Local Pattern

In the field of fingerprint matching, an algorithm named as k-plet presents excellent performance, which consists a fixed-length vector by quantizing the distance and orientation difference between a central minutia and its neighboring minutiae, Fig. 1.

Local pattern for indexing fingerprints is characterized based on the k-plet to get the local structural information in fingerprints. The main idea of k-plet is to represent a central minutia by using the nearest k neighbors of the central minutia. To solve the problem that minutiae are clustered and to maintain high connectivity in fingerprint image, k/4 nearest neighbors are sequentially selected in each of the four quadrant in local coordinate system of minutia m_i. It is noteworthy that the resulting k neighbors need not necessarily be the k closest neighbors of minutia m_i. The k-plet representation is invariant under translation and rotation since it is defined with its own local coordinate system. The advantage of these k neighbors is that it still performs well in the situation where the fingerprints exist missing or spurious minutiae. In this study, we considered 8 neighbors empirically (*i.e.*, k = 8), Fig. 2(a).

Fig. 1. K-plet local patterns of minutiae defined in a fingerprint [10]

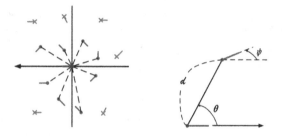

Fig. 2. (a). An example of k-plet (**b**). IFV formed by a neighbor minutia and central minutia

The k-plet consists of a central minutia m_i and other k neighborhood minutiae defined as $\{m_1, m_2, m_3, \ldots, m_k\}$. Each k-plet owns its local coordinate in which the central minutia m_i is the original point and the direction of the central minutia m_i is chosen as the positive direction of the X axis. Each neighboring minutia $m_j (j = 1, 2, \ldots, k)$ is defined with its local radial coordinates $(d_{ij}, \varphi_{ij}, \theta_{ij})$, where d_{ij} represents the Euclidean distance between minutia m_i and neighboring minutia m_j; φ_{ij} is the relative orientation of minutia m_j with respect to the central minutia m_i; θ_{ij} represents the direction of the edge connecting the two minutiae and θ_{ij} is also measured relative to orientation of minutia m_i. The translation-invariant and rotation-invariant vector $(d_{ij}, \varphi_{ij}, \theta_{ij})$ formed by minutia m_i and minutia m_j is represented as $A_j = (d_{ij}, \varphi_{ij}, \theta_{ij})$, named as *IFV* (Invariant Feature Vector), Fig. 2(b). For each k-plet, the k neighboring minutiae $\{m_1, m_2, m_3, \ldots, m_k\}$ and the central minutia m_i form k invariant feature vectors $\{A_1, A_2, A_3, \ldots, A_k\}$. Thus, the k-plet local pattern is defined in terms of $LP = \{A_1, A_2, A_3, \ldots, A_k\}$.

3 Indexing Approach

In the proposed approach, it consists of two stages, known as enrollment of template fingerprints T and indexing of query fingerprints I. Figure 3 illustrates a general flow chart of the proposed approach.

At the stage of enrollment, k index trees are constructed for storing the k-plet local patterns orderly and per tree has 3 levels representing d, φ and θ respectively. Finally,

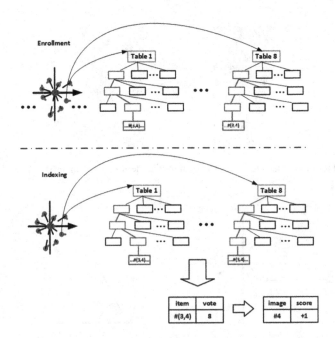

Fig. 3. The flow chart of the proposed indexing approach

the items # of *LP* ID and image ID is enrolled in the corresponding leaf bins. At the stage of indexing, the corresponding branches are selected. In the leaf bins, the items which are hit will be cast one vote. If votes are more than a certain threshold, the two *LP*s are regarded as matched successfully. According to the number of matched *LP*s, the similarity scores *S* between input fingerprint *I* and template fingerprint *T* can be acquired by Eq. (1).

$$S = \frac{\sum_m N_m}{N_i + N_t} \tag{1}$$

Where N_i is the number of *LP* in input fingerprint *I*, N_t is the number of LP in template fingerprint *T*, N_m is the number of matched *LP*s between input fingerprint *I* and template fingerprint *T*. Finally, the output result can be achieved by sorting all template fingerprints with the scores in descending order through the whole database.

3.1 Enrollment of Template Fingerprints

An off-line enrollment is made with the purpose of filling all *LPt* of template finger-prints into k index trees before indexing query fingerprints. Firstly, k duplicate index trees are constructed respectively. Each tree consists of 3 levels, which are presented by: d, φ, θ and there are numbers of fixed-length bins in every level. A $LP_t = (A_1^t, A_2^t, A_3^t, \ldots, A_k^t)$ need to be filled into the $j_{th}(j = 1, 2, \ldots, k)$ index tree depending on the jth invariant feature vector $A_j^t = (d_j^t, \varphi_j^t, \theta_j^t)$. That is, a *LPt* need to be stored totally k times. In the j_{th} index tree, the hash function (2), (3) and (4) at the corresponding level is used to obtain the sequence number n_j^d, n_j^φ and n_j^θ of the hit bin respectively.

$$n_j^d = H_d(d_j^t) = d_j^t/L_d + 1 \tag{2}$$

$$n_j^\varphi = H_\varphi(\varphi_j^t) = \varphi_j^t/L_\varphi + 1 \tag{3}$$

$$n_j^\theta = H_\theta(\theta_j^t) = \theta_j^t/L_\theta + 1 \tag{4}$$

Here, L_d, L_φ and L_θ is the length of one bin at d, φ and θ level and this operator / represents the quotient of the integer division. After the three levels browsed, along with the route in all levels, the item # of *LP* ID as well as fingerprint ID is registered into the last hit bin at the last level.

For example, suppose that we are going to fill the 3rd *LPt* in the 4th template fingerprint in database into the first index tree, which is instantiated as $LP_{3,4} = (A_1^t, A_2^t, A_3^t, \ldots, A_8^t)$, where the 1st invariant feature vector $A_1^t = (5, 8, 15)$. In addition, we assume that $L_d = L_\varphi = L_\theta = 10$. The enrollment process seems to be regarded as tracing the branches of the 3 levels in the 1st index tree. As $d = 5$, then $n_1^d = 5/10 + 1 = 1$, it need to be fell into the first bin of the first level, and all the rest branches should be traced along with $A_1^t = (5, 8, 15)$. At the end, the item #(3, 4) of *LPt* ID and fingerprint ID is enrolled into the second bin of the last level, Fig. 4.

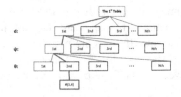

Fig. 4. An example of enrolling an *LPt* into the 1st hash table

Similarly, depending on the following $A_j^t = (j = 2, 3, \ldots, 8)$, the item $\#(3, 4)$ should be enrolled into the rest index trees by order (*i.e.*, the 2nd, 3rd,...,8th index tree). Summary of the off-line enrollment process is given in Algorithm 1.

```
Algorithm 1: Enrollment Algorithm
Input: the whole fingerprint database, DB
Output: K index trees
for all template fingerprints T in DB do
  for all minutiae m_t in T do
    Construct k-plet local patterns, get LPt
    for K neighboring minutiae in LPt do
      Select corresponding index tree by order
      for K index trees do
        Calculate sequence number n_d , n_φ , n_θ respectively
        Choose corresponding bin in different levels
        Register item # of LPt ID and image ID in leaf
      end for
    end for
  end for
end for
```

3.2   Indexing of a Query Fingerprint

Indexing of a query fingerprint aims to obtain the similar template fingerprints by looking up k established index trees. It is similar to the process of enrollment. A $LP_i = (A_1^i, A_2^i, A_3^i, \ldots, A_k^i)$ is need to be browsed into $j_{th}(j = 1, 2, \ldots, k)$ index tree depending on the jth invariant feature vector $A_j^i = (d_j^i, \varphi_j^i, \theta_j^i)$. Within the jth index tree, each sequence number $n_j^d, n_j^\varphi, n_j^\theta$ will be got by the hash function (2), (3), (4) separately. One bin in the last level will be searched after three levels' routing. All the template items filled in this bin will be recorded into a vote box, moreover the number of votes of these items will plus one. After searching all k index trees, the template items whose the number of votes is more than T will be treated as successfully matched with this *LPi*. According to the number of matched LPs, the similarity scores S between input fingerprint I and template fingerprint T can be calculated by Eq. (1). Finally, the output result can be achieved by sorting all template fingerprints with the scores in descending order through the whole database.

Summary of the on-line indexing process is shown in Algorithm 2.

```
Algorithm 2: Indexing Algorithm
Input: query image, I
Output: list of top N template fingerprints
for all minutiae mᵢ in I do
  Construct k-plet local patterns, get LPi
  for K neighboring minutiae in LPi do
    Select corresponding index tree by order
    for K index trees do
      Calculate sequence number n_d , n_φ , n_θ respectively

      Select corresponding bin in different levels
      Vote item # of LPi ID and image ID in leaf
    end for
  end for
  if votes of one LPi>T, then
    label LPi with LPt pairs
    for all labeled pairs do
      Calculate scores of query I and T by equation (1)
    end for
  end if
end for
Rearrange all templates in descending order with scores
Output list of top N candidates
```

3.3   Multipath Indexing Strategy

In practice, there are some inherent variations in the fingerprints, including translation, rotation, distortion and other noise. They get reflected in the indexing procedure and thus reduce the overall accuracy of the system. To account for errors that might be caused due to such variations, we implement the multipath techniques in our indexing scheme.

At the stage of indexing, we select the hit bin and also its brother bins according to (5), (6) and (7) at different levels.

$$|n_d - n_{d0}| \leq T_{nd} \tag{5}$$

$$|n_\varphi - n_{\varphi 0}| \leq T_{n\varphi} \tag{6}$$

$$|n_\theta - n_{\theta 0}| \leq T_{n\theta} \tag{7}$$

Where T_{nd}, $T_{n\varphi}$ and $T_{n\theta}$ are some certain thresholds. That is to say, if bin n_{d0} is hit at level d, all its brother bins between $(n_{d0} - T_{nd}, n_{d0} + T_{nd})$ are also selected. Finally, in this way, a LPi will search certain numbers of leaf bins so that it increasing the probability of indexing the correct item. Figure 5 gives a general example, where $T_{nd} = T_{n\varphi} = T_{n\theta} = 1$.

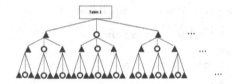

Fig. 5. An example of multipath indexing scheme. Circles are hit bins and triangles are their brother bins.

4 Experimental Results

This section describes experiments carried out to evaluate the proposed algorithm (*LP + Idx*) and to compare it with the algorithm (*OrigLP*) of original k-plet local pattern without indexing. As well, the time analysis of the proposed algorithm are given at the end.

4.1 Datasets

The experimental verification of *LP + Idx* and its comparison with *OrigLP* are performed based on three standard databases:

- FVC2002 DB1 database [11]: it contains 800 fingerprints, eight impressions for each of 100 distinct fingers, with the images of 388*374 pixels.
- NIST DB4 database [12]: it contains 4000 fingerprints, two different fingerprint instances (F and S) for each of 2000 distinct fingers, with the images of 512*512 pixels.
- NIST DB14 database [13]: we choose the first 2000 fingerprints, two impressions for each of 1000 distinct fingers, with the images of 832*768 pixels.

4.2 Evaluation Indicators and Setup

Typically, Correct Index Power (*CIP*) and Penetration Rate are adopted to evaluate the accuracy and efficiency. Here $CIP = N_{ci}/N_d$, where N_{ci} is the number of correctly retrieved input query fingerprints, N_d is the number of all input query fingerprints. Obtaining a higher *CIP* and a lower Penetration Rate is the common target of fingerprint indexing algorithm.

In the experiments, each database is equally divided into two parts: input fingerprints and template fingerprints. In FVC2002 DB1 database, the first 4 impressions of each individual finger are selected as the input fingerprints while the rest 4 impressions are regarded as the template fingerprints to build the index trees. Similarly, the first and second impressions of each individual finger in NIST DB4 and DB14 database are respectively chosen as input and template fingerprints.

For the feature extraction, an open investigation named as FM3 [14] is adopted to provide the reliable manual-marked minutiae in FVC2002 DB1 database. Meanwhile, a commercial automatic fingerprint identification system GAFIS [15] is used for the minutiae extraction in NIST DB4 and DB14, which is widely employed in Chinese criminal investigation departments.

4.3 Results and Discussions

Figures 6 and 7 show the trade-off between Penetration Rate and *CIP* in FVC 2002 DB1, NIST DB4 and NIST DB14 where k = 4 and 8, respectively. Table 1 compares the algorithm *LP + Idx* with *OrigLP* at some certain *CIP* in FVC 2002 DB1. In addition, Table 2 lists the time factors in FVC2002 DB1, NIST DB4 and NIST DB14.

From Figs. 6 and 7, it is noteworthy that the indexing performance in FVC2002 DB1, NIST DB4 and DB14 is quite outstanding. Moreover, it indicates the performance in FVC2002 DB1 looks much better than the one in NIST DB4 and DB14. It is mainly induced by two factors: (1) the image in database of FVC2002DB1 is more

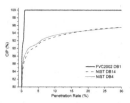

Fig. 6. Indexing performance in FVC 2002 DB1, NIST DB4 and NIST DB14, where k = 4

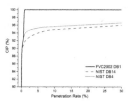

Fig. 7. Indexing performance in FVC 2002 DB1, NIST DB4 and NIST DB14, where k = 8

Table 1. Indexing performance of *LP + Idx* and *OrigLP* at certain *CIP* in FVC 2002 DB1

Algorithms \ Penetration Rate \ CIP	99%	97%	95%	90%
OrigLP	*0.84%*	*0.77%*	*0.71%*	*0.58%*
LP+Idx(k=8)	*0.85%*	*0.78%*	*0.71%*	*0.58%*
LP+Idx(k=4)	*0.86%*	*0.78%*	*0.71%*	*0.58%*

Table 2. Indexing time in FVC2002 DB1, NIST DB4 and NIST DB14

Algorithms \ Time \ Database	FVC02DB1	NIST DB4	NIST DB14
OrigLP	0.159ms	1.106ms	1.540ms
LP+Idx(k=8)	0.004ms	0.017ms	0.022ms
LP+Idx(k=4)	0.003ms	0.011ms	0.014ms

clear and smaller than the one in NIST DB4 and DB14. (2) Comparing with the only one template fingerprint for each individual input fingerprint in the database of NIST DB4 and DB14, there are four template fingerprints for each one in the database of FVC2002 DB1, which largely increases the probability of matching one input query fingerprint successfully.

In addition, with the increasing of the number of neighboring minutiae in *LP*, the discrimination of *LP* becomes stronger than before. Indeed, it achieves higher correct index power with a lower penetration rate. However, it will slow down the runtime and consume more memory resource. Consequently, we consider 8 neighbors empirically to achieve a balance between the two aspects (i.e., k = 8).

Furthermore, Table 1 compares the algorithm *LP + Idx* with *OrigLP* at some certain *CIP* in FVC 2002 DB1. The comparison demonstrates that the performance of *LP + Idx* is almost the same to *OrigLP* and verifies the robustness of *LP + Idx*.

For real-time fingerprint recognition system, time is another critical factor related to the efficiency. Consequently, the time factors in FVC2002 DB1, NIST DB4 and NIST DB14 are investigated and listed in Table 2. The time tests running on 2.40 GHz Intel Core CPU are implemented in C++ without particular optimizations. It is obvious to reveal that proposed algorithm *LP+ Idx* is very fast and efficient since these time factors are much faster than *OrigLP*.

5 Conclusion

In summary, we creatively adopt the k-plet local pattern for fingerprint indexing, which is translation-invariant and rotation-invariant. Then, according to these local patterns, k efficient index trees are constructed based on enrollment and indexing stages. The multipath indexing strategy ensures the accuracy and robustness while simple look-up operation makes it quite fast and effective. The performance testing proves that the proposed algorithm is beneficial for fingerprint indexing as it achieves high correct index power with a low penetration rate.

As our future focus, a completely hierarchical fingerprint indexing with other effective features, is expected to develop for large scale fingerprint database.

Acknowledgments. This work was funded by the Chinese National Natural Science Foundation (11331012, 71271204, 11101420)

References

1. Maltoni, D., Maio, D., Jain, A.K., Prabhakar, S.: Handbook of Fingerprint Recognition. Springer, New York (2009)
2. Henry, E.R.: Classification and Uses of Finger Prints. HM Stationery Office, London (1905)
3. Liu, T., Zhu, G., Zhang, C., Hao, P.: Fingerprint indexing based on singular point correlation. In: IEEE International Conference on Image Processing, vol. 3, pp. II-293–6 (2005)
4. Liu, M., Jiang, X., Kot, A.: Fingerprint retrieval by complex filter responses, In: 18th IEEE International Conference on Pattern Recognition, vol. 1, p. 1042 (2006)
5. Wang, Y., Hu, J., Phillips, D.: A fingerprint orientation model based on 2D fourier expansion (FOMFE) and its application to singular-point detection and fingerprint indexing. IEEE Trans. Pattern Anal. Mach. Intell. **29**, 573–585 (2007)
6. Liu, M., Jiang, X., Kot, A.C.: Efficient fingerprint search based on database clustering. Pattern Recogn. **40**(6), 1793–1803 (2007)
7. Bhanu, B., Tan, X.: Fingerprint indexing based on novel features of minutiae triplets. IEEE Trans. Pattern Anal. Mach. Intell. **25**, 616–622 (2003)
8. Iloanusi, O., Gyaourova, A., Ross, A : Indexing fingerprints using minutiae quadruplets. In: IEEE Conference on Computer Vision and Pattern Recognition Workshops, pp. 127–133 (2011)
9. Cappelli, R., Ferrara, M., Maltoni, D.: Fingerprint indexing based on minutia cylinder-code. IEEE Trans. Pattern Anal. Mach. Intell. **33**, 1051–1057 (2011)
10. Chikkerur, S., Cartwright, A.N., Govindaraju, V.: K-plet and coupled BFS: a graph based fingerprint representation and matching algorithm. In: Zhang, D., Jain, A.K. (eds.) ICB 2005. LNCS, vol. 3832, pp. 309–315. Springer, Heidelberg (2005)
11. Maio, D., Maltoni, D., Cappelli, R., Wayman, J.L., Jain, A.K.: FVC2002: second fingerprint verification competition. IEEE Int. Conf. Pattern Recogn. **3**, 811–814 (2002)
12. Watson, C., Wilson, C.: Nist Special Database 4. Fingerprint Database, National Institute of Standards and Technology (1992)
13. Watson, C.: Nist special database 14. Fingerprint Database, National Institute of Standards and Technology (1993)
14. Kayaoglu, M., Topcu, B., Uludag, U.: Standard fingerprint databases: Manual minutiae labeling and matcher performance analyses, CoRR, vol.3, abs/1305.1443 (2013)
15. GAFIS7.0, 2012. http://www.etgoldenfinger.com/

Color Characterization Comparison
for Machine Vision-Based Fruit Recognition

Farid García-Lamont[1(✉)], Jair Cervantes[1], Sergio Ruiz[1],
and Asdrúbal López-Chau[2]

[1] Centro Universitario UAEM Texcoco,
Universidad Autónoma del Estado de México, Av. Jardín Zumpango s/n,
Fraccionamiento El Tejocote, 56259 Texcoco-Estado de México, Mexico
fgarcial@uaemex.mx, {chazarral7,jsergioruizc}@gmail.com
[2] Centro Universitario UAEM Zumpango,
Universidad Autónoma del Estado de México,
Camino viejo a Jilotzingo continuación Calle Rayón,
55600 Zumpango-Estado de México, Mexico
alchau@uaemex.mx

Abstract. In this paper we present a comparison between three color characterizations methods applied for fruit recognition, two of them are selected from two related works and the third is the authors' proposal; in the three works, color is represented in the RGB space. The related works characterize the colors considering their intensity data; but employing the intensity data of colors in the RGB space may lead to obtain imprecise models of colors, because, in this space, despite two colors with the same chromaticity if they have different intensities then they represent different colors. Hence, we introduce a method to characterize the color of objects by extracting the chromaticity of colors; so, the intensity of colors does not influence significantly the color extraction. The color characterizations of these two methods and our proposal are implemented and tested to extract the color features of different fruit classes. The color features are concatenated with the shape characteristics, obtained using Fourier descriptors, Hu moments and four basic geometric features, to form a feature vector. A feed-forward neural network is employed as classifier; the performance of each method is evaluated using an image database with 12 fruit classes.

Keywords: Color characterization · Fruit classification · RGB images

1 Introduction

The automated recognition of different classes of fruits using artificial vision, has been a few studied. Related works have focus mainly on to evaluate the ripeness or quality of fruits [1–4] or to recognize varieties of the same fruit class [5–8]; that is, they do not recognize different classes of fruits. Reviewing the state of the art, we have found the proposal presented in references [9–11], where fruit recognition is performed by extracting color, shape and texture features, but texture is not always employed. Color is an important feature for fruit recognition, especially when the shapes of fruits are very similar. For instance, the shapes of tomato and orange are almost the same;

© Springer International Publishing Switzerland 2015
D.-S. Huang et al. (Eds.): ICIC 2015, Part I, LNCS 9225, pp. 258–270, 2015.
DOI: 10.1007/978-3-319-22180-9_26

Fig. 1. Tomatoes (a) and (b) with the same chromaticity but different intensity, onion (c) with different chromaticity but with the same intensity as the tomato (b)

however, the color feature becomes a discriminative data to classify the fruits successfully.

About color characterization, in reference [9] the color is extracted by discretizing each color channel in 4 intensity levels, then a histogram is computed where the occurrences of the $4 \times 4 \times 4 = 64$ possible colors are counted. In [10] the color is modeled by computing the mean of each color channel; these values are concatenated to form the feature vector of color. Betul et al. [11] characterize the color with fuzzy logic, where the inference system models the colors employing the intensity values of the color channels. In these works the feature vector of the fruit is obtained by concatenating the color and shape features; in [9] texture features are also included.

The drawback with the color extraction of these three approaches is that they depend on the intensity of colors. For instance, tomatoes (a) and (b) shown in Fig. 1 have the same red hue, but their intensities are different. Humans can associate this red hue to tomatoes, independently of the intensity; but in the RGB space, because of the intensity difference, they represent different colors. On the other hand, the onion (c) can be recognized due to its characteristic red hue, despite the intensity of the onion is similar to tomato (b).

The contribution of this paper consists on to show that fruit recognition can be more precise if color chromaticity is employed to model the color of fruits. Fruits have a characteristic color; due to the fruits can have different levels of ripeness, the intensity is different between fruits of the same class but the chromaticity changes just a little. With our approach it is possible to maintain the data of chromaticity despite intensity changes.

The rest of the paper is organized as follows: in Sect. 2 the methods for color and shape characterization are introduced, where our proposal is presented. Experiments with a fruit image database are performed in Sect. 3. In Sect. 4 the results are discussed, conclusions and future work in Sect. 5 close the paper.

2   Feature Extraction

In this section we present the methods employed to extract color and shape features of fruits. Figure 2 shows the steps we propose to characterize and recognize fruits.

First, the image of the fruit is acquired in the RGB space and then the image is mapped to the HSV space. It is important to remark that the background of the acquired images must be white, and in the scene there is just one kind of fruit. By using data about the saturation of colors, the fruit is segmented from the background. After

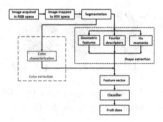

Fig. 2. Flow chart of the proposed approach

segmentation, the color and shape features are extracted. In Sect. 2.1 we explain our proposal for color characterization; in Sect. 2.2 the Fourier descriptors, Hu moments and basic geometric feature extraction methods are explained for shape characterization. The color and shape characteristics are concatenated to form a feature vector, which is feeded to a classifier and finally we obtain the class of the fruit.

2.1   Color Characterization

Before we introduce our proposal, it is convenient to give a short explanation of color representation in the RGB and HSV spaces. The RGB (Red, Green, Blue) space is based in a Cartesian coordinate system where colors are points defined by vectors that extend from the origin, where black is located in the origin and white in the opposite corner to the origin, see Fig. 3.

The color of a pixel p is written as a linear combination of the basis vectors red, green and blue, that is:

$$\phi_p = r_p \hat{i} + g_p \hat{j} + b_p \hat{k} \tag{1}$$

Where r_p, g_p and b_p are the red, green and blue components, respectively. The orientation and magnitude of a color vector defines the chromaticity and the intensity of the color, respectively [12].

The HSV space is cone shaped, see Fig. 4, the representation of the color of a pixel p in the HSV space is written as [13]:

$$\varphi_p = \left[h_p, s_p, v_p \right] \tag{2}$$

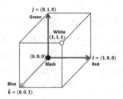

Fig. 3. RGB color space

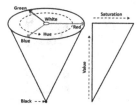

Fig. 4. HSV color space

Where h_p, s_p and v_p are the hue, saturation and value components, respectively. The hue is the chromaticity, saturation is the distance to the glow axis of black-white, and value is the intensity. The real ranges of the hue, saturation and value components are [0, 2π], [0, 1] and [0, 255], respectively.

As mentioned previously, we assume that in the image there is one kind of fruit and the background is white. The acquired image is mapped to the HSV space, the pixel's intensity is set to zero if its saturation is lower than a specific threshold.

As shown in Fig. 4, the white color is located in the axis of the cone; that is, white is a low saturated color. Thus, the color of the pixels with low saturation is white; therefore they are part of the image background.

The segmentation process is not performed in the RGB space because the computational load may be huge; to establish if a color vector represents white color, its orientation must be computed and compared to a specific reference value. By comparing the white color by its saturation, the computational load lies just in the image mapping from RGB space to HSV space; however, the computational load is low. The color extraction we propose consists on the following steps:

1. Let $\{\phi_1, ..., \phi_m\} \subset \mathbf{R}^3$ be the set of color vectors of the image's pixels, represented in the RGB space. The color vectors are mapped to the HSV space and we obtain the set of vector $\{\varphi_1, ..., \varphi_m\} \subset \mathbf{R}^3$. The segmentation is performed by comparing the saturation of the colors. If the saturation is lower than a threshold, then the color vector of the corresponding pixel in the RGB space is set to zero. In other words:

$$\phi_p^* = \begin{cases} \phi_p, & s_p \geq th \\ \vec{0} & s_p < th \end{cases} \qquad (3)$$

Where th is the threshold value and $\vec{0} = [0, 0, 0]$. In this step the fruit is segmented from the background. In this study $th = 0.2$. Figure 5 shows the images of two fruits segmented from the background using this segmentation approach.

2. The set S is built as follows:

$$S = \left\{ \phi_p^* | \phi_p^* \neq \vec{0} \right\} \qquad (4)$$

Fig. 5. Examples of segmented images: (a) and (c) input images, (b) and (d) images obtained after segmentation

3. The vectors of set S are sum, i.e.,

$$R = \sum_{\phi_p^* \in S} \phi_p^* \tag{5}$$

4. The vector R includes data of all the colors of the object, its orientation is the same or almost the same of the color vectors of the dominant color. It also includes data of the other colors of the fruit. The magnitude of vector R is very large; however, what we are interested on is the orientation of vector R because it provides data about the chromaticity of the color obtained from the fruit. Hence, the vector is normalized with:

$$u_R = \frac{R}{\|R\|} = [r_u, g_u, b_u] \tag{6}$$

Thus, the vector u_R characterizes the color of the fruit; the direction cosines of this vector are $\cos \alpha_R = r_u/\|u_R\|$, $\cos \beta_R = g_u/\|u_R\|$ and $\cos \theta_R = b_u/\|u_R\|$. But $\|u_R\| = 1$, therefore, the components of the vector u_R are the cosines of the angles between the vector and the basis vectors. Hence, the orientation of R is implicit in u_R. Therefore, the vector u_R characterizes the fruit's color by its chromaticity and the effect of intensity is reduced.

2.2 Shape Features

The shape features are characterized using Fourier descriptors, Hu moments and geometrical features. In order to compute the shape features we build the set $O = \{(x_0, y_0), \ldots, (x_{n-1}, y_{n-1})\}$, whose elements are the coordinates of the pixels of the segmented area. The shape features are computed as follows.

Fourier Descriptors: The Fourier descriptors compute the set of points of the object's contour as a sequence of complex numbers. With this data is built a periodic unidimensional function f that describes the object's contour as a sequence of complex numbers; the coefficients of the Fourier transform of function R characterizes the contour of an object in the frequency domain [14]. Let $C \subset O$ be the set whose elements are the coordinates of the contour's pixels of the segmented area. The coordinates in $C = \{(x_0, y_0), \ldots, (x_{M-1}, y_{M-1})\}$ are indexed according to the sequence they are located in the contour at a specific direction and starting at an arbitrary point. Thus, being $i = \sqrt{-1}$ and $\mathbf{C}$ the set of complex numbers, let $f: C \rightarrow \mathbf{C}$ be a function defined as follows:

$$f((x_k, y_k)) = x_k + iy_k \tag{7}$$

The Fourier transform of function f is computed with:

$$F(u) = \frac{1}{M} \sum_{k=0}^{M-1} f((x_k, y_k)) \exp\left(\frac{-i2\pi uk}{M}\right) \tag{8}$$

The complex coefficients $F(u)$ are known as the contour's Fourier descriptors. In this work we employ eight coefficients, that is, $F(u)$ is computed for $u = 0, 1, \ldots, 7$.

Hu moments: Hu moments compute the features of object's shape, considering the shape as a distribution of coordinates of 2D points [15, 16]. It is based in the central moments of the coordinates' distribution, which express characteristic parameters respect to a centroid point. Thus, let O be the set defined previously, the Hu moments are computed with:

$$
\begin{aligned}
H_1 &= \eta_{20} + \eta_{02} \\
H_2 &= (\eta_{20} - \eta_{02})^2 + 4\eta_1^2 \\
H_3 &= (\eta_{30} - 3\eta_{12})^2 + (3\eta_{21} - \eta_{03})^2 \\
H_3 &= (\eta_{30} - 3\eta_{12})^2 + (3\eta_{21} - \eta_{03})^2 \\
H_4 &= (\eta_{30} + \eta_{12})^2 + (\eta_{21} + \eta_{03})^2 \\
H_5 &= (\eta_{30} - 3\eta_{12})(\eta_{30} + \eta_{12})\left[(\eta_{30} + \eta_{12})^2 - 3(\eta_{21} + \eta_{03})^2\right] \\
&\quad + (3\eta_{21} - \eta_{03})(\eta_{21} + \eta_{03})\left[3(\eta_{30} + \eta_{12})^2 - (\eta_{21} + \eta_{03})^2\right] \\
H_6 &= (\eta_{20} - \eta_{02})\left[(\eta_{30} + \eta_{12})^2 - (\eta_{21} + \eta_{03})^2\right] + 4\eta_{11}(\eta_{30} + \eta_{12})(\eta_{21} + \eta_{03}) \\
H_7 &= (3\eta_{21} - \eta_{03})(\eta_{30} + \eta_{12})\left[(\eta_{30} + \eta_{12})^2 - 3(\eta_{21} + \eta_{03})^2\right] \\
&\quad - (\eta_{30} - 3\eta_{12})(\eta_{21} + \eta_{03})\left[3(\eta_{30} + \eta_{12})^2 - (\eta_{21} + \eta_{03})^2\right]
\end{aligned}
\tag{9}
$$

The centralized and normalized moments η_{pq} are obtained with:

$$\eta_{pq} = \frac{\mu_{pq}}{\mu_{00}^{c+1}} \tag{10}$$

Where $c = (p + q)/2$. The centralized moments are computed with:

$$\mu_{pq}(O) = \sum_{(x,y)\in O} (x - \bar{x})^p (y - \bar{y})^q \tag{11}$$

Where $\bar{x}$ and $\bar{y}$ are the mean values of the x and y values of coordinates of set O.

Basic geometric features: In this work we extract four features: eccentricity, solidity, compaction and roundness. Eccentricity is the ratio of the distance between the foci of the ellipse and its major axis length. Solidity is the proportion of the pixels in the object that are also in the convex hull.

Compaction is the relation between the area of an object and its perimeter:

$$Cp(O) = \frac{n}{P(O)} \tag{12}$$

Roundness is computed with:

$$R(O) = \frac{4\pi n}{P^2(O)} \tag{13}$$

Where $P(O)$ is the perimeter of object O. All these features are concatenated to form the feature vector whose dimension is 22. It is important to mention that all the shape features extracted are invariant to rotation, translation and scale.

3   Experiments and Results

The experimental set up consists on to classify twelve classes of fruits: red onion, green chili, corn, melon, orange, potato, pear, cucumber, pineapple, banana, tomato and carrot; hand labeled as classes 1, 2,...,12, respectively. Figure 6 shows images of the

Fig. 6. Samples of the twelve fruit classes employed for experiments

fruit classes recognized in this study. In order to appreciate easily the influence of the color characterizations, we select fruits with similar appearance. For instance, red onion and tomato, green chili and cucumber, melon and potato, banana and corn.

The image database employed has 720 images, 60 per fruit class; half of the image database is employed for training and the other half of the image database is used for testing. The image database was built by downloading images from internet and by photos acquired using a digital camera. The images were acquired such that in the scene there is only one fruit and the color of background is white. The fruit is acquired in the same position and angle in order to ease the shape feature extraction due to what we are interested on is the color characterization. The size of the images varies from 160×159 to 1000×965 pixels.

A feed-forward neural network (NN) is employed to recognize the fruits. The NN has three layers; one input layer, one hidden layer and one output layer. The first and second layers have 25 neurons and the third layer has 12 neurons. The activation function for the first and second layers is hyperbolic tangent sigmoid, while the activation function of the third layer is log-sigmoid. Table 1 shows the recognition rates obtained with our proposal and the methods presented in references [9, 10].

The highest recognition rate is obtained with our proposal; it is important to know between which fruit classes there are misclassifications, so as to analyze which color characterization works better. Thus, we present the confusion matrixes obtained with each of the three color characterizations implemented in this study. Table 2 shows the confusion matrix using the proposal of Zhang et al. [9].

The fruits best classified are red onion, melon, orange, potato, cucumber and tomato; the fruit worst classified is pineapple.

Table 1. Recognition rates obtained using the approaches presented in references [9, 10], and by the authors

Fruit class	Recognition rate		
	Zhang et al. [9]	Chaw and Hadi [10]	*Authors' proposal*
Red onion	93.34	96.67	*100*
Green chili	86.67	83.33	*86.67*
Corn	86.67	90	*93.34*
Melon	93.34	90	*93.34*
Orange	93.34	93.34	*93.34*
Potato	93.34	80.00	*93.34*
Cucumber	86.67	90.00	*90.00*
Pear	93.34	83.34	*96.67*
Pineapple	80.00	73.34	*86.67*
Banana	90.00	96.67	*96.67*
Tomato	93.34	93.34	*100*
Carrot	90.00	86.67	*96.67*
Average	90.01	88.06	*93.89*

Table 2. Confusion matrix obtained employing the proposal of reference [9]

Class						Target class						
	1	2	3	4	5	6	7	8	9	10	11	12
1	28	0	0	0	0	0	0	0	0	0	2	0
2	0	26	0	0	0	0	3	1	2	0	0	0
3	0	0	26	0	0	0	0	0	0	2	0	1
4	0	0	0	28	2	1	0	0	0	0	0	0
5	0	0	0	0	28	1	0	0	0	0	0	0
6	0	0	0	2	0	28	0	0	0	0	0	0
7	0	2	0	0	0	0	26	0	0	0	0	0
8	0	2	0	0	0	0	0	28	3	0	0	0
9	0	0	0	0	0	0	1	1	24	0	0	0
10	0	0	0	0	0	0	0	0	1	27	0	2
11	2	0	0	0	0	0	0	0	0	0	28	0
12	0	0	4	0	0	0	0	0	0	1	0	27

(Output class — left vertical label)

Table 3 shows the confusion matrix employing the color characterization used by Chaw and Hadi [10]. The fruits best classified are red onion and banana; the worst classified fruit is pineapple.

Table 4 shows the confusion matrix using the color characterization proposed by the authors.

The fruits best classified are red onion and tomato, the fruits worst classified are green chili and pineapple. In the following Sect. 4 we analyze and discuss the results shown in the current section.

Table 3. Confusion matrix obtained using the approach of reference [10]

Class						Target class						
	1	2	3	4	5	6	7	8	9	10	11	12
1	29	0	0	0	0	0	0	0	0	0	2	0
2	0	25	0	0	0	0	2	3	2	0	0	0
3	0	0	27	0	0	0	0	0	0	1	0	3
4	0	0	0	27	2	3	0	0	0	0	0	0
5	0	0	0	1	28	2	0	0	0	0	0	0
6	0	0	0	1	0	24	0	0	2	0	0	0
7	0	3	0	0	0	0	27	0	0	0	0	0
8	0	2	0	0	0	0	0	25	3	0	0	0
9	0	0	0	0	0	1	0	2	22	0	0	1
10	0	0	0	0	0	0	0	0	1	29	0	0
11	1	0	0	1	0	0	0	0	0	0	28	0
12	0	0	3	0	0	0	1	0	0	0	0	26

(Output class — left vertical label)

Table 4. Confusion matrix obtained using the authors' proposal

Class		Target class										
		1	2	3	4	5	6	7	8	9	10	11	12
Output class	1	30	0	0	0	0	0	0	0	0	0	0	0
	2	0	26	0	0	0	0	3	1	0	0	0	0
	3	0	0	28	0	0	0	0	0	0	1	0	1
	4	0	0	0	28	2	1	0	0	0	0	0	0
	5	0	0	0	2	28	1	0	0	0	0	0	0
	6	0	0	0	0	0	28	0	0	0	0	0	0
	7	0	0	0	0	0	0	27	0	0	0	0	0
	8	0	4	0	0	0	0	0	29	4	0	0	0
	9	0	0	0	0	0	0	0	0	26	0	0	0
	10	0	0	0	0	0	0	0	0	0	29	0	0
	11	0	0	0	0	0	0	0	0	0	0	30	0
	12	0	0	2	0	0	0	0	0	0	0	0	29

4 Discussion

According to the results of Table 1 the highest recognition rate is 93.89 %, which is obtained with our proposal; the second and third recognition rates are 90.01 % and 88.06 %, obtained with the methods presented in references [9, 10], respectively. The fruit worst classified by the three methods is pineapple. The cause may be due to the shape of pineapples is more complex than the other fruits, thus, it may be necessary to extract more shape features. Red onion is the best classified fruit by the three methods.

As stated in Sect. 1, color and shape features are extracted for fruit recognition. Because of the shape features, some fruits are misclassified, not by the color characterization; for instance pineapples and carrots. Hence, it is necessary to analyze the confusion matrixes built with the data obtained with the three methods.

From the results of Tables 2, 3 and 4, most of the red onions are recognized correctly, the misclassified onions are recognized as tomatoes; except with our proposal, where all the onions are recognized successfully.

The misclassified chilies are recognized as pears or cucumbers. With our method, all the chili misclassifications are assigned to cucumbers because the shape and chromaticity of both fruits are akin. With the other methods, the chilies misclassified as pears occur because the intensity of these chilies is as large as the pears. In other words, by employ intensity data of color, some chilies have the same or almost the same intensity of pears color; thus, these chilies are misclassified as pear.

With the three methods, the corn misclassifications are assigned to the carrot class. A plausible explanation is that, besides more shape features are needed, the color combination of carrots is a yellow-like hue; due to carrots usually have green and orange hues.

Misclassifications of melon class are with potato and orange. The color between potato and orange may be similar; the color of melons tends to the color of oranges, depending on the ripeness of the fruit.

Similarly to the melon class case, the orange class is misclassified melon. The color of the misclassified oranges is akin to the color of melon. The misclassification happens not only because of the color feature, but also the shape of both fruits is almost the same. To overcome the misclassification it may be necessary to extract more shape features or to obtain texture features.

Potato class is misclassified as melon and/or orange because the color of the misclassified potatoes is similar to some melons and oranges, not to mention that the shape is alike, to some extent.

With the methods of references [9, 10], some pears are misclassified as green chili and pineapple. As explained before, the color extracted from these misclassified pears has almost the same intensity of the chili color. Misclassifications with pineapples occur because pineapples have some parts with green color.

Cucumber is misclassified with green chili and pineapple by the methods of [9, 10], because the color and shape of both chili and cucumber are resembled; while cucumber is misclassified as pineapple due to pineapples have green hues. With our method, only one cucumber is classified as green chili.

Pineapple is classified as cucumber and banana; despite the complex shape of pineapple, due to the green and yellow hues and brightness of these colors, the pineapples are misclassified as cucumber and banana, respectively.

Banana is classified as corn and carrot, in this case the shape is not characterized correctly, so, the color of the fruit is the discriminant data for classification. In this case, yellow is the characteristic color of both banana and corn; thus, corn is misclassified as banana.

Tomato is misclassified as red onion; the brightness of those misclassified tomatoes is as low as the red onions, thus, they are classified as if they were the same fruit. Note that with our proposal all the tomatoes are classified successfully.

Carrot is classified as corn and banana, the color of carrots is a combination between orange and green, so, the resulting color is a color with yellow hue, to some extent. Therefore, carrot is misclassified as corn or banana.

It is important to remark that the dimension of the color feature vectors computed using the proposal of Zhang et al. [9] and ours is 64 and 3, respectively, The color features are concatenated with the 19 shape characteristics extracted in this work; thus, the dimension of the feature vector using the approach of reference [9] and ours is 83 and 22, respectively. Although the dimensionality of the feature vector can be performed using principal component analysis, computing the color feature vector as proposed in [9] can be a huge computational load.

The dimension of the color feature vector obtained with our proposal the approach presented in [10] is the same. But, as mentioned before, our proposal works better for fruit recognition because we employ the chromaticity data of fruits. While in [10] include the intensity data of colors; but in the RGB space despite the colors with the same chromaticity and with different intensities are classified as different colors.

5 Conclusions and Future Work

In this work we have introduced an approach to characterize color by its chromaticity in the RGB space, applied for fruit recognition. The color features extracted with our proposal are concatenated with the shape features extracted using Fourier descriptors, Hu moments and basic geometric features to form a feature vector that characterizes the fruits. Our approach is compared with two methods of color characterization for fruit recognition of related works, which employ intensity data of colors. The three methods are tested with an image database containing 720 fruit images of 12 fruit classes. According to results, our proposal obtained the highest recognition rate, 93.89 %.

The current approach is planned to be applied on supermarkets, to help the cashiers to identify the different fruit classes; fruits do not have bar codes print on them that eases the automated classification and for cashiers it is difficult to memorize all the fruit codes. Hence, the future work is to strength the performance of this proposal by improving or adding the following points. Include intensity data of colors as a feature separated from the chromaticity. The HSV color space can be employed for this purpose because, in this space, the intensity is decoupled from the chromaticity.

To improve the accuracy by collecting more fruit images per class, acquired at different angles, positions and sizes. Increase the amount of fruit classes to recognize. Include more shape features, such as, increase the number of Fourier descriptors and/or basic geometric features. In this work we do not extract texture features; this is an important characteristic [17, 18] of fruits that will be employed in future studies.

References

1. Bhatt, A., Pant, D.: Automatic apple grading model development based on back propagation neural network and machine vision, and its performance evaluation. AI Soc. **30**(1), 45–56 (2015)
2. Zhang, B., Huang, W., Li, W., Zhao, J., Fan, S., Wu, J., Liu, C.: Principles, developments and applications of computer vision for external quality inspection of fruits and vegetables: a review. Food Res. Int. **62**, 326–343 (2014)
3. Sekhar, N.C., Tudu, B., Koley, C.: A machine vision-based maturity prediction system for sorting of harvested mangoes. IEEE Trans. Instrum. Meas. **63**(7), 1722–1730 (2014)
4. Rodríguez-Pulido, F.J., Gordillo, B., González-Miret, M.L., Heredia, F.J.: Analysis of food appearance properties by computer vision applying ellipsoids to colour data. Comput. Electron. Agric. **99**, 108–115 (2013)
5. van Henten, E.J., Hemming, J., Van Tuijl, B.A.J., Kornet, J.G., Meuleman, J., Bontsema, J., Van Os, E.A.: An autonomous robot for harvesting cucumbers in greenhouses. Auton. Robot. **13**(3), 241–258 (2002)
6. Manickavasagan, A., Al-Mezeini, N., Al-Shekaili, H.: RGB color imaging technique for grading of dates. Sci. Hortic. **175**, 87–94 (2014)
7. Chen, X., Yang, S.: A practical solution for ripe tomato recognition and localization. J. Real-Time Image Process. **8**(1), 35–51 (2013)
8. Gatica, G., Best, S., Ceroni, J., Lefranc, G.: Olive fruits recognition using neural networks. Proc. Comput. Sci. **17**, 412–419 (2013)

9. Zhang, Y., Wang, S., Ji, G., Phillips, P.: Fruit classification using computer vision and feedforward neural network. J. Food Eng. **143**, 167–177 (2014)
10. Chaw, S.W., Hadi, M.S.: A new method for fruits recognition system. In: International Conference on Electrical Engineering and Information, pp. 130–134 (2009)
11. Bostanci, B., Hagras, H., Dooley, J.: A neuro fuzzy embedded agent approach towards the development of an intelligent refrigerator. In: IEEE International Conference on Fuzzy Systems, pp. 1–8 (2013)
12. Gonzalez, R.C., Woods, R.E.: Digital Image Processing, 2nd edn. Prentice Hall, Upper Saddle River (2002)
13. Rotaru, C., Graf, T., Zhang, J.: Color image segmentation in HSI space for automotive applications. J. Real-Time Image Process. **3**(4), 311–322 (2008)
14. Zahn, C., Roskies, R.: Fourier descriptors for plane closed curves. IEEE Trans. Comput. **C-21**(3), 269–284 (1972)
15. Hu, M.K.: Visual pattern recognition by moment invariants. IRE Trans. Inf. Theory **8**(2), 179–187 (1962)
16. Wang, X., Huang, D.S., Du, J.X., Xu, H., Heutte, L.: Classification of plant leaf images with complicated background. Appl. Math. Comput. **205**(2), 916–926 (2008)
17. Yuan, J.H., Huang, D.S., Zhu, H.D., Gan, Y.: Completed hybrid local binary pattern for texture classification. In: International Conference on Neural Networks, pp. 2050–2057 (2014)
18. Zhao, Y., Huang, D.S., Jia, W.: Completed local binary count for rotation invariant texture classification. IEEE Trans. Image Process. **21**(10), 4492–4497 (2012)

Real-Time Human Action Recognition Using DMMs-Based LBP and EOH Features

Mohammad Farhad Bulbul, Yunsheng Jiang, and Jinwen Ma[✉]

Department of Information Science,
School of Mathematical Sciences and LMAM,
Peking University, Beijing 100871, China
jwma@math.pku.edu.cn

Abstract. This paper proposes a new feature extraction scheme for the real-time human action recognition from depth video sequences. First, three Depth Motion Maps (DMMs) are formed from the depth video. Then, on top of these DMMs, the Local Binary Patterns (LBPs) are calculated within overlapping blocks to capture the local texture information, and the Edge Oriented Histograms (EOHs) are computed within non-overlapping blocks to extract dense shape features. Finally, to increase the discriminatory power, the DMMs-based LBP and EOH features are fused in a systematic way to get the so-called **DLE features**. The proposed DLE features are then fed into the l_2-regularized Collaborative Representation Classifier (l_2-CRC) to learn the model of human action. Experimental results on the publicly available Microsoft Research Action3D dataset demonstrate that the proposed approach achieves the state-of-the-art recognition performance without compromising the processing speed for all the key steps, and thus shows the suitability for real-time implementation.

Keywords: Human action recognition · Depth Motion Maps · Local Binary patterns · Edge Oriented Histograms

1 Introduction

In the area of computer vision, human action recognition is a process of detecting and labeling the humans in the video, and it has drawn much attention to researchers due to the growing demands from vast applications, such as visual surveillance, video retrieval, health monitoring, fitness training, human-computer interaction and so on (*e.g.*, [1, 2]). In the past decade, research has mainly focused on learning and recognizing actions from image sequences captured by traditional RGB video cameras [3, 4]. However, these types of data source have some inherent limitations. For example, they are sensitive to color and illumination changes, occlusions, and background clutters.

With the recent release of the cost-effective depth sensors, such as Microsoft Kinect, numerous research works on human action recognition have been carried out based on the depth maps. An example of depth map sequence is shown in Fig. 1. Compared with conventional RGB cameras, the depth camera has many advantages. For instance, the outputs of depth cameras are insensitive to changes in lighting

© Springer International Publishing Switzerland 2015
D.-S. Huang et al. (Eds.): ICIC 2015, Part I, LNCS 9225, pp. 271–282, 2015.
DOI: 10.1007/978-3-319-22180-9_27

conditions. Moreover, depth maps can provide 3D structural information for distinguishing human actions, which is difficult to characterize by using RGB video sequences. On the other hand, human skeleton information can be extracted from depth maps [5]. Specifically, 3D positions and rotation angles of the body joints can be estimated by using the Kinect Windows SDK [6].

In this paper, we present a computationally efficient and effective human action

Fig. 1. A depth map sequence for the *High throw* action

recognition method by utilizing Depth Motion Maps (DMMs) based Local Binary Patterns (LBPs) and Edge Oriented Histograms (EOHs) to the l_2-regularized Collaborative Representation Classifier (l_2-CRC). Specifically, first, for each depth video, all its video frames are projected onto three orthogonal Cartesian planes to generate the projected maps corresponding to three projection views (*front*, *side* and *top*). For each projection view, the accumulation of absolute differences between consecutive projected maps forms the corresponding DMMs (*i.e.*, DMM_f, DMM_s and DMM_t) [7]. Next, from these three DMMs, three LBP feature vectors are calculated with overlapping blocks, and three EOH feature vectors are calculated with non-overlapping blocks. After that, these six feature vectors are concatenated through a sequential way to form the so-called **DLE features** (since it contains DMMs-based LBP and EOH features). Finally, the dimension of DLE feature vector is reduced by Principal Component Analysis (PCA), and it is fed into the l_2-CRC to recognize human actions.

The rest of this paper is organized as follows. Section 2 reviews the related work. The whole approach (including the DLE features and l_2-CRC) is presented in Sect. 3. The experimental results are demonstrated in Sect. 4. Finally, Sect. 5 contains a brief conclusion of this work.

2   Related Work

In this section, we review the recent related work for human action recognition from depth video sequences, including low-level features and high-level skeleton information.

In the last few years, a lot of low-level features for recognizing human actions from depth video sequences have been introduced. In 2010, Li *et al.* [8] presented a framework to recognize human actions from sequences of depth maps. They employed an action graph to model the temporal dynamics of actions, and utilized a collection of 3D points to characterize postures. The loss of spatial context information between interest points and computational inefficiency were considered as limitations of this approach. To improve recognition rates, Vieira *et al.* [9] introduced the Space Time Occupancy Patterns (STOP) feature descriptor. Furthermore, Wang *et al.* [10] considered 3-dimensional action sequences as 4-dimensional shapes and proposed Random Occupancy Pattern (ROP), and sparse coding was utilized to further improve the robustness of the proposed approach.

Following another way, Yang *et al.* [11] computed Depth Motion Maps (DMMs) based Histogram of Oriented Gradients (HOG) features and fed them into SVM classifier to recognize human actions. In 2013, Chen *et al.* [7] utilized the DMMs as feature descriptor and l_2-CRC as classifier to recognize human actions. In that year, Oreifej and Liu [12] presented a new descriptor called histogram of oriented 4D surface normals (HON4D). To improve recognition accuracy, some researchers proposed methods based on features extracted from depth and RGB video sequences simultaneously. For instance, Luo *et al.* [13] extracted 3D joint features for each depth video, and utilized Centre-Symmetric Motion Local Ternary Pattern (CS-Mltp) to extract both the spatial and temporal features of the RGB sequences. Besides, some researchers are still working to improve the robustness as well as recognition rate of the action recognition methods. For example, binary range sample feature descriptor was proposed by Lu *et al.* [14]. In 2015, Chen *et al.* [15] proposed another DMMs method, where DMMs- based LBP features coupled with the Kernel-based Extreme Learning Machine (KELM) classifier was used to recognize human actions. Recently, inspired by the DMMs-based works in [7, 11, 15], Farhad *et al.* [16] proposed another method by designing an effective feature descriptor, called DMM-CT-HOG for short. More precisely, HOG was employed to DMMs-based contourlet sub-bands to compactly represent the body shape and movement information toward distinguishing actions.

There are also many skeleton based algorithms by utilizing high-level skeleton information extracted from depth maps. In 2012, Yang and Tian [17] proposed a human action recognition framework by using Eigen joints (position difference of points) and Naive-Bayes-Nearest-Neighbor (NBNN) classifier. In the same year, Xia *et al.* [18] proposed Histogram of 3D Joints Locations (HOJ3D) in depth maps. They applied Hidden Markov Model (HMM) to model the dynamics and action recognition. On the other hand, Wang *et al.* [19] utilized both skeleton and point cloud information and introduced an actionlet ensemble model to represent each action and capture the intra-class variance via occupancy information. In 2013, Luo *et al.* [20] proposed group sparsity and geometry-constraint dictionary learning (DL-GSGC) algorithm for recognizing human actions from skeleton data. Recently, body part based skeleton representation was introduced to characterize the 3D geometric relationships between different body parts by utilizing rotations and translation in [21]. In this approach, human actions were characterized as curves. In [22] the skeleton joint features were extracted from skeleton data and an evolutionary algorithm was utilized for feature selection. However, we have to point out that, though some of the skeleton-based methods show high recognition performance, they are not suitable in the case where skeleton information is not available.

3 Our Approach

In this section, DMMs and LBP are briefly reviewed, and our proposed DLE features (DMMs-based LBP and EOH features) are then introduced.

3.1 DMMs Construction

In the feature extraction stage, DMMs for each depth video sequence are first computed by using approaches demonstrated in [7]. Specifically, given a depth video sequence with K depth maps, each depth map is projected onto three orthogonal Cartesian planes (from front, side and top views) to get three projected maps. The accumulation of absolute differences between consecutive projected maps from front view constructs a DMM, which is referred to as DMM_f. Similarly, DMM_s and DMM_t can also be constructed for side and top views. The following equation is used to form each DMM:

$$DMM_v = \sum_{j=1}^{K-1} \left| map_v^{j+1} - map_v^j \right|, \tag{1}$$

where j is the frame index, $v \in \{f, s, t\}$ denotes the projection view, and map_v stands for the projected map. Figure 2 illustrates an example of DMMs for a High throw action video sequence.

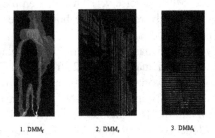

1. DMM_f 2. DMM_s 3. DMM_t

Fig. 2. DMMs for a *High throw* action video sequence

3.2 Overview of LBP

Local Binary Pattern (LBP) operator [23] is a powerful tool for texture description, and it has become very popular in various applications due to its discriminative power and computational simplicity. The original version of LBP operator works in 3×3 pixel blocks. All circumjacent pixels in the working block are thresholded by the center pixel and weighted by powers of 2 and then summed to label the center pixel. The LBP operator can also be extended to neighborhood with different sizes (see Fig. 3) [23]. To do this, consider a circular neighborhood denoted by (N, R), where N is the number of sampling points and R represents the radius of the circle. These sampling points lie around the center pixel (x, y) and at coordinates $(x_i, y_i) = (x + R\cos(2\pi i/N),$ $y - R\sin(2\pi i/N))$. If a sampling point does not fall at integer coordinates, then the

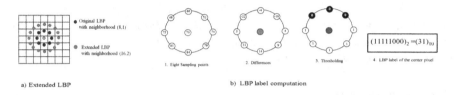

a) Extended LBP   b) LBP label computation

Fig. 3. Extended LBP and LBP label computation for the neighborhood (8, 1)

pixel value is bilinearly interpolated. Specifically, the LBP label for pixel (x, y) can be calculated as follows:

$$LBP_{N,R}(x,y) = \sum_{i=0}^{N-1} Th(f(x_i, y_i) - f(x, y)).2^i, \tag{2}$$

where $Th(x) = 1$ if $(x \geq 0)$ and $Th(x) = 0$ if $x < 0$. An example of LBP label computation is shown in Fig. 3.

Another extension to the original operator is so called *uniform patterns* [23]. A local binary pattern is considered as uniform if the binary pattern has at most 2 bitwise transitions from 0 to 1 or vice versa when the bit pattern is in circular form. For example, the patterns 00000000 (0 transitions) and 01110000 (2 transitions) are uniform whereas the patterns 11001001 (4 transitions) and 01010011 (6 transitions) are not uniform pattern.

After calculating the LBP codes for all pixels in an image, an occurrence histogram is computed for the image or an image region to represent the texture information. Though the original LBP operator in Eq. 2 provides 2^N different binary patterns, only uniform patterns are sufficient to describe image texture [23]. Therefore, only uniform patterns are used to compute the occurrence histogram, *i.e.*, the histogram has an independent bin for each uniform pattern.

3.3 DLE Features Extraction

To construct DLE features, we use the computed DMMs, LBPs and EOHs with the following three stages.

In the first stage, LBP coded images corresponding to DMM_f, DMM_s and DMM_t are divided into 4×2, 4×3 and 3×2 overlapped regions respectively [15]. For all the DMMs, histograms are computed for each block by setting 50 % overlap between two consecutive blocks. Finally, all histograms of all the blocks are merged to compute three feature vectors for the three DMMs. The LBP feature vectors for DMM_f, DMM_s and DMM_t are referred as LBP_f, LBP_s and LBP_t.

In the second stage, to compute EOHs [24], each DMM is first split into 4×4 non-overlapping rectangular regions. In each region, 4 directional edges (horizontal, vertical, 2 diagonals) and 1 non-directional edge are computed using Canny edge detector. Histograms are then computed for each block. The concatenation of all histograms for all regions yields a feature vector for the DMM. Thus, three feature vectors for the three DMMs are generated for each action video sequence. An example of EOHs computation is shown in Fig. 4, where DMM_f of *Horizontal wave* action video sequence is set as an image. The EOH feature vectors for DMM_f, DMM_s and DMM_t are labeled as EOH_f, EOH_s and EOH_t.

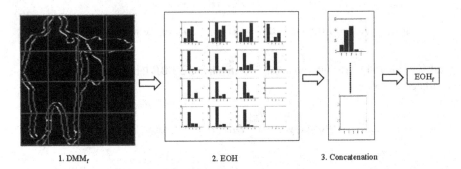

1. DMM_f   2. EOH   3. Concatenation

Fig. 4. EOH computation for DMM_f of *Horizontal wave* action sequence

In the third stage, the computed EOH_f and LBP_f are fused to get a feature vector that describes texture and edge features compactly for DMM_f and this feature vector is mentioned as $Feat_f$. In the same way, $Feat_s$ and $Feat_t$ can be calculated and their concatenation with $Feat_f$ provides a feature vector for the relevant action video sequence. Figure 5 illustrates our feature extraction approach for the *High throw* action video sequence.

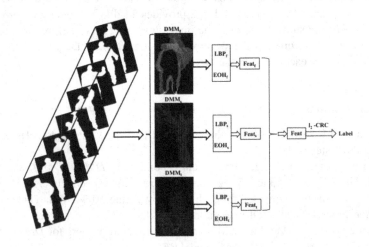

Fig. 5. Architecture of the proposed action recognition method

3.4 L_2-CRC

Motivated by the success of l_2-CRC in human action recognition [7, 16], we use this classifier to model human actions. For a better explanation of the l_2-CRC, let us consider a data set with C classes. By arranging the training samples column wisely, we can obtain an over-complete dictionary $B = [B_1, B_2........, B_C] = [b_1, b_2,........, b_n \in R^{d \times n}$, where d is the dimensionality of samples, n is the total number of training samples, $\mathbf{B}_j \in R^{d \times m_j}, (j = 1, 2,......, C)$ is the subset of the training samples belonging to the jth class and $\mathbf{b_i} \in R^d (i = 1, 2,......, n)$ is the single training sample.

Let us express any unknown sample $\mathbf{V} \in R^d$ using matrix $\mathbf{B}$ as follows:

$$v = \mathbf{B}\gamma, \tag{3}$$

Here γ is a $n \times 1$ vector associated with coefficients corresponding to the training samples. Practically, one cannot solve Eq. 3 directly as it's typically underdetermined [25]. Usually we obtain the solution by solving the following norm minimization problem:

$$\hat{\gamma} = \arg \min_{\gamma} \{\| \mathbf{v} - \mathbf{B}\gamma \|_2^2 + \alpha \| \mathbf{M}\gamma \|_2^2\}, \tag{4}$$

where $\mathbf{M}$ stands for the Tikhonov regularization matrix [26] and α represents the regularization parameter. The term involved with $\mathbf{M}$ permits the imposition of prior knowledge of the solution by utilizing the approach described in [27–29], where the training samples that are highly dissimilar from a test sample are provided less weight than the training samples that are highly similar. Specifically, the form of the matrix $\mathbf{M} \in R^{d \times n}$ is considered as follows:

$$\mathbf{M} = \begin{pmatrix} \|v - \mathbf{b_1}\|_2 & 0 & \cdots & 0 \\ 0 & \|v - \mathbf{b_n}\|_2 & \cdots & 0 \\ \vdots & \vdots & \ddots & \vdots \\ 0 & 0 & \cdots & \|v - \mathbf{b_n}\|_2 \end{pmatrix} \tag{5}$$

According to [30] the coefficient vector $\hat{\gamma}$ is calculated as follows:

$$\hat{\gamma} = (\mathbf{B}^T\mathbf{B} + \alpha\mathbf{M}^T\mathbf{M})^{-1}\mathbf{B}^T\mathbf{V}. \tag{6}$$

By using the class labels of all the training samples, $\hat{\gamma}$ can be partitioned into C subsets $\hat{\gamma} = [\hat{\gamma}_1; \hat{\gamma}_2;........; \hat{\gamma}_C]$ with $\hat{\gamma}_j (j = 1, 2, 3,......, C)$. After portioning $\hat{\gamma}$ the class label of the unknown sample $\mathbf{v}$ is then calculated as follows:

$$class(\mathbf{v}) = \arg \min_{j \in \{12,....C\}} \{e_j\}, \tag{7}$$

where $e_j = \| \mathbf{v} - \mathbf{B}_j\hat{\gamma}\|_2$.

4   Experimental Results

In this section, we first evaluate our proposed method on publicly available MSR-Action3D dataset and then compare our recognition results with other methods.

4.1   MSR-Action3D Dataset and Setup

MSR-Action3D dataset [8] contains 20 actions, where each action is performed by 10 different subjects 2 or 3 times facing the RGB-D camera. The list of 20 actions is: *high wave, horizontal wave, hammer, hand catch, forward punch, high throw, draw x, draw tick, draw circle, hand clap, two hand wave, side boxing, bend, forward kick, side kick, jogging, tennis swing, tennis serve, golf swing, and pickup throw.* This dataset is a challenging dataset as it contains many actions with similar appearance. In order to have a fair evaluation of the proposed method, we follow the same experimental settings as described in [8]. Specifically, we divide the 20 actions into three action subsets (AS1, AS2 and AS3), which are shown in Table 1. For each action subset, five subjects (1, 3, 5, 7 and 9) are used for training and the rest for testing. Usually, this type of experimental setup is known as **cross subject test**.

Table 1. Three subsets of the MSR-Action3D dataset

Label	Action set 1 (AS1)	Label	Action set 2 (AS2)	Label	Action set 3 (AS3)
2	Horizontal wave	1	High wave	6	High throw
3	Hammer	4	Hand catch	14	Forward kick
5	Forward punch	7	Draw x	15	Side kick
6	High throw	8	Draw tick	16	Jogging
10	Hand clap	9	Draw circle	17	Tennis swing
13	Bend	11	Two hand wave	18	Tennis serve
18	Tennis serve	14	Forward kick	19	Golf swing
20	Pickup throw	12	Side boxing	20	Pickup throw

In all the experiments, for each action video sequence, the first/last four frames are removed due to two reasons. Firstly, at the beginning/end of an action video sequence, the subjects are mostly at stand-static position with a small body movement, which is not necessary for the motion characteristics of the involved action. Secondly, in our approach of computing *DMMs*, small movements at the beginning/end result in a stand-still body shape with large pixel values along the DMM edges, which leads to a large amount of recognition error.

To find an appropriate value for the parameter N(number of sampling points) and R (radius) in LBP computation, we carry out experiments on different values of (N, R). Specifically, for each value of R in $\{1, 2, \ldots, 6\}$, we select four values $\{4, 6, 8, 10\}$ for N. We observe the promising result at pair $(4, 1)$. Moreover, the computational complexity of the uniform LBP features depends on the number of sampling points (*i.e.*, N), because the dimensionality of the LBP histogram feature based on uniform patterns is $N(N - 1) + 3$. [23] Since the pair $(4, 1)$ makes low computational complexity and high recognition accuracy, we set $N = 4$ and $R = 1$ for the entire experiment. Besides, in the $l2$-CRC, the key parameter α is set as $\alpha = 0.0001$ according to the five-fold cross-validation. In order to improve the computational efficiency in the classification step, Principle Component analysis (PCA) is utilized to reduce the dimension of the DLE feature vector. The PCA transform matrix is calculated using the training feature set and then applied to the test feature set. The principle components that account for 99 % of the total variation are retained.

4.2 Comparison with Other Methods

We compare the performance of our proposed method with several other competitive methods that were conducted on MSR-Action3D dataset with the same experimental setup. The comparison of the average recognition accuracy is shown in Table 2. It can be seen that, the recognition result of our method outperforms all the other methods listed in the table. Overall, our recognition result indicates that the fusion of DMMs-based LBP and EOH features obtains higher discriminatory power. On the other hand, the block-based dense features generated by LBPs and EOHs provides effective texture and edge information. Figure 6 shows the confusion matrices for the three action subsets separately. The confusion matrices state that actions with high similarities get relatively low accuracies. For example, action *draw x* is confused with *draw tick* due to the similarities of their *DMMs* and therefore action *draw x* is classified with low accuracy (see confusion matrix for the action subset AS2). The assigned label for each action (see Table 1) is used as axes label in the matrices to understand the classification accuracy and error of the corresponding action.

Table 2. Comparison of average recognition accuracies (%) corresponding to the three action subsets for the *cross subject test*

Method	Average accuracy (%)
Li *et al.* [8]	74.7
Yang *et al.* [11]	91.6
Chen *et al.* [7]	90.5
Farhad *et al.* [16]	92.3
Xia *et al.* [18]	79.0
Chaaraoui *et al.* [22]	93.2
Vemulapolli *et al.* [21]	92.5
Chen *et al.* [15]	94.9
Luo *et al.* [13]	93.8
Ours	**95.8**

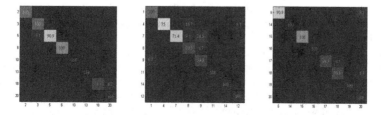

Fig. 6. Confusion matrices for the subset AS1, AS2 and AS3 (from left to right)

Table 3. Processing times (*mean ± STD*) associated with different components of our framework

Components	Average processing time (ms)
DMMs calculation	2.8 ± 0.34/frame
EOH feature extraction LBP feature extraction Dimension reduction (PCA) $l2$-CRC	16.20 ± 6.40/action sample
	12.0 ± 0.15/action sample
	0.13 ± 0.05/action sample
	0.41 ± 0.20/action sample

4.3 Computational Time

There are five main components in our human action recognition framework: DMMs calculation, EOHs computation, and LBPs extraction, dimensionality reduction (PCA) and action recognition (l_2-CRC). Table 3 shows the average processing times of the five components for each depth video sequence with 33 frames (in our experiments each video sequence has 33 frames on average). Our code is written with MATLAB and the processing time is obtained on a PC with 3.20 GHz Intel Core i5-3470 CPU. Noted that, according to the processing time, the proposed method can process 30 frames per second, and thus it is compatible for the real-time operation.

5 Conclusion

This paper has presented a computationally efficient and effective human action recognition method for the depth video sequences by applying the fused version of DMMs-based LBP and EOH features to the l2-CRC classifier. Experimental results on the public domain datasets have revealed that our method provides higher action recognition accuracy compared to the existing methods and allows to recognize actions in real-time.

Acknowledgment. This work was supported by the Natural Science Foundation of China for Grant 61171138

References

1. Chen, C., Kehtarnavaz, N., Jafari, R.: A medication adherence monitoring system for pill bottles based on a wearable inertial sensor. In: Proceedings of the 36th Annual International Conference of the IEEE Engineering in Medicine and Biology Society. pp. 4983–4986 (2014)
2. Chen, C., Jafari, R., Kehtarnavaz, N.: Improving human action recognition using fusion of depth camera and inertial sensors. IEEE Trans. Hum.-Mach. Syst. **45**(1), 51–61 (2015)
3. Laptev, I.: On space-time interest points. Int. J. Comput. Vision **64**(2/3), 107–123 (2005)
4. Niebles, J., Wang, H., Fei-Fei, L.: Unsupervised learning of human action categories using spatial-temporal words. Int. J. Comput. Vision **79**(3), 299–318 (2008)

5. Shotton, J., Fitzgibbon, A., Cook, M., Sharp, T., Finocchio, M., Moore, R., Kip- man, A., Blake, A.: Real-time human pose recognition in parts from single depth images. In: Proceedings of IEEE Conference on Computer Vision and Pattern Recognition, pp. 1297–1304 (2011)
6. Microsoft: kinect for windows. http://www.microsoft.com/en-us/kinectforwindows/
7. Chen, C., Liu, K., Kehtarnavaz, N.: Real-time human action recognition based on depth motion maps. J. Real-Time Image Process. (2013)
8. Li, W., Zhang, Z., Liu, Z.: Action recognition based on a bag of 3d Points. In: Proceedings of IEEE Conference on Computer Vision and Pattern Recognition Workshops, pp. 9–14 (2010)
9. Vieira, A., Nascimento, E., Oliveira, G., Liu, Z., Campos, M.: Space-time occupancy patterns for 3d action recognition from depth map sequences. In: Proceedings of Progress in Pattern Recognition, Image Analysis, Computer Vision, and Applications, pp. 252–259 (2012)
10. Wang, J., Liu, Z., Chorowski, J., Chen, Z., Wu, Y.: Robust 3d action recognition with random occupancy patterns. In: Proceedings of European Conference on Computer Vision, pp. 872–885 (2012)
11. Yang, X., Zhang, C., Tian, Y.: Recognizing actions using depth motion maps-based histograms of oriented gradients. In: Proceedings of ACM International Conference on Multimedia, pp. 1057–1060 (2012)
12. Oreifej, O., Liu, Z.: Hon4d: histogram of oriented 4d normals for activity recognition from depth sequences. In: Proceedings of IEEE Conference on Computer Vision and Pattern Recognition, pp. 716–723 (2013)
13. Luo, J., Wang, W., Qi, H.: Spatio-temporal feature extraction and representation for RGB-D human action recognition. Pattern Recogn. Lett. **50**, 139–148 (2014)
14. Lu, C., Jia, J., Tang, C.K.: Range-sample depth feature for action recognition. In: Proceedings of IEEE Conference on Computer Vision and Pattern Recognition, pp. 772–779 (2014)
15. Chen, C., Jafari, R., Kehtarnavaz, N.: Action recognition from depth sequences using depth motion maps-based local binary patterns. In: Proceedings of IEEE Winter Conference on Applications of Computer Vision, pp. 1092–1099 (2015)
16. Farhad, M., Jiang, Y., Ma, J.: Human action recognition based on DMMs, HOGs and Contourlet transform. In: Proceedings of IEEE International Conference on Multimedia Big Data, pp. 389–394 (2015)
17. Yang, X., Tian, Y.: Eigen joints-based action recognition using Naive-Bayes-Nearest-Neighbor. In: Proceedings of IEEE Conference on Computer Vision and Pattern Recognition Workshops, pp. 14–19 (2012)
18. Xia, L., Chen, C., Aggarwal, J.: View invariant human action recognition using histograms of 3d joints. In: Proceedings of Workshop on Human Activity Under- standing from 3D Data, pp. 20–27 (2012)
19. Wang, J., Liu, Z., Wu, Y., Yuan, J.: Mining actionlet ensemble for action recognition with depth cameras. In: Proceedings of IEEE Conference on Computer Vision and Pattern Recognition, pp. 1290–1297 (2012)
20. Luo, J., Wang, W., Qi, H.: Group sparsity and geometry constrained dictionary learning for action recognition from depth maps. In: Proceedings of the 14th IEEE International Conference on Computer Vision, pp. 1809–1816 (2013)
21. Vemulapalli, R., Arrate, F., Chellappa, R.: Human action recognition by representing 3d human skeletons as points in a lie group. In: Proceedings of IEEE Conference on Computer Vision and Pattern Recognition, pp. 588–595 (2014)

22. Chaaraoui, A.A., Padilla-Lopez, J.R., Climent-Perez, P., Florez-Revuelta, F.: Evolutionary joint selection to improve human action recognition with RGB-D devices. Expert Syst. Appl. **41**(3), 786–794 (2014)
23. Ojala, T., Pietikainen, M., Maenpaa, T.: Multiresolution gray-scale and rotation invariant texture Classification with local binary patterns. IEEE Trans. Pattern Anal. Mach. Intell. **24**(7), 971–987 (2002)
24. http://clickdamage.com/sourcecode/code/edgeOrientationHistogram.m
25. Wright, J., Ma, Y., Mairal, J., Sapiro, G., Huang, T., Yan, S.: Sparse representation for computer vision and pattern recognition. Proc. IEEE **98**(6), 1031–1044 (2010)
26. Tikhonov, A., Arsenin, V.: Solutions of ill-posed problems. Math. Comput. **32**(144), 1320–1322 (1978)
27. Chen, C., Tramel, E.W., Fowler, J.E.: Compressed-sensing recovery of images and video using multi-hypothesis predictions. In: Proceedings of the 45th Asilomar Conference on signals, Systems, and Computers, pp. 1193–1198 (2011)
28. Chen, C., Li, W., Tramel, E.W., Fowler, J.E.: Reconstruction of hyperspectral imagery from random projections using multi-hypothesis prediction. IEEE Trans. Geosci. Remote Sens. **52**(1), 365–374 (2014)
29. Chen, C., Fowler, J.E.: Single-image Super-resolution Using Multi-hypothesis Prediction. In: Proceedings of the 46th Asilomar Conference on Signals, Systems, and Computers, pp. 608–612 (2012)
30. Golub, G., Hansen, P.C., O'Leary, D.: Tikhonov-regularization and total least squares. SIAM J. Matrix Anal. Appl. **21**(1), 185–194 (1999)

Joint Abnormal Blob Detection
and Localization Under Complex Scenes

Tian Wang[1], Keyu Lai[1], Ce Li[2,3(✉)], and Hichem Snoussi[4]

[1] School of Automation Science and Electrical Engineering,
Beihang University, Beijing 100191, China
wangtian@buaa.edu.cn, keyulai@outlook.com
[2] College of Electrical and Information Engineering, Lanzhou University
of Technology, No. 287, Langongping Road, Lanzhou 730050, Gansu, China
xjtulice@gmail.com
[3] The School of Electronic and Information Engineering, Xi'an Jiaotong
University, No. 28, Xianning West Road, Xi'an 710049, Shaanxi, China
[4] Institut Charles Delaunay-LM2S-UMR STMR 6279 CNRS,
University of Technology of Troyes, 10004 Troyes, France
hichem.snoussi@utt.fr

Abstract. In this paper, an algorithm is proposed to detect the abnormal event in the form of rectangular blob in global images. Observing the status of the varying blobs, unusual behavior can be monitored and alarmed. A method extracting blobs from crowded video scenes is proposed, the covariance matrix descriptor fuses the image intensity and the optical flow to encode moving information and image characteristics of a blob. After characterizing normal behaviors of blobs or frames in a learning period, the nonlinear one-class SVM algorithm locates the abnormal blobs intra frame. The method is applied to detect abnormal events on several video surveillance datasets, and get promising results.

Keywords: Blob extraction · Abnormal blob localization · Covariance matrix descriptor · One-class SVM

1 Introduction

To be aware of the behavior of individuals or of group of people in video sequence obtaining by a visual sensor is a challenging task. Particularly, detection of abnormal movements requires sophisticated visual intelligence. Several normal and abnormal scenes are shown in Fig. 1. In Fig. 1(a), all the people are walking, this scene is considered as normal. In Fig. 1(b) and (c), an unusual group of movements is detected, the people are suddenly running to escape from dangers. Another abnormal example is shown in Fig. 1(d), a man in black suit is running across the walking pedestrians. The purpose of an intelligent surveillance system is to detect an abnormal video frame, or to locate abnormal blobs in the scene.

The local abnormal event detection is usually related to patch event analysis. In [1–6], a motion vector extracted over a region of interest or spatio-temporal points was

© Springer International Publishing Switzerland 2015
D.-S. Huang et al. (Eds.): ICIC 2015, Part I, LNCS 9225, pp. 283–292, 2015.
DOI: 10.1007/978-3-319-22180-9_28

Fig. 1. Examples of the normal and abnormal scenes. (a) All the people are walking, the normal indoor scene. (b) All the people are running, the abnormal indoor scene. (c) All the people are running, the abnormal plaza scene. (d) One person is running and the others are walking, the normal and abnormal blobs.

used as a feature descriptor. These methods tend to be dominated by local information, but not the larger visual structures, such as a global person or a global moving object. In [7], the histogram of optical flow orientation was used to detect abnormal frame event. In this paper, we address the abnormal *blob* detection and localization problem. These two conceptions depend on how you tackle the problem. Firstly, if the blobs of moving objects are provided, the abnormal action of the objects can be detected. On the other hand, the position of the object yielding an abnormal behavior in crowded scenes can be localized. The target that causes the abnormal event is labeled automatically without human intervention, and thus it can be tracked.

State event models were described in [8–10]. Bayesian network event model utilizes probability as a mechanism for handling the uncertainty of observations and interpreting existing events in a video. Because the model is difficult to be obtained generally, a model free method is more suitable for the abnormal detection problem. In the abnormal event detection problem, it is assumed that the samples from the normal class are obtainable. Thus, one-class SVM algorithm is used to deal with this situation is this paper.

The rest of the paper is organized as follows. In Sect. 2, a blob extraction method in a crowd scene is proposed. In Sect. 3, the proposed covariance matrix descriptor encoding motion feature is proposed. In Sect. 4, nonlinear one-class SVM for covariance feature is introduced. In Sect. 5, we present the abnormal blob localization and abnormal frame detection results on benchmark datasets. Finally, Sect. 6 concludes the paper.

Algorithm 1. Blob extraction.

Require:

Foreground image FG, optical flow OP

1: **Label the separate blobs in FG, the blob of the foreground image** $B_{FG}^k \leftarrow$ **is obtained.**

2: **if Blob size in** $\leftarrow \mathcal{FG} \leftarrow$ ⌉ $\leftarrow$**presetting size** $\leftarrow T_{blb} \psi$ **then** ψ

3: **Compute optical flow in this blob.**

4: **The optical flows with similar magnitudes and directions are clustered by mean-shift algorithm.**

5: **Delete redundant cluster by non-maximum suppression (NMS) algorithm, the blob of optical flow image** $\psi \{B_{OP}^i\}$ **is obtained. The remaining part of the blob is** $B_{RM} = B_{FG} \setminus B_{OP} \triangleright \psi$

6: **Traverse** $\psi B_{RM} \psi$ **by a rectangle template to find the blobs overlapped with the foreground. NMS algorithm is used to delete the redundant templates. Blob** $\psi \{B_{RM}^j\}$ **of** B_{RM} **is obtained.**

7: **Change foreground blob** $\psi \{B_{FG}^k\}$ **to** $\{B_{OP}^i\} \cup \{B_{RM}^j\}$.

8: **end if**

9: **The blobs of the image are extracted.**

---

2 Presentation of Event with Blob

Moving objects are usually conflicted with others. As shown in Fig. 2(a), the running person on the upper half in the 1-st rectangle is overlapped with another walking person. We present a method to improve the blob extraction performance by adopting optical flow, which presents the moving information. The method is summarized in Algorithm 1. The background subtraction algorithm used in this paper is presented in [11] to segment the moving object.

(a)   (b)   (c)

Fig. 2. The blobs before and after the proposed blob extraction method. (a) 2 extracted blobs based on the foreground template. (b) The optical flow image of Figure (a). The T_{blb} depends upon the size of individual. (c) 3 extracted blobs via the proposed blob extraction method.

3   Covariance Descriptor of Event

The optical flow describes the movement information intra-frame, and the frame intensity shows the appearance characteristic. We propose to construct covariance matrix descriptor based on the optical flow and the intensity to encode movement features both in a blob and in a global image. The covariance descriptor is defined as:

$$F(\text{x}, \text{y}, \ell) = \phi_\ell(\text{I}, \text{x}, \text{y}), \tag{1}$$

where I is an image, F is a $W \times H \times \text{d}$ dimensional feature of image I, W and H are the image width and height, d is the dimension of the used feature, ϕ_ℓ is a mapping relating to the image with the ℓ-th feature from the image I. For a given rectangular region R, the feature points can be represented as $\text{d} \times \text{d}$ covariance matrix:

$$C_R = \frac{1}{n-1} \sum_{k=1}^{n_p} (z_k - \mu)(z_k - \mu)^T, \tag{2}$$

where μ is the mean of the points, C_R is the covariance matrix of the feature vector F, z_k is the feature vector of pixel k, n_p pixels are chosen in region R. 13 different feature vectors F shown in Table 1 are proposed to construct the covariance descriptor, feature vector F_1 to F_6 contains the movement information, F_7 to F_{13} also takes the appearance of a frame into account. The optical flow information can be considered as an alternate form of a video clip with two frames. In Table 1, I is the intensity of the gray image, u is the horizontal optical flow, v is the vertical optical flow; $I_x, \text{u}_x, \text{v}_x, I_y, \text{u}_y, \text{v}_y$ are the first derivatives; $I_{xx}, \text{u}_{xx}, \text{v}_{xx}, I_{yy}, \text{u}_{yy}, \text{v}_{yy}, I_{xy}, \text{u}_{xy}, \text{v}_{xy}$ are the second derivatives.

4   One-Class SVM

One-class SVM is used to detect abnormal events by classifying the covariance matrix descriptor. If the whole image is taken as one big blob, the strategy of the abnormal blob detection method can also detect the abnormal frame event. One-class SVM has

Table 1. Features F used to form the covariance matrices

	Feature Vector F
F_1	$[y \quad x \quad u \quad v]$
F_2	$[y \quad x \quad u \quad v \quad u_x \quad u_y]$
F_3	$[y \quad x \quad u \quad v \quad v_x \quad v_y]$
F_4	$[y \quad x \quad u \quad v \quad u_x \quad u_y \quad v_x \quad v_y]$
F_5	$[y \quad x \quad u \quad v \quad u_x \quad u_y \quad v_x \quad v_y \quad u_{xx} \quad u_{yy} \quad v_{xx} \quad v_{yy}]$
F_6	$[y \quad x \quad u \quad v \quad u_x \quad u_y \quad v_x \quad v_y \quad u_{xx} \quad u_{yy} \quad v_{xx} \quad v_{yy} \quad u_{xy} \quad v_{xy}]$
F_7	$[y \quad x \quad u \quad v \quad I]$
F_8	$[y \quad x \quad u \quad v \quad u_x \quad u_y \quad v_x \quad v_y \quad I]$
F_9	$[y \quad x \quad u \quad v \quad v_x \quad v_y \quad u_{xx} \quad u_{yy} \quad v_{xx} \quad v_{yy} \quad I]$
F_{10}	$[y \quad x \quad u \quad v \quad u_x \quad u_y \quad v_x \quad v_y \quad u_{xx} \quad u_{yy} \quad v_{xx} \quad v_{yy} \quad u_{xy} \quad v_{xy} \quad I]$
F_{11}	$[y \quad x \quad u \quad v \quad u_x \quad u_y \quad v_x \quad v_y \quad I \quad I_x \quad I_y]$
F_{12}	$[y \quad x \quad u \quad v \quad u_x \quad u_y \quad v_x \quad v_y \quad u_{xx} \quad u_{yy} \quad v_{xx} \quad v_{yy} \quad I \quad I_x \quad I_y \quad I_{xx} \quad I_{yy}]$
F_{13}	$[y \quad x \quad u \quad v \quad u_x \quad u_y \quad v_x \quad v_y \quad u_{xx} \quad u_{yy} \quad v_{xx} \quad v_{yy} \quad u_{xy} \quad v_{xy} \quad I \quad I_x \quad I_y$ $I_{xx} \quad I_{yy} \quad I_{xy}]$

been used for abnormal detection in audio or video data in [12–14]. The non-linear one-class SVM [15, 16] can be presented as:

$$\min_{\omega,\xi,\rho} \frac{1}{2}\|\omega\|^2 + \frac{1}{vn}\sum_{i=1}^{n}\xi_i - \rho, \text{ subject to } \langle\omega,\Phi(x_i)\rangle \geq \rho - \xi_i, \xi_i \geq 0, \quad (3)$$

where $x_i \in X$, $i \in [1 \ldots n]$ are training samples in the original data space X, $\Phi : X \to H$ maps datum x_i into the feature space H, $\langle\omega,\Phi(x_i)\rangle - \rho = 0$ is the maximum margin decision hyperplane, ξ_i is the slack variable for penalizing the outliers. The hyperparameter $v \in (0,1]$ is the weight for restraining slack variable, it tunes the number of acceptable outliers.

The covariance matrix is an element in a Lie Group G, the Gaussian kernel in a Lie Group is:

$$\kappa(X_i, X_j) = \exp(-\frac{\|\log(X_i^{-1}X_j)\|}{2\sigma^2}), \quad (X_i, X_j) \in G \times G. \quad (4)$$

By using the first term of Baker Campbell Hausdorff formula [17], the approximate form of the Gaussian kernel is:

$$\kappa(X_i, X_j) = \exp(-\frac{\|\log(X_i) - \log(X_j)\|^2}{2\sigma^2}), \quad (X_i, X_j) \in G \times G. \quad (5)$$

where $\log(X)$ is a symmetrical matrix. The covariance descriptor C_R is a symmetry matrix of size $d \times d$ it has only $\frac{d^2+d}{2}$ different features. By choosing the upper triangular and the diagonal elements of the matrix log(X) to construct a vector x to replace log(X) in Eq. (5), we have:

$$\kappa(X_i, X_j) = \exp\left(-\frac{\left\|\bar{x}_i - \bar{x}_j\right\|^2}{2\sigma^2}\right) \tag{6}$$

5   Test Validation

This section presents the results of experiments conducted to analyze the performance of the one-class SVM classification method for abnormal blob localization, and abnormal global frame event detection. The receiver operating characteristic (ROC) curve and the area under the ROC curve (AUC) of the detection results are shown for each experiment.

The abnormal *blob* event detection results of the lawn scene and the plaza scene in UMN dataset [18] are shown in Fig. 3. All the blobs in the abnormal frame are labeled as abnormal. The maximum AUC value of lawn scene and plaza scene are 0.9721 and

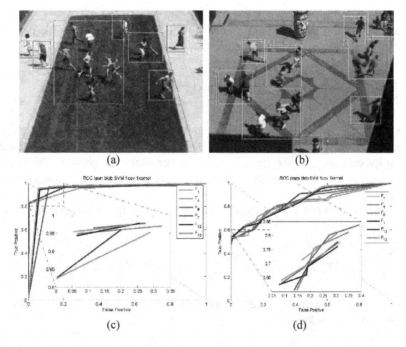

(a)   (b)

(c)   (d)

Fig. 3. The abnormal *blob* event localization results of the lawn scene and the plaza scene. (a) The abnormal detection results of lawn scene. The red rectangles label the abnormal running blobs. (b) The abnormal detection results of plaza scene. (c) ROC curve of different feature F results of the lawn scene results by using *"1 covariance 1 kernel"*. The maximum AUC value is 0.9721. (d) ROC curve of the plaza scene by using *"1 covariance 1 kernel"*. The maximum AUC value is 0.8523 (Color figure online)

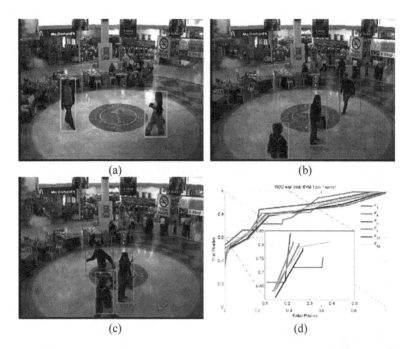

Fig. 4. The abnormal *blob* event localization results of the mall scene. (a) The normal blobs for training, two people are walking. (b) The detection result. The red rectangles label the abnormal blobs, the people are running. The blue rectangles label the normal blobs, the people are walking. (c) Another abnormal *blob* event detection result. (d) ROC curve by using *"1 covariance 1 kernel"*. The maximum AUC value is 0.8583 (Color figure online)

0.8523. The detection results of mall scenes [19] are shown in Fig. 4. The maximum AUC value is 0.8583. The results show that the abnormal detection algorithm of the blob covariance feature can obtain satisfactory detection results.

Taking the whole *frame* as one *blob*, the abnormal *blob* detection method can be adjusted to detect an abnormal *frame*. The detection results of UMN dataset are introduced below. The UMN dataset includes eleven video sequences of crowded escape events in three different scenes: lawn, indoor and plaza. The training samples and normal testing samples are the frames where the people are walking in different directions. The abnormal testing samples are the frames where the people are running. The results are shown in Table 2. The indoor scene is more difficult than the other two scenes, due to the instable illumination situation and the gloom circumstance. Our proposed abnormal detection method can handle this bad illumination scene. The abnormal frame detection results of indoor scene is shown in Fig. 5.

The performances of the covariance matrix descriptor based method and the state-of-the-art methods are shown in Table 3. The covariance matrix based abnormal frame detection method obtains competitive performance.

Table 2. AUC of abnormal *frame* event detection method of the UMN dataset

Features	Lawn	Indoor	Plaza
$F_1(4 \times 4)$	0.9382(12)	0.7359(13)	0.9103(13)
$F_2(6 \times 6)$	0.9474(11)	0.8381(10)	0.9148(12)
$F_3(6 \times 6)$	0.9583(10)	0.8410(9)	0.9192(11)
$F_4(8 \times 8)$	0.9656(7)	0.8483(8)	0.9367(9)
$F_5(12 \times 12)$	0.9798(2)	0.8477(6)	0.9782(2)
$F_6(14 \times 14)$	**0.9803(1)**	0.8752(5)	**0.9790(1)**
$F_7(5 \times 5)$	0.9337(13)	0.8314(11)	0.9220(10)
$F_8(9 \times 9)$	0.9617(8)	0.8529(7)	0.9219(8)
$F_9(13 \times 13)$	0.9786(4)	0.8797(4)	0.9421(4)
$F_{10}(15 \times 15)$	0.9789(3)	0.8145(12)	0.9734(3)
$F_{11}(11 \times 11)$	0.9583(9)	0.9000(3)	0.9472(7)
$F_{12}(17 \times 17)$	0.9758(6)	**0.9291(1)**	0.9549(6)
$F_{13}(20 \times 20)$	0.9767(5)	0.9253(2)	0.9580(5)

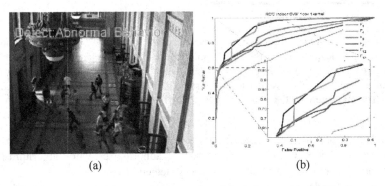

(a)   (b)

Fig. 5. The abnormal *frame* event detection results of the indoor scene. (a) The detection result of one abnormal panic frame. (c) ROC curve The maximum AUC value is 0.9291.

Table 3. The comparison of our proposed method with the state-of-the-art methods for global abnormal events detection in the UMN dataset. NN, nearest neighbor. SRC, sparse reconstruction cost. STCOG, spatial-temporal co-occurrence Gaussian mixture models. COV, covariance matrix descriptor

Method	Area under ROC		
	Lawn	Indoor	Plaza
Social Force [20]	0.96		
Optical Flow [20]	0.84		
NN [21]	0.93		
SRC [21]	0.995	0.975	0.964
STCOG [22]	0.9362	0.7759	0.9661
COV SVM (Ours)	0.9803	0.9291	0.9790

6 Conclusions

A method for abnormal blob localization and abnormal frame event detection is proposed in this paper. The blob is extracted based on the foreground and the optical flow in a crowd scene. The covariance matrix constructed by different features of the intensity and the optical flow information, is computed as the descriptor to encode the moving information of a blob or a frame. The influence of the different features is analyzed by experiments. Several benchmark datasets have been tested to demonstrate the effectiveness of the proposed technique.

Acknowledgment. This work is partially supported by the SURECAP CPER project and the Platform CAPSEC funded by REGION CHAMPAGNE-ARDENNE and FEDER, the Fundamental Research Funds for the Central Universities and the National Natural Science. Foundation of China (Grant No. U1435220, No. 61365003).

References

1. Tziakos, I., Cavallaro, A., Xu, L.-Q.: Event monitoring via local motion abnormality detection in non-linear subspace. Neurocomputing **73**(10), 1881–1891 (2010)
2. Saligrama, V., Chen, Z.: Video anomaly detection based on local statistical aggregates. In: Proceedings of IEEE Conference on Computer Vision and Pattern Recognition (CVPR), pp. 2112–2119 (2012)
3. Zhang, Z., Wang, C., Xiao, B., Zhou, W., Liu, S.: Action recognition using context-constrained linear coding. IEEE Sig. Process. Lett. **19**(7), 439–442 (2012)
4. Li, H., Achim, A., Bull, D.: Unsupervised video anomaly detection using feature clustering. IET Sig. Process. **6**(5), 521–533 (2012)
5. Biswas, S., Venkatesh Babu, R.: Sparse representation based anomaly detection with enhanced local dictionaries. In: Proceedings of IEEE International Conference on Image Processing (ICIP), pp. 5532–5536 (2014)
6. Cho, S.-H., Kang, H.-B.: Abnormal behavior detection using hybrid agents in crowded scenes. Pattern Recogn. Lett. **44**, 64–70 (2014)
7. Wang, T., Snoussi, H.: Detection of abnormal visual events via global optical flow orientation histogram. IEEE Trans. Inf. Forensics Secur. **9**(6), 988–998 (2014)
8. Wang, X., Ma, X., Grimson, W.E.L.: Unsupervised activity perception in crowded and complicated scenes using hierarchical bayesian models. IEEE Trans. Pattern Anal. Mach. Intell. **31**(3), 539–555 (2009)
9. Jiménez-Hernández, H., González-Barbosa, J.-J., Garcia-Ramírez, T.: Detecting abnormal vehicular dynamics at intersections based on an unsupervised learning approach and a stochastic model. Sensors **10**(8), 7576–7601 (2010)
10. Arbab-Zavar, B., Carter, J.N., Nixon, M.S.: On hierarchical modelling of motion for workflow analysis from overhead view. Mach. Vis. Appl. **25**(2), 345–359 (2014)
11. Tuzel, O., Porikli, F., Meer, P.: A Bayesian approach to background modeling. In: Proceedings of IEEE Conference on Computer Vision and Pattern Recognition Workshops (CVPR Workshops), pp. 58–58 (2005)
12. Lecomte, S., Lengellé, R., Richard, C., Capman, F., Ravera, B.: Abnormal events detection using unsupervised one-class SVM-application to audio surveillance and evaluation. In: Proceedings of IEEE International Conference on Advanced Video and Signal-Based Surveillance (AVSS), pp. 124–129 (2011)

13. Piciarelli, C., Micheloni, C., Foresti, G.L.: Trajectory-based anomalous event detection. IEEE Trans. Circ. Syst. Video Technol. **18**(11), 1544–1554 (2008)
14. Charalambous, P., Karamouzas, I., Guy, S.J., Chrysanthou, Y.: A data-driven framework for visual crowd analysis. Comput. Graph. Forum **33**, 41–50 (2014). Wiley Online Library
15. Schölkopf, B., Platt, J.C., Shawe-Taylor, J., Smola, A.J., Williamson, R.C.: Estimating the support of a high-dimensional distribution. Neural Comput. **13**(7), 1443–1471 (2001)
16. Canu, S., Grandvalet, Y., Guigue, V., Rakotomamonjy, A.: SVM and kernel methods matlab toolbox. Perception Systèmes et Information, INSA de Rouen, Rouen, France (2005)
17. Hall, B.C.: Lie Groups, Lie Algebras, and Representations: An Elementary Introduction, vol. 222. Springer, Heidelberg (2003)
18. UMN: Unusual Crowd Activity Dataset of University of Minnesota, Department of Computer Science and Engineering (2006). http://Mha.Cs.Umn.Edu/Movies/Crowd-Activity-All.Avi
19. Adam, A., Rivlin, E., Shimshoni, I., Reinitz, D.: Robust real-time unusual event detection using multiple fixed-location monitors. IEEE Trans. Pattern Anal. Mach. Intell. **30**(3), 555–560 (2008)
20. Mehran, R., Oyama, A., Shah, M.: Abnormal crowd behavior detection using social force model. In: Proceedings of IEEE Conference on Computer Vision and Pattern Recognition (CVPR), Miami, FL, USA, pp. 935–942, June 2009
21. Cong, Y., Yuan, J., Liu, J.: Sparse reconstruction cost for abnormal event detection. In: Proceedings of IEEE Conference on Computer Vision and Pattern Recognition (CVPR), Colorado Springs, CO, USA, pp. 3449–3456, June 2011
22. Shi, Y., Gao, Y., Wang, R.: Real-time abnormal event detection in complicated scenes. In: Proceedings of International Conference on Pattern Recognition (ICPR), Istanbul, Turkey, pp. 3653–3656, August 2010

Graph Based Kernel *k*-Means Using Representative Data Points as Initial Centers

Wuyi Yang$^{(\boxtimes)}$ and Liguo Tang

Key Laboratory of Underwater Acoustic Communication
and Marine Information Technology of the Minister of Education,
Xiamen University, Xiamen, China
{wyyang,liguotang}@xmu.edu.cn

Abstract. The *k*-means algorithm is undoubtedly the most widely used data clustering algorithm due to its relative simplicity. It can only handle data that are linearly separable. A generalization of *k*-means is kernel *k*-means, which can handle data that are not linearly separable. Standard *k*-means and kernel *k*-means have the same disadvantage of being sensitive to the initial placement of the cluster centers. A novel kernel *k*-means algorithm is proposed in the paper. The proposed algorithm uses a graph based kernel matrix and finds *k* data points as initial centers for kernel *k*-means. Since finding the optimal data points as initial centers is an NP-hard problem, this problem is relaxed to obtain *k* representative data points as initial centers. Matching pursuit algorithm for multiple vectors is used to greedily find *k* representative data points. The proposed algorithm is tested on synthetic and real-world datasets and compared with kernel *k*-means algorithms using other initialization techniques. Our empirical study shows encouraging results of the proposed algorithm.

Keywords: Kernel *k*-means · Graph Laplacian · Data clustering · Initial centers

1 Introduction

Clustering is one of the most important tasks in a variety of fields, such as data mining, statistical data analysis, compression, vector quantization, etc. Clustering is based on a similarity criteria and the aim of clustering is that of partitioning data into groups in which data points are similar to each other. Clustering algorithms can be roughly divided into two categories: hierarchical and partitioning [1].

The most widely used clustering method is undoubtedly the *k*-means algorithm [2]. The time and storage complexities of *k*-means are linear in the number of data points, the dimension of the data, and the number of cluster centers [3]. In addition, *k*-means is invariant to data ordering. However, *k*-means has two significant disadvantages [1, 3]. First, *k*-means can only find linear separators in the input space [4]. Second, *k*-means is highly sensitive to the selection of initial centers and often converges to a local minimum of its criterion function [4]. To counter the first disadvantage, kernel *k*-means [5] uses a nonlinear function to map data points to a higher-dimensional feature space.

© Springer International Publishing Switzerland 2015
D.-S. Huang et al. (Eds.): ICIC 2015, Part I, LNCS 9225, pp. 293–304, 2015.
DOI: 10.1007/978-3-319-22180-9_29

Then, the standard k-means algorithm is applied in this feature space. However, kernel k-means is still sensitive to the initial placement of the cluster centers.

Many works greedily find k data points as initial centers based on some criteria. The farthest-first method [6] chooses the first center randomly, and the i-th ($i \in \{2, \cdots, k\}$) center is chosen to be the point that has the largest minimum distance to the previously selected centers. Ball and Hall's method [7] takes the centroid of all data points as the first center and then traverses the points. The i-th ($i \in \{2, \cdots, k\}$) center is chosen to be a data point if it is at least T units apart from the previously selected centers. The k-means++ method [8] chooses the first center randomly and takes a data point $\mathbf{x}$ as a center with a probability of $md(\mathbf{x}) / \sum_{j=1}^{N} md(\mathbf{x}_j)$ until k centers are obtained, where md ($\mathbf{x}$) denotes the minimum-distance from a point $\mathbf{x}$ to the previously selected centers. The ROBIN method [9] uses a local outlier factor (LOF) to avoid selecting outlier points as centers. The method randomly takes a data point that has an LOF value close to 1 as the first center. In iteration i, the method first sorts the data points in decreasing order of their minimum-distance to the previously selected centers. It then traverses the points in sorted order and selects the first point that has an LOF value close to 1 as the i-th center.

To improve the performance of data clustering, we propose a novel graph based kernel k-means algorithm using representative data points (GKKMRDP) as initial cluster centers. GKKMRDP uses a graph based kernel matrix. A nearest neighbor graph is constructed to model the local geometric structure of data points. Graph based kernel matrix is constructed from the normalized graph Laplacian. It is NP-hard to find the optimal data points as initial centers for kernel k-means. So, this problem is relaxed and GKKMRDP greedily finds k representative data points as initial centers using the matching pursuit algorithm for multiple vectors. We compare GKKMRDP with kernel k-means algorithms using other initialization techniques and present experimental results on synthetic and real-world datasets. GKKMRDP outperforms or is competitive with other methods in terms of accuracy.

The rest of this paper is organized as follows: In Sect. 2, we give a brief review of kernel k-mean. Section 3 introduces the proposed algorithm. Extensive experimental results on synthetic and real-world datasets are presented in Sect. 4. Finally, conclusions are provided in Sect. 5.

2   A Brief Review of Kernel k-Means

Given a set of data points $\mathbf{X} = \{\mathbf{x}_1, \mathbf{x}_2, ..., \mathbf{x}_N\}$, the k-means algorithm seeks to divide $\mathbf{X}$ into k exhaustive and mutually exclusive clusters P_1, P_2, ..., P_k ($\cup_{i=1}^{k} P_i = \mathbf{X}, P_i \cap P_j = \varnothing$ for $1 \leq i \neq j \leq k$) that minimize the Sum of Squared Error (SSE) given by:

$$J(\{P_c\}_{c=1}^{k}) = \sum_{c=1}^{k} \sum_{\mathbf{x}_i \in P_c} ||\mathbf{x}_i - \mathbf{m}_c||^2, \tag{1}$$

where $|| \cdot ||_2$ denotes the Euclidean norm and $\mathbf{m}_c = (\sum_{\mathbf{x}_i \in P_c} \mathbf{x}_i)/|P_c|$ is the centroid or mean of cluster P_c whose cardinality is $|P_c|$.

Since the squared Euclidean distance is used as the distance measure, k-means can only find linear separators in the input space [3, 4]. To counter this disadvantage, kernel k-means [5] uses a nonlinear function to map data points to a higher-dimensional feature space. Then, the standard k-means algorithm is applied in this feature space, and the linear separators in the feature space correspond to nonlinear separators in the input space.

Let H be a reproduction kernel Hilbert space (RKHS) with a kernel function

$$k(\mathbf{x}, \mathbf{v}) = \, <\varphi(\mathbf{x}), \varphi(\mathbf{v}) > , \mathbf{x}, \mathbf{v} \in \Re^d, \tag{2}$$

where $\varphi : \Re^d \to H$ is a feature map from the input space to H. Using the nonlinear function φ, the objective function of kernel k-means is the SSE defined as:

$$J(\{P_c\}_{c=1}^k) = \sum_{c=1}^k \sum_{\mathbf{x}_i \in P_c} ||\varphi(\mathbf{x}_i) - \mathbf{m}_c||^2 \tag{3}$$

where $\mathbf{m}_c = (\sum_{\mathbf{x}_i \in P_c} \varphi(\mathbf{x}_i))/|P_c|$. The distance computation $||\varphi(\mathbf{x}_i) - \mathbf{m}_c||^2$ in the objective function can be expanded as:

$$\varphi(\mathbf{x}_i) \cdot \varphi(\mathbf{x}_i) - 2 \frac{\sum_{\mathbf{x}_j \in P_c} \varphi(\mathbf{x}_i) \cdot \varphi(\mathbf{x}_j)}{|P_c|} + \frac{\sum_{\mathbf{x}_j, \mathbf{x}_l \in P_c} \varphi(\mathbf{x}_j) \cdot \varphi(\mathbf{x}_l)}{|P_c|^2}. \tag{4}$$

Thus, to compute the distance between a point and a centroid, only inner products are used. If we compute a kernel matrix $\mathbf{K} = [\mathbf{K}_{ij}]_{N \times N}$ using the kernel function in (2), where $\mathbf{K}_{ij} = \varphi(\mathbf{x}_i) \cdot \varphi(\mathbf{x}_j)$, distances between points and centroids can be computed without explicitly knowing the mapping of $\mathbf{x}_i$ and $\mathbf{x}_j$ to representations of $\varphi(\mathbf{x}_i)$ and $\varphi(\mathbf{x}_j)$ respectively. Using the kernel matrix $\mathbf{K}$, the distance computation $||\phi(\mathbf{x}_i) - \mathbf{m}_c||^2$ can be rewritten as:

$$||\phi(\mathbf{x}_i) - \mathbf{m}_c||^2 = \mathbf{K}_{ii} - 2 \frac{\sum_{\mathbf{x}_j \in P_c} \mathbf{K}_{ij}}{|P_c|} + \frac{\sum_{\mathbf{x}_j, \mathbf{x}_l \in P_c} \mathbf{K}_{jl}}{|P_c|^2}. \tag{5}$$

At the beginning of kernel k-means, if initial clusters are not given, initial clusters are randomly generated. Since kernel k-means is often non-deterministic, it is common to perform multiple runs of kernel k-means and take the output of the run that produces the least SSE.

3 Graph Based Kernel k-Means Using Representative Data Points as Initial Centers

By selecting initial centers carefully, kernel k-means can converge to a better minimum of its objective function. In this section, we introduce our graph based kernel k-means algorithm which uses representative data points as initial centers.

3.1 Graph Based Kernel Matrix

Since any positive definite matrix can be thought of as a kernel matrix, a positive definite matrix $\mathbf{K} = [\mathbf{K}_{ij}]_{N \times N}$ used for clustering is derived from the data distribution. Spectral graph theory and manifold learning theory [10, 11] are increasingly being used for data modeling and prediction, due to their good performance on many tasks, such as dimensionality reduction, semi-supervised learning, etc. A natural assumption is that if two data points are close in the intrinsic geometry of the data distribution, then they are likely to belong to the same cluster. Local geometric structure can be effectively modeled through a nearest neighbor graph on a scatter of data points [11].

Consider a graph $G = (Q, E)$ with nodes Q corresponding to the N data points. For a data point $\mathbf{x}_j$, its p nearest neighbors are found and edges are put between $\mathbf{x}_j$ and its neighbors. The closeness of data points is measured by a weight matrix $\mathbf{W} = [\mathbf{W}_{ij}]_{N \times N}$ on the graph. There are many different methods to define the weight matrix $\mathbf{W}$. To differentiate the p nearest neighbors of a given data point, the heat kernel weighting is used to measure the closeness of nearby data points. If $\mathbf{x}_j$ and $\mathbf{x}_j$ are connected by an edge, the closeness of $\mathbf{x}_j$ and $\mathbf{x}_j$ is $\mathbf{W}_{ij} = \exp\{-||\mathbf{x}_i - \mathbf{x}_j||^2/2\sigma^2\}$, where $\sigma(\sigma > 0)$ is a parameter. The graph Laplacian [10] is defined as $\mathbf{L} = \mathbf{D} - \mathbf{W}$, where $\mathbf{D}$ is a diagonal matrix whose entries are column sums of $\mathbf{W}$, $\mathbf{D}_{ii} = \sum_j \mathbf{W}_{ij}$. The normalized graph Laplacian is defined as $\tilde{\mathbf{L}} = \mathbf{D}^{-\frac{1}{2}} \mathbf{L} \mathbf{D}^{-\frac{1}{2}}$. Graph based kernel matrix is constructed from the spectral decomposition of $\tilde{\mathbf{L}}$. Let us denote $\tilde{\mathbf{L}}$'s eigen-decomposition by $\{\lambda_i, \mathbf{v}_i\}$, so that $\tilde{\mathbf{L}} = \sum_{i=1}^N \lambda_i \mathbf{v}_i \mathbf{v}_i^T$. A graph based kernel matrix $\mathbf{K}$ can be constructed by transforming the spectrum of $\tilde{\mathbf{L}}$ as [12]

$$\mathbf{K} = \sum_{i=1}^N r(\lambda_i) \mathbf{v}_i \mathbf{v}_i^T, \tag{6}$$

where $r(\lambda)$ is a non-negative and decreasing function. Plugging $r(\lambda) = (\beta + \lambda)^{-1}$ into (6), we get the regularized Laplacian kernel matrix

$$\mathbf{K} = \sum_{i=1}^N \frac{1}{\beta + \lambda_i} \mathbf{v}_i \mathbf{v}_i^T = (\beta \mathbf{I} + \mathbf{D}^{-\frac{1}{2}} \mathbf{L} \mathbf{D}^{-\frac{1}{2}})^{-1}, \tag{7}$$

where $\beta(\beta > 0)$ is a regularization parameter. Since $\beta \mathbf{I} + \mathbf{D}^{-\frac{1}{2}} \mathbf{L} \mathbf{D}^{-\frac{1}{2}}$ is a symmetric, positive definite and sparse matrix, the kernel matrix $\mathbf{K}$ can be computed efficiently by Gaussian elimination.

3.2 Representative Data Points

If k data points, which are close to the optimal cluster centroids, can be chosen as initial centers, kernel k-means is more likely to converge to the global minimum of its objective function. However, this is an NP-hard problem. So, we try to select k representative data points (RDP) and then use these data points as initial centers for kernel k-means.

Suppose we have a set of M candidate initial centers $\mathbf{V} = \{\mathbf{x}_{t_1}, \mathbf{x}_{t_2}, \cdots, \mathbf{x}_{t_M}\}$, which are data points randomly selected from the data set $\mathbf{X}$, i.e. $\mathbf{V} \subseteq \mathbf{X}$, and $T = \{t_1, \cdots, t_M\}$ is the index set. k representative data points $\mathbf{Z} = \{\mathbf{x}_{s_1}, \mathbf{x}_{s_2}, \cdots, \mathbf{x}_{s_k}\}$ which are close to the optimal cluster centroids can be obtained by minimizing the objective function:

$$\min_{\mathbf{Z}} \sum_{i=1}^{N} ||\phi(\mathbf{x}_i) - \varphi^*(\mathbf{x}_i)||^2 \text{ subject to } \mathbf{Z} \subset \mathbf{V} \subseteq \mathbf{X}, |\mathbf{Z}| = k, \tag{8}$$

where $\varphi^*(\mathbf{x}_i) = \min_{\phi(\mathbf{x}_{s_j})} ||\phi(\mathbf{x}_i) - \phi(\mathbf{x}_{s_j})||^2$ subject to $\mathbf{x}_{s_j} \in \mathbf{Z}$. If $\mathbf{V} = \mathbf{X}$, (8) is the objective function of kernel k-medoids [13]. Let $\mathbf{a}_i = (a_{ij})_{k \times 1}$ be the cluster indictor vector of $\mathbf{x}_i$, where

$$a_{ij} = \begin{cases} 1, j = \arg\min_l ||\phi(\mathbf{x}_i) - \phi(\mathbf{x}_{s_l})||^2 \\ 0, \text{ otherwize} \end{cases}$$

Then, it is easy to see that

$$\sum_{i=1}^{N} ||\phi(\mathbf{x}_i) - \varphi^*(\mathbf{x}_i)||^2 = \sum_{i=1}^{N} ||\phi(\mathbf{x}_i) - \Phi(\mathbf{Z})\mathbf{a}_i||^2 \tag{9}$$

where $\Phi(\mathbf{Z}) = [\phi(\mathbf{x}_{s_1}), \phi(\mathbf{x}_{s_2}), \cdots, \phi(\mathbf{x}_{s_k})]$. (8) is equivalent to

$$\min_{\mathbf{Z},\mathbf{A}} \sum_{i=1}^{N} ||\phi(\mathbf{x}_i) - \Phi(\mathbf{Z})\mathbf{a}_i||^2 \text{ subject to } \mathbf{Z} \subset \mathbf{V}, |\mathbf{Z}| = k, \tag{10}$$

where $\mathbf{A} = [\mathbf{a}_1, \mathbf{a}_2, \cdots, \mathbf{a}_N] \in \Re^{k \times N}$. This is a difficult combinatorial optimization problem and is NP-hard. Ignoring the special structure of $\mathbf{a}_i$ and let it be an arbitrary vector, we obtain a relaxed minimization problem

$$\min_{\mathbf{Z},\mathbf{B}} \sum_{i=1}^{N} ||\phi(\mathbf{x}_i) - \Phi(\mathbf{Z})\mathbf{b}_i||^2 \text{ subject to } \mathbf{Z} \subset \mathbf{V}, |\mathbf{Z}| = k, \tag{11}$$

where $\mathbf{B} = [\mathbf{b}_1, \cdots, \mathbf{b}_N] \in \Re^{k \times N}$ is an arbitrary matrix.

From the above derivation, the problem of finding representative data points is transformed into a least squares formalism. The problem of (11) is equivalent to find the optimal set of basis vectors to approximate the whole set of vectors $\Phi(\mathbf{X}) \equiv$

$\{\phi(\mathbf{x}_i)\}$ by $\hat{\phi}(\mathbf{x}_i) = \Phi(\mathbf{Z})\mathbf{b}_i$. Since the approximations can be seen as projections of $\Phi(\mathbf{X})$ into the linear subspace spanned by $\Phi(\mathbf{Z})$, the problem tends to find representative data points $\Phi(\mathbf{Z})$ that span a linear space to retain most of the information of $\Phi(\mathbf{X})$ in the feature space. The objective function (11) is very similar to the objective function of transductive experimental design (TED) [14] for active learning. The optimization problem in (11), which seeks to approximate multiple vectors using sparse basis, is NP-hard. Matching pursuit algorithm [14, 15] for multiple vectors is used to greedily find k representative data points.

3.3   The Algorithm

In summary of the discussion so far, given a set of N data samples $(\{\mathbf{x}_1,\ldots,\mathbf{x}_n\})$, the number of clusters (k), the number of nearest neighbors (p), the regularization parameter (β), and the number of candidate initial centers (M), the steps of the proposed algorithm are given below:

1. Construct an adjacency graph to model the relationship between nearby data points. Calculate the weight matrix $\mathbf{W}$, the graph Laplacian $\mathbf{L}$, and the diagonal matrix $\mathbf{D}$.
2. Based on $\mathbf{L}$ and $\mathbf{D}$, calculate a kernel matrix $\mathbf{K} = (\beta\mathbf{I} + \mathbf{D}^{-\frac{1}{2}}\mathbf{L}\mathbf{D}^{-\frac{1}{2}})^{-1}$.
3. Randomly select M data points from the data set X as candidate initial centers.
4. Use the matching pursuit algorithm to select k representative data points from the candidate initial centers. The index set of these k data points is $S = \{s_1, \cdots, s_k\}$.
5. Initialize the clusters: $P_1^{(0)}\{\mathbf{x}_{S_1}\}, P_2^{(0)} = \{\mathbf{x}_{S_2}\}, \cdots, P_k^{(0)} = \{\mathbf{x}_{S_k}\}$, and set $t = 0$.
6. For each data point $\mathbf{x}_i$, and every cluster $c(c\in\{1,\ldots,k\})$, compute the distance

$$d(\mathbf{x}_i, \mathbf{m}_c) = \mathbf{K}_{ii} - 2\frac{\sum_{\mathbf{x}_j\in P_c^{(t)}}\mathbf{K}_{ij}}{|P_c^{(t)}|} + \frac{\sum_{\mathbf{x}_j,\mathbf{x}_l\in P_c^{(t)}}\mathbf{K}_{jl}}{|P_c^{(t)}|^2}.$$

7. For each sample $\mathbf{x}_i$, find its new cluster index as $c^*(\mathbf{x}_i) = \arg\min_c d(\mathbf{x}_i, \mathbf{m}_c)$.
8. Compute the updated clusters as $P_c^{(t+1)} = \{\mathbf{x}_i : c^*(\mathbf{x}_i) = c\}$.
9. If not converged, set $t = t + 1$ and go to Step 6.
10. Output the final clusters $(\{P_c^{(t)}\}_{c=1}^k)$.

In the proposed algorithm, M data points are randomly selected from the data set X as candidate initial centers. If $M < N$, k representative points are non-deterministic. If multiple runs are performed, the output of the run that produces the least SSE is taken.

4   Experiments

To demonstrate the effectiveness of the proposed algorithm, we evaluate and compare five data clustering algorithms:

1. Kernel k-means based on the Gaussian kernel function (KKMGKF). The elements of a Gaussian kernel matrix $\mathbf{K}$ are determined by virtue of kernel tricks:

$\mathbf{K}_{ij} = k(\mathbf{x}_i, \mathbf{x}_j) = \exp(-||\mathbf{x}_i - \mathbf{x}_j||^2/2\sigma^2)$. Initial cluster centers are initialized by k randomly selected data points.

2. Kernel k-means using the graph based kernel matrix (KKMGKM). k data points are randomly chosen as initial cluster centers.

3. Kerenl k-means++ (KKM++) (k-means++ in kernel space). A data point is randomly chosen as the first cluster center. Then, a data point $\mathbf{x}$ is taken as a cluster center with a probability of $\mathrm{md}(\mathbf{x})/\sum_{j=1}^{N} \mathrm{md}(\mathbf{x}_j)$ until k centers are obtained, where $\mathrm{md}(x)$ denotes the minimum-distance from a point x to the previously selected cluster centers.

4. Kerenl k-means using farthest-first initialization (KKMFFI). A data point is randomly chosen as the first cluster center. Then, the i-th ($i \in \{2, \cdots, k\}$) center is chosen to be the point that has the largest minimum distance to the previously selected centers until k centers are obtained.

5. GKKMRDP. M candidate initial centers are randomly selected from the data set. Then, k representative points are selected as initial centers for kernel k-means.

KKMGKM, KKM++, KKMFFI, and GKKMRDP use the graph based kernel matrix at each run of the test.

4.1   Datasets

Three data sets are used in the experiment. The first one is a synthetic data set and the other two are image data sets. The important statistics of these data sets are summarized below (see also Table 1).

The first data set is synthetic. We generate a mixture of k Gaussian components in a 2-D space. Each component has 60 samples. The mean vector of the k-th Gaussian components is $[\mathrm{rem}(k, 5) \cdot 5 + 10, \mathrm{floor}(k/5) \cdot 5 + 10]^T$, where $\mathrm{floor}(x)$ rounds x to the nearest integer towards minus infinity and $\mathrm{rem}(x, 5)$ is $x - 5 \cdot \mathrm{floor}(x/5)$. The covariance matrix of each Gaussian components is $\begin{pmatrix} 1 & 0 \\ 0 & 1 \end{pmatrix}$.

The second data set is the COIL20 image library from Columbia, which contains 20 objects viewed from varying angles. Each object has 72 images which were taken 5° apart as the object is rotated on a turntable. The size of each image is 32 × 32 pixels, with 256 gray levels per pixel. Thus, each image is represented by a 1,024-dimensional vector in the image space.

The third data set is the CMU PIE face data set which contains facial images of 68 people. The face images were captured under different light and illumination

Table 1. Statistics of the three data sets

Dataset	Size (N)	Dimensionality (d)	# of classes (k)
Synthetic	60k	2	k
COIL20	1440	1024	20
PIE	2856	1024	68

conditions, and each person has 42 facial images. Original images were normalized (in scale and orientation) so that the two eyes were aligned at the same position. Then, the facial areas were cropped into the final experimental images. The size of each cropped image is 32×32 pixels, with 256 gray levels per pixel. Thus, each face image is represented by a 1,024-dimensional vector in image space.

4.2   Evaluation Metric

The clustering result is evaluated by comparing the obtained label of each data point with that provided by the ground truth. The clustering performance is measured by the accuracy (AC).

The AC discovers the one-to-one relationship between clusters and classes. Given a point x_i, let c_i and t_i be the obtained cluster label and the label provided by the ground truth, respectively. The AC is defined as:

$$AC = \frac{\sum_{i=1}^{N} \delta(t_i, \mathrm{map}(c_i))}{N},$$ (12)

where N denotes the total number of data points in the test, $\delta(x, y)$ is the delta function that equals one if $x = y$ and equals zero otherwise, and $\mathrm{map}(c_i)$ is the permutation mapping function that maps each cluster label c_i to the equivalent label from the data set.

4.3   Experimental Setting

There is a parameter σ in the heat kernel weighting and selecting σ is very crucial to the performance. In the experiments, the parameter σ is set to the mean of the Euclidean distances between data points. The parameter β is empirically set to 0.0005. The number of candidate initial centers (M) in GKKMRDP is set to $5 \times k$.

In order to randomize the experiments, we conduct the evaluations with different cluster numbers in the experiments. At each run of the test, for a given cluster number k, the synthetic test data points are randomly generated according to the mean vectors and covariance matrices of the Gaussian components. On the COIL20 and PIE data sets, the test images are formed by mixing images from k clusters randomly selected from the data sets. At each run of the test, the mixed images, along with the cluster number k, are provided to the clustering algorithms. For each given cluster number k, 20 test runs are conducted on different randomly chosen clusters except the case when the entire data set is used. The final performance scores are obtained by averaging the scores from the 20 test runs. Each test run consists of 10 sub-runs among which the result of the best sub-run in terms of the corresponding clustering criterion is selected.

Table 2. Clustering results

Method	Synthetic data						
	$k = 5$	$k = 10$	$k = 15$	$k = 20$	$k = 25$	$k = 30$	$k = 35$
KKMGKF	0.9790	0.9341	0.9114	0.9190	0.9096	0.9057	0.8971
KKMGKM	0.9723	0.9117	0.8849	0.8634	0.8507	0.8436	0.8371
KKM++	0.9722	0.9467	0.8993	0.8814	0.8680	0.8631	0.8753
KKMFFI	0.9817	0.9524	0.9464	**0.9230**	**0.9305**	**0.9303**	**0.9238**
GKKMRDP	**0.9818**	**0.9627**	**0.9467**	0.9328	0.9212	0.9144	0.9134
Method	COIL20						
	$k = 8$	$k = 10$	$k = 12$	$k = 14$	$k = 16$	$k = 18$	$k = 20$
KKMGKF	0.6061	0.6180	0.5844	0.5549	0.5477	0.5374	0.5326
KKMGKM	0.8834	0.8218	0.8183	0.8030	0.7727	0.7672	0.8097
KKM++	0.9214	0.8694	0.8517	0.8666	0.8447	0.8242	0.8111
KKMFFI	0.8900	0.8519	0.8541	0.8498	0.8303	0.8061	0.8090
GKKMRDP	**0.9351**	**0.8800**	**0.8777**	**0.8820**	**0.8613**	**0.8375**	**0.8431**
Method	PIE						
	$k = 10$	$k = 20$	$k = 30$	$k = 40$	$k = 50$	$k = 60$	$k = 68$
KKMGKF	0.4644	0.4027	0.3892	0.3696	0.3593	0.3485	0.3407
KKMGKM	0.8908	0.8341	0.7806	0.7832	0.7587	0.7275	0.7227
KKM++	0.9045	0.8330	0.8164	0.7934	0.7738	0.7559	0.7094
KKMFFI	0.8775	0.8267	0.8164	0.8086	0.7999	0.7866	0.7770
GKKMRDP	**0.9452**	**0.9011**	**0.8933**	**0.8771**	**0.8707**	**0.8664**	**0.8715**

4.4 Clustering Results

Table 2 shows the clustering results on the synthetic, COIL20, and PIE data sets, respectively. On most case, GKKMRDP, KKM++ and KKMFFI outperform KKMGKM. This shows that finding data points as initial centers based on some

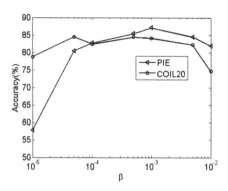

Fig. 1. The performance of GKKMRDP versus parameter β. The performance of GKKMRDP is stable with respect to the parameter β. It achieves consistently good performance when β varies from 5×10^{-5} to 10^{-2}.

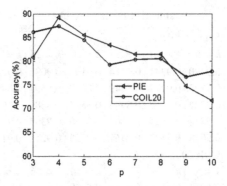

Fig. 2. The performance of GKKMRDP decreases as p increases.

criteria, kernel k-means can achieve a better clustering result. On the synthetic dataset, KKMFFI and GKKMRDP achieve comparative clustering results. On the COIL20 and PIE datasets, GKKMRDP achieves better performance than KKMFFI and KKM++, which suggests the effectiveness of GKKMRDP.

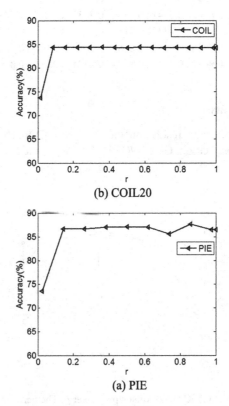

(b) COIL20

(a) PIE

Fig. 3. The performance of GKKMRDP versus the different number of candidate initial centers, $M = rN$.

4.5   Parameters Selection

GKKMRDP has three essential parameters: the number of candidate initial centers M, the number of nearest neighbors p, and the regularization parameter β. Figures 1, 2, and 3 show how the average performance of GKKMRDP varies with the parameters M, p, and β, respectively.

As we can see, the performance of GKKMRDP is very stable and achieves consistently good performance when β varies from 5×10^{-5} to 10^{-2}.

The performance of GKKMRDP decreases as p increases, as shown in Fig. 2. Based on the assumption that two neighboring data points share the same label, a p-nearest graph is used by GKKMRDP to capture the local geometric structure of the data distribution. Obviously, this assumption is more likely to fail as p increases.

The performance of GKKMRDP is very stable with respect to the parameter M. GKKMRDP achieves consistently good performance when M varies from $0.2N$ to N. When $M = k$ and k randomly chosen data points are used for initial centers, the performance of GKKMRDP is the same as KKMGKM.

5   Conclusion

A novel kernel k-means algorithm called GKKMRDP is presented. GKKMRDP uses a graph based kernel matrix which is derived from the geometrical structure of the data distribution. Since finding the optimal data points as initial centers is an NP-hard problem, this problem is relaxed to find k representative data points as initial centers. The problem of finding k representative data points can be transformed into a least squares formalism, which finds the optimal set of k data points to approximate the whole set of data points. Matching pursuit algorithm for multiple vectors is used to greedily find k representative data points. Our empirical study shows encouraging results of the proposed algorithm in comparison with kernel k-means algorithms using other initialization techniques on synthetic and real-world image clustering problems.

Acknowledgments. This work was supported by the National Natural Science Foundation of China (11374245).

References

1. Jain, A.K.: Data clustering, 50 years beyond K-means. Pattern Recogn. Lett. **31**, 651–666 (2010)
2. Macqueen, J.: Some methods for classification and analysis of multivariate observations. In: Fifth Berkeley Symposium on Mathematical Statistics and Probability, pp. 281–296 (1967)
3. Celebi, E.E., Kingravi, H.A., Vela, P.A.: A comparative study of efficient initialization methods for the K-means clustering algorithm. Expert Syst. Appl. **40**, 200–210 (2013)
4. Dhillon, I.S., Guan, Y., Kulis, B.: Weighted graph cuts without eigenvectors: a multilevel approach. IEEE Trans. Pattern Anal. Mach. Intell. **29**(11), 1944–1957 (2007)

5. Schölkopf, B., Smola, A., Müller, K.R.: Nonlinear component analysis as a kernel eigenvalue problem. Neural Comput. **10**, 1299–1319 (1998)
6. Hochbaum, D., Shmoys, D.: A best possible heuristic for the K-center problem. Math. Oper. Res. **10**(2), 180–184 (1985)
7. Ball, G.H., Hall, D.J.: A clustering technique for summarizing multivariate data. Behav. Sci. **12**(2), 153–155 (1967)
8. Arthur, D., Vassilvitskii, S.: K-means++: the advantages of careful seeding. In: Proceedings of the 18th Annual ACM-SIAM Symposium on Discrete Algorithms, pp. 1027–1035 (2007)
9. Hasan, M.A., Chaoji, V., Salem, S., Zaki, M.: Robust partitional clustering by outlier and density insensitive seeding. Pattern Recogn. **30**(11), 994–1002 (2009)
10. Belkin, M., Niyogi, P.: Laplacian eigenmaps and spectral techniques for embedding and clustering. In: Advances In Neural Information Processing Systems, pp. 585–591 (2001)
11. Roweis, S., Saul, L.: Nonlinear dimensionality reduction by locally linear embedding. Science **290**(5500), 2323–2326 (2000)
12. Smola, A.J., Kondor, R.: Kernels and regularization on graphs. In: Schölkopf, B., Warmuth, M.K. (eds.) COLT/Kernel 2003. LNCS (LNAI), vol. 2777, pp. 144–158. Springer, Heidelberg (2003)
13. Velmurugan, T., Santhanam, T.: A survey of partition based clustering algorithms in data mining: an experimental approach. Inf. Technol. J. **10**(3), 478–484 (2011)
14. Yu, K., Bi, J., Tresp, V.: Active learning via transductive experimental design. In: International Conference on Machine Learning (2006)
15. Mallat, S.G., Zhang, Z.: Matching pursuits with time-frequency dictionaries. IEEE Trans. Sig. Process. **41**(12), 3397–3415 (1993)

Palmprint Recognition Based on Image Sets

Qingjun Liang[1], Lin Zhang[1,2(✉)], Hongyu Li[1], and Jianwei Lu[1,3]

[1] School of Software Engineering, Tongji University, Shanghai, China
{13_qingjunliang, cslinzhang,
hyli, jwlu33}@tongji.edu.cn
[2] Jiangsu Key Laboratory of Image and Video Understanding for Social Safety,
Nanjing University of Science and Technology, Nanjing, Jiangsu, China
[3] The Advanced Institute of Translational Medicine,
Tongji University, Shanghai, China

Abstract. In recent years, researchers have found that palmprint is quite a promising biometric identifier as it has the merits of high distinctiveness, robustness, user friendliness, and cost effectiveness. Nearly all the existing palmprint recognition methods are based on one-to-one matching. However, recent studies have corroborated that matching based on image sets can usually lead to a better result. Consequently, in this paper, we present a novel approach for palmprint recognition based on image sets. In our approach, each gallery and query example contains a set of palmprint images captured from a same individual. Competitive code is used for palmprint feature extraction. After the feature extraction process, we use the method of sparse approximated nearest points (SANP) for palmprint image set classification. By calculating the minimum between-set distance, we can set the label of each testing palmprint set as that of the nearest training set. Effectiveness of the proposed approach has been corroborated by the experiments conducted on PolyU palmprint database.

Keywords: Palmprint recognition · Image set classification · Competitive code · Sparse approximated nearest points

1 Introduction

Recent years have witnessed a lot of effective biometrics authentication methods for recognizing a person's identity automatically [1]. Due to the uniqueness and inherent characteristics of biometric identifiers, more interests have been attracted in iris, fingerprint, finger-knuckle-print, and palmprint [2]. As an important member of the biometrics family, palmprint based personal authentication techniques have drawn much attention from researchers in recent years. A palmprint image has many unique features that can be used for identification, such as principal lines, wrinkles, valleys, ridges and minutiae points [3], as illustrated in Fig. 1. Palmprint recognition is traditionally regarded as the problem of single image comparison and matching, and many eminent algorithms have been proposed for palmprint matching.

PalmCode proposed in [3] uses a single Gabor filter to extract the local phase information. Its computational architecture is the same as the IrisCode [4]. In [5], Kong and Zhang proposed the competitive code (CompCode) scheme, which encodes the

© Springer International Publishing Switzerland 2015
D.-S. Huang et al. (Eds.): ICIC 2015, Part I, LNCS 9225, pp. 305–315, 2015.
DOI: 10.1007/978-3-319-22180-9_30

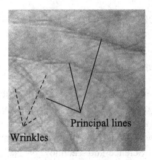

Fig. 1. Principal lines and wrinkles contained in a typical palmprint image.

local orientation field of a palmprint using symmetric Gabor filters along six different orientations. In [6], Jia *et al.* proposed a different coding method to extract the local orientation information of palmprints, namely robust line orientation code (RLOC), which is based on a modified finite Radon transform. In [7], Sun *et al.* used differences between two orthogonal Gaussians to extract the local ordinal measures from palmprints. In [8], Guo *et al.* proposed another coding method by binarizing a palmprint image's responses to the real Gabor filters along six different orientations and they named their method as binary orientation co-occurrence vector (BOCV).

The abovementioned methods actually all make use of a single image for matching. However, information contained in a single image is limited to some extent for matching. For instance, a single palmprint image cannot cover different illuminations and poses. Image set can solve this problem since each set can provide us more information about the interested object.

Image set classification is a very challenging topic. Unlike the verification and the identification, which are considered to be a one-to-one and one-to-many problem respectively, image set classification is a many-to-many problem. Generally, there are two main aspects merit attention: the representation of an image set, and the distance metrics to match the image sets.

Several approaches of image set classification have been proposed over past two decades [9–16]. In [12], Shakhnarovich *et al.* classify sets of images using the relative entropy between the estimated density of the input set and that of stored collections of images for each class. In [13], Arandjelovic *et al.* proposed a semi-parametric model for learning probability densities confined to highly non-linear but intrinsically low-dimensional manifolds. These methods use parametric representations to model an image set. When the test sets have weak statistical correlation to the training ones, the parameters may not characterize the image sets well. In [14], Fukui *et al.* presented a method to classify sets based on the canonical angles between the nonlinear subspaces, namely *kernel orthogonal mutual subspace method* (KOMSM). In [15], Nishiyama *et al.* used a hierarchical image-set matching (HISM) for face recognition. In [16], Nishiyama *et al.* applied ensemble learning to the Constrained Mutual Subspace Method (CMSM) and they named their new method as Multiple Constrained Mutual Subspace Method (MCMSM). These nonparametric methods can represent an image set without any assumption on data set distribution. In [10, 11], Hu *et al.* proposed a novel sparse formulation for image set classification by jointly minimizing the distance and maximizing the sparsity of the nearest points and achieved best performance compared with others. Hence, we choose Hu *et al.*'s method in palmprint image set classification. The method proposed in [10, 11] outperform those based on individual images, because we can synthesize the most nearest points from gallery and query sets to do the matching and ignore the unrelated outliers.

Since an image set taken from an object usually contains much more complementary information for matching, in this paper, we propose a novel approach for palmprint recognition based on image-set matching. With such a scheme, each gallery sample contains a set of images taken from the same palm at the registration stage and each probe sample also contains a set of images captured from the probe palm at the identification (or verification) stage. In this way, the identity of the probe palm can be determined by matching the probe image set with the gallery image sets with proper designed image-set metrics. Particularly, in our approach, for feature extraction and representation we make use of CompCode [5], which has been proved to be quite effective and efficient for palmprint representation in previous works. With respect to the scheme to compute the between-set distances, we resort to the approach Sparse Approximated Nearest Points (SANP) [10, 11], which is a state-of-the-art image-set matching method. We conducted experiments on the PolyU palmprint database [17] and the experimental results have corroborated the effectiveness of our proposed method. To our knowledge, our present work is the first work investigating the palmprint recognition problem in the context of image-set matching. This work may encourage colleagues to pay more attention to applications of image-set matching in the fields of biometrics.

The rest of the paper is arranged as follows. In Sect. 2, we will give a general review on competitive code. In Sect. 3, more details on palmprint image set classification will be given. Section 4 is about the experimental results and some conclusions will be drawn in Sect. 5.

2 Competitive Code

In our work, we will use CompCode [5] to represent each palmprint image. Thus, in this section, we will firstly give a brief review of CompCode [5].

CompCode has several advantages: (1) less space occupation for the features, (2) high matching speed and (3) high matching accuracy. The method has three core parts: (1) the filters, (2) the coding rules, and (3) the matching function. Since we will perform the image-set matching, the matching function for CompCode will be ignored here. We will only focus on the feature representation part: the Gabor filters and the coding rules.

2.1 Gabor Filters

As for the filters, the competitive coding scheme simply selects six different orientations $\theta_j = j\pi/6$, j = {0,1,...,5} of Gabor filters. The real parts of Gabor filters $\Psi_R(x,$

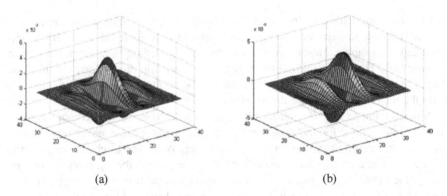

Fig. 2. The real part (a) and the imaginary part (b) of a Gabor filter.

y, ω, θ_j) with six different directions θ_j are applied to a preprocessed palmprint image $I(x, y)$. Figure 2 illustrates the real part and the imaginary part of a Gabor filter.

The circular Gabor filter proposed by Lee [18] has a general form as follows:

$$G(x, y, \theta, \mu, \sigma) = \frac{1}{2\pi\sigma^2} \exp\left\{-\frac{x^2 + y^2}{2\sigma^2}\right\} \exp\{2\pi i(\mu x \cos \theta + \mu y \sin \theta)\} \quad (1)$$

where $i = \sqrt{-1}$, μ is the frequency of the sinusoidal wave, θ controls the orientation of the function, and σ is the standard deviation of the Gaussian envelop.

2.2 Coding Rules

The Winner-take-all rule is applied to obtain the orientation of a local region:

$$k = \arg \min_j \left(I(x, y) * \psi_R\left(x, y, \omega, \theta_j\right)\right) \quad (2)$$

where $j = \{0, 1, 2, 3, 4, 5\}$ corresponds to the orientation index and k is called the winning index.

The competitive coding scheme adopts the Winner-take-all policy. It will only take the orientation that corresponds to the minimum value of the filtering result into account. The orientation will be encoded in 3 bits (there are only six possible values that the dominant orientation can take and thus 3 bits are enough to represent it).

2.3 Down-Sampling

Figure 3(a) and (b) shows an original 128 × 128 palmprint image and its corresponding competitive code where different gray level lines represent different orientations. Compared with the original image, its competitive code depicts the palm line more clearly. The size of the competitive code is a bit smaller because we exclude the boundary part in Gabor filtering process.

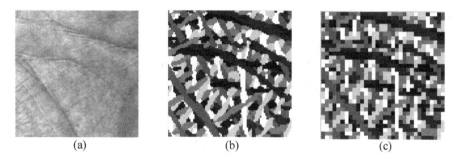

(a)   (b)   (c)

Fig. 3. (a) Original palmprint; (b) Competitive code; (c) Down-sampling competitive code.

Down-sampling is an indispensable process in our experiment. After Gabor filtering, the feature size is still too big for a single palmprint image. We can get a low-resolution image after down sampling (see Fig. 3(c)) and it will still contain enough information.

3 Palmprint Image Set Classification

In this section, we will describe our image-set matching based palmprint recognition scheme. For image-set matching, we resort to SANP [10, 11].

We model each image set as a single affine subspace, in which the palmprint image samples are sparsely scattered. To compare the similarity of two palmprint image sets, we use the geometric distance (distance between the closest pair of points lied on the convex surface) between their convex models. If two image sets come from the same palm, due to the illuminations and poses variance, they may not overlap everywhere, but they will close to each other at least some points. That's why the geometric distance is a sensible metric. In fact, since it is permissible to synthesize new palmprint example within each set by a sparse combination of sample images, finding the geo-metric distance also means synthesize the closest pair of palmprint examples.

3.1 Palmprint Image Set Representation

We first extract the feature of a palmprint image sample using the competitive code [5]. Let a feature vector be $x_{ci} \in R^d$ where c denotes the cth palmprint set and i denotes the ith palmprint in the cth palmprint image set. Thus a palmprint image set which contains N_c palmprint samples can be represented as:

$$X_c = [x_{c1}, x_{c2}, \ldots, x_{cN_c}] \tag{3}$$

We can model a palmprint image set as an affine hull of the data set [9]:

$$AH_c = \left\{ x = \sum_{i=1}^{N_c} \alpha_{ci} \cdot x_{ci} \middle| \sum_{i=1}^{N_c} \alpha_{ci} = 1 \right\} \tag{4}$$

where $a_{ci} \in R$ for all $i = 1, 2, \ldots, N_c$. A palmprint image set can be described as any affine combination of an individual's palmprint feature vectors. From this affine hull model, we can see that it is a rather loose approximation to the data set.

This affine hull model has another equivalent parametric form, which may look quite different. We choose the sample mean $\mu_c = \frac{1}{N_c} \sum_{i=1}^{N_c} x_{ci}$ as a reference point and rewrite the hull as:

$$AH_c = \left\{ x = \mu_c + U_c v_c \middle| v_c \in \mathcal{R}^l \right\} \tag{5}$$

where $U_c \in R^{d \times l}$ is an orthonormal basis obtained by applying the thin Singular Value Decomposition (SVD) to the centered feature matrix $\overline{X}_c = [x_{c1} - \mu_c, x_{c2} - \mu_c, \ldots, x_{cNc} - \mu_c]$ and v_c is a vector of coefficients with regard to U_c.

3.2   Convex Formulation and Optimization

Here we adopt Hu's method to establish the convex formulation [10, 11]. Given two palmprint image sets X_i and X_j, their affine hulls can be represented as $AH_i = \{x = \mu_i + U_i v_i \mid v_i \in R^l\}$ and $AH_j = \{x = \mu_j + U_j v_j \mid v_j \in R^l\}$ accordingly. In [10, 11], there are several functions needed to pay attention to and we list them below:

$$
\begin{aligned}
F_{v_i,v_j} &= \left| (\mu_i + U_i \cdot v_i) - (\mu_j + U_j \cdot v_j) \right|_2^2 \\
G_{v_i,\alpha} &= \left| (\mu_i + U_i \cdot v_i) - X_i \cdot \alpha \right|_2^2 \\
Q_{v_j,\beta} &= \left| (\mu_j + U_j \cdot v_j) - X_j \cdot \beta \right|_2^2
\end{aligned}
\tag{6}
$$

Where $\{v_i, v_j, \alpha, \beta\}$ are the unknowns and the rest are the known arguments. F_{v_i,v_j} defines the geometric distance between two palmprint image sets X_i and X_j using the parametric form. $G_{v_i,\alpha}$ defines the difference between two representation forms for image set X_i. Similarly, $Q_{v_j,\beta}$ defines the difference for image set X_j. The convex formulation can be set up as follows [10, 11]:

$$\min_{v_i,v_j,\alpha,\beta} \left(F_{v_i,v_j} + \lambda_1 \left(G_{v_i,\alpha} + Q_{v_j,\beta} \right) + \lambda_2 |\alpha|_1 + \lambda_3 |\beta|_1 \right) \tag{7}$$

where the minimization of F_{v_i,v_j} can ensure the inter-set distance is small (using the nearest pair of points). The second term is supposed to minimize the inner-set representation difference. The last two terms are proposed here to ensure the sparsity of the affine hull approximations.

In [9], the affine hull model can be used to synthesize new example from any affine combinations of the sample features. However, it is proved in [10, 11] that the sparse linear combination performs much better than the dense linear combination. There for a new term: Sparse Approximated Nearest Points (SANP) is put forward to calculate the distance between different image sets. The strategy of sparse combination will make SANPs from different identities further apart than SANPs from the same identity, which contributes to the classification accuracy. That's the reason for introducing the last two sparsity constraints.

Equation (7) can be proved as a convex formulation. The convex optimization problem in Eq. (7) can be solved by many algorithms, e.g., gradient-descent. In [19, 20], an efficient method Accelerated Proximal Gradient (APG) is adapted to optimize Eq. (7). Due to the high convergence rate among first order methods, the optimal solution $\{v_i, v_j, \alpha, \beta\}$ will be got at the end.

3.3 Distance Measurement

For the given two palmprint image sets X_i and X_j, the between-set distance based on SANPs can be represented as follows [10, 11]:

$$D(X_i, X_j) = (d_i + d_j) \cdot [F_{min} + \lambda_1 (G_{min} + Q_{min})] \tag{8}$$

where F_{min}, G_{min}, Q_{min} are the optimal values obtained by the optimal coefficients $\{v_i, v_j, \alpha, \beta\}$ (optimal solution of Eq. (7)). d_i and d_j are the dimensions of the affine hulls of X_i and X_j, respectively.

3.4 Algorithm

Summary of palmprint recognition based on image sets are listed in Table 1.

Table 1. The overall processes for palmprint image sets classification

Input: The gallery image set S_i and the query image set S_j.
Output: The distance d_{ij} between S_i and S_j.
1. Extract the features of the palmprint images in S_i and S_j by competitive code method. Arrange the features vectors of the palmprint images within a palmprint image set into a data matrix, and form X_i and X_j respectively.
2. Compute the necessary parameters: (1) mean vectors μ_i and μ_j, (2) orthonormal bases U_i and U_j from centralized X_i and X_j.
3. Establish the convex formulation Eq. (7) using Hu's method and find the optimal solution $\{v_i, v_j, \alpha, \beta\}$ of it by APG algorithm.
4. Take the optimal coefficients $\{v_i, v_j, \alpha, \beta\}$ back to Eq. (6) and get the optimal values F_{min}, G_{min}, Q_{min}.
5. Calculate the between-set distance d_{ij} of X_i and X_j using Eq. (8).
6. Classify S_j based on the between-set distance in step 5.

4 Experiments and Results

To evaluate the performance of the proposed method, we use the PolyU palmprint dataset [17] for the experiments.

4.1   Datasets

PolyU Palmprint Database [17] is used for the evaluation of the recognition performance. In the database, there are 386 different palms collected from 193 individuals at the Hong Kong PolyU. Each person provides his/her left palm and right palm. In addition, for each palm, there are about 20 samples collected on two separate occasions, the interval of which is at least two months. A subject is asked to provide about 10 samples for each of his/her palms on each occasion. Thus, 40 palmprint image samples are provided by an individual approximately. The palmprint database has 7,752 image samples in total.

The light source and the focus of the camera are changed in the second collecting occasion, which makes the samples collected on the first and second occasions look different (see Fig. 4).

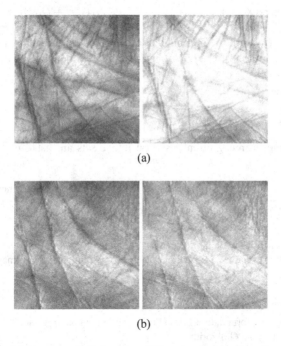

(a)

(b)

Fig. 4. Palmprint images collected on two different occasions from the same person.

4.2   Experimental Settings

All the palmprint images are normalize into the size of 128×128. After the process of the feature extraction using competitive code, the size of samples become 94×94 due to the exclusion of the boundary part in Gabor filtering process. To further reduce the feature dimension, we down sample each competitive code into the size of 32×32. Then the pixel values of the down sampled competitive code are vectorized to form the columns of data matrix X.

We separate the image samples of each palmprint into two image sets by the collecting occasion. The first collections of the palmprints are arranged as the gallery sets and the second collections of the palmprints are arranged as the query sets. Since there are 386 different palms, there can be totally 772 image sets: 386 image sets are used for training and the remaining 386 image sets are used for testing.

The lengths of the sets vary from 10 to 17. In the configuration script, we set the upper bound of maximum set length to 17. λ_1 in Eq. (7) is fixed to 0.01 in all the experiments.

We conduct experiments three times by selecting different number of image sets: (1) using 50 gallery sets and 50 query sets, (2) using 100 gallery sets and 100 query sets, (3) using 386 gallery sets and 386 query sets. For clear illustration, we adopt the second situation (100 gallery sets and 100 query sets) for example. For each query set, a geometric distance between this query set and a gallery set in linear subspace is calculated. Thus each query set needs 100 times distance calculation for matching. In total, 10,000 times distance calculation are needed for matching for whole query sets.

4.3   Experimental Results

Table 2 shows the classification accuracies of the three conditions with different number of image sets.

Table 2. Classification accuracies for different number of image sets

Number of gallery sets	Number of query sets	Recognition Rate
50	50	100 %
100	100	99 %
386	386	96.63 %

When the number of gallery sets and query sets are both 20, we illustrate the SANPs of a given query set to all gallery sets in Fig. 5. We can see that when two palmprint image sets are matched with each other, the distance between them is the smallest. Other distances that correspond to the wrong gallery sets are much larger than the correct gallery set.

5   Limitations

Table 2 shows that when the number of gallery and query sets increase, the recognition rate will drop down. That's mainly because the image set classification method we adopted here [10, 11] relies on the convex region in feature space. However, in our testing dataset, the difference between each palmprint sample within an image set (inner-set difference) is small, which makes the subspace over small and limits the construction of the convex model for distance measure.

Another underlying reason is that the set lengths of each gallery and query set in palmprint dataset (whose average set length is 10) are much shorter than those in the

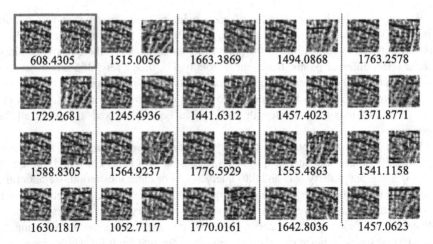

608.4305	1515.0056	1663.3869	1494.0868	1763.2578
1729.2681	1245.4936	1441.6312	1457.4023	1371.8771
1588.8305	1564.9237	1776.5929	1555.4863	1541.1158
1630.1817	1052.7117	1770.0161	1642.8036	1457.0623

Fig. 5. SANPs between a query set and 20 gallery sets based on the respective distance. Notice that the matching subjects which have the smallest distance are highlighted with a red box. Each palmprint image set is represented as the competitive code form in size of 32×32 in this figure (Color figure online).

face video related datasets, such as the Honda/USCD (whose maximum set length is 645) and CMU Mobo dataset.

6   Conclusion and Future Work

In this work, we propose a new approach for palmprint recognition based on image sets classification (many-to-many). Our method first extracts the features of the palmprint images and vectorizes the features. Next, we arrange those feature vectors from the corresponding image set into a data matrix. Then we adopt Hu's method-SANPs to characterize a palmprint image set into the forms a convex region (a joint representation) in feature space. Recognition is based on the geometric distance between the gallery region and the query region (finding the nearest points on the convex sets). In experiments, we evaluate our methods on PolyU palmprint dataset. It turns out a satisfactory recognition rate for the testing palmprint image sets.

For future work, we are interested in applying image set classification methods to more complex visual recognition problems, e.g. the 3D face recognition, where the inner-set difference could be much larger due to different face expressions.

Acknowledgement. This work was supported in part by the Natural Science Foundation of China under Grant 61201394, in part by the Shanghai Pujiang Program under Grant 13PJ1408700 and Grant 14PJ1408100, and in part by the Jiangsu Key Laboratory of Image and Video Understanding for Social Safety, Nanjing University of Science and Technology, Nanjing, China, under Grant 30920140122007.

References

1. Jain, A.K., Flynn, P.J., Ross, A.: Handbook of Biometrics. Springer, New York (2007)
2. Li, S.Z. (ed.): Encyclopedia of Biometrics. Springer, New York (2009)
3. Zhang, D., Kong, W.K., You, J., Wong, M.: On-line palmprint identification. IEEE Trans. PAMI **25**, 1041–1050 (2003)
4. Daugman, J.G.: High confidence visual recognition of persons by a test of statistical independence. IEEE Trans. PAMI **15**, 1148–1161 (1993)
5. Kong, A.W.K., Zhang, D.: Competitive coding scheme for palmprint verification. In: Proceedings of ICPR, pp. 520–523 (2004)
6. Jia, W., Huang, D., Zhang, D.: Palmprint verification based on robust line orientation code. Pattern Recogn. **41**, 1504–1513 (2008)
7. Sun, Z., Tan, T., Wang, Y., Li, S.Z.: Ordinal palmprint representation for personal identification. In: Proceedings of CVPR, pp. 279–284 (2005)
8. Guo, Z., Zhang, D., Zhang, L., Zuo, W.: Palmprint verification using binary orientation co-occurrence vector. Pattern Recogn. Lett. **30**, 1219–1227 (2009)
9. Cevikalp, H., Triggs, B.: Face recognition based on image sets. In: Proceedings of CVPR, pp. 2567–2573 (2010)
10. Hu, Y., Mian, A.S., Owens, R.: Face recognition using sparse approximated nearest points between image sets. IEEE Trans. PAMI **34**, 1992–2004 (2012)
11. Hu, Y., Mian, A.S., Owens, R.: Sparse approximated nearest points for image set classification. In: Proceedings of CVPR, pp. 121–128 (2011)
12. Shakhnarovich, G., Fisher III, J.W., Darrell, T.: Face recognition from long-term observations. In: Heyden, A., Sparr, G., Nielsen, M., Johansen, P. (eds.) ECCV 2002, Part III. LNCS, vol. 2352, pp. 851–865. Springer, Heidelberg (2002)
13. Arandjelovic, O., Shakhnarovich, G., Fisher, J., Cipolla, R., Darrell, T.: Face recognition with image sets using manifold density divergence. In: Proceedings of CVPR, pp. 581–588 (2005)
14. Fukui, K., Yamaguchi, O.: The kernel orthogonal mutual subspace method and its application to 3D object recognition. In: Yagi, Y., Kang, S.B., Kweon, I.S., Zha, H. (eds.) ACCV 2007, Part II. LNCS, vol. 4844, pp. 467–476. Springer, Heidelberg (2007)
15. Nishiyama, M., Yuasa, M., Shibata, T., Wakasugi, T., Kawahara, T., Yamaguchi, O.: Recognizing faces of moving people by hierarchical image-set matching. In: Proceedings of CVPR, pp. 1–8 (2007)
16. Nishiyama, M., Yamaguchi, O., Fukui, K.: Face recognition with the multiple constrained mutual subspace method. In: Kanade, T., Jain, A., Ratha, N.K. (eds.) AVBPA 2005. LNCS, vol. 3546, pp. 71–80. Springer, Heidelberg (2005)
17. PolyU palmprint database. http://www.comp.polyu.edu.hk/~biometrics/
18. Lee, T.S.: Image representation using 2D Gabor wavelets. IEEE Trans. PAMI **18**, 959–971 (1996)
19. Beck, A., Teboulle, M.: A fast iterative shrinkage-thresholding algorithm for linear inverse problems. SIAM J. Imaging Sci. **2**, 183–202 (2009)
20. Nesterov, Y.: Gradient methods for minimizing composite objective function. Technical report 2007076, Universite´ Catholique de Louvain, Center for Operations Research and Econometrics (CORE) (2007)

Face Recognition Using SURF

Raúl Cid Carro[1], Juan-Manuel Ahuactzin Larios[2],
Edmundo Bonilla Huerta[1(✉)], Roberto Morales Caporal[1],
and Federico Ramírez Cruz[1]

[1] Instituto Tecnológico de Apizaco, 90300 Apizaco, Tlaxcala, Mexico
`edbonn@hotmail.com, edbonn@hotmail.fr`
[2] Probayes Americas SA de CV, Vía Atlixcáyotl 2301,
Parque Tecnológico de Puebla, 72453 Puebla, Puebla, Mexico

Abstract. In recent years, several scale-invariant features have been proposed in literature, this paper analyzes the usage of Speeded Up Robust Features (SURF) as local descriptors, and as we will see, they are not only scale-invariant features, but they also offer the advantage of being computed very efficiently. Furthermore, a fundamental matrix estimation method based on the RANSAC is applied. The proposed approach allows to match faces under partial occlusions, and even if they are not perfectly aligned. Thus based on the above advantages of SURF, we propose to exploit SURF features in face recognition since current approaches are too sensitive to registration errors and usually rely on a very good initial alignment and illumination of the faces to be recognized.

1 Introduction

Face recognition has been an active area of research over the last two decades due to wide applications. It involves computer recognition of personal identity based on geometric or statistical features derived from face images. Nowadays, illumination invariance, facial expressions, and partial occlusions are one of the most challenging problems in face recognition [1], where face images are usually analyzed locally to cope with the corresponding transformations. Local feature descriptors describe a pixel in an image through its local neighborhood content, their purpose is to provide a representation that allows to efficiently match local structures between images. They should be distinctive and at the same time robust to changes in viewing conditions. Many different descriptors and interest-point detectors have been proposed in the literature, and the descriptor performance often depends on the interest point detector [2].

The Scale Invariant Feature Transform (SIFT) proposed by David G. Lowe [3] has been widely used in object detection and recognition. Nonetheless, despite the variety of works on the use of SIFT features in face recognition, these methods still cannot satisfy the speed requirement of on-line applications. On the other hand, SURF suggested by Herbert Bay [4] is a scale and in-plane rotation invariant detector and descriptor with comparable or even better performance than SIFT. Its feature is also personal specific. Just like SIFT, in SURF, detectors are first employed to find the interest points in an image, and then the descriptors are used to extract the feature vectors at each interest point. However, instead of Difference of Gaussians (DoG) filter

© Springer International Publishing Switzerland 2015
D.-S. Huang et al. (Eds.): ICIC 2015, Part I, LNCS 9225, pp. 316–326, 2015.
DOI: 10.1007/978-3-319-22180-9_31

used in SIFT, SURF uses Hessian-matrix approximation operating on the integral image to locate the interest points, which reduces the computation time drastically.

The benefits of facial recognition are that it is not intrusive, can be done from a distance even without the user being aware they are being scanned. What sets apart facial recognition from other biometric techniques is that it can be used for surveillance purposes; as in searching for wanted criminals, suspected terrorists, and missing children. Facial recognition can be done from far away so with no contact with the subject so they are unaware they are being scanned.

2   Speed-Up Robust Features (SURF)

SURF is a scale and in-plane rotation invariant feature. It contains interest point detector and descriptor. The detector locates the interest points in the image, and the descriptor describes the features of the interest points and constructs the feature vectors of the interest points. Conceptually similar to the SIFT descriptor, the 64-dimensional SURF descriptor also focusses on the spatial distribution of gradient information within the interest point neighborhood. The SURF descriptor is invariant to rotation, scale, brightness and, after reduction to unit length, contrast. It is outperforms to existing schemes in terms of repeatability, distinctiveness and robustness, with a faster performance.

2.1   Interest Point Detection

In computer vision, the concept of interest points, also called keypoints or feature points, has been largely used to solve many problems in object recognition, image registration, visual tracking, 3D reconstruction, and more. It relies on the idea that instead of looking at the image as a whole, it could be advantageous to select some special points in the image and perform a local analysis on these ones. The main difference between SURF and SIFT descriptors is mainly speed and accuracy. Since SURF descriptors are mostly based on intensity differences, they are faster to compute. However, SIFT descriptors are generally considered to be more accurate in finding the right matching feature [3]. In order to detect characteristic points on a scale invariably SIFT approach it uses cascaded filters, where the difference Gaussian, DoG, is calculated on rescaled images progressively.

2.2   Integral Image

The detection is speeded up by the summed area table known as an integral image, which is an algorithm for generating the sum of values in a rectangular sum of grid. The integral image can be given as:

$$I(x) = \sum_{i=0}^{i \le x} \sum_{j=0}^{j \le y} I(x, y)$$

2.3 Points of Interest in the Hessian Matrix

Different from SIFT using DoG to detect interest points, The SURF detector focuses its attention on blob-like structures in the image. These structures can be found at corners of objects, but also at locations where the reflection of light on specular surfaces is maximal. Integral images are used in Hessian matrix approximation, which reduce computation time drastically. Given a point $X = (x, y)$ in an image I, the Hessian matrix $H(x, \sigma)$ in X at scale σ is defined as follows:

$$H(X, \sigma) = \begin{bmatrix} L_{xx}(X, \sigma) & L_{xy}(X, \sigma) \\ L_{xy}(X, \sigma) & L_{yy}(X, \sigma) \end{bmatrix}$$

where $L_{xx}(X, \sigma)$, $L_{xy}(X, \sigma)$ and $L_{yy}(X, \sigma)$ are the convolutions of the second order Gaussian derivatives $\partial^2 g(\sigma)/\partial x^2$ with the image at point $X = (x, y)$. These derivatives are known as the Laplacian of Gaussians. The local maxima of this filter response occurs in regions where both L_{xx} and L_{yy} are strongly positive, and where L_{xy} is strongly negative. Therefore, these extrema occur in regions in the image with large intensity gradient variations in multiple directions, as well as at saddle points. Visually, this means that blob-like structures refer to corners and speckles. The calculation of all of these derivatives at different scales is computationally costly. The objective of the SURF algorithm is to make this process as efficient as possible.

Bay and many others have found that the number of results in scale-space decreases exponentially with increasing scale size. One potential explanation is that large amounts of Gaussian blurring can average out nearly all useful information in images. Therefore, searching through the scale space linearly can be very wasteful computationally. As an alternative, SURF introduces the notion of scale octaves: each octave linearly samples a small region of the scale space, but this region size is proportional to the scale magnitudes used. Each adjacent octave doubles the region size and sampling increment used in the previous one, to reduce the amount of search necessary at larger scales. The only negative side-effect of using octaves is that results found at larger scales can potentially have more error, due to the larger increments used in their corresponding octaves. To compensate, the SURF detector interpolates the coordinates of any local maxima found into the sub-pixel and sub-scale range.

Finally, Bay propose to make the distinction between bright blobs found on a dark background, versus dark blobs found on a bright background. This property can be represented by the sign of the Laplacian, as shown below:

$$sgm\{L_{xx}(X, \sigma) + L_{yy}(X, \sigma)\} = \begin{cases} +1 \Rightarrow bright\ blob\ over\ dark\ background \\ +1 \Rightarrow dark\ blob\ over\ bright\ background \end{cases}$$

To find blobs, we need to convolve the source image with various Gaussian-related filters. Because continuous Gaussian filters have non-integer weights, these convolutions require floating-point operations, which can severely hamper processing speed. Thus, SURF's authors propose to approximate these filters by using a set of 9×9 box filters (Fig. 1) this is used as the approximations of a Gaussian with $\sigma = 1.2$ and

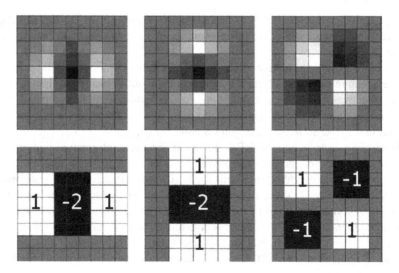

Fig. 1. Top row (left to right): The Gaussian second order partial derivative in L_{xx}, L_{yy}, and L_{xy}. Bottom row (left to right): The approximation for the second order Gaussian partial derivative D_{xx}, D_{yy} and D_{xy}

represents the lowest scale for computing the blob response maps. We will denote them by D_{xx}, D_{yy} and D_{xy}. The weights applied to the rectangular are kept simple for computational efficiency. This yields:

$$\det(H_{approx}) = D_{yy}D_{xx} - (wD_{xy})^2$$

where w is a weight for the energy conservation between the Gaussian kernels and the approximated Gaussian kernels, and

$$w = \frac{|L_{xy}(1.2)|_F |D_{yy}(9)|_F}{|L_{yy}(1.2)|_F |D_{xy}(9)|_F} = 0.912 \approx 0.9$$

$|X|_F$ is the Frobenius norm.

In order to detect keypoints using the determinant of Hessian it is necessary to introduce the notion of a scale space. The SURF constructs a pyramid scale space, like the SIFT. Different from the SIFT to repeatedly smooth the image with a Gaussian and then subsample the image, the SURF directly changes the scale of box filters to implement the scale space due to the use of the box filter and integral image [7].

2.4   Features Descriptors

In general, feature descriptors describe a pixel (or a position) in an image through its local content. They are supposed to be robust to small deformations or localization errors, and give us the possibility to find the corresponding pixel locations in images which capture the same amount of information about the spatial intensity patterns under different conditions.

In feature matching, feature descriptors are usually N-dimensional vectors that describe a feature point, ideally in a way that is invariant to change in lighting and to small perspective deformations. In addition, good descriptors can be compared using a simple distance metric (for example, Euclidean distance). Therefore, they constitute a powerful tool to use in feature matching algorithms [5].

To describe each feature, SURF summarizes the pixel information within a local neighborhood. The first step is determining an orientation for each feature, by convolving pixels in its neighborhood with the horizontal and the vertical Haar wavelet filters. Shown in Fig. 2, these filters can be thought of as block-based methods to compute directional derivatives of the image's intensity. By using intensity changes to characterize orientation, this descriptor is able to describe features in the same manner regardless of the specific orientation of objects or of the camera. This rotational invariance property allows SURF features to accurately identify objects within images taken from different perspectives.

In fact, the intensity gradient information can also reliably characterize these pixel regions. By looking at the normalized gradient responses, features in images taken in a dark room versus a light room, and those taken using different exposure settings will all have identical descriptor values. Therefore, by using Haar wavelet responses to generate a unit vector representing each feature and its neighborhood, the SURF feature framework inherits two desirable properties: lighting invariance and contrast invariance.

For the extraction of the descriptor, the first step consists of constructing a square region centered at the interest point and oriented along the orientation decided by the orientation selection method introduced by Bay. The region is split up equally into smaller 4×4 square sub-regions (as shown in Fig. 3). This preserves important spatial information.

For each sub-region, we compute Haar wavelet responses at 5×5 equally spaced sample points. For simplicity, we call d_x the Haar wavelet response in horizontal direction and d_y the Haar wavelet response in vertical direction. To increase the robustness towards geometric deformations and localization errors, the responses d_x and d_y are first weighted with a Gaussian centered at the interest point. These values are computed (with a kernel size of 4σ) within a circular neighborhood of radius 6σ at locations regularly spaced by intervals of σ. For a given orientation, the responses inside a certain angular interval $\pi/3$ are summed, and the orientation giving the longest vector is defined as the dominant orientation.

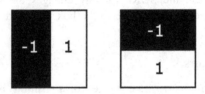

Fig. 2. Horizontal and vertical Haar wavelet filters

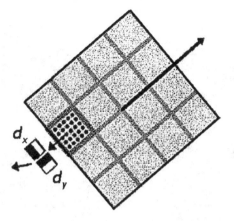

Fig. 3. Structure breakdown of each feature's neighborhood

Then, the wavelet responses d_x and d_y are summed up over each sub-region and form a first set of entries in the feature vector. In order to bring in information about the polarity of the intensity changes, we also extract the sum of the absolute values of the responses, $|d_x|$ and $|d_y|$. Hence, each sub-region has a four-dimensional descriptor vector v for its underlying intensity structure $v = \sum d_x, \sum d_y, \sum |d_x|, \sum |d_y|$. Concatenating this for all 4×4 sub-regions, this results a descriptor vector of length 64. The wavelet responses are invariant to a bias in illumination (offset). Invariance to contrast (a scale factor) is achieved by turning the descriptor into a unit vector.

2.5   Recognition by Matching

With the SURF features and descriptors, scale-invariant matching can be achieved. For example, (Fig. 4) shows two faces of the same person in different moments where the lines indicate the corresponding matched interest points. This is accomplished by first detecting features on each image, and then extracting the descriptors of these features. Each feature descriptor vector in the first image is then compared to all feature descriptors in the second image. The pair that obtains the best score (that is, the lowest distance between the two vectors) is then kept as the best match for that feature. This process is repeated for all features in the first image. Good feature descriptors must be invariant to small changes in illumination, in viewpoint, and to the presence of image noise. Therefore, they are often based on local intensity differences.

The matching is carried out by a nearest neighbor matching strategy $m(X, Y)$: the descriptor vectors $X = \{x_1, \ldots, x_I\}$ extracted at keypoints $1, \ldots, I$ in a test image X are compared to all descriptor vectors $Y = \{y_1, \ldots, y_J\}$ extracted at keypoints $1, \ldots, J$ from the reference images Y_n, $n = \{1, \ldots, N\}$ by the Euclidean distance. Additionally, a ratio constraint is applied only if the distance from the nearest neighbor descriptor is less than α times the distance from the second nearest neighbor descriptor, a matching pair is detected. Finally, the classification is carried out by assigning the class

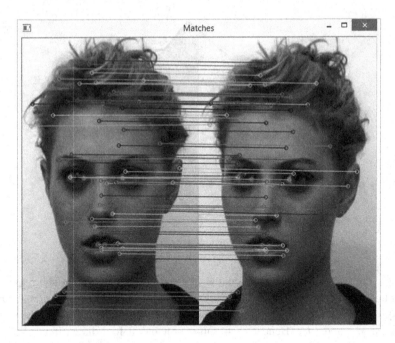

Fig. 4. Most of the matches correctly link a point on the left with its corresponding image point on the right, despite some differences in head position and facial expression between the images

$c = \{1, \ldots, C\}$ of the nearest neighbor image $Y_{n,c}$ which achieves the highest number of matching correspondences to the test image X.

2.6   Matching Images Using RANSAC

The use of the spatial information about matching points can help to reduce the amount of falsely matched correspondences, i.e. outliers. With the assumption that many parts of a face nearly lie on a plane, with only small viewpoint changes for frontal faces, a given homography (transformation) between the test and train images can reject outlier matches which lie outside a specified inlier radius.

The RANSAC (Random Sample Consensus) algorithm aims at estimating a given mathematical entity from a data set that may contain a number of outliers. The idea is to randomly select some data points from the set and perform the estimation only with these. The number of selected points should be the minimum number of points required to estimate the mathematical entity. Once a transformation has been estimated, all points of a test image will be projected to the train image. If a projected point lies in a given radius to its corresponding point it is classified as inlier for that particular homography, otherwise it is declared as an outlier. After a given number of iterations the maximum amount of inliers of all estimated homographies will be used as a measurement to determine the likelihood of the similarity between the test and the train image.

The central idea behind the RANSAC algorithm is that the larger the support set is, the higher the probability that the computed matrix is the right one. Obviously, if one (or more) of the randomly selected matches is a wrong match, then the computed fundamental matrix will also be wrong, and its support set is expected to be small. This process is repeated a number of times, and at the end, the matrix with the largest support will be retained as the most probable one [5].

2.7   Face Recognition Algorithm

The first step is detect the feature point and computing their descriptors. Next, we proceed to feature matching in order to find the best matching points for each feature. We now have two relatively good match sets, one from the first image to second image and the other one from second image to the first one. From these sets, we will now reject matches that do not obey the homography constraint. This test is based on the RANSAC (Fig. 5).

3   Experimental Results

We have used the openCV framework, this is a C++ framework for computer vision. It ships with its own implementation of SURF and several other computer vision algorithms. It was chosen as it provides different good low level routines for working with images and easy loading and saving of different image formats.

The algorithm was tested using two free face database images that contains about ten different images of each of 79 distinct subjects [8, 9]. For some subjects, the images

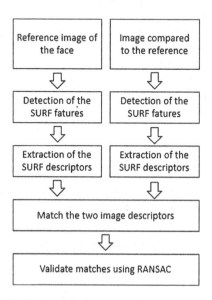

Fig. 5. Overview of the proposed approach

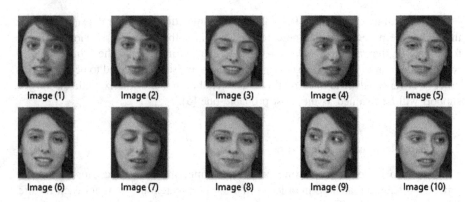

Image (1)   Image (2)   Image (3)   Image (4)   Image (5)

Image (6)   Image (7)   Image (8)   Image (9)   Image (10)

Fig. 6. Example of the images under different conditions

were taken at different times, varying the lighting, facial expressions (open, closed eyes, smiling, not smiling) and facial details (glasses, no glasses). All the images were taken against a dark homogeneous background with the subjects in an upright, frontal position and its total number of keypoints was calculated (Fig. 6).

Metrics like Correct Recognition Rate (*CRR*) for identification and the Equal Error Rate (*EER*) for recognition are used to measure the performance of SURF against SIFT. *CRR* of the system is defined as

$$CRR = \frac{NC}{TN} \times 100$$

where *NC* denotes the number of correct (Non-False) recognitions of face images and *TN* is the total number of face images in the testing set. At a given threshold, the probability of accepting the imposter, known as False Acceptance Rate (*FAR*) and the probability of rejecting the genuine person, known as False Rejection Rate (*FRR*) are obtained. Equal Error Rate (*EER*) is the error where *FAR = FRR* (Table 1).

As it can see, the algorithm SURF evinces a better performance and accuracy as expected. On the other hand SIFT turned out to be more time consuming. The range for matching points between each of the images of the same person is 7 to 45. Therefore, by the number of matching points we can determine a certain degree of similarity between two images. The Fig. 7 shows the comparison and matching between two images having a minimal difference. As can be seen, all points are matched correctly, therefore we can conclude with a high degree of certainty that both images are from the same person or someone very similar.

The Fig. 8 shows an example of the effectiveness in both algorithms at the time to matching the keypoints correctly that allow us to identify the resemblance or similarity

Table 1. Comparative results between SIFT and SURF

	CRR	ERR
SIFT	0.8734	0.3170
SURF	0.9897	0.2962

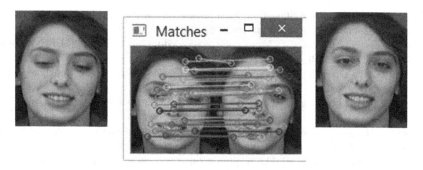

Fig. 7. Image(3) and Image(6) have the highest number of matched points. The difference between the two is minimal

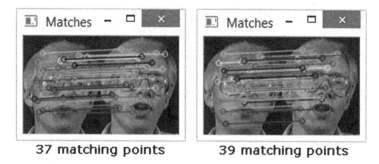

Fig. 8. Comparison between SIFT (left image) and SURF (right image)

present in images despite the different circumstances. Nonetheless, SURF obtained a greater number of correct matching points than SIFT in a shorter period of time.

4 Conclusions

Both algorithms can be used for face recognition, but nevertheless SURF algorithm is proven to be more suitable. However, we believe that facial recognition is more beneficial to use for facial authentication than for identification purposes, as it is too easy for someone to alter their face, features with a disguise or mask, etc. Environment is also a consideration as well as subject motion and focus on the camera.

References

1. Ekenel, H.K., Stiefelhagen, R.: Why is facial occlusion a challenging problem? In: International Conference on Biometrics, Sassari, Italy (2009)
2. Mikolajczyk, K.: Schmid.: A performance evaluation of local descriptors. IEEE Trans. Pattern Anal. Mach. Intell. **27**, 1615–1630 (2005)

3. Lowe, D.G.: Object recognition from local scale-invariant features. In: International Conference on Computer Vision, Corfu, Greece pp. 1150–1157 (1999)
4. Bay, H., Ess, A., Tuytelaars, T., Van-Gool, L.: Speeded-up robust features (SURF). Comput. Vis. Image Underst. **110**, 346–359 (2008)
5. Lagani-Re, R.: OpenCV 2 Computer Vision Application Programming Cookbook. Packt Publishing, Birmingham, Birmingham UK (2011)
6. Li, J., Zhang, Y.: Learning SURF Cascade for Fast and Accurate Object Detection. Intel Labs China (2013)
7. Hung-Fu, H., Shen-Chuan, T.: Facial expression recognition using new feature extraction algorithm. Electron. Lett. Comput. Vis. Image Anal. **11**, 41–54 (2012)
8. Lee, K.C., Ho, J., Kriegman, D.: Acquiring linear subspaces for face recognition under variable lighting. IEE Trans. Pattern Anal. Mach. Intell. **27**, 684–698 (2005)
9. Samaria, F., Harter, A.: Parameterisation of a stochastic model for human face identification. In: Proceeding of 2nd IEEE Workshop on Applications of Computer Vision, Sarasota FL (1994)

Research on an Algorithm of Shape Motion Deblurring

Xia Wu$^{(\boxtimes)}$, Hongzhe Xu, Xiaolin Gui, Wen Li, and Zhihai Yao

Shaanxi Key Lab of Computer Network, Xi'an Jiaotong University, Xi'an, China
wuxia900717@yeah.net,
{xuhz,leewhen,xlgui}@mail.xjtu.edu.cn

Abstract. According to the feature of fuzzy and the quality reduce background of plate image, this paper presents a Spectrum Estimation Partition Deblur (SEPD) algorithm to restore complex blur image. This algorithm is based on spectrum estimation and extracts the blur kernel information from the dark strips in frequency spectrum, and makes the traditional algorithm better to improve the algorithm's accuracy and noise capacity. Through the partition recovery strategy, each piece of image is deblurred separately, and then are integrated into a whole image by edge fitting. Through the analysis of actual measurements, the algorithm performs high accuracy and better noise immunity, and also improves the quality of the shape image obviously.

Keywords: Motion blur · Blur kernel · Spectrum estimation · Plate movement · Motion deblur

1 Introduction

Straightening machine is a main part of the sheet straightening industry, and in the intelligent straightening process, the measurement of sheet's flatness is one of the key factors of straightening quality. In laser sheet intelligent system, the phenomena of motion blur seriously impacts the measurement of sheet configuration. Because of the convolution of motion blurring, the images collected will has the phenomenon of incline, hauling tail, ghosting, and blur, which makes it hard to obtain accurate information from images and impacts the straightening effect [1].

There are mostly two aspects of motion deblur: the first one is stabilization, which is implemented by installing stabilization equipment on original ones, to mitigate motion blur of images collected by decreasing relative movement between CCD equipment and target object. However, stabilization doesn't work everywhere because of the increase of weight and cost. The other aspect is motion deblur. In which by researching image degradation or properties of blurred image, as well as acquiring the information of blur kernel by strategies, to recover clear image by deconvolution restoration method [2].

D.-S. Huang et al. (Eds.): ICIC 2015, Part I, LNCS 9225, pp. 327–338, 2015.
DOI: 10.1007/978-3-319-22180-9_32

2 Solution of Blur Kernel Based on Spectrum Estimation

The relation between spatial domain of motion blur image and original image is shown as Eq. (1):

$$g(x,y) = \frac{1}{T}\int_0^T f(x - \partial(t), y - \beta(t))dt \tag{1}$$

Transform Eq. (1) to frequency domain by Fourier transform:

$$G(u,v) = F(u,v)H(u,v) \tag{2}$$

In which,

$$H(u,v) = T\frac{\sin[c\pi(u\cos\theta + v\sin\theta)]}{\pi c(u\cos\theta + v\sin\theta)}e^{-j\pi c(u\cos\theta + v\sin\theta)} \tag{3}$$

$H(u,v)$ is the frequency-domain expression of blur kernel function $h(x,y)$, which is an expression of complex number, and its frequency spectrum (modulus of complex number) is acquired by square rooting the quadratic sum of its real part and imaginary part.

Equation of frequency spectrum is shown as Eq. (4):

$$|H(u,v)| = T\frac{\sin[c\pi(u\cos\theta + v\sin\theta)]}{\pi c(u\cos\theta + v\sin\theta)} \tag{4}$$

For the equation above, when $\theta = 0$, $|H(u,v)| = T\frac{\sin(c\pi u)}{\pi c(u)}$, and when u = n/c (n = 0, 1, 2...), $H(u,v) = 0$. In other words, the grey value of the special position on frequency spectrum ($n = u/c$, v is random) is 0. So there are some dark strips that are perpendicular to axis u existing in spectrum of motion blur image.

2.1 Calculation of Blurring Length Based on Vertical Projection

Assume u0 and u1 are positions of two dark strips that are next to each other:

$$cu_0 + cu_1 = 1 \tag{5}$$

Assume the pixel number on axis u of image is N, and the pixel on length c in Eq. (5) is a, so $c = a/n$, a is the blurring length to be solved. Formula of blurring length is as follows:

$$a_0 = \frac{N}{|u_0 - u_1|} \tag{6}$$

We can calculate plenty of blurring length a, to obtain a more accurate one by obtaining the arithmetic mean value of them, as which is shown in Eq. (7):

$$a = \frac{\sum\limits_{i=1}^{n} \frac{N}{|u_i - u_{i-1}|}}{n} = \frac{N}{n} \sum\limits_{i=1}^{n} \frac{1}{|u_i - u_{i-1}|} \tag{7}$$

In Eq. (7):

N—Width of frequency spectrum;
n—number of dark strips;
u_i—abscissa value of position of dark strip i

It is found from Eq. (7) that only when acquiring the abscissa value of dark strip position, the length of blur kernel can be solved. And we can obtain the position of dark strip u_i by vertical projecting the frequency spectrum. Projection formula is shown below:

$$S(v) = \sum\limits_{u=1}^{M} |G(u, v)| \tag{8}$$

After solving S(v), the one-dimensional vector is graphed in curve, which is a Bzier curve as Fig. 1. And then the positions of all dark strips can be found by solving all the minimums on the curve. Finally we substitute these minimums into the formula above to obtain length of blur kernel.

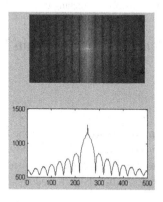

Fig. 1. Bzier curve of grayscale image

2.2 Calculations of Blur Direction Based on Hough Transform

Hough transform is a method to detect straight line came up with by Hough. It can transform coordinate space of image to parameter space, to realize the detection and fitting of straight line or curve [3].

In x-y coordinate space of image, the straight line that passes point (X_j, Y_j) is expressed as equation below:

$$y_i = ax_i + b \qquad (9)$$

Transform it into parameter space a-b as:

$$b = -ax_i + y_i \qquad (10)$$

The transform above is Hough transform, which is shown in Fig. 2.

In parameter space all straight lines that intersect at the same point have the corresponding collinear points in image coordinate space. Based on this feature, if all the intersections in parameter space can be found, then all the straight lines in image coordinate space can be found.

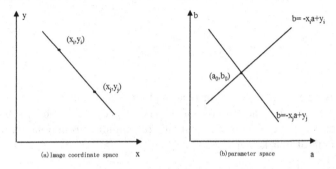

Fig. 2. Hough transform in rectangular coordinate system

3   Improvement of Algorithm for Spectrum Estimation Blur Kernel Solution

3.1   Preprocess

Logarithmic Transformation. In spectrum, because the high values are much bigger than low values, it becomes very hard to display the low-value part in spectrum. To enhance the low-value part, we can logarithmic transform the spectrum. The general expression of logarithmic transformation is as follows:

$$S = log(1 + r) \qquad (11)$$

In Eq. (11), r is grey value of original image; s is greyscale image after logarithmic transformation. After logarithmic transformation, a low-grey spectrum can be mapped as an even-grey one. Logarithmic transform a spectrum can compress the grey value from [0, 255] to [0, 5.5452], as shown in Fig. 3.

It can be found from Fig. 4 the pixel that has higher grey value is compressed in a smaller range, while the pixel that has lower grey value is distributed in a larger range. After logarithmic transformation, the logarithmic transformed grey value is homogenized from [0, 5.5452] to [0, 255], so that the low-value part in spectrum can be effectively enhanced.

Equation for greyscale homogenization is as Eq. (12):

$$b = \frac{(a-min) \times 255}{max-min} \tag{12}$$

In which:

a—grey value after logarithmic transformation;
b—grey value after homogenization;
max—maximum grey value after logarithmic transformation;
min—minimum grey value after logarithmic transformation;

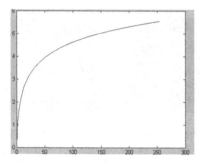

Fig. 3. Before and after image of logarithmic transformation

Centralization. The spectrum data obtained by Fourier transform is arranged in order calculated originally rather than zero frequency centered. Which makes zero frequency display at the four corners of the spectrum. Centralization utilizes the periodicity of spectrum to exchange the upper left part and lower right part, also the upper right part and lower left part. So that the high-frequency part can be moved to the center of spectrum, which makes the follow-up work easier. The method of centralization is shown in Fig. 5.

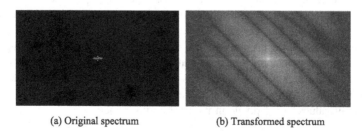

(a) Original spectrum   (b) Transformed spectrum

Fig. 4. Before and after image of logarithm transformation

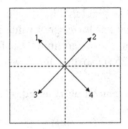

Fig. 5. Method of centralization

Figure 6 is the before and after image of centralization, from which it is seen that the higher grey value part is moved to the center of spectrum after centralization. This not only helps observe spectrum more conveniently, but also benefits the calculation of blur length.

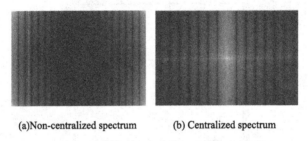

(a)Non-centralized spectrum   (b) Centralized spectrum

Fig. 6. Spectrum before and after effect of centralization

3.2 Improvement of Anti-noise Property

Algorithm for blur kernel solution is an algorithm based on spectrum characteristics, which is easily affected by noise [4]. This is because that in the Fourier transform process, the details of image and noise are both transformed into high-frequency area concurrently. The former is made use of for blur kernel calculation, while the latter will impact the performance of the former, making it impossible to show its features, as shown in Fig. 7.

In Fig. 7, it can be seen that the fuzzy image is seriously polluted by noise, and it becomes hard to tell the dark and light strips on the spectrum, which also means it is impossible to solve the blur kernel.

To improve anti-noise property of the algorithm, filtering and denoising can be conducted in spatial domain of image, so that the influence of noise on spectrum can be decreased. In this paper median filtering is used for denoising. Median filtering is a nonlinear smoothing technique, in which the grey value of a pixel is set as the median of all the grey values [5] of pixels in its neighboring window.

Steps:

(1) For a random point in the image, we pick up odd numbers of grey values in its neighboring window, and then rank them from small to large.
(2) We substitute the grey value of this random point with the median of the all the ranked grey values in step 1.

Median filtering is very effective in eliminating and impulse noise. Because the difference of grey values between these noise points and their neighboring area is significant, taking median of ranked values can enforce the grey value of the noise point to be as the same as the grey value of other pixels and this is how noise is eliminated. Before blur kernel solution algorithm, median filtering can be utilized to denoise image, to enable the algorithm to have anti-noise property and improve the application range as well as accuracy of algorithm.

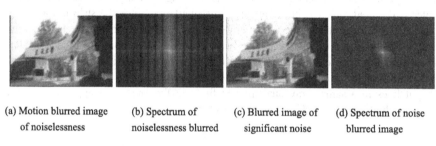

(a) Motion blurred image   (b) Spectrum of   (c) Blurred image of   (d) Spectrum of noise
      of noiselessness   noiselessness blurred   significant noise   blurred image

Fig. 7. Spectrum of noise blurred image

3.3   Improvement of Algorithm Accuracy

Weighted Projection. In the rotation process of spectrum that is inclined along blur direction, image becomes larger than original one. After saving rotated image, the redundant area is filled by zero-grey-value points, which is shown in Fig. 8.

It is found that the Bezier curve is not very ideal if directly projecting the rotated image, since it cannot obviously display the position of dark strips. In the middle of rotated image, because there are plenty of light points while rare zero points, a large value is projected. On two sides of the value, there are very few points that have larger than zero grey value, while there are many zero points, so the value projected is very small. As what is shown in (c) of Fig. 8, the minimum calculated from this projected curve is not accurate, so weighted projection is adopted in the algorithm. When projecting all the points on one vertical axis, we divide the summation by the number of non-zero grey values, and the projection result is shown in (d) of Fig. 8. It can be seen from (d) that the curve run by weighted projection is compact, and also the minimums corresponded by each dark strip has very clear position, which makes calculation easier and the result more accurate.

Selection of Dark Straight Strip. By making use of Hough transfer, all he dark straight strips in spectrum can be detected. Based on one of the strips, blur angle can be

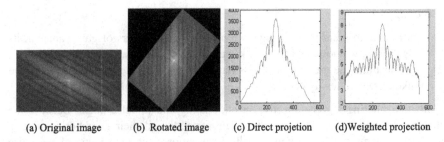

(a) Original image   (b) Rotated image   (c) Direct projetion   (d)Weighted projection

Fig. 8. Effect of weighted projection

calculated. To improve the accuracy of algorithm, this paper detects the ten longest straight lines in the spectrum as shown in (b) of Fig. 9. It is found from (b) that not all the straight lines are applicable for calculating blur angle. Firstly, the cross light strip in center is detected, but it is not applicable for calculation since the strip is formed by four edges of image that move to center. Secondly, due to error in each aspect, some straight lines that are not dark can also be detected, if these lines are used for calculation, then the error can be large. Therefore, it is very necessary to select the most applicable straight lines. The method adopted here is to put the inclination angles of all straight lines in a same array, and then to find out the angel that appears most frequently, which can be used for calculating blur angle.

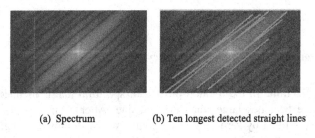

(a) Spectrum   (b) Ten longest detected straight lines

Fig. 9. Noiselessness motion blur image

4 SEPD Algorithm Flow

SEPD algorithm (Spectrum Estimation Partition Deblurring) is based on the following thought: since the motion mode of motion blur of the whole image is too complicated, so it is impossible to use a uniform blur kernel. Then partition the pending blurred image into many small pieces, and the motion blur mode of each piece can be approximately expressed as uniform linear motion. By solving the blur length and angle of each piece, blur kernel can be calculated, and then restore by convolution to acquire the clear image of each piece. In the end, all the images of each piece are fused to a whole image, to obtain an integrated clear image. So this algorithm is implemented in three steps: partition, recovery, and fusion.

Step1 Partition

This paper designs an algorithm based on partition of pixel colors. In which the pixel is determined as red or not based on the red proportion of the colors. After that, in accordance with the distribution of the red pixels, to divide the image on vertical direction in six parts, and then we calculate the width of image, so the width of every small image is acquired by dividing the overall width by ten.

Step2 Recovery

This paper uses three image recovery methods based on frequency domain: inverse filtering, Wiener filtering, and RL iterative algorithm. Because at present no ideal noise detection method has been found yet, it is hard to determine which recovery method is the most appropriate for which occasion. Therefore, the solution used here is: we recover a blurred image by three recovery methods respectively, and then estimate the quality of three recovered images, finally we select one of them as the ultimate recovered image according to the assessment result.

Step3 Fusion

After recovering the pieces separately, the partitions are fused, namely integrate all the small images into a big one. Generally since the strips are distributed continuously, the image has good consistency in small distance, all the images can be fused pretty well after combining the recovered partitions. When the joint of edges is not ideal, the least-square curve fitting can be performed for edges [6].

Fig. 10. Partition of sheet

5   Experiment Conclusions and Analysis

To examine the property of improved blur kernel solution algorithm that is based on frequency domain features, this paper makes use of image blurring model, and blurring processes multiple angles and lengths of an image. For the blurred image, we make estimation of blurred angles and lengths of blur kernel by algorithm proposed in this paper. The experimental results are shown in Tables 1 and 2.

From the experimental results above, the error of blurred angel is no more than 2 degree, and the error of blurred length is no more than 2 pixels, which satisfy the requirements of practical calculation.

When blurred angel is 45 degree and blurred length is 21 pixel, the performance of three filtering algorithm is analyzed with noise as well as without noise, respectively. Which is shown in Tables 3 and 4.

It can be seen from two tables above that when there is no noise at all, inverse filtering has the best recovery effect, however, when the nose exists, it becomes almost

Table 1. Error of blurred angle

Actual blurred angel (degree)	15	30	45	60	75	90	105	120	135	150
Algorithmic blurred angel (degree)	15	31	43	59	78	90	104	120	138	149
Error (degree)	0	1	2	1	1	0	1	0	1	1

Table 2. Error of blurred length

Actual blurred length (pixel)	8	10	13	15	17	20	22	25	27	30
Algorithmic blurred length (pixel)	8	9	12	15	19	20	23	25	29	31
Error (pixel)	0	1	1	0	2	0	1	0	2	1

Table 3. Recovery performance of three filtering without noise

	Ideal algorithm	Inverse filtering	Wiener filtering	Lucy-Richardson
Absolute mean square error	0	0.000000	0.036589	0.039626
Signal to noise ratio	high	241.900193	21.498953	21.068266
Maximum signal to noise ratio	high	245.304794	24.903555	24.472867
Energy normalization mean square error	0	0.000000	0.007081	0.007819
Mean square error	0	0.000000	0.015095	0.035995

Table 4. Recovery performance of three filtering with noise

	Ideal algorithm	Inverse filtering	Wiener filtering	Lucy-Richardson
Absolute mean square error	0	174.531945	0.037255	0.039747
Signal to noise ratio	high	−50.032784	21.377703	21.045744
Maximum signal to noise ratio	high	−46.628183	24.82304	24.450345
Energy normalization mean square error	0	100757.73	0.007282	0.007860
Mean square error	0	45365.80474	0.015272	0.0360460

useless. While the other two algorithms both have certain anti-noise ability in two conditions. From the results of plenty of experiments, when the influence of noise is reinforced, the quality of images recovered by three algorithms is shown in Fig. 11.

SEPD algorithm in this paper is selected from one of inverse filtering, Wiener filtering, and RL iterative algorithm based on their performance. So the recovery effect of SEPD algorithm will always be the best of these three algorithms.

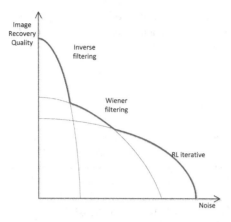

Fig. 11. Anti-noise ability of three recovery algorithms

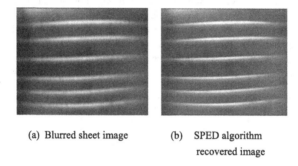

(a) Blurred sheet image   (b)   SPED algorithm
                                        recovered image

Fig. 12. Comparison of sharpness effect

Deconvolution recovery in SEPD algorithm uses complexity of algorithm to exchange for performance. It recovers the image by three different algorithms, and then selects the one with best quality as the final image. So the final image not only has best quality, but also the best anti-noise property. The performance of recovered image is as the thick line in Fig. 11.

Figure 12 displays the SEPD algorithm before and after effect of image sharpness. Figure 12(a) shows the effect before process, in which there is motion blur on the edges. After recovering Fig. 12(a) by SEPD algorithm, the recovered image obviously becomes clearer than before as shown in Fig. 12(b). By comparing the effect, it can be found that SEPD algorithm helps reduce the motion blur of sheet image to some extent, which evidently improve the image quality.

6   Conclusion

Motion blur is a universal phenomenon in imaging process, so that's why recover techniques for motion blur image becomes an important research topic. The motion blur image studied in this article has very complicated blur kernel, and it is impossible to deblur using common motion deblur methods. Though the image studied here has only six laser rays, other kinds of images can also be recovered by reforming algorithm in the future.

References

1. Li, Z., Huang, D.: Study on continuous strip tension leveling mechanism. Cooper Eng. **2005** (02), 29–33 (2005). doi:10.3969/j.issn.1009-3842.2005.02.011
2. Huang, C., Wang, J., Gao, X., Ding, S.: Study on video stabilization performance assessment in electronic image stabilization. Infrared Laser Eng. **43**(5), 477–481 (2013). doi:10.3969/j. issn.1001-5078.2013.05.001
3. Duan, D., Xie, M., Mo, Q., et al.: An improved Hough transform for line detection. In: 2010 International Conference on Computer Application and System Modeling (ICCASM), vol. 2, pp. V2-354–V2-357. IEEE (2010)
4. Cho, S., Matsushita, Y., Lee, S.: Removing non-uniform motion blur from images. In: Proceedings of the IEEE International Conference on Computer Vision (2007)
5. Yin, L., Yang, R., Gabbouj, M., et al.: Weighted median filters: a tutorial. IEEE Trans. Circ. Syst. II: Analog Digit. Sig. Proc. **43**(3), 157–192 (1996)
6. Yang, A., Ai-ling, W., Chang, J.: The research on parallel least squares curve fitting algorithm. In: International Conference on Test and Measurement, ICTM 2009, pp. 201–204. IEEE (2009)

A New Approach for Greenness Identification from Maize Images

Wenzhu Yang[✉], Xiaolan Zhao, Sile Wang, Liping Chen,
Xiangyang Chen, and Sukui Lu

School of Computer Science and Technology,
Hebei University, Baoding 071002, Peoples Republic of China
wenzhuyang@163.com

Abstract. Greenness identification from crop growth monitoring images is the first and important step for crop growth status analysis. There are many methods to recognize the green crops from the images, and the visible spectral-index based methods are the most commonly used ones. But these methods can not work properly when dealing with images captured outdoors due to the high variability of illumination and the complex background elements. In this paper, a new approach for greenness identification from maize images is proposed. Firstly, the crop image was converted from RGB color space to HSV color space to obtain the hue and saturation value of each pixel in the image. Secondly, most of the background pixels were removed according to the hue value range of greenness. Then, the green crops were identified from the processed image using the excess green index method and the Otsu method. Finally, all noise objects were removed to get the real crops. The experimental results indicate that the proposed approach can recognized the green plants correctly from the maize images captured outdoors.

Keywords: Greenness identification · Images captured outdoors · HSV color space · Hue · Excess green index

1 Introduction

To get better crop yields, precision agriculture is considered by more and more countries. It can take different forms, such as estimation physiological status (Sakamoto et al. 2012), crop disease detection (Jun et al. 2011, Pugoy and Mariano 2011) and plant or weed identification (M. Montalvo et al. 2013). To realize the philosophy of precision farming, getting the information of the crops during each growing stage is the most important task, where plants must be identified as a previous step.

This paper proposed a new method for greenness identification. It consists of four main steps: (1) Color space transformation. (2) General background removal. Based on the Hue analysis and the excess green method, some pixels belong to the background are removed. (3) Greenness identification. After the above step, a threshold is needed to separate the object and the background. (4) Small objects removal. Some small objects which belong to the noise are needed to remove.

© Springer International Publishing Switzerland 2015
D.-S. Huang et al. (Eds.): ICIC 2015, Part I, LNCS 9225, pp. 339–347, 2015.
DOI: 10.1007/978-3-319-22180-9_33

2 Related Works

Different methods and strategies for plant identification have been applied. One is the visible spectral-index base method: Woebbecke et al. (1995) studied excess green index method (ExG), which was partly useful for identifying plant regions. However, bright soil or residue pixels which contained high green content (although not appearing green to the human eye) tended to provide false plant information (George. E. Meyer, et al. 2004). Kataoka et al. (2003) analyzed the intensity of the RGB in the images and then gave a method called color index of vegetation extraction (CIVE) to enhance the green information in the images. Marchant and Onyango (2002) proposed the vegetative index method (VEG), which makes a good contrast between plant material and soil background. Hague, Tillet and Wheeler (2006) use this method to assessment crop or weed. Neto et al. (2004) proposed the excess green minus excess red index method (ExGR). Considering all the above indices, Guijiaro et al. (2011) proposed a combined index method (COM). All the visible spectral-index based methods are based on the assumption that plants display a clear high degree of greenness and the soil contains low green content. However the images captured from outdoor environments are easily affected by the variable illumination which may result in a failure when separating the green plant from the background. The other is the learning-based method: Ruiz-Ruiz et al. (2009) applied the EASA under the HIS (hue-saturation-intensity) color space to deal with the illumination variability. Zheng et al. (2010) use a supervised mean-shift algorithm to identify greenness. J. Romeo et al. (2013) applied the fuzzy clustering method for greenness identification. Zhenghong Yu et al. (2013) approached the AP-HI method which combined the hue-intensity color model and affinity propagation clustering algorithm.

Several strategies have been proposed for segmenting crop images, especially for greenness identification. Color is a human visual perception concept. Red, green, and blue are primary colors with associated intensity image subsets. The visible spectral-index based methods are the most commonly used ones for greenness identification. They include the excess green index (ExG), the excess green minus excess red index (ExGR), the vegetative index (VEG), the color index of vegetation extraction (CIVE), the combined index (COM), etc. All these methods are based on the fact that green plants have larger green indexes than others in the normalized RGB color space.

The normalized color r, g and b in RGB color space are defined as follows, where R, G and B are the color components of the RGB crop image.

$$r = \frac{R}{R+G+B}, g = \frac{G}{R+G+B}, b = \frac{B}{R+G+B} \tag{1}$$

The ExG, ExGR, VEG, CIVE, and COM are defined as follows.
Excess green index (ExG):

$$ExG = 2g - r - b \tag{2}$$

Excess green minus excess red (ExGR):

$$ExGR = ExG - (1.4r - g) \tag{3}$$

Color index of vegetation extraction (CIVE):

$$CIVE = 0.441r - 0.811g + 0.385b + 18.78745 \tag{4}$$

Vegetative index (VEG):

$$VEG = \frac{g}{r^{0.667}b^{0.333}} \tag{5}$$

The combined method (COM):

$$COM = 0.25 * ExG + 0.30 * ExGR + 0.33 * CIVE + 0.12 * VEG \tag{6}$$

3 Materials and Methods

3.1 Color Images Captured in Different Illumination Conditions

The color images used for this study are captured in the maize field in different illumination conditions, as shown in Fig. 1. We selected different illumination conditions for image capturing to check the robustness of the greenness identification methods proposed in this paper. The images were acquired under a high variety of illumination conditions, including morning, midday and afternoon. The illumination of

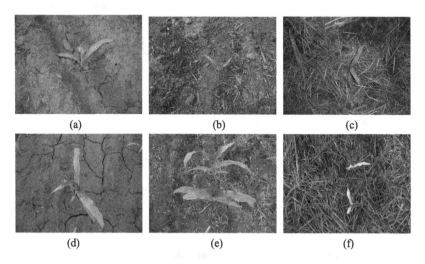

Fig. 1. Color image under different illumination conditions. (a) Sunny with clean soil, (b) sunny with straw ash, (c) sunny with wheat straw, (d) cloudy with clean soil, (e) cloudy with straw ash, (f) cloudy with corn straw(Color figure online)

the maize images are in different conditions which ranges from low to high. The leaves of the maize seedlings are in different green depth which range from dark green to bright green.

The maize seedling images were stored as 24-bit color images with resolutions of 3264*2448 pixels and saved in RGB color space in the JPEG format.

3.2 Hue Distribution Analysis

Color is an effective and essential visual cue to distinguishing an object from the others. RGB color space is the mostly used color space for segmentation. The visible spectral-index based methods failed when dealing with images with high or low contrast. So we choose the HSV (Hue, Saturation and Value) color space to void the influence of variable illumination.

In our research, the maize seedling images are captured in different illumination conditions, so the color of the maize leaves range from dark green to light green. To distinguish the leaves and the background, colors of all elements in the image should be analyzed. So we select six samples which are made of 8*8 pixels to do the hue analysis. They are the dark green leave, the bright green leave, the soil, the straw ash, the corn straw and the wheat straw respectively. Figure 2 shows the hue distribution. We can see that the hue values of green leaf range from about 50 to 150. But the hues of the straw ashes and the wheat straws are very close to the green leaf. So we need combine some other method such as excess green to identify crops.

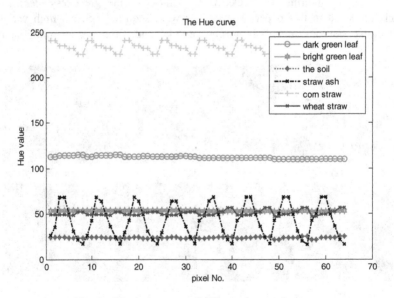

Fig. 2. The hue curves of different image elements

3.3   Greenness Identification

(1) Color Space Transformation

Let I be a RGB image of maize seedling, and R, G and B be the three color components of I. Let M be the corresponding image in HSV color space of this RGB image, and H be the hue component of M.

$$H = \begin{cases} \frac{G-B}{\max(R,G,B)-\min(R,G,B)} & \text{if } R = \max(R,G,B) \\ 2 + \frac{B-R}{\max(R,G,B)-\min(R,G,B)} & \text{if } G = \max(R,G,B) \\ 4 + \frac{R-G}{\max(R,G,B)-\min(R,G,B)} & \text{if } B = \max(R,G,B) \end{cases} \tag{7}$$

(2) Background Removal

Most of the background pixels can be eliminated by judging their corresponding hue values according to the common hue value range of greenness.

$$\text{R} = \begin{cases} 0 & \text{if } H < h_1 \ or \ H > h_2 \\ R & \text{otherwise} \end{cases} \tag{8}$$

$$G = \begin{cases} 0 & \text{if } H < h_1 \ or \ H > h_2 \\ G & \text{otherwise} \end{cases} \tag{9}$$

$$B = \begin{cases} 0 & \text{if } H < h_1 \ or \ H > h_2 \\ B & \text{otherwise} \end{cases} \tag{10}$$

where h_1 is the smallest hue value of greenness and h_2 is the largest hue value of greenness. After this step, most of the background elements are removed.

(3) Greenness Identification

The excess green index metrics of the RGB image was calculated after it processed by step (2).

$$ExG = 2G - R - B \tag{11}$$

Taking the metrics ExG as a grey image, we can identify the green crops from it by applying the thresholding method.

$$BW = \begin{cases} 1 & \text{if } ExG > T \\ 0 & \text{otherwise} \end{cases} \tag{12}$$

where T is the threshold.

(4) Small Objects Removal

In the binarized image, most of the small objects are noises. Therefore, all small objects with area less than S should be removed, where S is the smallest area of the real object of green plant.

4   Results and Discussion

To verify whether the proposed method is robust to illumination variations, we applied it to recognize the green plants in the maize images captured under various outdoor natural illumination conditions in the field. The proposed method is also compared with other crop segmentation methods, including the ExG, the ExGR, the CIVE, the VEG and the COM. A personal computer with Intel Core i5 CPU and 4 GB SDRAM was chosen as the test environment, and Windows 7 was selected as the operation system. Matlab 7.0 was used to implement and validate the algorithm.

Totally 73 maize images are employed to compare different segmentation approaches. They are taken under different whether conditions, i.e. sunny and cloudy, in Baoding, Hebei Province, China. The background elements in the captured images mainly include soil, straw and ash. The images shown in Fig. 1 are the representative examples of the set of images processed. The segmentation results by our method, by the ExG, the ExGR, the CIVE, the VEG and the COM were shown in Figs. 3, 4, 5, 6, 7 respectively.

From Figs. 4 and 5 we can see that the results of the ExG and the ExGR method are over segmented. The VEG method does not work properly when dealing with the maize images. The CIVE and the COM method also failed in result of bringing insufficient segmentation results under complex outdoor environment as shown in Figs. 6 and 7. The result of the proposed method in Fig. 3 is most close to the ground truth. The results indicate that it is robust in different illumination conditions.

From Table 1 we can see the proposed method is better than the other five approaches and it reached to 95 % of correctness in high or low illumination conditions. The ExG method only reaches 70 % and 85 % of correctness in segmenting the

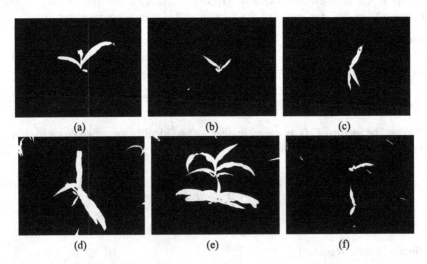

Fig. 3. The greenness identification results by the proposed method. (a) Sunny with clean soil, (b) sunny with straw ash, (c) sunny with wheat straw, (d) cloudy with clean soil, (e) cloudy with straw ash, (f) cloudy with corn straw

(a)   (b)   (c)

Fig. 4. The greenness identification results by the ExG method. (a) sunny with straw ash, (b) sunny with wheat straw, (c) cloudy with corn straw

(a)   (b)   (c)

Fig. 5. The greenness identification results by the ExGR method. (a) sunny with straw ash, (b) sunny with wheat straw, (c) cloudy with corn straw

(a)   (b)   (c)

Fig. 6. The greenness identification results by the CIVE method. (a) sunny with straw ash, (b) sunny with wheat straw, (c) cloudy with corn straw

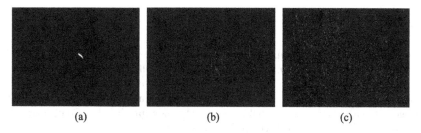

(a)   (b)   (c)

Fig. 7. The greenness identification results by the COM method. (a) sunny with straw ash, (b) sunny with wheat straw, (c) cloudy with corn straw

Table 1. Recognition accuracy of different methods when processing images of field maize captured in sunny or cloudy

	Recognition accuracy of different methods (%)					
	Our method	ExG	ExGR	CIVE	VEG	COM
Sunny	95	70	65	75	0	55
Cloudy	95	85	95	90	0	75

sunny and cloudy images respectively. The recognition accuracy of the ExGR method is near to 95 % when processing images captured in low illumination conditions, but only 65 % in high illumination conditions. The COM method can not work well when processing images captured in sunny days. The VEG method can't do the identification of the greenness completely. So the proposed segmentation method can be used to crop recognition effectively.

5 Conclusion

The visible spectral-index based method including the ExG, the ExGR, the CIVE, the VEG and the COM method were commonly employed for identifying green plants from crop images. But neither of them works very well when dealing with images captured outdoors with complex background. This is because all these visible spectral-index based methods are working on the assumption that plants display a clear high degree of greenness, and soil is the only background element. A HSV-based approach was proposed to improve the robustness of the greenness identification for processing crop images captured in different illumination conditions with complex background. The method works well not only in the normal illumination condition but also in the low and high illumination conditions. It can also process the images with complex background. The results indicate it is robust when processing images captured in different illumination conditions.

Acknowledgements. The authors thank The Ministry of Science and Technology of the People's Republic of China (2013DFA11320), The Natural Science Foundation of Hebei Provence (F2015201033), for their financial support.

References

Sakamoto, T., Gitelson, A.A., Nguy-Robertson, A.L., Arkebauer, T.J., Wardlow, B.D., Suyker, A.E., Verma, S.B., Shibayama, M.: An alternative method using digital cameras for continuous monitoring of crop status. Agric. Forest Meteorol. **154–155**, 113–126 (2012)

Jun, P., Bai, Z.-Y., Jun-chen, L., Li, S.-K.: Automatic segmentation of crop leaf spot disease images by integrating local threshold and seeded region growing. Int. Conf. Image Anal. Signal Process. **2011**, 590–594 (2011)

Pugoy, R.A.D.L., Mariano, V.Y.: Automated rice leaf disease detection using color image analysis. In: 3rd International Conference on Digital Image Processing (ICDIP 2011), pp. 80090F–80097. Chengdu (2011)

Pugoy, R.A.D.L., Mariano, V.Y.: Automated rice leaf disease detection using color image analysis. 3rd International Conference on Digital Image Processing (ICDIP 2011), pp. 80090F–80097. Chengdu, China (2011)

Montalvo, M., Guerrero, J.M., Romeo, J., Emmi, L., Guijarro, M., Pajares, G.: Automatic expert system for weeds/crops identification in images from maize fields. Expert Syst. Appl. **40**, 75–82 (2013)

Kataoka, T., Kaneko, T., Okamoto, H., & Hata, S. (2003). Crop growth estimation system using machine vision. In: the 2003 IEEE/ASME International Conference On Advanced Intelligent Mechatronics

Woebbecke, D.M., Meyer, G.E., Von Bargen, K., Mortensen, D.A.: Color indices for weed identification under various soil, residue, and lighting conditions. Trans. ASAE **38**(1), 259–269 (1995)

Meyer, G.E., Camargo-Neto, J., Jones, D.D., Hindman, T.W.: Intensified fuzzy clusters for classifying plant, soil, and residue regions of interest from color images. Comput. Electron. Agric. **42**, 161–180 (2004)

Marchant, J.A., Onyango, C.M.: Shadow-invariant classification for scenes illuminated by daylight. J. Opt. Soc. Am. A: **17**(11), 1952–1961 (2002)

Hague, T., Tillet, N., Wheeler, H.: Automated crop and weed monitoring in widely spaced cereals. Precis. Agric. **1**(1), 95–113 (2006)

Neto, J.C.: A Combined Statistical – Soft Computing Approach For Classification And Mapping Weed Species In Minimum Tillage Systems. University of Nebraska, Lincoln (2004)

Guijarro, M., Pajares, G., Riomoros, I., Herrera, P.J., Burgos-Artizzu, X.P., Ribeiro, A.: Automatic segmentation of relevant textures in agricultural images. Comput. Electron. Agric. **75**, 75–83 (2011)

Ruiz-Ruiz, G., Gómez-Gil, J., Navas-Gracia, L.M.: Testing different color spaces based on hue for the environmentally adaptive segmentation algorithm (EASA). Comput. Electron. Agric. **68**, 88–96 (2009)

Zheng, L., Shi, D., Zhang, J.: Segmentation of green vegetation of crop canopy images based on meanshift and Fisher linear discriminate. Pattern Recogn. Lett. **31**(9), 920–925 (2010)

Romeo, J., Pajares, G., Montalvo, M., Guerrero, J.M., Guijarro, M., de la Cruz, J.M.: A new expert system for greenness identification in agricultural images. Expert Syst. Appl. **40**, 2275–2286 (2013)

Zhenghong, Yu., Cao, Z., Xi, W., Bai, X., Qin, Y., Zhuo, W., Xiao, Y., Zhang, X., Xue, H.: Automatic image-based detection technology for two critical growth stages of maize: emergence and three-leaf stage. Agric. For. Meteorol. **174–175**, 65–84 (2013)

Image Super-Resolution Reconstruction Based on Sparse Representation and POCS Method

Li Shang[1(✉)], Shu-fen Liu[1], and Zhan-li Sun[2]

[1] Department of Communication Technology, College of Electronic Information Engineering, Suzhou Vocational University, Suzhou 215104, Jiangsu, China
{sl0930, sfliu}@jssvc.edu.cn
[2] School of Electrical Engineering and Automation, Anhui University, Hefei 230039, Anhui, China
zhlsun2006@126.com

Abstract. The traditional projection onto convex set (POCS) algorithm can reconstruct a low resolution (LR) image, but it is contradictory in retaining image detail and denoising, so the quality of a reconstructed image is limited. To avoid defects of POCS and obtain higher resolution, the image denoising idea based on sparse representation is led into this paper. Sparse representation can learn well the optimized overcomplete sparse dictionary of an image, which has self-adaptive property to image data and can describe image essential features so as to implement the goal of denoising efficiently. At present, K- singular value decomposition (K-SVD) is the emerging image processing method of sparse representation and has been used widely in image denoising. Therefore, combined the advantages of K-SVD and POCS, a new image ISR method is explored here. In terms of signal noise ratio (SNR) values and the visual effect of reconstructed images, simulation results show that our method proposed has clear improvement in image resolution and can retain image detail well.

Keywords: Sparse representation · Image super-resolution reconstruction · Sparse dictionary · K-SVD algorithm · POCS algorithm

1 Introduction

Image super-resolution (ISR) is such a method that combines some similar but not identical low resolution (LR) images into a higher resolution (LR) single image [1–3]. For some digital imaging system, using ISR algorithms can help to improve images' quality without modifying hardware system so as to reduce funds, therefore ISR techniques have been developed in image processing fields. Currently, ISR methods can be divided into two main categories, one is frequency domain methods and another is spatial domain ones [3]. Frequency methods are earlier ISR ones, which can only deal with image sequences that only translational motions are allowed. Spatial methods use general observation models, which have better adaptability and performance and they are used generally now [3]. To this day, some spatial methods used widely contain iterative back projection (IBP) [4, 5], non-uniformly interval sample interpolation [5],

© Springer International Publishing Switzerland 2015
D.-S. Huang et al. (Eds.): ICIC 2015, Part I, LNCS 9225, pp. 348–356, 2015.
DOI: 10.1007/978-3-319-22180-9_34

constrained least square (CLS) [6], self-adaptive filter [3], project onto convex set (POCS) [2, 3, 7, 8], maximum a posteriori probability estimation (MAP) and so on [8]. Among these methods, POCS method can keep well edge features of an image and be flexibly added prior information, so it is utilized widely in image processing fields. However, usual POCS methods are contradictory in the aspect of retaining image detail and denoising, so it is essential to explore some ISR methods combined POCS with denoising techniques. In view of reasons above-mentioned, the denoising method based on sparse representation is introduced in this paper. Sparse representation's theoretic proof is that a clear image with smooth property behaves the sparse decomposition under the overcomplete dictionary. Sparse representation can trained a self-adaptive dictionary of any image, which can describe well image's essential features. And if a noise image can be represented utilizing this dictionary learned, then it can be viewed as denoising. Currently, K- singular value decomposition (K-SVD) based sparse representation [9–11] is used widely in image denoising. This algorithm not only can reduce well Gauss additive noise, but also can keep edge and texture information [9]. Usually, K-SVD algorithm uses orthogonal matching pursuit (OMP) algorithm to realize sparse optimization of image feature coefficients, however, when image is too large, the effectiveness of OMP becomes limited [11]. In order to avoid this defect of K-SVD, the fast sparse coding (FSC) algorithm is used to implement sparse optimization learning. Considered artificial LR image and real LR image (i.e. millimeter (MMW) wave image), and utilized the criterion of signal noise ratio (SNR) to estimate results of image reconstructed, the image ISR task is successfully implemented by our method. Compared with POCS and usual K-SVD, the simulation results show that our method indeed has remarkable improvement in SNR and visual effect.

2 POCS Algorithm

POCS method can reconstruct a super-resolution image form LR images, and not only restore the details, but also amplify the dimension of a LR image [7, 8]. It is an iteration and projection process, which starts from a given or random initial image in the Hilbert space. The projection process is defined as Eq. (1)

$$X_{k+1} = T_m T_{m-1} \cdots T_1 \{X_k\} \ (k = 0, 1, \cdots, m), \tag{1}$$

where X denotes an image with $N_1 \times N_2$ pixels, $T_i = (1 - \beta_i)I + \beta_i P_i \ (0 < \beta_i < 2)$, and the parameter β_i is a relaxed projection operator, which can adjust the convergence rate of iteration. If $\beta_i = 1$, then $T_i = P_i$, and P_i can adjust the step length and direction of projection. Thus, the process of projection is defined as Eq. (2)

$$X_{k+1} = P_m P_{m-1} \cdots P_1 \{X_k\} \ (k = 0, 1, \cdots, m). \tag{2}$$

Based on above definitions, many convex sets of an image can be defined :

$$
\begin{cases}
C_k(m_1, m_2) = \left\{ X(i_1, i_2) : \left| r_k^{(X)}(m_1, m_2) \right| \leq \delta_0 \right\} \\
r_k^{(X)}(m_1, m_2) = g_k(m_1, m_2) - \sum\limits_{i_1=0}^{M_1-1} \sum\limits_{i_2=0}^{M_2-1} X(i_1, i_2) h_k(m_1, m_2; i_1, i_2)
\end{cases}, \quad (3)
$$

where the area of parameter k is from 1 to m. C_k denotes prior information, $X(i_1, i_2)$ is an estimation of ideal super-resolution image, and $h_k(m_1, m_2; i_1, i_2)$ is the PSF of pixel (i_1, i_2). Utilized the projection operator P, the step length and direction of projection can be adjusted. And $P_{(m_1, m_2)}\{X(i_1, i_2)\}$ can be defined as the estimation of $X(i_1, i_2)$, namely the following formula can be obtained:

$$
\hat{X}(i_1, i_2) = X(i_1, i_2) + \begin{cases}
\dfrac{r_k^{(X)}(m_1, m_2) - \delta_0}{\sum\limits_{i_1}\sum\limits_{i_2} h_k^2(m_1, m_2; j_1, j_2)} h_k(m_1, m_2; j_1, j_2) & r_k^{(X)}(m_1, m_2) > \delta_0 \\
0 & -\delta_0 < r_k^{(X)}(m_1, m_2) < \delta_0 \\
\dfrac{r_k^{(X)}(m_1, m_2) + \delta_0}{\sum\limits_{i_1}\sum\limits_{i_2} h_k^2(m_1, m_2; j_1, j_2)} h(m_1, m_2; j_1, j_2) & r_k^{(X)}(m_1, m_2) < -\delta_0
\end{cases}. \quad (4)
$$

In Eq. (2), all parameters and convex sets define a boundary corresponding to high resolution image. The parameter δ_0 reflects a prior confidence interval of original image, which is decided by the statistical characteristic of noise process, and for any pixel coordinate (m_1, m_2), the difference between observed image $g_k(m_1, m_2)$ and super-resolution estimation will be decided by δ_0. Further, another constraint of convex set named C_A is considered, which a range constraint of pixels is.

$$
C_A = \{ X(i_1, i_2) : 0 \leq X(i_1, i_2) \leq 255 \} \ (0 \leq i_1 \leq N_1 - 1, 0 \leq i_2 \leq N_2 - 1), \quad (5)
$$

where N_1 and N_2 is the row number and column number of the original image. And the projection operator of C_A is defined as follows

$$
P_A[X(i_1, i_2)] = \begin{cases}
0 & X(i_1, i_2) < 0 \\
X(i_1, i_2) & 0 \leq X(i_1, i_2) \leq 255 \\
255 & X(i_1, i_2) > 255
\end{cases}. \quad (6)
$$

All in all, the POCS method can ensure the monotonicity of convergence and an optimal solution can be obtained in theory, which will satisfy all constraints of original prior knowledge.

3   Sparse Representation and K-SVD Algorithm

3.1   Sparse Representation of Images

Let the matrix $D = \{ d_k \in \Re^{N \times K}, \|d_k\| = 1 \}$ $(N < < K, 1 \leq k \leq K)$ be an over-complete dictionary of K prototype atoms, and supposed an image $u \in \Re^N$ with the size of

$\sqrt{N} \times \sqrt{N}$ can be represented as a sparse linear combination of d_k, thus, this sparse representation model of image u can be defined as follows

$$\hat{s} = \arg \min_{s} \|s\|_0 \; s.t. \, u = Ds, \tag{7}$$

where $\|s\|_0$ is the L_0 norm of sparse coefficient vector, which denotes the number of non-zero elements of feature coefficients of an image. Commonly, even for a high resolution (HR) image, it is also difficult to solve the optimization problem of Eq. (7), at the same time, for a noise image, generally, it is not necessary to obtain absolutely precise reconstruction result. Therefore, the approximate process is usually required. In application, an approximated sparse model is often used the following formula:

$$\hat{s} = \arg \min_{s} \|s\|_0 \; s.t. \, \|u - Ds\|_2^2 \leq \varepsilon, \tag{8}$$

where ε is a positive constant, which approximate zero value. In Eq. (8), $\|\cdot\|_0$ and $\|\cdot\|_2$ denote respectively the l_0 and l_2 norm. Let u_h and u_l denote respectively the HR and LR image, according to Eq. (1), the equation of $u_h = D_h s$ can be obtained. And u_l can be approximated by the form of $u_l = T u_h \approx T D_h s$, where T is the mapping matrix. Thus, the LR over-complete dictionary D_l can be calculated by $D_l = T D_h$. Namely, corresponding respective dictionary, the high and LR image patches have the same sparse representation.

3.2 K-SVD Denoising Model

Assume that x is an image patch (a column vector), and consider its noisy version $y = x + n$, which is contaminated by the additive zero-mean white Gaussian noise with standard deviation $\sigma(n \in N(0, \sigma))$. In order to denoise x, the maximum a posteriori (MAP) estimator, namely, common K-SVD denoising object model, is often used, which is defined as follows

$$\hat{s} = \arg \min_{s} \|Ds - y\|_2^2 + \alpha \|s\|_0. \tag{9}$$

Noted that the optimization of Eq. (9) is a NP problem, to obtain unique solution, some algorithms, such as basis pursuit (BP), OMP, regularized OMP (ROMP) [16], Stage-wise OMP (StOMP) [7] and so on. Once $\hat{s}$ estimated is enough to be sparse, and then $\hat{s}$ is thought to approach the exact solution, thus, the estimation of x can be obtained by the formula of $\hat{x} = D\hat{s}$.

For a large image, let U denote a clear image patch set, namely the HR image patch set, and Y denote the noise version of U, namely a LR patch image set. Assumed that the dictionary $D \in \Re^{N \times K}$ is known in advance, and R_{ij} is the extraction mark of lapped image patches, thus, each image patch $U_{ij} = R_{ij} U$ with $p \times p$ pixels in every location has a sparse representation with bounded error. In other words, R is the extraction matrix of size $p \times N$, which extracts the (i, j) block from an image of size $N \times N$. And then, the denoising model of Eq. (10) is rewritten as follows

$$J(\hat{D}_{ij}, \hat{U}) = \arg\min_{s_{ij}, U} \left[\lambda \|U - Y\|_2^2 + \sum_{i,j} \mu_{ij} \|S_{ij}\|_0 + \sum_{i,j} \|D S_{ij} - R_{ij} U\|_2^2 \right]. \quad (10)$$

In Eq. (10), the first term controls the degree of approximation of U and Y by controlling the relational expression of $\|U - Y\|_2^2 \leq Const \cdot \sigma^2$. And the larger the parameter σ is, the smaller the parameter λ is. In common, the optimal relation between σ and λ is thought to be $\lambda_{optimal} = 30/\sigma$. The coefficients μ_{ij} must be location dependent so as to comply with a set of constraints of the form $\|D S_{ij} - U_{ij}\|_2^2 \leq \varepsilon$. Further, to assure the maximum sparsity, on the base of Eq. (10), a constraint term is considered additional in the K-SVD denoising model shown in Eq. (10), which is described as the following Eq. (11)

$$J(\hat{D}_{ij}, \hat{U}) = \arg\min_{s_{ij}, U} \left[\lambda \|U - Y\|_2^2 + \sum_{i,j} \mu_{ij} \|S_{ij}\|_1 + \gamma \sum_{i,j} \left(D_{ij}^T D_{ij} \right) + \sum_{i,j} \|D S_{ij} - R_{ij} U\|_2^2 \right]. \quad (11)$$

And utilizing the dictionary matrix $\hat{D}$ and sparse coefficient matrix $\hat{S}$ learned by Eq. (11), the estimation of noise data Y, which is denoted by $\hat{Y}$, can be obtained by the formula of $\hat{Y} = \hat{D}\hat{S}$. Furthermore, super-resolution reconstruction result, denoted by $\hat{U}$, can be calculated by the following formula:

$$\hat{U} = \left(\lambda Y + \sum_{i,j} R_{ij}^T \right)^{-1} \left(\lambda \hat{Y} + \sum_{i,j} R_{ij}^T \hat{D} \hat{S}_{ij} \right). \quad (12)$$

4   Our ISR Method

To reduce convergence time, in the sparse coding step of K-SVD algorithm, we use fast sparse coding (FSC) algorithm to replace OMP algorithm. FSC algorithm is based on iteratively solving well two convex optimization problems, the L_1-regularized least squares problem and the L_2-constrained least squares problem [9]. This algorithm updates feature bases (i.e. dictionary) using the Lagrange Dual and learns sparse coefficient vectors considering the sign of feature coefficient elements. For a LR image, first fixed the dictionary D, the estimated sparse coefficient matrix $\hat{S}$ is trained by FSC in the sparse coding process of K-SVD. Then utilized $\hat{S}$ to update the k column of D, the estimated dictionary $\hat{D}$ can be obtained. Limited by the paper's length, the detail of FSC algorithm is neglected, which was described in detail in the document [11]. Here, much known noise is denoised by FSC based K-SVD algorithm for LR images, furthermore, denoised results are super-resolution restored by POCS algorithm, and then the reconstruction method is generalized as follows:

Step 1. Denoted U to be a HR image, $\hat{U}$ to be the estimation of U, Y, to be the LR version of U, and $\hat{Y}$ to be the estimation of Y. In test, a LR image Y is sampled randomly L times with an image patch with $p \times p$ pixels. Each image patch is converted into a column vector to save, so, a HR image and its LR version is divided into a image patch set with $p^2 \times L$ pixels denoted respectively by $\hat{U}$ and $\hat{Y}$

Step 2. Utilized the basic K-SVD denoising model shown in Eq. (10) to learn the HR image patch set $\hat{U}$, the sparse dictionary $\hat{D}$ can be learned.

Step 3. Viewed $\hat{D}$ to be fixed, utilized FSC algorithm to update sparse coefficient vectors of $\tilde{Y}$, the sparse coefficient matrix $\hat{S}$ can be obtained. Thus, the denoised image $\hat{Y}$ can be calculated by the estimation of $\hat{Y} = \hat{D}\hat{S}$.

Step 4. Finally, according to Eq. (12), the super-resolution reconstruction image $\hat{U}$ can be obtained.

5   Experimental Results and Analysis

5.1   Denoising Results by the FSC Based K-SVD Model

In test, an artificial LR image and a real LR image, namely MMW image, were used. The artificial LR was from a degenerated version of the corresponding HR image. First, this HR image was contaminated severally by adding Gaussian white noise and further

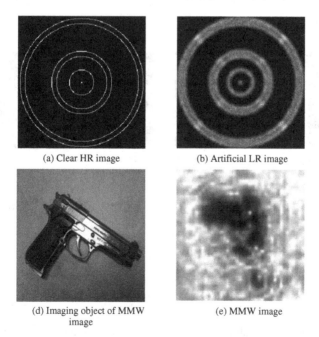

(a) Clear HR image   (b) Artificial LR image

(d) Imaging object of MMW   (e) MMW image
image

Fig. 1. The original images and corresponding low-resolution versions as well as the corresponding denoised results by our K-SVD model.

blurred by the simulated point spread function (PSF) filter. Each image was sampled randomly with 8 × 8 image patch 50000 times, and each image patch was converted into a column, thus the training image patch set with 64 × 50000 size was obtained. Here, noted that the MMW image was generated by the State Key Lab. of Millimeter Waves of Southeast University. The HR images and their corresponding LR versions were shown in Fig. 1. Utilized the FSC based K-SVD algorithm to train LR image set, the dictionary updated with 256 atoms were shown in Fig. 2, here the iteration time was 100 and the redundant degree of the dictionary was 4.

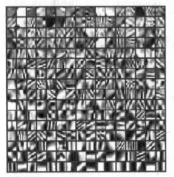
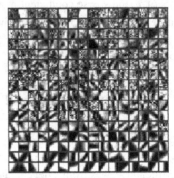

(a) Dictionary of artificial noise image (256 atoms)   (b) Dictionary of the MMW iamge (256 atoms)

Fig. 2. LR dictionary of artificial LR image and real MMW image.

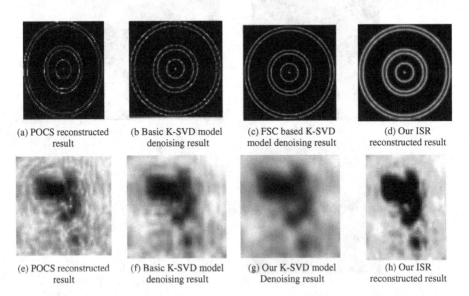

(a) POCS reconstructed result   (b Basic K-SVD model denoising result   (c) FSC based K-SVD model denoising result   (d) Our ISR reconstructed result

(e) POCS reconstructed result   (f) Basic K-SVD model denoising result   (g) Our K-SVD model Denoising result   (h) Our ISR reconstructed result

Fig. 3. ISR reconstructed results by POCS and our method, as well as denoised results of basic K-SVD denoising model and FSC based K-SVD denoising model. The topmost row is processed results of artificial image, and the second row is processed results of the MMW image.

5.2   ISR Reconstructed Results by the K-SVD and POCS Method

According to our ISR method's steps described in Sect. 4, the ISR results were obtained. Considered different noise variance σ, such as 10, 20, 30, 50, etc., the denoised results can be obtained by using the basic K-SVD denoising model shown in Eq. (10) and our K-SVD model shown in Eq. (11). Here, considered the paper's length, only the denoised result and ISR reconstructed results obtained by different algorithm under $\sigma = 30$ was given out, as shown in Fig. 3. Clearly, compared with LR images, denoised results of two K-SVD models and ISR reconstructed results of POCS and our ISR method have better visual effect. Otherwise, for artificial LR image, using signal noise ratio (SNR) to measure processed results of different algorithms, the SNR values from Fig. 3 (a) to (d) are respectively 12.35, 15.22, 17.83 and 21.36. This also proves that our ISR reconstruction method outperforms POCS and K-SVD denoising methods.

6   Conclusions

A novel image super-resolution reconstruction method using the combination method of FSC based K-SVD denoising model and POCS algorithm is discussed in this paper. The FSC algorithm used in the K-SVD denoising model can optimize well the sparse coding process and obtain better sparse dictionary than the OMP algorithm. At the same time, the maximum sparse constraint of dictionary is also considered. In test, the restoration effect of our method is testified by using five highly degraded natural images, which simulate some artificial low-resolution images, and a real MMW image. Using our FSC based K-SVD denoising model, much unknown noise is reduced clearly. Further, denoised results are super-resolution restored by POCS algorithm. This method not only reduces noise efficiently but also retains well image edge detail information. Experimental results show that, compared with methods of the basic K-SVD denoising model, FSC based K-SVD denoising model and POCS, our method indeed behaves clear improvement in SNR and better visual effect.

Acknowledgements. This work was supported by the grants from National Nature Science Foundation of China (Grant No. 61373098 and 61370109), the grant from Natural Science Foundation of Anhui Province (No. 1308085MF85).

References

1. Dong, W.S., Zhang, L., Shi, G.M., Wu, X.L.: Image deblurring and super-resolution by adaptive sparse domain selection and adaptive regularization. IEEE Trans. Image Process. **20**(7), 1838–1857 (2011)
2. Salehin, S.M.A., Abhayapala, T.D.: Constrained total variation minimization for photoacoustic tomograhy. In: Proceedings of 2012 IEEE International Conference on Acoustics, Speech and Signal Processing (ICASSP), Vancouver, Canada, pp. 561–564. IEEE Press (2012)

3. Le Montagner Y., Angelini E., Olivo-Marin J.: Comparison of reconstruction algorithms in compressed sensing applied to biological imaging. In: Proceedings of IEEE International Symposium on Biomedical Imaging: From Nano to Macro, Chicago, pp. 105–108. IEEE Press (2011)

4. Zhi-ming, P., Hong-kai, X.: Sparse representation over adaptive dictionaries for super-resolution. Inf. Technol. **4**(3), 73–76 (2012)

5. Sun, J., Xu, Z.B., Shum, H.Y.: Gradient profile prior and its applications in image super-resolution and enhancement. IEEE Trans. Image Process. **20**(6), 1529–1542 (2011)

6. Jin-zheng, L., Qi-heng, Z., Zhi-yong, X., Zhen-ming, P.: Image super-resolution reconstruction algorithm using over-complete sparse representation. Syst. Eng. Electron. **34**(2), 403–408 (2012)

7. Wang, S.Q., Zhang, J.H.: Fast image inpainting using exponential threshold POCS plus conjugate gradient. Imaging Sci. J. **62**(3), 161–170 (2014)

8. Salehin, S.M.A., Huang, S., Abhayapala, T.D.: Projection onto convex sets (POCS) method for photoacoustic tomography with a nonnegative constraint. In: 35th Annual International Conference of the IEEE EMBS, Osaka, Japan, pp. 3–7. IEEE Press (2013)

9. Elad, M.: Sparse and Redundant Representation: From Theory to Applications in Signal and Image Processing. Springer, New York (2010)

10. Yang, J.C., Wright, J., Huang, T., Ma, Y.: Image super-resolution via sparse representation. IEEE Trans. Image Process. **19**(11), 2861–2873 (2010)

11. Zhu, Z., Guo, F., Yu, H., Chen, C.: Fast single image super-resolution via self-example learning and sparse representation. IEEE Trans. Multimedia **16**(8), 2178–2190 (2014)

An Improved Denoising Method
Based on Wavelet Transform
for Processing Bases Sequence Images

Ke Yan[✉], Jin-Xing Liu, and Yong Xu

Bio-Computing Research Center, Shenzhen Graduate School,
Harbin Institute of Technology, Shenzhen, China
yanke401@163.com, sdcavell@126.com, yongxu@ymail.com

Abstract. In this article, we present an improved images denoising method for base sequence images. It is based on the multiscale analysis of the images resulting from the à trous wavelet transform decomposition. We define a new thresholding function and use it to improve the denoising performance of the isotropic undecimated wavelet transform (IUWT). The proposed method selects the best suitable wavelet function based on IUWT. The advantages of the new thresholding function are that it is more robust than previous thresholding function, and the convergence of function is more efficient. The experimental results indicate that the proposed method can obtain higher signal-to-noise ratio (SNR) and mean squared error ratio (MSE) than conventional wavelet thresholding denoising methods.

Keywords: Bases sequence images · Image denoising · Isotropic undecimated wavelet transform

1 Introduction

Second generation sequencing technologies such as Genome analysis tools (Illumina, San et al) have increased the production of sequence data by several orders of magnitude. The major technology of the second generation sequencing technology is the high- throughput genome sequencing technology, and a large number of base sequence images will be observed and captured from the tool in the process of sequencing, such as fluorescence microscopy, CCD and so on. Fluorescence microscopy is broadly used in the biological imaging, and the corresponding images allow us to detect the presence of particular biological or to examine the distribution of antigens in cellular systems by molecular medicine [1, 2]. In the fluorescence microscopy, a fluorescence protein label object is usually displayed as a blurred spots in the image, which consist of a few pixels, and have no clear borders [3]. The phenomenon of blurring is caused by some reasons, and the mainly source is caused by imperfections of the optical system. When the photon emission is randomly nature, the base sequencing images are polluted by photon noise (Poisson noise).

Currently, there are several automatic methods to noise reduction [4–8]. Those classical methods such as morphological top-hat [4], adaptive thresholding, wiener

© Springer International Publishing Switzerland 2015
D.-S. Huang et al. (Eds.): ICIC 2015, Part I, LNCS 9225, pp. 357–365, 2015.
DOI: 10.1007/978-3-319-22180-9_35

filter do not have satisfactory results on biological images. This can be explained by two reasons: first, the quality of the image is rather low, due to the limitations in the image acquisition process [3]; second, the grey level distributions of spots are not homogeneous, and the images may have an uneven background. In other words, some spots in one part don't have the same pixel levels with the spots in another part [9]. By using an global operator, we will get over- or sub- detectivity depending on the local properties of the image [9]. As a result, those traditional methods cannot achieve the best performance of denoising on the dataset of bases sequence images.

Xu and co-workers use the properties of the wavelet coefficients' direct correlation to denoise images in an edge- oriented manner [10]. In order to detect the edges of the image and reconstruct the denoising image, they employ the wavelet transform proportional to the first derivation of a smoothing function. Mallet and Zhong use wavelet proportional to a first derivative to provide an edge-oriented approach to image filtering [11]. Bijaoui and co-workers use the multiresolution support to the filter and reconstruct images by the scheme of the iterative regularization in the à trous wavelet domain [12–14]. Recently, Sadler and Swami detect the estimation of the edge by the wavelet transform multiscale products [15].

Isotropic Undecimated Wavelet Transform (IUWT) [9] introduces a multiresolution algorithm for detecting bright spots. The feature detection is the process of extracting and combining multilevel elements of response, with each element coming from successive resolution level. To retain the significant responses of the filter to the desired feature, the step of extracting features is accomplished by a denoising technique using a hard thresholding value. In the end, the new selected coefficients allow us to combine multiscale information and to detect the spots. IUWT is effective for detecting and counting the spots but may perform badly in the case of the low quality of image. Due to the discontinuous of the shrink function, the wavelet transform based on the hard thresholding will yields abrupt artefacts in recovered image To avoid the disadvantages of the hard threshold in the denoising stage, a number of wavelet transform choose the soft threshold, such as the threshold based on Bayesian estimation [16–18]. The disadvantages of the soft threshold shrinkage function tend to have bigger bias, due to the large coefficients.

In this paper, we introduce a new method to improve Isotropic Undecimated Wavelet Transform (IIUWT) for denoising in fluorescence microscopy images. To overcome the drawbacks of the conventional IUWT, we address the problem by choosing an improved soft thresholding by ℓ_1 norm robustly. In the wavelet thresholding denoising method, we should choose an optimal threshold value beforehand. The wavelet coefficients whose magnitudes are smaller than the threshold are set to zero. And the other coefficients should be shrunk (the soft-thresholding case) [19–22]. To avoid the drawback of traditional soft thresholding shrinkage function, the new proposed method proposed a new function to reconstruct the wavelet coefficients. When the scale increases, the distance between original wavelet coefficients and reconstruction coefficients are decrease at each scale.

This paper is organized as follows. A brief introduction to the algorithm of IUWT, including its process and mainly properties are presented in Sect. 2. The proposed algorithm of IIUWT is introduced in Sect. 3. Experimental results are provided in Sect. 4.

2   A Brief Introduction of IUWT

The wavelet transform of an image captures a set of wavelet coefficients values $\{\omega_j\}$, the mean of which is zero [23]. IUWT uses the à trous algorithm [24] and its B3-spline version proposed by Bijaoui [25]. The original image M_0 can be decomposed as a series of wavelet components ω_i and the smoothed array M_k.

$$M_0(x, y) = M_K(x, y) + \sum_{i=1}^{K} \omega_i(x, y) \tag{1}$$

The process of the wavelet transform is as follows:

1. Initialize i to 0, starting with the original image $M_0(x, y)$.
2. Increase the value of i, the data $M_i(x, y)$ is convolved row by row, and then column by column with a 1-D kernel h, and the result is $M_{i+1}(x, y)$. The kernel h is $\left[\frac{1}{16}, \frac{1}{4}, \frac{3}{8}, \frac{1}{4}, \frac{1}{16}\right]$, and is modified in term of the scale i by inserting $(2^{i-1} - 1)$ zeros between two taps.
3. Compute the discrete wavelet transform: $\omega_i(k) = M_{i-1}(k) - M_i(k)$.
4. Return to step 2 till scale i equals to the number k which is the deepest resolution level.

Spots in the image can be distinguished by the large value of the coefficients which are correlated across levels. If the image is polluted by additive correlated Possion noise, the value of local maxima in the wavelet plane will propagate across scales when they are caused by significant discontinuities in the image such as spots, otherwise they do not propagate across scales if caused by noise [26]. So we should choose those special wavelet coefficients, which give a coarse estimation of spot positions.

IUWT chooses the spatial hard thresholding strategy to distinguish the spot from the noise. So we can define the denoising process as follows:

$$\tilde{\omega}_{i,hard} = \begin{cases} 0, \omega_i < t_i \\ \omega_i, \omega_i \geq t_i \end{cases} \tag{2}$$

IUWT introduces a special spatial filtering which sets the hard thresholding as $t_i = k\sigma_i$. In the formula, σ_i represents the standard deviation of the noisy wavelet coefficients at scale i, and a usual choice is $k = 3$ [25].

When we use the wavelet coefficients to distinguish spots from the noise, the wavelet representation as a set of $(k + 1)$ images $\tilde{\omega}_1, \tilde{\omega}_2, \ldots, \tilde{\omega}_K, M_K(x, y)$. The reconstruction can be easily performed as:

$$M(x, y) = M_K(x, y) + \sum_{i=1}^{K} \tilde{\omega}_i \tag{3}$$

The high values of $\tilde{\omega}_i(x, y)$ indicate the presence of the spot and the others represent the background or large structures [26, 27].

3 Methodology

Using the IUWT based à trous algorithm and its B3-spline version to decompose the original image, we can decompose the noised signal. As a result, the approximated wavelet coefficients ω_i s and the smoothed image M_k are obtained.

In the denoising context, the decorrelation property suggests the independence of the wavelet coefficients. And the sparseness property suggests the way to estimate those wavelet coefficients with small amplitudes is feasible. The conventional threshold functions are the hard or soft threshold functions defined below.

$$\tilde{\omega}_i^{hard} = \begin{cases} 0, |\omega_i| \leq \lambda \\ \omega_i, |\omega_i| > \lambda \end{cases} \tag{4}$$

$$\tilde{\omega}_i^{soft} = \begin{cases} 0, |\omega_i| \leq \lambda \\ \mathrm{sign}(\omega_i)(|\omega_i| - \lambda), |\omega_i| > \lambda \end{cases} \tag{5}$$

Function $\mathrm{sign}(x)$ is defined as: if $x \geq 0$, then $\mathrm{sign}(x) = 1$; otherwise $\mathrm{sign}(x) = -1$, and λ is the threshold level. One of the alternative thresholds is $k\delta_i(\delta_i$ is the standard deviation of ω_i, and $k \in R)$. In our study, the improved threshold is represented as:

$$\tilde{\omega}_i = \begin{cases} 0, |\omega_i| \leq \lambda \\ \mathrm{sgn}(\omega_i)\left(|\omega_i| - \alpha^{\frac{1}{\|\omega_i - \lambda\|_2}} * \lambda\right), |\omega_i| > \lambda \end{cases} \tag{6}$$

where $\alpha > 1$ and $\alpha \in R$. λ is represented as:

$$\lambda = \frac{k}{N} \|\omega_i - \mathrm{median}(\omega_i)\|_1 \tag{7}$$

where $\mathrm{median}(\omega_i)$ denotes the median value of the values of the wavelet coefficients ω_i, and N stands for the size of the origin image.

If the absolute value of ω_i is larger than λ, the deviation between $\tilde{\omega}_i$ and ω_i decreases with the increase of ω_i when parameter α is fixed. It makes the reconstruction image remains the characteristic of the original image mostly. Given an original image in the form of:

$$Y = y + n \tag{8}$$

where Y is the observation, and y is the noise-free data, and n is the noise. We want to calculate the wavelet transformation of Y, i.e. $\omega^Y = \omega^y + \omega^n$. To distinguish the image data ω^y from the wavelet coefficient ω^Y, we define the threshold λ which is associated with noise data, and the value of $\|\omega_i - \lambda\|_2$ defines the intensity of the noise-free data. As a result, the formula $\alpha^{\frac{1}{\|\omega_i - \lambda\|_2}} * \lambda$ is used to estimate the value of the noise data from the wavelet coefficients ω_i. The estimation of $\tilde{\omega}$ is replaced by zero due to the noise.

The value of threshold λ is calculated by the ℓ_1 norm of the difference value between ω_i and the median of ω_i. In the traditional method, the value of threshold λ is $k\delta_i$. ω_i is the standard deviation of the wavelet coefficients ω_i. Parameter δ_i can be calculated by the formula: $\delta_i = \frac{1}{N}\|\omega_i - \text{mean}(\omega_i)\|_2$, where $\text{mean}(\omega_i)$ is the mean value of the wavelet coefficient ω_i. Compared with ℓ_2 regularization, ℓ_1 regularization is more robust. We can see ℓ_2 regularization spreads the error throughout vector $(\omega_i - \text{median}(\omega_i))$, whereas ℓ_1 regularization is stable with a sparse vector. Some values of vector $(\omega_i - \text{median}(\omega_i))$ are exactly zero whereas others may be relatively large.

4   Experiments

4.1   Evaluation Criteria

We utilize both the mean square error (MSE) and signal-to-noise (SNR) [17] values of the reconstruction image and original image to evaluate the performance of our method.

MSE provides a quantitative evaluation of the performance. The MSE provides a quality of an estimator error between the denoising image and origin image. Minimizing MSE is a key criterion in denoting the denoising performance. We calculate the MSE using the following formula:

$$MSE = \frac{1}{n_1 * n_2} \sum_{i=1}^{n_1} \sum_{j=1}^{n_2} \left(M(i,j) - M'(i,j)\right)^2 \tag{9}$$

where $M(i,j)$ denotes the original image, $M'(i,j)$ denotes the reconstruction image, i.e. the resultant image of wavelet transform denoising, and $n_1.n_2$ is the total number of the pixels.

The output SNR is given by

$$SNR = \frac{\sum_{i=1}^{n_1}\sum_{j=1}^{n_2} M'(i,j)^2}{\sum_{i=1}^{n_1}\sum_{j=1}^{n_2}\left(M(i,j) - M'(i,j)\right)^2} \tag{10}$$

4.2   Evaluation on Synthetic Image Data

The proposed IIUWT method is first evaluated using synthetic 2-D image (the sizes are 128*128 pixels) containing intensity profile of round objects modeled using $\delta_{max} = \delta_{min} = 2$, and different levels of Poisson noise. To simulate the real image data which is acquired in our biological applications and is lower than the critical level of SNR=4 ~ 5 [3], SNRs of our synthetic images are in the range of 5,8,10. The maximum amplitude of some spots is 2530, and the maximum amplitude of the others spots is 1275.

To validate the superiority of the proposed improvement of Isotropic Undecimated Wavelet Transform (IIUWT), we compare the proposed method with the others popular methods. The IUWT wavelet denoising is performed for hard-thresholding, soft-thresholding, an improved thresholding denoising methods(ITDM) [19], and IIUWT respectively to denoise the noisy synthetic image data. In my IIUWT method, the parameter $\alpha = 2, k = 3$, and the thresholding of λ in the soft-thresholding is $2\delta_i$ (δ_i is the standard deviation of ω_i). The parameter k in the hard-thresholding is 3. Table 1 lists the SNR of the different methods in the database of synthetic image data. Table 2 lists the MSE of the different methods in the database of synthetic image data.

Table 1. SNR values of the different methods in the images of Type A, Type B, Type C

Image	Type A	Type B	Type C
IUWT(hard threshold)	28.36	26.86	28.26
IUWT(soft threshold)	26.05	24.23	25.23
ITDM	28.86	27.52	29.14
Proposed method	31.45	30.12	31.66

Table 2. MSE values of the different methods in the images of Type A, Type B, Type C

Image	Type A	Type B	Type C
IUWT(hard threshold)	81.98	76.59	68.11
IUWT(soft threshold)	87.92	82.49	74.48
ITDM	80.52	74.86	65.84
Proposed method	70.26	64.96	57.24
Proposed method	70.26	64.96	57.24

From Table 1, we can know that the values of SNR of the proposed method are the highest among all methods. One can see that on some images, our method achieves an improvement of more than 2 dB. As a result, our proposed method can improve the quality of the noisy signal efficiently. From Table 2, we can know that the MSE values of the proposed method are lower than those of all other method. The denoising performance of our method is better than the other methods from the results.

4.3   Evaluation on the Real Image Data

The described denoising method is also tested by real microscopy image data from several biological studies. We can get images from an open source, which is named as Swift. Swift is the first freely available solution to the primary data analysis problem "from image to base calls" and is available under the LGPL3 [28]. The sizes of the biological image are 1794*2048, and the total number of the biological image is 148. An image whose sizes are 150*150 pixels is segmented from the real image as the experiment data. Due to the noise intensity of image is low, we will add the Possion noises of different levels. There are three images which is named as type A, type B, type C. The SNR of those images are 4,7,12, respectively. To firm the superiority of the

Table 3. SNR values of different methods in the images of Type A, Type B, Type C

Image	Type A	Type B	Type C
IUWT(hard threshold)	19.73	17.56	15.31
IUWT(soft threshold)	18.5	15.8	13
ITDM	21.45	19.17	16.78
Proposed method	22.49	20.41	17.95

Table 4. MSE values of different methods in the images of Type A, Type B, Type C

Image	Type A	Type B	Type C
IUWT(hard threshold)	303.15	267.66	271.39
IUWT(soft threshold)	302.94	268.42	272.29
ITDM	301.02	265.55	268.67
Proposed method	294.37	260.23	263.35

proposed method, we compared with other popular methods. In my IIUWT method, the parameter $\alpha = 2, k = 3$, and the thresholding of λ in the soft-thresholding is $2\delta_i$ (δ_i is the standard deviation of ω_i). The parameter of k in the hard thresholding is 3. Table 3 shows the SNRs of different methods in the database of real data. Table 4 shows the MSE of different methods in the database of real data.

From the Table 3, we can know that our proposed method's SNR is higher than the other methods in the database of the real data. Compared with the SNR values of the IUWT (hard thresholding), IUWT(soft thresholding), ITDM, the proposed method's SNR values can improve the 3 dB, 4 dB, 1 dB respectively. Comparing with the SNR of the noisy signal, our proposed method can improve the quality of the image in the database. It is clearly that our proposed method is better than the state-of-art algorithm in the Table 4. The MSE values of our method are lower than the other methods. It can be seen from the table, decrease of the MSE of residuals and increase of the SNR are achieved simultaneously.

5 Conclusion

In this paper, we proposed the improvement of the isotropic undecimated wavelet transform (IIUWT) which can improve the performance of the conventional isotropic undecimated wavelet transform with a hard threshold. In the proposed method, we redefine the expression of the wavelet denoising function, and the threshold is calculated by the ℓ_1 norm. To validate the superiority of the proposed method, we compared it with other conventional methods. The values of the SNR and the MSE of our proposed method are more competitive than these conventional algorithms.

Acknowledgement. This work was supported by Shenzhen Municipal Science and Technology Innovation Council (Grant No. JCYJ20130329151843309, Grant No. CXZZ20140904154910774, Grant No.JCYJ20140417172417174) and China Postdoctoral Science Foundation funded project (Grant No.2014M560264).

References

1. Hémar, A., Olivo, J., Williamson, E., Saffrich, R., Dotti, C.: Dendroaxonal transcytosis of transferrin in cultured hippocampal and sympathetic neurons. J. Neurosci. **17**(23), 9026–9034 (1997)
2. Dumenil, G., Olivo, J., Pellegrini, S., Fellous, M., Sansonetti, P., Van Nhieu, T.G.: Interferon α inhibits a Src-mediated pathway necessary for shigella-induced cytoskeletal rearrangements in Epithelial cells. J. Cell Biol. **143**(4), 1003–1012 (1998)
3. Smal, I., Loog, M., Niessen, W., Meijering, E.: Quantitative comparison of spot detection methods in fluorescence microscopy. IEEE Trans. Med. Imaging **29**(2), 282–301 (2010)
4. Matheron, G., Serra, J.: Image Analysis And Mathematical Morphology. Academic Press, London (1982)
5. Bright, D., Steel, E.: Two-dimensional top hat filter for extracting spots and spheres from digital images. J. Microsc. **146**(2), 191–200 (1987)
6. Liang, L., Xu, Y., Shen, H., Camilli, P.D., Toomere, D., Duncan, J.: Automatic detection of subcellular particles in fluorescence microscopy via feature clustering and bayesian analysis. In: 2012 IEEE Workshop on Mathematical Methods in Biomedical Image Analysis (MMBIA), pp. 161–166 (2012)
7. Eils, R., Athale, C.: Computational imaging in cell biology. J. Cell Biol. **161**(3), 477–481 (2003)
8. Zhou, X., Wong, S.T.C.: Informatics challenges of high-throughput microscopy. IEEE Sig. Proc. Mag. **23**(3), 63–72 (2006)
9. Olivo-Marin, J.: Extraction of spots in biological images using multiscale products. Pattern Recogn. **35**(2), 1989–1996 (2002)
10. Xu, Y., Weaver, J., Healy Jr, D.M., Lu, J.: Wavelet transform domain filters: a spatially selective noise filtration technique. IEEE Trans. Image Proc. **3**(6), 747–758 (1994)
11. Mallat, S., Zhong, S.: Characterization of signals from multiscale edges. IEEE Trans. Pattern Anal. Mach. Intell. **14**(7), 710–732 (1992)
12. Starck, J., Murtagii, F., Bijaoui, A.: Multiresolution support applied to image filtering and restoration. Graph. Models Image Proc. **57**(5), 420–431 (1995)
13. Murtagh, F., Starck, J.: Image processing through multiscale analysis and measurement noise modeling. Stat. Comput. **10**(2), 95–103 (2000)
14. Starck, J., Murtagh, F., Bijaoui, A.: Image Processing and Data Analysis: The Multiscale Approach. Cambridge University Press, Cambridge (1998)
15. Sadler, B.M., Swami, A.: Analysis of multiscale products for step detection and estimation. IEEE Trans. Inf. Theo. **45**(3), 1043–1051 (1999)
16. Robert, C.: The Bayesian Choice: A Decision Theoretic Motivation. Springer Texts in Statistics, New York (1994)
17. Smal, I., Draegestein, K., Galjart, N., Niessen, W., Meijering, E.: Particle filtering for multiple object tracking in dynamic fluorescence microscopy images: application to microtubule growth analysis. IEEE Trans. Med. Imaging **27**(6), 789–804 (2008)
18. Smal, I., Meijering, E., Draegestein, K., Galjart, N., Grigoriev, I., Akhmanova, A., Royen, M., Houstsmuller, A., Niessen, W.: Multiple object tracking in molecular bioimaging by rao-blackwellized marginal particle filtering. Med. Image Anal. **12**(6), 764–777 (2008)
19. Reddy, G., Muralidhar, M., Varadarajan, S.: ECG de-noising using improved thresholding based on wavelet transforms. IJCSNS **9**(9), 221 (2009)
20. Donoho, D., Johnstone, J.: Ideal spatial adaptation by wavelet shrinkage. Biometrika **81**(3), 425–455 (1994)

21. Chen, G., Qian, S.: Denoising of hyperspectral imagery using principal component analysis and wavelet shrinkage. IEEE Trans. Geosci. Remote Sens. **49**(3), 973–980 (2011)
22. Parrilli, S., Poderico, M., Angelino, C., Verdoliva, L.: A nonlocal SAR image denoising algorithm based on LLMMSE wavelet shrinkage. IEEE Trans. Geosci. Remote Sens. **50**(2), 606–616 (2012)
23. Murtagh, F., Starck, J.: Image processing through multiscale analysis and measurement noise modeling. Stat. Comput. **10**(2), 95–103 (2000)
24. Holschneider, M., Kronland-Martinet, R., Morlet, J., Tchamitchian, Ph: A real-time algorithm for signal analysis with the help of the wavelet transform. In: Combes, J.-M., Grossmann, A., Tchamitchian, P. (eds.) Wavelets. Inverse Problems and Theoretical Imaging, pp. 286–297. Springer, Heidelberg (1990)
25. Starck, J., Murtagii, F., Bijaoui, A.: Multiresolution support applied to image filtering and restoration. Graph. Models Image Proc. **57**(5), 420–431 (1995)
26. Jansen, M., Bultheel, A.: Multiple wavelet threshold estimation by generalized cross validation for images with correlated noise. IEEE Trans. Image Proc. **8**(7), 947–953 (1999)
27. Ma, Z., Wen, J., Liu, Q., Tuo, G.: Near-infrared and visible light image fusion algorithm for face recognition. J. Modern Optics. **62**(9), 745–753 (2015)
28. Whiteford, N., Skelly, T., Curtis, C., Ritchie, M., Lohr, A., Zaranek, A., Abnizova, I., Brown, C.: Swift: primary data analysis for the illumina Solexa sequencing platform. Bioinf. **25**(7), 2194–2199 (2009)

License Plate Extraction Using Spiking Neural Networks

Qian Du, LiJuan Chen, RongTai Cai[✉], Peng Zhu, TianShui Wu,
and QingXiang Wu

College of Photonic and Electronic Engineering, Fujian Normal University,
Fuzhou 350108, Fujian, China
624426409@qq.com, gjrtcai@163.com

Abstract. In this paper, we present an algorithm for license plate detection and extraction using spiking neural networks (SNNs). We propose an SNN for the detection of license plate by simulating the color perception principle in human beings' visual system, where synchronization of spiking trains are employed as a color detection function and used to detect the license plate according to the difference of color in the license plate's patch and those in the other image patches. By doing so, we can extract those image regions that are likely to be license plates. And then we use another SNN to produce the edge images of these candidates by simulating the receptive field of orientation in human beings' visual cortex. Finally, we extract the license plate from these candidates according to the texture difference between a real license plate image and the distracters, where the numbers of strokes in image rows are served as cues for the texture difference. The experimental results show that the proposed biological inspired SNNs are valid in the detection and extraction of license plate.

Keywords: License plate recognition · OCR · Spiking neural networks · Object detection

1 Introduction

License plate recognition system (LPRS) is an important part of intelligent transportation system. As an essential part of license plate recognition system, the localization of license plate is of great importance. There are many features in the license plate that can be used as the cues for the localization and extraction of license plate. In [1] X. Wan and J. Liu use a Color Barycenters Hexagon (CBH) model to localize the license plate which is based on the color feature. In [2] F. Wang, L. Man extract the color features in the HSV space and then three components of the HSV space are firstly mapped to fuzzy sets according to different membership functions, lastly use a the fuzzy classification function for color recognition. In [3] a geometry-based method is used to locate the position of the license plate. In [4] Jinn-Li Tan, Abu-Bakar introduce the idea of edge-geometrical features in detecting these plates. In [5] R. Bremananth, A. Chitra, V. Seetharaman and in [6] C.-T. Hsieh, Y.-S. Juan use texture features to acquire the license plate. In [7] global image features are used to gain the license plate. In [8] a new class of locally threshold separable detectors based on external regions are

© Springer International Publishing Switzerland 2015
D.-S. Huang et al. (Eds.): ICIC 2015, Part I, LNCS 9225, pp. 366–377, 2015.
DOI: 10.1007/978-3-319-22180-9_36

used to localize the license plate. In [9] H. W. Lim and Y. H. Tay present a license plate detector using a fusion of Maximally Stable External Regions and SIFT-based unigram classifier trained with Core Vector Machine to extract the license plate. In [10] B. K. Cho, S. H. Ryu use the character features to extract the license plate. In general cases, one feature is not enough for the extraction of license plate in practical applications, thus multiple cues would be employed in the task [11, 12].

It is well known that producing computational models by means of mimicking human beings' perception principle is promising to advance the development of computer vision. Recent research reveals that biological neurons use spike trains to encode information. Meanwhile, researchers have recently come to realize that the spike-based computation can be performed by spiking neural networks. The computation of spike trains was implemented on hardware by Harris in University of Florida in 2008 [13] and Kit Cheung in 2012 [14]. When it comes to the application of SNN in the field of visual information processing, we can find that it was widely used in the field of image feature extraction or target detection. For examples, Meigui chen and Qingxiang Wu extracted multiple features of breast cancer areas in mammography using SNN [15], Rongtai Cai et al. extracted infrared target from image sequences using SNN [16]. Except for the feature or target extraction, it was also used in the classification of visual target, such as target detection and classification in Rongtai Cai's work [17] and action recognition in Maria Jose Escobar's work [18]. Following this line, we proposed a SNN model that can be used in the detection and extraction of license plate by simulating the visual functions in human being's visual perception function in this paper.

2   The Procedure of License Plate Extraction

The procedure of the extraction of license plate is shown in Fig. 1. After acquisition of the image that contains license plate patch, based on the color features of the license plate, we use a SNN to acquire the candidates, details of which are given in Sect. 3. Considering that there are many noises, we then go on with the work of noise removals. After that, we can obtain the location of the license plate. Because there may still be image region, that are distracters, we can employ other features of the license plate to exclude the distracters, such as the number of strokes in image rows in the candidates. In this stage, we use another SNN to extract the edges of the candidates, details of which are presented in Sect. 4. By doing so, we can obtain an edge image. Then we produce binary image from the edge image. In the following steps, we project the binary image rows onto containers that record the number of stokes horizontally. By calculating the number of strokes,we can exclude the false plate area and extract the license plate accurately.

Fig. 1. The flow chart of extracting license plate

3 The Detection of License Plate Using SNN

3.1 Spiking Neuron Model

Many spiking neuron models have been proposed in the literature, which differ in biological plausibility and their computational efficiency [19]. In this paper, we employ a conductance-driven integrate-and-fire neuron model to construct SNNs. Considering a neuron i, defined by its membrane potential $u_i(t)$, the integrate-and-fire equation is

$$\frac{d u_i(t)}{dt} = G_i^{exc}(t)(E^{exc} - u_i(t)) + G_i^{inh}(t)(E^{inh} - u_i(t)) + g_L(E_L - u_i(t)) + I_i(t), \quad (1)$$

where $u_i(t)$ is the membrane potential, $G_i^{exc}(t)$ is the normalized excitatory conductance directly associated with the pre-synaptic neurons connected to neuron i. The conductance g_L is the passive leaks in the membrane. $I(t)$ is an external input current, $G_i^{inh}(t)$ is an inhibitory normalized conductance dependent on lateral connections or feedbacks from upper layers. E_L, E^{exc}, E^{inh} are reverse potential.

Here is the spike emission process: the neuron i will generate a spike when the membrane potential of the cell reaches the threshold value $u_i(t) = v_{th}$, then the $u_i(t)$ is reset to its resting potential E_L. The neuron membrane potential $u_i(t)$ evolves according to inputs through either conductance's ($G_i^{exc}(t)$ or $G_i^{inh}(t)$) or external currents($I_i(t)$).

Here we use $S_{N1}(t)$ to represent the spike train of neuron $N1$, the expression of which is

$$S_{N1}(t) = \begin{cases} 1 & \text{if neuron } N1 \text{ fires at time } t. \\ 0 & \text{if neuron } N1 \text{ does not fire at time } t. \end{cases} \quad (2)$$

The firing rate of the neuron $N1$ is calculated according to

$$F_N(t) = \frac{1}{T}\sum_{t}^{t+T} S_{N1}(t) \quad (3)$$

where $S_{N1}(t)$ is the spike train, T is a period of time, in which we calculate the firing rate, t is time variable, F is the firing rate, N_i indicates the N_i neuron.

3.2 The Detection of License Plate Using SNN

By simulating the retinal cone cell's color perception function, we designed a spiking neural network, using the color features to detect the license plate candidates. The three kinds of cone cell and their color perception characteristics and presented in Table 1. The Illustrations of SNN composed of retinal cone cells are depicted in Fig. 2.

Figure 2 shows the color detection principles using synchronization of spiking trains in a SNN in the red channel. The first layer is the photonic receptor layer, neurons in this layer convert input pixels into currents. The second layer is composed of color sensors corresponding to the receptors in the input layer. In this layer, the input

Table 1. Cone cells in the human eye [21]

Cone type	Range	Peak wavelength	Color
Short (S)	400–500 nm	420–440 nm	B
Medium (M),	450–630 nm	534–555 nm	G
Long (L)	500–700 nm	564–580 nm	R

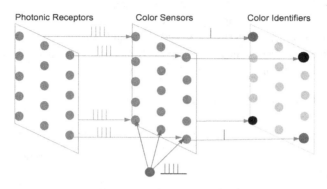

Fig. 2. Color detection using synchronization of spiking trains in a SNN

spike trains are compared with the reference red spike train and judged whether they are similar to the reference color. If they are similar with the reference color, the neurons will produce spikes and send them to the corresponding neurons in the output layer. The neurons in the output layer determined whether the corresponding pixels belong to license plate candidates or not according to whether they receive spikes or not.

A Chinese license plate usually have four types of background color, that is blue background, yellow background, white background, and black background. According to the perception feature of cone cell in retina, if the three components R, G, and B of a color pixel $P(x,y)$ satisfy $R < R_r$, $G < G_r$ and $B > B_r$, we can conclude that $P(x, y)$ belong to blue background. Here R_r, G_r, and B_r are three threshold, which are specified through experiments. Similarly, if R, G, and B satisfy $R < R_r$, $G < G_r$ and $B < B_r$, we can conclude that $P(x,y)$ belong to black background; if R, G, and B satisfy $R > R_r, G > G_r$, and $B > B_r$, we can conclude that $P(x,y)$ belong to white background; if $R > R_r, G > G_r$, and $B < B_r$, we can conclude that $P(x,y)$ belong to yellow background. Based on these color features of license plate, we can obtain the candidates of the license plate, thus detecting the candidates of license plate from an image. Considering a color image $I(x,y)$, whose dimension is $P \times Q \times 3$. Its component images are labeled as $I_R(x,y)$, $I_G(x,y)$, and $I_B(x,y)$. Firstly, we transform each pixel of I_R, I_G, and I_B into the corresponding current, then feed into the neurons in the first layer of the SNN, which will convert the input currents into spike trains. Then we compute the firing frequency of each spike train corresponding to each pixel in the three component images. We can use three arrays to represent the firing frequency, labeled as *spikeR, spikeG* and *spikeB*, the dimension of which are $P \times Q$. For example,$spikeR(3,4)$ correspond to the firing frequency of the pixel in the coordinate $(3, 4)$ of component image IR. We transform three reference pixel R_r,G_r,and B_r to spike trains and compute their firing rate respectively. Their firing rate are labeled as $spikeR_{rr}$,$spikeG_{rg}$,$spikeB_{rb}$.Then comparing each value of $spikeR$ with $spikeR_{rr}$, each value of $spikeG$ with $spikeG_{rg}$, each value of $spikeB$ with $spikeB_{rb}$, if they meet the requirements, we label the corresponding position as 1. Thus we obtain three models labeled as ModelR, ModelG, ModelB, the dimension of which are $P \times Q$. For example, we assume that the background of the license plate is blue and $spikeR(x,y)$ $spikeG(x,y)$ $spikeB(x,y)$ are their firing rates corresponding to pixel (x,y). If $spikeR(x,y) <$ $spikeR_{rr}$, $spikeG(x,y) < spikeG_{rg}$, and $spikeB(x,y) > spikeB_{rb}$, we conclude that the pixel in (x,y) has a blue color. By analogy, this procedure is available to the other three types of background. The specific algorithm is as following:

Algorithm 1: Detection of license plate using color cues

```
Step 1. Read in a color image.
Step 2. Convert pixels in images I_R, I_G and I_B into spike
        trains.
Step 3. Compute the firing times of each spike trains
        using 100 sampling points. Obtain three arrays
        SpikeR, SpikeG, SpikeB, the dimension of which are
        pxQ.
Step 4. Transform three reference pixel R_r, G_r and B_r into
        spike trains,labeled as SpikeR_rr, SpikeG_rg and
        SpikeB_rb respectively.
Step 5.
  for x=1 : P
      for y=1:Q
          If SpikeR(x,y)< SpikeR_rr & spikeG(x,y)< SpikeG_rg
             & spikeB(x,y)> SpikeB_rb
```

```
                ModelR(x,y)=1;
          Else
                ModelR(x,y)=0;
          end
      end
  end
Step 6. If Model(x,y)=1, I(x,y) belongs to the blue
        background.
```

---

4   Extraction of License Plate Based on SNN

By using the SNN in Sect. 3, we can preliminarily detect a license plate in an image. In the section we will use the SNN proposed in [20] to obtain the edge of the license plate for the extraction of license plate.

4.1   Edge Detection Using Spiking Neural Network

Edge detection can be performed efficiently by the human beings' visual system. Various receptive fields from simple cells in the striate cortex to those of the retina and LGN have been found by the Neuroscientists. Besides, the neurons can be simulated by the integrate-and-fire model. Here, we use a network model to detect edges in a visual image, which is based on these receptive fields of orientation and the spiking neuron model. The structure of the network is shown in Fig. 3.

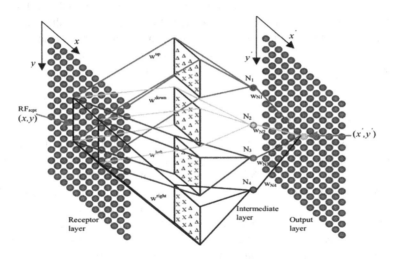

Fig. 3. The SNN model for image edge extraction [20]

Suppose that the first layer represents photonic receptors. Each pixel corresponds to a receptor. The intermediate layer consists of four types of neurons corresponding to four different receptive fields respectively. `X' in the synapse connections represents an excitatory synapse and `Δ' represents an inhibitory synapse. Each neuron in the output layer integrates four corresponding outputs from intermediate neurons. The firing rate map of the output layer produce an edge image corresponding to the input image.The intermediate layer includes four parallel arrays of neurons and each array is the same dimensions as the output layer. These arrays are flagged as N1,N2,N3 and N4 and only one neuron in each array is shown in·Fig. 3 for simplicity. Each of these layers perform the function of up, down, left and right edges detection and are connected to the receptor layer by different weight matrices. For example neuron N1 connects to receptive field RF_{rept} in the receptor layer through the weight matrices w^{up} and this receptive field can respond to a up-edge. The principle of up-edge detection using the above structure are as follow. If a uniform image is within the RF_{rept}, through the weight matrices the potential caused by the excitatory synapses are the same as that caused by the inhibitory synapses but the signs are reverse. Therefore the membrane potential of Neuron N1 has not been changed, and no spikes are generated by Neuron N1. However, if an edge image is within the RF_{rept} and the low-half of the edge has a strong signal and the upper-half of the edge has a weak signal namely this edge is a up-edge, then the strong signal will potential the Neuron N1, but the weak signal will not depress the membrane significantly. Therefore the membrane potential of Neuron N1 rise up fast and generates spikes frequently to respond to an up-edge within its receptive field. The synapse distribution matrix w^{up} acts as a filter for up-edge within the receptive field. By analogy, neuron N2 with synapse strengthen distribution w^{down} can best respond to a down-edge within the receptive field; neuron N3 with synapse strength distribution w^{left} can best respond to a left-edge within the receptive field; neuron N4 with synapse strength distribution w^{right} can best respond to a right-edge within the receptive field; Neuron (x', y') in the output layer integrate the outputs from these four neurons from the neuron arrays in the intermediate layer and can respond to any direction edge within receptive field RF_{rept}. The firing rate map of the output layer produce an edge image corresponding to the input image.

4.2 Extraction of License Plate from Texture Cues

Here we use the texture features in the license plate as the cue to exclude the distracters from the real license plate. The edge image produced by the SNN contains a lot of texture information, one of which is unique in the license plate, that is the stokes in the image rows in the license plate are larger than 12 in Chinese license plate results from the arrangement of the characters in the license plate. This unique texture feature can be obtained by horizontal projection of the pixels into a container, an accumulative value mathematically. By determining the columns that composed of consecutive containers with values larger than 12, we can extract the license plate accurately, while exclude the distracters from the list of candidates. All of the projections and computation of accumulative values are based on the edge image produced by the SNN.

5 Experimental Results and Discussions

After acquisition of the image, which contains license plate patch, based on the color feature, we obtain the candidate area using the SNN introduced in Sect. 3.We run the experiment on the Matlab 2012a. Here are the parameters for this network: $v_{reset} = -65mv, E_L = -65mv, R = 3.5, \tau = 5\,ms, \quad dt = msV_{th} = -60mv, n = 100$. By doing so, we may obtain one, two or more candidate areas. Therefore we should use other features of license plate to achieve accurate extraction. In this paper we exploit the number of strokes in image rows to exclude the pseudo-plate region. Firstly, we extract the edge of the gray scale image corresponding to the candidate area by using the SNN presented in Sect. 4. Then we use the edge image to produce a binary image. Next we calculate the horizontal projection of the binary image. According to the number of strokes, we can extract the license plate accurately. The parameters in the SNN for the extraction of image edge are as follows: $v_{th} = -60mv$ $E_{ex} = 0mv$, $E_{ih} = -75mv$, $\tau_{ex} = 4ms, \tau_{ih} = 10ms, \tau_{ref} = 6ms, A_{ih} = 0.028953\,mm^2$, $A_{ex} = 0.014103\,mm^2$.

Figure 4 is a process of detecting the license plate using the proposed method in our experiments.

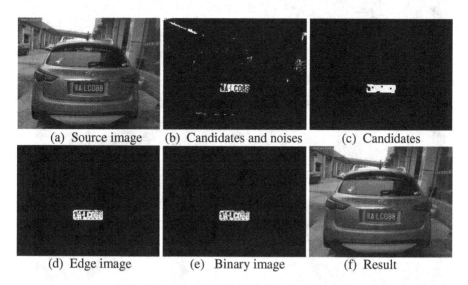

(a) Source image   (b) Candidates and noises   (c) Candidates

(d) Edge image   (e) Binary image   (f) Result

Fig. 4. A process of detecting the license plate

Figure 4(a) is a original license plate with blue background; Fig. 4(b) is the image after the extraction of color features using the SNN method. We use the SNN method which based on the color feature to segment the license plate. By doing so we can get the candidate area. From the picture, we can see that the blue background can be preliminarily extracted and the blue background are represented by the white area. We can also see that there are many noisy area in the picture. Therefore in the next step we consider remove the noisy area; Fig. 4(c) is the image which is processed by removing the noise area. By doing so we can obtain the coarse location of the license plate. Figure 4(d) is the edge image which is obtained by using the SNN method mentioned in this chapter. Figure 4 (e) is the binary image. We binarize the edge image and thus get the binary image. Then we compute the horizontal projection and use the character jump times to exclude the false license plate area. Figure 4(f) is the accurate image after excluding the false license plate area.

We run the experiments on 150 samples using the proposed algorithm. Some of the results are showed in Fig. 5, where we use red line to identify the license plate region.

In order to access the performance of the license plate extraction. We summarize the results of the experiment in Table 2.

Fig. 5. Some results in the experiments

Table 2. Performance of the proposed algorithm

Classes	# of samples	# of correctly detected	Success rate
Blue	100	94	94 %
Yellow	50	46	92 %
Total	150	140	93.3 %

It is clear from the experimental results that the proposed algorithm can detect the license plate effectively. Firstly, it can detect and extract all kind of Chinese license plates, that is license plate with blue,yellow,white, black background. Secondly, our method can detect the license plates with two lines of characters. See the first row of Fig. 5 for examples. Thirdly, It is robust to certain degree of blur and stain of image. From the first image of Fig. 5, we can see that although the image is stained, our method can detect the license plate correctly. Fourthly, it is robust to certain degree of perspective distortion, see the second image and third image in the first row of Fig. 5 for examples. Fifthly, it is robust for license plate detection in general cases by employing color information and texture information. At last, compared with other traditional methods, it is a more biologically plausible by simulation the retinal cone cell's perception function and edge detection receptive fields in the VI cortex.

Figure 6 shows some failure cases of the experiments. There are mainly three reasons that may cause this unsatisfied results. (i) Similar details like the color of the scene happen to appear around the license plate region, which may result in the false location. (ii) License plate is heavily stained. As showed in Fig. 6(a),the last character is such heavily stained that our algorithm fail to identify the region through color information correctly. (iii) the color of the license plate is far from the standard color. As shown in Fig. 6 (b) and (c),the color of the license plate is far from the standard color of blue, that our algorithm is unable to find the candidate regions correctly.

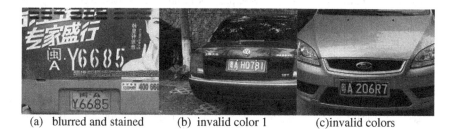

(a) blurred and stained   (b) invalid color 1   (c)invalid colors

Fig. 6. Failure examples

Since we use detail color and texture information extracted by SNN to locate the license plate in our work. If the color and texture information can be extracted effectively, the proposed method can find the license plate successfully. As shown in Table 2, the proposed SNN methods can locate the license plate effectively. Only in some cases, we cannot tell the difference of the distracters and the license plate region from color and texture feature that our method may fail to extract the correct license plate regions. Though there are some failure cases, from the experimental results, we found that the proposed SNN can detect and extract Chinese license plate effectively by using the color feature and texture features in the license plate. The success rate of license plate extraction is comparable to those reported in literature [11], which means that the SNN-based visual computation principles in the human being's cortex are applicable in the field of computer vision. And with further study, this kind of biological inspired computation maybe outperforms those traditional algorithm in computer vision.

6 Conclusions

In this paper, we proposed a SNN to detect the license plate and use another SNN to extract the edge of the candidates. The major contribution of this paper is that we propose the SNN methods to locate the license plate. The proposed SNN which simulates the color perception principle of human beings' visual cortex can detect the candidate area effectively. The SNN which simulates the receive fields of orientation in human beings' visual cortex can effectively extract the edges of candidates. The experimental results show that the proposed SNN is valid in the license plate extraction, and it is a promising intelligent computation method which can be extended into other visual information processing fields.

Acknowledgement. This work was supported by the Science-Technology Project of Education Bureau of Fujian Province, China (Grant No. JA13073), the Natural Science Foundation of Fujian Province, China (Grant No. 2014J01224), and the National Natural Science Foundation of China (Grant No. 61179011).

References

1. Wan, X., Liu, J., Liu, J.: A vehicle license plate localization method using color barycenters hexagon model. In: 3rd International Conference on Digital Image Processing, vol. 8009, pp. 80092O-1–80092O-5 (2011)
2. Wang, F., Man, L., Wang, B., Xiao, Y., Pan, W., Lu, X.: Fuzzy-based algorithm for color recognition of license plates. Pattern Recogn. Lett. **29**(7), 1007–1020 (2008)
3. Babu, C.N.K., Nallaperumal, K.: An efficient geometric feature based license plate localization and recognition. Int. J. Imaging Sci. Eng. **2**(2), 189–194 (2008)
4. Tan, J. L., Abu-Bakar, S. A. R., Mokji, M. M.: License plate localization based on edge-geometrical features using morphological approach. In: 2013 20th IEEE International Conference on Image Processing (ICIP), pp. 4549–4553 (2013)
5. Bremananth, R., Chitra, A., Seetharaman, V., Nathan, V. S. L.: A robust video based license plate recognition system. In: Proceedings of 2005 International Conference on Intelligent Sensing and Information Processing. IEEE, pp. 175 – 180 (2005)
6. Hsieh, C. T., Juan, Y. S., Hung, K. M.: Multiple license plate detection for complex background. In: Proceedings of the 19th International Conference on Advanced Information Networking and Applications, vol. 2, pp. 389–392. IEEE Computer Society (2005)
7. Wu, B.F., Lin, S.P., Chiu, C.C.: Extracting characters from real vehicle licence plates out-of-doors. Comput. Vis. Iet **1**(1), 2–10 (2007)
8. Matas, J., Zimmermann, K.: Unconstrained license plate and text localization and recognition. In: Proceedings of IEEE Conference on Intelligent Transportation Systems, pp. 225 – 230 (2005)
9. Hao, W. L., Tay, Y. H.: Detection of license plate characters in natural scene with MSER and SIFT unigram classifier. In: 2010 IEEE Conference Sustainable Utilization and Development in Engineering and Technology (STUDENT), pp. 95–98 (2010)
10. Cho, B.K., Ryu, S.H., Shin, D.R., Jung, J.I.: License plate extraction method for identification of vehicle violations at a railway level crossing. Int. J. Automot. Technol. **12**(2), 281–289 (2011)

11. Du, S., Ibrahim, M., Shehata, M., Badawy, W.: Automatic license plate recognition (ALPR): a state-of-the-art review. IEEE Trans. Circ. Syst. Video Technol. **23**(2), 311–325 (2013)
12. Mao, S., Huang, X., Wang, M.: An adaptive method for chinese license plate location. In: 2010 8th World Congress on IEEE Intelligent Control and Automation (WCICA), vol. 20, pp. 6173–6177 (2010)
13. Harris, J. G., Xu, J., Rastogi, M., Alvarado, A. S., Garg, V., Principe, J. C., et al.: Real time signal reconstruction from spikes on a digital signal processor. In: IEEE International Symposium on Circuits and Systems ISCAS, pp. 1060–1063 (2008)
14. Cheung, K., Schultz, S.R., Luk, W.: A large-scale spiking neural network accelerator for FPGA systems. In: Villa, A.E., Duch, W., Érdi, P., Masulli, F., Palm, G. (eds.) ICANN 2012, Part I. LNCS, vol. 7552, pp. 113–120. Springer, Heidelberg (2012)
15. Chen, M., Wu, Q., Cai, R., Ruan, C., Fan, L.: Extraction of breast cancer areas in mammography using a neural network based on multiple features. In: Deng, H., Miao, D., Lei, J., Wang, F.L. (eds.) AICI 2011, Part III. LNCS, vol. 7004, pp. 228–235. Springer, Heidelberg (2011)
16. Cai, R.T., Wu, Q.X.: Target extraction in infrared image based on spiking neural networks. J. Comput. Appl. **30**(12), 3327–3330 (2010)
17. Cai, R., Wu, Q., Wang, P., Sun, H., Wang, Z.: Moving target detection and classification using spiking neural networks. In: Zhang, Y., Zhou, Z.-H., Zhang, C., Li, Y. (eds.) IScIDE 2011. LNCS, vol. 7202, pp. 210–217. Springer, Heidelberg (2012)
18. Escobar, M.J., Masson, G.S., Vieville, T., Kornprobst, P.: Action recognition using a bio-inspired Feedforward spiking network. Int. J. Comput. Vis. **82**(3), 284–301 (2009)
19. Shu, N., Tang, Q., Liu, H.: A bio-inspired approach modeling spiking neural networks of visual cortex for human action recognition. In: 2014 International Joint Conference on Neural Networks (IJCNN), pp. 3450–3457. IEEE (2014)
20. Wu, Q.X., Mcginnity, M., Maguire, L., et al.: Edge detection based on spiking neural network model. In: Advanced Intelligent Computing Theories and Applications. With Aspects of Artificial Intelligence. LNCS, pp. 26–34. Springer, Heidelberg (2007)
21. Wyszecki, G., Stiles, W.S.: Color Science: Concepts and Methods, Quantitative Data and Formulae, 2nd edn, p. 950. Wiley, New york (1982). Billmeyer, F.W.: Color Research and Application, **8**(4), 262–263 (1983)

Pedestrian Detection and Counting Based on Ellipse Fitting and Object Motion Continuity for Video Data Analysis

Yaning Wang[1,2] and Hong Zhang[1,2(✉)]

[1] College of Computer Science and Technology,
Wuhan University of Science and Technology, Wuhan, China
[2] Hubei Province Key Laboratory of Intelligent Information Processing
and Real-Time Industrial System, Wuhan, China
yaningyaluo@sina.com, 46476522@qq.com

Abstract. In order to detect and count pedestrians in different kinds of scenes, this paper put forward a method of solving the problem on video sequences captured from a fixed camera. After preprocessing operations on the original video sequences (Gaussian mixture modeling, three-frame-differencing, image binaryzation, Gaussian filtering, dilation and erosion) we extract the relatively complete pedestrian contours. Then we use the least square ellipse fitting method on those contours that has been extracted, the center of the ellipse is undoubtedly regarded as the tracking point of a pedestrian. With those points, a pedestrian matching pursuit and counting algorithm based on object motion continuity is used for tracking and counting pedestrians, this method can be better used in those scenes which are sparse and rarely obscured. Experiments validate that our pedestrian matching pursuit and counting algorithm has obvious superiorities: good real-time performance and high accuracy.

Keywords: Human detection · Pedestrian counting · Matching pursuit algorithm · Ellipse fitting

1 Introduction

The detection and counting of pedestrians is an important research direction of video surveillance which has wide applications such as tourists flow estimation and people flow monitoring. However, due to illumination, occlusion, the complexity and irregularity of pedestrian motion and the influence of other targets such as animals and cars, the problem is still far from being solved. Recent years, research on pedestrian detection and counting has made great progress: [1, 2] use regression techniques to learn a map between features and the number of people in the training set and then use the map to estimate the number of people in novel test images or videos; In [3, 4], multi-scale windows slide over the whole image and a binary classifier is adopted to determine whether there is a people within the window; literature [5] put forward a method based on facial feature description and SVM (support vector machine) to count pedestrians; literature [6] adopts background subtraction based on threshold to extract

© Springer International Publishing Switzerland 2015
D.-S. Huang et al. (Eds.): ICIC 2015, Part I, LNCS 9225, pp. 378–387, 2015.
DOI: 10.1007/978-3-319-22180-9_37

object information, then uses connected component detection algorithm, setting the object feature and shape judgment condition and marking object region, finally count the number of people, but it can't remove the influences by some problem such as illumination, rapidly changing weather conditions, people head which are covered completely etc.; literature [7] construct a novel system that uses vertical Kinect sensor for people counting which equals to find the suitable local minimum regions, then propose a novel unsupervised water filling method that can find these regions with the property of robustness, but it can't handle the situation where some moving object is closer to the sensor than head; literature [8] uses Bayesian Gaussian process to learn the map between holistic features and the number of people. In [9], KLT tracker and agglomerative clustering were used. An unsupervised Bayesian approach was further proposed in [10]. These researches can achieve some success in specific situations, even then, on the account of occlusion, variation of illumination and other complex reasons, there is still more to do to solve the problem.

In this paper, we adopt pedestrian matching pursuit and counting algorithm based on object motion continuity to solve the problem of pedestrian detection and counting. As shown in Fig. 1, this algorithm is mainly divided into two parts. The objective of the first part is to segment the pedestrian objects from the video and then obtain the contour of each pedestrian. So in the first part, Gaussian mixture modeling is combined with three-frame-differencing to extract the foreground object firstly. After choosing appropriate threshold, filtering operation, erosion and dilation, we get relatively complete and smooth binary image of the pedestrian secondly. Thirdly, through the binary image we've got we can easily obtain the contour of each pedestrian. Considering as a decisive factor of pedestrian counting, contours need to be screened which means we need to estimate the scope of the contours in terms of the distance between pedestrians and camera. Fourthly according to the scope we should remove those contours which

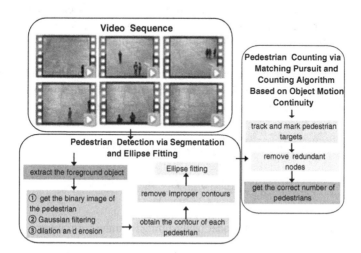

Fig. 1. The flowchart of the pedestrian matching pursuit and counting algorithm

are too big or too small. Fifthly, we use the least square ellipse fitting method to get the ellipse which closely matches the contours, and the center of the ellipse is undoubtedly regarded as the tracking point of a pedestrian. In the second part, considering the fact that the intervals between consecutive frames are very short, we take the movement of pedestrians between consecutive frames as uniform motion. By using the pedestrian matching pursuit and counting algorithm based on object motion continuity, we achieve the aim to track and mark pedestrian targets, at the same time we put the former and present coordinates and other information into a dynamic linked list. At last, according to the coordinates and flags we get the correct number of pedestrians after removing redundant nodes in the dynamic linked list.

2 Pedestrian Detection via Segmentation and Ellipse Fitting

As described in Sect. 1, we should segment the pedestrian objects from the video at the first step. It is well known that there are some common segmentation methods for moving targets: background subtraction, frame subtraction and optical flow method. In the practical application, due to the occlusion, illumination, transparency, noise and etc., we can't get the right flow field threshold, besides, the operation process is complex and the real-time performance is poor [11–14], therefore, optical flow method will not be considered here. Frame subtraction method extract the target by temporal difference and thresholding between two or three consecutive frames of the video sequence, it is easy to implement and has strong adaptability to the dynamic environment, but it is sensitive to illumination variant and will bring forth cavitation. Background subtraction get the object target via the difference of current frame and background frame, it can better adapt to the changes of illumination but cannot deal with the problem of scene light rapid change [15–19]. So in this paper, we combine Gaussian mixture modeling which is an effective method of background subtraction [20] with three-frame-differencing to extract the foreground object. This way of combining has better control to the interference of illumination and other external unrelated events [21, 22].

As for Gaussian mixture modeling, the gray value of each pixel is in line with the Gaussian distribution, $\mu_n(x,y)$ signify the mean value of the n frames in front and the σ_n^2 stands for the mean-square deviation [23–25], we can calculate:

$$\mu_n(x,y) = \frac{1-n}{n}\mu_{n-1}(x,y) + \frac{1}{n}f(x,y) \tag{1}$$

$$\sigma_n^2 = \frac{\sum_{i=1}^{n}(f_i(x,y) - \mu_n(x,y))^2}{n-1} \tag{2}$$

We can build a background model using limited number of frames:

$$B_0(x,y) = \frac{\sum_{k=1}^{m} f_{ik}(x,y)}{m} \tag{3}$$

$f_{ik}(x, y)$ meet the conditions of $|f_{ik}(x, y) - \mu_n(x, y)| \leq \beta \alpha_n(x, y)$, in this formula β is a previously set constant. In the process of obtaining background, the background model must be updated timely because of the changes caused by illumination and other factors, therefore we use the following formula to update background:

$$B_k(x, y) = \alpha B_{k-1}(x, y) + (1 - \alpha)f_k(x, y) \tag{4}$$

$B_k(x, y)$ represent the background model of the k_{th} frame, $0 \leq \alpha \leq 1$ stands for updating rate. After all, we get the $B_k(x, y)$ as the foreground image [26, 27].

Next comes to the three-frame-differencing. Firstly according to the formula (5) we can get difference images $D_{k-1}(x, y)$ and $D_k(x, y)$ of the three consecutive frames $I_{k-1}(x, y)$, $I_k(x, y)$ and $I_{k+1}(x, y)$, secondly by using and operation of $D_{k-1}(x, y)$ and $D_k(x, y)$, we get the final result $FD_k(x, y)$ of the three-frame-differencing [28, 29].

$$D_k(x, y) = |I_{k+1}(x, y) - I_k(x, y)| \tag{5}$$

After obtaining the threshold T by using the OTSU method, conduct the two images ($B_k(x, y)$ and $FD_k(x, y)$) with binary operation respectively and then implement or operation on the two images, then we've segmented pedestrian objects from the video. However after this step we can see not only noise but also burr in the foreground image which will bring great influence to later work, so Gaussian filtering, dilation and erosion is used to eliminate these noises.

The second step is to detect pedestrian targets. We divide this part into three steps: Firstly, we should find the contours of each pedestrian and it is not very difficult since we've got the binary image of the foreground image. Secondly, we'll remove those contours which are too big or too small. Because the video is captured from a fixed monocular camera, the size of the pedestrian targets in the monitored area is controlled within a certain range. We will estimate the size of pedestrian in every part of the image previously by collecting filed data and get the scope of the contours, then while finding a contour, we ascertain whether the size of the contour is in the range and remove those contours which are too big or too small. Thirdly, we use the least square ellipse fitting method to get the ellipse which closely matches the contours [30], and the center of the ellipse is undoubtedly regarded as the tracking point of a pedestrian. The general equation of planar curve of second order can be expressed as:

$$F(m, n) = m \cdot n = ax^2 + bxy + cy^2 + dx + ey + f = 0 \tag{6}$$

where m $= [a, b, c, d, e, f]^T$, $n_i = [x_i^2, x_iy_i, y_i^2, x_i, y_i, 1]$, B $= [n_1^T, n_2^T, \ldots n_i^T]^T$. $F(m, n_i)$ is the algebraic distance between point (x_i, y_i) in the plane and the curve $F(m, n) = 0$. From geometry knowledge we can know that when the curve equation coefficient meets $b^2 - 4ac = -1$, the quadratic curve equation described in (6) is an ellipse.

The least square ellipse fitting method adopts generalized characteristic root method for solving, the algorithm is not only simple, efficient but also has a good robustness.

3 Pedestrian Counting via Matching Pursuit and Counting Algorithm Based on Object Motion Continuity

After ellipse fitting for each pedestrian contour in the foreground image, we get the tracking point of each pedestrian in Sect. 2. Considering the speed and direction of the same target centroid's movement between two consecutive frames, in this section we propose a novel matching pursuit and counting algorithm based on object motion continuity.

3.1 Pedestrian Matching Algorithm

Considering the fact that the intervals between consecutive frames is very short, some parameters of pedestrians won't mutate, so we take the movement of pedestrians between consecutive frames as uniform motion.

Suppose in the n_{th} frame I_n of image sequence $I = (I_1, I_2, I_3 \cdots, I_n, \cdots)$, m moving targets $p_1, p_2, \cdots, p_m$ are detected which are recorded as $T = (T_1, T_2, T_3, \cdots, T_m)$. In this dynamic linked list T_i = (f_x, f_y, l_x, l_y, f_frame, l_frame, flag, p_entry, p_center, p_departure), among which (f_x, f_y), f_frame respectively mean the coordinate of the tracking point and the frame number which is detected at the first time, (l_x, l_y), l_frame respectively stand for the coordinate of the tracking point and the frame number which is matched with (f_x, f_y) of T_i. p_entry, p_center, p_departure take the charge of marking the position of pedestrians. The pedestrian matching algorithm is shown as Algorithm 1.

Algorithm 1. Pedestrian Matching Algorithm

```
    Input : the dynamic linked list which store infor-
mation of pedestrians T, the maximum of lateral and
longitudinal  movement  of  pedestrians  xspeed  and
yspeed, the current detected point coor, the current
frame number framenum.
    Output: complete pedestrian target dynamic linked
list T
1.  r=T;
2.  while(r->next!=NULL)
3.      r=r->next;
4.      if(|coor->x-r->f_x|<xspeed*(framenum-r->f_frame)
        &&|coor->y-r->f_y|<yspeed*(framenum-r->f_frame))
5.          if(|coor->x-r->l_x|!=0&&|coor->y-r->l_y|!=0)
6.              r->l_x=coor->x, r->l_y=coor->y,
                r->l_frame=framenum, r=T;
                break;
7.          end if;
8.      end if;
9.  end while;
10. if(r->next==NULL)
11.     r->next=coor;
12.     give the corresponding value to flag, p_entry,
        p_center, p_departure.
13. end if;
```

3.2   Pedestrian Pursuit and Counting Algorithm

In Sect. 3.1 we have put the coordinates and other information of the pedestrians into a dynamic linked list, so in this section we'll use those coordinates and information to draw a rectangle which completely cover the pedestrian, and judge the moving direction of pedestrians then count in the corresponding direction. In order to do this we divide the monitor area into two parts by a center line which is shown in Fig. 2.

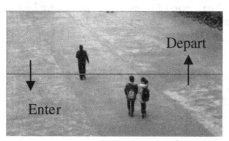

Fig. 2. Counting line set

When pedestrians first appear on the monitored area as well as has a tendency to enter, *flag*, *p_entry*, *p_center*, *p_departure* will be respectively marked as 2, 0, 0, 1. If pedestrians first appear on the monitored area and has a tendency to depart, *flag*, *p_entry*, *p_center*, *p_departure* will be marked as 1, 1, 0, 0.

After the pedestrian matching algorithm we've got many nodes in the dynamic linked list, however some of them are not pedestrian tracking points at all, even some are almost repetitive due to the irregularity of pedestrians' movement, therefore we need to remove those redundant nodes. Specific practice is: Traverse the dynamic linked list T in sequence, try to find those nodes which meet the conditions that if the difference of frame numbers is so small that can be negligible, at the same time the difference of x-coordinates l_x and y-coordinates l_y are all within the scope that is set in advance. We can judge out that those nodes are repetitive with each other, so we only need to leave one of them and then remove the rest of the nodes. After completing this step, the result of pedestrian counting will be more accurate.

Finally we comes to pedestrian counting. In this paper, we realize it according to the tag values of pedestrian's location and the change of coordinates of each pedestrian's tracking point. *p_entry* is the tag which stands for entering the monitor area or not, *p_center* means whether the pedestrian is crossing the counting line or not, *p_departure* is the tag which stands for departing the monitor area or not. We only count in the corresponding direction when all of the three flags equals 1. The pedestrian pursuit and counting algorithm is shown in Algorithm 2 (we only give the algorithm of pedestrians entering the monitored area, the situation that pedestrians depart is similar to this algorithm):

Algorithm 2. Pedestrian Pursuit and Counting Algorithm

Input: the dynamic linked list which store information of pedestrians T, current image pfr.

```
1.  c_p=T;
2.  while(c_p->next!=NULL)
3.     c_p=c_p->next;
4.     if(c_p->flag=1)
5.         if(c_p->p_entry=1&c_p->p_center=0&
            c_p->departure=0)
6.             if the center coordinate of pedestrian is
               quite near to the counting line
7.                 make c_p->p_center=1
8.             end if;
9.         end if;
10.        if(c_p->p_entry=1&c_p->p_center=1&
           c_p->departu-re=0)
11.            if( c_p->f_y <pfr.rows/2 &
               c_p->l_y >pfr.row-s/2)
12.                make c_p -> p_departure=1;
                   entrynumber+1;
                   totalnumber+1;
                   remove and free c_p from T;
13.            end if;
14.        end if;
15.    end if;
16. end while;
```

---

4 Experiment Result

In this section we perform several videos to test the performance of the proposed method. The videos is shot at the gate of our teaching build. The pedestrian detection results are shown in Fig. 3. The image shown in (a) is the source video sequence, (b) is the contour and ellipse we've extracted, (c) is the foreground image.

The pedestrian counting result is shown in Table 1. There are three people who depart from normal behavior in the first video shown in Table 1. In the second video two pedestrians left and one pedestrian entered the monitored area. At the beginning of the third video, three pedestrians have crossed the counting line and are leaving the monitored area, so the three pedestrians could not be contained in the departure number. In similar circumstances, at the beginning of the fourth video, two pedestrians have crossed the counting line and are entering the monitored area, so the two pedestrians could not be contained in the enter number thus there are only three pedestrians left and two pedestrians entered the monitor area. In the fifth video, four pedestrians entered the monitored area and one of the four pedestrians turn around when he just crossed the counting line to leave. In the sixth video, at the beginning there are four pedestrians who have a tendency to enter the monitored area but only three pedestrians crossed the counting line. Experiment results show that the algorithm shown in this paper has a good real-time performance and robustness. What's more, the algorithm have a higher accuracy in counting.

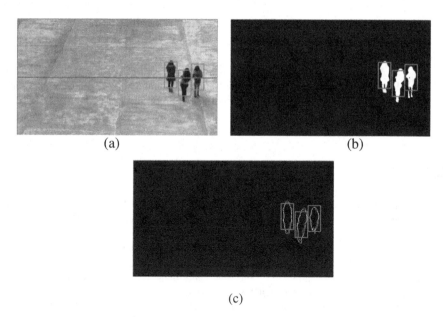

Fig. 3. (a) (b) (c) Pedestrian detection results

Table 1. Pedestrian counting results.

	Actual value		Measured value	
	Enter	Departure	Enter	Departure
Video 1	0	3	0	3
Video 2	1	2	1	2
Video 3	0	2	0	2
Video 4	2	3	2	3
Video 5	4	1	4	1
Video 6	3	0	3	0

5   Conclusion

In this paper, we construct a novel pedestrian counting system via pedestrian matching pursuit and counting algorithm based on object motion continuity which is robust to appearance variations. Seeing from this algorithm, pedestrian counting problem equals to finding all the pedestrian contours and then get the number of all the contours. We use the least square ellipse fitting method that can closely matches the contours and give a rectangle completely cover the pedestrian as well as runs at real-time. Experiment results show the algorithm have a higher accuracy in counting even when several pedestrians crossing the counting line at the same time. However, according to the limitation of the algorithm, the pedestrian matching pursuit and counting algorithm cannot handle the situation where occlusion is very serious such as multiple people have interactions with each other. Besides, the system is sensitive to the accuracy of the extracted pedestrian contours. In our future work, we will consider to give a perfect solution to this situation.

Acknowledgments. This work is supported by National Natural Science Foundation of China (No. 61373109, No. 61003127), State Key Laboratory of Software Engineering (SKLSE2012-09-31).

References

1. Dong, L., Parameswaran, V., Ramesh, V., Zoghlami, I.: Fast crowd segmentation using shape indexing. In: IEEE 11th International Conference on Computer Vision (ICCV), pp. 1–8 (2007)
2. Kong, D., Gray, D., Tao, H.: Counting pedestrians in crowds using viewpoint invariant training. In: British Machine Vision Conference (BMVC) (2005)
3. Viola, P., Jones, M., Snow, D.: Detecting pedestrians using patterns of motion and appearance. Int. J. Comput. Vis. (IJCV) **63**(2), 153–161 (2005)
4. Dalal, N., Triggs, B.: Histograms of oriented gradients for human detection. In: International Conference on Computer Vision and Pattern Recognition (CVPR) (2005)
5. Shimosaka, M.: Counting pedestrians in crowded scenes with efficient sparse learning. In: First Asian Conference on Pattern Recognition (ACPR), pp. 27–31 (2011)
6. Ye, Q.: A robust method for counting people in complex indoor spaces. In: International Conference on Education Technology and Computer (ICETC), pp. 450–454 (2012)
7. Zhang, X., Yan, J., Feng, S., Lei, Z., Yi, D., Li, S.Z.: Water filling, unsupervised people counting via vertical kinect sensor. In: IEEE Ninth International Conference on Advanced Video and Signal-Based Surveillance, pp. 215–220 (2012)
8. Chan, A.B., Vasconcelos, N.: Counting people with low-level features and bayesian regression. Trans. Image Process. (TIP) **21**(4), 2160–2177 (2012)
9. Rabaud, V., Belongie, S.: Counting crowded moving objects. In: International Conference on Computer Vision and Pattern Recognition (CVPR) (2006)
10. Brostow, G., Cipolla, R.: Unsupervised bayesian detection of independent motion in crowds. In: International Conference on Computer Vision and Pattern Recognition (CVPR) (2006)
11. Barron, J.L., Fleet, D.J., Beauchemin, S.S.: Performance of optical flow techniques. In: Computer Vision and Pattern Recognition (CVPR), pp. 236–242 (1992)
12. Glocker, B., Paragios, N., Komodakis, N., Tziritas, G.: Optical flow estimation with uncertainties through dynamic MRFs. In: IEEE Conference on Computer Vision and Pattern Recognition (CVPR), pp. 1–8 (2008)
13. Chan, V.W.S., Zheng, L., Huang, H., Weichenberg, G.: Optical flow switching: an end-to-end "UltraFlow" architecture. In: 15th International Conference on Transparent Optical Networks (ICTON), pp. 1–4 (2013)
14. Sidram, M.H., Bhajantri, N.U.: Exploitation of regression line potentiality to track the object through color optical flow. In: Third International Conference on Advances in Computing and Communications (ICACC), pp. 181–185 (2013)
15. Benezeth, Y., Jodoin, P.M., Emile, B.: Review and evaluation of commonly-implemented background subtraction algorithms. In: 19th International Conference on Pattern Recognition (ICPR), pp. 1–4 (2007)

16. Marie, R., Potelle, A., Mouaddib, E.M.: Dynamic background subtraction using moments. In: IEEE International Conference on Image Processing (ICIP) (2011)

17. Dharamadhat, T., Thanasoontornlerk, K., Kanongchaiyos, P.: Tracking object in video pictures based on background subtraction and image matching. In: IEEE International Conference on Robotics and Biomimetics (ROBIO), pp. 1255–1260 (2008)

18. Bayona, A., SanMiguel, J.C., Martinez, J.M.: Comparative evaluation of stationary foreground object detection algorithms based on background subtraction techniques. In: IEEE International Conference on Advanced Video and Signal Based Surveillance (AVSS), pp. 25–30 (2009)

19. Mohamed, S.S., Tahir, N.M., Adnan, R.: Background modelling and background subtraction performance for object detection. In: 6th International Colloquium on Signal Processing and Its Applications (CSPA), pp. 1–6 (2010)

20. Zivkovic, Z.: Improved adaptive Gaussian mixture model for background subtraction. In: International Conference on Pattern Recognition (ICPR), pp. 28–31 (2004)

21. Lan, Y.: Robot fish detection based on a combination method of three-frame-difference and background subtraction. In: Control and Decision Conference (CCDC), pp. 3905–3909 (2014)

22. Zhang, H., Wu, K.: A vehicle detection algorithm based on three-frame-differencing and background subtraction. In: Computational Intelligence and Design (ISCID), pp. 148–151 (2012)

23. Flam, J.T., Jalden, J., Chatterjee, S.: Gaussian mixture modeling for source localization. In: IEEE International Acoustics, Speech and Signal Processing (ICASSP), pp. 2604–2607 (2011)

24. Allili, M.S., Bouguila, N., Ziou, D.: Online video foreground segmentation using general Gaussian mixture modeling. In: IEEE International Conference on Signal Processing and Communications (ICSPC), pp. 959–962 (2007)

25. Soh, Y., Hae, Y., Kim, I.: Spatio-temporal Gaussian mixture model for background modeling. In: IEEE International Symposium on Multimedia, pp. 360–363 (2012)

26. Allili, M.S., Bouguila, N., Ziou, D.: A robust video foreground segmentation by using generalized Gaussian mixture modeling. In: Computer and Robot Vision (CRV), pp. 503–509 (2007)

27. Zhuang, S., Huang, Y., Palaniappan, K., Zhao, Y.: Gaussian mixture density modeling, decomposition, and applications. IEEE Trans. Image Process. **5**(9), 1293–1302 (1996)

28. Lian, X., Zhang, T., Liu, Z.: A novel method on moving-objects detection based on background subtraction and three frames differencing. In: Measuring Technology and Mechatronics Automation (ICMTMA), pp. 252–256 (2010)

29. Mendi, E., Bayrak, C., Milanova, M.: A video quality metric based on frame differencing. In: IEEE International Conference on Information and Automation (ICIA), pp. 829–832 (2011)

30. Liu, P., Dong, X.: Constrained least squares fitting of ellipse. In: Computational Intelligence and Software Engineering (CISE), pp. 1–3 (2010)

Regularized Level Set Method
by Incorporating Local Statistical Information
and Global Similarity Compatibility
for Image Segmentation

Yu Haiping$^{(\boxtimes)}$ and Zhang Huali$^{(\boxtimes)}$

Faculty of Information Engineering,
City College Wuhan University of Science and Technology,
Wuhan 430083, China
seapingyu@163.com, zhanghuali@foxmail.com

Abstract. This paper presents a regularized level set method for image segmentation, where the local statistical information and global similarity compatibility are both incorporated into the construction of energy functional. By considering the image local statistical information, the proposed model can efficiently segment images with intensity inhomogeneity. To improve the convergence speed, an adaptive stop strategy is proposed. In addition, the distance regularization term is defined with a five power of polynomial function for maintaining the stability during the curve evolution. Finally, experimental results show that our proposed model is efficient for segmenting noisy images, texture images and images with intensity inhomogeneity.

Keywords: Image segmentation · Level set · Intensity inhomogeneity

1 Introduction

Image segmentation is one of the most crucial and fundamental task in the field of image processing and computer vision. It plays an important role in many practical applications, such as computer vision, artificial intelligence, medical images analysis and tracking. The general segmentation task involves separating two sets, which are referred to as "Object" and "Background". To solve the problem, an integrated framework for ACMS has been proposed over the last few decades [1, 2].

Active contour models have been extensively applied to image segmentation [3, 4]. There are several desirable advantages of active contour models over classical image segmentation methods [5], such as edge detection, threshold, and region grow. The existing active contour models can be categorized into two classes: edge-based models and region-based models. These two types of models both have their pros and cons, and the choice of them in applications depends on different characteristics of images.

Edge-base models utilize the image gradient to stop the curve evolving process near the object boundaries. Osher and Sethian built the LSM equation by Hamilton-Jacobi, which was the former proposed. In 1993 Caselles et al. set up the LSM formula by utilizing the curves' mean curvature and the image gradient [6]. In order to keep the

© Springer International Publishing Switzerland 2015
D.-S. Huang et al. (Eds.): ICIC 2015, Part I, LNCS 9225, pp. 388–399, 2015.
DOI: 10.1007/978-3-319-22180-9_38

property of a signed distance in the re-initialization process, Li et al. added the penalty term into the standard energy functional [7] and its proposed method [8]. These typical LSMs can segment successfully for some kinds of homogenous images. However, It usually makes the model sensitive to noise and the curve initial position, in addition, it maybe detect the false boundaries or boundary leakage in images with weak object boundaries. Recently, region-based level set methods and its variant methods have been proposed for image segmentation by using the global region information into the energy functional. For instance, Chan and Vese proposed a piecewise constant approximation model (named CV model) [9], which was derived from the Mumford functional [10]. The model is insensitive to noise for performing the regional statistical information. However, the setting of initializing correctly contour remains unavoidable problem. To solving the limitations of the CV model, many researchers have extended the CV model by different methods [11–13].

The rest paper is organized as follows: We briefly review four well-known methods for image segmentation in Sect. 2. In Sect. 3, We describe the proposed method with three subsections in details. Experimental results are validated to be effective by comparing with the representative LSMs on synthetic and real images in Sect. 3.7. Finally, research conclusions are included in Sect. 4.

2 Related Works

2.1 The Chan-Vese Model (CV)

Based on the special case of MS problem [10], Chan and Vese proposed an active contour approach named CV model for two phase image segmentation. The CV model is an alternative solution to the MS problem which solves the minimization of Eq. (1) by minimizing the following energy functional [9, 14]:

$$E^{CV}(c_1, c_2, C) = \mu Length(C) + vArea(in(C)) + \lambda_1 \int_{in(C)} |I(x,y) - c_1|^2 dxdy$$

$$+ \lambda_2 \int_{out(C)} |I(x,y) - c_2|^2 dxdy \tag{1}$$

Where C represents the curve, the constants c_1 and c_2 denote the average intensities inside and outside of the curve, respectively, and the coefficients λ_1 and λ_2 are fixed parameters. The first is a regularizing term and the second term is a smoothness of the boundary. Where H (f) and d (f) are Heaviside function and Dirac function, respectively. Generally, the regularized versions are selected as:

$$\delta_\varepsilon(x) = \begin{cases} \frac{1}{2}[1 + cos(\frac{\pi x}{\varepsilon})], & |x| \leq \varepsilon \\ 0, & |x| > \varepsilon \end{cases} \qquad H_\varepsilon(x) = \begin{cases} \frac{1}{2}(1 + \frac{x}{\varepsilon} + \frac{1}{\pi}sin(\frac{\pi x}{\varepsilon})), & |x| \leq \varepsilon \\ 1, & x > \varepsilon \\ 0, & x < -\varepsilon \end{cases} \tag{2}$$

The curve C is the zero level set, so, we can replace the curve C by the unknown variable $\phi(x)$, so Eq. (1) replace by Eq. (3):

$$
E^{CV}(c_1, c_2, \phi) = \mu \int_\Omega \delta(\phi)|\nabla\phi|dxdy + v \int_\Omega H(\phi)dxdy
$$

$$
+ \lambda_1 \int_{inside(C)} |I(x,y) - c_1|^2 dxdy + \lambda_2 \int_{outside(C)} |I(x,y) - c_2|^2 dxdy \tag{3}
$$

2.2 The Region-Scalable Fitting Model (RSF)

To overcome the difficulty caused by intensity inhomogeneities, Li et al. proposed the region-scalable fitting model, which can utilize the local regions at a controllable scale [15, 16].

$$
E^{RSF}(\phi, f_1, f_2) = \sum_{i=1}^{2} \lambda_i \int \left(\int K_\sigma(x-y)|I(y) - f_i(x)|^2 M_i^\varepsilon(\phi(y))dy \right) dx
$$

$$
+ v \int |\nabla H_\varepsilon(\phi(x))|dx + \mu \int \frac{1}{2}(|\nabla\phi(x)| - 1)^2 dx \tag{4}
$$

where the first term is referred to as the data fitting term; this term plays a key role in the RSF model; the second term is called the length term which maintain the smoothness of the boundary; the third one is called a level set regularization term, which serves to maintain the regularity of the level set function. And K is a Gaussian Kernel function.

2.3 The Local Chan–Vese (LCV) Model

Wang et al. proposed a local Chan–Vese (LCV) model which can utilize both global and local image information for image segmentation [11]. For the LCV model, the energy functional consists of three parts: global term E^G, local term E^L, and regularization term E^R, and it is defined as following:

$$
E^{LCV}(c_1, c_2, d_1, d_2, \phi) = \alpha \cdot E^G(c_1, c_2, \phi) + \beta \cdot E^L(d_1, d_2, \phi) + E^R(d_1, d_2, \phi) \tag{5}
$$

Where α and β are two positive parameters which govern the tradeoff between the global term and the local term. For inhomogeneous images, the value of α should be selected less than that of β so as to restrict the intensity inhomogeneity. On the contrary, for images without intensity inhomogeneity, the value of α is equal to that of β. The formula of E^G can been rewritten the first term of E^{RSF}; the formula of E^L and E^R refer to the local information and the regularize term respectively.

3   Proposed Model of Intensity Inhomogeneity

In this section, we shall address our proposed hybrid level set model (HLSM) by combining global and local statistical information to segment the images with inhomogeneous intensity distribution. Particularly, in order to maintain the signed distance property, we incorporating a so-called double-well potential polynomial function into the proposed HLSM. Therefore, the overall energy functional in HLSM is made up of three parts: global term E^G, local term E^L and double-well penalty term E^R. Our proposed model ALSM can be described as:

$$E^{HLSM} = \alpha \cdot E^G + \beta \cdot E^L + E^R \tag{6}$$

3.1   The Description of Intensity Inhomogeneity Images

Because of the poor radio-frequency coil uniformity, static field inhomogeneity and the illumination, the images have the property of intensity inhomogeneity, which usually make it difficult to analyze, e.g., in registration, quantification, or segmentation. In [17], An observed image can be decomposed as two components: the reflectance image and the illumination image in computer vision. Different from it, in this paper, for images with intensity inhomogeneity, they manifest themselves as a smooth spatially varying function. The most general model in describing the acquired images with intensity inhomogeneity effect is described as follows:

$$I(x) = u(x)b(x) + n(x) \tag{7}$$

Where $u(x)$ is the inhomogeneity free image; $b(x)$ denotes the bias field that accounts for the intensity inhomogeneity; $n(x)$ means the noise; and $I(x)$ is the acquired image.

3.2   Local Energy Term

The local statistical method assumed that the images are intensity homogeneous in some local regions and the noise is ignored, therefore, the intensity inhomogeneity is as the follows:

$$I(x) = u(x)b(x) \tag{8}$$

To simplify the computation, the local region in our method is defined in square window. Accordingly the average filtering is defined in the local region:

$$K(x - y) = \frac{1}{(2\pi)^{n/2}\sigma^n} e^{-|x-y|^2/2\sigma^2} \tag{9}$$

Where $K(x-y)$ is the Gaussian filtering with the scale parameter $\sigma > 0$, if the scale parameter is too small, only litter neighborhood pixel is analyzed for each pixel so that the local statistical information cannot be well included. On the contrary, if the scale parameter is too large, the assumption that intensity is homogeneous in a small region does not possess. Therefore, the Gaussian function decreases drastically to zero as y goes away from x.

Generally speaking, the Gaussian function is effectively zero when $|x - y| \geq 3\sigma$, so the value of scale parameter plays a very important role. In this paper, we fix $\sigma = 3$, we have the square window no less than 4σ, so we have the size of the mask is 13×13 for all the experiments.

In Subsect. 3.1, we regard the intensity inhomogeneity as a low frequency artifact in image; In addition, what is more important is that the Gaussian low-pass filter can filter the noise. So replacing $I(y)$ by $J'(y) = K(x-y)*I(y)-I(y)$, the contrast between foreground intensities and background ones can be significantly increased. Thus, the energy functional can be constructed as follows:

$$E^L(d_1, d_2, C) = \sum_{i=1}^{2} \int_{in(C)} |J'(y) - d_i|^2 dy \tag{10}$$

Where d_1 and d_2 are the intensity averages of inside and outside C, respectively. Then, we represent the evolving contour C by the zero level set of a Lipchitz function:

$$\phi(x) = \begin{cases} > 0 & x \text{ is inside } C \\ = 0 & x \in C \\ < 0 & x \text{ is outside } C \end{cases} \tag{11}$$

Accordingly, the local term in (11) can be rewritten in terms of ϕ on the domain Ω as follows:

$$E^L(d_1, d_2, \phi) = \int_\Omega |J'(x) - d_1|^2 H(x)dx + \int_\Omega |J'(x) - d_2|^2 (1 - H(x))dx \tag{12}$$

3.3   Global Energy Term

The global term is directly derived from Chan-Vese model. Obviously, the global term is defined based on the global properties, i.e., the averages of I inside and outside of C, which is stated as follows:

$$E^G(c_1, c_2, \phi) = \int_\Omega |I(x) - c_1|^2 H(x)dx + \int_\Omega |I - c_2|^2 (1 - H(x))dx \tag{13}$$

Where c_1, c_2 are the average intensity of inside and outside C. $H(.)$ is the Heaviside function.

3.4   Regularization Term

It is crucial to maintain the level set function in a stable property, which is well satisfied by signed distance functions during the evolution. Therefore, we add to the regularization term for keeping the signed distance property. Li et al. [8] provided a double-well potential to maintain the signed distance property near the zero level set. The double-well potential function is based on trigonometric function. Inspired by this, we construct the following double-well potential function by using polynomial function:

$$p_2(s) = \begin{cases} \dfrac{1}{2}s^2(s^2 - 1)^2, & \text{if } 0 \le s \le 1 \\ \dfrac{1}{2}(s - 1)^2, & \text{if } s \ge 1 \end{cases} \tag{14}$$

Obviously, the potential function $p_2(s)$ has two minimum points at s = 0 and s = 1, which has the same feature as the reference [8], the difference is that our proposed function has its good characteristics, i.e. the piecewise function has the same structure and its lower computation complexity. The function $p_2(s)$ and the corresponding function $d_p(s)$ are plotted in Fig. 1. It is easy to prove the function $d_p(s) = p'_2(s)/s$ satisfies:

$$|d_p(s)| \le 1, \text{ for } s \in [0, \infty) \tag{15}$$

Currently, we obtain our regularization term defined in the following:

$$R_p(\phi) = \int_\Omega p_2(|\nabla\phi|)dx \tag{16}$$

Where p_2 (.) is a double-well potential function. By using Gâteaux derivative, we can obtain the gradient flow of (19):

$$\frac{\partial\phi}{\partial t} = -\frac{\partial R_p}{\partial \phi} = div(d_{p2}(|\nabla\phi|)\nabla\phi) \tag{17}$$

Where div (.) is the divergence operator. To analyze the property of the potential function, we replace the above gradient flow by a standard form of a diffusion equation: $\partial\phi/\partial t = div(D\nabla\phi)$. Where $D = div(d_{p2}(|\nabla\phi|))$ is called diffusion rate. The property of the potential function, as plotted in Fig. 1(b), infers the following tree cases:

1. If $|\nabla\phi| > 1$, D is positive, and It will forward diffusion in order to decrease the value of $|\nabla\phi|$.
2. If $1/3 < |\nabla\phi| < 1$, D is negative, and it will backward diffusion in order to increase the value of $|\nabla\phi|$.
3. If $|\nabla\phi| < 1/3$, D is positive, and it will forward diffusion in order to decrease the value of $|\nabla\phi|$ down to zero.

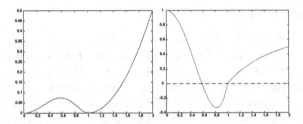

Fig. 1. The potential function $p(s)$ for the left and $d_p(s)$ for the right

The double-well potential function can maintain the signed distance property near the zero level set, Not only that, we have to consider the evolution of the smoothness of the curve in order to avoid the occurrence of isolated regions. The length penalty term should be taken into account.

$$L(\phi = 0) = \int_{\Omega} \delta(\phi(x))|\nabla \phi(x)|dx \qquad (18)$$

Through the above analysis, we can define the penalty term: $E^R = \mu \cdot L(\phi = 0) + R_p(\phi)$. Where μ is the penalty term control parameter.

3.5   Level Set Formulation

As a result, we propose the overall energy functional in details as follows:

$$\begin{aligned}
E &= \alpha \cdot E^L(d_1, d_2, \phi) + \beta \cdot E^G(c_1, c_2, \phi) + E^R \\
&= \alpha \cdot \int_{\Omega} |J'(x) - d_1|^2 H(x)dx + \alpha \cdot \int_{\Omega} |J'(x) - d_2|^2 (1 - H(x))dx \\
&\quad + \beta \cdot \int_{\Omega} |I(x) - c_1|^2 H(x)dx + \int_{\Omega} |I - c_2|^2 (1 - H(x))dx \\
&\quad + \mu \cdot \int_{\Omega} \delta(\phi(x))|\nabla \phi(x)|dx + \int_{\Omega} p_2(|\nabla \phi|)dx
\end{aligned} \qquad (19)$$

And the parameters are described as follows:

$$\begin{aligned}
d_1 &= \frac{\int_{\Omega} J'(x) H(\phi(x))dx}{\int_{\Omega} H(\phi(x))dx} & d_2 &= \frac{\int_{\Omega} J'(x)(1 - H(\phi(x)))dx}{\int_{\Omega} (1 - H(\phi(x)))dx} \\
c_1 &= \frac{\int_{\Omega} I(x) H(\phi(x))dx}{\int_{\Omega} H(\phi(x))dx} & c_2 &= \frac{\int_{\Omega} I(x)(1 - H(\phi(x)))dx}{\int_{\Omega} (1 - H(\phi(x)))dx}
\end{aligned} \qquad (20)$$

Keeping d_1, d_2, c_1, c_2 fixed and minimizing E with respect to ϕ, we can deduce the associated Euler-Lagrange equation for ϕ. The following variational formulations can be obtained:

$$\frac{\partial \phi}{\partial t} = \delta(\phi)[-(\alpha(J'(x) - d_1)^2 + \beta(I(x) - c_1)^2) + (\alpha(J'(x) - d_2)^2 + \beta(I(x) - c_2)^2)]$$
$$+ \mu\delta(\phi)\text{div}\left(\frac{\nabla\phi}{|\nabla\phi|}\right) + div(d_{p2}(|\nabla\phi|)\nabla\phi) \tag{21}$$

In the experiments, the initial level set function is defined as follows:

$$\phi_0 = \begin{cases} 2 & when\, x \in \Omega_0 - \partial\Omega_0 \\ 0 & when\, x \in \partial\Omega_0 \\ -2 & when\, x \in \Omega - \Omega_0 \end{cases} \tag{22}$$

In this paper, we use the finite difference scheme to solve the gradient flow equation in (21), and approximate the spatial partial derivatives by the central difference and the temporal partial derivatives by the forward difference.

3.6   Description of Our Proposed Model

Finally, we shall address the steps of our proposed model as follows:

Step 1: Input the original image u_0;
Step 2: the initial work:

(a) Place the initial contour C_0 inside u_0 or outside u_0 according to (22);
(b) Set the value of time step $\Delta t = 0.1$, the grid spacing h = 1 and ε = 1 for Heaviside function;
(c) Set the adaptive window size k of convolution operator in the local term, place the controlling parameter of global term α and local term β, respectively, set the length smoothness $\mu = \rho*255^2$, $\rho \in (0,1)$;

Step 3: Evolve the level set function according to (21);
Step 4: Extract the zero level set if the level set evolution terminates at time t

3.7   Experimental Results

In this section, we shall present the experimental results of our method on some synthetic and real images. Our proposed model was implemented by Matlab2013 on a computer with Intel Core i5 2.3 GHz CPU, 4G RAM, and Windows 7 operating system. In all experiments, we choose the time step $\Delta t = 0.1$, the difference spacing h = 1, the parameter for Heaviside ε = 1. We set the controlling parameters of local and global term α, β range (0, 1) according to the image intensity property. And we should adjust the value of length parameter μ according to practical situation. In our experiments we choose the two values: 0.01*2552, 0.001*2552. In this section, we shall present the experimental results of our methods (HLSM). Our method has been tested with synthetic and real images from different modalities comparing with the well-known method in reference [18] (M1). Firstly we show the results for three synthetic images in Figs. 2, 3 and 4. To be fairness, the initial position is the same in comparing with the two methods. Figure 2(b), (c) show the segmentation results of synthetic image with intensity homogeneous for the proposed ALSM and M1, respectively. It can be seen from Fig. 2(d), (e) that the two models are able to regularize the LSF, and can keep stable level set evolution. So the signed distance function can be revealed near the zero level set. Figure 3(b), (c) show the segmentation results of a texture image for the proposed HLSM and M1, respectively. Though M1 is able to segment the result, it is obvious that the convergence LSF (Fig. 3(e)) exhibits a sharp corner outside the vicinity of the zero level set, which implies the level set evolution unstable. Finally, our proposed HLSM can segment a synthetic image with noisy in Fig. 4(c), the results indicate that M1 cannot detect the real boundaries because the level set function cannot convergence. In order to quantitatively compare the results, their iteration times and elapse time for segmenting images in Figs. 2, 3 and 4 are presented in Table 1.

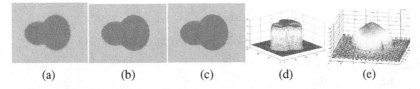

(a)   (b)   (c)   (d)   (e)

Fig. 2. The comparisons of M1 and our proposed ALSM on segmenting a synthetic image: (a) initial contour; (b) final segmentation result using HLSM; (c) final segmentation result using M1; (d) the convergence LSF of HLSM; (e) the convergence LSF of M1.

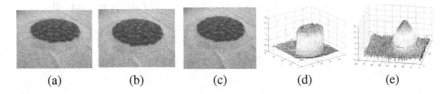

(a)   (b)   (c)   (d)   (e)

Fig. 3. The comparisons of M1 and our proposed ALSM on segmenting a texture image: (a) initial contour; (b) final segmentation result using HLSM; (c) final segmentation result using M1; (d) the convergence LSF of HLSM; (e) the convergence LSF of M1.

Let us then compare the segmentation results for real images with intensity inhomogeneity using M1 and our proposed HLSM. They have the same initial position shown in Fig. 5(b)–(d) show the segmentation result of iterative 10,20,40 times respectively using M1; (e)–(f) show the segmentation result of iterative 10,20 times respectively using HLSM. Obviously, segmentation result cannot be obtained from M1

(a)   (b)   (c)   (d)   (e)

Fig. 4. The comparisons of M1 and our proposed ALSM on segmenting a noisy image: (a) initial contour; (b) final segmentation result using HLSM; (c) final segmentation result using M1; (d) the convergence LSF of HLSM; (e) the convergence LSF of M1.

Table 1. Iteration times and elapse time for the M1 and our proposed HLSM in segmenting the image Figs. 2, 3 and 4.

	Image from Fig. 2		Image from Fig. 3		Image from Fig. 4	
	M1	HLSM	M1	HLSM	M1	HLSM
Iteration times	20	12	20	9	100	9
Elapsed time	5.11	4.09	6.35	4.68	49.79	4.73

from Fig. 5(b–d), and our proposed model can efficiently segment the object. Figure 6 shows the results for an X-ray image of vessel with intensity inhomogeneity. The M1 model fails to segment the image because of its intensity inhomogeneity. Our method successfully extracts the object boundary as shown in Fig. 6(f).

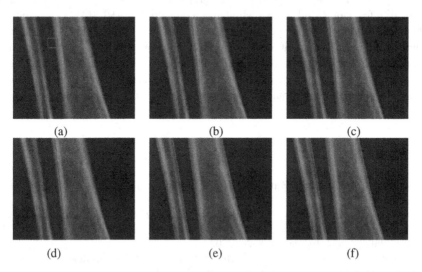

(a)   (b)   (c)

(d)   (e)   (f)

Fig. 5. The comparisons of M1 and our proposed HLSM on segmenting a X-ray image in the shin portion: (a) initial contour; (b)–(d) segmentation result of iterative 10,20,40 times respectively using M1; (e)–(f) segmentation result of iterative 10,20 times respectively using HLSM.

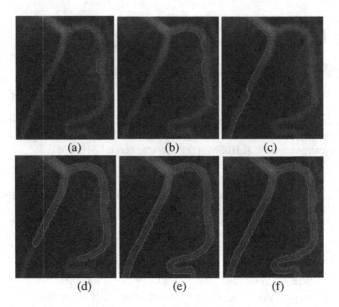

Fig. 6. The comparisons of M1 and our proposed HLSM on segmenting a X-ray image of vessel; (a) initial contour; (b–c) segmentation result of iterative 100,200 times respectively using M1; (d)–(f) segmentation result of iterative 50,100,150 times respectively using HLSM.

4 Conclusion

In this paper, we proposed a hybrid regularized level set method by Incorporating local statistical information and global similarity compatibility (HLSM). Our proposed method can segment the images correctly. Meanwhile, the HLSM has an essential feature of preserving regularity of the level set function with high speed computing performance. Most importantly, we also introduce the adaptive iteration strategy in the curve evolution. Finally, experimental results compared with the well known nonlinear adaptive level set show that our proposed model is effective and efficient.

References

1. Andrew, A.M.: Geometric Partial Differential Equations and Image Analysis, Kybernetes, vol. 31 (2002)
2. Suri, J.S., Liu, K., Singh, S., Laxminarayan, S.N., Zeng, X., Reden, L.: Shape recovery algorithms using level sets in 2-D/3-D medical imagery: a state-of-the-art review. IEEE Trans. Inf. Technol. Biomed. **6**, 8–28 (2002)
3. Paragios, N., Deriche, R.: Geodesic active regions and level set methods for supervised texture segmentation. Int. J. Comput. Vis. **46**, 223–247 (2002)
4. Trontis, A., Spathopoulos, M.P.: Hybrid control synthesis for eventuality specifications using level set methods. Int. J. Control **76**, 1599–1627 (2003)

5. Gout, C., Le Guyader, C., Vese, L.: Segmentation under geometrical conditions using geodesic active contours and interpolation using level set methods. Numer. Algorithms **39**, 155–173 (2005)

6. Caselles, V., Catté, F., Coll, T.: A geometric model for active contours in image processing. Numer. Math. **66**, 1–31 (1993)

7. Li, C., Xu, C., Gui, C., Fox, M.D.: Level set evolution without re-initialization: a new variational formulation. In: IEEE Computer Society Conference on Computer Vision and Pattern Recognition, 2005, CVPR 2005, pp. 430–436 (2005)

8. Li, C., Xu, C., Gui, C., Fox, M.D.: Distance regularized level set evolution and its application to image segmentation. IEEE Trans. Image Process. **19**, 3243–3254 (2010)

9. Chan, T.F., Vese, L.A.: Active contours without edges. IEEE Trans. Image Process. **10**, 266–277 (2001)

10. Mumford, D., Shah, J.: Optimal approximations by piecewise smooth functions and associated variational problems. Commun. Pure Appl. Math. **42**, 577–685 (1989)

11. Wang, X.-F., Huang, D.-S., Xu, H.: An efficient local Chan-Vese model for image segmentation. Pattern Recogn. **43**, 603–618 (2010)

12. Chan, T.F., Sandberg, B.Y., Vese, L.A.: Active contours without edges for vector-valued images. J. Vis. Commun. Image Represent. **11**, 130–141 (2000)

13. Xia, R., Liu, W., Zhao, J., Li, L.: An optimal initialization technique for improving the segmentation performance of Chan-Vese model. In: IEEE International Conference on Automation and Logistics, 2007, pp. 411–415 (2007)

14. Osher, S., Sethian, J.A.: Fronts propagating with curvature-dependent speed: algorithms based on Hamilton-Jacobi formulations. J. Comput. Phys. **79**, 12–49 (1988)

15. Li, C.M., Kao, C.Y., Gore, J.C., Ding, Z.H.: Implicit active contours driven by local binary fitting energy. In: IEEE Conference on Computer Vision and Pattern Recognition 2007, vol. 1–8, pp. 339–345 (2007)

16. Li, C.M., Kao, C.Y., Gore, J.C., Ding, Z.H.: Minimization of region-scalable fitting energy for image segmentation. IEEE Trans. Image Process. **17**, 1940–1949 (2008)

17. Likar, B., Viergever, M.A., Pernus, F.: Retrospective correction of MR intensity inhomogeneity by information minimization. IEEE Trans. Med. Imaging **20**, 1398–1410 (2001)

18. Wang, B., Gao, X., Tao, D., Li, X.: A nonlinear adaptive level set for image segmentation. IEEE Trans. Cybern. **44**, 418–428 (2014)

Sparse Learning for Robust Background Subtraction of Video Sequences

Yuhan Luo[1,2] and Hong Zhang[1,2(✉)]

[1] College of Computer Science and Technology,
Wuhan University of Science and Technology, Wuhan, China
[2] Hubei Province Key Laboratory of Intelligent Information Processing
and Real-time Industrial System, Wuhan, China
luoyuhan_1@163.com, 46476522@qq.com

Abstract. Sparse representation has been applied to background detecting by finding the best candidate with minimal reconstruction error using target templates. However most sparse representation based methods only consider the holistic representation and do not make full use of the sparse coefficients to discriminate between the foreground and the background. Learning overcomplete dictionaries that facilitate a sparse representation of the data as a liner combination of a few atoms from such dictionary leads to state-of-the-art results in image and video restoration and classification. To take these challenges, this paper proposes a new method for robust background detecting via sparse representation. Our method explores both the strength of the well-patch adaptive dictionary learning technique to video frame structure analysis and the robustness background detection by the l_l-norm data-fidelity term. By using linear sparse combinations of dictionary atom, the proposed method learns the sparse representations of video frame regions corresponding to candidate particles. The experiments show that the proposed method is able to tolerate the background clutter and video frame deterioration, and improves the existing detecting performance.

Keywords: Sparse representation · Background subtraction · Dictionary learning · Overcomplete dictionary

1 Introduction

Background detection is widely used in various computer vision tasks, in particular, automated video surveillance, crowd and traffic monitoring, to separate the foreground, i.e. objects of interest or movement, from the background. For a visual detecting algorithm to be useful in real-world scenarios, it should be designed to handle and overcome cases where the background changes due to foreground objects changes from frame-to-frame. Significant and rapid variation due to time, noise, occlusion, varying viewpoints, background clutter, and illumination and scale changes pose major challenges to any detecting method. Over the years, many methods have been proposed to overcome these challenges.

© Springer International Publishing Switzerland 2015
D.-S. Huang et al. (Eds.): ICIC 2015, Part I, LNCS 9225, pp. 400–411, 2015.
DOI: 10.1007/978-3-319-22180-9_39

Recently, sparse representation [3] has been successfully applied to visual detecting [1, 2]. In this situation, the detector represents each candidate as a sparse linear combination of sparse dictionary that can be dynamically updated to sustain an up-to-date target detect model. This representation has been shown to be robust against partial occlusions, which leads to improving the detecting performance. However, sparse coding based detector performs computationally expensive l1 minimization at each frame [4]. In a patch filter framework, computational cost grows linearly with the number of patches sampled. It is this computational bottleneck that obstructs the use of these detectors in real-time application. Consequently, very recent efforts have been made to speed up this detecting paradigm [12]. More importantly, these methods assume that sparse representations of patches are independent. Ignoring the relationships that ultimately constrain particle representations tends to make the detector more prone to drift away from the target in cases of significant changes in position.

In this paper, we propose an efficient detecting algorithm with adaptive sparse detecting model and adaptive dictionary update strategy. The proposed method detected non-overlapped local image patches within the background region. We observe that sparse coding of image patches with a dimensional layout contains both spatial and partial information of the background pixels. The similarity measure is obtained by proposed alignment-pooling method across the patches within one candidate region. This helps locate the background more accurately and handle partial occlusion. In addition, the dictionary for sparse coding is generated from the dynamic video frames, which are updated based on sparse representation. The update scheme facilitates the detector to account for appearance changes of the background. Due to the simplicity of detecting model and dictionary update strategy, our method can detect the background efficiently.

The rest of this paper is organized as follows. Section 2 discusses related works. Section 3 describes robust background detection of video sequences via sparse learning. The efficiency and accuracy of the proposed method are illustrated by some experiment tests in Sect. 4. Finally, the conclusion is given in Sect. 5.

2 Related Works

Sparse representation has been successfully applied in numerous computer vision applications [7, 10]. In this section, we discuss the most relevant methods to our work. Visual detecting methods can be categorized as generative and discriminative [8]. Generative detecting methods adopt detecting models to represent non-background objects and search for the most similar image regions. Discriminative methods formulate object detecting as a binary classification with local search which aims to find the foreground image region that best distinguishes from the background.

An adaptive appearance model that accounts for foreground appearance variation is proposed in the incremental visual detecting method. Although it has been shown to perform well when the background tolerates lighting and appearance variation, this method is less effective in handling heavy occlusion or non-rigid background as a result of the adopted holistic appearance model. The fragment detector addresses the partial occlusion problem by modeling object appearance with histograms of local patches. The detecting task is carried out by combining votes of matching local patches based on histograms. As the model is not updated, this method is less effective for handling large appearance changes. The visual detecting by decomposition method extends the conventional patch filter framework with multiple background and observation models to account for large appearance variation caused by change of pose, lighting and scale as well as partial occlusion. As the adopted generative representation scheme is not designed to distinguish between foreground and background patches, it is prone to drift in complex situations.

In 2006, a method based on online adaptive boosting was proposed to select discriminative features for object detecting. As each detecting result and model update is based on the object detection of each frame, detecting errors are likely accumulated and thereby causing drifts. To account for ambiguities in selecting the best background segment, a boosting approach that extends the multiple instances learning framework for online object detecting is developed. While it is able to reduce detecting drifts, this method does not handle large non-rigid shape deformation or scale well. A mixed approach that combines a generative model and a discriminative classifier is proposed to handle appearance changes.

Sparse representation has been successfully applied in numerous computer vision applications. With sparsity constraints, one signal can be represented in the form of linear combination of only a few basis vectors. In [9], the image patch candidate is sparsely represented as a linear combination of the atoms of a dictionary which is composed of dynamic foreground and trivial background. This sparse representation problem is then solved through l1 minimization with non-negativity constraints. In [6], dynamic group sparsity which includes both spatial and temporal adjacency is introduced into the sparse representation to enhance the robustness of the detector. In [11], a local sparse representation scheme is employed to model the non-background appearance and then represent the basis distribution of the foreground with the sparse coding histogram. Due to the representation of local patches, their method performs well especially in handling the partial occlusion. A mean-shift algorithm [5] and a sparse representation based voting map are used to better detect the foreground.

However, we sample non-overlapped local image patches with prepared spatial layout where there are more spatial structural information in them. In addition, we make full use of the sparse coding coefficients with the proposed method rather than histograms and kernel densities to measure the similarity. Instead of using fixed dictionary learned from the first frame, we update the dictionary adaptively using dynamic video frame. Object detecting with a static dictionary is likely to fail in dynamic scenes due to large position changes. In [16], the dictionary is updated according to both the weights assigned to each atom and the similarity between templates and current estimation of non-background candidate. Different from that dictionary update scheme, we adopt sparse representation to update the dictionary adaptively. This dictionary update method reduces the drifting problem and puts more weights on the important parts of the background. In addition, it reduces the influence of the atom with partial occlusion.

3   Consistent Background Detection with Adaptive Sparse Representation

In this section, we present the proposed detecting algorithm based on temporally consistent sparse representations of image patch samples.

In our detecting method, video frames are divided into patches at the same size. In the t-th frame, we consider each image patches, whose observations (pixel color values or gradients) are denoted in matrix form as: $X = [x_1, x_2, \cdots, x_n]$. We represent each observation as a linear combination of atoms from a dictionary $D_t = [d_1, d_2, \cdots, d_m]$, such that $X = D_t Z_t$. Here, the columns of $Z_t = [z_1, z_2, \cdots, z_m]$ denote the representations of patch observations with respect to D_t.

Besides predefined sparsifying transforms, sparse and redundant representations of image patches based on learned dictionaries have drawn considerable attention in recent years. Considering image patches set $R(u) = [R_1 u, R_2 u, \cdots, R_L u]$ consisting of L samples, $R_l u \in R^M$ denotes a vectored form of the $M^{1/2} \times M^{1/2}$ patch extracted from image u of size $N^{1/2} \times N^{1/2}$. The sparse land model for image patches suggests that every image patch $R_l u$ could be represented sparsely over a learned dictionary D i.e.

$$Z = \arg \min_Z \|DZ - R_l u\|_2^2 \tag{1}$$

The combination of sparse and redundant representation modeling of signals, together with a learned dictionary from signal examples, has shown its promise in our problem.

The dictionary columns comprise the atoms that will be used to represent each patch. These atoms include visual observations of the detected object and the background (non-object) possibly under a variety of position changes. Since our representation is constructed at the pixel level, misalignment between dictionary atoms and patches might lead to degraded performance. To alleviate this problem, one of two strategies can be employed. (1) D_t can be constructed from a dense sampling of the foreground and the background, which includes transformed versions of both. (2) Columns of X can be aligned to columns of D_t as in [14]. In this paper, we take the first strategy, which leads to a larger m but a lower overall computational cost. We denote D_t with a subscript because its atoms will be progressively updated to incorporate variations in object position due to changes in time, illumination, viewpoint, etc. How to update D_t systematically will be introduced later.

The dictionary columns contain atoms that are used to represent each patch including image observations of the detected object and the background. Since our representation is constructed on the pixel level, misalignment between dictionary atoms and patch observations may lead to detecting drifts. To alleviate this problem, the dictionary D_t can be constructed from an overcomplete set using the transformed dictionary atoms of the foreground and background part. In addition, this dictionary is progressively updated.

For efficient and effective detecting, we exploit temporal consistency to prune patches. A patch is considered temporally inconsistent if its observation is not linearly represented well by the dictionary D_t and the representation of the background in the previous frame, denoted as z_0. More specifically, if its l_2 reconstruction error is above a predefined threshold σ, then it is pruned in the current frame. Temporal consistency is exploited in this work as the appearances of the detected object and its representations do not vary much in a short time period. Consequently, this process effectively reduces the number of patches to be represented from n_0 to n, where $n_0 < n$ in most cases. In what follows, we denote the ones after pruning as candidate patches, their corresponding observations as $X \in R^{d \times n}$, and their representations as $Z \in R^{m \times n}$.

3.1   Sparse Representation of Image Patches

The representation of each candidate patch is based on the following observations. After pruning, the candidate patch observations can be modeled by a subspace and therefore Z (their representations with respect to D_t) is expected to be sparse. The observation x_i of a good candidate patch can be modeled by a small number of nonzero coefficients in its corresponding representation z_i. The aim of object detecting is to search patches which have a representation similar to previous detecting results. In other words, a "well" representation should be consistent over time.

We base the formulation of our detecting method on the following observations. Firstly because patches are intensively sampled around the current detected object state, most of them will have similar representations with respect to D_t. Therefore, the resulting representation matrix Z is expected to be sparse, even for large values of m and n. Then inspired by the l_1 detector [15], a good target candidate's (patch) observation x_i has only a limited number of nonzero coefficients in its corresponding representation z_i. In other words, only a few dictionary atoms are required to reliably represent a patch. Lastly in many visual detecting situations, background regions are often corrupted by noise or partially occluded. As in [1], this noise can be modeled as sparse additive noise that can take on large values anywhere in its support. We combine these to obtain the problem in Eq. (2), where E is the error due to noise or occlusion and λ are three parameters to balance the importance of each term.

In this work, we formulate the detecting problem by

$$\min_{Z,E} \|DZ - R_l u\|_2 + \lambda \|E\|_1 \tag{2}$$

so accordingly

$$X = D_t Z + E. \tag{3}$$

And E is the error due to noise or occlusion. In this formulation, λ is the weight that quantifies the balance between different terms. We denote the representation of the previous detecting result with respect to D_t as z_0.

To encourage temporal consistency in the representation of the detecting result, we compare the representations of the patches in the current frame to that in the previous frame z_0. This approach effectively enforces temporal consistency for visual detecting although more complicated methods can be adopted. In other words, the regularization norm encourages the representations of most patches that represented well by the current object representation in D_t of the current frame to be similar to that of the previous detecting result [13]. Equivalently, it allows only a small number of patches that observations not represented well by object patches to have representations different from the previous detecting result.

For robustness against sparse significant errors due to noise or occlusion, we seek to minimize the l_1 norm of each column of E. This sparse error assumption has been adopted in detecting and other applications. Unlike the some detector that incorporates sparse error by augmenting D_t with a large number of trivial patches and computing the corresponding coefficients, we obtain the reconstruction error $E \in R^{d \times n}$. Furthermore, the values and support of columns in E are informative since they indicate the presence of occlusion which is large values but sparse support and whether a candidate patch is sampled from the background those large values with non-sparse support.

3.2   Adaptive Dictionary Updating

To alleviate the problem of detecting drift, we gain D_t with representative patches of the background such that $D_t = [D_O D_B]$ where D_O and D_B represent the foreground object and background patches respectively. The dictionary D_t is initialized by sampling image patches around the initial foreground position. For accurate detecting, the dictionary is updated in successive frames to model appearance change of the non-background object. The detecting result y_t at instance t is the patch x_i such that

$$i = \arg \max_{k=1,2,\cdots,n} \left(\left\| Z_k^O \right\|_1 - \left\| Z_k^B \right\|_1 \right) \tag{4}$$

This encourages the detecting result to be modeled well by object representation and not background representation. We also exploit discriminative information to design a systematic procedure for updating D_t.

Updating the Gradient Image Variables $u^i, i = 1, 2$ in the horizontal and vertical directions: the minimization with respect to u^1 and u^2 is decoupled, and then can be solved separately. It yields:

$$u_t^i = \arg \min_{u^i} \sum_I D_t Z_t - R_l u^i \tag{5}$$

In this work, we propose to compute the final detection by iteratively reconstructing the gradients dictionary via dictionary learning and solving the detected image. The proposed method is based on the observation that the gradients are sparser than the image itself and therefore may have sparser representation with the learned dictionary than the pixel-domain image. This motivates us to learn the dictionary in the gradient domain. It is expected that such learning is more accurate and robust than that from the pixel domain based on the illustrations from [17, 18].

3.3 Consistent Background Detector with Adaptive Sparse Representation

In the previous detectors, the sparse representations of patches are learned for each patch independently. In our proposed detecting method, the patches are sampled with a Gaussian distribution cover the target image, and the observations of some patches may be different to each other. Experimental results in Sect. 4 provide evidence of this crucial algorithmic difference.

The use of gradients property facilitates learning effective sparse representation for background detecting. In the l_1 detector, the representation of each patch is learned independently. Due to the gradients property, our detector learns the representations of all patches. In other words, the sparse representations of observations are learned by considering all patches and the gradients property. However, in the l_1 detector, each patch is processed independently without considering the other patches. Thus, the learned sparse representation by our algorithm is more compact and robust for background detecting.

Algorithm 1 shows the main steps of our method, given observations of all pruned patches X and the current dictionary D, we learn the representation matrix Z by solving (1). Clearly, Z is sparse. The patch observation x_i is selected as the current detecting result y_t as its difference is largest among all patches. Since the patch observation can be considered as a misaligned representation of the target, it is not modeled well by the foreground object dictionary D_O. On the other hand, the patch observation is represented well by the background dictionary D_B. As illustrated in this example, the detecting drift problem is alleviated by the proposed formulation.

Algorithm 1. The SparseBackground Detector Algorithm

Input:	Output:
– Current frame at t	– Detected result y_t
– Gradients dictionary D_{t-1}	– Updated gradients dictionary D_t
– Representation of previous detecting result Z	– Updated representation Z

1: Create the image patches

2: According to observations corresponding to all patches to get X_0

3: Use consistency property to prune and obtain candidate patch observations X based on the reconstruction error

4: Compute sparse representation Z for X

5: Select the patch with the highest value of $\left\| Z^O - Z^B \right\|_1$ as the current detecting result y_t

6: Update z_0 based on the detecting result

7: Update dictionary atoms via Algorithm 2

It is known that detecting algorithms with a fixed dictionary of atoms is not effective to account for position change in complex scenes. However, small errors are likely to be introduced and accumulated if the atoms are updated too frequently, thereby making the detector drift away from the target. Numerous approaches have been proposed for dictionary update to alleviate the detecting drift problem. In this work, we address this issue by dynamically updating the dictionary in D_t.

We shift the initial bounding box by 1 to 3 patches in each dimension, and thus obtain object patches for the object dictionary D_O. In addition, we initialize the background dictionary D_B, with image patches sampled at a sufficient distance from the surrounding background based on the initial detecting result and obtain background representation. All dictionary atoms are normalized to fit the size of the patches which have been manually initialized.

Sparse representation for gradient patches with respect to dictionary variables D_t and Δz. The minimization with respect to dictionary and coefficient variables of the gradient images in the horizontal and vertical directions is also decoupled, and thus can be solved separately. It yields:

$$\{D_t\} = \arg\max_{D_t} \left\| Z^O - Z^B \right\|_1 \tag{6}$$

Algorithm 2. Dictionary Update

Input:	Output:
– Foreground object dictionary D_O .	– Dictionary $D_t = [D_O D_B]$
– Background dictionary D_B .	
– y_t , which is the newly chosen detecting patch.	

1: Initialize predefined parameters at $t = 1$.

2: z is the solution of (1). Set $\Delta z = \left\| Z^O - Z^B \right\|_1$

3: Update weights according to the coefficients of the target templates.

4: Update D_B based on the current detecting result.

5: Update D_O based on $\{D_o\} = \max_{D_o,\Delta z} \left\| X - D_o Z^o \right\|_2$.

Each object patch in D_O is associated with a weight proportional to the frequency that it is selected for detecting. The weight of an object patch in D_O is updated based on how frequently that dictionary is used in representing the current detecting result z. If Z is adequately represented (based on a predefined threshold) by the current dictionary, then there is no need to update it. Otherwise, the object patch with the smallest weight is replaced by the current detecting result, and its weight is set to the median of the current normalized weight vector ω. The main steps of the dictionary update for dictionary are summarized in Algorithm 2. On the other hand, the background dictionary D_B is updated at every frame by re-sampling patches at a sufficient distance from the detecting result.

4 Experimental Results

In this section, we present experimental results that validate the effectiveness and efficiency of our method. All our experiments are carried out in MATLAB on a 3.40 GHz Intel Core i3 machine with 4 GB RAM.

To evaluate our method, we take a set of 16 challenging video sequences (in 4 categories) that are publicly available online, each sequence contains 1000 frames. Due to space constraints, we will only show results on two of these sequences. These videos are recorded in indoor and outdoor environments and include challenging variations due to changes in appearance, illumination, scale, and the presence of occlusion.

The first column (a, d) of Fig. 1 shows two frames of a typical sequence. The third columns (c, f) are detecting results by the proposed algorithm, respectively. This experiment shows good results can be obtained by the proposed algorithm. However, there is some difference between the images in the second (b, e, the ground truth) and third column. Almost all background object pixels are rightly classified according to the ground truth image, but some non-background object pixels are still classified into background in the detected image. This phenomenon is caused by the model inaccuracy due to the fast movement of the object.

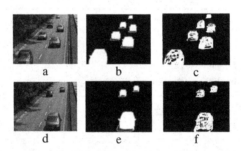

Fig. 1. Detecting result of the highway sequences. The original frames are in the first column, ground truth are displayed in the middle column, our detecting result in the third column.

Another example of background detection is shown in Fig. 2. The first column (a) of Fig. 2 shows the test image of the second sequence. A pedestrian (left) dressed in a white coat is moving in front of a similarly colored background in this sequence. Detected results by the proposed algorithm are shown in the second column (b). Good

Fig. 2. Detecting result of the pedestrians sequence. The original frame on the left, our detecting result on the right.

Table 1. Average performance comparisons of 4 methods

Methods	Recall	Specificity	FPR	FNR	PWC	F-1
Proposed algorithm	0.7103	**0.9799**	**0.0201**	0.2897	**2.8450**	**0.6606**
KDE	0.7375	0.9519	0.0481	0.2625	5.6262	0.5688
GMM	0.6604	0.9725	0.0275	0.3396	3.9953	0.5566
Euclidean distance	0.6803	0.9449	0.0551	0.3197	6.6542	0.5161

results can be obtained by the proposed algorithm, but many target pixels still cannot be detected, especially in the part of shoulder. In this example, the shape of the moving object changes quickly and periodically.

We also compared the performance of our method to the widely used method by using the 15 test sequences [20–22]. The sequences include both indoor and outdoor scenes. For all sequences, our method gave less false negatives than the comparison method. In the case of false positives, our method was better in five cases. For the rest of the sequences, the difference is very small. It should be noticed that, for the proposed method, most of the false positives occur on the contour areas of the moving objects. This is because the features are extracted from the pixel neighborhood.

Six measurements, including Recall, Specificity, False Positive Rate, False Negative Rate, Percentage of Wrong Classifications (PWC) and F-Measure (also referred as F-1), are used to determine comprehensive evaluations on different features of each method. Average performance of the proposed method comparing with three other methods is shown in Table 1. It can be seen that the proposed method outperforms the other methods in most measurements.

5 Conclusion

In this paper we proposed a new compressed sensing model and an efficient algorithm for background subtraction, based upon an adaptive dictionary learning strategy and a penalized splitting approach. The model aims to learn dictionary in the sparse gradient domain for better detection. Consequently, an efficient splitting method was presented to decouple the difference operators, dictionary and sparse coefficients, which constitutes our algorithm. The results of several numerical experiments demonstrate the superiority of our method in terms of detection accuracy and convergence. The superior detection performance of the proposed method opens a new avenue to study how the sparse representation model can be particularized to background detecting and processing framework. Further studies will be conducted to improve the proposed algorithm, such as introducing some other filter outputs [19], or learning dictionaries in both pixel and gradient domains.

Acknowledgments. This work is supported by National Natural Science Foundation of China (No. 61373109, No.61003127), State Key Laboratory of Software Engineering (SKLSE2012-09-31).

References

1. Pique-Regi, R., Monso-Varona, J., Ortega, A., et al.: Sparse representation and Bayesian detection of genome copy number alterations from microarray data. Bioinformatics **24**(3), 309–318 (2008)
2. Agarwal, S., Roth, D.: Learning a sparse representation for object detection. In: Heyden, A., Sparr, G., Nielsen, M., Johansen, P. (eds.) ECCV 2002, Part IV. LNCS, vol. 2353, pp. 113–127. Springer, Heidelberg (2002)
3. Aharon, M., Elad, M., Bruckstein, A.: k-svd: An algorithm for designing overcomplete dictionaries for sparse representation. IEEE Trans. Sig. Process. **54**(11), 4311–4322 (2006)
4. Yang, J., Yu, K., Gong, Y., et al.: Linear spatial pyramid matching using sparse coding for image classification. In: IEEE Conference on Computer Vision and Pattern Recognition, CVPR 2009, pp. 1794–1801. IEEE (2009)
5. Zhou, H., Yuan, Y., Shi, C.: Object tracking using SIFT features and mean shift. Comput. Vis. Image Underst. **113**(3), 345–352 (2009)
6. Liu, B., Huang, J., Yang, L., et al.: Robust tracking using local sparse appearance model and k-selection. In: 2011 IEEE Conference on Computer Vision and Pattern Recognition (CVPR), pp. 1313–1320. IEEE (2011)
7. Bruckstein, A.M., Donoho, D.L., Elad, M.: From sparse solutions of systems of equations to sparse modeling of signals and images. SIAM Rev. **51**(1), 34–81 (2009)
8. Felzenszwalb, P.F., Girshick, R.B., McAllester, D., et al.: Object detection with discriminatively trained part-based models. IEEE Trans. Pattern Anal. Mach. Intell. **32**(9), 1627–1645 (2010)
9. Xu, R, Zhang, B., Ye, Q., et al.: Cascaded L1-norm minimization learning (CLML) classifier for human detection. In: 2010 IEEE Conference on Computer Vision and Pattern Recognition (CVPR), pp. 89–96. IEEE (2010)
10. Wright, J., Ma, Y., Mairal, J., et al.: Sparse representation for computer vision and pattern recognition. Proc. IEEE **98**(6), 1031–1044 (2010)
11. Liu, B., Yang, L., Huang, J., Meer, P., Gong, L., Kulikowski, C.: Robust and fast collaborative tracking with two stage sparse optimization. In: Daniilidis, K., Maragos, P., Paragios, N. (eds.) ECCV 2010, Part IV. LNCS, vol. 6314, pp. 624–637. Springer, Heidelberg (2010)
12. Newcombe, R.A., Davison, A.J., Izadi, S., et al.: KinectFusion: real-time dense surface mapping and tracking. In: 2011 10th IEEE International Symposium on Mixed and Augmented Reality (ISMAR), pp. 127–136. IEEE (2011)
13. Liu, Q., Wang, S., Luo, J.: A novel predual dictionary learning algorithm. J. Vis. Commun. Image Represent. **23**(1), 182–193 (2012)
14. Peng, Y., Ganesh, A., Wright, J., et al.: RASL: robust alignment by sparse and low-rank decomposition for linearly correlated images. IEEE Trans. Pattern Anal. Mach. Intell. **34**(11), 2233–2246 (2012)
15. Mei, X., Ling, H.: Robust visual tracking and vehicle classification via sparse representation. IEEE Trans. Pattern Anal. Mach. Intell. **33**(11), 2259–2272 (2011)
16. Wright, J., Yang, A.Y., Ganesh, A., et al.: Robust face recognition via sparse representation. IEEE Trans. Pattern Anal. Mach. Intell. **31**(2), 210–227 (2009)
17. Mei, X., Ling, H.: Robust visual tracking using $\ell 1$ minimization. In: 2009 IEEE 12th International Conference on Computer Vision, pp. 1436–1443. IEEE (2009)

18. Mairal, J., Leordeanu, M., Bach, F., Hebert, M., Ponce, J.: Discriminative sparse image models for class-specific edge detection and image interpretation. In: Forsyth, D., Torr, P., Zisserman, A. (eds.) ECCV 2008, Part III. LNCS, vol. 5304, pp. 43–56. Springer, Heidelberg (2008)
19. Liu, Q., Luo, J., Wang, S., et al.: An augmented Lagrangian multi-scale dictionary learning algorithm. EURASIP J. Adv. Signal Process. **2011**(1), 1–16 (2011)
20. Wang, X., Ma, X., Grimson, W.E.L.: Unsupervised activity perception in crowded and complicated scenes using hierarchical bayesian models. IEEE Trans. Pattern Anal. Mach. Intell. **31**(3), 539–555 (2009)
21. Elgammal, A., Harwood, D., Davis, L.: Non-parametric model for background subtraction. In: Vernon, D. (ed.) ECCV 2000. LNCS, vol. 1843, pp. 751–767. Springer, Heidelberg (2000)
22. Zivkovic, Z.: Improved adaptive gaussian mixture model for background subtraction. In: International Conference on Pattern Recognition, vol. 2 (2004)
23. Benezeth, Y., Jodoin, P.-M., Emile, B., Rosenberger, C., Laurent, H.: Comparative study of background subtraction algorithms. J. Electron. Imaging **19**(3), 033003–033003-12 (2010)

A New Microcalcification Detection Method in Full Field Digital Mammogram Images

Xiaoming Liu[1,2(✉)], Ming Mei[1], Weiwei Sun[1], and Jun Liu[1,2]

[1] College of Computer Science and Technology, Wuhan University of Science and Technology, Wuhan 430065, China
lxmspace@gmail.com
[2] Hubei Province Key Laboratory of Intelligent Information Processing and Real-time Industrial System, Wuhan, China

Abstract. Breast cancer is a great threat for women around the world. Mammography is the main approach for early detection and diagnosis. Microcalcification (MC) in mammograms is one of the important early signs of breast cancer. Their accurate detection is important in computer-aided detection (CADe). In this paper, we proposed a new Microcalcification detection method for full field digital mammograms (FFDM) by integrating Possibilistic Fuzzy c-Means (PFCM) clustering algorithm and weighted support vector machine (WSVM). The method includes a training process and a testing process. In the training process, possible microcalcification regions are located and extracted. Extracted features are selected with mutual information based technique. Positive and negative samples are weighted with PFCM and used to train a weighted SVM. A similar procedure is performed on test images. The proposed method is evaluated on a database of 410 clinical mammograms and compared with a standard unweighted support vector machine classifier.

Keywords: Microcalcification detection · Computer aided diagnosis · Possibilistic fuzzy c-means · Support vector machine

1 Introduction

Breast cancer is the most frequent form of cancer in women, and is also the leading cause of mortality in women each year. According to the statistics from the World Health Organization, 521,907 women worldwide died in 2012 due to breast cancer [1]. Studies have indicated that early detection and treatment improve the survival chances of the patients. Unfortunately, only 20 % of the breast cancer patients are diagnosed at the early stage. Among all the diagnostic methods currently available for detection of breast cancer, mammography is regarded as the only reliable and practical method capable of detecting breast cancer in its early stage [2].

Various types of abnormalities can be observed in mammograms, such as microcalcification clusters and mass lesion, distortion in breast architecture, and asymmetry. Microcalcification clusters and mass [3, 4] are the most common signs of breast cancer, microcalcification (MC) clusters appear in 30 %–50 % diagnosed cases. MCs are

© Springer International Publishing Switzerland 2015
D.-S. Huang et al. (Eds.): ICIC 2015, Part I, LNCS 9225, pp. 412–420, 2015.
DOI: 10.1007/978-3-319-22180-9_40

calcium deposits of very small dimension and appear as a group of granular bright spots in a mammogram.

The detection of abnormalities in mammogram has attracted researcher's great interest for decades, and various methods have been proposed. While most of them have been targeted on screen-film mammography. With the advance of imaging techniques, now Full Field Digital Mammography (FFDM) has been proved by FDA, and become widely available. Compared to traditional screen-film mammography, FFDM can provide higher spatial resolution and better contrast. Thus, the computer aided detection and diagnosis on FFDM is an important problem.

A typical mammogram with microcalcification clusters is shown in Fig. 1(a) and the full view of a cluster of microcalcifications in Fig. 1(b).

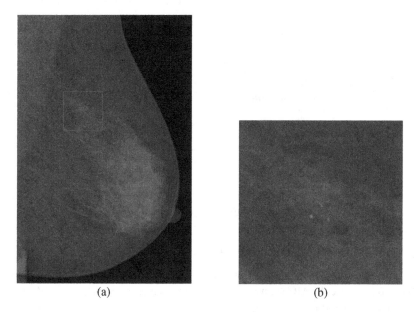

(a)   (b)

Fig. 1. (a) Mammogram with a microcalcification cluster, (b) enlarged view of red part in (a) (Color figure online)

In this paper, we proposed a new weighted support vector machine based microcalcification detection method for FFDM images. Inspired by the work in [5], possibilistic fuzzy c-means (PFCM) [6] clustering algorithm is used to derivative weights for the samples. Several features are extracted and used to train SVM. The proposed method is evaluated on a FFDM dataset [7], consisting of 410 images.

2   Proposed Method

The workflow of the proposed method is shown in Fig. 2. For each image, suspicious MC regions are extracted with active contour segmentation, then geometry and texture features are extracted for each suspicious MC. A mutual information based supervised criterion is used to select important features, and PFCM is applied to cluster the samples. Weights of the samples are calculated based on possibilities and typicality values from the PFCM. A weighted nonlinear SVM is trained. During the test process, when an unknown image is presented, a similar process is performed. The suspicious MC regions are classified by the powerful weighted nonlinear SVM.

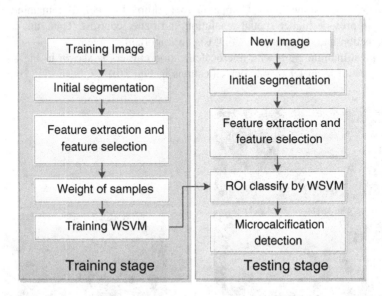

Fig. 2. Workflow of the proposed method

2.1   Level Set Based MC Segmentation

The segmentation of MC consists of two steps, firstly several edge points are detected and used to initialize the MC segmentation, and then an active contour is used to refine the initial segmentation.

The initial step follows the method proposed in [8]. Since microcalcifications on image are with high grayscale values. For a given image $f(x, y)$, the edge of a microcalcification to be segmented is a closed contour around a known pixel (x_0, y_0), which is the location of local highest grayscale value pixel. For each pixel, a slope value $s(x, y)$ with respect to $f(x_0, y_0)$ is defined as [8]

$$s(x, y) = \frac{f(x_0, y_0) - f(x, y)}{d(x_0, y_0, x, y)} \tag{1}$$

where $d(x_0, y_0, x, y)$ is the Euclidean distance between the local maximum pixel (x_0, y_0) and pixel (x, y). A pixel is considered on the edge if $s(x, y)$ is maximal along a line segment originating from (x_0, y_0). The length of the considered line segment is chosen as 15. The line search is applied in 16 equally spaced directions originating from the seed pixel. Thus, with each local maximal, 16 edge points are located. Note that segmentation step will encounter some difficulties for small MC, the approach used in [9] is adopted to overcome the problem.

For a given local maximal with 16 edge points, a circle is fitted to the points. And the circle is used as the initialization of the level set based segmentation. Level set method achieves the segmentation by solving an energy optimization problem. We used the segmentation method proposed in our previous work by optimizing the following energy [3, 10]:

$$
\begin{aligned}
E(\phi, f_1, f_2) = {} & \lambda_1 \int \left(\int K_\sigma(x - y)|I(y) - f_1(x)|^2 H(\phi(y)) dy \right) dx \\
& + \lambda_2 \int \left(\int K_\sigma(x - y)|I(y) - f_2(x)|^2 (1 - H(\phi(y))dy \right) dx \\
& + \gamma_1 \int |I(x) - c_1|^2 (1 - H(\phi(x))) dx + \gamma_2 \int |I(x) - c_2|^2 H(\phi(x)) dx \\
& + \mu \int |\nabla H(\phi(x))| dx + v \int g\delta(\phi)|\nabla\phi| dx + w \int \frac{1}{2} \left(|\nabla\phi(x)|^2 - 1 \right) dx
\end{aligned}
\tag{2}
$$

2.2 Feature Extraction from ROI

Fourteen geometry features are considered in the study, including area (denoted as GF1), perimeter (GF2), compactness (C, GF3), Normalized Distance Moment (NDM2, NDM3, NDM4, GF4-F6), Fourier Feature (FF, GF7), Normalized Radial Length (NRL) based features (μ_{NRL}, σ_{NRL}, E_{NRL}, AR_{NRL}, GF8-GF11), relative gradient orientation (RGO) based features (μ_{RGO}, σ_{RGO}, E_{RGO}, GF12-GF14). For more details about geometry features, please see our previous work [3].

For each suspected MC, a patch with size 16×16 is extracted to calculate texture features, whose center is determined by the center of the suspected MC. Besides the average grayscale in the segmented region in a block (denoted as TF1, where TF means texture feature), the grayscale difference between average suspicious region and background (TF2) is also used.

We used several GLCM features, including autocorrelation (TF3), contrast (TF4), correlation (TF5), Cluster Prominence (TF6), Cluster Shade (TF7), Energy (TF8), Entropy (TF9), Homogeneity (TF10), Maximum probability (TF11), Sum of squares (TF12), Sum average (TF13), Sum variance (TF14), Sum entropy (TF15), Difference variance (TF16), Difference entropy (TF17), information measure of correlation (TF18, TF19), Inverse difference normalized (TF20), Inverse difference moment normalized (TF21).

We also used undecimiated wavelet transform with the Daubechies 4 filter for each suspicious MC patch (a 16×16 window) in the paper. The entropy and energy of each sub-band are used as features. Twelve sub-images are generated for each ROI with a three-level wavelet decomposition, as the first level decompose consists of mostly noise, features are extracted from the level 2 and 3 subbands. Thus, 16 wavelet features are extracted for each ROI. Denote them with TF22-TF37.

2.3   Feature Selection Based on Mutual Information

With the above procedure, a lot of features are extracted to represent the possible MC. However, not every feature is useful to discriminate non-MC and MC. In this paper, we used the mRMR (a mutual information MI based filter feature selection) method [11]. For the details of mRMR, please see reference [11].

2.4   Clustering with PFCM and Weight Samples

With the obtained MC and non-MC samples, we selected the features by mRMR criterion. PFCM is applied to cluster the samples. Each sample with selected feature values is regarded as a data point. As shown above, for each sample after PFCM clustering, it has a probability and a typicality value.

Let y_i denote the label of sample i, and let $y_i \in \{+1, -1\}$ denote the class variable (MC or non-MC) which we can obtain by the doctor's manual annotation. Let MU_i denote the probability of sample i belonging to calcification, let MT_i^{+1} denote the typicality value of sample i belongs to MC, and let MT_i^{-1} denote the typicality value of it belongs to non-MC. Both MU_i, MT_i^{+1} and MT_i^{-1} can be get by PFCM clustering, and their value ranges are between 0 and 1.

We want to give more weights to the samples with higher confidence, and define the weight $W1_i$ as

$$W1_i = \frac{1 + y_i}{2} \times MU_i + \frac{1 - y_i}{2} \times (1 - MU_i) \tag{3}$$

Besides the confidence value, we also used the typicality values outputted by PFCM. The weight term considering the typicality value is defined as follows:

$$W2_i = \frac{1 + y_i}{2} \times MT_i^1 \times \left(1 - MT_i^{-1}\right) + \frac{1 - y_i}{2} \times \left(1 - MT_i^1\right) \times MT_i^{-1} \tag{4}$$

For typical sample belongs to MC or non-MC, its weight $W2_i$ is high. We take both possibility information and typical information into consideration, and the final weight of sample i we defined is

$$W3_i = W1_i * W2_i \tag{5}$$

2.5   Weighted SVM Based Classification

The weighted SVM optimization problem can be formulated as follows: Given a training sample set $S = ((x_1, y_1, v_1), \ldots, (x_n, y_n, v_n))$, where v_i is the weight of the sample x_i (calculated with Eq. (5)), find the hyperplane (w, b) that solves the following optimization problem:

$$\text{minimize } \langle w \cdot w \rangle + C \sum_{i=1}^{n} g(v_i)\xi_i$$

$$s.t. \, y_i(\langle w \cdot x_i \rangle + b)f(v_i) \geq 1 - \xi_i, i = 1, \cdots, n \tag{6}$$

$$\xi_i \geq 0, i = 1, \cdots, n$$

realizes the maximal weighted soft margin hyperplane. If both function f and g are setted to be constant 1, then the WSVM coincides with standard SVM.

In the above formulation, the final decision plane will be less affected by those margin violating samples with low confidence, and samples with high confidence have higher impact on the final decision plane.

3 Experimental Results

3.1 Mammogram Database

The proposed method was tested on the INbreast database [7]. The database was acquired at the Breast Centre in CHSJ, Porto. The database has a total 115 cases (410 images), from which 90 cases are from women with both breasts affected (4 images per case) and 25 cases are from mastectomy patients (two images per cases). The pixel size is 70 μm (microns), with 14-bit resolution. The image matrix was 3328×4084 or 2560×3328 pixels. Among the 410 images, calcifications are present in 301 images. A total of 6880 microcalcifications were individually identified in 299 images (≈ 23.0 calcifications per image).

3.2 Detection Results

Our method first extracted suspicious MC regions and then used WSVM to reduce the false positives. If a MC is missed in the segmentation step, it will not show up in the final detection. With visual inspection of the output images and their corresponding annotations showed that all the MCs had been detected in the first segmentation stage. Figure 3 shows the segmentation stage of an image.

With the extracted features and known class label, MI is used to select the important features. We have extracted 51 features (14 geometry features, 2 grayscale features, 19 GLCM features, and 16 wavelet features) to represent MC and non-MC.

Five-fold cross-validation method was used to select the number of features. The averaged performances were recorded to set parameter values. For the classifier here, we used a standard SVM (with radial basis function kernel) without weights, more specifically, the LIBSVM toolbox [12] was used. The most suitable number of feature used here is 22 (the top 10 features ranked by MI is shown in Table 1).

The receiver operating characteristic (ROC) curves are used to evaluate the performance of MC detection. The standard SVM without weighting is used to evaluate the effect of the PFCM based weighting scheme.

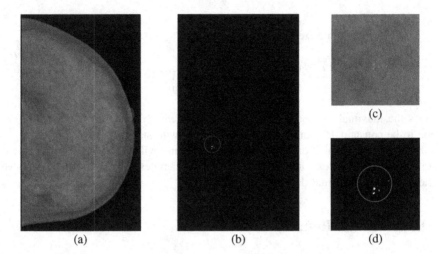

(a)   (b)   (d)

Fig. 3. Segmentation of suspected of MC, (a) original image, red circle is the annotation of MC (b) segmentation mask of possible MC, (c) enlarged part of MC in (a), (d) enlarged part of segmentation mask of (b) (Color figure online).

Table 1. Top 10 features ranked by mutual information filter

Order	Feature no. and name	Order	Feature no. and name
1	TF2, grayscale difference	6	GF1, area of the MC
2	GF3, compactness	7	GF7, Fourier feature
3	TF17, difference entropy with GLCM	8	TF1, the average grayscale in segmented region
4	TF23, entropy in approximate matrix of level 2 with wavelet	9	GF12, RGO mean
5	TF4, contrast with GLCM	10	GF2, perimeter of the MC

We compared the performance of the standard unweighted SVM and the proposed PFCM clustering based weighted SVM on MC classification. The test set contains 1366 true MCs, we selected 2732 non-MC samples from the segmentation on test images, similar to the training step. The performance of standard unweighted SVM and our weighted SVM are shown in Fig. 4 with ROC curve. The AUC (A_z), which is the area under the ROC curve, is used to compare the performance of two classification methods. The AUC for standard unweighted SVM is 82.68 % and the AUC of the proposed PFCM based weighted SVM is 86.76 %. We can see that with the same training samples and test samples, the proposed weight SVM achieved better performance than standard unweighted SVM.

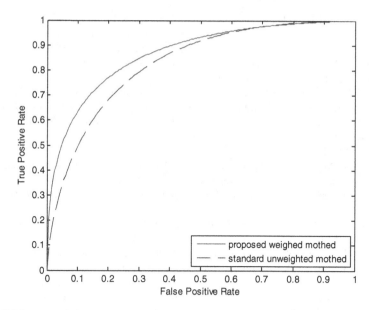

Fig. 4. ROC curve of the standard unweighted SVM and the proposed PFCM based weighted SVM.

4   Conclusion

In this paper, we proposed a weighted SVM technique for detection of MC in FFDM. In this approach, suspicious MC regions are first segmented with active contour, and then the regions are classified by a trained weighted SVM. The non-MC training samples are selected from the segmented regions, which can be better tuned to the whole procedure than random sampling. Mutual information criterion was used to select important features. The training samples are weighted with the possibility and typicality value of a sample belongs to MC output by the novel introduction of possibilistic fuzzy c-means (PFCM) clustering. Experimental results with ROC analysis using a set of 410 FFDM mammograms demonstrated that the proposed method outperformed standard unweighted SVM.

Acknowledgments. This work is partially supported by the National Natural Science Foundation of China (NO. 61403287, NO. 61273303, NO. 31201121), the Natural Science Foundation of Hubei Province, China (No. 2014CFB288), the Educational Commission of Hubei Proince, China (NO. D20131101), and the China Postdoctoral Science Foundation (NO. 2014M552039).

References

1. GLOBOCAN 2012 v1.0, Cancer Incidence and Mortality Worldwide: IARC CancerBase No. 11 [Internet]. International Agency for Research on Cancer, Lyon (2013). http:// globocan.iarc.fr. Accessed 4 May 2015

2. Lee, C.H., Dershaw, D.D., Kopans, D., Evans, P., Monsees, B., Monticciolo, D., Brenner, R. J., Bassett, L., Berg, W., Feig, S.: Breast cancer screening with imaging: recommendations from the Society of Breast Imaging and the ACR on the use of mammography, breast MRI, breast ultrasound, and other technologies for the detection of clinically occult breast cancer. J. Am. Coll. Radiol. **7**, 18–27 (2010)

3. Liu, X., Tang, J.: Mass classification in mammograms using selected geometry and texture features, and a new SVM-based feature selection method. IEEE Syst. J. **8**, 910–920 (2014)

4. Tang, J., Rangayyan, R.M., Xu, J., El Naqa, I., Yang, Y.: Computer-aided detection and diagnosis of breast cancer with mammography: recent advances. IEEE Trans. Inf. Technol. Biomed. **13**, 236–251 (2009)

5. Quintanilla-Domínguez, J., Ojeda-Magaña, B., Marcano-Cedeño, A., Cortina-Januchs, M. G., Vega-Corona, A., Andina, D.: Improvement for detection of microcalcifications through clustering algorithms and artificial neural networks. EURASIP J. Adv. Sig. Process. **2011**, 91 (2011)

6. Pal, N.R., Pal, K., Keller, J.M., Bezdek, J.C.: A possibilistic fuzzy c-means clustering algorithm. IEEE Trans. Fuzzy Syst. **13**, 517–530 (2005)

7. Moreira, I.C., Amaral, I., Domingues, I., Cardoso, A., Cardoso, M.J., Cardoso, J.S.: INbreast: toward a full-field digital mammographic database. Acad. Radiol. **19**, 236–248 (2012)

8. Bankman, I.N., Nizialek, T., Simon, I., Gatewood, O.B., Weinberg, I.N., Brody, W.R.: Segmentation algorithms for detecting microcalcifications in mammograms. IEEE Trans. Inf. Technol. Biomed. **1**, 141–149 (1997)

9. Soltanian-Zadeh, H., Rafiee-Rad, F., Pourabdollah-Nejad, D.S.: Comparison of multiwavelet, wavelet, Haralick, and shape features for microcalcification classification in mammograms. Pattern Recognit. **37**, 1973–1986 (2004)

10. Tang, J., Liu, X.: Classification of breast mass in mammography with an improved level set segmentation by combining morphological features and texture features. In: El-Baz, A.S., Rajendra, A.U., Laine, A.F., Suri, J.S. (eds.) Multi Modality State-of-the-Art Medical Image Segmentation and Registration Methodologies, pp. 119–135. Springer, New York (2011)

11. Peng, H., Long, F., Ding, C.: Feature selection based on mutual information criteria of max-dependency, max-relevance, and min-redundancy. IEEE Trans. Pattern Anal. Mach. Intell. **27**, 1226–1238 (2005)

12. Chang, C.-C., Lin, C.-J.: LIBSVM: a library for support vector machines. ACM Trans. Intell. Syst. Technol. (TIST) **2**, 1–27 (2011)

Forensic Detection of Median Filtering in Digital Images Using the Coefficient-Pair Histogram of DCT Value and LBP Pattern

Yun-Ni Lai[1(✉)], Tie-Gang Gao[1], Jia-Xin Li[1], and Guo-Rui Sheng[2]

[1] College of Software, Nankai University, Tianjin, China
laiyunni_nk@163.com,
gaotiegang@nankai.edu.cn, vivianink@126.com,
shengguorui@hotmail.com
[2] College of Information and Electric Engineering,
Ludong University, Yantai, China

Abstract. Looking for modification traces of digital media is of great value for forensic analysis. The median filter can be used to remove the fingerprints left by other image operations, and the detection of median filtering has become more and more significant. In this paper, a new detector for median-filtering operation is proposed. In the method, the image features combined by LBP (Local Binary Pattern) and coefficient-pair histogram in DCT (Discrete Cosine Transform) domain are firstly extracted; then classifier SVM is used to train the authentic and median-filtered image; lastly, some suspicious images are used to test the effectiveness of the proposed scheme. Large amounts of experiments show that the proposed method can detect median filtering under a variety of scenarios, and further more it has letter robustness against JPEG post-compressed image, this outperforms the existing state-of-the-art method.

Keywords: Median filter · Forensic analysis · Steganalysis

1 Introduction

Due to the rapidly development of image processing tools, such as Photoshop, image can be easily modified, and people can hardly find any differences in the new modified image from the original one, so, digital image forensic has become a hot research area. Among all the methods of image forgery, median filter and JPEG compression are two of the most popular post-processing method for image forgery [1]. It may not change the visual expression of image, but it has a great influence on steganalytic algorithms. Median filter is widely used as a nonlinear denoising operator for hiding the traces of resampling. And it can also preserve image content and leads to the invalidation of image forensic technology. In the meanwhile, low-quality and JPEG compression will also greatly influence the forgery analysis. Therefore, the detection of median filtering remains a challenge.

Recently, a few of researches have been conducted on the detection of median filtering [1–6]. Kirchner and Fridrich extract first-order differences SPAM (subtractive pixel adjacency matrix) feature and ratio of first-order difference-histogram bins to detect non-linear median filtering [1]. It constructs a complex detector with a false

© Springer International Publishing Switzerland 2015
D.-S. Huang et al. (Eds.): ICIC 2015, Part I, LNCS 9225, pp. 421–432, 2015.
DOI: 10.1007/978-3-319-22180-9_41

positive rate as low as 1.8 % (3×3 median filtering) and 1.1 % (5×5 median filtering). Cao *et al.* use the probability of zero values on the first order difference map in texture regions to detect median filtering (MF) [3]. The median-filtered image is identified by perform thresholding adjudication on the fingerprint metric. For arbitrary images, a novel approach (MFF) based on the new feature set and the scalar feature is proposed by Hai-Dong Yuan [4]. It even reliably detects tempering when part of a median-filtering image is inserted into a non-median-filtered image. Xiangui Kang [5] introduced a robust median filtering scheme based on the autoregressive model of median filtered residual. The median filtered residual (MFR) is the difference between the initial image and the filtered output image. It is used as the forensic fingerprint, which has the features of low dimension. A novel local texture operator named the second-order local ternary pattern (LTP) is proposed by Yujin Zhang et al. [6]. The operator encodes the local derivative direction variations by using a 3-valued coding function. It has shown that LTP performs better than several state-of-the-art approaches investigated.

This paper proposed a method for detecting median filter based on the coefficient-pair histogram of DCT (Discrete Cosine Transform) value and LBP (Local Binary Pattern) pattern [7, 8]. The proposed detector extracts LBP value of image and conduct DCT translation on images. Then, a parameter T is chosen to find feature pairs and the coefficient-pair histogram is introduced. It combines LBP features in space domain with DCT value in frequency domain. The proposed detector takes frequency domain coefficients into account for the first time and outperforms some existing work in the detection of median filtering.

The rest of this paper is organized as follows. In Sect. 2, the coefficient-pair histogram is introduced. In Sect. 3, the proposed detector is compared with the state-of-the-art methods [2, 4] and the experiment results are reported. Finally, the conclusion is drawn in Sect. 4.

2 Detection of Median Filtering

2.1 LBP Feature

LBP (Local Binary Pattern) was introduced by Ojala to analysis the texture features of the image. It was firstly introduced in 1996 [7]. LBP is theoretically simple and widely used in human face identification [9]. The main algorithm of LBP is to extract LBP feature by comparing 3×3-neighborhood pixel to a center pixel. A pixel in the center generally has circularly symmetric P neighbors within a certain radii R.

If the gray value of the local neighborhood $(g_1 \cdots g_p)$ is larger than the gray value of the center pixel (g_c), it is labeled as 1 and otherwise it is labeled as 0. Thus a binary number is formed and it has 2^P different binary patterns. The algorithm is given by formula (1)

$$LBP_{P,R} = \sum_{P=0}^{P-1} s(g_P - g_c)2^P, \; s(x) = \begin{cases} 1, x \geq 0 \\ 0, x < 0 \end{cases}. \tag{1}$$

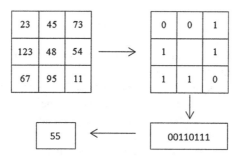

Fig. 1. Computing process for LBP operator

Based on the 3×3-neighborhood of P = 8 with R = 1, the LBP operator has $2^8 = 256$ types of different binary patterns. And the computational process is shown in Fig. 1.

Although LBP operator is scale invariant in certain cases but it is not rotation invariant. With rotation invariant, the derivation of custom LBP operator is proposed and the effect of rotation of images can be removed [10].

Another extension of original operator is uniform pattern LBP. The uniform pattern LBP is proposed to decrease the dimensionality of the feature vector, because some LBP patterns occur more frequently than others. The uniform pattern LBP classifies some binary patterns as the same pattern.

The uniform LBP has been limited at most two bitwise transitions from 0 to 1 such as 11011111. A uniformity measure U was introduced to define uniform patterns:

$$U\left(LBP_{P,R}\right) = \sum_{i=1}^{P}|s(g_i - g_c) - s(g_{i-1} - g_c)|. \tag{2}$$

The uniform LBP pattern $LBP_{P,R}^{riu2}$ is given by:

$$LBP_{P,R}^{riu2} = \begin{cases} \sum_{P=0}^{P-1} s(g_p - g_c), \mathrm{U}(\mathrm{LBP}_{P,R} \leq 2) \\ P + 1, otherwise \end{cases}. \tag{3}$$

The basic idea of median filtering is to replace each pixel with the median of neighboring pixel and affects the correlation of pixel values. In order to find the texture change of median-filtered images, the LBP features are taken into account. LBP pattern is a powerful feature for texture classification and express the correlation of pixel in 3×3-neighborhood. LBP can be employed in finding the traces of median filtering, since median-filtering images obtain less content and introduce the correlation among the pixels.

2.2 The Coefficient-Pair Histogram

The coefficient-pair histogram comes from the pixel-pair histogram. For an $m \times n$ grayscale image, every value in (i, j) means the frequency of pixel values converting i into j in the histogram of pixel pairs. Therefore:

$$P(i,j) = \sum A(\omega). \tag{4}$$

In order to generate the pixel-pair histogram, the grayscale image is converted to a 1-D vector with column ordering, row ordering or zigzag ordering. The pixel-pair histogram is an image of size 256×256 (for grayscale image) where the intensity of each location (i,j) represents the number of times that the pixel pairs with the intensities i and j occurs in the generated vector. The computational process of pixels-pair histogram is shown in Fig. 2.

The pixel-pairs histogram is often used for steganalysis and application. Recently, people are trying to use it for identifying contrast tamper. For example, Mahmoo Shabanifard computes the pixel-pair histogram in the space domain for integrity verification or forgery detection of image [11]. The absolute value of the first 36 Zernike moments of the pixel-pair histogram is calculated to identify contrast enhancement.

In Fig. 2, the 3×3 grayscale image is converted to a 1D-vector using column ordering. The pixel-pair histogram is a matrix of the size 256×256 (for grey-level images) where each location represent the intensity, (4) is used to compute the pixel-pairs histogram.

In this paper, the coefficient-pair histogram is proposed to replace pixel points with coefficients. In the research of median filtering detection, directly using the LBP pattern features has a bad performance for JPEG compression (detail of discussion shown in Sect. 3.4). Therefore, the coefficient-pair histogram of LBP pattern is computed, and it takes the correlation of LBP pattern into account. The coefficient-pair histogram is employed to identify the contrast enhancement fingerprint, since median filtering can change the contrast of images. In order to make the method of detection for median filtering robust against compression, the coefficient-pair histogram of LBP pattern has a perfect performance.

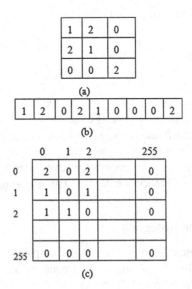

Fig. 2. (a) Image (b) 1D-vector by column (c) Pixel-pair histogram.

2.3   The Coefficient-Pair Histogram of DCT Value and LBP Pattern

JPEG (Joint Photographic Experts Group) compression provides good compression performance for raw images. But the existing median-filtering detector is not robust to JPEG compression. By researching the computational process of JPEG compression, it is obvious that JPEG compression is based on the discrete cosine transform (DCT).

DCT expresses a finite sequence of image points in terms of a sum of cosine functions oscillating at different frequencies. For a 2D-discrete image, DCT coefficients of image are:

$$C(u, v) = a(u)a(v) \sum_{x=0}^{M-1} \sum_{y=0}^{N-1} f(x, y) \cos[\frac{(2x + 1)u\pi}{2M}] \cos[\frac{(2y + 1)v\pi}{2N}]. \tag{5}$$

Among this,

$$a(u) = \begin{cases} \sqrt{1/M} & u = 0 \\ \sqrt{2/M} & u = 1, 2, \cdots M - 1 \end{cases}. \tag{6}$$

$$a(v) = \begin{cases} \sqrt{1/N} & v = 0 \\ \sqrt{N} & v = 1, 2, \cdots N - 1 \end{cases}. \tag{7}$$

The main steps of JPEG compression are computing 2-D DCT of 8×8-block and the results are quantized and entropy coded. This is the main step of DCT transformation. Since DCT value can express the change of JPEG compressed images in frequency domain, the DCT transform coefficients are taken into account for forensic analysis of median filtering.

The experiments have indicated that LBP pattern features have a bad performance in the detection of median filtering in compressed images. In this paper, the coefficient-pair histogram of LBP pattern is computed to express the correlation of LBP pattern. To improve the robustness for JPEG compression, the coefficient-pair histogram of DCT value is combined with the coefficient-pair histogram of LBP pattern which called HDL. The combined coefficient-pair histogram can express the structure features in the space domain and frequency domain. The main steps of computation process of HDL are shown as follows:

(1) In the first step, the uniform pattern LBP (P = 8, R = 1) is selected for median-filtering detection to avoid the high-dimension. For an m × n grayscale image, the uniform pattern LBP is computed. And it results in an m × n matrix in the ranges of $\{0 \sim 59\}$. Then, LBP matrix is divided by a nonnegative integer T and rounded up. So the LBP values range in $\{0 \sim \lceil 59/T \rceil\}$. At last, LBP matrix is reordering to a row vector (V_{LBP}) by column ordering.

(2) 2-D DCT of 8×8-block is computed and forms a DCT coefficient matrix. The DCT coefficients can be positive or negative. In order to ignore the influence of sign, DCT coefficients are changed into absolute values. Same as above, DCT

coefficients matrix is rounded up and reordering to a row vector (V_{DCT}) by column ordering.

At last two row vectors are formed (V_{LBP}, V_{DCT}). The parameter T can be various parameters. In the pixel-pair histogram, parameter T is in the range of $\{0 \sim 255\}$. The parameter T determines the dimension of the coefficient-pair histogram. Different from the pixel-pair histogram, LBP vector ranges between $\{0 \sim \lceil 59/T \rceil\}$. The dimension of V_{LBP} is smaller than the pixel-pair histogram of greyscale images. The histogram of LBP vector pairs (h^{LBP}) and DCT coefficient pairs (h^{DCT}) are computed in the range of $\{0 \sim \lceil 59/T \rceil\}$. In this way, the dimension of h^{LBP} is $(\lceil 59/T \rceil + 1)^2$ and so as of h^{DCT}. Every histogram is reshaped to be a row feature vector. And a direct combination of the two feature vectors results in a $2 \times (\lceil 59/T \rceil + 1)^2$-dimensional feature vector:

$$h^{HDL} = (h^{DCT}, h^{LBP}). \tag{8}$$

It is obvious that the dimension of coefficient-pair histogram is much smaller than the pixel-pair histogram. The new improvement of pixel-pair histogram greatly reduces the complexity of operator. In order to take the influence of JPEG post-compression into account, the space domain features (LBP pattern values) are combined with frequency domain features (DCT values) for the first time.

3 Experiment Results

3.1 Experiment Setup

To indicate the performance of our proposed median filtering detector and to compare the performance of the existing median filtering detector, the image database needs to be prepared before conducting the experiment.

The images used in our experiment are taken from UCID [12] image database. UCID is often used for image forensics and steganalysis. Therefore, it is hard to identify the tampering of median filtering. The database has 1338 types of RGB images of size 512×384 or 384×512. 1000 images are randomly selected from database and converted to 8-bit grayscale images in the first step. Then, the following datasets are prepared:

- Since the different size of images differs from each other, 256×256, 128×128, 64×64 center portions are cropped from each image of UCID database, to obtain the dataset S^{ORI}.
- 3×3 and 5×5 median filtering are performed on database S^{ORI} to obtain the median-filtered dataset $\{S^{MF3}, S^{MF5}\}$.
- JPEG post-compression is introduced in all the image dataset $\{S^{ORI}, S^{MF3}, S^{MF5}\}$, with the JPEG compression quality factor (QF) in the range $\{70, 80, 90\}$.

Before testing, two types of training-testing pairs are formed based on the UCID, including $\{S^{ORI}, S^{MF3}\}, \{S^{ORI}, S^{MF5}\}$. The size of the training set is set to be equal to 75 % of the corresponding database size (750 images). And the rest of the database (250 images) constitutes the testing set.

3.2   Classification

LIBSVM is proposed by Lin Chih-Jen from Taiwan University. LIBSVM has 5 kinds of kernels and the experiment is conducted with RBF kernel:

$$K(x_i, x_j) = \exp\left(\frac{-x_i - x_j^2}{2\sigma^2}\right). \tag{9}$$

as the classifier in our experiment. For each training-testing pair, 50 % of the median-filtered images and 50 % of the non-median-filtered images are randomly selected. The rate of the positive samples and negative samples is 1:1. Then, 3/4 of samples are randomly selected for training and the rest for testing. And finding the best parameter C and g in the parameter grid:

$$(C, g) \in \left\{ (2^i, 2^j) \big| i \in \{0, 0.5, \ldots, 6\}, j \in \{-5, -4.5, \ldots, 0\} \right\}. \tag{10}$$

by a five-fold cross-validation on the training set. The classifier model is achieved on the training set after the best parameter C and g are achieved. Then, the classification is performed on the testing set. The performance of our proposed method is compared by the area under ROC curve (AUC) and the minimal average decision error under the assumption of equal priors and equal costs [2]:

$$P_e = \min \frac{P_{FP} + (1 - P_{TP})}{2}. \tag{11}$$

where P_{FP} and P_{TP} denote the false positive and true positive rates, respectively.

3.3   Selection of T

The dimension of h^{DCT} is decided by T. The dimension will influence the algorithm complexity. Image recognition rate and algorithm complexity is important for median-filtering detection techniques. To further reduce the dimensionality of our detector, the following experiments firstly choose the appropriate T and then conduct a series of contrast experiment.

To observe how the parameter affects the performance of detection, the T varies in the range of {3, 4, 5} in the experiments. Table 1 shows the results in terms of the

Table 1. Minimal average decision error P_e of median filtering detectors based on HDL for filter size $\{3 \times 3, 5 \times 5\}$. All results are conducted on the images of size 256×256. The parameter $T \in \{3, 4, 5\}$.

	JPEG 90			JPEG 80			JPEG 70		
	$T = 3$	$T = 4$	$T = 5$	$T = 3$	$T = 4$	$T = 5$	$T = 3$	$T = 4$	$T = 5$
MF3	0	0.004	0.016	0.032	0.018	0.07	0.024	0.038	0.054
MF5	0	0.004	0.01	0.004	0.016	0.024	0.008	0.026	0.028

minimal average decision error P_e for varying JPEG quality factor in the range of {70, 80, 90}.

The minimal average decision error drops rapidly with $T = 5$. While the results for $T = 3$ and $T = 4$ are compared, the results for $T = 3$ have a better performance. The parameter T determines the scale of LBP pattern value. With $T = 3$, the results are perfect and P_e drops to zero with JPEG quality QF = 90 for filter size 3×3 and 5×5. The operator results in a $2 \times (\lceil 59/T \rceil + 1)^2$-dimension feature vector. When parameter T is chosen to be 3, the dimension of feature vectors is 882 and it is acceptable for complexity.

3.4   Detection of Median Filtering Using LBP Values

Combining the coefficient pair histogram of LBP values and the coefficient pair histogram of DCT values is of great value. Compare directly using LBP value with our propose methods can highlight the performance. The experiment is conducted in images set with size of ($64 \times 64, 128 \times 128$ and 256×256). The images set with median filter size. And JPEG compression QF ranges in {70, 80, 90}.

The results of detecting 3×3 median filtering are shown in Table 2. The minimal average decision error of LBP is larger than HDL in any case. Directly using LBP values has a poor performance on the JPEG compressed images.

3.5   Detection of Median Filtering in Uncompressed Images

In order to investigate the performance of HDL, two methods (SPAM and MFF) are used to compare with the HDL in the same dataset. SPAM is an earlier method for median-filtering detector. But when the size of images is reduced, the detector rate will decrease quickly. For the SPAM detector, the parameter T is set to 3, which leads to the 686-D SPAM features. For the uncompressed images extracted from UCID, The classification results of the three detectors are shown in Fig. 3.

The three methods have a good performance for filter size 3×3 and 5×5 in uncompressed images. The AUC of HDL and MFF are up to 1. Because the three detectors achieve perfect detection accuracy, the post-compression images must be taken into account to conduct contrast effect.

Table 2. Minimal average decision error P_e of median filtering detectors based on LBP and HDL. All results are conducted on the different size ($64 \times 64, 128 \times 128$ and 256×256) and different QF {70, 80, 90}.

	64×64			128×128			256×256		
	70	80	90	70	80	90	70	80	90
LBP	0.208	0.22	0.192	0.088	0.16	0.172	0.108	0.168	0.144
HDL	0.081	0.052	0.034	0.05	0.034	0.01	0.024	0.032	0

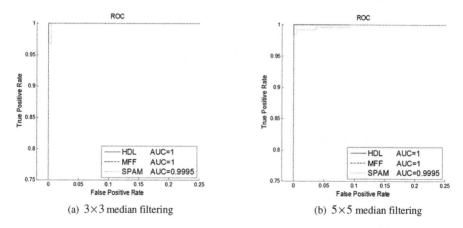

(a) 3×3 median filtering   (b) 5×5 median filtering

Fig. 3. ROC curves of detecting different filter size on uncompressed images

3.6   Detection of Median Filtering in JPEG Post-compressed Images

JPEG post-compressed images are widely used in the real world. But JPEG post-compressed will affect the detection accuracy of median filtering. In our experiment, the quality factor (QF) of JPEG post-compressed is decreased to indicate the robustness of HDL. To highlight the performance of our proposed method, the experiment is conducted on image dataset of small size (128×128) with quality factor in the range $\{70, 80, 90\}$.

The Fig. 4 shows that the classification performance of the three method decreases as QF decreases. Because higher QF will bring less distortion to the image, it is difficult to hide median-filtering traces. The HDL outperforms the other two methods in any case. Our proposed method has a reliable detection of median filtering even for QF as low as 70 (for 3×3 median, $P_e = 5.6\%$; for 5×5 median, $P_e = 1.8\%$).

For the three methods, the performance of 3×3 median filtering is better than 5×5 median filtering especially for SPAM. But the classification results of HDL for different size of filter are very close (with QF = 80, for 3×3 median, $P_e = 3.4\%$; for 5×5 median, $P_e = 1.4\%$).

The classification performance of the three methods are compared at lower quality factor (with QF = 70, for HDL, $P_e = 5.6\%$; for MFF, $P_e = 14\%$; for SPAM, $P_e = 34.8\%$) in the detection of 3×3 median filtering and the same as in detection of 5×5 median filtering. The results indicate the HDL is robust for JPEG compression.

3.7   Detection of Median Filtering in Different-Size Images

Kirchner and Fridrich indicated that low-size images will be hard to detect median-filtering traces [2]. Therefore, the scale of images must be taken into account. In our experiment, the experiment is conducted on the three sizes (64×64, 128×128 and 256×256) of images. Previous research indicated that the

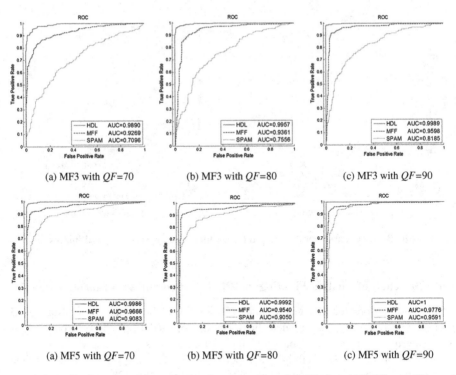

(a) MF3 with QF=70　　　　(b) MF3 with QF=80　　　　(c) MF3 with QF=90

(a) MF5 with QF=70　　　　(b) MF5 with QF=80　　　　(c) MF5 with QF=90

Fig. 4. ROC curves of detecting 3×3 and 5×5 median filtering with different QF

performance of 3×3 median filtering is better than 5×5 median filtering. In order to simplify the experimental procedure, the three methods are only compared for filter size 3×3.

The conclusion is shown in Fig. 5. It is obvious that the performance of HDL is much better than the other two methods in different-size images. The three methods all have a perfect performance for the big size of images (256×256).

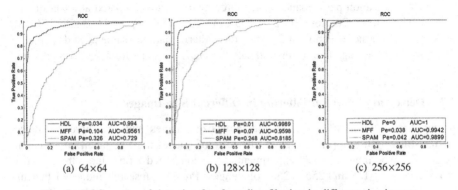

(a) 64×64　　　　　　(b) 128×128　　　　　　(c) 256×256

Fig. 5. ROC curves of detecting 3×3 median filtering in different-size images

For size of images as low as 64×64, the results is compared (for HDL, $P_e = 3.4\%$; for MFF, $P_e = 10.4\%$; for SPAM, $P_e = 32.6\%$). The results show that HDL is robust for small-size images.

4 Conclusion

In this paper, a novel detection of median-filtered based image operation is made. The proposed method extract image features of the histogram of the LBP pattern and DCT domain coefficient pairs (HDL), then these two features are merged into one, as the features includes spatial domain (LBP features) and frequency-domain features, so the algorithm has better robustness for detection of median-filtered image. Large amounts of experiment results have indicated the efficiency of HDL.

In the future, some works will be made to test the usability of the scheme for average filter and Gaussian filter. This is a very attractive subject for future research.

References

1. Cao, G., Zhao, Y., Ni, R.R., Yu, L.F., Tian, H.W.: Forensic detection of median filtering in digital images. In: 2010 IEEE International Conference on Multimedia and Expo (ICME), pp. 89–94 (2010)
2. Kirchner, M., Fridrich, J.: On detection of median filtering in digital images. In: Proceedings SPIE, Electronic Imaging, Media Forensics and Security II, vol. 7541, pp. 1–12 (2010)
3. Cao, G., Zhao, Y., Ni, R., Yu, L., Tian, H.: Forensic detection of median filtering in digital images. In: IEEE International Conference on Multimedia and Expo, pp. 89–94 (2010)
4. Yuan, H.D.: Blind forensics of median filtering in digital images. IEEE Trans. Inf. Forensics Secur. 6(4), 1335–1345 (2011)
5. Kang, X.G., Stamm, M.C., Peng, A.J., Liu, K.J.R.: Robust median filtering forensics based on the autoregressive model of median filtered residual. In: Signal & Information Processing Association Annual Summit and Conference (APSIPA ASC), 2012 Asia-Pacific, pp. 1, 9, 3–6 December 2012
6. Zhang, Y.J., Li, S.H., Wang, S.L., Shi, Y.Q.: Revealing the traces of median filtering using high-order local ternary patterns. IEEE Signal Process. Lett. 21(3), 275–279 (2014)
7. Ojala, T., PietikaÈinen, M., Harwood, D.: A comparative study of texture measures with classification based on feature distributions. Pattern Recogn. 29(1), 51–59 (1996)
8. Ahmed, N., Natarajan, T., Rao, K.R.: Discrete cosine transform. IEEE Trans. Comput. C-23 (1), 90–93 (1974)
9. Ren, J.F., Jiang, X.D., Yuan, J.S., Wang, G.: Optimizing LBP structure for visual recognition using binary quadratic programming. IEEE Signal Process. Lett. 21(11), 1346–1350 (2014)
10. Ojala, T., Pietikainen, M., Maenpaa, T.: Multiresolution gray-scale and rotation invariant texture classification with local binary patterns. IEEE Trans. Pattern Anal. Mach. Intell. 24 (7), 971–987 (2002)

11. Mahmood, S., Mahrokh, G.S., Mohammad, A.A.: Forensic detection of image manipulation using the Zernike moments and pixel-pair histogram. Image Process. IET **7**(9), 817–828 (2013)
12. Schaefer, G., Stich, M.: An uncompressed color image database. Storage Appl. Multimedia **5307**, 472–480 (2003)

An Image Enhancement Method Based on Edge Preserving Random Walk Filter

Zhaobin Wang$^{(\boxtimes)}$, Hao Wang, Xiaoguang Sun, and Xu Zheng

School of Information Science and Engineering,
Lanzhou University, Lanzhou 730000, China
{wangzhb,wangh2013}@lzu.edu.cn

Abstract. Some previous edge preserving smoothing methods suffer from halo artifacts when they are applied for image enhancement. In this paper, an edge preserving random walk filter is proposed, our method suffers free from artifacts. Unlike previous methods, the proposed method is able to obtain a smoothing result by just solving a system of linear equation. The proposed filter is then adopted to design an image enhancement algorithm. By just amplifying and adding the detail layer to the base layer, the algorithm can produce a satisfactory result. The simulation results demonstrate that our approach performs much better than other existing techniques.

Keywords: Random walk · Image enhancement · Edge preserving · Image decomposition

1 Introduction

Image enhancement can be achieved by a two-step process, decomposing an image into base layer and detail layer, and combining base layer and detail exaggerated layer. Edge-preserving smoothing method plays an important role in the first step, it is a popular technology in image decomposition. Edge preserving smoothing may be viewed as a compromise between smoothing and edge preserving. It wipes out tiny details and noise, and at the same time, preserves salient important edges. During the development of past decades years, it is applied in the field of image processing such as edge detection, image restoration, image enhancement, and many other high-level image processing tasks. As an integral step of many computer vision problem, the results of image smoothing influence the performance of the whole vision system.

Anisotropic diffusion model introduced by Perona and Malik is widely used in practice and extensively studied in theory [1], it is modeled using partial differential equations and implemented as an iterative process, it uses an edge-stopping function of local gradient to make smoothing take place only in the interior of regions without crossing edges. It has been demonstrated to be able to achieve a good trade-off between noise removal and edge preservation. There has been a large number of work aimed at optimizing and extending the idea [2, 3]. Rudin et al. propose to regularize total variation, which utilizes the gradient sparsity enforced by an L1 penalty term to do edge-preserving smoothing [4]. It has been successful in denoising problems. It is the inspiration source of many work. By now, the algorithm has also been used for other

© Springer International Publishing Switzerland 2015
D.-S. Huang et al. (Eds.): ICIC 2015, Part I, LNCS 9225, pp. 433–442, 2015.
DOI: 10.1007/978-3-319-22180-9_42

restoration tasks such as deblurring, blind deconvolution and inpainting [5]. The bilateral filter computes a smoothed output with a weighted average of neighboring pixels intensities, it takes into account spatial and color distances [6]. It is an extension of typical Gaussian smoothing. There exists a number of methods to boost the algorithm [6, 7]. Because of its simplicity and effectiveness, it is one of the most popular smoothing methods. In recent years, many exciting new techniques have been invented to solve this problem. All the methods have demonstrated satisfying results with good performance. Farbman et al. propose to perform the edge-preserving smoothing using the weighted least square (WLS) framework [8, 9], edge preserving smoothing is viewed as a compromise between data term and regularization term. By minimizing the proposed energy functional, image smoothing result can be obtained by solving a large linear system. Xu et al. present L0 gradient minimization for image smoothing, it minimizes a specific objective function [10]. It can remove low-amplitude structures and globally preserve and enhance salient edges. Subr et al. consider edge preserving smoothing as interpolation between local signal extremes [11]. The method defines detail as oscillations between local minima and maxima. It smoothes high contrast texture while preserving salient edges. Guided image filter assume there is a local linear model between the guidance image and the filtering output, and they seek a solution that minimizes the difference between the filtering input and output while maintaining the linear model [12].

In this paper, an edge preserving filter for image enhancement is proposed, unlike previous approaches, our proposed method has several advantages: 1. Smoothing scale can be easily controlled by changing a parameter; 2. Our method has a clear physical meaning; 3. Our method is based on random walk algorithm, the result can be obtained by solving a system of linear equation; 4. When applied to image enhancement, our method suffers free from halo artifacts.

The rest of the paper is organized as follows. Section 2 introduces traditional random walks algorithm; Sect. 3 provides implementation details of our method and theoretical connections with previous random walks algorithm and anisotropic diffusion algorithm; Sect. 4 discusses experimental results and performance, along with comparisons with other smoothing methods. Finally, Sect. 5 gives the conclusions.

2   Traditional Random Walk Algorithm

The main idea of the random walks algorithm is simple [13], given a set of n data points. Based on the data points, construct a connected graph G (V, E, W) with n vertices $V\{v_1, v_2, \ldots, v_n\}$ and l edges $E\{e_1, e_2, \ldots, e_l\}$. An edge connecting two neighboring vertices, v_i and v_j, is denoted by e_{ij}. The weight of an edge e_{ij} is denoted by w_{ij},

$$w_{ij} = e^{-\beta(I_i - I_j)^2} \tag{1}$$

I_i and I_j are intensity at position i and j. Degree d_i of vertex i is defined as the sum of the weights of the edges, that connect i with its neighboring pixel node.

Define the combinatorial Laplacian matrix as L, L is a symmetric positive semi-definite sparse matrix of size $n \times n$.

$$L = \begin{cases} d_i & \text{if } i = j \\ -w_{ij} & \text{if } v_i, v_j \text{ are connected by an edge} \\ 0 & \text{else} \end{cases} \qquad (2)$$

A combinatorial formulation of the Dirichlet integral is E, which is the sum of the weights on the boundary of the segmentation. E can be denoted as follows:

$$E = \frac{1}{2} \sum_{e_{ij} \in E} w_{ij}(f_i - f_j)^2 = \frac{1}{2} f^T L f \qquad (3)$$

To minimize the Dirichlet integral, the random walks algorithm requires some pre-labeled data points, suppose there are v pre-labeled points of k classes, and u unlabeled points remains to be labeled.

So the data points can be divided into labeled points V_L and unlabeled points V_U, the Dirichlet integral can be represented as:

$$\begin{aligned} E &= \frac{1}{2} [f_L^T f_U^T] \begin{bmatrix} L_L & B \\ B^T & L_U \end{bmatrix} \begin{bmatrix} f_L \\ f_U \end{bmatrix} \\ &= \frac{1}{2} (f_L^T L_L f_L + 2 f_U^T B^T f_L + f_U^T L_U f_U) \end{aligned} \qquad (4)$$

The minimization of above integral with respect to f_U is given by the system:

$$L_U f_U = -B^T f_L, \qquad (5)$$

where L_U is a matrix of size $u \times u$, which can be seen as closeness between u unlabeled points to be classified. B^T is a matrix of size $u \times v$, which can be seen as closeness between v labeled points and u unlabeled points. x_L is a matrix of size $v \times k$, which is pre-labeled information, it can be seen as relationship between v labeled points and k labels.

After solving above equations, f_U can be obtained, which is a matrix of size $u \times k$, indicating closeness between u unlabeled points and k labels. So u unlabeled points each have a k-tuple of closeness number indicating their closeness with k labels.

The combinatorial Dirichlet problem has the same solution as the random walker probabilities, so f_U can also be explained as the probabilities of random walks starting from u unlabeled points first reach each labeled points. The probabilities indicate closeness between labeled points and unlabeled points.

Random walk algorithm can be formulates on a weighted graph model as Fig. 1(a) shows. Grady used the idea for image segmentation [13, 14]. Each node or vertex represents a pixel and each edge spanning two vertices has a weight indicating the closeness of the two vertices. Given several marked pixels, let random walker starts at

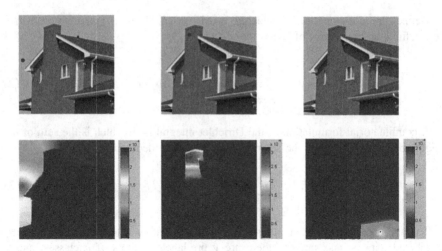

Fig. 1. Smoothing kernel are computed for the labeled pixel in the house image $\mu = 1E\text{-}4$, $\beta = 100$.

other unmarked pixels, by analyzing the probability random walkers will first arrive at each marked pixel, the algorithm is able to give a satisfactory segmentation result.

3 Proposed Method for Image Enhancement

3.1 Random Walk for Image Smoothing

In this section, we first describe a novel edge-preserving smoothing filter based on the random walk framework and then show why it works. Random walks algorithm for image segmentation minimize Eq. (3), to adapt the algorithm to image smoothing, we minimize the following energy functional,

$$E_{RW} = \frac{1}{2} \sum_{e_{ij} \in E} w_{ij}(f_i - f_j)^2 + \frac{\mu}{2} \sum_{i=1}^{n} d_i(f_i - I_i)^2, \tag{6}$$

The first term comes from traditional random walk algorithm, it stives to achieve smoothness of the output image f, while the second term is the data term, the goal of the term is to minimize the distance between f and I, the requirement is enforced via the degree of each pixel. Finally, μ is tradeoff between the two terms.

Using matrix notation we may rewrite Eq. (6) as:

$$\begin{aligned} E_{RW} &= \frac{1}{2}f^T L f + \frac{\mu}{2}(f - I)^T D(f - I) \\ &= \frac{1}{2}f^T L f + \frac{\mu}{2}f^T D f - \mu f^T D I + const \end{aligned} \tag{7}$$

Where *const* means a constant number which is independent of f, f is the smoothed pixel intensity. In fact, f is the only critical point that minimizes the E_{RW}, so it can be obtained by differentiating E_{RW} to f. So the final question goes to solve the equation:

$$((1 + \mu)D - W)f = \mu DI, \tag{8}$$

The Eq. (8) can be solved using matrix inversion,

$$f = \mu \times [(1 + \mu)E - D^{-1}W]^{-1} I. \tag{9}$$

So the method can be thought of as applying a filter $\left(\mu \times [(1 + \mu)E - D^{-1}W]^{-1} \right)$ on the input image I. Each row of the operator matrix may be thought as a kernel that determines the weight value of the corresponding pixel, output smoothing result can be thought as a weighted combination of other pixels in the input image. Figure 2 shows filter kernels of some position.

However, solving Eq. (9) costs $O(n^3)$ operations, where n is the number of pixels. And applying the inverse matrix require $O(n^2)$ operations [8]. Solving the sparse linear equation system Eq. (8) is more computational efficient. Above equations is a system of linear equations with n unknowns. The resulting linear system is a form of $Ax = b$, and the above equation is nonsingular [17], it is easy to be solved using modern numerical linear algebra. Many good methods exist on the solution to such large, sparse, symmetric systems. This particular type of linear system of equations has been well studied and can be solved quickly by many methods such as the conjugate gradient and the algebraic multigrid method [18, 19]. An appropriate preconditioner or a multi-grid

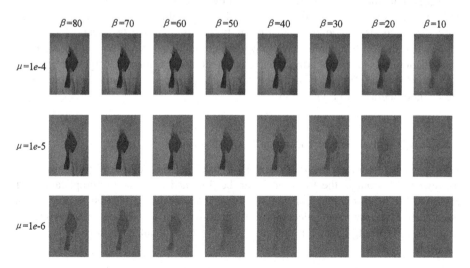

Fig. 2. Effect of varying parameters. μ controls the extent of smoothing; β controls edge preserving property.

(a) Input image   (b) μ=0.5   (c) μ=0.1   (d) μ=0.01   (e) μ=0.001   (f) μ=0.0001

Fig. 3. Gaussian filter effect. Set $\beta = 0.1$, the left free parameter μ in weighting function controls smoothing scale.

solver requires only $O(n)$ operations. Using MATLAB's direct solver, solution of Eq. (12) for a 512 × 512 image and 256 × 256 image requires only 0.97 and 0.18 s respectively on an Intel i5 3.2 GHZ CPU and 4 GHZ of memory.

In fact, solving a system of Eq. (8) using Jacobi iteration [20] is given by:

$$I(k + 1) = \frac{1}{1 + \mu} D^{-1} W I(k) + \frac{\mu}{1 + \mu} I(0). \tag{10}$$

Where $I(0)$ is the initial state of iteration process (i.e. the input image). It can be thought of as anisotropic diffusion. However, compared with traditional anisotropic diffusion [1], our algorithm converges to a non-trivial solution, rather than to a constant image.

In this paper, we use a typical Gaussian function for edge weight calculation. However it is more appropriate to modify the function to texture information, filter coefficients of other image features when applying our methods to other specific problems. For image smoothing. Figure 2 shows the effect of varying parameters. μ controls the extent of smoothing; When μ decreases, the smoothing scale gradually improved. β controls edge preserving property. Figure 3 shows the situation when we set $\beta = 0.1$, the left free parameter μ in weighting function controls smoothing scale. In this situation, β is small, all edge weights calculate using Gaussian weighting function are basically equal. Proposed method has no edge preserving property, in which case the method degrades to a Gaussian filter.

3.2   Image Enhancement by Image Decomposition

A simple and commonly used framework is used in our method, Fig. 4 shows the framework. Of course, the framework can be designed to be more complicated and effective. The proposed filter can smooth image at any scale, so it is easy to construct a multi-scale edge preserving decomposition. However in the following experiments, detail layer is multiplied by a constant 3 to create a detail enhancement layer for simplicity.

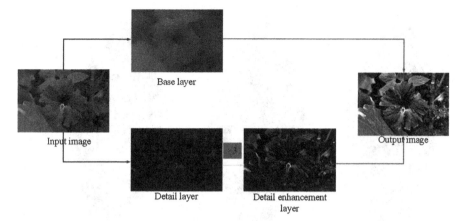

Fig. 4. Image enhancement framework. First, base layer is calculated using above random walk based filter; second, detail layer is multiplied by a constant; third, detail enhancement layer is added to the base layer and enhancement result is obtained.

4 Experimental Results

As edge preserving image smoothing is one of the most fundamental work in image processing area, it finds lots of interesting applications in field of computer vision. For some problems, such as image enhancement, edge detection, image denoising and HDR tone mapping, our problem helps to make a satisfactory result. We show in this section how flattening, enhancement can be effectively addressed using the proposed method.

The proposed filter may be used in place of the edge preserving filter based applications. Figure 5 gives an example, (a) is the original input image, (b) is the smoothed image with our proposed edge preserving filter, Note that foreground (leaf) and background (white board) are smoothed in their own regions respectively, the proposed filter can wipe out tiny details and noise, and at the same time, preserves salient important edges. (c) shows the result of image enhancement with above simple framework.

Some image decomposition based image enhancement algorithm suffers from halos and artifacts effect, along many of the edges, their result exhibits gradient reversals, while our method does not suffer from the drawback, the edges in our result are much better. In order to evaluate the performance of the proposed method, we conduct experiments on a commonly used flower image. We compare our method with other popular methods including: L0 regularization [10] and local extrema [11]. Figure 6 shows the magnification results of L0, local extreme and our method. (a) is the widely used flower image for image enhancement, (b)–(d) show similar results of the three methods. For the two existing methods, we use the source code or executable codes provided by the authors or other researchers. Figure 7 gives a comparison with a close examination. As shown in (b)–(c), results obtained from L0 regularization and local

(a) input image   (b) smoothed image   (c) image enhancement result

Fig. 5. Base-detail separation and enhancement. (a) shows the input image, (b) shows the base layer, the smoothed image is computed using our proposed random walk filter, (c) shows the image enhancement result.

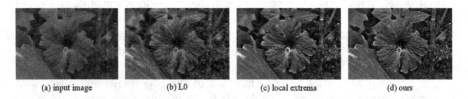

(a) input image   (b) L0   (c) local extrema   (d) ours

Fig. 6. Detail exaggeration using our tool. The detail enhanced results (3 ×). Parameters: L0 ($\lambda = 3E$-2), local extrema (17), and ours ($\mu = 0.005, \beta = 9$).

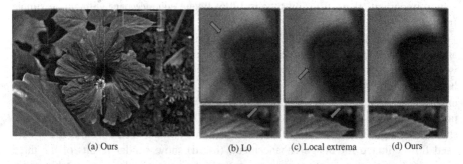

(a) Ours   (b) L0   (c) Local extrema   (d) Ours

Fig. 7. Detail enhancement. L0 and local extrema based method results in halos (b) and artifacts (c) along the high-contrast edges, while our method (d) suffers free from these problems.

extreme may suffer from artifacts along some of the edges, while our method suffers free from halos and artifacts effects.

5   Conclusion

In this paper, a random walk based filter is proposed for image smoothing. The filter has a clear physical meaning, it can be thought of as a random walk process on a specified graph or anisotropic diffusion. The proposed filter is able to get an edge preserving smoothed image by solving a system of linear equation. When applied to image enhancement, our method suffers free from halo and artifacts effects, the simulation results demonstrate that our approach performs much better than other existing techniques in terms of visual quality.

In future work we would like to investigate more sophisticated schemes for image enhancement and apply the random walk filter for more applications.

Acknowledgments. The authors would like to thank L. Grady for providing the source code of the original random walks algorithm. The authors would also like to thank the reviewers for their valuable comments. This work was jointly supported by National Science Foundation of China (Grant No. 61201421), China Postdoctoral Science Foundation (Grant No. 2013M532097), Fundamental Research Funds for the Central Universities (lzujbky-2014-52 & lzujbky-2015-197), and Science Foundation of Gansu Province of China (Grant No. 1208RJYA058).

References

1. Perona, P., Malik, J.: Scale-space and edge detection using anisotropic diffusion. IEEE Trans. Pattern Anal. Mach. Intell. **12**(7), 629–639 (1990)
2. Black, M.J., Sapiro, G., Marimont, D.H., Heeger, D.: Robust anisotropic diffusion. IEEE Trans. Image Process. **7**(3), 421–432 (1998)
3. Lopez-Molina, C., Galar, M., Bustince, H., De Baets, B.: On the impact of anisotropic diffusion on edge detection. Pattern Recogn. **47**(1), 270–281 (2014)
4. Rudin, L.I., Osher, S., Fatemi, E.: Nonlinear total variation based noise removal algorithms. Physica D **60**, 259–268 (1992)
5. Chan, T., Esedoglu, S., Park, F., et al.: Recent developments in total variation image restoration. Math. Models Comput Vis. **24**(8), 19–22 (2005)
6. Elad, M.: On the origin of the bilateral filter and ways to improve it. IEEE Trans. Image. Process. **11**(10), 1141–1151 (2002)
7. Sanun, S.: Bilateral filtering as a tool for image smoothing with edge preserving properties. In: 2014 IEEE International Conference on Electrical Engineering Congress (iEECON) (2014)
8. Farbman, Z., Fattal, R., Lischinski, D., Szeliski, R.: Edge-preserving decompositions for multi-scale tone and detail manipulation. ACM Trans. Graph. **27**(3), 1 (2008)
9. Min, D., Choi, S., Lu, J., Ham, B., Sohn, K., Do, M.: Fast global image smoothing based on weighted least squares. IEEE Trans. Image Process. **7149**(c), 1–15 (2014)
10. Xu, L., Lu, C., Xu, Y., Jia, J.: Image smoothing via L0 gradient minimization. In: Proceedings of 2011 SIGGRAPH Asia Conference- SA 2011, 30(60), 1 (2011)

11. Subr, K., Soler, C., Durand, F.: Edge-preserving multiscale image decomposition based on local extrema. ACM Trans. Graph. (TOG) **28**(5), 147 (2009)
12. He, K., Sun, J., Tang, X.: Guided image filtering. IEEE Trans. Pattern Anal. Mach. Intell. **35** (6), 1397–1409 (2013)
13. Grady, L.: Random walks for image segmentation. IEEE Trans. Pattern Anal. Mach. Intell. **28**, 1768–1783 (2006)
14. Grady, L., Funka-Lea, G.: Multi-label image segmentation for medical applications based on graph-theoretic electrical potentials. In: Sonka, M., Kakadiaris, I.A., Kybic, J. (eds.) CVAMIA/MMBIA 2004. LNCS, vol. 3117, pp. 230–245. Springer, Heidelberg (2004)
15. Dodziuk, J.: Difference equations, isoperimetric inequality and transience of certain random walks. Trans. Am. Math. Soc. **284**(2), 787 (1984)
16. Luxburg, U.: A tutorial on spectral clustering. Stat. Comput **17**(4), 395–416 (2007)
17. Biggs, N.: Algebraic potential theory on graphs. Bull. London Math. Soc. **29**, 641–682 (1997)
18. Hazra, S.B.: Introduction. In: Hazra, S.B. (ed.) Large-Scale PDE-Constrained Optimization in Applications. LNACM, vol. 49, pp. 1–4. Springer, Heidelberg (2010)
19. Falgout, R.D.: An introduction to algebraic multigrid computing. Comput. Sci Eng. **8**(6), 24–33 (2006)
20. Saad, Y.: Iterative methods for sparse linear systems. IEEE Comput. Sci. Eng. **3**, 88–90 (1996)

Latent Fingerprint Segmentation Based on Sparse Representation

Kaifeng Wei[(⊠)], Xiaoying Chen, and Manhua Liu

Department of Instrument Science and Engineering, School of EIEE,
Shanghai Jiao Tong University, Shanghai 200240, China
mhliu@sjtu.edu.cn

Abstract. Latent fingerprints are the finger skin impressions which are left at the scene of a crime by accident. They are usually of poor quality with weak fingerprint ridge flows and various overlapping irrelevant patterns. It is still a challenging problem for automatic latent fingerprint processing and recognition. Latent fingerprint segmentation, which segments the fingerprint ridge area from complex backgrounds, is an important preprocessing step for latent fingerprint recognition. This paper proposes a latent fingerprint segmentation algorithm based on sparse representation. First, the total variation (TV) model is used to decompose a latent image into two components: texture and cartoon. The texture component, which contains the weak fingerprint ridge and valley structures, is used for further processing, while the cartoon component mainly consisting of the irrelevant information is discarded as noises. Then, we compute the sparse representation of the texture image against the dictionary constructed by a set of Gabor elementary functions. Since the sparse coefficients measure the weights of the basis atoms in fingerprint representation, an image quality measure is computed from the sparse coefficients, which evaluate how well the texture image can be sparsely reconstructed from the basis atoms. Finally, this image quality measure is used for fingerprint segmentation. We test the proposed method on the NIST SD27 latent fingerprint database. Experimental results and comparisons demonstrate the effectiveness of the proposed method.

Keywords: Latent fingerprint segmentation · Sparse representation · Orientation consistency · Gabor dictionary

1 Introduction

Latent fingerprints are the finger skin impressions inadvertently touched or handled by a person typically at crime scenes [1]. Comparing with the rolled and plain fingerprint images, latent fingerprints are usually of low image quality, caused by unclear ridge structure, uneven image contrast, and various overlapping patterns such as lines, printed letters, handwritings or even other fingerprints, etc. [2]. Figure 1(a), (b) and (c) show three image samples of rolled, plain and latent fingerprints, respectively. Obviously, the rolled and plain fingerprints are both typically of good quality and are rich in information content [2]. Latent fingerprint images are usually of poor quality as shown in Fig. 2. For example, some latent fingerprint images are partly overlapped by letters, lines and other fingerprints (see Fig. 2a, b and c). In addition, the latent

© Springer International Publishing Switzerland 2015
D.-S. Huang et al. (Eds.): ICIC 2015, Part I, LNCS 9225, pp. 443–453, 2015.
DOI: 10.1007/978-3-319-22180-9_43

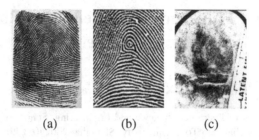

(a)   (b)   (c)

Fig. 1. Three types of fingerprint images: (a) rolled, (b) plain and (c) latent

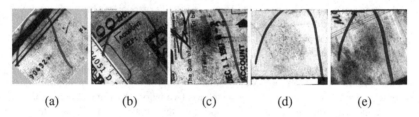

(a)   (b)   (c)   (d)   (e)

Fig. 2. Different kinds of latent fingerprint images: (a) overlapped by lines, (b) overlapped by letters, (c) overlapped by fingerprints, (d) blurry and (e) stain

fingerprints are usually blurred like Fig. 2(d), corrupted with noise like Fig. 2(e). These problems cause the difficulties in segmenting the fingerprint ridge area from the latent image for reliable feature extraction and recognition.

Fingerprint segmentation is one of the first and most important pre-processing steps for fingerprint identification which affects the results of fingerprint analysis and recognition [3]. The object of fingerprint segmentation is to extract the region of interest (ROI) which contains the desired fingerprint texture [4]. Nowadays, great progress has been made in feature extraction and matching but the performance mainly depends on the quality of the image. That is to say, the low match performance with latent images is due to the high number of unreliable extracted minutiae [5]. Due to the poor image quality, it is necessary to do segmentation on latent fingerprint images before feature extraction and feature matching. Segmentation on rolled and plain fingerprint images has been well-studied and significant efforts have been made in developing algorithms for these fingerprints [6]. Nevertheless, these methods cannot be directly used for latent fingerprint segmentation. The main difficulties are weak fingerprint ridge structure and the complex overlapping irrelevant patterns in latent images. Ratha et al. considered the orientation consistency of the local ridge in a fingerprint image as the key basis for segmentation [7]. Zhang et al. [6] proposed an adaptive directional total variation (ADTV) model for latent fingerprint segmentation. This model incorporates both the anisotropic directional TV term and the spatially-adaptive fidelity weight into the model formulation. Short et al. [5] proposed an algorithm that uses an adaptable ridge template to determine a goodness of fit score with a local region of the fingerprint. Choi et al. [8] utilizes both ridge orientation and frequency features to segment the fingerprint region from background. Karimi et al. [9] proposed a robust algorithm which does not rely on the information of local gradients and provides robust estimates to orientations and frequencies of fingerprints in a local region. Liu et al. [10] proposed an automatic segmentation of useful ridge-like fingerprint region based on orientation coherence. In this method, they compute orientation coherence with various gradients to measure the fingerprint image quality for segmentation.

Different from the segmentation method based on orientation coherence in [10], this paper proposes a latent fingerprint segmentation algorithm based on sparse coefficients. First, a total variation (TV) model is used to decompose a latent image into two components: texture and cartoon. Second, we compute the sparse representation of the texture image against the dictionary constructed by a set of Gabor elementary functions, which have both good frequency and orientation selective properties and characterize the fingerprint ridge structure well. Third, since the sparse coefficients measure the weights of the basis atoms in fingerprint representation, an image quality measure is computed from the sparse coefficients, which evaluate how well the texture image can be sparsely reconstructed from the basis atoms. Finally, this image quality measure is used for fingerprint segmentation. Figure 3 shows the flowchart of the proposed latent fingerprint segmentation algorithm based on sparse representation. The paper is organized as follows. Section 2 presents in detail the proposed latent fingerprint segmentation algorithm. In Sect. 3, experimental results are presented and analyzed. Finally, the Sect. 4 concludes the paper and last section is the acknowledgment.

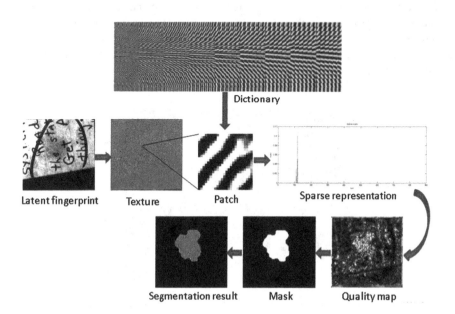

Fig. 3. The flowchart of proposed latent fingerprint segmentation algorithm

2   Latent Fingerprint Segmentation

In this section, we present the proposed latent fingerprint segmentation algorithm based on sparse representation. To begin with, we introduce a TV image model and use it to decompose the latent image into texture and cartoon components. Then, sparse representation is applied on the texture component to segment the fingerprint ridge structure area from the background.

2.1   Latent Fingerprint Decomposition with TV Model

An image can be decomposed into texture and cartoon by using total variation (TV) model. The texture component is often considered as repeated and meaningful structure of small patterns, while the cartoon component is consist of sharp edges and piecewise-smooth patterns [10]. Generally, given an image y, the decomposition can be obtained by minimization of total variation [11]:

$$\min_{(u,v)\in L^2(\Omega)}\left\{TV(u)\lambda\|u-y\|_2^2\right\} \tag{1}$$

Where $y = u + v$ with u and v denoting the image cartoon and texture components, respectively; $TV(u)$ denotes the total variation of u and $\|.\|_2$ is a two norm. The parameter λ, which ranges from 0 to 1, can be used to adjust the contents of texture and cartoon components on the results of decomposition. The smaller λ will produce weak texture component and strong cartoon component and vice versa.

In the latent fingerprint image, the fingerprint pattern is composed of parallel ridges and valley flows and the non-fingerprint pattern usually contains the smooth inner surface and sharp edges. Figure 4(b) and (c) are the texture and cartoon components of the latent fingerprint image shown in Fig. 4 (a), respectively. In our experiment, the parameter λ is set as 0.5 to balance the proportion of the cartoon and texture components. After the cartoon-texture decomposition, the region of interest will be easily extracted from the texture component.

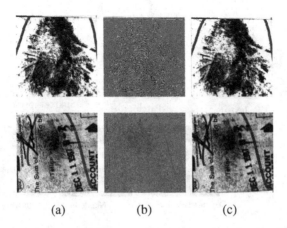

(a)   (b)   (c)

Fig. 4. (a) The latent fingerprint image and its TV decomposed components: (b) texture and (c) cartoon

2.2   Fingerprint Segmentation Based on Sparse Representation

A signal can be usually decomposed into a combination of elementary signals or dictionary atoms [12]. This dictionary-based approach is also widely used in image processing. Research on image statistics considers that image patches can be well-represented as a sparse linear combination of elements from an appropriate dictionary [13]. As a powerful statistical image modeling technique, sparse representation on redundant dictionary has been successfully used in image processing [14]. Generally, a real-valued signal $y \in R^N$ can be represented by a linear combination of a set of N-dimensional basis atoms [10]:

$$y = \sum_i \alpha_i \varphi_i + e = \Phi\alpha + e \tag{2}$$

In this formula, $\Phi = [\varphi_1, \varphi_2, ..., \varphi_M] \in R^{N \times M}$ is a dictionary and it is composed of a set of basis atoms. $\alpha = [\alpha_1, \alpha_2, ..., \alpha_M]$ represents the basis coefficient vector and e is the additive noise imposed on the signal. For the redundant dictionary, the representation coefficients are sparse which means that the signal can be represented by a small number of dictionary atoms. Calculating sparsity by l_0- norm is a nonconvex problem, so it cannot be efficiently solved. Therefore, l_1- norm is more suitable for computing sparsity and a l_1- norm regularization is shown as follows:

$$\min\|\alpha\|_1 \text{s.t.} \|y - \Phi\alpha\|_2^2 < \in \tag{3}$$

where $\|\alpha\|_1$ is the l_1-norm of α and $\in > 0$ is a threshold to control the extent of the model approximation to y.

To compute the spare representation of a local fingerprint patch, a dictionary needs to be constructed firstly. As a Gabor function is defined with a sinusoidal plane wave and the local fingerprint patch usually has a specific ridge frequency and orientation, we take use of Gabor functions to build dictionary. Gabor functions have both good frequency and orientation selective properties and have optimal joint resolution in both spatial and frequency domains [15, 16]. To characterize the fingerprint ridge structure, we propose to build the basis atoms of dictionary with a set of Gabor elementary functions with different parameters in this paper.

The Gabor function has the general form:

$$h(x, y, \theta, f) = \exp\left\{-\frac{1}{2}\left[\frac{x_\theta^2}{\delta_x^2} + \frac{y_\theta^2}{\delta_y^2}\right]\right\} \cos(2\pi f x_\theta + \varphi_0) \tag{4}$$

$$x_\theta = x\cos\theta + y\sin\theta \tag{5}$$

$$y_\theta = -x\sin\theta + y\cos\theta \tag{6}$$

where δ_x and δ_y are the space constants of the Gaussian envelope along x and y axes, respectively. In this paper, we set $\delta_x = \delta_y$ equal to the patch size. f represents the frequency of a Gabor filter and it ranges from 5 to 21 at a step of 2. θ is the orientation of Gabor filter and it ranges from 0 to $15\pi/16$ at a step of $\pi/16$. φ_0 is the initial phase of Gabor filter and it ranges from 0 to $5\pi/16$ at a step of $\pi/6$. Figure 5 presents the dictionary consisting of 864 Gabor atoms.

After dictionary construction, we compute the sparse representation of the texture image against the Gabor dictionary. Since a local fingerprint patch usually has a specific frequency and orientation, the representation coefficient is usually sparse which means that only a few dictionary atoms can be used to represent the fingerprint patch. As for other patches in the non-fingerprint area, such as character or other noises, which have no specific ridge frequency and orientation, the representation coefficients are more evenly distributed. Therefore, different image patches of the texture have quite different sparse representation of the Gabor dictionary (as shown in Fig. 6). Figure 6(a) shows the decomposed texture images of a latent fingerprint. Figure 6(b) shows four different image patches extracted from Fig. 6(a), where the blue rectangles are noise patches and the red rectangles are fingerprint patches (see Fig. 6a). Figure 6(c) presents the reconstructed results by combination of the Gabor dictionary and sparse representation. Figure 6(d) shows sparse coefficients of the Gabor dictionary. From these figures, we can see that the sparsity or the coefficient concentration is larger in the fingerprint patch than in the noise patch. Thus we propose a fingerprint image quality measure based on the sparse representation of the image patch as:

$$Q = \frac{\max|\alpha_i|}{\sum_{i=1}^{N} |\alpha_i|} \tag{7}$$

Algorithm 1. Fingerprint Segmentation Based on Sparse Representation

1 **Input:** Φ: Gabor dictionary; Y: the original fingerprint.

2 **First step:** for each 24×24 image patch y of Y, starting from the upper-left corner with 8 pixel forward step. Then, repeating the following steps:

— Normalize the image patch y to have unit l_2 norm.

— Compute the sparse representation α in Equation (3) by using Φ and y.

— Reconstruct the image patch $\hat{y} = \Phi\alpha$.

— Compute the fingerprint image quality measure Q with Equation (7).

— For an image patch, if its image quality is below a predefined threshold (empirically set as 0.28 in our experiment), it is considered as background or noise, otherwise, this patch is the valid fingerprint.

— Finally, we get a binary segmentation map M.

3 **Second step**

— Morphological processing is applied to the binary segmentation map M and get the biggest and continuous mask of the fingerprint foreground.

4 **Output:** Fingerprint segmentation mask

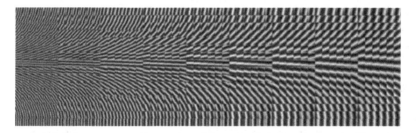

Fig. 5. The Gabor dictionary used for sparse fingerprint representation, which consists of 864 basis atoms at 16 orientations, 9 frequencies and 6 initial phases

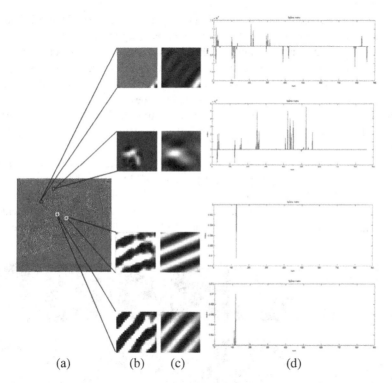

Fig. 6. The sparse representation of different image patches: (a) The texture of a latent fingerprint image, (b) the extracted image patches, (c) reconstructed image patches and (d) the plots of sparse representation coefficients where the horizontal ordinate denotes the number of the dictionary atoms and the vertical ordinate is the coefficient value of the corresponding dictionary atom

Finally, we can generate a fingerprint image quality map with Eq. (7) on the texture image and the parameter N is 864 in our experiment. Thresholding can be applied to the quality map to generate a binary segmentation map. To fill the gaps and eliminate the islands, morphological dilation and erosion processing are applied to get the biggest and continuous mask of the fingerprint foreground. The proposed latent fingerprint automatic segmentation algorithm based on sparse representation is outlined in Algorithm 1.

3   Experimental Result

To show the performance of the proposed algorithm, we test it on the NIST SD27 latent fingerprint database firstly and then compare the proposed algorithm with other methods. NIST SD27 database has 258 latent fingerprint images and all of them are divided into three categories: "good", "bad" and "ugly". Their numbers are 88, 85 and 85, respectively.

First, we perform the proposed algorithm on some sample latent fingerprint images. Figure 7 shows the results of each step by the proposed algorithm. Figure 7(a) includes three latent fingerprint images which are "bad", "good" and "ugly" of the NIST SD27 database from top to bottom, respectively. Figure 7(b) shows their decomposed texture images. Figure 7(c) presents the fingerprint image quality maps of textures. Figure 7(d) contains the eroded results and Fig. 7(e) shows the biggest and continuous region of the fingerprint chosen from Fig. 7(d). The dilation results of Fig. 7(e) are presented in Fig. 7(f). The biggest area in the eroded result is most likely the foreground part of fingerprint.

Second, we compare the proposed algorithm to the segmentation method based on orientation consistency (i.e., coherence) [10] and the ADTV method as shown in Fig. 8. The results of NIST SD27 database by ADTV method are downloaded from their website http://www-scf.usc.edu/~jiangyaz/files/. Figure 8(a) shows three latent fingerprint images which are selected from the "bad", "good" and "ugly" of the NIST

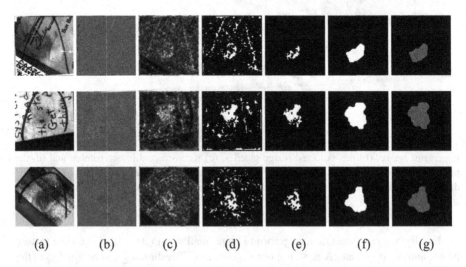

(a)   (b)   (c)   (d)   (e)   (f)   (g)

Fig. 7. (a) The latent fingerprint images, (b) the textures of latent fingerprint images, (c) the quality maps, (d) the eroded results, (e) the most likely part of fingerprints, (f) the dilated results and (g) the final segmented results

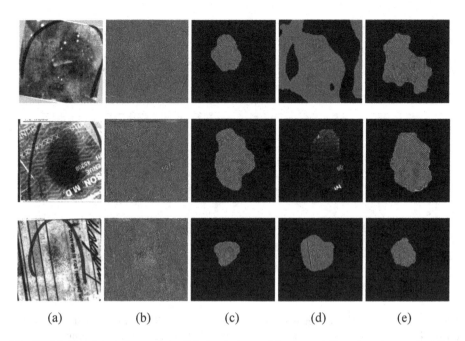

(a)   (b)   (c)   (d)   (e)

Fig. 8. (a) The latent fingerprints, (b) the textures, (c) segmentations based on orientation consistency, (d) the ADTV method and (e) the proposed algorithm

SD27 database, respectively. The decomposed texture images are shown in Fig. 8(b). Figure 8(c), (d) and (e) show the segmentation results of the algorithm based on orientation consistency, the ADTV and the proposed algorithm. The algorithm based on orientation consistency cannot segment the fingerprint well (see the first one and third one from the top in Fig. 8(c). The ADTV method cannot eliminate the noise well comparing to the other two methods (see Fig. 8(d)). The proposed algorithm can achieve better performance for segmentation than other two methods (see Fig. 8(e)).

Finally, we perform the latent fingerprint identification to test the effectiveness of the proposed algorithm. For fair comparison, the only difference is the segmentation methods. In addition to the 258 fingerprint images of NIST SD27, the template database includes 27,000 fingerprint images of NIST SD14. The feature extraction and matching process are conducted by the commercial software VeriFinger SDK 4.3 and it will produce a matching score between a latent and a template fingerprint. The matching scores are sorted in descending order and they can be evaluated by Cumulative Match Characteristic (CMC) curve. A CMC curve draws the identification rate with different rank -k and it indicates the proportion of times that the mated rolled fingerprint appears in the top k matches. A CMC curve presents the probability that a given latent fingerprint appears in the template candidate lists of different sizes.

Figure 9 shows the CMC curves of latent fingerprint identification by using different segmentation results. The "proposed" means segmentation based on sparse representation and the "orientation consistency" means segmentation based on orientation consistency. We don't show the results of ADTV for comparison because these

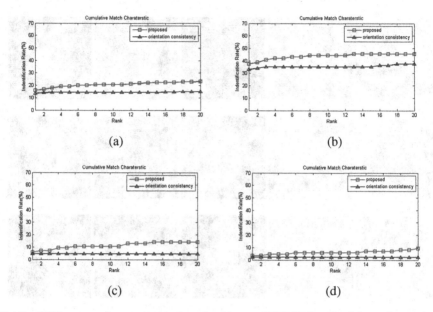

Fig. 9. CMC curves of two segmentation algorithm on NIST SD27: (a) all latent fingerprints, (b) good latent fingerprints, (c) bad latent fingerprints and (d) ugly latent fingerprints

downloaded results of NIST SD27 database by ADTV method can't produce comparable CMC curves. Figure 9(a) shows the CMC curves on all latent fingerprint images of NIST SD27. Obviously, the proposed algorithm performs better than the orientation consistency method. Figure 9(b), (c) and (d) present the CMC curves on "good", "bad" and "ugly" quality fingerprint images of NIST SD27 database, respectively. From these results, we can get that the proposed algorithm performs better than the algorithm based on orientation consistency.

4   Conclusion

In this paper, we propose a latent fingerprint segmentation algorithm based on sparse representation to extract the foreground fingerprint area from the complex background. To remove the various image noises, the TV model is used to decompose the latent image into texture and cartoon components. Then, we compute the sparse representation of the texture image against the Gabor dictionary. An image quality measure is computed from the sparse coefficients, which evaluate how well the texture image can be sparsely reconstructed from the basis atoms. Finally, this image quality measure is used to segment the fingerprint foreground area from the background. The proposed algorithm is tested on the NIST SD27 latent fingerprint database and compared with the existing method. Experimental results and comparisons demonstrate the effectiveness of the proposed latent segmentation method.

Acknowledgment. This work was supported by National Natural Science Foundation of China under the grants No. 61005024 and No. 61375112.

References

1. Zhang, J., Lai, R., Kuo, C.C.J.: Latent fingerprint segmentation with adaptive total variation model. In: 5th IAPR International Conference on Biometrics (ICB), pp. 189–195. IEEE (2012)
2. Jain, A.K., Farrokhnia, F.: Unsupervised texture segmentation using Gabor filters. In: IEEE International Conference on Systems Man and Cybernetics, pp. 14–19. IEEE (1990)
3. Ma, J., Jing, X., Zhang, Y. et al.: Simple effective fingerprint segmentation algorithm for low quality images. In: 3rd IEEE International Conference on Broadband Network and Multimedia Technology (IC-BNMT), pp. 855–859. IEEE (2010)
4. Akram, M.U., Nasir, S., Tariq, A. et al.: Improved fingerprint image segmentation using new modified gradient based technique. In: Canadian Conference on IEEE Electrical and Computer Engineering, CCECE 2008, pp. 001967–001972 (2008)
5. Short, N.J., Hsiao, M.S., Abbott, A.L. et al.: Latent fingerprint segmentation using ridge template correlation (2011)
6. Zhang, J., Lai, R., Kuo, C.J.: Adaptive directional total-variation model for latent fingerprint segmentation. IEEE Trans. Inf. Forensics Secur. **8**(8), 1261–1273 (2013)
7. Ratha, N.K., Chen, S., Jain, A.K.: Adaptive flow orientation-based feature extraction in fingerprint images. Pattern Recognit. **28**(11), 1657–1672 (1995)
8. Choi, H., Boaventura, M., Boaventura, I.A.G., et al.: Automatic segmentation of latent fingerprints. In: IEEE Fifth International Conference on Biometrics: Theory, Applications and Systems (BTAS), pp. 303–310. IEEE (2012)
9. Karimi-Ashtiani, S., Kuo, C.C.J.: A robust technique for latent fingerprint image segmentation and enhancement. In: 15th IEEE International Conference on Image Processing, ICIP 2008, pp. 1492–1495. IEEE(2008)
10. Liu, M., Chen, X., Wang, X.: Latent Fingerprint Enhancement via Multi-scale Patch Based Sparse Representation (2013)
11. Buades, A., Le, T.M., Morel, J.M., et al.: Fast cartoon + texture image filters. IEEE Trans. Image Process. **19**(8), 1978–1986 (2010)
12. Rubinstein, R., Bruckstein, A.M., Elad, M.: Dictionaries for sparse representation modeling. Proc. IEEE **98**(6), 1045–1057 (2010)
13. Yang, J., Wright, J., Huang, T.S., et al.: Image super-resolution via sparse representation. IEEE Trans. Image Process. **19**(11), 2861–2873 (2010)
14. Elad, M., Aharon, M.: Image denoising via sparse and redundant representations over learned dictionaries. IEEE Trans. Image Processing **15**(12), 3736–3745 (2006)
15. Daugman, J.G.: Uncertainty relation for resolution in space, spatial frequency, and orientation optimized by two-dimensional visual cortical filters. JOSA A **2**(7), 1160–1169 (1985)
16. Jain, A.K., Feng, J.: Latent fingerprint matching. IEEE Trans. Pattern Anal. Mach. Intell. **33**(1), 88–100 (2011)

Pose Estimation for Vehicles Based on Binocular Stereo Vision in Urban Traffic

Pengyu Liu, Fei Wang$^{(\boxtimes)}$, Yicong He, Hang Dong, Haiwei Yang, and Yang Yang

Xi'an Jiaotong University, Xi'an, Shaanxi, China
{liu.pengyu,heyicong2013,yanghw.2005,
dhunter}@stu.xjtu.edu.cn,
{wfx,yyang}@mail.xjtu.edu.cn

Abstract. Extensive research has been carried out in the field of driver assistance systems in order to increase road safety and comfort. We propose a pose estimation algorithm based on binocular stereo vision for calculating the pose of on-road vehicles and providing reference for the decision of driving assistant system, which is useful for behavior prediction for vehicles and collision avoidance. Our algorithm is divided into three major stages. In the first part, the vehicle is detected and roughly located on the disparity map. In the second part, feature points on the vehicle are extracted by means of license plate detection algorithm. Finally, pose information including distance, direction and its variation is estimated. Experimental results prove the feasibility of the algorithm in complex traffic scenarios.

Keywords: Pose estimation · Behavior prediction · Binocular stereo vision

1 Introduction

The automotive industry is constantly evolving as there is a growing interest in the field of safety. In order to improve security in driving, several applications of Advanced Driving Assistance Systems (ADAS) were proposed. When a potential safety problem occurs, this system can timely alerts the driver. If the driver does not respond to warnings, measures will be taken by the system automatically so as to avoid the accident. This paper is aimed at the Forward Collision Avoidance System (FCAS), which is designed for assisting the driver to maintain a safe stopping distance related to the vehicle ahead in order to avoid collision when accident occurs or at least reduce the severity of the collision [1]. However, in many cases the safe distance is not maintained because of the driver's occasional distraction or other negligence. Besides, some emergencies, such as sudden changing of lanes, may cause accidents as the drivers don't have enough time for response (cf. Fig. 1). The algorithm we propose is based on binocular stereo vision algorithm which can detect vehicles and calculate its pose information including distance, direction and its variation. So that behaviors of the vehicles can be predicted at the same time. To date, RADAR-based system becomes a very popular solution for FCAS. However, most of them are used in the high class cars due to the exorbitant cost of laser equipment. One of the main goals of our research is

© Springer International Publishing Switzerland 2015
D.-S. Huang et al. (Eds.): ICIC 2015, Part I, LNCS 9225, pp. 454–465, 2015.
DOI: 10.1007/978-3-319-22180-9_44

to provide cars with a not only cheap but also reliable forward collision avoidance system with speed, distance as well as pose information. Due to the fact that vision system can provide depth information as well as texture, it is still the first choice considering both cost and efficiency.

Fig. 1. Accident due to sudden changing of lanes

Our paper is organized as follows: the second part is the recent research work. The third part describes the algorithm in detail, including the coarse localization of the vehicles forward, vehicle feature extraction and pose estimation. Experimental results are displayed in the fourth part. The fifth part is the conclusion and prospect of our work.

2 Related Work

Many approaches to assist drivers in avoiding collisions have been developed based on different technologies such as RADAR [2], LASER scanner [3], ultrasounds [4] or vision systems [5–7]. As for drawbacks, the vision field of RADAR can be narrow, and mutual interference may occur when other vehicles on road are installed with same equipment due to radar's active sensing characteristic. Cheap LASERs are easily affected by fog and rain. Ultra-sounds easily fail because of lateral wind.

To date, many researches aims at vehicle detection. In [8], scene depth is deduced from texture gradients, defocus, color and haze using a multi-scale Markov Random Field (MRF). However, this algorithm may fail when the actual scene differs greatly with the training scene because of the algorithm's need for pre-training. In [9], the vehicle's distance measurement is based on the contact between the vehicle's tires and the road. This method may easily overestimate because of the difficulty of obtaining the exact tire-asphalt contact point which in addition does not correspond to the real back of the vehicle.

For number plate detection, there are methods based on Hough transform [10], color SVM classifiers to model character appearance variations [11] and methods based on top-hat transform [12]. The number plate location approach proposed adapts the widely employed morphological top-hat method to vehicles in motion where the position of the vehicle is unknown and therefore the dimensions of the number plate in the image are, in principle, unknown.

For pose estimation, the aim of the Perspective-n-Point problem (PnP) is to determine the position and orientation of a camera given its intrinsic parameters and a set of n correspondences between 3D points and their 2D projections. It is widely used in computer vision, Robotics, Augmented Reality and has received much attention in both the Photogrammetry [13] and Computer Vision [14] communities. A robust pose estimation algorithm is proposed by Yang [15]. He divides approximate coplanar points into coplanar points and non-coplanar points and uses non-coplanar points for final pose calculating.

In the field of behavior prediction, Toru Kumagai [16] uses Bayesian Dynamic Models as well as driving status to predict the parking intention at the intersection. Tesheng Hsiao [17] obtains parameters of turning model based on maximum a posterior estimation. However, the methods above require long time for modeling or training, which can't meet the real-time requirements essential for collision avoidance.

3   Algorithm Description

3.1   Method Overview

Our algorithm is divided into three parts: vehicle detection, vehicle feature extraction and pose estimation. The flowchart of our algorithm is shown in Fig. 2.

Fig. 2. Flowchart of our algorithm

In order to calculate the pose information of the vehicle, four feature points are needed according to the pose estimate algorithm. Unfortunately, different kinds of vehicles have different shapes, which makes it difficult to find fixed feature points. However, a common element on vehicle is the license plate which has strict regulations in every country. As a result, the vehicle's pose information can be calculated directly by localizing the front vehicle's license plate and establishing the relationship between the license plate's size in the image and the 3D coordinates in space in advance.

Nevertheless, as the front vehicle is far from the camera, the license plate occupies only a small proportion in the image, which makes it difficult to detect and locate. Thus coarse localization of the vehicles is needed previously. We calculate the bounding-box of the front vehicle in image and set it as ROI. License plate detection and pose estimation can be realized subsequently.

The license detection algorithm requires the plates being installed according to the regulations of China. Vehicles without plates or installed with twisted plates are not discussed in this paper.

3.2   Vehicle Detection

The vehicle detection procedure is based on two parts: the disparity map generation based on binocular stereo vision and front vehicle localization.

3.2.1   Scene Disparity Map Calculating

The model of binocular stereo vision observes the same scene from different viewpoints using two fixed cameras with given intrinsic parameters. Correspond pixels are calculated by stereo matching of the left and right images. In the structure of standard epipolar geometry, the epipolar of space pointare parallel and aligned in left and right image planes. Suppose the corresponding points of p are p_1 and p_2 in left and right images, the base line is b, focus is f, and the distance from p to base line is z. (cf. Fig. 3).

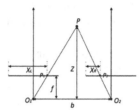

Fig. 3. Parallax theory based on epipolar geometry

According to geometry we have:

$$\frac{b}{Z} = \frac{(b + x_R) - x_L}{Z - f} \tag{1}$$

Then we get z from point p to base line:

$$z = \frac{bf}{x_R - x_L} = \frac{bf}{d} \tag{2}$$

where $d = x_R - x_L$

From the analysis above, the depth z of point p depends on the parallax d only. Thus the location of p in space can be uniquely identified by calculating the corresponding point p_2 of p_1. We use block matching algorithm based on Graph Cuts.

The disparity map and pseudo-color image calculated by the model of binocular stereo vision are shown in Fig. 4, which is prepared for vehicle localization in the next section.

3.2.2   Vehicle Localization

After generating disparity map, more work is done to locate the vehicle in the image. The algorithm is described in [18] as follows.

1. Count u-disparity on map to detect horizontal line using Hough transform. The detected lines are the position information in x-axis of the vehicle.

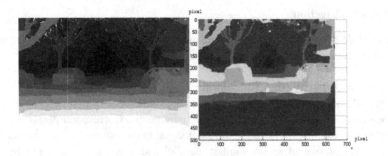

Fig. 4. Disparity map and pseudo-color image (Color figure online)

2. Count v-disparity on map to detect vertical line using Hough transform. The crossing points of road and vehicle are the position information in y-axis of the vehicle.
3. Sort the coordinates of vehicles by y-axis and match the vehicles after sorting. Then the (x, y) information of vehicles can be got.
4. Calculate the 3D position of vehicles according to the binocular stereo vision model.

The detected vehicle is marked with red box as Fig. 5 shows. After calculating the bounding-box of the vehicle forward using binocular stereo vision algorithm, we set the bounding-box area as ROI. We detect the license plate and calculate the coordinates of the feature points in ROI.

Fig. 5. Vehicle detection (Color figure online)

Considering the real condition, there may be several vehicles in the visual field at the same time. Thus several ROIs may be detected. We choose three nearest ones from the camera, if any, to calculate the pose of vehicles on current lane as well as the neighbor lanes.

3.3 Vehicle Feature Extraction

In order to calculate pose information, four feature points are required at least. However, different kinds of vehicles have unique shapes, which makes it difficult to look for fixed feature points. Fortunately, the standard of license plate is fixed in every country. In China, for instance, according to regulations based on the public security industry standards of People's Republic of China, we summarize some characteristics of license plate designed for cars which are most common on roads. As Fig. 6 shows.

Fig. 6. Sample of license plate of China (Color figure online)

1. The background color of the license plate is quite different from the vehicle and characters.
2. The contour of the license plate is continues or interrupted due to abrasion.
3. The characters on the plate are on the same horizontal line. Much boundary information can be found on the plate.
4. The size of plate is fixed.

On the basis of the characteristics above, neighboring pixels on plate's boundary will vary frequently from 0 to 1 or 1 to 0, and the sum of changes will be larger than a threshold, which can be used for detecting the license plate.

The algorithm for license plate detection in this paper is based on grey level transformation of plate's region. Due to the contrast of gray level between the plate and the vehicle, binaryzation is achieved with a threshold calculated by local histogram. The small areas which are not plates are removed. Finally, in all connected domains, the one with max likelihood is marked as the plate.

The process of license plate detection is shown in Fig. 7. Figure 7-a is the RGB image of one ROI. Binary image with mark points of likelihood areas of license plate is shown in Fig. 7-b. The final plate is marked on Fig. 7-c and the four vertexes are used for pose estimation algorithm in next section.

a   b   c

Fig. 7. Results for plate detection (Color figure online)

3.4   Vehicle Pose Estimation

PnP problem is the estimation of the pose of a calibrated camera from n 3D-to-2D point correspondences. The central idea of the algorithm is to express the n 3D points as a weighted sum of four virtual control points and estimate the coordinates of these control points in the camera referential. At last, the rotation matrix R and translation matrix T are calculated. The algorithm is described in detail as follows:

Suppose reference points of n points with 3D coordinates known in the world coordinate system be:

$$p_i, \quad i = 1, \ldots, n \tag{3}$$

Similarly, let the 4 control points we use to express their world coordinates be

$$c_j, \quad j = 1, \ldots, 4 \tag{4}$$

Then the references can be expressed by control points uniquely:

$$p_i^w = \sum_{j=1}^{4} \alpha_{ij} c_j^w, \quad with \sum_{j=1}^{4} \alpha_{ij} = 1 \tag{5}$$

Where α_{ij} are homogeneous barycentric coordinates.

The same relation holds in the camera coordinate system:

$$p_i^c = \sum_{j=1}^{4} \alpha_{ij} c_j^c \tag{6}$$

Suppose A be the camera internal calibration matrix and $\{u_i\}_{i=1,\ldots,n}$ are the projections of reference points expressed by $\{p_i\}_{i=1,\ldots,n}$. We have:

$$\forall i, \quad w_i \begin{bmatrix} u_i \\ 1 \end{bmatrix} = A p_i^c = A \sum_{j=1}^{4} \alpha_{ij} c_j^c \tag{7}$$

Where w_i is scalar projective parameter. Matrix A consists of the f_u, f_v focal length coefficients and the (u_c, v_c) principal point.

We now consider the specific 3D coordinates $[x_j^c, y_j^c, z_j^c]^T$ of each control point c_j^c, the 2D coordinates $[u_i, v_i]^T$ of the u_i projections. Then Eq. (7) becomes:

$$\forall i, \quad w_i \begin{bmatrix} u_i \\ v_i \\ 1 \end{bmatrix} = \begin{bmatrix} f_u & 0 & u_c \\ 0 & f_v & v_c \\ 0 & 0 & 1 \end{bmatrix} \sum_{j=1}^{4} \alpha_{ij} \begin{bmatrix} x_j^c \\ y_j^c \\ z_j^c \end{bmatrix} \tag{8}$$

Expand Eq. (8), we have:

$$Mx = 0 \tag{9}$$

Where $x = [c_1^{cT}, c_2^{cT}, c_3^{cT}, c_4^{cT}]^T$ is a 12-vector with the unknowns, and M is a $2n \times 12$ matrix. The solution can be expressed as:

$$x = \sum_{i=1}^{N} \beta_i v_i \tag{10}$$

Where the set v_i are the columns of the right-singular vectors of M.

Finally, we compute solutions for all four values of N and keep the one that yields the smallest re-projection error:

$$res = \sum_i dist^2 \left(A[R|t] \begin{bmatrix} p_i^w \\ 1 \end{bmatrix}, u_i \right) \tag{11}$$

Where the rotation matrix R and the translation matrix T represent the direction and distance of the vehicle respectively.

4   Experimental Results

The experiments are implemented with a desktop PC with processor of i7-2.80 GHz, RAM 32 GB. The videos were recorded by two cameras made by Imagesource which are mounted in a car.

4.1   Pose Estimation of Vehicles Forward

In order to prove the robustness of our algorithm, we choose the image with a vehicle at corner. The results are shown in Fig. 8.

Figure 8-a shows the original image. Figure 8-b shows the coarse detection of vehicle according to disparity map. Figure 8-c shows the plate's vertexes detection result. Finally, as proposed in Sect. 3.4, we calculate the pose of vehicle forward and re-project the outlines of the vehicle using a virtual cube with axes for verification, as Fig. 8-d shows. The axes x, y and z are represented with green, yellow and blue respectively.

4.2   Pose Estimation of Vehicles While Changing Lanes

One of our goals is to predict vehicle's behavior, especially the intention of changing lanes, which may cause a collision. We calculate pose variations from consecutive frames. The re-projection results are shown in Fig. 9.

Fig. 8. Pose estimation of vehicle forward (Color figure online)

Fig. 9. Continuously pose estimation

The pose estimating result is expressed as:

$$pose = [x, y, z, \alpha, \beta, \gamma] \qquad (12)$$

Where x, y, z represents translation in relevant axis direction and α, β, γ represents the rotation around x axis, y axis and z axis. The pose information in the 6 frames above are displayed in Table 1 and Fig. 10. The last column of Table 1 represents the error of pose estimation algorithm according to Eq. (11).

Table 1. Pose estimation results

Frame	x(cm)	y(cm)	z(cm)	α(°)	β(°)	γ(°)	err
1	205.6	405.2	15.0	0.1	−0.2	8.6	0.2
2	180.2	320.6	14.2	−0.2	0.1	9.9	0.3
3	160.3	250.6	16.1	0.1	0	10.2	1.5
4	140.6	365.2	14.5	−0.3	0.3	9.6	2.2
5	80.8	1050.3	15.3	0	−0.2	7.5	1.6
6	30.5	1500.3	14.6	0.5	−0.3	4.3	1.2

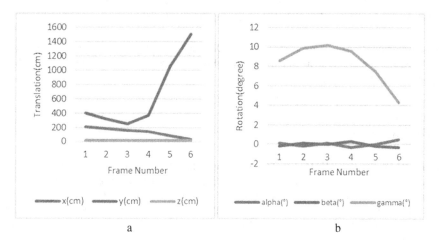

Fig. 10. Line chart of pose estimation results

From Fig. 10-a, the x value is continually decreasing when the vehicle is approaching to our car in horizontal direction. The y value increases first and then decreases as the vehicle comes close and then move away. The z value maintains around zero because the road is relatively flat.

From Fig. 10-b, the angle of gamma, which represents rotation around z axis, increases first and then decreases. It indicates the turning action of the vehicle. The angle of beta and gamma undulate around zero due to the flat road.

The results and analysis above exactly match the vehicle's behavior in the picture, which proves that the pose we get can be used in vehicle behavior prediction as a reliable reference.

5   Conclusion and Forecast

This paper puts forward an innovative pose estimation method for vehicles' behavior prediction based on binocular stereo vision. Using the binocular camera, we get pose information of other vehicles including distance, speed and direction, which is useful for assisting drivers to keep a safe distance and predicting the behavior of other vehicles.

The method proposed in this paper uses two cameras as sensors, which cost much less than the traditional laser equipment. In addition, the vehicle detection based on disparity map gives coarse localization of vehicle and thus greatly reduces the searching space and increases the correct rate of license plate detection. We get vehicle's pose information with only four points based on the proposed pose estimation algorithm gets, which is accurate enough and provides sufficient reference to the ADAS. In road tests, pose information including distance, direction and its variation is estimated, and the analysis proves the feasibility of the algorithm in complex traffic scenarios.

Although the pose information can be calculated with four feature points in most cases using the algorithm we propose, more feature points are needed to reduce the error in pose estimation. In the future, more work are needed for feature point extraction and error analysis [19] to improve the accuracy.

Acknowledgment. This work is sponsored by National Natural Science Foundation of China (No.612173366) and National High Technology Research and Development Program of China (No.2013AA014601).

References

1. Bertolazzi, E., Biral, F., Da Lio, M., et al.: Supporting drivers in keeping safe speed and safe distance: the saspence subproject within the European framework programme 6 integrating project Prevent. IEEE Trans. Int. Transp. Syst. **11**(3), 525–538 (2010)
2. Skutek, M., Mekhaiel, M., Wanielik, G.: A Precrash system based on radar for automotive applications. In: Proceedings of IEEE of Intelligent Vehicles Symposium, Columbus, pp. 37–41 (2003)
3. Velupillai, S., Guvenc, L.: Laser scanners for driver-assistance systems in intelligent vehicles. IEEE Control Syst. Mag. **29**(2), 17–19 (2009)
4. Alonso, L., Pérez-Oria, J., Fernández, M., Rodríguez, C., Arce, J., Ibarra, M., Ordoñez, V.: Genetically tuned controller of an adaptive cruise control for urban traffic based on ultrasounds. In: Diamantaras, K., Duch, W., Iliadis, L.S. (eds.) ICANN 2010, Part II. LNCS, vol. 6353, pp. 479–485. Springer, Heidelberg (2010)
5. Nedevschi, S., Danescu, R., Frentiu, D., et al.: High accuracy stereovision approach for obstacle detection on non-planar roads. In: Proceedings of IEEE Intelligent Engineering Systems (INES), Cluj Napoca, Romania, pp. 211–216 (2004)
6. Broggi, A., Bertozzi, M., Fascioli, A., Guarino, C., Piazzi, A.: Visual perception of obstacles and vehicles for platooning. IEEE Trans. Intell. Transp. Syst. **1**(3), 164–176 (2000)
7. Sun, Z., Bebis, G., Miller, R.: On-road vehicle detection: a review. IEEE Trans. Pattern Anal. Mach. Intell. **28**(5), 694–711 (2006)

8. Michels, J., Saxena, A., Ng, A.Y.: High speed obstacle avoidance using monocular vision and reinforcement learning. In: 22nd International Conference on Machine Learning, Bonn, Germany, pp. 593–600 (2005)

9. Gat, I., Benady, M., Shashua, A.: A monocular vision advance warning system for the automotive aftermarket. SAE Technical Paper (2005)

10. Yanamura, Y., Goto, M., Nishiyama, D., Soga, M., Nakatani, H., Saji, H.: Extraction and tracking of the license plate using Hough transform and voted block matching. In: Proceedings of IEEE Intelligent Vehicles Symposium, pp. 243–246 (2003)

11. Chengwen, H., Yannan, Z., Jiaxin, W., Zehong, Y.: An improved method for the character recognition based on SVM. In: Proceedings of the IASTED International Conference on Artificial Intelligence and Applications, pp. 457–461 (2006)

12. Hung, K.M., Hsieh, C.T.: A real-time mobile vehicle license plate detection and recognition. J. Sci. Eng. **13**, 433–442 (2010)

13. McGlove, C., Mikhail, E., Bethel, J.: Manual of Photogrametry. American Society For Photogrammetry and Remote Sensing, New York (2004)

14. Hartley, R., Zisserman, A.: Multiple View Geometry In Computer Vision. Cambridge University Press, Cambridge (2003)

15. Yang, H., Wang, F., Chen, L., He, Y., He, Y.: Robust pose estimation algorithm for approximate coplanar targets. In: Huang, D.-S., Jo, K.-H., Wang, L. (eds.) ICIC 2014. LNCS, vol. 8589, pp. 350–361. Springer, Heidelberg (2014)

16. Kumagai, T., Sakaguchi, Y., Okuwa, M. et al.: Prediction of driving behavior through probabilistic inference. In: Proceedings of 8th International. Conference on Engineering Applications of Neural Networks, pp. 117–123 (2003)

17. Hsiao, T.: Time-varying system identification via maximum a posteriori estimation and its application to driver steering models. In: American Control Conference, pp. 684–689. IEEE (2008)

18. Hu, Z., Uchimura, K.: UV-disparity: an efficient algorithm for stereovision based scene analysis. In: Proceedings of IEEE on Intelligent Vehicles Symposium, pp. 48–54. IEEE (2005)

19. Song, Y., Wang, F., Gao, S., Yang, H., He, Y.: Error tracing and analysis of vision measurement system. In: Huang, D.-S., Bevilacqua, V., Premaratne, P. (eds.) ICIC 2014. LNCS, vol. 8588, pp. 729–740. Springer, Heidelberg (2014)

Robust Segmentation of Vehicles Under Illumination Variations and Camera Movement

Zubair Iftikhar$^{(\boxtimes)}$, Prashan Premaratne, Peter Vial, and Shuai Yang

School of Electrical Computer and Telecommunications Engineering,
University of Wollongong, North Wollongong, NSW, Australia
zi770@uowmail.edu.au

Abstract. Vision-based vehicle detection and segmentation in intelligent transportation systems, particularly under outdoor illuminations, camera vibration, cast shadows and vehicle variations is still an area of active research for analysis and processing of traffic data. This paper proposes an effective scheme that improves Gaussian mixture model (GMM) for non-stationary temporal distributions through dynamically updating the learning rate at each pixel. In this proposed technique, sleeping foreground pixels and slow moving vehicles cannot become the part of background model that also does not lead to extra computational cost as compare to other methods that are proposed in the literature. Sudden illumination change is also captured in this technique. Vision based system cannot be efficient without fixing of camera vibration, so movement of camera is adjusted based on clues from background model. At the end, shadows are removed from detected vehicles through applying a new recursive method in dark regions. Experimental results demonstrate the robustness and high level performance of the proposed adaptive foreground extraction algorithm under illumination variations compared to state-of-the-art methods.

Keywords: Intelligent transportation systems · Gaussian mixture model · Sleeping foreground pixels

1 Introduction

A robust system should be capable to detect and segment the moving or stopped object under outdoor illumination changes, camera vibrations, shadows appearance and motions in background. Gaussian Mixture Models (GMM) [1] is one of the widely-used parametric technique for background modelling which can cope with non-consistent environmental and spectral conditions. GMM method obtains the adaptive model of each pixel in an image plane through mixture of several Gaussians that are based on motions in background [2]. Robustness against gradual illumination changes (GIC), quasi-periodic and multi-model [3–5] appraises the GMM as a competitive method. Thus, it is enormously recognized and many new versions of this techniques have been suggested in literature [6–9]. Despite of all these features, GMM method causes a number of limitations, it is unable to cope with sudden illumination change [6, 10] and parametric behavior of this modelling leads to tricky and tedious

© Springer International Publishing Switzerland 2015
D.-S. Huang et al. (Eds.): ICIC 2015, Part I, LNCS 9225, pp. 466–476, 2015.
DOI: 10.1007/978-3-319-22180-9_45

task for parameters tuning [11, 12]. In GMM, learning rate determines that how much influence of an intensity value on specific location can cause on distribution that is tuned through mean and standard deviation. Low learning rate does not provide the fast recovery of a pixel that is wrongly considered to be foreground pixel. It takes many frames to become the background pixel due to low learning rate which slowly increases the weight of that distribution so it takes time to reach the threshold τ. High learning rate quickly increases the weight of slow moving object and that object becomes the part of background model. Improvement was made in [13] which provides the high learning rate in initial frames and after that provides the low learning rate. This technique works well for sudden illumination change in start but fails to adopt this sudden environmental change in late coming frames. These problems of low and high learning rates are addressed in proposed algorithm. Generally, GMM methods only depend on temporal information for updating the parameters but our method assigns the weight of a foreground distribution at frame level which shows the strong relation between temporal and spectral domains. A vision system can not provide the accurate results without the fixing of camera movement. In [14], horizontal and vertical histograms of an image plane are calculated in temporal domain to find the displacement for particular frame in both directions. However, camera vibration does not provide the same displacement at each point of a particular direction. Usually, high displacement is appeared on boundary of an image plane and low displacement is seen on middle part of that image. We measured the camera vibration by using the prominent edges information from image background. Sometimes sudden illumination change (SIC) appears so fast due to clouds movements in outdoor environment that it becomes impossible to deal them through GMM. A SIC provides the same increase or decrease level of intensity on a particular location and neighbors so the difference between the intensity values of a pixel and its neighbor are not changed. This idea is utilized in [15] which applies the gradient of Gaussian as a pre-processing step to avoid the SIC and then GMM is performed for background modelling. This technique does not work well in case of camera vibration because the displacements in x and y coordinates change the pixels locations and their neighbors. This [15] for SIC also fails for background motions such as waving leaves of trees or flags. Our proposed algorithm handles SIC at frame level.

2 Algorithm Description

This proposed algorithm extracts the vehicles from image plane through a new advanced Gaussian mixture model Fig. 1.

It consists of three modules. The first module performs the preprocessing and fixes the camera vibration. The second module learns the background through following steps: learning rate is determined for both Gaussians and for every pixel which provides the speed of adaptation to illumination changes; then, mean, variance and weight of each pixel are updated; last step of this module classifies each pixel as a foreground pixel or a background pixel. The third module works on frame level, clusters of foreground pixels are segmented; after that, detection of sudden illumination change

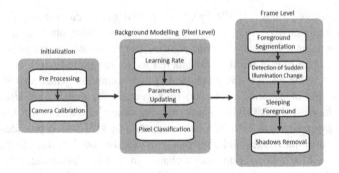

Fig. 1. Flow diagram

(SIC) is applied; recognizes the sleeping foreground vehicles and removes the shadows from vehicles.

3 Background Modeling

RGB color space requires extra computational task so gray level color space is used for this system. The concept of Gaussian mixture model is to manifest the intensity level of each pixel point with K status, which are all approximated by normal distribution. So, a mixture of K Gaussians is defined for each pixel. GMM defined by Stauffer and Grimson [1] observes the probability of each pixel value through following equation:

$$P(X_t) = \sum_{i=1}^{K} \omega_{i,t} . \eta(X_t, \mu_{i,t}, \sum_{i,t}) \tag{1}$$

$$\eta(X_t, \mu, \Sigma) = \frac{1}{(2\pi)^{n/2} |\sum|^{1/2}} e^{-\frac{1}{2}(X_t-\mu)\Sigma^{-1}(X_t-\mu)} \tag{2}$$

K is the number of distributions, for the ith Gaussian at time t, $\omega_{i,t}$ is a weight with mean $\mu_{i,t}$ and η is a Gaussian probability density function. K value between 3 to 5 is suggested in [1]. Gaussians which satisfies following condition, are considered to background. Remaining distributions which do not reach the certain threshold T are retained for foreground distributions.

$$B = \arg\min_b = (\sum_{i=1}^{b} \omega_{i,t} > T) \tag{3}$$

$$sqrt((X_{t+1} - \mu_{i,t})^T . \sum_{i,t}^{-1} . (X_{t+1} - \mu_{i,t})) < 2.5\sigma \tag{4}$$

σ is standard deviation, if difference value of pixel from Gaussian exists in one of the distribution from K, and that distribution is a background distribution then the pixel is classified as background pixel otherwise it will be a foreground pixel. In one other case, pixel is again considered to be foreground if difference value is not part of any distribution. [1] proposed the following rules for updating the background model, if (4) is a segment of one of the Gaussians.

$$\omega_{i,t+1} = (1 - \alpha)\omega_{i,t} + \alpha \tag{5}$$

$$\mu_{i,t+1} = (1 - \rho)\mu_{i,t} + \rho X_{t+1} \tag{6}$$

$$\sigma^2_{i,t+1} = (1 - \rho)\sigma^2_{i,t} + p(X_{t+1} - \mu_{i,t+1}).(X_{t+1} - \mu_{i,t+1})^T \tag{7}$$

Where, α is a constant learning rate and ρ ($\alpha / \omega_{i,t+1}$) is a learning factor of the distribution. If (4) is not matched in a Gaussian then only weight is updated:

$$\omega_{i,t+1} = (1 - \alpha)\omega_{i,t} \tag{8}$$

Further, if difference value (4) is failed to adopt any Gaussian, then:

$\omega_{k,t+1}$ = Low Prior Weight, $\mu_{k,t+1} = X_{t+1}$, $\sigma^2_{k,t+1}$ = Large Initial Variance K is three, two are allocated for background and one for foreground. Generally, α also incorporates that how fast an object stops as explained in [17]. Here, α does not perform the extra duty of determining the sleeping pixels. Selection of α is very critical, it should deal with the slow process of storing a temporal history and quickly adaptation of illumination changes Fig. 3.

α is introduced for each Gaussian that is calculated on each pixel. 1st Gaussian, is fast enough to adopt the GIC and 2nd Gaussian is slow enough to store the temporal history. GIC on particular location moves from the corners to outward as seen in Fig. 2. So, corners of a Gaussian are more sensitive as compare to other regions. If the current

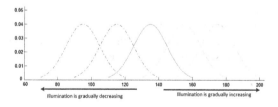

Fig. 2. Sliding of gaussian

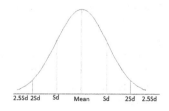

Fig. 3. α of a gaussian

value is within the σ then it is considered that it is not moving towards illumination change. If value is not in σ but exists in the 2 σ then pixel value on this location is slightly moving outside the Gaussian. Finally, if value is outside of 2 σ and within range of 2.5 σ then GIC is more expected. In this case, future value can be outside of Gaussian and will be considered as foreground pixel which is not true information. These facts are utilized in 1st Gaussian which cope the speed of GIC. Gaussian is divided into three regions as shown in figure. Learning rates in region A to C, are 0.2, 0.3 and 0.5 respectively of total values in that Gaussian. If α is high then μ and σ quickly moves toward new location as can see in (6) and (7). If α is slow then new value does not influence the existing μ and σ of Gaussian which indicates stationary illumination conditions. 2nd Gaussian uses only one α. Initially, both distributions start with same μ which is pixel value of 1st frame and same σ = 7. After that, a race is started between distributions. If (4) is matched for a distribution then μ and σ are updated. If value of (4) is only exist in one distribution then values of μ and σ of matched distribution are also applied to unmatched distribution. In 1st case, if GIC is continuously happening then 1st Gaussian is adopting that change but 2nd Gaussian moves slowly and fails to consider the pixel value in distribution after some frames. It is happened because of 2nd Gaussian does not compete with the GIC. In 2nd case, if GIC is increased for certain time and then starts to decrease, at that point, 1st Gaussian does not adopt that condition but 2nd Gaussian sustains that condition. High Threshold T fails to adopt the SIC and low T considers slow moving objects as background. Our method does not require T which makes the modeling efficient as compare to [1]. If a pixel value is not exist in any Gaussian according to (4) then it is considered as foreground pixel. Current value of foreground pixel is assigned to μ of both Gaussians and σ is initialized to 7. Any foreground pixel can become the part of background model if ω reaches to 10. ω of foreground pixels are updated after segmentation. If a location of foreground pixel is the part of moving vehicle according to procedure of vehicle tracking described in Sect. 4 then ω with 0 is assigned to that foreground pixel. All the other foreground pixels will get the value of 10 for ω so that quickly become the background pixel. Therefore, a slow moving vehicle is not included in background. Repetitive motions in background is also handled in this updating procedure of ω. In this way, no separated Gaussian distribution is required for background motion compare to [1, 14, 16].

4 Tracking and Sudden Illumination Change

Aforementioned steps perform the operations on individual pixels and finally each pixel is classified either foreground pixel (fg = 1) or background pixel (fg = 0). At frame level, all the connected foreground pixels are grouped into region through Matlab function "bwlabel(fg,4)". Properties of these regions such as corners, widths and centroids are determined through "regionprops(labelimg,['basic'])". Then, bubble sort is performed on regions to rank them and those regions are discarded which have less than 300 pixels. Top eight regions are selected. These regions are determined based on traffic density. GMM can only capture the gradual shifting of a pixel's intensity and it is not capable to deal with sudden illumination change (SIC) due to unexpected shift of the pixel's intensity at new location. SIC is generally happened due to clouds motion. A pixel value in SIC goes out of Gaussian and considers as foreground. This problem is solved by using the vehicle's tracking. Firstly, determined the all possible entrance points for vehicles that are three in our experiment, as can see in Fig. 4. So, a label is assigned to each vehicle when it enters to the scene through these defined entering points. At next frame, if a segmented region is partially or fully occluded to any previously region then label of previous region is assigned to this new region at current frame, as seen in Fig. 11. In this way, all the other foreground regions which are not occluded to any of the previous regions but except for three entrance points are discarded. Weights of foreground pixels in these discarded regions are increased to 10 so that they can be included in background model. If SIC is occurred at entrance then if the segmented region are within the lane then will be considered as an object. However, if the foreground region are not within the one lane at entrance points then considers it as background. If region is fully occluded to any of the previously detected region which means that current region has same width, height as in previous. Then, ω of foreground pixels in this region are not updated. So, pixels do not reach the threshold ($\omega >= 10$). Therefore, a slow moving or stopped vehicle does not consider as background region.

5 Camera Vibration Control System

A vision based system is ineffective unless camera vibration is not being control. Surveillance camera may vibrate due to wind or in case of movements on bridge if camera is installed on bridge. GMM is a parametric and temporal algorithm that applies the parameters which are updated in previous frame to extract the information in current frame. Locations are wrongly interpreted in case of camera vibration so parameters like μ and ω which are associated to a specific location are mismatched. A traffic based algorithm generally deals with a view of ground that has road surface markings for defining lanes, as shown in Fig. 4. White solid lines on borders of lanes are used in this step.

$$Starting\_point\_of\_1_{st}line@i_{th}row = Starting\_point\_of\_line@31_{th}row + \Delta y_{start} \quad (9)$$

$$\Delta y_{start} = i_{th}row/2.1882 \quad (10)$$

$$Ending\_point\_of\_1_{st}line@i_{th}row = Ending\_point\_of\_line@31_{th}row + \Delta y_{end} \quad (11)$$

$$\Delta y_{end} = i_{th}row/2.3866 \quad (12)$$

$$Width\_of\_line@i_{th}row = Starting\_point\_of\_1_{st}line@i_{th}row$$
$$: Ending\_point\_of\_1_{st}line@i_{th}row \quad (13)$$

Positions of these lines in x and y directions are determined before starting the process by continually evaluating using virtual red lines on edges as shown in Fig. 4. Rate of change for each white line in both directions are measured as described in (10) and

(12). On each iteration, if 90 % pixels which are bounded in white region for particular row (13) satisfy the condition of (4) and exists in any Gaussian then it is considered that no vibration is happened. Otherwise, parameters will be shifted in y direction in a sequence of [+ 1-1... + 20-20] and x direction in a sequence of [+ 1-1... + 30-30] according to (4) to find the best match. So, if more than 90 % pixels which are bounded in white region for particular row satisfy the condition of (4) for a particular shift of parameters in both directions and exists in any Gaussian then shifting process in both directions are stopped. So, parameters of pixels are shifted to these new locations. These steps are repeated at each row for all lines which exist at that row. Frame is divided into five zones based on lines so shifting values which are determined from line are also applied on its particular zone.

6 Shadows Removal

A searching process is applied with intensity level of 70 is used as threshold for shadow detection. This threshold is selected from observing the maximum value in shadows which appear due to trees. Maximum intensity in shadows can vary according to lighting conditions. Searching of shadow starts from edge of region in y direction and stops at that point where searching location is reached 25 % away from centroid in y direction, as described in (14).

$$Expected\_shadow\_length = (Vehicle\_Width/2) \times 0.75 \qquad (14)$$

This limit is selected due to fact that shadow will not cover the more than half area of a detected vehicle. Shadows which appear perpendicular to motion are determined and removed. Instead of using the all locations in a row at y location, here, we use three points and they represent the middle, left and right. So, following two conditions are applied for these cases.

$$At\,Center,\,Left\,and\,Right\,point <= Threshold \qquad (15)$$

$$At\,Center,\,Left\,and\,Right\,point <= Threshold\,or\,background\,pixel \qquad (16)$$

These conditions are verified on every third pixel in y direction and if there will be no condition is satisfied then it means that vehicle boundary has started. Shadows which appear parallel to vehicle are discarded using white lines on road surface which separates two lanes. If region of any vehicle covers more than one lane then white line used as reference.

7 Experiment Results and Conclusion

This algorithm is applied on a freeway which remains under GIC due to sunlight and also effected from SIC due to movement of clouds. Figure 4. shows the five white lines along the y direction which are used for measurement of displacement that is caused by

Fig. 4. Lines for measurements

movement of camera. Figure 5. describes the many ghost foreground regions which are appeared due to movement of camera. In all these foreground regions, diagonals and center values remain along the white line of image plane because the intensity values of these narrow strips of white lines remain more sensitive compare to dark background in neighboring region under camera vibration. As a result, dark part of background occupies the location of white lines so their values are not satisfied for μ and σ at these locations and wrongly considered as foreground. These ghost foreground regions are vanished through applying the proposed method of measurement and adjusting of displacements as shown in Fig. 6. Results of shadows removal technique can be seen in Figs. 7, 8, and 9. It is important to understand that our method also works well for black cars, as shown in Fig. 8. Intensity level of a of black car and its shadows is quite close, however, when the region of car from posterior side starts for shadows searching according to our approach then initially number plate appears which usually have different color as compare to vehicle body. So this information makes it possible to separate the black car from shadows and also same results are got for big vehicle as shown in Fig. 9. Convergence and adaption of Gaussian distribution is enhanced in our

Fig. 5. Segmentation

Fig. 6. Segmentation after of displacements without displacements displacements

Fig. 7. Shadows removal

Fig. 8. Shadows removal of a ` black car

Fig. 9. Shadows removal of a black car, white car and truck

method but sudden change in illumination is difficult to adjust at pixel level as discussed in Sect. 1. So, an approach to handle the SIC at frame level is defined. A ghost foreground region is shown in Fig. 10 which does not provide the any relationship to true foreground region based on previous and next frames so that region is discarded through our approach as shown in Fig. 11. Numerous challenges which can affects the true segmentation of the vehicles are dealt in our suggested approach. Requirements and utilization of this system are independent which makes it reliable and robust for all kinds of traffic roads.

Fig. 10. Ghost foreground region due to SIC

Fig. 11. Removal of ghost foreground

References

1. Stauffer, C., Grimson, W.: Learning patterns of activity using real-time tracking. IEEE Trans. Pattern Anal. Mach. Intell. **22**, 747–757 (2000)
2. Lin, H., Chuang, J., Liu, T.: Regularized background adaptation: a novel learning rate control scheme for gaussian mixture modeling. IEEE Trans. Image Process. **20**, 822–836 (2011)
3. Huang, T., Qiu, J., Ikenaga, T.: A Foreground extraction algorithm based on adaptively adjusted gaussian mixture models. In: Proceedings of the 5th International Joint Conference on INC, IMS and IDC, pp. 1662–1667. IEEE Computer Society, New York (2009)
4. Piccardi, M.: Background subtraction techniques: a review. In: Proceedings of the IEEE International Conference on Systems, Man And Cybernetics, vol. 4, pp. 3099–3104. IEEE, New York (2004)
5. Stauffer, C., Grimson, W.: Adaptive background mixture models for real-time tracking. In: Proceedings of the IEEE Computer Society Conference On Computer Vision And Pattern Recognition, Cvpr 1999, vol. 2, pp. 246–252. IEEE Computer Society, Los Alamitos (1999)
6. Bouwmans, T., Baf, F.E., Vachon, B.: Statistical background modeling for foreground detection: a survey. In: Chen, C.H. (ed.) Handbook of Pattern Recognition and Computer, pp. 181–199. World Scientific Publishing, Singapore (2010)
7. Lee, D.S.: Effective gaussian mixture learning for video background subtraction. IEEE Trans. Pattern Anal. Mach. Intell. **27**(5), 827–832 (2005)
8. Shimada, A., Arita, D., Ichiro Taniguchi, R.: Dynamic control of adaptive mixture-of-gaussians background model. In: Proceedings of the IEEE Conference on Advanced Video And Signal Based Surveillance, p. 5. IEEE Computer Society, Los Alamitos (2006)
9. Zivkovic, Z.: Improved adaptive gaussian mixturemodel for background subtraction. In: Proceedings of the 17th International Conference on Pattern Recognition, ICPR 2004, vol. 2, pp. 28–31. IEEE Computer Society, New York (2004)

10. Cheng, J., Yang, J., Zhou, Y., Cui, Y.: Flexible background mixture models for foreground segmentation. Image Vis. Comput. **24**(5), 473–482 (2006)
11. Figueiredo, M.A.T., Jain, A.K.: Unsupervised learning of finite mixture models. IEEE Trans. Pattern Anal. Mach. Intell. **24**(3), 381–396 (2000)
12. White, B., Shah, M.: Automatically tunings background subtraction parameters using particle swarm optimization. In: Proceedings of the IEEE International Conference on Multimedia And Exposition, pp. 1826–1829. IEEE, New York (2005)
13. Lee, D.S.: Effective gaussian mixture learning for video background subtraction. IEEE Trans. Pattern Anal. Mach. Intell. **27**, 827–832 (2005)
14. Nguyen, T.T.: Compensating background for noise due to camera vibration in uncalibrated-camear-based vehicle speed measurement system, IEEE Trans. Veh. Technol. **60**(1) (2011)
15. Xiangdong, Y., Jie, Y., Na, W.: Removal of disturbance of sudden illumination change based on color gradient fusion gaussian model. Int. J. Advancements Comput. Technol. **5**(2), 86–92 (2013)
16. Chen, Z., Ellis, T.: A self-adaptive gaussian mixture model. Comput. Vision Image Underst. **122**, 35–46 (2014)

Binarization Chinese Rubbing Images Using Gaussian Mixture Model

Zhi-Kai Huang[1]([⊠]), Fang Wang[1], Jun-Mei Xi[1], and Han Huang[2]

[1] Nanchang Institute of Technology,
College of Mechanical and Electrical Engineering,
Nanchang 330099, Jiangxi, China
huangzhik2001@163.com
[2] School of Mechanical and Engineering,
Harbin Institute of Technology,
Harbin 150080, Heilongjiang, China
863503892@qq.com

Abstract. Rubbings are important components of ancient Chinese books, and are the main source for people to learn, study, and research history. Image segmentation plays a crucial role in extracting useful information and characteristics of Chinese character from the rubbing images. In this paper, binarization using a Gaussian Mixture Model (GMM) with 2 components for representation of background and foreground distribution in a Chinese rubbing image has been proposed. To model the likelihood of each pixel belonging to foreground or background, a foreground and background color model are learned from three color bands samples that using RGB color space. The standard Expectation-Maximisation (EM) algorithm had been used to estimate the GMM parameters. Experimental results on real rubbing images validate the effectiveness of the model when working with Chinese rubbing images.

Keywords: Chinese rubbing images · Gaussian Mixture Model (GMM) · Expectation-Maximisation (EM) algorithm · Binarization

1 Introduction

What is rubbing? If you look for this question in the internet, you could find that a rubbing, by accurately reproducing every line of the texture of a surface created by placing a piece of paper or similar material over the subject and then rubbing the paper with something to deposit marks, most commonly charcoal or pencil, but also various forms of blotted and rolled ink, chalk, wax, and many other substances as well [1]. There is no civilization that has relied as much as the Chinese on carving inscriptions into stone as a way of preserving the memory of its history and culture.

It is very important for transformation of history Chinese rubbing into electronic form that digital libraries broaden access to ancient rubbing image for other research. There are some famous museum such as Harvard library, East Asian library in university of California and Field Museum, etc., which stored a large number Chinese rubbing image [2]. At the same time, the digitization of Chinese rubbing image is a

D.-S. Huang et al. (Eds.): ICIC 2015, Part I, LNCS 9225, pp. 477–482, 2015.
DOI: 10.1007/978-3-319-22180-9_46

valuable tool not only to preserve and spread their content, but also to extract information from such documents that could be used for retrieval and classification. It is difficult to extracted Chinese characters from rubbing image, because most rubbings contain graphics and images in addition to text. In the field of document image processing, the text/graphic separation is a major step that conditions the performance of the recognition and indexing systems. That involves identifying and separating the graphical and textual components of a document image. By focusing on segmentation algorithms, binarization process involves a pre-processing step that aims to improve recognition performance for rubbing image had been considered. Most of the document image binarization methods rely on global or local discriminating thresholds. These thresholds are determined according to some statistical analysis of the luminance or chrominance distribution generally based on histograms, without taking into account the characters shapes [3, 4]. Global techniques attempt to statistically define a global threshold. These methods show good results only when there is a clear separation between the two classes of background and text. Lee et al. [5] conducted a comparative analysis of five global thresholding methods and advanced several useful criteria for thresholding performance evaluation. In most cases, threshold is computed for every element (i.e. pixel or region) based on local statistics [6]. For example, the approach proposed by Niblack [7] attempts to vary the threshold over the image based on the local mean and standard deviation computed in a small neighborhood of each pixel. Another binarization approach using text contours and a local thresholding method was proposed by Zhou et al. [12]. Saïdane and Garcia [8] introduced an automatic binarization step based on aConvNet particularly robust to complex background and low resolution. Mishra et al. [9] presented a Markov Random Field (MRF)-based technique of binarization adapted to scene text images. Existing foreground–background separation-based systems for enhancing degraded historical documents include the work done by Gatos et al. in [10] and Agam et al. in [11, 15].

In this paper, we propose a novel automatic binarization scheme handles color Chinese rubbing documents using adaptive threshold segmentation based on Gaussians mixture models (GMM) utilizing the expectation maximization (EM) algorithm that GMM based on unsupervised learning, without making any assumptions or using tunable parameters.

2 Gaussian Mixture Model (GMM)

Suppose that we have obtained a set of N independent and identically distributed vectors $X = \{x_1, \ldots, x_j, \ldots, x_N\}$, where each D-dimensional random vector $X_i = \{x_1, \ldots, x_j, \ldots, x_D\}$ is assumed to be distributed according to a finite Gaussian Distribute mixture model with M components as [16]:

$$p(X|\Theta) = \sum_{i=1}^{M} \alpha_i p_i(X|\theta_i) \tag{1}$$

where α_i refers to the priority probability of each component.

The parameters are

$$\Theta = \{\alpha_1, \ldots, \alpha_M; \theta_1, \ldots, \theta_M\} \text{ and } \sum_{i=1}^{M} \alpha_i = 1, \theta_i = \left(\mu_i \sum_i \right) \qquad (2)$$

μ_i is the mean and Σ_i is the covariance matrix, $i = 1, \ldots, \ldots M$

Let $y_j (j = 1, \ldots, N)$ refers to the Gaussian from which x_j has come. The probability of x_j coming from the i-th Gaussian is shown in Eq. 3.

$$P(x_j | y_j = i, \theta_i) = \frac{\exp\left(-\frac{1}{2} (x_j - \mu_i)^T \sum_i^{-1} (x_j - \mu_i) \right)}{(2\pi)^{\pi/2} |\sum_i|^{1/2}} \qquad (3)$$

The aim of GMM is to estimate the hidden distribution from the image data and in order to maximise Eq. 3, the estimation of the unknown parameter Θ is needed. A GMM is usually computed in an unsupervised manner using the Expectation Maximization (EM) algorithm. The EM algorithm may estimate the GMM parameters. It works by repeating the E-step and M steps until convergence. The detail of EM algorithm for GMM could be reference [14].

3   The Results of Experiments

To verify the performance of our method, a set of various images which has stored in Harvard library (could be found in reference [15]) with resolution of 2107 * 1781 (another rubbing image size is 2108 * 1781) pixels and 256 intensity levels was tested by our methods and Ostu's methods respectively. The results of rubbing image segmentation are shown in Figs. 1, 2 and 3. For all the tested images, the images labeled (a) are original images, these label (b) are the histogram of original images, and these labeled (c), (d) are thresholding images of our method and Ostu's method, respectively. We compare our method to the well-known binarization methods that is the Otsu method. For the purpose of comparison, the results of Peak Signal to Noise Ratio (PSNR) and Mean Square Error (MSE) for two algorithms have been shown in Table 1 also. The algorithm execution times were measured in this comparison.

From the experiments results, we can find that GMM method gives more litter SNR results. From the view of PSNR, two methods show the similar value. At the same time, Execution times of two algorithms have been recorded that GMM method take on an average 900 s to produce the segmentation result for images of an average size of 2100 × 1780 pixels. The programs which had been written in Matlab runs on a PC with Intel® core(TM) 1.8 GHz clock speed and 4.0 GB of RAM.

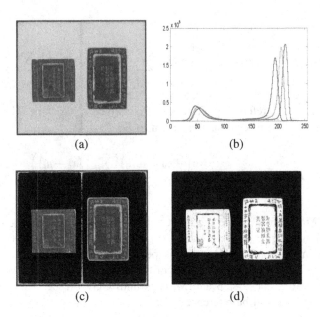

Fig. 1. Thresholding results of rubbing image which number is 12443743 [13].

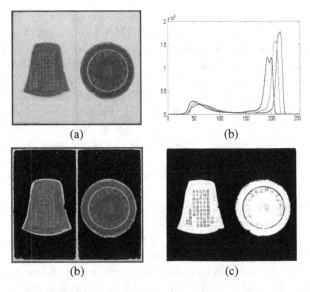

Fig. 2. Thresholding results of rubbing image which number is 12443743 [13].

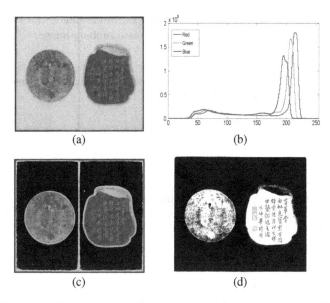

(a) (b)

(c) (d)

Fig. 3. Thresholding results of rubbing image which number is 12443743 [13].

Table 1. Thresholding evaluation of rubbing images.

No.	Method	MSE	PSNR	CPU time (s)
Name.12443743	GMM method	163.73	25.99	907.17
Size of image: 2107 * 1781	Ostu's method	179.52	25.59	0.6947
Name. 12443745	GMM method	154.53	26.24	951.89
Size of image: 2108 * 1781	Ostu's method	176.64	25.66	0.7693
Name. 12443746	GMM method	148.84	26.40	869.63
Size of image: 2108 * 1781	Ostu's method	181.87	25.53	0.7381

4 Conclusions

In this paper, we proposed a Gaussian Mixture Model (GMM) based adaptive thresholding segmentation algorithm for color rubbing image. It only calculates the threshold values automatically with the help of merging process. The first step of the method is that construct the histograms for each color channel. With this aim, information based histogram of the color intensities have been obtained. In the second step of the method, GMM is used on each of the three histograms in R (red), G (green) and B (blue) dimensions, rubbing image segmentation is obtained the performance of the GMM thresholding for each color channel. The performance of the algorithm is demonstrated by real rubbing image segmentation experiments. Experiment results show that this method can determine automatically the number of the thresholds levels and achieves good results for color rubbing images. In the future, text extraction and character recognition would be test on more Chinese rubbing image.

Acknowledgements. This work was supported by the National Natural Science Foundation of China (Grant No. 61472173), the grants from the Science and Technology Planning Project of Jiangxi Province of China, No. 20111BBG70032-2.

References

1. http://en.wikipedia.org/wiki/Rubbing
2. http://archive.fieldmuseum.org/chineserubbings/introduction_2.asp
3. Roy, B., Chatterjee, R.K.: Historical handwritten document image segmentation using morphology. In: Sengupta, S., Das, K., Khan, G. (eds.) ETCC 2014. LNEE, vol. 298, pp. 123–131. Springer, Heidelberg (2014)
4. Huang, Z.-K., Chau, K.-W.: A new image thresholding method based on Gaussian mixture model. Appl. Math. Comput. **205**(2), 899–907 (2008)
5. Le, S.U., Chung, S.Y., Park, R.H.: A comparative performance study of several global thresholding techniques for segmentation. Graph. Models Image Process. **52**, 171–190 (1990)
6. Kittler, J., Illingworth, J.: Threshold selection based on a simple image statistic. Comput. Vis. Graph. Image Process. CVGIP **30**, 125–147 (1985)
7. Niblack, W.: An Introduction to Digital Image Processing, pp. 115–116. Prentice-Hall, Englewood Cliffs (1986)
8. Saïdane, Z., Garcia, C.: Robust binarization for video text recognition. In: International Conference on Document Analysis and Recognition, vol. 2, pp. 874–879 (2007)
9. Mishra, A., Alahari, K., Jawahar, C.: An MRF model for binarization of natural scene text. In: International Conference on Document Analysis and Recognition, pp. 11–16 (2011)
10. Gatos, B., Pratikakis, I., Perantonis, S.J.: Adaptive degraded document image binarization. Pattern Recogn. **39**(6), 317–327 (2006)
11. Agam, G., Bal, G., Frieder, G., Frieder, O.: Degraded document image enhancement. In: Lin, X., Yanikoglu, B.A. (eds.) Document Recognition and Retrieval XIV. Proceeding of the SPIE, vol. 6500, pp. 65000C-1–65000C-11 (2007)
12. Zhou, Z., Li, L., Tan, C.: Edge based binarization for video text images In: International Conference on Pattern Recognition, pp. 133–136 (2010)
13. http://pds.lib.harvard.edu/pds/view/10401767?n=1&imagesize=1200&jp2Res=.25&print Thumbnails=no
14. Redner, R.A., Walker, H.F.: Mixture densities, maximum likelihood and the EM algorithm. SIAM Rev. **26**(2), 195–239 (1984)
15. Garain, U., Paquet, T., Heutte, L.: On foreground—background separation in low quality document images. Int. J. Doc. Anal. Recogn. (IJDAR) **8**(1), 47–63 (2006)

Locally Linear Representation Manifolds Margin

Bo Li[1,2(✉)], Yun-Qing Wang[1,2], Lei Lei[1,2], and Zhang-Tao Fan[1,2]

[1] School of Computer Science and Technology, Wuhan University of Science and Technology, Wuhan 430081, China
liberol@126.com
[2] Hubei Province Key Laboratory of Intelligent Information Processing and Real-Time Industrial System, Wuhan, China

Abstract. In this paper, a novel supervised multiple manifolds learning method is presented for dimensionality reduction, which is titled locally linear representation manifold margin (LLRMM). In the proposed LLRMM, both an inter-manifold graph and intra-manifold graph are constructed, where any point in the inter-manifold graph must select neighbors from other manifolds while the neighborhood in the intra-manifold graph are composed of samples from the same manifold. Then the least locally linear representation technique is introduced to optimize the reconstruction weights as well as the corresponding inter-manifold scatter and intra-manifold scatter, based on which manifolds margin can be reasoned. At last, a discriminant subspace is explored. Experiments on some benchmark face data sets have been conducted and experimental results show that the proposed method outperforms some related state-of-the-art dimensionality reduction methods.

Keywords: Dimensionality reduction · Supervised learning · Manifolds margin

1 Introduction

In many pattern recognition problems, especially in face classification, patterns are usually represented to high dimensional vectors. So it will contribute to accomplish the task of classification at low computational by mapping those data into a low dimensional space with dimensionality reduction techniques. Currently, researchers have developed many dimensionality reduction methods.

Principal component analysis (PCA) is a widely used linear dimensionality reduction method, which projects the original data into a subspace spanned by the eigenvectors associated to the leading eigenvalues of sample covariance [1]. However, PCA is completely unsupervised with regard to class information, which may result in useful information loss and recognition performance weakening [2]. Unlike PCA, linear discriminant analysis (LDA) takes full consideration of class labels of data. Generally speaking, LDA seeks an optimal subspace by a linear transformation, where the ratio of between-class scatter to within-class scatter can be maximized [3].

Unfortunately, both linear methods fail to detect the essential structure of nonlinear data in practical applications. So many nonlinear methods have to be developed such as

© Springer International Publishing Switzerland 2015
D.-S. Huang et al. (Eds.): ICIC 2015, Part I, LNCS 9225, pp. 483–490, 2015.
DOI: 10.1007/978-3-319-22180-9_47

kernel approaches and manifold learning ones. In Kernel methods, observations are implicitly mapped using some kernels into a space with much higher dimensions, where data can be well classified linearly. Kernel principal component analysis (KPCA) [4] and kernel linear discriminant analysis (KLDA) [5] are the kernel extensions of PCA and LDA, respectively.

Instead of concentrating on global Euclidean structure like KPCA and KLDA, manifold learning methods pay more attentions to local geometry of original data. In tradition manifold learning methods, k nearest neighbors (KNN) is introduced to approach the local patch of manifold distributed data, where any point and its k nearest neighbors are viewed on a super-plane and the locality of manifolds can be well constructed with linear tricks. Moreover, in order to improve their classification accuracy, some supervised modifications to these methods are also made to take advantage of class information to direct construction of local patch. Combining to data labels, marginal Fisher analysis (MFA) [6] and discriminant multi-manifold learning (DMML) [7] model an intrinsic graph representing intra-class compactness and a penalty graph characterizing inter-class separability, respectively. Similar to MFA, locality sensitive discriminant analysis (LSDA) [8] and local discriminant embedding (LDE) [9] define a within-class graph and a between-class graph, based on which the local margin defined by the corresponding graph Laplacian spectrum can be maximized. Both local Fisher discriminant analysis (LFDA) [10] and nonparametric discriminant analysis (NDA) [11] apply the traditional LDA to any local patch, which can minimize the trace of local within-class graph scatter and maximize the trace of local between-class graph scatter, simultaneously.

In this paper, we propose a locally linear representation manifolds margin (LLRMM) method for dimensionality reduction, where the proposed manifolds margin can be reasoned to be difference between the reconstruction errors in an inter-manifold graph to the reconstruction errors in an intra-manifold graph. In addition, unlike the supervised manifold learning methods mentioned above, not intra-manifold locality but all manifolds locality is well preserved in LLRMM.

2   Method

2.1   Motivation

The original LLE succeeds in data visualization. However, LLE can not efficiently extract features for classification. The reason probably lies in that LLE nonlinearly mines high dimensional data by preserving local manifold structure without considering the class information. Therefore, both manifold locality preserving and class information should be integrated to improve the recognition accuracy of LLE. On the one hand, the locality should be preserved, which helps to mine the local structure of manifold distributed data; on the other hand, the class information should be introduced to improve the classification performance. Thus a manifolds margin is globally modeled based on an inter-manifold graph and an intra-manifold graph. In the inter-manifold graph, any point in the inter-manifold graph must select neighbors from manifolds with different labels. While in the intra-manifold graph, any patch is composed of points sampled from one manifold. Moreover, a total-manifold graph is also constructed to approach the locality of all manifolds, where the k nearest neighbors of any point can be determined just by the sorted Euclidean distances without taking into account label information of manifolds.

2.2 Intra-manifold Graph, Inter-manifold Graph and Total-Manifold Graph

An intra-manifold graph can be constructed according to both class labels and local information. For any data in the intra-manifold graph, those both from the same manifold and with k bottom Euclidean distances to it are selected as its neighbors. In the local patch of the intra-manifold graph, the sample can be reconstructed by its k intra-manifold nearest neighbors with the optimal weights, which is stated below.

$$\varepsilon\big((w_{\text{intra}})_i\big) = \min \left\| X_i - \sum_{j=1}^{k} (w_i)_j X_j \right\|^2 \tag{1}$$

Similarly, we can also establish an inter-manifold graph where data must be sampled from the different manifolds with k bottom Euclidean distances to it. Moreover, a total-manifold graph is also constructed, where we only select those with k bottom Euclidean distances to point X_i as its neighbors. Thus the weights between any two points in both inter-manifold graph and total-manifold graph can be computed below, respectively.

$$\varepsilon\big((w_{\text{inter}})_i\big) = \min \left\| X_i - \sum_{j=1}^{k} (w_i)_j X_j \right\|^2 \tag{2}$$

$$\varepsilon\big((w_{Total})_i\big) = \min \left\| X_i - \sum_{j=1}^{k} (w_i)_j X_j \right\|^2 \tag{3}$$

2.3 Intra-manifold Scatter, Inter-manifold Scatter and Total-Manifold Scatter

As mentioned above, an intra-manifold graph, an inter-manifold graph and a total-manifold graph can be constructed with the corresponding optimal weights between nodes. In the following, we will quantify the intra-manifold scatter, the inter-manifold scatter and the total-manifold scatter, respectively.

For Eq. (1), a minimum reconstruction error between any point and its intra-manifold neighbors is involved to optimize the reconstruction weights. It can also be expressed with the quadratic form.

$$\varepsilon(W_{\text{intra}}) = tr(XU_{\text{intra}}X^T) \tag{4}$$

where $U_{\text{intra}} = (I - W_{\text{intra}})^T (I - W_{\text{intra}})$.

Thus the intra-manifold scatter and the total-manifold scatter can be defined in the corresponding matrix form.

$$S_{\text{inter}} = XU_{\text{inter}}X^T \tag{5}$$

where $U_{\text{inter}} = (I - W_{\text{inter}})^T (I - W_{\text{inter}})$

$$S_{total} = XU_{total}X^T \tag{6}$$

where $U_{total} = (I - W_{total})^T (I - W_{total})$

Moreover, in order to characterize the separability between different labeled manifolds, a novel manifolds margin can be modeled as follows.

$$S_M = \sum_i d(M_i, M) - \sum_i tr(M_i) \tag{7}$$

where M_i denotes the *ith* manifold and $d(M_i, M)$ means the distances between M_i and other manifolds. $tr(M_i)$ is employed to measure the clustering of M_i.

Based on Eq. (7), we introduce the sum of the minimum variance between all the points on manifold M_i to other manifolds as the expected distances $d(M_i, M)$. That is:

$$d(M_i, M) = \sum_i \{\min d(X_i, M)\} \tag{8}$$

Furthermore, the minimum distance between X_i and other manifolds M can be expressed by the variance between any point X_i to the weighted mean among X_i's k inter-manifold nearest neighbors. Thus Eq. (8) can be rewritten to:

$$d(M_i, M) = \sum_i \left\| X_i - \sum_{j=1}^k (w_i)_j X_j \right\|^2 \tag{9}$$

where $X_j(j = 1, 2, \ldots, k)$ must be the k nearest neighbors on other manifolds.

In the inter-manifold graph, $\sum_i \left\| X_i - \sum_{j=1}^k (w_i)_j X_j \right\|^2$ just characterizes the inter-manifold scatter, thus Eq. (7) can be represented to following form.

$$S_M = S_{\text{inter}} - \sum_i tr(M_i) \tag{10}$$

Equation (1) shows that any point can be reconstructed by its intra-manifold nearest neighbors and the least error $\varepsilon((w_{\text{intra}})_i)$ demonstrates data compactness sampled from one manifold. Thus the sum of all the reconstruction errors in the intra-manifold graph can be introduced to weigh the clustering the corresponding manifold. Thus, manifolds margin can be represented to:

$$S_M = S_{\text{inter}} - S_{\text{intra}} = X(U_{\text{inter}} - U_{\text{intra}})X^T \tag{11}$$

2.4   Manifolds Margin

For the task of classification, a low dimensional subspace will be explored where the points with different labels will be more separable and points from the same classes will be more clustered. In the inter-manifold graph, the reconstruction error represents the distance between a point on one manifold to the weighted mean of its nearest neighbors sampled from other manifolds. Meanwhile, for this point, it is also the minimum inter-manifold distance because its nearest neighbors are involved in calculating the weighted mean. In addition, the reconstruction error in the intra-manifold graph displays the shortest distance between a point to the weighted mean of its intra-manifold nearest neighbors, which can demonstrate the compactness of intra-manifold data. Thus it will contribute to find the optimal subspace for classification to maximize difference between the reconstruction errors in the inter-manifold graph to the reconstruction errors in the intra-manifold graph. Moreover, we also expect to well preserve the local structure of all manifolds instead of intra-manifold in the low dimensional space, which can be achieved by minimizing the reconstruction errors in the total-manifold graph. So the following objective function can be modeled in low dimensional space.

$$J(Y) = \max \frac{tr\{A^T X(U_{\text{inter}} - U_{\text{intra}})X^T A\}}{tr\{A^T X U_{total} X^T A\}} \tag{12}$$

Obviously, the solutions of Eq. (12) can be gained by solving the following eigen-decomposition problem.

$$S_M A_i = \lambda_i S_{total} A_i \tag{13}$$

3   Experiments

In this Section, the proposed LLRMM is compared with some related dimensionality reduction methods including kernel principal component analysis (KPCA), nonparametric discriminant analysis (NDA), reconstructive discriminant analysis (RDA) and discriminant multiple manifold learning (DMML).

3.1   Experiments on AR Face Data

AR face contains over 4,000 color face images of 126 people, including frontal views of faces with different facial expressions, lighting conditions, and occlusions. In this experiment, 14 grayscale face images (each session containing 7) of these 120 individuals are selected. The face portion of each image is manually cropped, which is displayed in Fig. 1.

Fig. 1. The sample images for one person in AR database.

Figure 2 shows the mean accuracies and the corresponding standard deviations by DMML, KPCA, NDA, RDA and LLRMM with 5, 6, 7, 8 training samples each class, where the experiments are repeated 10 times. It is disclosed that the proposed LLRMM acquires the best mean accuracy no matter how many training samples are chosen.

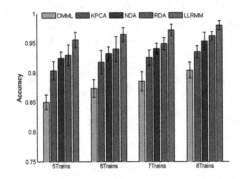

Fig. 2. The mean accuracy and standard deviation on AR face data with various training samples

3.2 Experiments on Yale Face Data

Yale face database was constructed at the Yale Center for Computation Vision and Control. There are 165 images about 15 individuals in YALE face data sets, where each person has 11 images. The images demonstrate variations in lighting condition, facial expression and with or without glasses. Figure 3 shows one object from Yale database.

We randomly select the 4, 5, 6 images as training sets and the rest 7, 6, 5 images as test sets for each class. Displayed in Table 1 are the best recognition rates and the corresponding dimensions. It can find that the proposed method has the comparably better recognition performance.

Fig. 3. Sample images of one person in Yale database

Table 1. The maximum accuracy on Yale face data with the corresponding dimensions

	4Trains	5Trains	6Trains
DMML	82.86 %(14)	87.78 %(14)	90.67 %(10)
KPCA	84 %(14)	91.11 %(14)	93.34 %(14)
NDA	85.71 %(14)	93.78 %(22)	96 %(12)
RDA	88.57 %(14)	94.44 %(8)	97.33 %(14)
LLRMM	89.52 %(14)	95.56 %(10)	97.33 %(14)

4   Conclusions

In this paper, we reviewed some extensions to LLE to solve the supervised learning problem, which either use class information to adjust distances between two nodes in KNN graph or combine LDA to LLE for features extraction discriminately. In LLRMM, labels are introduced to supervise the construction of the inter-manifold graph and the intra-manifold graph respectively. In the intra-manifold graph, a point and its KNN neighborhood points must be sampled from the same manifold. Oppositely, in the inter-manifold graph, any point should select its k nearest neighbors from other manifolds. Based on the inter-manifold graph and the intra-manifold graph, a novel manifolds margin is defined with the least locally linear reconstruction technique, which can be reasoned to be the difference between the total reconstruction errors in the inter-manifold graph to the total reconstruction errors in the intra-manifold graph. At last, a subspace is explored with the maximum manifolds margin and the minimum manifolds locality.

Acknowledgments. This work was partly supported by the grants of National Natural Science Foundation of China (61273303, 61273225 & 61373109).

References

1. Jolliffe, I.T.: Principal Component Analysis. Springer, New York (2002)
2. Yang, H., Zhang, C.: The research on two kinds of restricted biased estimators based on mean squared error matrix. Commun. Stat. Theory Methods **37**(1), 70–80 (2008)
3. Kim, T.K., Stenger, B., Kittler, J., Cipolla, R.: Incremental linear discriminant analysis using sufficient spanning sets and its applications. Int. J. Comput. Vision **91**(2), 216–232 (2011)
4. Martinez, A.M., Kak, A.C.: PCA versus LDA. IEEE Trans. Pattern Anal. Mach. Intell. **23**(2), 228–233 (2001)

5. Scholkopf, B., Smola, A., Muller, K.R.: Nonlinear component analysis as a kernel eigenvalue problem. Neural Comput. **10**(5), 1299–1319 (1998)

6. Xu, D., Yan, S., Tao, D., Lin, S., Zhang, H.J.: Marginal fisher analysis and its variants for human gait recognition and content- based image retrieval. IEEE Trans. Image Process. **16** (11), 2811–2820 (2007)

7. Lu, J., Tan, Y.P., Wang, G.: Discriminative multi manifold analysis for face recognition from a single training sample per person. IEEE Trans. Pattern Anal. Mach. Intell. **35**(1), 39–51 (2013)

8. Cai, D., He, X., Zhou, K., Han, J., Bao, H.: Locality sensitive discriminant analysis. In: Proceedings of the 20th International Joint Conference on Artificial Intelligence, pp. 708–713 (2007)

9. Chen, H.T., Chang, H.W., Liu, T.L.: Local discriminant embedding and its variants. In: Proceedings of the 2005 IEEE Computer Society Conference on Computer Vision and Pattern Recognition, pp. 846–853 (2005)

10. Sugiyama, M.: Local fisher discriminant analysis for supervised dimensionality reduction. In: Proceedings of International Conference on Machine Learning, pp. 905–912 (2006)

11. Li, Z., Lin, D., Tang, X.: Nonparametric Discriminant Analysis for Face Recognition. IEEE Trans. Pattern Anal. Mach. Intell. **31**(4), 755–761 (2009)

Modified Sparse Representation Based Image Super-Resolution Reconstruction

Li Shang[1(✉)], Pin-gang Su[1], and Zhan-li Sun[2]

[1] Department of Communication Technology, College of Electronic Information Engineering, Suzhou Vocational University, Suzhou 215104, Jiangsu, China
{sl0930, supg}@jssvc.edu.cn
[2] School of Electrical Engineering and Automation, Anhui University, Hefei 230039, Anhui, China
zhlsun2006@126.com

Abstract. A modified sparse representation based image super-resolution reconstruction (ISR) is discussed in this paper. The edge features of high resolution (HR) image patches and the gradient and texture features of low resolution (LR) image patches are considered in our method. Meanwhile, features of LR image patches are classified by extreme learning machine (ELM) classifier. Further, For image patches' features classified, the fast sparse coding (FSC) algorithm based K-SVD sparse representation is used to train sparse dictionaries. And utilized these dictionaries, LR images can be super-resolution reconstructed well. Simulation results show that our method has clear improvement in visual effect and retain well image detail.

Keywords: Sparse representation · Super-resolution reconstruction · Fast sparse coding · Extreme learning machine (ELM) classifier · K-SVD algorithm

1 Introduction

Sparse representation is an efficient image representation model and has been used widely in image super-resolution reconstruction (ISR) fields [1, 2]. The usual sparse representation method is the k-means based singular value decomposition (K-SVD) algorithm [3, 4], and the K-SVD dictionary is commonly learned by using algorithms of matching pursuit (MP) [3], orthogonal MP (OMP) [4], regularized OMP (ROMP) and so on. To an extent, these methods developed can obtain high image resolution and only need one input image. However, they can not capture many image details and still consume much calculation time [5]. To reduce convergent time and obtain better image structure and details, we propose a modified ISR method here, where spare dictionaries are trained by fast sparse coding (FSC) [6] based K-SVD algorithm [3] that can consume less time than common K-SVD algorithms above-mentioned. In this method, the reconstruction effect of utilizing features classified of LR image patches is considered. For high resolution (HR) image patches, their edge features are extracted by using the Canny algorithm. And for LR image patches, their first order and second order gradient images in horizontal direction and vertical direction, namely four gradient images, are considered. Then, utilized the center values of image patches as the

© Springer International Publishing Switzerland 2015
D.-S. Huang et al. (Eds.): ICIC 2015, Part I, LNCS 9225, pp. 491–497, 2015.
DOI: 10.1007/978-3-319-22180-9_48

criterion of image edge pixels, the LR edge and texture image patches can be solved. Further, the extreme learning machine (ELM) classifier is utilized to classify edge and texture features of LR images, thus, each class's cluster center value can be calculated. Then according to feature classification results of LR image patches, the HR image features are classified well corresponding to the classification sequence number of LR image patches' features. For classification samples, the (FSC) based K-SVD algorithm is used to train HR and LR sparse dictionaries. Further, utilized dictionaries learned, the LR image can be reconstructed well. Finally, utilized signal noise ratio (SNR) values to estimate the quality of reconstructed images, simulation results show that our method proposed here can obtain better visual effect and more details.

2 FSC Algorithm

2.1 The Objection Function of FSC Algorithm

The fast sparse coding (FSC) algorithm is based on iteratively solving two convex optimization problems, namely the L_1-regularized and the L_2-constrained least squares problem. And L_1 regularization is known to produce sparse coefficients and can be robust to irrelevant features. The maximum a posteriori estimate of bases and coefficients is the solution to the following optimization problem:

$$J = \frac{1}{2\sigma^2} \|X - DS\|_F^2 + \beta \sum_{i,j} \phi\left(s_j^i\right). \tag{1}$$

where s_j is the *jth* column vector of the coefficient matrix S and $\phi(\cdot)$ is a sparsity function and β is a constant.

2.2 Learning of FSC Feature Bases

In FSC algorithm, the basis feature vectors and sparse coefficients are updated in turn. Keeping the bases fixed, the sparse coefficient matrix S can be solved by optimizing each $s_j^{(t)}$ individually [6]:

$$J1 = \left\| x^{(i)} - \sum_j d_j s_j^{(i)} \right\|_2^2 + (2\sigma^2 \beta) \sum_j \left| s_j^{(i)} \right|. \tag{2}$$

The learning process of $s_j^{(i)}$ is called feature-sign search algorithm. This algorithm converges to a global optimum in a finite number of steps. Given fixed sparse coefficients S, the bases D can be learned by the following problem [6]:

$$\|X - DS\|_2^2. \tag{3}$$

Subject to $\sum_{i=1}^{k} D_{i,j}^2 \leq c$, $\forall j = 1, 2, 3, \ldots, n$. This is a least squares problem with quadratic constraints. Using a Lagrange dual, this problem can be much more efficiently solved. The Lagrangian form of Eq. (3) is written as follows [6]:

$$L(D, \lambda) = trace\left((X - DS)^T (X - DS)\right) + \sum_{j=1}^{n} \lambda_j \left(\sum_{l=1}^{K} D_{i,j}^2 - c\right). \qquad (4)$$

And the Lagrange dual can be obtained:

$$D(\lambda) = \min_{D} L(D, \lambda) = trace\left(X^T X - X S^T \left(S S^T + \Lambda\right)^{-1} \left(X S^T\right)^T - c\Lambda\right). \qquad (5)$$

where $\Lambda = diag(\lambda)$. The gradient and Hessian of $D(\lambda)$ are computed as follows:

$$\partial D(\lambda)/\partial \lambda_i = \left\| X S^T \left(S S^T + \Lambda\right)^{-1} e_i \right\|^2 - c. \qquad (6)$$

$$\partial^2 D(\lambda)/(\partial \lambda_i \partial \lambda_j) = -2\left(\left(S S^T + \Lambda\right)^{-1}\left(X S^T\right)^T X S^T \left(S S^T + \Lambda\right)^{-1}\right)_{i,j} \left(\left(S S^T + \Lambda\right)^{-1}\right)_{i,j}. \qquad (7)$$

where e_i is the i–th unit vector. Using Newton's method or conjugate gradient, the Lagrange dual Eq. (5) can be optimized. The optimal basis vectors are deduced as follows:

$$D^T = \left(S S^T + \Lambda\right)^{-1}\left(X S^T\right)^T. \qquad (8)$$

3   The Modified ISR Method

In the modified ISR method, the classed features of image patches are used as the input samples of sparse representation algorithm. The sparse dictionaries are learned by the FSC based K-SVD algorithm. Utilized dictionaries learned, the LR image can be reconstructed. And the reconstruction method is generalized briefly as follows:

Step 1.   Input data. The HR image and LR image are randomly divided into image patches with $p \times p$ pixels. Thus, the HR and LR image patch set can be obtained.

Step 2.   Extracting features of HR image patches by using the Canny algorithm.

Step 3.   Extracting edge and texture features of LR image patches. The four gradient images of a LR image are first calculated by convolution calculation. Each gradient image is randomly divided into patches. Used each image patch's center pixel value to be the edge pixel criterion, the edge and texture image patches are obtained.

Step 4.   Classifying features learned of LR image patches by using ELM classifier.

Step 5.   Learning the sparse dictionary of HR image features by using basic K-SVD model.

Step 6.   Learning the sparse dictionary of LR image features by using the FSC based K-SVD model.

Step 7.   Utilizing dictionaries trained to reconstructing the LR image by using the ordinary ISR idea.

4   Experimental Results

4.1   Feature Extraction of HR and LR Image Patches

In test, a clear image called Elaine (i.e. HR image) with the size of 512 × 512 pixels and its degenerated version (i.e. LR image) was used, as shown in Fig. 1(a)–(c). For the LR Elaine image, four gradient feature images of the first and second order in horizontal and vertical direction were deduced as shown in Fig. 2. At the same time, the 4 gradient images' real parts of texture images obtained by Gabor filter were also shown in Fig. 2. And then, each HR and LR gradient feature images were sampled randomly with 3 overlapped pixels. For HR image patches, the edge features were extracted by Canny algorithm, and some of edge feature image patches were shown in Fig. 3. For LR gradient image patches, utilized each image patch's center value as the threshold value to retain maximize feature information, then the edge and texture feature image patches were obtained, and some texture patches of gradient image patches were shown in Fig. 4. Further, using ELM algorithm to classify LR feature image patches, and the classified image patch set was used as K-SVD model's input set.

4.2   Feature Extraction of HR and LR Image Patches

Utilized our modified K-SVD method to train LR feature image patches classified, which were obtained according to the description in Subsect. 4.1, the corresponding LR and HR dictionaries with 256 atoms, denoted by $\hat{D}_{fl}$ and $\hat{D}_{fh}$, were obtained. And then combined the cluster center values of LR feature image patches classified and $\hat{D}_{fl}$,

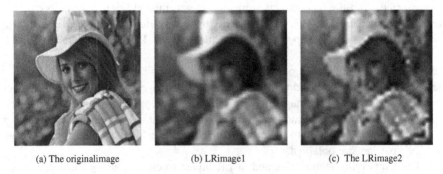

(a) The originalimage   (b) LRimage1   (c) The LRimage2

Fig. 1. The original Elaine image and its corresponding LR versions.

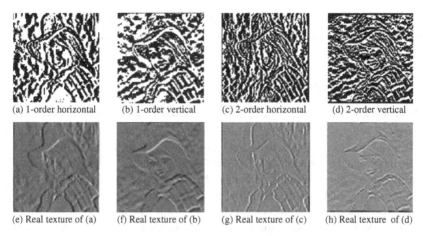

(a) 1-order horizontal   (b) 1-order vertical   (c) 2-order horizontal   (d) 2-order vertical

(e) Real texture of (a)   (f) Real texture of (b)   (g) Real texture of (c)   (h) Real texture of (d)

Fig. 2. LR Elaine image's 4 Gradient images and corresponding texture images.

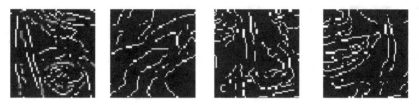

Fig. 3. Some edge feature patches randomly selected from Elaine's HR edge image patch set.

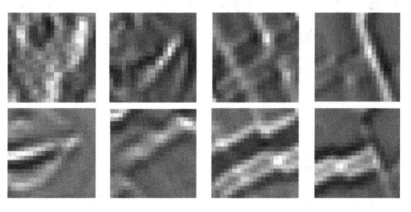

Fig. 4. Some 2-order gradient feature patches randomly selected from Elaine's LR gradient image patch set. The top line: 2-order vertical patches. The bottom line: 2-order horizontal.

the LR dictionary $\hat{D}_l$ was obtained. Utilized $\hat{D}_l$ and the LR feature image patch vector $\{x_l^k\}$, the LR coefficient vector $\{s_l^k\}$ can be learned by the following form

$$s_l^k = \arg\min_s \frac{1}{2}\left\|x_l^k - \hat{D}_l s_l^k\right\|_2^2 + \gamma \left\|s_l^k\right\|_1. \tag{9}$$

Utilizing LR coefficient vector s_l^k and the HR dictionary $\hat{D}_h$, the reconstructed HR image patch vector $\hat{x}_h^k = \hat{D}_h s_l^k$ can be calculated. Further, considered the mean values of HR image patches and the corresponding image patch location sampled randomly, the HR image can be restored well. Several LR Elaine images and their corresponding reconstructed results obtained by different algorithms were shown in Fig. 5. To compare different algorithms' ISR effect, the signal noise ratio (SNR) criterion was used to evaluate the quality of reconstructed images, as well as calculated SNR values was listed in Table 1. According to the visual effect and SNR values, it is clear to see that our method outperforms FSC and common K-SVD algorithm.

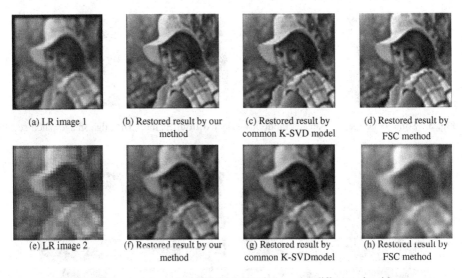

(a) LR image 1   (b) Restored result by our method   (c) Restored result by common K-SVD model   (d) Restored result by FSC method

(e) LR image 2   (f) Restored result by our method   (g) Restored result by common K-SVDmodel   (h) Restored result by FSC method

Fig. 5. The restored results of Elaine's LR images by different algorithms.

Table 1. SNR values of restored images using different algorithms.

Algorithms \ Images	Restored result of LR image 1	Restored result of LR image 2	LR image 1	LR image 2
Our method	12.72	9.42		
K-SVD model	11.27	6.13	7.76	4.77
FSC method	9.46	5.56		

5 Conclusions

A new image super-resolution method using the K-SVD model based on fast sparse coding is discussed in this paper. In this ISR method, edge features of HR image patches and the edge and texture features of LR image patches are considered. Further, the ELM classifier is used to classify LR gradient image patches. The classified feature image patches is used as the input set of K-SVD algorithm, thus the LR dictionary $\hat{D}_{fl}$ with classification is obtained. Combined the cluster center values of LR feature image patches classified and $\hat{D}_{fl}$, LR dictionary of LR images can be obtained. Further, utilized HR dictionary and LR sparse coefficients, at the same time, considered the classification label between HR and LR image patches, as well as the HR image patches' mean values, the LR image patches can be restored. And then, considered the sampled location of HR image patches, the LR image can be reconstructed well. Simulation results show that, compared with FSC and K-SVD model, our method indeed has better image structures and visual effect.

Acknowledgment. This work was supported by the grants from National Nature Science Foundation of China (Grant No. 61373098 and 61370109), the grant from Natural Science Foundation of Anhui Province (No. 1308085MF85).

References

1. Dong, W.S., Shi, G.M., Zhang, L., Wu, X.L.: Super-resolution with nonlocal regularized sparse representation. In: 2010 SPIE Conference on Visual Communication and Image Processing. IEEE Press, Huangshan, pp. 1–10 (2010)
2. Elad, M.: Sparse and Redundant Representation: From Theory to Applications in Signal and Image Processing, pp. 228–237. Springer, New York (2010)
3. Aharon, M., Elad, M., Bruckstein, A.: K-SVD: an algorithm for designing overcomplete dictionaries for sparse representation. IEEE Trans. Sig. Process. **54**(11), 4311–4322 (2006)
4. Tony, C.T., Wang, L.: Orthogonal matching pursuit for sparse signal recovery with noise. IEEE Trans. Inf. Theory **57**(7), 4680–4688 (2011)
5. Yang, J.C., Wright, J., Huang, T., Ma, Y.: Image super-resolution via sparse representation. IEEE Trans. Image Process. **19**(11), 2861–2873 (2010)
6. Shang, L., Cui, M., Chen, J.: Palm recognition using fast sparse coding algorithm. In: Huang, D.-S., Gan, Y., Gupta, P., Gromiha, M. (eds.) ICIC 2011. LNCS, vol. 6839, pp. 701–707. Springer, Heidelberg (2012)

Planning Feasible and Smooth Paths
for Simulating Realistic Crowd

Libo Sun, Lu Ding, and Wenhu Qin[✉]

School of Instrument Science and Engineering, Southeast University,
Nanjing 210096, China
{sunlibo, qinwenhu}@seu.edu.cn

Abstract. A very important challenge in many virtual applications is to plan feasible, smooth and congestion-free paths for virtual agents in dynamic and complex environments. The agents should move towards their destinations successfully while avoiding the collisions with other agents and static and dynamic obstacles. In this paper, we propose a novel approach for realistic path planning. We first create a navigation mesh for the walkable regions in a two-dimensional environment. Then an A* search on this graph determines a series of connected meshes for agents to go through from the start position to the goal position and furthermore, the walkable corridor whose radii equal to maximum clearance to the obstacles is built based on backbone path derived from the inflection point method and Catmull-Rom spline. Finally, a local collision avoidance algorithm is integrated to guarantee that agents navigating in the corridor do not collide with other agents and dynamic obstacles. Our experiments show that we can compute feasible, smooth and realistic paths for agents situated in dynamic environments in real time.

Keywords: Crowd simulation · Path planning · Navigation mesh

1 Introduction

Recently, real-time crowd simulation has been gaining considerable attention due to its applications in entertainment, education, architecture, training, urban engineering and virtual heritage. Crowd simulation consists of many different components, including perception, path planning, behavior, locomotion and how to integrate them effectively. Path planning can guarantee that agents reach their goals without colliding with obstacles and other agents and it is a very important aspect of crowd simulation that researchers should put great effort into.

Simulating crowd motions in a realistic manner is complex and difficult, especially when agents are moving in dynamic and complicated environments containing both static and moving obstacles. Existing approaches often simplify this problem by decoupling global planning from local collision avoidance. However, most of global planning can only deal with the static environments in which positions of obstacles cannot change at all while local collision avoidance gets stuck easily so that it may not find a global and existent path. Even the approach combined global planning with local collision avoidance may compute an aesthetically unpleasant and fixed path at high

D.-S. Huang et al. (Eds.): ICIC 2015, Part I, LNCS 9225, pp. 498–509, 2015.
DOI: 10.1007/978-3-319-22180-9_49

computation cost. Therefore, our aim is to provide an optimal and smooth path that not only can adapt to a dynamic environment and avoid local minima, but can also reduce the computation cost and guarantee the simulation is real time.

In this paper, we propose a novel approach to plan feasible, smooth and congestion-free paths for virtual agents in dynamic and complex environments. Our approach first creates a navigation mesh which partitions the two-dimensional environment into a collection of walkable regions. In a second stage, an A* search on this graph is used to guide each agent through the environment and furthermore, the walkable corridor, which allows agents to move free inside and show their social characteristics, is built according to backbone path derived from the inflection point method and Catmull-Rom spline so that a global and smooth path can be extracted for each agent. Finally, a local collision avoidance algorithm is combined with global planning to make agents not collide with dynamic obstacles and other agents.

The key contribution of this work can be summarized as follows:

- Our approach can reduce the computation complexity of running A* algorithm on navigation mesh than a low-resolution grid;
- Our approach can compute much smoother, more flexible and realistic paths for agents situated in dynamic environment;
- Our approach can provide a basis for simulating realistic crowd in various scenarios in real time.

2   Related Work

Virtual crowd simulation is a wide topic and this overview focuses on crowd motion simulation. In the crowd simulation research field, most researchers are concerned with the realism of the behaviors of each individual in the crowd. Thus, the perception, the memory, the planning, the psychology, and the emotion of every agent are taken into account and each agent could react differently to the same event as a result.

Since Reynolds [1] demonstrated that emergent flocking and other behaviors can be generated from simple local rules, the rule-based model has become popular and a considerable corpus of research has emerged focused on modeling human crowds with it. The most sophisticated representatives are artificial fish [2] and cognitive modeling [3]. Furthermore, Shao [4] extends the rule-based model with pedestrian visibility and path planning. Other further work has accounted for sociological factors [5], psychological effects [6], cognitive reasoning [7], geographically-based direction [8], and so on. However, human behaviors are so complex that it requires vast computational resources to use and many related factors to consider. Therefore, much work only deals with and focuses on navigation, which can be separated into local collision avoidance and global planning.

Collision avoidance methods have been proposed including geometrically-based algorithms [9], grid-based methods [10], force-based methods [11], Bayesian decision processes [12] and divergence-free flow tiles [13]. However, collision avoidance alone cannot properly model real crowds with specific goal due to the possibility of getting stuck. Therefore, local methods are often combined with global planning techniques.

In general, global planning has taken the form of graph-based techniques or static potential fields, such as navigation graphs [14], probabilistic maps [15], coarse graph-based roadmaps [16] and so on.

In recent years, the concept of path planning inside corridors has been introduced [17, 18]. Such a corridor is defined as a sequence of empty disks. Since the union of these disks is two-dimensional, corridors facilitate creating collision-free smooth paths for agents while avoiding other agents. This flexibility is difficult to obtain for approaches operating on one-dimensional graphs.

Our method is inspired by the concept of path planning inside corridors [18]. We create the navigation mesh and run A* algorithm to guide agents through the environment. However, we create the navigation mesh by adopting Constrained Delaunay triangulation method. We compute a global and smooth backbone path based on the results of the inflection point method and Catmull-Rom spline. Furthermore, the integration of the corridor concept can allow agents move flexibly inside corridors and show their social characteristics which guarantees the realism of the crowd.

3 Overview

The focus of this work is to plan realistic and smooth paths for the crowd. It should not only provide an optimal path for each agent, but should also show crowd characteristics at low cost. Our approach can be divided into three phases to achieve the aforementioned objectives. The first phase is to create a navigation mesh according to the environment the crowd situated in, which is described in Sect. 4.1. The second phase runs A* algorithm on navigation mesh to get a series of connected meshes for each agent to go through from start position to goal position and furthermore, the walkable corridor is constructed based on the backbone path derived from the inflection point method and Catmull-Rom spline to allow agents move inside freely and make their paths smooth and natural, which are described in Sect. 4.2. The third phase integrates local collision avoidance method to make agents not collide with other agents and dynamic obstacles, which is described in Sect. 4.3. In each time step, these three phases are executed successively until the simulation ends.

The framework architecture is illustrated in Fig. 1. The perception information is first acquired as an input so that the positions of static obstacles are determined and navigation mesh can be created for the two-dimensional environment. Then the global planning algorithm formulates an optimal and smooth path while the local collision avoidance algorithm makes agents not collide with each other as well as dynamic obstacles. That is, local collision avoidance algorithm predicts whether there is a collision going to happen every time step. When there is no collision, the global planning algorithm guides agents to achieve their goals. However, once there is a predicted collision, the local collision avoidance algorithm takes the priority to make agents adapt motions so that the collision is resolved. The loop continues until all agents achieve their goals.

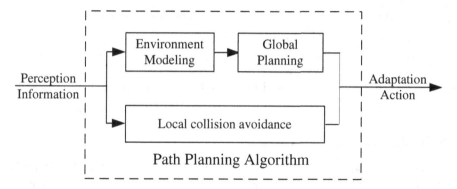

Fig. 1. The framework architecture of our approach

4 Path Planning Approach

4.1 The Generation of Navigation Mesh

The navigation mesh can be created off-line to improve the performance of our path planning algorithm. We first extract the vertexes of all static obstacles as well as those of outer boundary of the environment, and after that, Constrained Delaunay Triangulation method is adopted to partition the two-dimensional environment into non-overlapping walkable regions. The concrete algorithm is described as follows:

(1) Identify set of vertexes V and set of edges E for all static obstacles and outer boundary of the environment;
(2) Take any one edge from set of edges composing outer boundary of the environment and supposed it is called p_1p_2;
(3) Compute DT point called p_3 and construct a constrained triangle called $\Delta p_1p_2p_3$;
(4) If the new generated edge p_1p_3 is not a constrained edge and it hasn't already existed in the stack, then push it on the stack; else delete it. The same operation can be done for the new generated edge p_3p_2;
(5) Popup an edge and go to Step 3 until the stack is empty.

- **Computation of DT Point**

   We define the vertex opposite one edge as DT point of this edge in constrained Delaunay Triangulation. We call point p_3 a visible point for edge p_1p_2 if it satisfies the following conditions at the same time:

(1) Point p_3 is on the right side of edge p_1p_2 (Note that the vertex order is clockwise);
(2) Point p_3 is visible to point p_1, which means for any edge $e \in E$, $e \cap (p_1p_2) = \phi$;
(3) Point p_3 is visible to point p_2.

The computation of DT point of edge p_1p_2 can be described as follows:

(1) Find a visible point for edge p_1p_2, which is called p_3;
(2) Search another visible point p for edge p_1p_2 with $p \in V$ and check whether $\angle p_1pp_2 > p_1p_3p_2$, if it is, then $p_3 = p$;
(3) Go to step 2 until there is no visible point for edge p_1p_2 belonging to V.

4.2   Global Planning Using Corridors

In general, A* algorithm is used to plan a global path on a low-resolution grid since it is simplest and always finds a shortest path if one exists. However, it is time consumed when the virtual environment becomes very large since many paths have to be planned simultaneously. In addition, paths resulting from A* algorithm tend to have little clearance to the obstacles and can be aesthetically unpleasant. Therefore, we only run A* algorithm searching on navigation mesh to compute the connected meshes for agents to go through from start position to goal position. We construct the walkable corridor for agents to move inside freely based on the backbone path derived from the inflection point method and Catmull-Rom spline. As a result, a flexible, smooth and natural path can be extracted for each agent.

- **A* Algorithm Searching on Navigation Mesh**

   After the start position s and the goal position g are determined, our approach firstly finds mesh M_s with the start position s situated in and mesh M_g with the goal position g situated in respectively, then performs A* algorithm between mesh M_s and mesh M_g on navigation mesh to get a sequence of meshes for agents to go through. Note that for A* algorithm searching on navigation mesh,

$$f(n) = g(n) + h(n)$$

$g(n) = g(n-1) +$ Euclidean distance between the center of mesh M_{n-1} and the center of mesh M_n

$h(n) =$ Euclidean distance between the center of mesh M_n and the goal position g

- **The Inflection Point Method**

   In our approach, we adopt the inflection point method to compute the positions of inflection points and furthermore, construct the corridor in which agents can move inside freely. The concrete steps for determining the inflection points are described as follows:

(1) Find all outlet edges of a sequence of meshes for agents to go through from M_s to M_g;
(2) Connect the start position s with each endpoint of the nearest outlet edge respectively. We call one endpoint Point-Left and the other one Point-Right where Point-Left and Point-Right are determined by the sequence for representing

triangles. As a result, two segments are formed, in which one connecting the start position *s* with Point-Left is called Line-Left and the other one connecting the start position *s* with Point-Right is called Line-Right;

(3) Find two endpoints of the next nearest outlet edge continually. Check whether new Point-Left is between Line-Left and Line-Right, if it is, update Line-Left to be the segment connecting the start position *s* with new Point-Left; else, do nothing on Line-Left. Similarly, the same operations are done on Line-Right;

(4) Go to Step 3 until both new Point-Left and new Point-Right are on the right side of Line-Right or new Point-Left is on the right side of Line-Right while new Point-Right stays the same, then the endpoint of Line-Right (not the start position) is the inflection point. The way for dealing with Line-Left is the same as Line-Right.

Once the inflection point is determined, we could view it as a new start position and do the same operations to find other inflection points if there are until the last outlet edge to the goal position *g* has been checked.

- **The Construction of Corridor**

The inflection points have little clearance to the obstacles and global backbone path is not pleasant if it is generated by directly connecting start position with goal position through the inflection points in sequence. Therefore, we adopt two different methods to improve a global backbone path from start position *s* to goal position *g* according to the existence of inflection points:

(1) There are inflection points. We firstly find all outlet edges between Line-Left and Line-Right with inflection points, and then compute each midpoint for every outlet edge respectively. We view these midpoints as waypoints and connect start position *s* with goal position *g* through a series of midpoints sequentially to get a global backbone path;

(2) There are no inflection points. We firstly check whether goal position *g* is between Line-Left and Line-Right, if it is, a global backbone path is obtained by connecting start position *s* with goal position *g* directly; else find two intersecting point of last outlet edge with Line-Left and Line-Right, called point L_1 and point L_2. Compute the midpoint of segment L_1L_2, called L_{mid}, and connect start position *s* with goal positionthrough L_{mid} to get a global backbone path.

After we obtain a global backbone path, we use Catmull-Rom spline to make it smoother and furthermore, we construct discrete corridors based on smoothed backbone paths to guarantee that paths are not fixed for every agent and the realism of global planning can be improved. We define a discrete corridor $B = (B_i, R_i)$ as a sequence of largest empty balls with radii R_i whose center points B_i lie along its backbone path B. To ensure that a path can be found inside the corridor, every two adjacent balls, $\mathbf{B_i} = (\mathbf{B_i}, \mathbf{R_i})$ and $\mathbf{B_{i+1}} = (\mathbf{B_{i+1}}, \mathbf{R_{i+1}})$, should satisfy the following conditions: $d(\mathbf{B_i}, \mathbf{B_{i+1}}) + \mathbf{r} \leq \mathbf{R_{i+1}}$ where $d(B_i, B_{i+1})$ is the Euclidean distance between B_i and B_{i+1}, r is the radius of the agent, R_{i+1} is the radius of ball B_{i+1}.

4.3   Local Collision Avoidance Algorithm

Agents should avoid collisions with other agents and dynamic obstacles after their global paths are determined. In our paper, we adopt RVO [9] local collision avoidance algorithm to guarantee that the path of the agent is collision-free, smooth and natural. Compared with Velocity Obstacles concept whose idea is to choose a velocity that lies outside any of the velocity obstacles induced by the moving obstacles [19], the idea of Reciprocal Velocity Obstacles is to choose a new velocity that is the *average* of its current velocity and a velocity that lies outside the other agent's velocity obstacle. It is demonstrated that RVO local collision avoidance algorithm could resolve the problem of undesirable oscillations in multi-agent navigation.

5   Results

All the experiments were run on a PC with a NVIDIA GeForce 9800GT graphics card and an Intel Core 2 E8400 CPU (3.0 GHZ) with 4 GB memory. Our virtual scenes are constructed through the Unity3D Engine.

Figure 2 shows the navigation mesh of 2D environment we create. The frame in blue represents the outer boundary of the environment. The squares in red represent static obstacles the agents should avoid collisions when they move towards their goals. The lines in blue partition our two-dimensional environment into a collection of walkable regions.

Figure 3 shows the process of computing a global backbone path and constructing corresponding corridor for agents moving from start position to goal position. In Fig. 3 (a), start position is labeled as S circled in blue and goal position is labeled as G circled in red. A* algorithm searching on navigation mesh is performed to get a series of connected meshes for agents to go through from mesh with start position S situated in to mesh with goal position G situated in, which are shown in navy-blue arrows. After

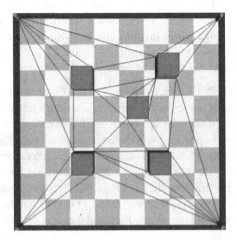

Fig. 2. The navigation mesh of 2D environment

that, the inflection point method is adopted to check whether inflection point exists. Figure 3(b), (c), (d) and (e) show the process of updating Line-Left and Line-Right since new Point-Left or new Point-Right of outlet edge is between Line-Left and Line-Right. In Fig. 3(f), we find that no inflection point exists and goal position G is not

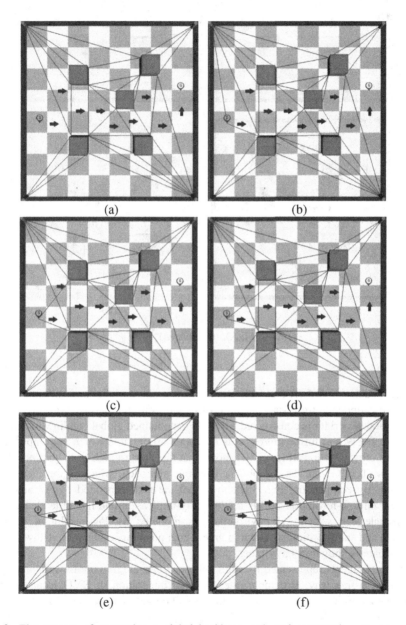

(a)   (b)

(c)   (d)

(e)   (f)

Fig. 3. The process of computing a global backbone path and constructing corresponding corridor for agents moving from start position S to goal position G

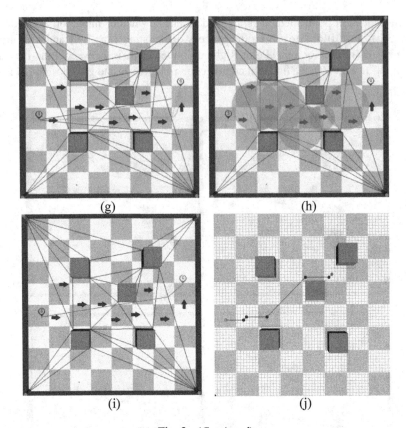

Fig. 3. (*Continued*)

between updated Line-Left and Line-Right. As a result, a global and smoothed back-bone path is obtained from start position S to goal position G in Fig. 3(g). Figure 3(h) shows that the corridor is constructed based on the obtained global path in which agents can move freely and furthermore, paths of agents are flexible which can improve the realism of crowd motions. Figure 3(i) shows that the shortest path from start position S to goal position G can also be obtained by our approach, which is the idea of A* algorithm. By comparison, Fig. 3(j) shows that the path generated by A* algorithm searching on low-resolution grid is aesthetically unpleasant and has little clearance to the obstacles, in which 2D environment is modeled by low-resolution grids in light-blue and start position is shown in green, goal position is shown in yellow and the planned path is shown in navy-blue.

In Fig. 4, agents represented by solid yellow circles are moving towards their goal in a collision-free way. It demonstrates that the integration of RVO local collision avoidance algorithm can make agents do not collide with each other and show social characteristics of the crowds.

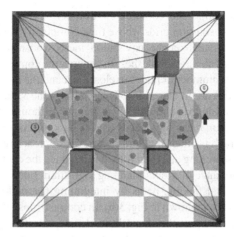

Fig. 4. The agents avoid collisions successfully when they move towards their goal

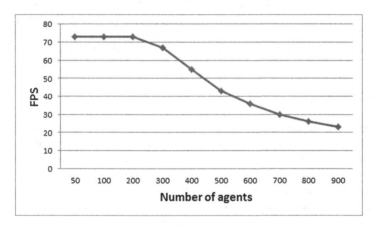

Fig. 5. The relationship between number of agents and FPS

Figure 5 shows the relationship between the number of agents and the frame rate. We can see that the frame rate decreases as the number of agents increases. When there are 900 agents in 2D scenario, the frame rate of our approach is still 24 fps (frame per second) although Unity 3D engine also uses some computation cost for rendering the scene with agents and obstacles in. It demonstrates that our approach can compute feasible, smooth and realistic paths for agents in real time.

6   Comparisons and Discussions

Many prior works have addressed the problem of path planning to simulate more realistic crowds, either by global planning or local collision avoidance or combining above two methods. We present a novel approach to plan feasible, smooth and congestion-free paths for virtual agents in dynamic and complex environments. Compared with traditional global planning on low-resolution grids whose time complexity depends on the number of grids, we search the paths on the navigation mesh whose time complexity only depends on the number of navigation meshes. In general, the complexity of A* algorithm is closely related with the number of nodes in closed list and the more the grids are in the scene, the more the nodes are in closed list. That is, when the virtual scene is large, A* algorithm on low-resolution grid shown in Fig. 3(j) needs thousands of searches while our approach on navigation mesh only needs tens or hundreds of searches at most since our navigation mesh graph is much smaller. That is, our approach reduces the number of vertexes and edges for searching and therefore, it can reduce the computation complexity and improve the performance of global planning algorithm. Furthermore, our approach can produce smoother and flexible paths by integrating Catmull-Rom spline and the concept of corridors, which is shown in Fig. 3(h). In contrast, most paths that are searched by A*algorithm in many approaches have little clearance to the obstacles and can be aesthetically unpleasant and always fixed once start position and goal position are determined, which can be seen in Fig. 3(j). Even some work such as [18] produces flexible paths with variable clearance to the obstacles by using corridors, it does not take smoothing paths into account or leaves it to local collision avoidance algorithm. Finally, our approach can show social characteristics of crowds when simulating crowd motions while other algorithm has to integrate behavior model to reflect this feature.

7 Conclusion

We propose a novel approach to plan feasible, smooth and congestion-free paths for virtual agents in dynamic and complex environments. We state that it is time consumed for traditional global planner to search on the low-resolution grids and the computed paths are almost fixed and cannot reflect the characteristics of human motions. Therefore, our approach combines a fast global planner with an effective local planner to provide real-time performance and the flexibility to model natural motion of agents. We evaluate our approach and compare it with previous techniques. We demonstrate the achieved improvements: our approach can find smoother, variable and collision-free paths for agents to move towards their goals at a low computation cost by using navigation mesh and corridors. Furthermore, our approach can display more realistic planning behaviors and well reflect social characteristics of crowds. The main objective of future work is to integrate our approach with behavior model to develop a more general and improved model for crowd simulation.

Acknowledgments. This work is supported by a grant from the National Natural Science Foundation of China (No. 61300101) and a grant from the Natural Science Foundation of Jiangsu Province (No. BK20130638).

References

1. Reynolds, C.W.: Flocks, herds and schools: a distributed behavioral model. ACM Comput. Graph. **21**(4), 25–34 (1987)
2. Tu, X., Terzopoulos, D.: Artificial fishes: physics, locomotion, perception, behavior. In: Proceedings of SIGGRAPH 1994, Computer Graphics Proceedings, pp. 43–50 (1994)
3. Funge, J., Tu, X., Terzopoulos, D.: Cognitive modeling: knowledge, reasoning and planning for intelligent characters. In: Proceedings of SIGGRAPH 1999, Computer Graphics Proceedings, pp. 29–38 (1999)
4. Shao, W., Terzopoulos, D.: Autonomous pedestrians. In: Proceedings of the 2005 ACM SIGGRAPH/Eurographics Symposium on Computer animation, pp. 19–28 (2005)
5. Musse, S.R., Thalmann, D.: A model of human crowd behavior: group inter-relationship and collision detection analysis. In: Thalmann, D., van de Panne, M. (eds.) Computer Animation and Simulation 1997, pp. 39–51. Springer, Vienna (1997)
6. Pelechano, N., O'Brien, K., Silverman, B., Badler, N.: Crowd simulation incorporating agent psychological models, roles and communication. In: First International Workshop on Crowd Simulation (2005)
7. Yu, Q., Terzopoulos, D.: A Decision network framework for the behavioral animation of virtual humans. In: Proceedings of the 2007 ACM SIGGRAPH/Eurographics Symposium on Computer Animation, pp. 119–128 (2007)
8. Sung, M., Gleicher, M., Chenney, S.: Scalable Behaviors for Crowd Simulation. Comput. Graph. Forum **23**(3), 519–528 (2004)
9. Van Den Berg, J., Lin, M.C., Manocha, D.: Reciprocal velocity obstacles for real-time multi-agent navigation. In: Proceedings of IEEE Conference on Robotics and Automation, pp. 1928–1935 (2008)
10. Narain, R., Golas, A., Curtis, S., Lin, M.C.: Aggregate dynamics for dense crowd simulation. In: ACM SIGGRAPH Asia 2009 Papers, pp. 1–8 (2009)
11. Sud, A., Gayle, R., Andersen, E., Guy, S., Lin, M., Manocha, D.: Real-time navigation of independent agents using adaptive roadmaps. In: Proceedings of the ACM Symposium on Virtual Reality Software and Technology, pp. 99–106 (2007)
12. Metoyer, R.A., Hodgins, J.K.: Reactive pedestrian path following from examples. Visual Comput. **20**(10), 635–649 (2004)
13. Chenney, S.: Flow tiles. In: ACM SIGGRAPH/ Eurographics Proceedings of Symposium on Computer Animation, pp. 233–242 (2004)
14. Pettre, J., Laumond, J.-P., Thalmann, D.: A navigation graph for real-time crowd animation on multilayered and uneven terrain. In: First International Workshop on Crowd Simulation (2005)
15. Sung, M., Kovar, L., Gleicher, M.: Fast and accurate goal-directed motion synthesis for crowds. In: Proceedings of the 2005 ACM SIGGRAPH/Eurographics symposium on Computer animation, SCA 2005, pp. 291–300 (2005)
16. Bayazit, O.B., Line, J.-M., Amato, N.M.: Better group behaviors in complex environments with global roadmaps. In: International Conference on the Simulation and Synthesis of Living Systems (Alife), pp. 362–370 (2002)
17. Geraerts, R., Overmars, M.: The corridor map method: a general framework for real-time high-quality path planning. Comput. Anim. Virtual Worlds **18**(2), 107–119 (2007)
18. Geraerts, R.: Planning short paths with clearance using explicit corridors. In: IEEE International Conference on Robotics and Automation, pp. 1997–2004 (2010)
19. Fiorini, P., Shiller, Z.: Motion planning in dynamic environments using velocity obstacles. Int. J. Rob. Res. **17**(7), 760–772 (1998)

A Comparative Investigation of PSG Signal Patterns to Classify Sleep Disorders Using Machine Learning Techniques

Thakerng Wongsirichot[(✉)] and Anantaporn Hanskunatai

Department of Computer Science Faculty of Science,
King Mongkut's Institute of Technology Ladkrabang (KMITL),
Bangkok, 10520, Thailand
thakerng.w@gmail.com, ksananta@kmitl.ac.th

Abstract. Patients with Non-Communicable Diseases (NCDs) are increasing around the globe. Possible causes of the NCDs are continuously being investigated. One of them is a sleep disorder. In order to detect specific sleep disorders, the Polysomnography (PSG), is necessary. However, due to the lack of the PSG in many hospitals, researchers attempt to discover alternative approaches. This article demonstrates comparisons of sleep disorder classifications using machine learning techniques. Three main machine learning techniques have been compared including Classification And Regression Tree (CART), k-Mean Clustering (KMC) and Support Vector Machine (SVM). The SVM achieves the best classification results in NREM-1 and NREM-2. The CART performs superior in NREM-3 and REM. Implications in terms of medical diagnosis, there are two main selected features, SaO2 and Pulse, based on the CART in all of the sleep stages. The features may be pieces of evidences to predict various types of sleep disorders.

Keywords: Sleep disorders · Classification · CART · K-Mean clustering · SVM

1 Introduction

One third of our lifetimes are allocated for one of the most vital activities, sleeping. Wakefulness has been subsided during sleep intervals with minimum apparent physical activities. The Central Nervous System (CNS) plays a significant role in controlling all human's activities even in the sleeping periods. Quality of sleep is considered as a vital property of human beings unless some symptoms may be recognized, especially the Non-Communicable Disease (NCD). Statistically, almost 1 of 2 adults (more than 133 million) in the United States live with at least one NCD. There are a number of clinical studies show that adults with NCDs such as hypertension and cardiovascular diseases are dramatically increasing worldwide [1]. The NCDs are caused by various factors such as lacks of exercises, genetic inheritances and mutations, prolonged use of some medications, etc. NCD patients' conditions are steadily degrading if medications or treatments are not properly provided. In order to discover appropriate medications or treatments, deepen diagnostic techniques are

© Springer International Publishing Switzerland 2015
D.-S. Huang et al. (Eds.): ICIC 2015, Part I, LNCS 9225, pp. 510–521, 2015.
DOI: 10.1007/978-3-319-22180-9_50

possibly appointed. With advancements of medical diagnostic technologies, a number of NCDs and symptoms are deeply studied in order to accomplish their causes and discover suitable treatments.

According to the American Academy of Sleep Medicine (AASM) standards, a sleep cycle contains four stages that are separated into two phrases namely, the Rapid Eye Movement (REM) and the Non-Rapid Eye Movement (NREM). There is only one stage in REM and three stages in NREM including stage 1, 2 and 3 [2]. A sleep cycle initiates when eye lids are closed (NREM-1), follows by a light sleep episode (NREM-2), and a deep sleep episode (NREM-3). The REM phrase is entered immediately after the NREM-3. An identification of sleep stages is a key main factor to detect sleep disorders [3].

Clinically, there are eight types of sleep disorders and may be classified into two main categories so called Dyssomnias and Parasomnias [4]. The sleep disorders may develop to serious sleep disorders that is the sleep apnea. One of the most encountered sleep apnea is Obstructive Sleep Apnea (OSA). The OSA is caused by the obstructions of the upper airway and series of repetitive pauses of breathing (apnea) during sleep. Possible causes of the OSA include overweight, short jaw structure, etc. Another type of sleep disorder, the Central Sleep Apnea (CSA), is similar to the OSA. However, the CSA is caused by abnormal brain activities [5]. The OSA and CSA are obviously found in severe patient cases. Most of the researchers attempt to identify only the OSA-related problems. However, there are other related sleep disorders that also require in-depth attentions and researches, especially the potential OSA/CSA patients.

2 Related Works

Researchers in various research fields including medicines, biomedicines and computer sciences attempt to investigate information beneath gigantic amount of recorded data from the PSG test. A group of researchers performed analyses using statistical analytical techniques such as regression analyses, autoregressive techniques, etc. On the other hand, others utilise machine mining techniques to conduct the analysis such as the Artificial Neuron Network (ANN), the Support Vector Machine (SVM), etc. Due to the vast amount of raw data have been collected in each PSG test, practically only subsets of potential variables, which are in EEG, EKG, ECG or a combination of these signals, are selected for analyses. Additionally, the PSG test is one of the dedicated systems that has to be performed in a formal sleep laboratory. Due to high setup and maintenance costs, the PSG test is rarely available in many hospitals.

Over the past few decades, there are a number of targeted investigations of the OSA and the CSA. A research work performed a pattern analysis in order to discriminate the CSA and the OSA using only the EEG recorded data. The researchers selected series of signals from C3-A2 and C4-A1 nodes wherein the sleep stage 2 per se. The Feed-Forward Neural Network (FFNN) was used to classify three studied groups, which have been preclassified by sleep specialists. Specifically, there are ten subjects with the CSA, ten subjects with the OSA, and ten healthy subjects. The EEG synchronisation methods, both the Coherence Function (CF) and the Mutual Information (MI), have been formulated to discriminate the CSA and the OSA cases. The classification result is 93.3 % accuracy [6].

Rather than using the partial EEG signal to classify the OSA and the healthy subjects, a group of researchers gathered and analysed the ECG signals. Clinically, there are common irregular ECG patterns, specifically based on the Heart Rate Variability (HRV), that are associated with apnea. A presence of bradycardia, the heart rate is below 60 bpm, follows by tachycardia, the heart rate is over 100 bpm. In order to conduct the HRV analysis, which is non-stationary, the wavelet decomposition has been engaged. In addition to the analysis of the HRV, the QRS complex also represents a correlation between the apnea and its pattern. The ECG-derived respiration (EDR) is retrieved in order to investigate the attenuation of respiratory effort. Therefore, the combination of HRV and EDR have been selected as classifying parameters using the SVM technique. 83 subjects have been used to develop the classification algorithm in this study together with 42 test cases from three sources. The collected data has been initially processed using the Wavelet Decomposition technique. It generated 14 levels with the Daubechies wavelets. The SVM has been performed in order to conduct the binary classification. The final result showed a promising result of 92.85 % accuracy on the independent tests with Cohen's K value of 0.85 [7].

In general, the OSA subjects usually have daytime sleepiness occurrences even they had full night sleeps. The researchers observe the OSA subjects that are related to daytime sleepiness. A research work collected data from three different groups. There are five untreated subjects, four narcoleptic subjects, and six healthy subjects. The interesting point of this research work is pupillometry data have been considered as a classifier along with the EEG signal. Specifically, the EEG signals have been selected only the signals from C3-A2, O1-A2 and P3-O1, which relate to eye movements. The NN technique, specifically the ART2 NN algorithm, has been selected as a main classifier between the OSA and the healthy subjects. The result shows 91 % classification accuracy [8].

Since 2002, there are a number of researchers interested in studies of snoring sound may relate to the NCDs. A report showed approximately 20–40 % of adults snore whilst asleep [9] However, not all of the snorers have or will eventually encounter the NCDs. Out of the medical observations and diagnostics, our previous work attempted to find other possible variables that are able to predict sleep disorder episodes. We also attempted to analyse snoring sound patterns with mobile devices using the modified k-Mean Clustering technique. With our experiment test, 74.70 % instances has been correctly classified. It implies that only the snoring sound patterns may not be sufficient for the classifications [10]. According to the reviewed related works, the sample sizes of the studied subjects are limited due to the availabilities of the data, which are collected from actual clinical sleep test laboratories. All of the studied subjects are extracted from the full PSG recordings. In addition to the sample sizes, most of the researches targeted to only severe cases of the OSA or the CSA. However, people with apnea risks are not thoroughly studied.

3   Methods

Our study investigates not only severe sleep apnea cases per se. It also includes mild or non-sleep apnea cases with anonymity. A set of full PSG recordings from five OSA cases were digitally extracted from a PSG recording software application, which installed at the Songklanagarin Hospital, Thailand. The study has been approved by the hospital's director with explicit attentions to the patient confidentiality. The main objective is to apply the DT, specifically the Classification and Regression Trees (CART), and the k-Mean Clustering (KMC) techniques to classify frequently found sleep disorder patterns. A set of 19 original explanatory variables has been stamped, which are predefined by the PSG machine [10]. (Table 1).

The variables are categorised into EEG, ECG, EKG, and other movement detections. The recorded data are separated according to sleep stages and excluded the normal sleep patterns. Four sleep disorders are manually identified by sleep technicians including Oxygen Desaturation (D), Hypopnea (H), PLMD (P), and IPLMD (I) episodes. Specific skills are required for the identification processes. Due to the gigantic amount of data, a simple technique that is able to represent possible decision paths and chance events. Moreover, the CART technique is able to act as a preliminary feature selector choosing possible nominated variables. On the other hand, the KMC technique is selected, which has been used in our previous research work, for a comparison purpose [11]. Additionally, the SVM, a well-known classification technique [7] has been tested on this clinical dataset.

```
Algorithm 1 CART (rpart function in R) [12]
Require: D is a set of PSG records.
          attr_list is a set of selected attributes.
          rpart() is the attribute selection method in the
          R (rattle package).

 1: create a node N //as an initial node
 2: if tuples in D are all the same as C, then
 3:    return N as a leaf node labelled with the class C;
 4: if attr_list is empty then
 5:    return N as a leaf node labelled with the majority
       class in D;
 6: apply rpart(D, attr_list) to find the "best"
    splitting_criterion;
 7: label Node N with splitting_criterion;
 8: if splitting_atrribute is discreate-valued and
 9:    Multiway splits allowed then
10:      attr_list ← attr_list - splitting_attribute
11: for each outcome j of splitting_criterion
12:    let Dj be the set of data records in D satisfying
       outcome j
13:    if Dj is empty then
14:       attach a leaf labelled with D to node N
15:    else attach the node returned by
       Generate_CART_decision_tree(Dj, attr_list) to node N
16: endfor
17: return N
```

Table 1. Variables from the PSG Signals

Variable	Description	Variable	Description
C3-A2	Monopolar EEG at C3-A2	FLOW	Mouth Airflow
C4-A1	Monopolar EEG at C4-A1	CHEST	Chest movement
F3-A2	Bipolar EEG at F3-A2	ABDOMEN	Abdomen movement
F4-A1	Bipolar EEG at F4-A1	LAT	Left Anterior Tibialis
O1-A2	Monopolar EEG at O1-A2	RAT	Right Anterior Tibialis
O2-A1	Monopolar EEG at O2-A1	EKG	Electrocardiography
CHIN	Chin Movement	Pulse	Pulse
LOC	Left Outer Canthus	SaO2	Saturation level of oxygen in haemoglobin
ROC	Right Outer Canthus	Snore	Amplitude of snoring sounds
canular	Nasal Cannula		

The CART technique is employed to classify the dataset into meaningful tree-like structures. There are three main parameters including D, *attr_list*, and *rpart()*. Specifically, the D is a data partition of the extracted PSG dataset. The *attr_list* is a set of selected attributes of the dataset. The *rpart()* is an attribute selection method in R. It selects a set of best discriminated attributes in order to classify the dataset into corresponding classes [12]. With the same dataset, the KMC has been performed to partition observations into potential sleep disorder classes (k). In general, the initial dataset has been loaded for a training purpose. According to Algorithm 2, a number of sleep disorder classes in each of the sleep stages are predefined by sleep technicians. The classes are the number of clusters k in the KMC. An initial set of centroids is calculated with continuously adjustments throughout the process. The final set of centroids are referenced in order to evaluate the testing dataset. The Euclidean distance (*argminDistance(D_i, C_k)*) is used to measure the differences between an element with the closest centroid. Final decisions of classifications are represented with the minimum distance between the element and its corresponding centroid [12].

Algorithm 2 k-Mean Clustering [12]
Require: D **is a set of PSG records.**
         k **is a number of clusters.**

```
1:  set k based on predefined sleep disorder classes
2:  let C_k is a centroid of each clusters
3:  repeat
4:      assign D_i to a cluster to which the record is the
        most similar using argminDistance(D_i, C_k)
5:      update C_k
6:      MaxIter++
7:  until MaxIter is 500 // a number of iterations
```

4   Results

All of the computational analyses have been successfully performed by the Rattle package in R. The dataset has been divided into two portions, training and testing datasets, for our analyses with the ratio of 70:30, respectively. Table 2 represents the CART analysis results.

Four main performance evaluation of the measures have been selected to evaluate each of the techniques according to the sleep stages. The measures include Accuracy, Precision, Sensitivity, Specificity, and F-Measure. The measures are calculated from incremental counts of True Positive (TP), True Negative (TN), False Positive (FP) and False Negative (FN) in confusion matrices. The followings are the formulae of the measures [13].

$$Accuracy = \frac{|TP| + |TN|}{|TP| + |TN| + |FP| + |FN|} \tag{1}$$

$$Precision = \frac{|TP|}{|TP| + |FP|} \tag{2}$$

$$Sensitivity = \frac{|TP|}{|TP| + |FN|} \tag{3}$$

$$Specificity = \frac{|TN|}{|TN| + |FP|} \tag{4}$$

$$F - Measure = \frac{(2 \times |TP|)}{(2 \times |TP|) + |FP| + |FN|} \tag{5}$$

Specifically, confusion matrices are constructed according to sleep stages on both of the analysis methods, the CART, the KMC and the SVM. D, H, I, and P are the classes of the Oxygen Desaturation, Hypopnea, IPLMD, and PLMD, respectively. Furthermore, $\hat{D}, \hat{H}, \hat{I}$ and $\hat{P}$ are the representations of classified classes corresponding to D, H, I, and P, respectively. Due to the confusion matrices are based on multi-classes classification, each of the measures is compared a class with the rest of the classes. Tables 3 and 4 represent the confusion matrices, the sleep disorder episode classification results, of NREM-1, NREM-2, NREM-3 and REM using the CART technique. In order to clarify the evaluation measures based on the confusion matrices, Table 4 in

Table 2. Features Selected by the CART technique

Sleep Stage	Observation (n)	Tree Construction Variables
NREM-1	86072	Canular, Pulse, SaO2
NREM-2	83274	Abdomen, Chest, Flow, Pulse, SaO2
NREM-3	85164	Abdomen, LOC, Pulse, ROC, SaO2
REM	14958	O2.A1, Pulse, SaO2

Table 3. Confusion Matrices of the NREM-1 and NREM-2 using the CART Technique

NREM-1	$\hat{D}$	$\hat{H}$	$\hat{I}$	$\hat{P}$	NREM-2	$\hat{D}$	$\hat{H}$	$\hat{P}$
D	5457	11398	5399	5331	D	19903	7442	0
H	3359	24801	11474	15522	H	5603	47402	0
I	0	80	0	258	P	330	2594	0
P	129	429	137	2298				

Table 4. Confusion Matrices of the NREM-3 and REM using the CART Technique

NREM-3	$\hat{D}$	$\hat{H}$	REM	$\hat{D}$	$\hat{P}$
D	37473	84	D	5537	0
H	77	84030	P	9	9389

the REM stage, the TP value is the corrected classified D class as the $\hat{D}$, class, which is 5537. The FP value is the P class that has been classified as $\hat{P}$, which is 9 (Bottom left in the matrix). The FN value is the D class that has been classified as, which is 0 (Top right in the matrix). The TN value is the P class that has been classified as $\hat{P}$, which is 9389. Additionally, the calculations of TP, TN, FP and FN are performed separately according to each of the base class. For example, if class D is considered, the TP, TN, FP and FN are based on only class D.

Tables 5 and 6 show the performance evaluation results in each of the sleep stages. In NREM-1, there are four determined classes, D, H, I and P. The classification of the $H - \hat{H}$ achieves highest F-Measure value of 0.54. In NREM-2, there are three determined classes, D, H and P. The classification of the $H - \hat{H}$ achieves highest F-Measure value of 0.86. However, in the NREM-3 and REM, all of the F-Measure values reach the maximum of 1.00, which mean the majority of the instances are correctly classified.

Figure 1 represents a tree structure of the REM stage using the CART technique. The CART technique implements a binary classification [11] There are three selected variables including O2.A1, Pulse and SaO2, which have been mentioned in Table 2. In this sleep stage, there are two classes, D and P. At the beginning of the tree, the condition of SaO2 > = 94 is set based on the root node calculation. If the SaO2 > = 94

Table 5. Performance Evaluation Results of NREM-1 and NREM-2 using the CART Technique

Performance Measure	NREM-1				NREM-2		
	D-$\hat{D}$	H-$\hat{H}$	I-$\hat{I}$	P-$\hat{P}$	D-$\hat{D}$	H-$\hat{H}$	P-$\hat{P}$
Accuracy	0.702	0.509	0.798	0.747	0.839	0.812	0.965
Precision	0.610	0.676	0.000	0.098	0.770	0.825	–
Sensitivity	0.198	0.450	0.000	0.768	0.728	0.894	0.000
Specificity	0.941	0.615	0.801	0.746	0.894	0.668	1.000
F-Measure	0.30	0.54	0.00	0.00	0.75	0.86	0.00

Table 6. Performance Evaluation Results of NREM-3 and REM using the CART Technique

Performance Measure	NREM-3		REM	
	D-$\hat{D}$	H-$\hat{H}$	D-$\hat{D}$	P-$\hat{P}$
Accuracy	0.999	0.999	0.999	0.999
Precision	0.998	0.999	0.998	1.000
Sensitivity	0.998	0.990	1.000	0.999
Specificity	0.999	0.998	0.999	1.000
F-Measure	1.00	1.00	1.00	1.00

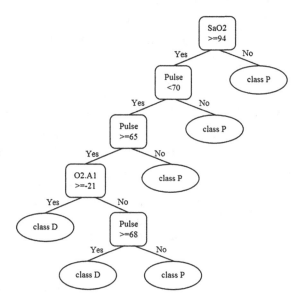

Fig. 1. A Tree Structure of the REM Stage using the CART Technique

is true, the latter condition is Pulse < 70. If the SaO2 >= 94 is false, the element is classified to the class P. The process is continuously performed until all of the instances are classified into classes.

On the other hand, the KMC has been performed with the k clusters based on actual known cluster sizes in each of the sleep stages. For example, the NREM-1 has four types of presented sleep disorders so that the k is set to four.

Tables 7 and 8 represent the confusion matrices, classifications of the sleep order episode results, of the NREM-1, NREM-2, NREM-3 and REM using the KMC technique. Overall performance evaluation results show that the KMC technique also performs relatively acceptable in the REM stage, in Tables 9 and 10. Specifically, the $P - \hat{P}$ and $H - \hat{H}$ are best in REM and NREM-1 in some classified classes, respectively. On the other hand, the $I - \hat{I}$ in NREM-1, is totally incorrectly classified.

Table 7. Confusion Matrices of NREM-1 and NREM-2 using the KMC Technique

NREM- 1	$\hat{D}$	$\hat{H}$	$\hat{I}$	$\hat{P}$	NREM-2	$\hat{D}$	$\hat{H}$	$\hat{P}$
D	8272	19049	0	264	D	13037	14308	0
H	409	54747	0	0	H	13321	36479	3205
I	0	0	0	338	P	2441	177	306
P	129	629	0	2235				

Table 8. Confusion Matrices of the NREM-3 and REM using the KMC Technique

NREM-3	$\hat{D}$	$\hat{H}$	REM	$\hat{D}$	$\hat{P}$
D	15170	22387	D	5560	0
H	18353	65754	P	2568	6830

Table 9. Performance Evaluation Results of NREM-1 and NREM-2 using the KMC Technique

Performance Measure	NREM-1			NREM-2			
	D-$\hat{D}$	H-$\hat{H}$	D-$\hat{D}$	H-$\hat{H}$	D-$\hat{D}$	H-$\hat{H}$	D-$\hat{D}$
Accuracy	0.769	0.767	0.996	0.984	0.639	0.628	0.930
Precision	0.939	0.736	–	0.788	0.453	0.716	0.087
Sensitivity	0.299	0.993	0.00	0.747	0.477	0.688	0.105
Specificity	0.991	0.364	100.00	0.993	0.718	0.522	0.960
F-Measure	0.45	0.84	0.00	0.77	0.46	0.70	0.10

Table 10. Performance Evaluation Results of NREM-3 and REM using the KMC Technique

Performance Measure	NREM-3		REM	
	D-$\hat{D}$	H-$\hat{H}$	D-$\hat{D}$	H-$\hat{H}$
Accuracy	0.665	0.665	0.828	0.828
Precision	0.453	0.746	1.000	0.727
Sensitivity	0.404	0.782	0.684	1.000
Specificity	0.782	0.404	1.000	0.684
F-Measure	0.43	0.76	0.81	0.84

One of the standard classification algorithm is the Support Vector Machines (SVMs) [9]. The SVM is selected to perform on the same collected clinical dataset. The performance evaluation results are shown in Tables 11 and 12.

Tables 13 and 14 show a comparison summary of F-Measures in all of the selected techniques. The best classifying techniques are marked according to the sleep stages and the specified sleep disorder classes.

Table 11. Performance Evaluation Results of NREM-1 and NREM-2 using the SVM Technique

Performance Measure	NREM-1				NREM-2		
	D-D̂	H-Ĥ	D-D̂	H-Ĥ	D-D̂	H-Ĥ	D-D̂
Accuracy	0.816	0.811	0.999	0.992	0.875	0.873	0.996
Precision	0.956	0.775	1.000	0.991	0.899	0.857	0.995
Sensitivity	0.447	0.992	0.893	0.764	0.697	0.961	0.897
Specificity	0.990	0.487	1.000	0.999	0.962	0.719	0.999
F-Measure	0.61	0.87	0.94	0.86	0.79	0.91	0.94

Table 12. Performance Evaluation Results of NREM-3 and REM using the SVM Technique

Performance Measure	NREM-3		REM	
	D-D̂	H-Ĥ	D-D̂	H-Ĥ
Accuracy	0.825	0.825	0.992	0.992
Precision	0.951	0.804	0.979	0.999
Sensitivity	0.458	0.989	0.999	0.987
Specificity	0.989	0.458	0.987	0.999
F-Measure	0.62	0.89	0.99	0.99

Table 13. A Comparison Summary of F-Measures in NREM-1 and NREM-2

F-Measure	NREM-1				NREM-2		
	D-D̂	H-Ĥ	D-D̂	H-Ĥ	D-D̂	H-Ĥ	D-D̂
CART	0.30	0.54	0.00	0.00	0.75	0.86	0.00
KMC	0.45	0.84	0.00	0.77	0.46	0.70	0.10
SVM	0.61*	0.87*	0.94*	0.86*	0.79*	0.91*	0.94*

Table 14. A Comparison Summary of F-Measures in NREM-3 and REM

F-Measure	NREM-3		REM	
	D-D̂	H-Ĥ	D-D̂	H-Ĥ
CART	1.00*	1.00*	1.00*	1.00*
KMC	0.43	0.76	0.81	0.84
SVM	0.62	0.89	0.99	0.99

5 Discussions and Implications

According to the classification results, the CART technique classification results are superior in the NREM-3 and the REM. There are a number of variables that have been selected as tree construction variables. For example, the SaO2 and Pulse variables are significantly nominated in all of the sleep stages, according to Table 2. However, there are a number of additional variables such as Abdomen that also engages into two of the sleep stages, the NREM-2 and the NREM-3. Specially, there is only a variable, O2.A1, which is in the category of EEG. The SVM achieves the best classification performance in the NREM-1 and the NREM-2. The KMC technique has performed with lower accuracy. One of the possible explanations is the KMC technique utilises all of the variables for the classification without pre-feature selection stages.

Implications in terms of medical diagnosis, the distinction sleep disorder diagnostic is able to be performed by the PSG test in modern hospitals. On the other hand, the PSG test is not applicable in some areas due to its high costs. Our initial research discovers only a small set of variables that are main classifiers in the CART technique. The determined nominated variables may be pieces of evidences to classify various types of sleep disorders. Specifically, a traditional ECG and an oximeter may be able to use for overnight sleep tests. Additionally, other machine learning technique may be selected sequentially in order to construct a hybrid machine learning technique.

6   Conclusions

The sleep disorders are hidden symptoms that can provoke potential NCDs. The PSG test is recently the best accurate diagnostic tool per se in order to detect the sleep disorders. However, due to its high costs and availabilities, some researchers are seeking for alternative diagnostic tools and techniques that are able to partially or fully substitute the ordinary PSG test. Researchers attempt to select some properties from other measurements such as ECG, EEG, and EKG. Alternatively, a combination of signals may also be considered. The machine learning techniques have devoted as key mechanisms to classify sleep disorders. Most of the studies investigated only moderate to severe cases of sleep apnea. Based on our research works, other less severe sleep disorder subjects, which may lead to severe sleep apnea disorders, have been thoroughly studied. The selected data set is an original extracted dataset from the PSG test, which has not been altered or rescaled. The SVM achieves the best classification results in NREM-1 and NREM-2. The CART performs superior in NREM-3 and REM. The utmost goal is to minimise the number of nominated variables. The traditional ECG and the oximeter may be able to use for overnight sleep tests. It benefits rural hospitals or medical centres to achieve acceptable sleep disorder diagnosis results with standard medical devices. In terms of sleep disorder classifications, hybrid machine learning techniques will be investigated.

References

1. Centers for Disease Control and Prevention Information. http://www.cdc.gov/
2. Moser, D., Anderer, P., Gruber, G., Paraptics, S., Loretz, E., Boeck, M., Kloesch, G., Heller, E., Schmidt, A., Danker-Hopfe, H., Saletu, B., Zeitlhofer, J., Dorffner, G.: Sleep classification according to AASM and Rechtschaffen & Kales: effects on sleep scoring parameters. In: Sleep, vol.32, pp. 139–149 (2009)

3. Ruehland, W., Rochford, P., O'Donoghue, F., Pierce, R., Singh, P., Thornton, A.: The new AASM criteria for scoring hypopneas: impact on the apnea hypopnea index. Sleep 32, 150–157 (2009)
4. Kocak, O., Bayrak, T., Erdamar, A., Ozparlak, L., Telatar, Z., Erogul, O.: Automated detection and classification of sleep apnea types using electrocardiogram (ECG) and electroencephalogram (EEG) features. In: Advances in Electrocardiograms – Clinical Applications, pp. 211–230 (2012)
5. Azarbarzin, A.Z.: Snoring sounds' statistical characteristics depend on anthropometric parameters. J. Biomed. Sci. Eng. 5, 245–254 (2012)
6. Aksahin, M.F., Aydin, S., Firat, H., Erogul, O., Ardic, S.: Classification of sleep apnea types using EEG synchronization criteria. In: The 15th National Biomedical Engineering Meeting (BİYOMUT), pp. 1–4 (2010)
7. Khandoker, A.H., Palaniswami, M., Karmarkar, C.K.: Support vector machines for automated recognition of obstructive sleep apnea syndrome from ECG recordings. IEEE Trans. Inf. Technol. Biomed. 13, 37–48 (2009)
8. Liu, D., Pang, Z., Lloyd, S.: A neural network method for detection of obstructive sleep apnea and narcolepsy based on pupil size and EEG. IEEE Trans. Neural Netw. 19, 308–318 (2008)
9. Hoffstin, V.: Apnea and snoring: state of the art and future direction. Acta Otorhinolaryngol Belg 56, 205–236 (2002)
10. Berry, R.B.: Fundamentals of Sleep Medicine. Elsevier Saunders, Philadelphia (2012)
11. Wongsirichot, T., Iad-ua, N., Wibulkit, J.: A snoring sound analysis application using k-mean clustering method on mobile devices. In: Springer Series Advances in Intelligent Systems and Computing (2015)
12. Han, J., Kamber, M.: Data Mining: Concepts and Techniques. Waltham, MA (2012)
13. Costa, E.P., Lorena, A.C., Carvalho, A.C., Freitas, A.A.: A review of performance evaluation measures for hierarchical classifiers. In: Evaluation Methods for Machine Learning II, AAAI Press (2007)

A Multi-valued Coarse Graining of Lempel-Ziv Complexity and SVM in ECG Signal Analysis

Deling Xia[1,2], Qingfang Meng[1,2(✉)], Yuehui Chen[1,2],
and Zaiguo Zhang[3]

[1] The School of Information Science and Engineering,
University of Jinan, Jinan 250022, China
[2] Shandong Provincial Key Laboratory of Network Based
Intelligent Computing, Jinan 250022, China
ise_mengqf@ujn.edu.cn
[3] CET Shandong Electronics Co., Ltd., Jinan 250101, China

Abstract. Lempel-Ziv (LZ) complexity method has been widely applied to detection ventricular tachycardia (VT) and ventricular fibrillation (VF). The coarse-graining process (Quantization levels, L) plays an important role in the LZ complexity measure analysis. In this paper, we present a multi-valued coarse-graining process approaches ($L > 2$), our test shows that this algorithm is superior to the two-valued coarse-graining of LZ complexity approaches ($L = 2$) in VT and VF separation. Furthermore, we used support vector machine (SVM) classifier to discriminate VF and VT. Using the complexity as a feature to input classifiers can significantly improve the classification results. Particularly, optimum performance is achieved at a 4-second length.

Keywords: Lempel-Ziv (LZ) complexity · Quantization levels · Multi-valued coarse-graining of Lempel-Ziv complexity (MLZ) · SVM

1 Introduction

The Lempel-Ziv (LZ) complexity for sequences of finite length was suggested by Lempel and Ziv [1]. It is a nonparametric, simple-to-calculate measure of complexity in a one-dimensional signal that does not require long data segments to compute [2]. With larger values corresponding to more complexity in the data, LZ complexity is related to the number of distinct substrings and the rate of their recurrence along the given sequence. It has been applied to study electroencephalographic [3, 12], electromyography (EMG) [4], DNA sequences analysis [5] and classification of ECG signal [6, 7].

Sudden cardiac death (SCD) is a major problem in the worldwide. It is most frequently caused by ventricular tachycardia (VT) and ventricular fibrillation (VF). VF has been considered a random and irregular process. VT is a periodic motion exhibiting a rapid heart rate; it's giving rise to a diminished cardiac output when it occurs. In non-linear signal processing, the LZ complexity measure can cope with a dynamical system entering a chaotic state by quantifying the rate of new pattern occurrences along given finite symbolic sequences [8]. Therefore, the LZ complexity measure has been adopted to classify the VT and VF.

© Springer International Publishing Switzerland 2015
D.-S. Huang et al. (Eds.): ICIC 2015, Part I, LNCS 9225, pp. 522–528, 2015.
DOI: 10.1007/978-3-319-22180-9_51

SVM classification algorithms have been used in a wide number of practical applications due to its good properties of regularization, maximum margin, and robustness with data distribution and with input space dimensionality [9]. SVM is based on structural risk minimization principle, and could construct an Optimal Separating Hyper plane (OSH) in the feature space. With the minimum risk of misclassification, the OSH can classify both the training samples and the unseen samples in the test set [9–11].

In the binarization processing of the LZ complexity method ($L = 2$), many of the original sequence information details were not be reflected and saved. In this paper, we proposed a multi-valued coarse graining of LZ complexity (MLZ, $L > 2$) arithmetic. The multistate LZ index can quantify the impact of complexity changes on amplitude variations of the signals. The paper is organized as followed. In Sect. 2 we explain the MLZ complexity arithmetic and SVM. Section 3 presents the results of our study. Finally, conclusions are given in Sect. 4.

2 Methods

2.1 Data Selection

The data were selected from MIT-BIH Malignant Ventricular Ectopy Database (MIT-BIH Database) and Creighton University Ventricular Tachyarrhythmia Database (CU Database). 100 VF episodes and 100 VT episodes are respectively extracted from CU Database and MIT-BIH Database. The data length is four-second times. Then all samples are normalized before using it.

2.2 Multi-valued Coarse Graining of LZ Complexity Arithmetic

The Lempel-Ziv (LZ) complexity algorithm is a measure that with the increase of the length of the sequence, the new model also increases [12]. This method was put forward by Lempel and Ziv in 1979.

LZ complexity analysis is based on a coarse-graining of the measurements, so before calculating the complexity measure $C(n)$ (n is the length of the sequence), the signal must be transformed into a finite symbol sequence. In this study we have used two different sequence conversion methods:

(a) *0–1 sequence conversion*. This method is also called the LZ complexity. The median value is estimated as a threshold Td, as partitioning about the median is robust to outliers.

(b) *0–(L-1) sequence conversion* (*L* is the number of different symbols in coarse graining, defined as different quantization levels (QLs)). This method is called MLZ complexity. For each of the ECG segments, given sample *L*, the lower (*L*) and upper (*L*) bounds are fixed.

In the *0–1 sequence conversion*, many of the original sequence information details were not be reflected and saved. Obviously, this method is a kind of rules form and not totally with the actual dynamic characteristics of sequence itself. In order to overcome this phenomenon, we used 0–(L-1) *sequence conversion*. But how to determine the value of L, we introduced the MLZ index f [13] that was used for estimation of muscle force to assess the complexity of VT and VF.

(1) Give a discrete-time signal X, Define $X_M = \{x_{(N-1)(n+1)}, \cdots, x_{(N-1)(n+N)}\}$, ($M = 1, 2, \cdots N$), n represents the length of each sample.
(2) According to the following formula, make these samples be normalized.

$$X'_M = X_M - \bar{x}/\sigma \tag{1}$$

where $\bar{x}$ and σ represent the mean and the standard deviation of the sample X_M.

(3) Repeat step (2) until all samples is normalized.
(4) For each normalized sample, calculate the maximum absolute value X_{abs}, the standard deviation of the signal X_{std}.
(5) Finish step (4) until all the f the samples are calculated.

The rest of the algorithm is expressed in formal terms [7, 14, 15].

2.3   Statistical Analysis

SVM is a machine learning technique usually for solving two-class problem. Due to the SVM classifiers good properties of regularization, robustness with data distribution, maximum margin, and with input space dimensionality; we selected SVM classifiers to detect VF and VT.

After all the samples are calculated by the MLZ complexity, the MLZ complexity feature is combined to form a feature vector. Then to get the optimal SVM parameters, the feature vector of training data are fed into a SVM classifier for training. Finally, using the trained SVM classifier to detect VT and VF in the testing set. The sensitivity, the specificity, and the accuracy (ACC) were used to assess.

Figure 1 shows a block diagram with the different steps followed in this study.

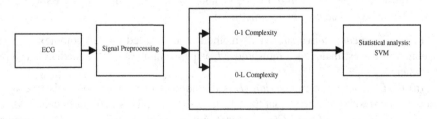

Fig. 1. Block diagram with the different steps in this study

3 Simulation and Results

The recorded ECG signal were analyzed using the MLZ algorithm with different quantization levels (QLs, is an even number, defined L). Specifically, we selected three VF and three VT samples and the QLs was used 2, 4, 6, 8, 10, 20, 40, 60, 80, 100, 120, 140, 160, 180 and 200. The values of BB, x_{abs}, x_{std} of each sample were shown in Tables 1 and 2.

Table 1. The parameter values of the six samples of VF

samples / parameters	BB	x_{abs}	x_{std}
VF$_1$	900	-872	310.5611
VF$_2$	800	-796	312.0254
VF$_3$	850	-840	308.7302

Table 2. The parameter values of the six samples of VT

samples / parameters	BB	x_{abs}	x_{std}
VT$_1$	780	-775	124.0643
VT$_2$	800	-775	155.0369
VT$_3$	850	-848	134.2531

Tables 3 and 4 shows the variation of f for all the L used with the VF and VT signals. We can draw that for a large f the information provided by the MLZ index will be more affected by changes in signal complexity, while for a small f the information will be less affected by changes in signal complexity. In general, when the $L \geq 60$, the value f changes small. So in this paper, we selected $L = 60$ for the MLZ complexity method.

After that we used the LZ complexity method described previously to calculate the values of each episode. The results calculated from an episode of VF and an episode of VT is presented in Table 5 and Fig. 2. By examining probability density function, a threshold for distinguishing between VT and VF is found. The signal is considered to be VF if $c(n)$ is less than the threshold, otherwise it is classified as VT. So we can draw that the classification results for VF and VT are 63.10 % of accuracy only using the LZ complexity. While the classification results using the MLZ complexity method for VF and VT are 88.50 %, respectively.

Table 3. The relationship between the MLZ index f and the L for three samples of VF

QLs	2	4	6	8	10	20	40	*60*	80	100	120	140	160	180	200
F1	2.89	1.44	0.96	0.72	0.57	0.28	0.14	*0.09*	0.07	0.05	0.04	0.04	0.03	0.03	0.02
	80	90	60	45	96	98	49	*66*	24	80	83	14	62	22	90
F2	2.56	1.28	0.85	0.64	0.51	0.25	0.12	*0.08*	0.06	0.05	0.04	0.03	0.03	0.02	0.02
	39	19	46	10	28	64	82	*55*	41	13	27	66	20	85	56
F3	2.75	1.37	0.91	0.68	0.55	0.27	0.13	*0.09*	0.06	0.05	0.04	0.03	0.03	0.03	0.02
	32	66	77	83	06	53	77	*18*	88	51	59	93	44	06	75

Table 4. The relationship between the MLZ index f and the L for three samples of VT

QLs	2	4	6	8	10	20	40	*60*	80	100	120	140	160	180	200
T1	6.28	3.14	2.09	1.57	1.25	0.62	0.31	*0.20*	0.15	0.12	0.10	0.08	0.07	0.06	0.06
	71	35	57	18	74	87	44	*96*	72	57	48	98	86	99	29
T2	5.16	2.58	1.72	1.29	1.03	0.51	0.25	*0.17*	0.12	0.10	0.08	0.07	0.06	0.05	0.05
	01	00	00	00	20	60	80	*20*	90	32	06	37	45	73	16
T3	6.33	3.16	2.11	1.58	1.26	0.63	0.31	*0.21*	0.15	0.12	0.10	0.09	0.07	0.07	0.06
	13	57	04	28	63	31	66	*10*	83	66	55	04	91	03	33

Table 5. The results of the classification for VF and VT

Component	TH	VF		VT		ACC
		Sensitivity	Specificity	Sensitivity	Specificity	
LZ complexity ($L = 2$)	0.1813	62.09	64.11	64.11	62.09	63.10
MLZ complexity ($L = 60$)	0.3687	86.00	91.00	91.00	86.00	*88.50*

In order to improve the detection precision, we used SVM classifiers to discriminate VF and VT. To evaluate the generalization capability of the SVM models, data consisted of training and testing. From these 200 ECG episodes, 150 episodes were allocated to training phase, while another 50 episodes were allotted to testing phase. From Table 6, we can see that using the complexity as a feature to input the SVM classifier to classify can significantly improve the classification results. This demonstrates that SVM constitute an adequate tool for developing VF detection algorithms.

4 Discussion

In this paper, the MLZ complexity measure shows itself possessing good advantages in time domain analysis and is particularly useful in describing the complexity of random processes. We used the information theory to analyze the chaotic ECG signal. In order to improve the detection precision, we used SVM classifier to discriminate VF and VT. In this experiment, we analyzed the performances of the SVM when using the LZ complexity ($L = 2$) and the MLZ complexity ($L = 60$) as the feature respectively. Our results not only show that a proper coarse-graining technique is essential in the success

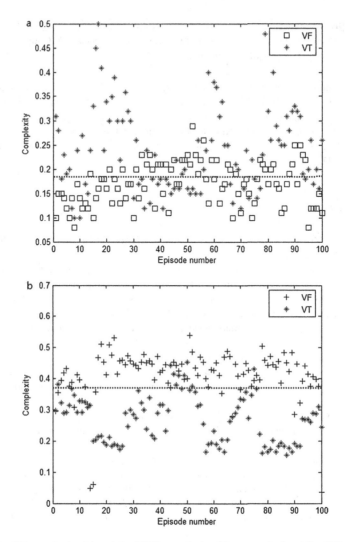

Fig. 2. The LZ complexity (a) and the MLZ complexity (b) were calculated for different VF and VT episodes. The dotted horizontal line is the threshold.

Table 6. SVM classifier performance for VF and VT detection using this two method

Component	Parameters	VF		VT		ACC
		Sensitivity	Specificity	Sensitivity	Specificity	
LZ complexity ($L = 2$)	Ker = 'poly' P1 = 1.9 C = 940	63.93	91.92	91.92	63.93	77.93
MLZ complexity ($L = 60$)	Ker = 'poly' P1 = 1.9 C = 940	99.92	91.92	91.92	99.92	**95.92**

of LZ complexity analysis in ECG signal classification, but also suggest that using the MLZ complexity as a feature to input the classifier to classify VT and VF can significantly improve the classification results.

Acknowledgements. This work was supported by the National Natural Science Foundation of China (Grant No. 61201428, 61302090), the Natural Science Foundation of Shandong Province, China (Grant No. ZR2010FQ020, ZR2013FL002), the Shandong Distinguished Middle-aged and Young Scientist Encourage and Reward Foundation, China (Grant No. BS2009SW003, BS2014DX015), the Graduate Innovation Foundation of University of Jinan (Grant No. YCX13011).

References

1. Lempel, A., Ziv, J.: On the complexity of finite sequences. IEEE Trans. Inform. Theory **22**, 75–81 (1976)
2. Zhang, X.S., Zhu, Y.S., Thakor, N.V., Wang, Z.Z.: Detecting ventricular tachycardia and fibrillation by complexity measure. IEEE Trans. Biomed. Eng. **46**, 548–555 (1999)
3. Zhang, X., Roy, R.J., Weber, E.: EEG complexity as a measure of depth of anesthesia for patients. IEEE Trans. Biomed. Eng. **12**, 1424–1433 (2001)
4. Nagarajan, R.: Quantizing physiological data with Lempel-Ziv complexity-certain issues. IEEE Trans. Biomed. Eng. **49**(11), 1371–1373 (2002)
5. Otu, H.H.: K. Sayood.: Sequence analyses. Bioinformatics **19**, 2122–2130 (2003)
6. Xu, Y., Ma, Q.D.Y., Schmitt, D.T., et al.: Effects of coarse-graining on the scaling behavior of long-range correlated and anti-correlated signals. Fuel Energy Abstr. **390**, 4057–4059 (2011)
7. Zhang, H.X., Zhu, Y.S., Wang, Z.M.: Complexity measure and complexity rate information based detection of ventricular tachycardia and fibrillation. Med. Biol. Eng. Comput. **38**, 553–557 (2000)
8. Zhou, S., Zhang, Z., Gu, J.: Interpretation of coarse-graining of Lempel-Ziv complexity measure in ECG signal analysis. In: 33rd Annual International Conference of the IEEE EMBS Boston, Massachusetts, USA, August 30–September 3 2011
9. Cortes, C.: V. Vapnik.: Support-vector networks. Mach. Learn. **20**(3), 273–297 (1995)
10. Pontil, M., Verri, A.: Support vector machines for 3D object recognition. IEEE Trans. Pattern Anal. Mach. Intell. **20**(6), 637–646 (1998)
11. Li, S., Zhou, W., QiYuan, S.G., Cai, D.: Feature extraction and recognition of ictal EEG using EMD and SVM. Comput. Biol. Med. **43**, 807–816 (2013)
12. Luo, Z., Cao, M.: EEG feature analysis of motor imagery based on Lempel-Ziv complexity at multi-scale. Chin. J. Sens. Actuators **24**(7), 1033–1037 (2011)
13. Sarlabous, L., Torres, A., Fiz, J.A., Morera, J., Jané, R.: Index for estimation of muscle force from mechanomyography based on the Lempel-Ziv algorithm. J. Electromyogr. Kinesiol. **23**, 548–557 (2013)
14. Aboy, M., Hornero, R., Abásolo, D., Álvarez, D.: Interpretation of the Lempel-Ziv complexity measure in the context of biomedical signal analysis. IEEE Trans. Biomed. Eng. **53**(11), 2282–2288 (2006)
15. Gómez, C., Hornero, R., Abásolo, D., Fernández, A., López, M.: Complexity analysis of the magnetoencephalogram background activity in Alzheimer's disease patients. Med. Eng. Phys. **28**, 851–859 (2006)

A Survey of Multiple Sequence Alignment Techniques

Xiao-Dan Wang$^{(\boxtimes)}$, Jin-Xing Liu, Yong Xu, and Jian Zhang

Bio-Computing Research Center, Shenzhen Graduate School,
Harbin Institute of Technology, Heilongjiang, China
{wangxiaodan0608, zpower007}@163.com,
sdcavell@126.com, yongxu@ymail.com

Abstract. Multiple sequence alignment (MSA) is a basic step in many bioinformatics analyses, and also a NP-hard problem. In order to improve the speed, accuracy and cater to the requirement of large-scale sequences alignment, a wide variety of MSA methods and softwares have been subsequently developed. In this article, we will systematically review the wildly used methods and introduce their practical results on the benchmark Balibase 3.0 references. We come to the conclusion that computational complexity still is the bottleneck of MSA. We also consider future development of MSA methods with respect to applying of more different technologies and the prospect of parallelization of MSA.

Keywords: Multiple sequence alignment · MSA techniques

1 Introduction

With the further rapid development of new sequencing technology, the biological applications become more and more widely, including exposition of relationship between nucleosome positioning and DNA methylation [1], prediction of missense mutation or protein functionality [2, 3], the assembly of new genomes [4], crop breeding [5], and so on. For most of these applications, multiple sequence alignments are fundamental.

For N sequences of length L, the exact way of computing an optimal alignment has a computational complexity of $O(N^L)$, which is excessive even for small number of sequences. Unfortunately, all sequencing technologies in production, such as Illumina, Helicos, SOLiD and Roche/454, can produce thousands or millions of sequences concurrently [6, 7]. In order to overcome this difficulty, many heuristic methods, including progressive methods [8] and iterative refinement methods [9] are developed.

This article aims to systematically review the recent advance of MSA methods. It is organized as follows. We first introduce the basic theory of heuristic methods and review the development of wildly used techniques, including Clustal, T-Coffee, MAFFT, MUSCLE and Kalign in Sect. 2, and then examine their programs on the benchmark Balibase 3.0 references [10], Oxbench [11] and Homestrad in Sect. 3. Finally, we discuss the future development of multiple sequence alignment in Sect. 4.

© Springer International Publishing Switzerland 2015
D.-S. Huang et al. (Eds.): ICIC 2015, Part I, LNCS 9225, pp. 529–538, 2015.
DOI: 10.1007/978-3-319-22180-9_52

2 Overview

2.1 Theory

Progressive Method. The progressive method is the first practical MSA construction strategy, and still composes the key of a majority of MSA programs by now. A progressive method usually is made up of four steps as follows [12]:

Step 1: Calculate a distance matrix for N input sequences. The element of this matrix is the distance of every pair of the input sequences, and there are many ways to messure distance, for example, angle cosine and Euclidean distance. In a exact way, $\binom{N}{2}$ pair-wise alignments are needed to count the numbers of matches, mismatches, and indels, which are then converted to the distance measures. This procedure is costly when N is large, as its time complexity is $O(N^2L^2)$;

Step 2: Construct a guide tree according to the distance matrix calculated in Step1 by a clustering analysis method. The most widely used method is UPGMA (Unweighted Pair-Group Method with Arithmetic means) [13] which takes computation time of $O(N^2)$ to construct the guide tree;

Step 3: In the guide tree, an external node represents each input sequence, while an internal node represents an MSA;

Step 4: Repeat Step1 and Step2 for the generated pair-wise alignments after construction of the initial MSA.

Iterative Refinement. The progressive method is implemented using a "greedy algorithm" by what mistakes made at the initial alignment stages cannot be corrected later [14]. To overcome this defect, an effective approach relies on post process known as iterative refinement, which also consists of four steps as follow [12]:

Step 1: Construct an initial MSA;

Step 2: Divide the MSA constructed in Step1 into two groups, then get rid of the columns made up of nulls from each of the two groups;

Step 3: Realign the two groups produced in Step2 by a pair-wise sequence-to-group or group-to-group alignment method;

Step 4: Repeat Step2 and Step3 until no gain in the alignment score or the iterative times exceeding a predefined number.

Scoring Function. A good scoring function is necessary to guarantee this procedure work accurately. The most widely used function is sum-of-pairs (SP) score [15] and weighted sum-of-pairs score (WSP) [16] with affine gaps.

For a sequence set A which is made up of N sequences of length L, we define WSP as follow:

$$WSP(A) = \sum_{1 \le i < j \le N} w_{i,j} H(a_i, a_j) = \sum_{1 \le l \le L} \sum_{1 \le i < j \le N} w_{i,j} [S(a_{i,l}, a_{j,l}) - v \cdot G(i,j,l)],$$

(1)

where $H(a_i, a_j)$ is the alignment score of a pair of sequences in A, $w_{i,j}$ is the weight corresponding to the pair sequences $[a_i, a_j]$ ($w_{i,j} = 1$ is an unweighted case), $S(a_{i,l}, a_{j,l})$ is the match score of the pair sequences $[a_i, a_j]$ at position l, $G(i,j,l)$ is a Boolean variable which is defined as follows, if a gap opens between a_i and a_j at position l, $G(i,j,l) = 1$, else $G(i,j,l) = 0$, and v is the penalty of gap.

2.2   Alignment Technique

Clustal. In 1988, the first Clustal program was written by Des Higgins [17], and a dynamic programming algorithm [18] and the progressive alignment strategy developed by Feng and Doolittle [8] were combined in this program. It used a word-based alignment algorithm [19] to calculate the distance matrix and UPGMA method was used to construct the guide tree. In 1992, ClustalV [20] implemented profile alignments to generate guide trees from the multiple alignment using the Neighbour-Joining (NJ) method [21]. In 1994, ClustalW [22] improved the sensitivity of progressive multiple sequence alignment through sequence weighting, position-specific gap penalties and weight matrix choice. In 1997, ClustalX [23] provided a visual interface, so that the multiple alignment can be displayed on the screen and all parameters were optional, which was a significant convenience to the user's of evaluation. The latest member of Clustal series program is Clustal Omega [14], which can align virtually any number of protein sequences quickly and delivers accurate alignments. For constructing a guide tree, Clustal Omega uses a modified version of mBed [24] which has complexity of $O(N \log N)$ and the guide tree is just as accurate as those from conventional methods. In Clustal Omega, the alignments are then computed using the very accurate HHalign package [25], which aligns two profile hidden Markov models [17].

T-Coffee. The first T-Coffee (Tree-based Consistency Objective Function for alignment Evaluation) [26] version can be track back to 2000. It implemented progressive alignment with a consistency-based objective function [27] and tried to maximize the score between the final multiple alignment and a library of pair-wise aligned residue scores which is derived from a mixture of local and global pair-wise alignments. M-Coffee [28] is an extension of T-Coffee and uses consistency to estimate a consensus alignment, and a meta-method for assembling multiple sequence alignments (MSA) by combining the output of several individual methods into one single MSA. TCS (Transitive Consistency Score) [29] is a new extended version of the T-Coffee scoring scheme for overcoming the problem that homology and evolutionary modeling are sensitive to the underlying MSA accuracy, and it also can improve phylogenetic tree reconstruction.

MAFFT. MAFFT [30] was a method for rapid multiple protein sequence alignment based on FFT (Fast Fourier Transform), first released in 2002. Homologous region were rapidly identified by the FFT. FFT converted an amino acid sequence to a sequence whose composition were volume and polarity values of each amino acid residue. The original MAFFT included two different heuristics, the progressive methods were FFT-NS-1 and FFT-NS-2 and the iterate refinement method was FFT-NS-i. In 2005, MAFFT version 5 [31] was released with improvement of accuracy by offering new iterative refinement options, H-INS-i, F-INS-i and G-INS-i. And MAFFT version 5 incorporated pair-wise alignment information into objective function. In 2007, MAFFT version 6 [32] improved accuracy of multiple ncRNA alignment with two techniques: the PartTree algorithm and the Four-way consistency objective function. In 2010, for speeding up program, two natural parallelization strategies (best-first and simple hill-climbing) were implemented for the iterative refinement stage based on MAFFT version 6, and a simple hill-climbing approach was selected as the default [33]. In 2012, two methods had been implemented as the '–add' and '–addfragments' options in the MAFFT package [34] for adding unaligned sequences into an existing multiple sequence alignment.

The newest version is MAFFT version 7 [35], it has options for adding unaligned sequences into an existing alignment, and beyond this, it has several new features, including adjustment of direction in nucleotide alignment, constrained alignment and parallel processing.

MUSCLE. MUSCLE (MUltiple Sequence Comparison by Log-Expectation) [36] is a multiple sequence alignment method of protein sequences. MUSCLE uses two distance measures for each pair of sequences: a kmer distance (for an unaligned pair) and the Kimura distance (for an aligned pair). Guide tree is constructed using UPGMA. MUSCLE uses a profile function called log-expectation (LE) score. And MUSCLE includes three stages as follow:

Stage 1: Draft progressive. This stage includes four steps (similarity measure, distance estimate, tree construction, progressive alignment) and produces a rapid multiple alignment, while de-emphasizing accuracy.

Stage 2: Improved progressive. This stage also includes four steps (similarity measure, tree construction, tree comparison, progressive alignment). In the stage1, the main source of error is the k-mer distance measure, which leads to a suboptimal tree. MUSCLE therefore re-estimates the tree using the Kimura distance, which is more accurate but requires an alignment.

Stage 3: Refinement. This stage is made up of four steps (choice of bipartition, profile extraction, re-alignment, accept/reject). The third stage performs iterative refinement using a approximate tree-dependent restricted partitioning [21].

Kalign. Kalign [31] was a MSA algorithm, which proposed in 2005. It also implemented progressive alignment. And unlike other progressive methods, Kalign employed Wu-Manber approximate string-matching algorithm [37] which made Kalign more accurate in aspect of distance estimation. In 2007, Emmanuelle Becher etc. proposed a tool called HMM-Kalign [38] for generating sub-optimal alignments. As the name implies, HMM-Kalign was based on original Kalign by implementing Hidden Markove Model. The newest inproved edition of Kalign was Kalign-LCS [39].

It applied the longest common subsequence (LLCS) in similarity measure step, and obtained a balance between accuracy and speed.

3   Practical Result

We examine ClustalW, Clustal Omega, T-Coffee, MAFFT:Auto, MAFFT:FFT-NS-1, MAFFT:G-INS-i, MUSCLE and Kalign on the benchmark Balibase 3.0 references, OXbench and Homestrad, respectively.

We evaluate the alignment results with BaliScore, including SP-score (Sum of Pairs score) which is the percentage of homologies in the reference alignment recovered in the estimated alignment and TC-score (Total column score) is the percentage of columns that are recovered entirely correctly in the estimated alignment (Tables 1, 2 and 3).

Table 1. Summary of the techniques described in the review

Name		Method	Guide tree	Sequence	Server
ClustalW [22]		Progressive	NJ	Protein DNA	http://www.clustal.org/ clustal2/ http://www.ebi.ac.uk/Tools/ msa/clustalw2/
Clustal Omega [14]		Progressive	mBed, PartTree	Protein DNA RNA	http://www.clustal.org/ omega/ http://www.ebi.ac.uk/Tools/ msa/clustalo/
T-Coffee [26]		Progressive	–	Protein DNA RNA	http://www.tcoffee.org/ http://www.ebi.ac.uk/Tools/ msa/tcoffee/
MAFFT [35]	*FFT-NS-1* *FFT-NS-2* *G-INS-1*	Progressive	PartTree	Protein DNA RNA	http://mafft.cbrc.jp/ alignment/server/ http://www.ebi.ac.uk/Tools/ msa/mafft/
	FFT-NS-i *E-INS-i* *L-INS-i* *G-INS-i* *Q-INS-i*	Iterative refinement			
MUSCLE [36]	*Step1* *Step2*	Progressive	UPGMA	Protein	http://www.drive5.com/ muscle/ http://www.ebi.ac.uk/Tools/ msa/muscle/
	Step3	Iterative refinement			
Kalign		Progressive	Wu-Manber	Protein DNA RNA	http://msa.sbc.su.se/cgi-bin/ msa.cgi http://www.ebi.ac.uk/Tools/ msa/kalign/

Table 2. The SP-score of various individual methods on the benchmark Balibase 3.0 references

SP-score	ClustalW	Clustal Omega	T-Coffee	MAFFT: Auto	MAFFT: FFT-NS-1	MAFFT: G-INS-i	MUSCLE	Kalign
BaliBase Set: 11	98.7 %	99.3 %	100 %	82.9 %	62.7 %	88.4 %	90.4 %	91.2 %
BaliBase Set: 12	97.3 %	100 %	100 %	93.1 %	87.8 %	94.4 %	90.3 %	90.5 %
BaliBase Set: 20	43.5 %	93.8 %	95.6 %	42.9 %	36.9 %	48.0 %	47.6 %	70.1 %
BaliBase Set: 30	61.4 %	70.7 %	94.4 %	63.7 %	63.6 %	66.4 %	60.8 %	65.0 %
BaliBase Set: 40	93.4 %	93.3 %	97.4 %	90.7 %	88.4 %	90.6 %	90.3 %	88.7 %
BaliBase Set: 50	69.1 %	71.4 %	84.7 %	63.5 %	58.3 %	66.5 %	58.7 %	61.7 %
Average of BaliBase	77.2 %	88.1 %	95.4 %	66.3 %	66.3 %	75.7 %	73.0 %	77.9 %
OXbench set: full	0	100 %	100 %	2.0 %	1.7 %	2.3 %	2.4 %	1.0 %
OXbench set: master	7.7 %	73.2 %	100 %	6.9 %	6.1 %	9.0 %	7.0 %	7.1 %
OXbench set: extended	8.1 %	11.7 %	96.2 %	7.4 %	7.3 %	8.3 %	8.8 %	7.8 %
Average of OXbench	5.3 %	61.6 %	98.7 %	5.4 %	5.0 %	6.5 %	6.1 %	5.3 %
Homestrad	96.9 %	95.1 %	99.1 %	82.7 %	75.1 %	83.9 %	77.9 %	77.3 %

Table 3. The TC-score of various individual methods on the benchmark Balibase 3.0 references

TC-score	ClustalW	Clustal Omega	T-Coffee	MAFFT: Auto	MAFFT: FFT-NS-1	MAFFT: G-NS-i	MUSCLE	Kalign
BaliBase Set: 11	97.4 %	98.7 %	100 %	76.3 %	42.1 %	80.3 %	85.5 %	84.2 %
BaliBase Set: 12	94.2 %	100 %	100 %	86.5 %	78.8 %	88.5 %	80.8 %	80.8 %
BaliBase Set: 20	0	85.9 %	88.5 %	0	0	0	0	0
BaliBase Set: 30	22.7 %	33.1 %	79.1 %	22.7 %	22.1 %	23.3 %	23.9 %	23.0 %
BaliBase Set: 40	61.5 %	58.1 %	80.8 %	50.6 %	38.5 %	48.3 %	45.3 %	40.4 %
BaliBase Set: 50	24.4 %	27.7 %	0	16.5 %	13.5 %	19.0 %	12.7 %	12.2 %
Average of BaliBase	50.3 %	67.3 %	74.7 %	42.1 %	32.5 %	43.2 %	41.4 %	40.1 %
OXbench set: full	0	100 %	100 %	0	0	0	0	0
OXbench set: master	0	50.0 %	100 %	0	0	0	0	0
OXbench set: extended	0	0	83.7 %	0	0	0	0	0
Average of OXbench	0	50.0 %	94.6 %	0	0	0	0	0
Homestrad	91.8	87.0 %	95.5 %	53.0 %	31.4 %	57.2 %	44.2 %	31.5

From the results of SP-score and TC-score, we can see that all programs we examined are not sensitive to divergence of sequence. All programs suffer by the impact of a highly divergent "orphan" sequence, residue difference between groups, N/C-terminal extensions, and internal insertions to varying degrees, respectively. And on the whole, Clustal Omega and T-Coffee perform well, especially the results corresponding to T-Coffee are the best.

4   Conclusion and Future Development

In the past years, MSA achieved great development, and obtained good effect which applied in many biological applications. But there still is plenty room to improve multiple sequence alignment, especially in the respect of robustness and accuracy. In order to solve these problems, in one hand, we should continue to develop recent efficient MSA techniques, such as T-Coffee, in other hand we should transform the way of thinking and apply more techniques which are not just heuristic methods, even not just biological informatics technology to improve MSA.

Happily, many researchers devote themselves to develop MSA method. Sabari Pramanik and S.K. Setua [40] define a new form of chromosome representation, and deploy it on steady state Genetic Algorithm, then get better results. Siavash Mirarab, Nam Nguyen, and Tandy Warnow propose an algorithm called PASTA [41] to realize estimation of large-scale multiple sequence alignment. And there is a interesting method called Phylo [42], which is a human-based computing framework applying "crowd sourcing" techniques to solve the Multiple Sequence Alignment (MSA) problem. The key idea of Phylo is to convert the MSA problem into a casual game that can be played by ordinary web users with a minimal prior knowledge of the biological context. Cactus [43] caters to the phenomenon that much attention has been given to the problem of creating reliable multiple sequence alignments in a model incorporating substitutions, insertions, and deletions while far less attention has been paid to the problem of optimizing alignments in the presence of more general rearrangement and copy number variation.

Another trend of development is parallelization of MSA. Because of that MSA is a NP-hard problem and the huge amount of data, the programs of MSA are costly in the respect of time. Hence, it's necessary to implement parallel solutions in MSA. Jucele F. A. et al. [44] present two parallel solutions using the BSP/CGM model, with MPI and CUDA implementations. And the results of this method show that the use of parallel processing allows the manipulation of more and larger sequences. Evandro A. Marucci et al. [45] propose a parallel algorithm for multiple sequence similarities calculation based on the k-mer counting method, and obtain a very good scalability and a nearly linear speedup.

Acknowledgement. This work was supported by Shenzhen Municipal Science and Technology Innovation Council (Grant No. CXZZ20140904154910774, Grant No.JCYJ20140417172417174, Grant No. JCYJ20140904154645958, Grant No. JCYJ20130329151843309) and China Post-doctoral Science Foundation funded project (Grant No. 2014M560264).

References

1. Chodavarapu, R.K., Feng, S., Bernatavichute, Y.V., Chen, P.-Y., Stroud, H., Yu, Y., et al.: Relationship between nucleosome positioning and DNA methylation. Nature **466**, 388–392 (2010)

2. Hicks, S., Wheeler, D.A., Plon, S.E., Kimmel, M.: Prediction of missense mutation functionality depends on both the algorithm and sequence alignment employed. Hum. Mutat. **32**, 661–668 (2011)

3. Wang, P., Hu, L., Liu, G., Jiang, N., Chen, X., Xu, J., et al.: Prediction of antimicrobial peptides based on sequence alignment and feature selection methods. PLoS one **6**, e18476 (2011)

4. Brenchley, R., Spannagl, M., Pfeifer, M., Barker, G.L., D'Amore, R., Allen, A.M., et al.: Analysis of the bread wheat genome using whole-genome shotgun sequencing. Nature **491**, 705–710 (2012)

5. Varshney, R.K., Terauchi, R., McCouch, S.R.: Harvesting the promising fruits of genomics: applying genome sequencing technologies to crop breeding. PLoS Biol. **12**, e1001883 (2014)

6. Li, H., Homer, N.: A survey of sequence alignment algorithms for next-generation sequencing. Briefings Bioinform. **11**, 473–483 (2010)

7. Zhou, X., Ren, L., Meng, Q., Li, Y., Yu, Y., Yu, J.: The Next-generation sequencing technology and application. Protein Cell **1**, 520–536 (2010)

8. Feng, D.-F., Doolittle, R.F.: Progressive sequence alignment as a prerequisitetto correct phylogenetic trees. J. Mol. Evol. **25**, 351–360 (1987)

9. Hogeweg, P., Hesper, B.: The alignment of sets of sequences and the construction of phyletic trees: an integrated method. J. Mol. Evol. **20**, 175–186 (1984)

10. Thompson, J.D., Koehl, P., Ripp, R., Poch, O.: BAliBASE 3.0: latest developments of the multiple sequence alignment benchmark. Proteins Struct. Funct. Bioinf. **61**, 127–136 (2005)

11. Raghava, G., Searle, S.M., Audley, P.C., Barber, J.D., Barton, G.J.: OXBench: a benchmark for evaluation of protein multiple sequence alignment accuracy. BMC Bioinf. **4**, 47 (2003)

12. Gotoh, O.: Heuristic Alignment Methods. Multiple Seq. Alignment Meth. **1079**, 29–43 (2014)

13. Kersters, K., De Ley, J., Sneath, P., Sackin, M.: Numerical taxonomic analysis of agrobacterium. J. Gen. Microbiol. **78**, 227–239 (1973)

14. Sievers, F., Wilm, A., Dineen, D., Gibson, T.J., Karplus, K., Li, W., et al.: Fast, scalable generation of high-quality protein multiple sequence alignments using clustal omega. Mol. Syst. Biol. **7**, 539 (2011)

15. Altschul, S.F.: Gap costs for multiple sequence alignment. J. Theor. Biol. **138**, 297–309 (1989)

16. Altschul, S.F., Carroll, R.J., DJ, L.: Weights for Data Related by a Tree. J. Mol. Biol. **207**, 647–653 (1989)

17. Eddy, S.R.: Profile hidden markov models. Bioinformatics **14**, 755–763 (1998)

18. Myers, E.W., Miller, W.: Optimal alignments in linear space. Comput. Appl. Biosci. CABIOS. **4**, 11–17 (1988)

19. Wilbur, W.J., Lipman, D.J.: Rapid similarity searches of nucleic acid and protein data banks. Proc. Natl. Acad. Sci. **80**, 726–730 (1983)

20. Higgins, D.G.: CLUSTAL V: multiple alignment of DNA and protein sequences. Comput. Anal. Seq. Data **25**, 307–318 (1994)

21. Saitou, N., Nei, M.: The neighbor-joining method: a new method for reconstructing phylogenetic trees. Mol. Biol. Evol. **4**, 406–425 (1987)

22. Thompson, J.D., Higgins, D.G., Gibson, T.J.: CLUSTAL W: improving the sensitivity of progressive multiple sequence alignment through sequence weighting, position-specific gap penalties and weight matrix choice. Nucleic Acids Res. **22**, 4673–4680 (1994)

23. Thompson, J.D., Gibson, T.J., Plewniak, F., Jeanmougin, F., Higgins, D.G.: The CLUSTAL_X windows Interface: Flexible Strategies for Multiple Sequence Alignment Aided by Quality Analysis Tools. Nucleic Acids Res. **25**, 4876–4882 (1997)

24. Blackshields, G.S.F., Shi, W., Wilm, A., Higgins, D.G.: Sequence embedding for fast construction of guide trees for multiple sequence alignment. Algorithms Mol Biol. **5**, 21 (2010)
25. Söding, J.: Protein homology detection by HMM–HMM comparison. Bioinformatics **21**, 951–960 (2005)
26. Notredame, C., Higgins, D.G., Heringa, J.: T-Coffee: a novel method for fast and accurate multiple sequence alignment. J. Mol. Biol. **302**, 205–217 (2000)
27. JD, K.: The maximum weight trace problem in multiple sequence alignment. In: Apostolico, A., Crochemore, M., Galil, Z., Manber, U. (eds.) CPM 1993. LNCS, vol. 684, pp. 106–119. Springer, Heidelberg (1993)
28. Wallace, I.M., O'Sullivan, O., Higgins, D.G., Notredame, C.: M-Coffee: combining multiple sequence alignment methods with t-coffee. Nucleic Acids Res. **34**, 1692–1699 (2006)
29. Chang, J.-M., Di Tommaso, P., Notredame, C.: TCS: A New Multiple Sequence Alignment Reliability Measure to Estimate Alignment Accuracy and Improve Phylogenetic Tree Reconstruction. Molecular Biology and Evolution. msu117(2014)
30. Katoh, K., Misawa, K., K.-I, K., Miyata, T.: MAFFT: a novel method for rapid multiple sequence alignment based on fast fourier transform. Nucleic Acids Res. **30**, 3059–3066 (2002)
31. Katoh, K., Kuma, K.-i, Toh, H., Miyata, T.: MAFFT Version 5: improvement in accuracy of multiple sequence alignment. Nucleic Acids Res. **33**, 511–518 (2005)
32. Katoh, K., Toh, H.: Improved accuracy of multiple ncRNA alignment by incorporating structural information into a MAFFT-based framework. BMC Bioinform. **9**, 212 (2008)
33. Katoh, K., Toh, H.: Parallelization of the MAFFT multiple sequence alignment program. Bioinform. **2**, 1899–1900 (2010)
34. Katoh, K., Frith, M.C.: Adding unaligned sequences into an existing alignment using MAFFT and LAST. Bioinform. **28**, 3144–3146 (2012)
35. Katoh, K., Standley, D.M.: MAFFT multiple sequence alignment software Version 7: improvements in performance and usability. Mol. Biol. Evol. **30**, 772–780 (2013)
36. Edgar, R.C.: MUSCLE: multiple aequence alignment with high accuracy and high throughput. Nucleic Acids Res. **32**, 1792–1797 (2004)
37. Wu, S., Manber, U.: Fast text searching: allowing errors. Commun. ACM **35**, 83–91 (1992)
38. Becker, E., Cotillard, A., Meyer, V., Madaoui, H., Guérois, R.: HMM-Kalign: a tool for generating sub-optimal HMM alignments. Bioinform. **23**, 3095–3097 (2007)
39. Deorowicz, S., Debudaj-Grabysz, A., Gudyś, A.: Kalign-LCS — a more accurate and faster variant of kalign2 algorithm for the multiple sequence alignment problem. In: Gruca, A., Czachórski, T., Kozielski, S. (eds.) Man-Machine Interactions 3. AISC, vol. 242, pp. 499–506. Springer, Heidelberg (2014)
40. Pramanik, S., Setua, S.: A steady state genetic algorithm for multiple sequence alignment. In: International Conference on Advances in Computing, Communications and Informatics (ICACCI), pp. 1095–1099. IEEE (2014)
41. Mirarab, S., Nguyen, N., Warnow, T.: PASTA: ultra-large multiple sequence alignment. In: Sharan, R. (ed.) RECOMB 2014. LNCS, vol. 8394, pp. 177–191. Springer, Heidelberg (2014)
42. Kawrykow, A., Roumanis, G., Kam, A., Kwak, D., Leung, C., Wu, C., et al.: Phylo: a citizen science approach for improving multiple sequence alignment. PLoS one **7**, e31362 (2012)
43. Paten, B., Earl, D., Nguyen, N., Diekhans, M., Zerbino, D., Haussler, D.: Cactus: algorithms for genome multiple sequence alignment. Genome Res. **21**, 1512–1528 (2011)

44. Vasconcellos, J.F., Nishibe, C., Almeida, N.F., Cáceres, E.N.: Efficient parallel implementations of multiple sequence alignment using BSP/CGM model. In: Proceedings of Programming Models and Applications on Multicores and Manycores, 103. ACM (2014)
45. Marucci, E.A., Zafalon, G.F., Momente, J.C., Neves, L.A., Valêncio, C.R., Pinto, A.R. et al.: An Efficient Parallel Algorithm for Multiple Aequence Aimilarities Calculation Using a Low Complexity Method. BioMed research international (2014)

Analyzing the Genomes of Coxsackievirus A16 and Enterovirus 71 in Relation to Hand, Foot and Mouth Disease(HFMD) Using Apriori Algorithm, Decision Tree and Support Vector Machine (SVM)

Chaeyun Jung[1]($\boxtimes$), Yonghyun Park[2], Seunghui Han[2],
and Taeseon Yoon[3]

[1] Department of International,
Hankuk Academy of Foreign Studies,
Yongin, Republic of Korea
dbsco3293@hanmail.net
[2] Department of Natural Science,
Hankuk Academy of Foreign Studies,
Yongin, Republic of Korea
{eastlifeyh,mrhan1998}@naver.com
[3] Faculty of Computer Science,
Hankuk Academy of Foreign Studies,
Yongin, Republic of Korea
tsyoon@hafs.hs.kr

Abstract. Hand, foot and mouth disease (HFMD), caused by highly infectious intestinal viruses of either coxsackievirus A16 (CVA16) or enterovirus 71 (EV71), is a common children syndrome featured by mild fever, spots and bumps that blister the skin of hands, oral cavity in mouth, feet and sometimes to the extent of buttocks and genitalia. Though CVA16 and EV71 both cause the HFMD, the intensity of each symptom is obviously different. Normal cases are HFMD by CVA16, which are typically characterized by mild symptoms usually treated in 6−8 days. Conversely, HFMD by EV71 results severe neural disorders and various influenza complications that even cause death. Currently, no vaccine is available to protect individuals from infection by the viruses that cause HFMD. In order to investigate why these two viruses have too much differences in degree of symptoms and compare the inner relationships to the medical effects, we analyzed the genomes of CVA16 and EV71 by using apriori algorithm, decision tree and support vector machine (SVM). Therefore, by comparing the genomes of each virus, we found out better results for analyzing the relationship between the two and state the potential of developing medical remedy in DNA point of view.

Keywords: Hand, Foot and Mouth Disease (HFMD) · Coxsackievirus A16 · Enterovirus 17 · Apriori Algorithm · Decision Tree · Support Vector Machine (SVM)

© Springer International Publishing Switzerland 2015
D.-S. Huang et al. (Eds.): ICIC 2015, Part I, LNCS 9225, pp. 539–545, 2015.
DOI: 10.1007/978-3-319-22180-9_53

1   Introduction

Hand, foot and mouth disease (HFMD), first described in New Zealand in 1957, tends to occur in spring, summer, and fall seasons for children under the age of 10. Common striking signs of the HFMD include fever, nausea, vomiting, spots followed by vesicular sores with blisters on palms, feet, buttocks and lips, painful facial ulcers, blisters, and lesions around the nose or mouth [2, 3]. Coxsackievirus A16 (CVA16) and Enterovirus 71 (EV71) are known as the primary causes of HFMD, highly contagious and frequently transmitted by direct contact of nasopharyngeal secretions such as saliva or nasal mucus or by fecal-oral transmission [12]. These symptoms rarely improve to severe complications. However, HFMD by EV71 is different. It also causes blisters and rash, but the symptoms appear much severer and are more likely to improve to neurologic or cardiac complications including death and viral or aseptic meningitis featured by fever, headache, stiff neck, and back pain. In rare situations, serious complications including swelling of the brain, defined as encephalitis, or flaccid paralysis are caused [10].

In this paper, we focused on investigating the comparison and contrast of the CVA16 and EV71 in genomic sequences since the differences between two viruses are clearly stated. Genomic similarities of two types proved by Apriori algorithm and decision tree prove the similar symptoms and fatal complications of different severity concerning with EV71 is shown in DNA point of view, as evidenced by decision tree and SVM. And in the process of analyzing the genomes of each type of virus, we can further bring up the potential of developing vaccine related with certain inner DNA amine proteins.

2   Related research

2.1   Comparison of Coxsackievirus A16 (CVA16) and Enterovirus 71 (EV71)

The use of CVA16 and EV71 as the research topics has been considered previously a lot because it is the major causative agent of HFMD. Coxsackie virus is a member of the Picornaviridae family of viruses in the genus termed Enterovirus [2]. In addition, EV71 infrequently causes polio-like syndrome permanent paralysis. In RNA perspective, EV71 infection leads to increase in the level of mRNAs encoding chemokines, complement proteins, and proapoptosis proteins [9]. Analysis of CVA16 and EV71 has been studied separately on various points of view.

Among them, the research on the role of each virus in a large outbreak of HFMD during 2012 in Changsha, China, was impressive. By using real-time RT-PCR and sequencing of the VP1 regions, the detection and genotyping of enterovirus were performed [6]. One study reveals differences of two viruses in binding cell surface heparan sulphate. To be specific, the involvement of HS (in mediating viral infection of isolates of human enteroviruses was investigated in Vero and human neural cells [4]. Another study about research on vaccines based on their similarities of both types shows how novel recombinant chimeric virus-like particle is immunogenic and protective against both EV71 and CVA16, using mice as experimental agent [5].

3   Method

Apriori is an algorithm for frequent item set mining and association rule learning over broad databases [11]. A decision tree can be used as a model for sequential decision problems under uncertainty [8]. A Support Vector Machine (SVM) is a non-probabilistic binary linear classifier formally defined by a separating hyperplane [8].

4   Results

We extracted the genome sequences of CVA16 and EV71 from the National Center for Biotechnology Information (NCBI). For the Apriori Algorithm, we extracted genomic databases of two viruses each and found out the rules of patterns in position number and certain amino acids. For the decision tree, we used the sequences for a 10-fold cross validation experiment held with 2 classes (EV71 for class 1, and CVA16 for class 2). Among the algorithm results, we first gathered the rules of patterns divided into each virus and extracted data with very high frequency rates; only data with frequency rates higher than 0.800 were chosen. We deleted the overlapped patterns except for maximum value. For SVM, we experimented in 10 folds cross validation and experimental results show the division probability of NORMAL, POLY and RBF. In all these procedures, for better experiments, we divided the experiments in three segments: 9 window, 13 window and 17 window.

4.1   Apriori Algorithm

We did this in order to find the certain features of amino acids in two of each virus: CVA16 and EV71. Rules of patterns are characterized by the position number of certain amino acids. In the output of the algorithms, specific sequences of amino acids are shown among whole twenty amino acids. As experimental result comes out, it is shown that both EV71 and CVA16 have rules of Serine, Leucine, and Argine mostly.

Table 1. Rule extraction under 9 windows a prior

EV71	CVA16
1. amino1 = S 33	1. amino4 = S 39
2. amino2 = L 32	2. amino1 = R 33
3. amino2 = S 32	3. amino8 = S 33
4. amino6 = S 32	4. amino9 = G 31
5. amino9 = R 32	5. amino2 = S 30
6. amino6 = R 30	6. amino3 = L 28
7. amino8 = G 30	
8. amino3 = S 29	

Results of rule extraction under 13 and 17 window are also shown in the similar patterns that both EV71 and CVA16 have similar inner components of amino acids in common.

542   C. Jung et al.

4.2   Decision Tree

After being clearly proved that both EV71 and CVA16 have certain amino acids in common, we conducted cross-validating experiment, not dividing the features of each virus, but tracking out rules of patterns with certain frequencies. By using two viruses as experiment variables in every fold, this experiment shows both similarities and overall differences. In conducting 10 experiments, class 1 is EV71 and class 2 is CVA16. Like Apriori, we divided 9, 13, and 17 windows for efficient rule extraction.

Among extraction results, we first gathered the rules of patterns divided into each two viruses and selected data with very high frequency rates; only data with frequency rates higher than 0.800 were chosen because it is considered as almost fixed patterns to influence the experiment results.

Table 2. Rule extraction unde 13 window decision tree

Virus	Rule		Frequency
EV71	pos1 = Q	pos6 = A	0.833
	pos5 = Q	pos6 = G	0.833
	pos1 = W		0.889
	pos11 = S	pos13 = A	0.833
CVA16	pos13 = D		0.857
	pos3 = S	pos13 = G	0.833
	pos6 = S	pos13 = R	0.833
	pos6 = L	pos13 = H	0.857
	pos1 = S	pos8 = L	0.833
	pos12 = Q	pos13 = V	0.833
	pos6 = E		0.800

Results of rule extraction under 9 and 17 window are also shown in the similar patterns. Because of the more specified rules of patterns with mostly two amino acids and high frequency rates, we found out EV71 and CVA16 is quite different. Rule extraction experiments by decision tree reveals whole kinds of amino acids. These results turned out to be different from Apriori that has large amounts of Leucine, Arginine and Serine in common since decision tree tracks out more specified inner segments of amino patterns with accurate rates.

As the number of window increases as 9 to 13 to 17, length of rule lists become longer. And unlike the 9 window and 13 window results, 17 window results show particularly higher frequency rates. Especially, CVA16's rule of pos2 = D and the rule of pos4 = L has frequency rates of each 0.900 and 0.929. These are strikingly high frequency that means except for very rare cases, these patterns are fixed. Apriori and Decision tree proves that though EV71 and CVA16 is similar as member of the Picornaviridae family of viruses in the genus termed Enteroviruss in overall point of view, when analyzed thoroughly, inner components that consist of these two viruses are different.

4.3   Support Vector Machine (SVM)

For Support Vector machine, we experimented in 10 folds cross validation, divided into three segments: 9 window, 13 window and 17 window. Experimental results show the division probability of NORMAL, POLY and RBF. Results of Table 4 to Table 6 is to divide sums of accuracy sets into 10 and calculate the average value. Since Precision/recall is possible with accuracy, we focused on accuracy to prove their nonlinear classification.

Table 3. Classification accuracy rate under SVM

	NORMAL	POLY	R.B.F.
9 window	50.39	59.18	100
13 window	51.94	61.62	100
17 window	54.59	66.10	100

Although results of decision tree shows seemingly similar EV71 and CVA16 are different, it doesn't confer any scientific accuracy of its clear difference. We used SVM as a tool because it has features of linear classification and non-linear classification so that these kinds of classification prove whether it is similar in more cases or it is rather different in rates. Table 3 approximately shows NORMAL function is divided into 50 %, linear POLYnomial function into 60 % and R.B.F into 100 % in all three cases of 9 window, 13 window and 17 window. Therefore, it is non-linear classification which accurately proves EV71 and CVA16 are different.

4.4   Analysis and Discussion

Hand-foot-and-mouth disease (HFMD) has been recognized as an important global public health issue, which is predominantly caused by EV71 and CVA16. There is no available vaccine against HFMD. An ideal HFMD vaccine should be bivalent against both EV71 and CVA16 [1]. By using Apriori algorithm, we found out EV71 and CVA16 has similar amino acids of Leucine, Serine and Argine. And with Decision tree, we were able to find several acceptable rules that exceed a frequency of 0.75. We were able to find over one hundred rules for each window, indicating that though similar amino acids are discovered in overall perspective, EV71 and CVA16 have many differences. As endpoint of thorough scientific analysis, SVM proves non-linear classification. Too many differences amongst the viruses that data shows make it impossible for immunity to deal with both viruses.

With results from Apriori, however, we can state the potential of vaccine with Leucine, Serine and Argine since these three amino factors are most common in both viruses. Although it is far to go to develop perfect vaccine satisfying the variety, the results of our Apriori experiment must be quite necessary in the process of development. How Leucine, Serine and Argine in two viruses affects the human body has not been conducted experimentally. Further investigation on the influence of certain amino acids in EV71 and CVA16 may lead to deeper research. Another notable results by decision tree that CVA16's rule of pos2 = D and the rule of pos4 = L have frequency rates of each 0.900 and 0.929 may be an important factor in determining standard to moderate differences in vaccine development. Further experiments on this specific position may allow scientists to find the binding differences in other factors between EV71 and CVA16.

5 Conclusions

Finding a vaccine for Hand, Foot and Mouth Disease (HFMD) and creating a treatment for it are arduous tasks since Enterovirus 71 and Coxsackievirus A16 have too many differences in inner amino elements, as is substantiated by the experimental result we extracted from the decision tree and SVM. However, we were also able to find certain similarities in the viruses and particular traits of amino acids that can further state the potential to take a significant role as a vaccine. Those components and certain positions are significant factors in adjusting appropriate levels to compare one another in higher frequency. We believe the experiments we conducted will provide help for further investigations and experiments on better analysis for Enterovirus 71 and Coxsackievirus A16 and for vaccine development.

References

1. Chen, X.P., Tan, X.J., Xu, W.B.: Research Advances in Molecular Epidemiology and Vaccines of Coxsackievirus A16. US National Library of Medicine National Institutes of Health. Bing Du Xue Bao, 30 July 2014
2. Hou, W., Yang, L., He, D., Zheng, J., Xu, L., Liu, J., Zhao, H., Liu, Y., Ye, X., Cheng, T., Xia, N.: Development of a Coxsackievirus A16 Neutralization Test Based on the Enzyme-linked Immunospot Assay. National Center for Biotechnology Information. U.S. National Library of Medicine
3. Lu, G., Qi, J., Chen, Z., Xu, X., Gao, F., Lin, D., Qian, W., Liu, H., Jiang, H., Yan, J., Gao, G.F.: Enterovirus 71 and Coxsackievirus A16 3C Proteases: Binding to Rupintrivir and Their Substrates and Anti-Hand, Foot, and Mouth Disease Virus Drug Design. Journal of Virology **85**, 10319–10331 (2011). American Society for Microbiology
4. Pourianfar, H.R., Kirk, K., Grollo, L.: Initial Evidence on Differences among Enterovirus 71, Coxsackievirus A16 and Coxsackievirus B4 in Binding to Cell Surface Heparan Sulphate. National Center for Biotechnology Information. U.S. National Library of Medicine, 4 December 2013
5. Zhao, H., Li, H.Y., Han, J.F., Deng, Y.Q., Zhu, S.Y., Li, X.F., Yang, H.Q., Li, Y.X., Zhang, Y., Qin, E.D., Chen, R., Qin, C.F.: Novel recombinant Chimeric Virus-like particle is immunogenic and protective against both Enterovirus 71 and Coxsackievirus A16 in Mice. National Center for Biotechnology Information, U.S. National Library of Medicine
6. Zhu, Z., Zhu, S., Guo, X., Wang, J., Wang, D., Yan, D., Tan, X., Tang, L., Zhu, H., Yang, Z., Jiang, X., Ji, Y., Zhang, Y., Xu, W.: Retrospective seroepidemiology indicated that human Enterovirus 71 and Coxsackievirus A16 circulated wildly in central and Southern China before large-scale outbreaks from 2008. J. Virol. **7**(1), 300 (2010)

7. Go, E., Lee, S., Yoon, T.: Analysis of Ebolavirus with decision tree and apriori algorithm. Int. J. Mach. Learn. Comput. **4**(6), 543–546 (2014)
8. Theodoridis, S., Koutroumbas, K.: Pattern Recognition, 4th edn. Academic Press, Burlington (2009). ISBN 978-1-59749-272-0
9. Shih, S.R., Stollar, V., Lin, J.Y., Chang, S.C., Chen, G.W., Li, M.L.: Identification of genes involved in the host response to enterovirus 71 infection. J. Neurovirol. **10**(5), 293–304 (2004). doi:10.1080/13550280490499551. PMID 15385252
10. Shimizu, H., Utama, A., Yoshii, K., et al.: Enterovirus 71 from fatal and nonfatal cases of hand, foot and mouth disease epidemics in Malaysia, Japan and Taiwan in 1997–1998. Jpn. J. Infect. Dis. **52**(1), 12–5 (1999). PMID 10808253
11. Agrawal, R., Srikant, R.: Fast algorithms for mining association rules, pp. 1–32. IBM Almaden Research Center
12. Enterovirus 71 Infection. Centre for Health Protection

A Utility Function Based Resource Allocation Method for LEO Satellite Constellation System

Fangfang Yuan[✉], Xingwei Wang, Fuliang Li, and Min Huang

College of Information Science and Engineering, Northeastern University,
Shenyang, China
dolphinNEUQ@163.com, {wangxw,isemhuang}@mail.neu.edu.cn,
lifuliang@ise.neu.edu.cn

Abstract. Low Earth Orbit (LEO) satellite constellation system has been regarded as a very promising satellite mobile communication system for its low propagation delay, global coverage for communication and mature technologies. However, considering its crucial but limited resources on satellites, efficient resource allocation methods are needed to guarantee the network carrying capability under specific requirements. In this paper, we put forward a utility function based resource allocation method for LEO satellite constellation system. We first utilize utility function to represent the resource acquisition satisfaction of adjacent satellites. The bigger the utility is the higher satisfaction the adjacent satellites can get. In addition, we adopt the improved Multiple Population Cloud Differential Evolution Algorithm (MPCDEA) to solve the resource allocation problem, which belongs to the nonlinear mixed integer programming problem. Finally, we evaluate the proposed utility function based resource allocation method according to the topologies of Iridium and Globalstar systems and quantify it with performance indexes like throughput capacity and network capacity. Evaluation results show that our method is feasible and effective.

Keywords: Low Earth Orbit (LEO) · Resource allocation · Utility function · MPCDEA

1 Introduction

Integration of the Internet and satellite communication in recent years makes satellite communication network applied widely, especially the LEO satellite constellation. LEO satellite constellation system is a popular satellite network with on-board processing technology and inter-satellite links (ISLs). Consequently, it has high communication quality, large bandwidth and nearly no effect from terrain feature, weather or other factors. For LEO satellites belonging to small satellites with limited energy and storage space, when faced with increasing number of Internet users and demand of faster Internet access, higher network share bandwidth and so on, its resource-constrained disadvantages become more evident. Thus the goal of maximizing the carrying capability of the whole network under specific environment through resource allocation methods is most significant. Network capacity is an important metric of network carrying capability, which refers to the specific maximum data transfer capability of the

© Springer International Publishing Switzerland 2015
D.-S. Huang et al. (Eds.): ICIC 2015, Part I, LNCS 9225, pp. 546–557, 2015.
DOI: 10.1007/978-3-319-22180-9_54

entire network. But related work about network capacity or network capacity oriented resource allocation seldom concentrates on ISLs of LEO constellation system.

In this paper, we put forward a utility function based resource allocation method. In the beginning, we introduce utility function to represent the adjacent satellites' satisfaction of resource acquisition based on the analysis of network capacity and consideration of the actual demand as well as the fairness between adjacent satellites of LEO satellite constellation system. Then we adopt the improved MPCDEA to conduct resource allocation, which actually belongs to one kind of nonlinear mixed integer programming problems. We accurately calculate throughput capacity and network capacity so as to quantitatively describe network performances through dividing time slices into smaller time intervals.

The remainder of the paper is organized as follows. The related work is presented in Sect. 2. We construct three models to help understand our problems in Sect. 3. Our proposed resource allocation method is described in Sect. 4. Section 5 presents our simulation and evaluation results including throughput capacity and network capacity. Our conclusions and future remarks are presented in Sect. 6.

2　Related Work

As LEO satellite communication network gets more widely applied, how to maximize network carrying capability within limited resources under specific demand environments becomes urgent. Network carrying capability can be measured through network capacity. Current studies on network capacity and allocation of resources can be summarized as follows.

Studies on network capacity are mainly for ground network. The literature [1] built wireless Ad hoc network model of static nodes, proposed the concept of network capacity and analyzed the relation between network carrying capability and node number via mathematical derivation. The path loss and absorption attenuation have an influence on network capacity [2]. The research of network capacity brought in transmitting power and power control [3, 4]. The literature [5] put forward an optimization model for bending tube satellite communication network based on FDMA and digital channelized technology with the goal of maximizing all uplink channel Shannon capacity and adopted multi-stage optimization algorithm to get a solution.

Studies about network capacity oriented resource allocation methods are focused on ground networks as well. The literature [6] proposed PAUS algorithm based on resource allocation and user scheduling by studying cognitive radio network resource allocation. In terms of maximizing channel capacity, the literature [7] did a survey on resource allocation such as time and power by considering upper and lower bounds of the orthogonal relay channel capacity. To maximize the satisfaction of all participants, an effective mechanism was put forward for allocating resources [8]. For satellite communication, the literature [9] proposed a novel subchannel allocation algorithm aimed at spectral transmission and compression transmission to assure the effective utilization of satellite bandwidth and power. Considering QoS requirements, the literature [10] put forward a resource allocation method based on utility, which can be solved by utility based Lagrange heuristic algorithm.

From above analysis, we find that current studies about network capacity or resource allocation methods of network capacity mainly concentrate on the ground or satellite network of bending cube. In contrast, resource allocation methods of satellite communication networks with ISLs oriented network capacity are relatively little, so that there is still a lot of room to grow. We allocate certain resources to satellites in LEO environment according to their actual needs and adopt improved multiple population cloud differential evolution algorithm to get resource allocation matrices.

3   Problem Description

Here, we model LEO satellite constellation network, analyze power loss in the transmission process and give a kind of analysis method of network capacity.

3.1   Network Model

Network model can be simplified as $G(t) = (S, E, D(t))$, $S = \{s_{ij}\}$ is the set of satellite nodes; i stands for the number of orbit planes; j is the satellite's number in one orbital plane; $E \subseteq S \times S$, $E = \{e_1, \ldots e_n\}$ is set of ISLs. T stands for the cycle of network model, $D(t) = \{d_{i_2j_2}^{i_1j_1}(t)\}$ stands for the distance from satellite node $s_{i_1j_1}$ to $s_{i_2j_2}$ at the moment t and $t \in [0, T]$. Here, for simplification, we use s_A and s_B to represent $s_{i_1j_1}$ and $s_{i_2j_2}$ respectively. Then the length of ISL between s_A and s_B is as Eq. (1). R stands for the orbit radius of s_A and s_B, φ_A and φ_B represent latitudes of s_A and s_B, λ_A and λ_B stands for longitudes of s_A and s_B.

$$d_{AB} = R\sqrt{2(1 - (\sin \varphi_A \sin \varphi_B + \cos \varphi_A \cos \varphi_B \cos(\lambda_A - \lambda_B)))} \tag{1}$$

3.2   Transmission Model

Satellite communication waves are affected by various losses. Signal-to-noise ratio (SNR) is a main metric to measure the communication quality of satellite ISLs. Here, we obtain SNR via considering transmitting and receiving power as well as noise power at the receiving end. Figure 1 depicts the inter-satellite link (ISL) structure.

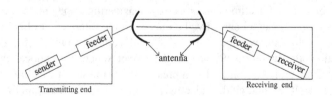

Fig. 1.   Structure of ISL

3.3 Analysis of Network Capacity

The limited resources of satellites mainly contain bandwidth and power. As a result, we consider network capacity under these two resource constraints. Channel capacity, node throughput capacity and network capacity are three key concepts to help quantitatively evaluate our method and their definitions are as follows. Among them, B stands for channel bandwidth measured in hertz, S represents signal average power measured in watt, N is noise power measured in watt, M stands for the number of channels and C_m presents the channel capacity of channel M.

Channel Capacity:

$$C(bit/s) = B\log_2(1 + S/N) \tag{2}$$

Node Throughput Capacity:

$$\lambda(bit/s) = \sum\nolimits_{m=1}^{M} C_m \tag{3}$$

Network Capacity:

$$NC(bit/s) = \sum\nolimits_{i=1}^{n} \lambda_i \tag{4}$$

Considering the dynamics of satellites' motion, network capacity can be classified into instantaneous network capacity and average network capacity. And we split entire network cycle into several time slices considering the ISLs' handover caused by satellites' periodic movement. So the average network capacity of one cycle can be obtained by calculating the average network capacity within one time slice. Equation (5) is the average network capacity of the time slice $[t_i, t_{i+1}]$. More detailed researches about how to divide time slices can be found in [11].

$$NC_{[t_i,t_{i+1}]} \approx \frac{1}{n}\sum\nolimits_{k=1}^{n} NC(t_{i_k}) \tag{5}$$

4 Utility Function Based Resource Allocation Method

After considering network model, transmission model and analyzing network capacity, starting from business needs, we put forward the utility function based resource allocation method [12]. Utility Function represents adjacent satellites' resource acquisition satisfaction degree and the bigger it is, the higher the satisfaction degree is. For each pair of adjacent nodes, the satisfaction degree is defined as Eq. (6). T_k^{link} depicts the ISL capacity between satellite s_{ij} and the t_{th} adjacent satellite. $Utility(\cdot)$ represents utility function which is convex aimed at maintaining fairness. $N_{subchannel}$ is the number of allocated subchannels. The mathematical model of utility function is as Eqs. (7) and (8).

$$Utility(T_k^{link}) = \ln\left(1 + \frac{N_{subchannel} \cdot T_k^{link}}{B}\right) \tag{6}$$

Objective Function:

$$\max \sum_{k=1}^{N_{Neighbor}} Utility(T_k^{link}) \tag{7}$$

Constraints:
$$\begin{cases} \sum_{i=1}^{N_{subchannel}} p_i \le p_{total} & (1) \\ SNR_i \ge SNR_{min} \quad i = 1, \cdots, N_{subchannel} & (2) \\ \sum_{k=1}^{N_{neighbor}} c_{ki} = 1 \quad i = 1, \cdots, N_{subchannel} & (3) \\ p_i \in [0, \ p_{total}] \quad i = 1, \cdots, N_{subchannel} & (4) \\ c_{ki} \in \{0, 1\} \quad k = 1, \cdots, N_{neighbor}; \ i = 1, \cdots, N_{subchannel} & (5) \end{cases} \tag{8}$$

In Eq. (8), (1) represents power constraint, (2) represents the normal communication conditions of ISLs, equation (3) stands for the subchannel-allocated constraints, (4) and (5) are the ranges of optimization variables. $N_{neighbor}$ is the number of neighbor satellites of $s_{ij}.p_{total}$ is the power resource of satellites.

Seeing from the above objective function and constraints, we can find that the utility function based resource allocation method needs to solve the nonlinear mixed integer programming problem which is born with non-convexity and have many local solutions. Traditional mathematics methods such as branch and bound method, cutting plane method and Lagrange relaxation method have certain limits on objective function and often can't get the global optimal solution. Naturally, the best choices to find optimal solution within a number of local solutions are heuristic and intelligent optimization algorithms except for their inability of dealing with constraints. Considering the above two limits, we improve MPCDEA by utilizing the most common method penalty function to handling constraints.

4.1   Constraints Handling

Using penalty function method converts the constraints in mathematical model in this paper into inequality and equality constraints, defined as Eqs. (9) and (10).

$$g_1(X) := \left(\sum_{i=1}^{N_{subchannel}} p_i - p_{total}\right) + \sum_{i=1}^{N_{subchannel}} (SNR_{min} - SNR_i) \le 0 \tag{9}$$

$$g_2(X) := \sum_{i=1}^{N_{subchannel}} \left(\sum_{k=1}^{N_{neighbor}} c_{ki} - 1\right) = 0 \tag{10}$$

Combing Eq. (9) with (10), the constraint-violated value equations is as (11) and (12). $G(X)$ is a result which has been standardized. Each X is corresponding to a type of resource allocation scheme including the allocation of subchannel $\vec{c}$ and power $\vec{p}$. $X = (\vec{c}, \vec{p})$, $\vec{c} = (c_{1,1}, \cdots, c_{1,N_{subchannel}}, \cdots, c_{N_{neighbor},1}, \cdots, c_{N_{neighbor} \cdot N_{subchannel}})$ is a vector of $N_{neighbor} \cdot N_{subchannel}$ dimensions with value of 0 and 1, $\vec{p} = (p_1, \cdots, p_{N_{subchannel}})$ is a real vector of $N_{subchannel}$ dimensions.

$$Con(X) = (g_1(X))^2 + (g_2(X))^2 \tag{11}$$

$$G(X) = \frac{Con_{max} - Con(X)}{Con_{max}} N_{population} \tag{12}$$

Equation (13) is our original objective function and Eq. (14) represents the standardized one.

$$Obj(X) = \sum_{k=1}^{N_{Neighbor}} Utility(T_k^{link}) \tag{13}$$

$$F(X) = \frac{Obj(X) - Obj_{min}}{Obj_{max} - Obj_{min}} N_{population} \tag{14}$$

Equation set (15) describes fitness function. Among it, $iter$ is the current iteration number and $N_{feasible}$ is the feasible solutions' number.

$$Fit(X) = \begin{cases} F(X) + 2N_{population} & X \in feasible\ solution\ set \\ F(X) + 2\left(1 - \dfrac{iter}{maxIter}\right)\left(1 - \dfrac{N_{feasible}}{N_{population}}\right)G(X) & else \end{cases} \tag{15}$$

4.2 MPCDEA

MPCDEA is based on cloud differential evolution algorithm [13] and splits the whole population into several subpopulations. After independent initialization, subpopulations will execute mutation, crossover, selection, exchange or expel operation.

(1) Initialization: Initiate individual $X = (\vec{c}, \vec{p})$ as Eqs. (16) and (17).

$$c_l = rand\{0,1\}, \quad l = 1, \cdots, N_{neighbor} \cdot N_{subchannel} \tag{16}$$

$$p_m = rand \cdot p_{total}, \quad m = 1, \cdots, N_{subchannel} \tag{17}$$

(2) Mutation: DE/rand/1, DE/best/1, DE/rand/2 and DE/best/2 are four main differential mutation ways [13] in continuous search domains. Considering our nonlinear mixed integer programming problem, we define a new mutation scheme called

DisOperation to generate variant individuals' subchannel distribution matrices for individuals $X_{rand1}, X_{rand2}, \ldots, X_{randn}$.

Step1: Calculate subchannel distribution matrices $C_{rand1}, C_{rand2}, \ldots, C_{randn}$.

Step2: Comparing each column of matrices $C_{rand1}, C_{rand2}, \ldots, C_{randn}$, if column l meets condition $col_{rand1,l} = col_{rand2,l} = \ldots = col_{randn,l}$, values of column l should be set to $col_l^{mutation} = col_{rand1,l}$; otherwise sum the values of column l;

Step3: If the above sum results have unique maximum value located in row r, column l, replace the maximum value with 1 and the remaining with 0, which means the subchannel l is assigned to adjacent satellite r. Otherwise, randomly select a location from those maximum elements, set the value to 1 and the values of other locations to 0 and do this until variant matrix's columns are all dealt with.

(3) Crossover: Let individual $X_i = (c_i, p_i)$ and $X_i^{mutation} = (c_i^{mutation}, p_i^{mutation})$ in the population do crossover operation as Eqs. (18) and (19) and produce $X_i^{cross} = (c_i^{cross}, p_i^{cross})$. $CR_i \in [0, 1]$ is the crossover factor of X_i. $r1$ and $r2$ are integers randomly produced from $[1, N_{neighbor} \cdot N_{subchannel}]$ and $[1, N_{subchannel}]$ respectively.

$$c_{i,l}^{cross} = \begin{cases} c_{i,l}^{mutation} & rand \leq CR_i \, or \, l = r1 \\ c_{i,l} & else \end{cases} \tag{18}$$

$$p_{i,l}^{cross} = \begin{cases} p_{i,l}^{mutation} & rand \leq CR_i \, or \, l = r2 \\ p_{i,l} & else \end{cases} \tag{19}$$

(4) Selection: Greedily select best surviving individuals according to Eq. (20).

$$X_i^{new} = \begin{cases} X_i^{cross} & Fit(X_i^{cross}) > Fit(X_i) \\ X_i & else \end{cases} \tag{20}$$

(5) Exchange: For diversity, after a fixed iteration *exchangeIter*, selecting an exchange subpopulation from each subpopulation and randomly exchange equal number of individuals between these two subpopulations.

(6) Expel: When $Diff = Fit(X^{best}) - Fit(X^{worst})$ is less than the predefined threshold value *DiffBound*, half of the current subpopulation except for the optimal are randomly replaced with new randomly generated individuals to avoid converging to local optimal solutions.

Next, we utilize MPCDEA to optimally conduct satellites' resource allocation and Algorithm 1 is its specific procedures. The inputs of this algorithm are satellites' power p_{total}, satellites' bandwidth B, $N_{subchannel}$, $N_{neighbor}$ of the satellite $S_{i_0 j_0}$, distances between adjacent satellites and $S_{i_0 j_0}$ as well as *maxIter*. The outputs are the allocation matrices c of subchannels and allocation vector P of power.

Algorithm 1. Multi-population Cloud Differential Evolution algorithm

1. Initiate all individuals of the population, set $iter = 1$;
2. For $iter \leq maxIter$
3. For $subpopulationID \leq 4$
4. Switch $subpopulationID$
5. Case 1: Do mutation $DE / rand / 1$ to the real vector p of one individual, do $DisOperation(X_{rand1}, X_{rand2}, X_{rand3})$ to integer vector C;
6. Case 2: Do mutation $DE / best / 1$ to the real vector P of one individual , do $DisOperation(X_{best}, X_{rand1}, X_{rand2})$ to integer vector C;
7. Case 3: Do mutation $DE / rand / 2$ to the real vector P of one individual , do $DisOperation(X_{rand1}, ..., X_{rand5})$ to integer vector C;
8. Case 4: Do mutation $DE / best / 2$ to the real vector P of one individual , do $DisOperation(X_{best}, X_{rand1}, ..., X_{rand4})$ to integer vector C;
9. End Switch
10. Calculate individuals' crossover factors and do crossover
11. Do selection operation
12. End For
13. If $iter$ mod $exchangeIter = 0$, Do exchange operation among subpopulations, End If
14. For $subpopulationID \leq 4$
15. Calculate fitness difference $Diff$ between the best and worst individual of subpopulation $subpopulationID$
16. If $Diff < DiffBound$, Do Expel to subpopulation $subpopulationID$, End If
17. End For
18. $iter + +$
19. End For
20. Calculate subchannels' allocation matrices C, vector P and return

5 Simulation and Evaluation

5.1 Experimental Platform

In order to verify the feasibility and effectiveness of this utility function based resource allocation method, we utilize Iridium and Globalstar these two classic LEO satellite constellation systems as simulation topologies and obtaining real-time data can be collected from STK. And Eclipse is our simulation software. Iridium has two kinds of satellites, called border satellites and non-border satellites respectively. Border satellites can build at most three ISLs including two intra-plane ISLs and one inter-plane ISL. While non-border satellites can build at most two inter-plane ISLs and two intra-plane ISLs. As to Globalstar System, we only use its topology structure, for it doesn't have ISLs. Main parameters of Iridium and Globalstar systems are listed in Table 1. For example, bandwidth is set to 200 Mbps, power is 20w for both 154 and 144 sampling time points are selected in Iridium and Globalstar respectively within a cycle.

Table 1. Main Simulation Parameters of LEO Satellite Constellation System

Parameters	Iridium	Globalstar
satellites' number	66	48
orbit planes' number	6	8
intra-plane satellites' number	11	6
orbit inclination angle (°)	86.4	52
phase factor	–	2
system cycle (min)	100.45	114.09
high latitude boundary values (°)	60	48
antenna diameter (m)	0.8	0.8
antenna efficiency (%)	0.6	0.6
bandwidth (Mbps)	200	200
sampling time points' number	154	144

5.2   Performance Indexes

Here, we choose Satellite Instantaneous Throughput Capacity (SITC), Satellite Average Throughput Capacity (SATC), Instantaneous Network Capacity (INC), and Average Network Capacity (ANC) to evaluate the effect of UFRAM on network capacity. Their equations are as follows.

$$INC = \sum\nolimits_{i-1}^{|S|} SITC_i(t) \tag{21}$$

$$SATC = (1/|S_{sample}|) \sum\nolimits_{t_i \in S_{sample}} SITC(t_i) \tag{22}$$

$$ANC = (1/|S_{sample}|) \sum\nolimits_{t_i \in S_{sample}} INC(t_i) \tag{23}$$

SITC represents the satellite's maximum data transfer speed at time t. INC is used to analyze the LEO satellite constellation system's carrying capability at time t. SATC describes a satellite's average maximum data transfer rate mainly during a system cycle. ANC represents LEO satellite constellation system's network capacity during a system cycle. S_sample represents the sampling time points within a system cycle.

5.3   Experiment Results

Bandwidth and power are two major components of network capacity. Here, we set bandwidth to 200 Mbps and powers to 20w, 50w, 100w. P(W) represents power.

From Fig. 2, we can know how satellites' latitudes or locations are changing with time. Figure 3 shows a contrast result of border and non-border satellites' SITC by utilizing utility function based resource allocation method. Apparently, the curves in Fig. 4 present periodicity caused mainly by satellites' periodic movement. And the SITC of non-border satellites fluctuates more widely for two inter-plane ISLs will disconnect when satellites move in high-latitude zones and when satellites move near the equator, the distance of two inter-plane ISLs will reach greatest (Fig. 2).

However, border-satellites have only one inter-plane ISL. There are 44 non-border satellites and 22 border satellites in Iridium. Therefore, we'll mainly analyze non-border ones. In Fig. 5, we can find that satellites' performance is positively proportional to powers. Seeing from Figs. 6, 7, compared with Iridium, INC in Globalstar system is more complex but SATC and ANC is relatively smaller.

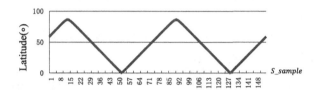

Fig. 2. Absolute Value of Latitude Curve Satellite of Iridium System

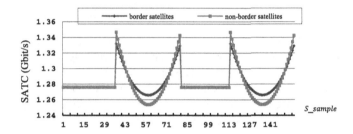

Fig. 3. Contrast of Border and Non-border satellites of Iridium System

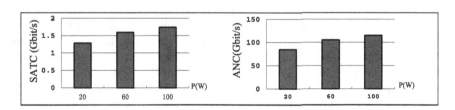

Fig. 4. SATC Curve and ANC Curve of Iridium System

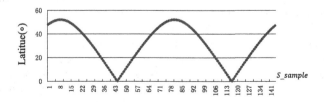

Fig. 5. Absolute Latitude Values of Satellites in Globalstar System

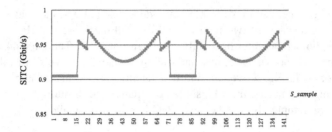

Fig. 6. SITC Curve of Globalstar System

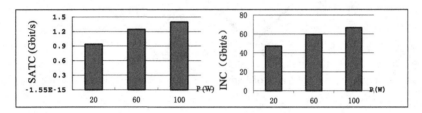

Fig. 7. SATC Curve and INC Curve of Globalstar System.

By obtaining data from the above figures and do calculating, SATC in Iridium is 1.28 Gbit/s and 0.935 Gbit/s in Globalstar with a power-set of 20 W and a bandwidth-set of 200 M. ANC in Iridium is 115.0 Gbit/s and 66.6 Gbit/s in Globalstar with power-set 100 W. The different results of Globalstar and Iridium demonstrate that the proposed resource allocation method is feasible.

6 Conclusions and Future Remarks

In this paper, based on the analysis of network capacity and considering the actual demand and fairness of adjacent satellites, we introduce utility function to represent satellites' satisfaction of obtaining resources and put forward a resource allocation method based on utility function with the goal of maximizing satisfaction degree of all adjacent nodes. And we adopt MPCDEA to get a relatively optimal solution. The simulation demonstrates that our resource allocation method is feasible and effective though the exploration is micro in the process of constructing satellite communication network which still has great room for improvement, such as bringing in optical communication [14] to construct green satellite network, considering malicious electromagnetic interference and sun transit outage interruption, etc.

Acknowledgment. This work is supported by the National Science Foundation for Distinguished Young Scholars of China under Grant No. 61225012 and No. 71325002; the Specialized Research Fund of the Doctoral Program of Higher Education for the Priority Development Areas under Grant No. 20120042130003; Liaoning BaiQianWan Talents Program under Grant No. 2013921068.

References

1. Gupta, P., Kumar, P.R.: The capacity of wireless networks. IEEE Trans. Inf. Theory **46**(2), 388–404 (2000)
2. Xie, L., Kumar, P.: A network information theory for wireless communication: scaling laws and optimal operation. IEEE Trans. Inf. Theory **50**(5), 748–768 (2004)
3. Behzad, A., Rubin, I.: High transmission power increases the capacity of ad hoc wireless networks. IEEE Trans. Wireless Commun. **5**(1), 156–165 (2006)
4. Liu, J., Jiang, X., Nishiyama, H., Kato, N.: Exact throughput capacity under power control in mobile ad hoc networks. In: IEEE INFOCOM 2012 - IEEE Conference on Computer Communications, pp. 1–9. Orlando (2012)
5. Liu, C., Yan, J., Chen, X., Mei, S.: Capacity and loading analysis of digital channelized SATCOM system. In: 2013 8th International ICST Conference on Communications and Networking, pp. 155–160. China (2012)
6. Li, J., Luo, T., Yue, G.: A resource allocation scheme jointly considering network capacity and user satisfaction in cognitive radio networks. Acta Electronica Sinica **40**(7), 1315–1322 (2012)
7. Wang, W., Wu, R., Liang, J.: Capacity of orthogonal relay channel and resource allocation. Acta Electronica Sinica **38**(4), 771–775 (2010)
8. Wang, X., Wang, X., Wang, C., Huang, M.: Resource allocation in cloud environment: a model based on double multi-attribute auction mechanism. In: 6th IEEE International Conference on Cloud Computing Technology and Science, pp. 15–18. Singapore (2014)
9. Nakahira, K., Abe, J., Mashino, J.: Novel channel allocation algorithm using spectrum control technique for effective usage of both satellite transponder bandwidth and satellite transmission power. IEICE Trans. Commun. **95**(11), 3393–3403 (2012)
10. Wu, X., Chen, Y., Gao, L., Meng, C.: A utility-based OFDM resource allocation scheme for LEO small satellite system. In: International Conference on Cyberspace Technology, pp. 68–73. Stevenage (2014)
11. Wang, J., Li, L., Zhou, M.: Topological dynamics characterization for LEO satellite networks. Comput. Netw. **51**(1), 43–53 (2007)
12. Sun, J., Wang, X., Gao, C., Huang, M.: Resource allocation scheme based on neural network and group search optimization in cloud environment. Journal of Software **25**(8), 1858–1873 (2014). (in Chinese)
13. Zhu, C., Ni, J.: Cloud model-based differential evolution algorithm for optimization problems. Internet Computing for Science and Engineering. In: 2012 Sixth International Conference Internet Computing for Science and Engineering (ICICSE), pp. 55–59. China (2012)
14. Wang, X., Hou, W., Guo, L., Cao, J., Jiang, D.: Energy saving and cost reduction in multi-granularity green optical networks. Comput. Netw. **55**(3), 676–688 (2011)

A Novel Naive Bayes Classifier Model
Based on Differential Evolution

Jun Li[1,2,3(✉)], Guokang Fang[1,2], Bo Li[1,2], and Chong Wang[1,2]

[1] College of Computer Science and Technology,
Wuhan University of Science and Technology, WUST,
Wuhan 430065, China
lijun@wust.edu.cn
[2] Hubei Province Key Laboratory of Intelligent Information
Processing and Real-time Industrial System,
WUST, Wuhan 430065, China
[3] State Key Laboratory of Software Engineering,
Wuhan University,
Wuhan 430072, China

Abstract. Naive Bayes (NB) classifier is a simple and efficient classifier, but the independent assumption of its attribute limits the application of the actual data. This paper presents an approach called Differential Evolution-Naive Bayes (DE-NB) which takes advantage of combining differential evolution with naive Bayes for attribute selection to improve naive Bayes classifier. This method applies DE firstly to search out an optimal subset of attributes reduction in the original attribute space, and then constructs a naive Bayes classifier on the gotten subset of the attributes reduction. Nineteen experimental results on UCI datasets distinctly show that compared with Cfs-BestFirst algorithm, NB algorithm, Support Vector Machine (SVM) algorithm, Decision Tree (C4.5) algorithm, K-neighbor (KNN) algorithm, the proposed algorithm has higher classification accuracy.

Keywords: Naive Bayes · Differential evolution · Feature selection · Attribute subset · Classification accuracy

1 Introduction

Classified prediction is an important branch of data mining. Classification is to identify a set of data collection that can describe typical characteristics of the model to make predictions of the unknown variables or categories. The core part of the classification algorithm is to construct a classifier. Because of the efficient calculation, the high accuracy and the solid theoretical foundation, naive Bayes classifier has been widely used. But naive Bayes classification is based on the assumption: under the given condition of the classification characteristics the attribute values are independent of each other. But in the real world, this independence assumption is not always satisfied. Therefore, for naive Bayes classification deficiency, many scholars study Bayes classification to improve its performance.

© Springer International Publishing Switzerland 2015
D.-S. Huang et al. (Eds.): ICIC 2015, Part I, LNCS 9225, pp. 558–566, 2015.
DOI: 10.1007/978-3-319-22180-9_55

Related work can be broadly divided into three main categories:

1. Structure extension: extending the structure of naive Bayes to represent the dependencies among attributes. Zhang and Ling [1] observed that the dependence among attributes tends to cluster into groups in many real world domains with a large number of attributes. Based on their observation, they presented an improved algorithm StumpNetwork. Their motivation is to enhance the efficiency of the Superparent algorithm while maintaining a similar predictive accuracy.
2. Attribute weighting: assigning different weights to attributes in building naive Bayes. Harry Zhang and Shengli Sheng [2] extended the naïve Bayes into weighted naive Bayes (WNB), in which attributes have different weights. They explored various methods and the experiments showed that a weighted naive Bayes is trained to produce accurate ranking outperformed naive Bayes.
3. Feature selection: selecting an attribute subset from the whole space of attributes. In the following part, this approach is introduced detailed.

The feature selection approach improves naive Bayes by removing redundant or irrelevant attributes from training data sets, and only selecting those that are the most informative in learning tasks [3]. In fact, any of this kind of improved algorithms is a variant of naive Bayes that only uses a subset of the given attributes in making predictions. Several feature selection algorithms have been presented and demonstrated considerable improvement over naive Bayes.

Feature selection involves the choice of a feature subset evaluator and a search method. Langley and Sage [4] presented an algorithm called selective Bayes classifiers (simply SBC). It uses a forward greedy search method to select an attribute subset through the whole space of attributes. More specifically, it uses naive Bayes' classification accuracy to evaluate alternative subsets of attributes. S. Casale and A. Russo [5] used an algorithm called CFSSubsetEval-BestFirst in feature selection. CfsSubsetEval means that it evaluates the worth of a subset of features by considering the individual predictive ability of each feature along with the degree of redundancy between them; subsets of features that are highly correlated with the class while having low inter-correlation are preferred. BestFirst means that it searches the space of feature subsets by greedy hill-climbing augmented with a backtracking facility. The CFSSubsetEval-BestFirst method is also embedded in the WEKA [6] (Waikato Environment for Knowledge Analysis) software. In addition, there are many general feature selection algorithms, which are not specifically for naive Bayes. Ratanamahatana and Gunopulos [7] proposed another feature selection approach by building decision trees.

Differential evolution algorithm is a kind of real number encoding optimization algorithm based on population evolution [8]. The algorithm can extract different information from the current population and to further guide the search. The principles are relatively simple, and there is a relatively small number of control parameters in the algorithm, while the global search ability is far stronger.

In the paper, the authors present an algorithm called DE-NB, which applies DE to NB. The remainder of this paper is organized as follows. The basic principle of the naive Bayes is given in Sect. 2. Section 3 introduces the differential evolution algorithm. Section 4 elaborates the proposed DE-NB method. Section 5 gives the experimental result of comparison of the proposed method with the five other methods. Section 6 summaries this paper.

2   Naive Bayes Classification

The naive Bayes classifier uses a probabilistic approach to assign each record of the data set to a possible class. It works as follows [9–11]:

1. Let T be a training set of samples, each with their class labels. There are m classes, $C_1, C_2, \ldots, C_m$. Each sample is represented by an n-dimensional vector, $X = \{x_1, x_2, \ldots, x_n\}$, respectively describe the n attributes $A_1, A_2, \ldots, A_n$.
2. Given a sample X, the classifier will predict that X belongs to the class having the highest a posteriori probability, conditioned on X. That is, X is predicted to belong to the class C_i if and only if $P(C_i|X) > P(C_j|X)$ for $1 \leq j \leq m$, $j \neq i$. Thus the class that maximizes $P(C_i|X)$ is found. The class C_i for which $P(C_i|X)$ is maximized is called the maximum posteriori hypothesis. By Bayes' theorem

$$P(C_i|X) = P(X|C_i)\ P(C_i)/P(X) \tag{1}$$

3. As P(X) is the same for all classes, only $P(X|C_i)P(C_i)$ need be maximized. If the priori probability of class C_i is not known, then it is usually assumed that the probabilities of these classes are equal, that is $P(C_1) = P(C_2) = \ldots = P(C_m)$, the question therefore is converted to maximize the $P(X|C_i)$. $P(X|C_i)$ is often called the likelihood of data X when it is given C_i, while the assumption of maximizing $P(X|C_i)$ is called the maximum likelihood. Otherwise, $P(X|C_i)P(C_i)$ need to be maximized. Note that the class a priori probabilities may be estimated by $P(C_i) = s_i/s$, where s_i is the number of training samples in the class C_i, while s is the total number of training samples.
4. Given data sets with many attributes, it would be computationally expensive to compute $P(X|C_i)$. In order to reduce the overhead of calculating $P(X|C_i)$, the naive assumption of class conditional independence is made. This presumes that the values of the attributes are conditionally independent of one another, given the class label of the sample. Mathematically this means that

$$P(X|C_i) \approx \prod_{k=1}^{n} P(x_k|C_i) \tag{2}$$

The probabilities $P(x_1|C_i)$, $P(x_2|C_i)$, ..., $P(x_n|C_i)$ can easily be estimated from the training set. Here x_k refers to the value of attribute A_k for sample X.

5. In order to predict the class label of X, $P(X|C_i)P(C_i)$ is evaluated for each class C_i. The classifier predicts that the class label of X is C_i if and only if it is the class that maximizes $P(X|C_i)P(C_i)$.

3 Differential Evolution

Differential evolution algorithm can be used to solve the global optimization problems which have N continuous variables. In the population initialization phase, individuals in a population randomly initialize space in the search space. In the evolutionary stage, mutation, hybridization and selection process are in-order until the stop condition is met [12].

A. Coding and initialized
Classical differential evolutionary algorithm uses real number coding, which makes the algorithm more suitable for real optimization problems. Provided the independent variable of the problem has D dimensions, then, in the population the ith individual X_i is shown by the following: $X_i = \{x_i(j), x_i(j), \ldots, x_i(j)\}$.

Among this, $x_i(j) \in [l_j, u_j]; i = 1, \ldots, NP; j = 1, \ldots, D; x_i(j)$ as a real number is initialized uniformly and randomly in the $[l_j, u_j]$ of the independent variable, that is, $x_i(j) = rndreal[l_j, u_j]$. Three control parameters of original differential evolution algorithm, they are respectively, population size NP, scaling factor F and the probability of hybridization CR.

B. The differential mutation
In the differential evolution algorithm, mutation operator is one of the most important operator. DE/rand /2 is used in this paper by the following:

$$V_i = X_{r_1} + F(X_{r_2} - X_{r_3}) + F(X_{r_4} - X_{r_5}) \tag{3}$$

Among this, X_{best} is the best individual in the current population, X_i is a father individual, $r_1 \neq r_2 \neq r_3 \neq r_4 \neq r_5 \neq i$ are five individuals randomly selected from population, V_i is a variation vector, $X_{r_2} - X_{r_3}$ is a difference vector, and $F \in [0, 1+)$ is a zoom factor, which is used for the difference vector to zoom in, thus to control the search step length.

C. Crossover operation
Differential evolution algorithm use discrete crossover operator, including the Binomial Crossover and Exponential Crossover. Crossover operator let variable vector V_i generated by mutation operator and father individual vector X_i do the discrete crossover, and acquire the try vector U_i. Binomial Crossover operator can also be expressed as:

$$U_i(j) = \begin{cases} V_i(j) & if\,(rndreal_j[0, 1] < CR \; or \; j = j_{rand}) \\ X_i(j) & otherwise \end{cases} \tag{4}$$

Among this, $j = 1, \ldots, D$; j_{rand} is a random integer in the range [1, D], which ensures that at least one dimension of the attempt vector U_i from the variation vector V_i, thus to avoid the same as the parent individual vector X_i. In each mutation, operator of differential evolution algorithm can be combined with index of crossover operator.

D. Select operation

Through mutation operator and crossover operator, differential evolution algorithm generates subgroup, there adopt the one-to-one selection operator to compare individual to corresponding parent individual, then the excellent individual can be saved to the next group. Optimization for minimizing the selection operator can be described as:

$$X_i = \begin{cases} U_i & if(f(U_i) \leq f(X_i)) \\ X_i & otherwise \end{cases} \qquad (5)$$

Among this, $f(X_i)$ is the adaptive value for X_i.

4 DE-NB

DE-NB selects an attribute subset through the whole space of attributes by carrying out a differential evolution search process. And it uses naive Bayes' classification accuracy as the fitness function to evaluate alternative subsets of attributes and selects the individual with the maximum classification accuracy after a fixed number of generations. Figure 1 shows the flow chart of the developed DE-NB model.

The outline of the proposed algorithm lists as follows:

Step1. Initialize population X by binary code, each individual is composed of a string of feature selection bit.

Step2. Remove attributes which are not selected from the training samples attribute to get the training data T, according to the feature of each individual selection bit.

Step3. Calculate priori probability $P(y_i)$ of each class of training data.

Step4. Calculate conditional probability $P(x|y_i)$ of each attribute' division.

Step5. Calculate priori probability $P(y_i)*P(x|y_i)$ of each class.

Step6. Select the maximum priori probability $P(y_i)* P(x|y_i)$ as the class x belongs to.

Step7. Calculate the entire sample classification accuracy as the classification accuracy BestAccuracy and its corresponding feature selection Bestf.

Step8. Determine whether the current accuracy and number of iterations reaches the end of the condition, if reached, go to step13, otherwise the next step go into the next generation iterative process.

Step9. Choose two individuals from population X. Execute differential mutation on each individual in population X to generate population Y.

Step10. Execute Crossover between individual in population X and individual in population Y to generate population Z.

Step11. Repeat step2 to step7 to get the current best accuracy BestAccuracy_temp and its corresponding subset of feature selection. If BestAccuracy_temp > BestAccuracy, then BestAccuracy = BestAccuracy _temp, Bestf = f_temp.

Step12. Choose the best one into population X in the next generation between individual in the population X and individual in the population Z in a way of one-to-one selection according to classification accuracy.

Step13. Repeat step2 to step7 according to training data selected by Bestf attributes to get final classified accuracy.

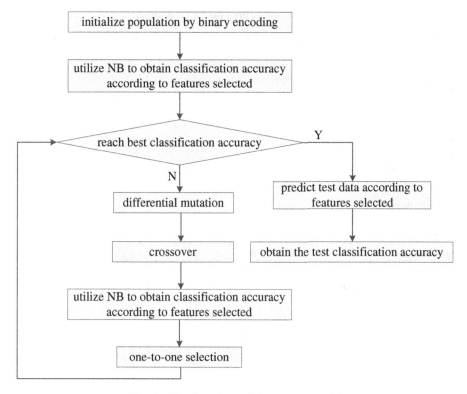

Fig. 1. The flow chart of the DE-NB model

5 Experiment Results

The platform adopted to develop the DE-NB approach is Dell desktop with the following features: Intel Pentium Cores duo CPU E7200 2.53 GHz main frequency, 2 GMB RAM, a Windows XP sp3 operating system. Our implementation is carried out on the Matlab2010a development environment.

In order to measure the performance of the developed DE-NB approach, the following datasets in UC Irvine Machine Learning Repository [13] are used: balance, blood, diabetes, glass, haberman, iris, kr_V_kp, Libras_Movement, liver, liverdisorder, p_gene, parkinsons, sonar_all_data, Soybean, splice, tic_tac_toc, wdbc, wine, zoo. Table 1 presents the properties of these datasets.

To guarantee valid results for making predictions regarding new data, the dataset is further randomly partitioned into training sets and independent test sets via a k-fold cross validation. Each of the k subsets acts as an independent holdout test set for the model trained with the remaining k-1 subsets. The advantages of cross validation are that all of the test sets were independent and the reliability of the results could be improved. This study used k = 10. Each of the 10 subset is used as test data sets in turn, so the program runs 10 times. The final classification accuracy is expressed in the form "mean ± standard deviation".

Table 1. Datasets from the UCI repository

Dataset	Number of instances	Number of features	Number of classes	Numeric
balance	625	4	3	Y
blood	748	4	2	Y
diabetes	768	8	2	N
glass	214	9	6	N
haberman	306	3	2	Y
iris	150	4	3	N
kr_V_kp	3196	36	2	Y
Libras_Movement	360	89	15	N
liver	327	6	2	N
liverdisorder	345	6	2	Y
p_gene	106	57	2	Y
parkinsons	195	22	2	N
sonar_all_data	208	60	2	N
Soybean	47	35	4	Y
splice	1000	60	2	Y
tic_tac_toc	958	9	2	Y
wdbc	569	30	2	N
wine	178	12	3	N
zoo	101	16	7	Y

In the DE-NB algorithm, the parameters are initially set as follows. Iterations GEN is 200, population size NP is 40, crossover probability CR is 0.5.

To verify the excellence of DE-NB for parameters optimization, the authors design an experiment to compare the DE-NB algorithm, NB algorithm, SVM algorithm, C4.5 algorithm, KNN algorithm and Cfs-BestFirst algorithm (that is an embedded feature selection algorithm in WEKA. In WEKA's AttributeSelectedClassifier, evaluator is CfsSubsetEval, search method is BestFirst. And classification algorithm is naive Bayes. So, it shorts for Cfs-BestFirst). The experimental results are shown in Table 2.

The experiment results are illustrated as follows:

It is obviously seen that the DE-NB classification performance is much better than Cfs-BestFirst algorithm, NB algorithm, SVM algorithm, C4.5 algorithm, KNN algorithm, etc. Among the 19 testing data sets, 11 test results of DE-NB rank first (* is put at the end of the test results). Its average classification accuracy is also significantly higher than that of other data sets.

When comparing DE-NB algorithm, Cfs-BestFirst algorithm, NB algorithm, it is found that the average classification accuracy of Cfs-BestFirst (77.78) is slightly less than that of NB (78.07). But when compared to DE-NB average classification accuracy, both of them are inferior, with the gap about 5 %. Among the whole 19 testing sets, 15 test results of DE-NB rank first, thus proves that its classification performance is much better than the former two.

Table 2. Experimental results

Dataset	DE-NB (%)	Cfs-BestFirst (%)	NB (%)	SVM (%)	C4.5 (%)	KNN (%)
balance	73.39 ± 3.51	63.52 ± 4.97	90.62 ± 1.34	89.92 ± 1.77	77.76 ± 3.85	86.99 ± 2.83
blood	68.38 ± 4.19	76.27 ± 3.08	75.28 ± 3.47	75.23 ± 2.65	78.20 ± 3.71	77.18 ± 3.45
diabetes	79.47. ± 3.41*	77.06 ± 4.70	75.75 ± 5.32	65.11 ± 0.34	74.49 ± 5.27	73.86 ± 4.61
glass	61.90 ± 11.45	48.11 ± 9.97	47.84 ± 8.74	58.00 ± 9.32	68.08 ± 9.28	66.18 ± 8.22
haberman	68.33 ± 5.72	74.21 ± 5.48	75.06 ± 5.42	73.42 ± 2.11	71.05 ± 5.20	70.49 ± 5.17
iris	97.33 ± 4.66*	96.20 ± 4.26	94.87 ± 5.26	95.60 ± 4.85	94.73 ± 5.30	95.73 ± 4.60
kr_V_kp	93.13 ± 0.87	92.97 ± 1.43	84.00 ± 1.96	96.43 ± 1.12	99.34 ± 0.41	95.83 ± 1.10
Libras_Movement	53.33 ± 8.86	66.94 ± 7.56	64.42 ± 7.67	85.08 ± 5.54	45.72 ± 7.92	63.56 ± 7.39
liver	72.50 ± 6.72*	62.54 ± 7.32	63.00 ± 7.75	63.28 ± 7.84	63.28 ± 7.57	65.32 ± 7.61
liverdisorder	70.59 ± 4.60*	56.15 ± 6.48	54.89 ± 8.83	59.37 ± 2.28	65.84 ± 7.40	60.48 ± 7.92
p_gene	100.00 ± 0.00*	85.35 ± 10.57	81.10 ± 11.76	77.35 ± 11.69	80.05 ± 12.04	72.34 ± 13.14
parkinsons	93.16 ± 4.33*	77.83 ± 8.92	70.14 ± 9.24	79.04 ± 4.72	84.69 ± 7.96	92.73 ± 5.27
sonar_all_data	83.50 ± 5.80*	67.62 ± 9.26	67.71 ± 8.66	64.99 ± 7.66	73.61 ± 9.34	82.28 ± 9.12
Soybean	100.00 ± 0.00*	100.00 ± 0.00	98.00 ± 6.03	100.00 ± 0.00	97.65 ± 7.12	100.00 ± 0.00
splice	93.40 ± 2.07	82.32 ± 3.61	83.48 ± 3.33	87.25 ± 3.54	94.44 ± 2.54	69.13 ± 3.78
tic_tac_toc	73.68 ± 4.15	65.40 ± 0.50	71.28 ± 2.45	87.64 ± 2.94	89.15 ± 3.93	83.69 ± 2.66
wdbc	96.96 ± 2.07*	94.55 ± 3.03	93.31 ± 3.30	62.74 ± 0.70	93.27 ± 3.55	96.88 ± 2.25
wine	97.65 ± 4.11*	94.78 ± 4.58	95.60 ± 4.29	89.37 ± 6.46	92.98 ± 5.77	94.21 ± 4.90
zoo	100.00 ± 0.00*	95.95 ± 5.09	96.95 ± 4.75	92.75 ± 5.64	92.61 ± 7.33	95.05 ± 6.70
mean	82.98 ± 4.03	77.78 ± 5.31	78.07 ± 5.77	79.08 ± 4.27	80.89 ± 6.08	81.05 ± 5.30

The obtained results clearly confirm the superiority of the DE-NB algorithm compared to Cfs-BestFirst algorithm, NB algorithm, SVM algorithm, C4.5 algorithm, KNN algorithm, etc.

6   Conclusions

In this paper, aiming at the shortcomings of the independent assumption of naive Bayes classification, combined with differential evolution, a novel DE-NB algorithm is proposed. It selects an attribute subset through the whole space of attributes by carrying out a de search process. Simulation results show that the proposed algorithm greatly enhances classification accuracy of naive Bayes.

Acknowledgements. This research is supported by National Natural Science Foundation of China under Grant No. 61273303. The authors thank professor Jiang Liangxiao for his very useful comments and suggestions.

References

1. Zhang, H., Ling, C.X.: An improved learning algorithm for augmented naive Bayes. In: Cheung, D., Williams, G.J., Li, Q. (eds.) PAKDD 2001. LNCS (LNAI), vol. 2035, pp. 581–586. Springer, Heidelberg (2001)
2. Zhang, H., Sheng, S.: Learning weighted naive Bayes with accurate ranking. In: Fourth IEEE International Conference on Data Mining, ICDM 2004, pp. 567–570. IEEE (2004)

3. Jiang, L., Zhang, H., Cai, Z.: A novel Bayes model: hidden naive Bayes. IEEE Trans. Knowl. Data Eng. **21**(10), 1361–1371 (2009)
4. Langley, P., Sage, S.: Induction of selective Bayesian classifiers. In: Proceedings of the Tenth International Conference on Uncertainty in Artificial Intelligence, pp. 399–406. Morgan Kaufmann Publishers Inc. (1994)
5. Casale, S., Russo, A., Serrano, S.: Analysis of robustness of attributes selection applied to speech emotion recognition. In: Proceedings of the 18th European Signal Processing Conference (EUSIPCO 2010). EURASIP, Aalborg, Denmark (2010)
6. Hall, M., Frank, E., Holmes, G., Pfahringer, B., Reutemann, P., Witten, I.H.: The WEKA data mining software: an update. ACM SIGKDD Explor. Newsl. **11**(1), 10–18 (2009)
7. Ratanamahatana, C.A., Gunopulos, D.: Scaling up the naive Bayesian classifier: using decision trees for feature selection (2002)
8. Storn, R., Price, K.: Differential evolution–a simple and efficient heuristic for global optimization over continuous spaces. J. Glob. Optim. **11**(4), 341–359 (1997)
9. Kantardzic, M.: Data Mining: Concepts, Models, Methods, and Algorithms. Wiley, New York (2011)
10. Han, J., Kamber, M., Pei, J.: Data Mining, Southeast Asia Edition: Concepts and Techniques. Morgan Kaufmann, San Francisco (2006)
11. Lin, J., Yu, J.: Weighted naive bayes classification algorithm based on particle swarm optimization. In: 2011 IEEE 3rd International Conference on Communication Software and Networks (ICCSN), pp. 444–447. IEEE (2011)
12. Jun, L., Lixin, D., Ying, X.: Differential evolution based parameters selection for support vector machine. In: 2013 9th International Conference on Computational Intelligence and Security (CIS), pp. 284–288. IEEE (2013)
13. Hettich, S., Blake, C.L., Merz, C.J.: UCI repository of machine learning databases, Department of Information and Computer Science, University of California, Irvine, CA (1998). http://www.ics.uci.edu/~mlearn/MLRepository.html

Chaotic Iteration Particle Swarm Optimization Algorithm Based on Economic Load Dispatch

Zhenghong Yu[(✉)] and Fengli Zhou

Faculty of Information Engineering,
Wuhan University of Science and Technology City College, Wuhan 430083, China
wtuyzh@126.com, thinkview@163.com

Abstract. To solve the non-convex and non-linear economic dispatch problem efficiently, a chaotic iteration particle swarm optimization algorithm is presented. In the global research of particle swarm optimization and local optimum, ergodicity of chaos can effectively restrain premature. To balance the exploration and exploitation abilities and avoid being trapped into local optimal, a new index, called iteration best, is incorporated into particle swarm optimization, and chaotic mutation with a new Tent map imported can make local search within the prior knowledge, a new strategy is proposed in iteration strategy. The algorithm is validated for two test systems consisting of 6 and 15 generators. Compared with other methods in this literature, the experimental result demonstrates the high convergency and effectiveness of proposed algorithm.

Keywords: Chaotic mutation · Iteration best · Particle swarm optimization · Power system · Premature

1 Introduction

As a typical optimization problem in power system, the target of economic load dispatch (ELD) is that minimize generating cost under the conditions of meeting the load and operation constraints. Power transmission capacity and stability of the system make the feasible region of the ELD problem is non-convex, and due to the impact of turbine valve point effect, consumption curve of unit is nonlinear. Therefore, ELD problem presents essential features of high-dimensional, non-convex, multi-constraint and nonlinear. Traditional way such as dynamic programming can solve ELD problem, but it needs the object to be optimized with good mathematical properties and the treatment effect is not ideal. In recent years, with the development of artificial intelligence technology and the integration of cross-curricular interests, the solution of ELD problem has a new breakthrough, some new methods such as genetic algorithm [1], chaos particle swarm optimization algorithm [2], differential evolution algorithm [3] and so on are used extensively, thus overcoming the shortcomings of traditional methods and improving the quality of solution.

Particle swarm optimization (PSO) algorithm uses memory and feedback mechanism together for high efficiency optimization, and can solve large-scale optimization problems quickly, but the algorithm efficiency is affected seriously by parameter setting and initial distribution of particle. To improve algorithm efficiency, literature [4]

© Springer International Publishing Switzerland 2015
D.-S. Huang et al. (Eds.): ICIC 2015, Part I, LNCS 9225, pp. 567–575, 2015.
DOI: 10.1007/978-3-319-22180-9_56

combines global search ability of PSO and local searching ability of tabu algorithm (TS), and uses this method for solving the combined heat and power economic load dispatch problem, it can effectively overcome premature convergence problem of PSO by TS algorithm's strong gradeability. Literature [5] embeds simplex operator in PSO and is used to update part of the particles that have the fitness value, it can improve the optimization ability and convergence speed of PSO. Literature [6] adopts the weight of inertia and renovates the best and worst particle to make the improving PSO algorithm possess high efficiency in searching for the best solution in the global area. Literature [7] solves multi-objective problem by improved PSO algorithm based on Pareto Dominant strategy and crowding distance ordering, and obtains more uniform and more complete Pareto optimal fronts. Literature [8] fine-tunes the global optimum solution by introducing sequential quadratic programming (SQP) into chaotic particle swarm optimization algorithm, solves economic load dispatch problem with valve-point effect, and obtains solution with faster convergence and better quality than the existing algorithm.

Based on the previous chaotic particle swarm algorithm, this paper introduces the optimal iteration factor to enhance the global search ability of particles [9], leads in a mutation operator that is guided by corresponding fitness value of the global optimal particle based on updating formula of particle swarm velocity, and imports chaotic mutation with a new Tent map to the priori solution, which can change particles' chaotic search mechanism and improve the particles' iterative strategy. Simulation example verifies that the algorithm can well balance the global and local search of particles, and avoid premature effectively.

2 Multi-constraint ELD Mathematical Model

ELD problem is that distributes unit load reasonably and minimizes the cost of electricity under the constraints of system power balance, ramp, dead-zone, etc.

2.1 Power Balance Constraint

The constraint based on the balance of system total power output $\sum\limits_{m=1}^{N} P_m$, total load P_D and total line loss P_L.

$$\sum_{m=1}^{N} P_m = P_D + P_L , \quad m = 1, 2, \ldots, N \tag{1}$$

P_L is calculated by $\boldsymbol{B}$ coefficient

$$P_L = P^T BP + P^T B_0 + B_{00} \tag{2}$$

Where P is a matrix of $1 \times N$, $\boldsymbol{B}$ is a $\boldsymbol{B}$ coefficient matrix of $N \times N$ that is used to calculate line loss.

2.2   Units Operation Constraint

$$P_m^{\min} \leqslant P_m \leqslant P_m^{\max}, \quad m = 1, 2, \ldots, N \tag{3}$$

Where P_m is the m-generator's output power, $P_m^{\min}$ is the minimum value of generator active power and $P_m^{\max}$ is the maximum value of generator active power.

2.3   Units Ramp Constraint

In practice, units ramp constraint limits the units run in two operating range.

$$\begin{cases} P_m - P_m^0 \leq UR_m; \\ P_m^0 - P_m \leq DR_m, \quad m = 1, 2, \ldots, N \end{cases} \tag{4}$$

Where P_m is the operating power after adjusting output power; P_m^0 is the m-generator's initial operating power; UR_m, DR_m is the upper and lower limit of unit m's power increased and decreased value respectively, if unit breaches the constraint, it will be limited to the nearest boundary value.

2.4   Dead-Zone Constraint

By the impact of unit's physical characteristics or instability such as steam valve bearing vibration, as well as saving running cost, unit m needs to be run in a certain feasible region A_{am}:

$$A_{am} = \begin{cases} P_m^{\min} \leq P_m \leq P_m^{1l}; \\ P_m^{(j-1)u} \leq P_m \leq P_m^{jl}; \\ P_m^{n_m u} \leq P_m \leq P_m^{\max}, \quad m = 1, 2, \ldots, N \end{cases} \tag{5}$$

Where P_m^{ju} and P_m^{jl} are upper and lower boundary of unit m's j-th work dead-zone respectively; n_m is the number of unit m's work dead-zone, if unit breaches the constraint, it will be limited to the nearest boundary value.

2.5   Line Power Flow Constraint

$$\left| P_{Lf,k} \right| \leq P_{Lf,k}^{\max}, \quad k = 1, 2, \ldots, L \tag{6}$$

Where $P_{Lf,k}$ is the real-time power flow on line k; $P_{Lf,k}^{\max}$ is the upper limit of power flow on line k; L is the transmission line number.

The objective function of ELD problem is as follows.

$$\text{minimize } F_t = \sum_{m=1}^{N} F_m(P_m) \tag{7}$$

Where F_t is total power cost for the system; N is the total number of generators; P_m is the m-th generator's active power; $F_m(P_m)$ is the m-th generator's consumption characteristics which is generally approximated by a quadratic function as follows.

$$F_m(P_m) = a_m P_m^2 + b_m P_m + c_m, \quad m = 1, 2, \ldots, N \tag{8}$$

Where a, b and c are the coal consumption parameters in the unit coal function.

3   Iterative PSO (IPSO) and Chaotic Mutation

3.1   Classical Particle Swarm Optimization (PSO) Algorithm

Particle swarm optimization (PSO) is a swarm intelligent optimization method that proposed by Eberhart and Kennedy in 1995, it came of the research on bird flock preying behavior. Because the algorithm has many features such as simplicity, less adjustable parameters, easy to implement and fast convergence, it is widely used in engineering fields. PSO algorithm is as follows:

$$\begin{cases} V_i(t+1) = \omega V_i(t) + c_1 r_1(P_{best i}(t) - X_i(t)) + c_2 r_2(g_{best}(t) - X_i(t)) \\ X_i(t+1) = X_i(t) + V_i(t+1) \end{cases} \tag{9}$$

Where the initial particle swarms is a group of random particles in d-dimensional space; t represents the current iterative times; $V_{id} = (v_{i1}, v_{i2}, \ldots, v_{id})$ is a d-dimensional vector which represents the i-th particle's velocity in d-dimension, d-dimensional vector $X_{id} = (x_{i1}, x_{i2}, \ldots, x_{id})$ is i-th particle's position in d-dimension. Among them; individual extremum P_{id} (P_{best}) is each particle's optimum solution in the search process; global extremum P_{gd} (g_{best}) is entire group's optimum solution. According to the two extremums, particle can find the optimal solution by updating iteration based on the above formula. c_1, c_2 are acceleration coefficients that can diverse velocity of group particles. r_1, r_2 are random values in the range of (0, 1).

3.2   Iterative Particle Swarm Optimization (IPSO) Algorithm

Iterative optimal value I_b has introduced to PSO that can enhance the diversity of algorithmic search and avoid falling into local minimum [9]. The expression is as follows:

$$\begin{cases} V_i(t+1) = \omega \times V_i(t) + c_1 \times r_1 \times (P_{id}(t) - X_{id}(t)) + c_2 \times r_2 \times (P_{gd}(t) - X_{id}(t)) + c_3 \times r_3 \times (I_b(t) - X_{id}(t)) \\ X_i(t+1) = X_i(t) + V_i(t+1) \end{cases}$$

$$\tag{10}$$

Where I_b is the fitness of optimal particle that searching in population; c_3 is particle's accelerate weight to I_b, here taking $c_3 = c_1 \times (1 - e^{-c_1 t})$. Through constantly iterating until close to the optimal particle's corresponding best fitness, particles' search is more robust and efficient.

Inertia weight ω adjusts dynamically with iteration number.

$$\omega = (\omega_{max} - \omega_{min}) \times e^{-\beta t} + \omega_{min} \tag{11}$$

Where ω_{max} and ω_{min} are the maximum and minimum of inertia weight, the initial search for larger ω is conducive to global search, and the smaller ω will limit search to local area on later ($\omega_{max} = 0.94$, $\omega_{min} = 0.4$). β is the constriction factor that can control speed changes of ω ($\beta = 0.001$), t is the current iteration.

3.3 Chaotic Mutation

Tent mapping that has strong ergodic search is introduced [10].

$$x^{k+1} = \begin{cases} x^k/0.7, 0 \le x^k < 0.7 \\ 10 x^k (1 - x^k)/3, 0.7 \le x^k \le 1 \end{cases} \tag{12}$$

Where k is the number of iterations; x^k is the chaotic variable ($x^k \in [0,1]$), x^0 has initialized in the range of $(0, 1)$ and $x^0 \notin \{0, 0.7, 1\}$.

Chaotic mutation process is as follows:

(1) Find each particle's individual extremum p_{besti} and set it as the initial position, then introduce chaos optimization into the neighborhood of the particle, the initial value of chaotic sequence is determined by following formula:

$$z_{i,j}^0 = \frac{p_{besti} - x_{i,min}}{x_{i,max} - x_{i,min}}$$

(2) Generate m chaotic sequence $z_{i,j}^k$ with different trajectory according to Eq. (12) based on the initial value, $k = 1, 2, ..., m$.

(3) Chaotic mutate to particle.

$$x_{i,j}^k = x_{i,j}^* + 2\alpha \left(z_{i,j}^k - 0.5 \right) \tag{13}$$

Where α is the factor that determines the variation range and defines the particle's scope, if it does not satisfy the condition, set boundary value to it; $x_{i,j}^*$ is the current optimal solution; $z_{i,j}^k$ is the chaotic variable in the range of $[0, 1]$.

(4) Calculate the fitness value of mutated particle $x_{i,j}$, if it is superior to the particle before mutation, replace it.

(5) If satisfy the iteration convergence conditions, output the optimum solution, otherwise return step (2) and continue to search.

4   Realization of Chaos Iterative Particle Swarm Optimization (CIPSO)

The algorithm is encoded based on Matlab 8.0 and all experiments are performed on a personal computer (Experiments run 30 times independently).

(1) Set system parameters: particle's maximum number of iteration is 100, the number of chaos local search is 30, chaotic sequence $m = 20$, the number of particles involved in the simulation $N = 30$, the initial iterative times $t = 0$, acceleration coefficients $c_1 = c_2 = 0.01$.

(2) Correspond to the generator output load $P_i = [p_{i1}, p_{i2}, \ldots, p_{ij}]$ in actual problem, represented by the particles' position $X_i = [x_{i1}, x_{i2}, \ldots, x_{ij}]$, initialize the particle velocity.

$$v_{ij} = \left(v_{jmax} - v_{jmin} \right) \times rand + v_{jmin}, j = 1, 2, 3, \ldots, D \qquad (14)$$

Speed limit: $V_{i,j}^{max} = - V_{i,j}^{min} = 0.1 \, P_{i,j}^{max}$
Initialize particles' position:

$$x_{ij} = \left(x_{jmax} - x_{jmin} \right) \times rand + x_{jmin} \qquad (15)$$

Which rand is a random number in the range of (0, 1).

(3) Initialization particle individual extremum P_{besti} and global extremum g_{best}. The initial particle in population is set to individual extremum, the smallest fitness value in the individual extremum is set to global extremum, and the corresponding fitness value of global extremum is set to I_b.

(4) Calculate the fitness value of each particle, then introduce the penalization function, the mathematical model is as follows.

$$PF_T = \sum_{i=1}^{N} F(P_i) + \xi \left| \sum_{i=1}^{N} P_i - (P_{load} + P_{loss}) \right| \qquad (16)$$

Where variable PF_T represents the final consumption cost; ξ is the penalty factor (The test have found that the punishment has good effect when $\xi = 100$, now the magnitude is same as the initial particle).

(5) Adjust the inertia weight ω according to Eq. (11).

(6) Update the speed and location according to Eq. (10), judge whether satisfy the constraints or not, limit the particle's scope to the boundary value if it does not meet the constraints.

(7) Select 20 % of particles in population that corresponding fitness value is better, and implement chaotic mutation introduced in 2.3 to the selected particles' individual extremum.

(8) Update population's individual extremum P_{besti}, global extremum g_{best} and fitness value I_b of optimal particle.

(9) Reduce search area of particles dynamically.

$$\begin{cases} p_i^{\min} = \max\left(p_i^{\min}, \; g_{\text{best}} -0.5 \bullet rand \bullet \left(p_i^{\max} - p_i^{\min}\right)\right) \\ p_i^{\max} = \min\left(p_i^{\max}, \; g_{\text{best}} -0.5 \bullet rand \bullet \left(p_i^{\max} - p_i^{\min}\right)\right) \end{cases}$$

(10) Generate 80 % of the particles randomly according to formula (14) (15) in the new search area and calculate its fitness value.

(11) Reassemble particles in step (7) (10) and replace the original population.

(12) Return to step (4) and iterate, until reach the given maximum times and output g_{best}.

Table 1. Simulation results and comparisons of 6 units

Unit Output	IPSO [9]	CPSO [12]	PSO [13]	SOH_PSO [14]	CIPSO
P1	440.57	434.43	447.81	438.21	429.84
P2	179.84	173.32	173.32	172.58	159.78
P3	261.38	274.47	263.47	257.42	271.87
P4	131.91	128.06	139.06	141.09	146.14
P5	170.98	179.48	165.48	179.37	151.03
P6	90.82	85.93	87.13	86.88	114.12
$\sum P_m$	1275.51	1276.00	1276.01	1275.55	1272.78
P_L	12.55	12.96	12.96	12.55	9.78
COST/$	15446.30	15446.00	15450.00	15446.00	15423.77

Table 2. Simulation results and comparisons of 15 units

Unit Output	IPSO [9]	CPSO [12]	PSO [13]	SOH_PSO [14]	CIPSO
P1	455.00	450.02	439.12	455.00	415.85
P2	380.00	454.06	407.97	380.00	411.00
P3	129.97	124.81	119.63	130.00	128.85
P4	130.00	124.81	129.99	130.00	126.19
P5	169.93	151.06	151.07	170.00	188.10
P6	459.88	460.00	460.00	459.96	427.70
P7	429.25	434.57	425.56	430.00	431.73
P8	60.43	148.46	98.57	117.53	99.80
P9	74.78	63.59	13.49	77.90	95.02
P10	158.02	101.12	101.11	119.54	117.73
P11	80.00	28.66	33.91	54.50	70.87
P12	78.57	20.91	79.96	80.00	52.74
P13	25.00	25.00	25.00	25.00	27.16
P14	15.00	54.41	41.41	17.86	35.76
P15	15.00	20.62	35.61	15.00	26.64
$\sum P_m$	2660.80	2662.10	2662.40	2662.29	2655.16
P_L	30.86	32.13	32.43	32.28	25.16
COST/$	32784.50	32834.00	33039.00	32878.00	32745.35

5 Simulation Results and Analysis

The algorithm is validated for two test systems consisting of 6 and 15 generators in literature [12], unit parameters, B coefficient and so on can be seen in literature [13]. The simulation results and the algorithms comparison are shown in Tables 1 and 2. Simulation result comparisons of the algorithm in this thesis and CPSO in literature [12] are shown in Figs. 1 and 2. The simulation results show that new algorithm converges faster, the optimal solution getting in the global scope is better, cost of electricity-generating is cheaper under the same condition and save the economic cost.

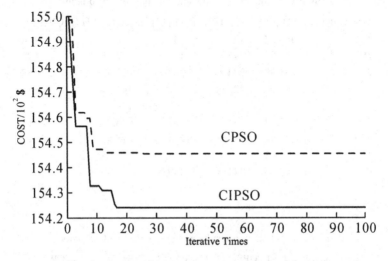

Fig. 1. Average fitness convergence curves of 6 units

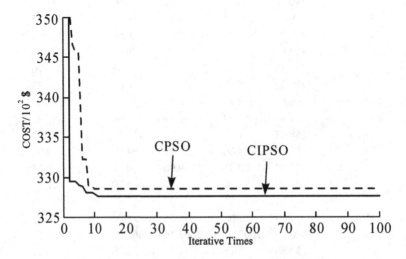

Fig. 2. Average fitness convergence curves of 15 units

6   Conclusion

This paper explores the economic load allocation problem of complex power system under multi-constraints, then proposes an improved chaotic particle swarm optimization algorithm and verifies it through simulation examples. This algorithm corrects the global search strategy of PSO by introducing optimal iterative factor, implements chaotic mutation to some better solution based on a new Tent mapping and improves algorithm iterative process. The experimental result demonstrates the proposed algorithm well balances global and local search of particles, avoids premature, and acquires more accurate global optimal solution while speeding up the convergence of particles.

References

1. Liu, L.H., Han, P., Wang, D.F.: A new chaos genetic algorithm and its application in function optimization. J. North China Electric Power Univ. **37**(3), 93–96 (2010)
2. Zou, E., Xin, J.T., Fang, S.Y., et al.: Improved chaotic particle swarm optimization algorithm and its application in economic load dispatch. Proc. Chin. Soc. Univ. Electr. Power Syst. Autom. **24**(4), 1942–1948 (2012)
3. Deng, Z.P.: Improved differential evolution algorithm for transmission network planning. Electr. Eng. **8**, 66–69 (2012)
4. Gu, H., Guo, Z.Y., Liu, W., et al.: Cogeneration economic dispatch based on taboo-particle swarm optimization algorithm. J. Southeast Univ. **43**(1), 83–87 (2013)
5. Tang, Y.G., Cui, Y.H., Qiao, L.J., et al.: Application of simplex search method and particle swarm optimization in economic dispatch. Proc. Chin. Soc. Univ. Electr. Power Syst. Autom. **21**(1), 20–26 (2009)
6. Wang, K., Wang, W., Zhang, Z.L., et al.: Improved particle swarm optimization algorithm for economic load dispatch of power system. J. Qingdao Univ. **24**(1), 79–84 (2009)
7. Liu, G., Peng, C.H., Xiang, L.Y.: Economic-environmental dispatch using improved multi-objective particle swarm optimization. Power Syst. Technol. **35**(7), 139–144 (2011)
8. Cai, J.J., Li, Q., Li, L.X., et al.: A hybrid CPSO-SQP method for economic dispatch considering the valve-point effects. Energy Convers. Manag. **53**(1), 175–181 (2012)
9. Safari, A., Shayeghi, H.: Iteration particle swarm optimization procedure for economic load dispatch with generator constraints. Expert Syst. Appl. **38**(1), 6043–6048 (2011)
10. Alatas, B., Akin, E., Ozer, A.B.: Chaos embedded particle swarm optimization algorithms. Chaos, Solitions Fractals **40**(4), 1715–1734 (2009)
11. Wang, Y., Zhou, J.Z., Qin, H., et al.: Improved chaotic particle swarm optimization algorithm for dynamic economic dispatch problem with valve-point effects. Energy Convers. Manag. **51**(1), 2893–2900 (2010)
12. Cai, J.J., Ma, X.Q., Li, L.X., et al.: Chaotic particle swarm optimization for economic dispatch considering the generator constraints. Energy Convers. Manag. **48**(2), 645–653 (2007)
13. Gang, Z.L.: Particle swarm optimization to solving the economic dispatch considering the generator constraints. IEEE Trans. Power Syst. **18**(3), 1187–1195 (2003)
14. Chaturvedi, K.T., Pandit, M., Srivastava, L.: Self-organizing hierarchical particle swarm optimization for nonconvex economic dispatch. IEEE Trans. Power Syst. **23**(3), 1079–1087 (2008)

Training Artificial Neural Network Using Hybrid Optimization Algorithm for Rainfall-Runoff Forecasting

Jiansheng Wu[1,2(✉)] and Chengdong Wei[3]

[1] Department of Mathematics and Computer,
Liuzhou Teacher College, Liuzhou, Guangxi 545004
People's Republic of China
Wjsh2002168@163.com

[2] School of Information Engineering, Wuhan University of Technology, Wuhan,
Hubei 430070, People's Republic of China

[3] School of Mathematics and Statistics, Guangxi Teachers Education University,
Nanning, Guangxi 530001, People's Republic of China
13077732268@163.com

Abstract. In this paper, a hybrid optimization algorithm is proposed to train the initial connection weights and thresholds of artificial neural network (ANN) by incorporating Simulated Annealing algorithm (SA) into Genetic Algorithm (GA), and then the Back Propagation (BP) algorithm is applied to adjust the final weights and biases, namely HGASA-ANN. Finally, a numerical example of daily rainfall-runoff data is used to elucidate the forecasting performance of the proposed HGASA-ANN model. The GASA is employed to accelerate the training speed and helps to avoid premature convergence and permutation problems. The HGASA-NN can make use of not only strong global searching ability of the GASA, but also strong local searching ability of the BP algorithm. The forecasting results indicate that the proposed model yields more accurate forecasting results than the back-propagation neural network and pure GA training artificial neural network. Therefore, the HGASA-ANN model is a promising alternative for rainfall-runoff forecasting.

Keywords: Artificial neural network · Simulated annealing algorithm · Genetic algorithm · Hybrid optimization · Rainfall-runoff forecasting

1 Introduction

Accurate and timely rainfall-runoff forecasting is a major challenge job for many catchment management applications, such as design of spillways and waterways, water quality modeling, urban planning, in particular for flood warning systems, because it can help prevent casualties and damages caused by natural disasters [1–3]. Artificial Neural Network (ANN) has been successful applied to modeling rainfall–runoff, due to the ability of extracting essential characteristics relationships through empirical data. Riad et al., used the multilayer perception (MLP) neural network to model the rainfall-runoff relationship in a catchment located in a semiarid climate in Morocco [4].

© Springer International Publishing Switzerland 2015
D.-S. Huang et al. (Eds.): ICIC 2015, Part I, LNCS 9225, pp. 576–586, 2015.
DOI: 10.1007/978-3-319-22180-9_57

Mutlu et al., applied the multilayer perception (MLP) and the radial basis neural network (RBFNN) to predict stream flow at four gauging stations in order to compare their performance prediction. The results show that ANN models are useful tools for forecasting in agricultural watersheds [5]. Asadiet et al., presented a hybrid intelligent model for runoff prediction combined of data preprocessing methods, genetic algorithms and levenberg-marquardt (LM) algorithm for learning feed forward neural networks [6].

Despite a lot of applications of ANN with great success on rainfall–runoff forecasting by a back–propagation (BP) learning algorithm based on the gradient descent searching technology (BP–ANN), there have several drawbacks, such as network easily over-fitting, difficult to determine network parameters and effects to depend on the user's experience, etc. [7–9]. Recently, Genetic Algorithm (GA) has been widely used for training and/or automatically designing neural networks. Heckerling et al. have used GA to determine the number of nodes, connection weights and training parameters for ANN for predicting [10]. Sedki et al. have designed and optimized the neural network by a real coded GA strategy and hybrid with a BP algorithm for rainfall–runoff forecasting [11]. Nambiar et al. have utilized a multi–population parallel genetic algorithm (GA) to simultaneously optimize the structure and system latency of evolvable BPNN [12]. Those papers shows GA can improve the performance of the forecasting ability of NN application because GA is heuristic optimization algorithms to acquire an optimal or suboptimal solution. However, the GA's convergent speed is very slow when the structures of NNs are complex and there are large training samples, so that it is easily trapped in local optimum [13]. Moreover, GA's ability of fine-tuning solution is not good, because the selected mutation probability may be too small to escape a trap. Anyway, GA is an important but not powerful optimization tools.

In order to overcome these drawbacks from BP-ANN and GA, it is necessary to find some effective approach to play the advantages of each algorithm, and avoid misleading in the local optimum. This paper propose a novel and specialized hybrid optimization strategy by incorporating SA into GA to search for optimal or approximate optimal beginning connection weights and thresholds for the network, then using the BP to adjust the final weights, namely HGASA-NN. As such, this paper aims at identifying an optimal Ann's rainfall-runoff forecasting model. The rest of this study is organized as follows. Section 2 describes the proposed HGASA-ANN, ideas and procedures. For further illustration, the method has been used to establish a prediction model for rainfall–runoff forecasting in Sect. 3 and conclusions are drawn in the final Section.

2 The Developed HGASA-ANN Approach

MLP-ANN is optimized by HGASA, which the network consists of an input layer, one hidden layer, and an output layer in this paper. The number of nodes of network input layer and output layer nodes is decided by the training data, the number of network hidden node equal to the number of input nodes.

2.1   Encoding Strategy of HGASA

In this paper, a real genetic algorithm is proposed to code processes. In this encoding strategy, every chromosome is encoded for real vectors, which represent each neuron's connection weights and bias to its correspondent gene segments. Each chromosome has four gene segments, which represent four parameters w_{ji}, w_{j0}, v_{kj} and v_{k0}. For TFNN neural network involved, each chromosome represents all weights of a network structure. For example, for the TFNN with the structure of 3-4-1, the corresponding encoding style for each individual can be represented as:

$$individual(i) = [w_{11}w_{12}w_{13}w_{21}w_{22}w_{23}w_{31}w_{32}w_{33}w_{10}w_{20}w_{30}v_{11}v_{21}v_{31}v_{10}] \quad (1)$$

$$population = \begin{bmatrix} individual(1) \\ individual(2) \\ \vdots \\ individual(M) \end{bmatrix} \quad (2)$$

where M is the number of the total population, i = 1,…, M. In this encoding strategy, every chromosome is encoded for a matrix. We also take the TFNN with the structure of 3-3-1 for an example, the encoding strategy can be written as

$$generation(:,:,:,:,i) = [W, W_0, V, V_0] \quad (3)$$

$$W = \begin{bmatrix} w_{11} & w_{12} & w_{13} \\ w_{21} & w_{22} & w_{23} \\ w_{31} & w_{32} & w_{33} \end{bmatrix} \quad w_0 = \begin{bmatrix} w_{10} \\ w_{20} \\ w_{30} \end{bmatrix} \quad V = \begin{bmatrix} v_{11} \\ v_{21} \\ v_{31} \end{bmatrix} \quad V_0 = [v_{10}] \quad (4)$$

Where w is the hidden layer weight matrix, v is the output layer weight matrix, while w_0 is the hidden layer bias matrix and of v_0 is the output layer bias matrix.

2.2   Methods and Steps HGASA-ANN

The hybrid GASA training ANN process consists of two stages: firstly employing GASA to search for optimal or approximate optimal connection weights and thresholds for the network, then using the BP to adjust the final weights. Figure 2 shows flowchart of the proposed algorithm. The major steps of the proposed algorithm are as follows:

1. Generate initial population. The connection weights and thresholds are encoded as float string, randomly generated within [-1,1].
2. Input training data and calculate the fitness of each chromosome according to Eq. (5).

$$f_{fitness\ function} = \frac{1}{[1 + \frac{1}{n}\sum_{i=1}^{n}(y_t - \hat{y}_t)^2]} \quad (5)$$

Where n is the number of training data samples; y_t is the actual value, and $\hat{y}_t$ is the network output.

3. Perform GA process. The selection operator of genetic algorithm is implemented by using the roulette wheel algorithm to determine which population members are chosen as parents that will create offspring for the next generation.

4. The paper uses arithmetical crossover which can ensure the offspring are still in the constraint region and moreover the system is more stable and the variance of the best solution is smaller. The crossover of connection weights and thresholds are operated with probability p_c at float string according to Eqs. (6) and (7)

$$x_i^{t+1} = \alpha x_i^t + (1 - \alpha) x_{i+1}^t \qquad (6)$$

$$x_{i+1}^{t+1} = (1 - \alpha) x_i^t + \alpha x_{i+1}^t \qquad (7)$$

where x_i^{t+1} stands for the real values of the ith individual of the $(t + 1)$th generation, x_i^t and x_{i+1}^t are a pair of individuals before crossover, x_i^{t+1} and x_{i+1}^{t+1} are a pair of individuals after crossover, a is taken as random value within [0, 1].

5. The mutation of connection weights and thresholds are operated with probability p_m at float string according to Eq. (8)

$$x_i^{t+1} = x_i^t + \beta \qquad (8)$$

where x_i^t stands for the real values of the ith individual of the tth generation, x_i^t is individual before mutation, x_i^t is individual after mutation, β is taken as random value within [0, 1].

6. Stop condition. If the number of generation is equal to a given scale, then the best chromosomes are presented as a solution, otherwise go to the step 1 of the SA part. GA will deliver its best individual to SA for further processing.

7. Perform SA operators. Generate initial current state. Receive values of the weights and thresholds from GA's. The value of fitness, shown as Eq. (5), is defined as the system state (E). Here, the initial state (E_0) is obtained.

8. Provisional state. The existing system state is denoted by S_{old}, Make a random move to change the existing system state to a provisional state, named S_{new}. Another set of weights and thresholds are generated in this stage S_{new}.

9. Metropolis criterion tests. The probability of accepting the new state is given by the following probability function

$$P = \begin{cases} 1 & \text{if } E(S_{new}) > E(S_{old}) \\ \exp\left[\dfrac{E(S_{new}) > E(S_{old})}{kT}\right] & \text{if } E(S_{new}) < E(S_{old}) \end{cases} \qquad (9)$$

T is the thermal equilibrium temperature, k represents the Boltzmann constant. If the provisional state is accepted, then set the provisional state as the current state.

10. Temperature reduction. After the new system state is obtained, reduce the temperature. The new temperature reduction is obtained by the Eq. (10)

$$T_{i+1} = \alpha T_i \qquad (10)$$

Where T_i is ith temperature stage and determines the gradient of cooling, α is set at 0.8 in this paper. If the pre-determined temperature is reached, then stop the algorithm and the latest state is an approximate optimal solution. Otherwise, go to step 8.

11. Perform GA process. Once the termination condition is met, output the final solution, obtain the appropriate connection weights and thresholds. Input validation data and compute fitness for all the individuals.

12. Input testing data and Output forecasting results by the GASA-ANN (Fig. 1).

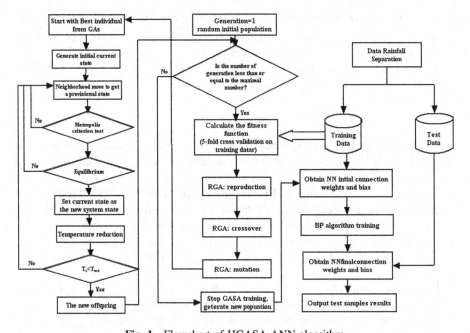

Fig. 1. Flowchart of HGASA-ANN algorithm.

3   Experiments Results and Analysis

To evaluate the purposed model performance, the obtained results compared with the other two model for rainfall–runoff prediction, such as BP-NN and pure GA-ANN. BP-ANN model was established by the trial-and-error approach following the previous studies [10], and GA-NN model bullied by the real code GA to evolve initial the values of weights and thresholds for MLP neural networks, which the method has presented by Sedki et al. [11]. The parameter settings of the proposed HGASA-NN model with other models are shown in Table 1.

Table 1. Results of the proposed HGASA-ANN model with other models.

Model	The description of parameter	Value
GA	Number of generations	100
	Population size	20
	Crossover probability	0.80
	Mutation probability	0.05
SA	Initial temperature	5000
	Termination temperature	0.9
BP-ANN	Architecture (input-hidden-output)	5-6-6-1
	Transfer function	Sigmoid-Sigmoid-Tanh

3.1 Study Area and Data

Daily flow as runoff data from Daqiao stream flow gauging stations, were obtained from the observation archives of Liuzhou Water Management Information System. The collected data were prepared for the period between 2006 and 2010 years, a total of 5 years with 1826 data points. The data were divided into two parts, the first fours years (2006–2009) samples with 1461 data points for model training and the remaining one year (2010) samples with 365 data points for model testing.

3.2 Input Variables

In general, the causal variables involved in rainfall-runoff relations are those associated with rainfall, previous water level, evaporation, temperature, etc. Most studies applied rainfall and previous flow (or water level) as inputs variables [14]. According to the literature (see Ref. [14]), runoff at time steps t-1($X_{(t-1)}$), t-2($X_{(t-2)}$), t-3($X_{(t-3)}$) and rainfall at time step t-1($R_{(t-1)}$) are considered as inputs to the model by the stepwise regression method to eliminate low impact factors and choose the most influential factors.

3.3 Performance Evaluation

There four indexes are applied for performance evaluation of rainfall-runoff forecasting model in this paper, such as root mean square error (RMSE), mean absolute percentage error (MAPE), coefficient of efficiency (CE), and coefficient of efficiency for peak values (CE_{peak}), which can be found in many paper [8].

3.4 Application HGASA-ANN for Rainfall-Runoff

Figure 2 shows the curve of fitness in the learning stage for GA-ANN approach arising from the iteration number. One can see that the maximum, average and the minimum fitness are unstable with increase of iteration 100. Figure 3 shows the curve of fitness in the learning stage for HGASA-ANN approach arising from the iteration number. One

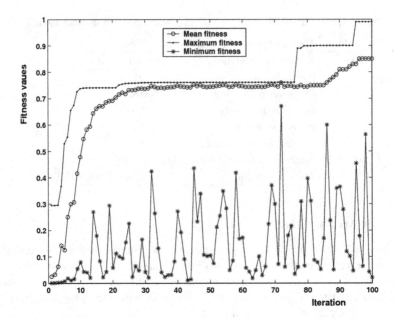

Fig. 2. Fitness function curve of GA-ANN approach in the learning stage

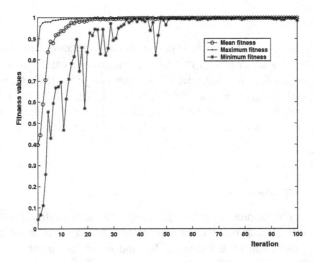

Fig. 3. Fitness function curve of HGASA-ANN approach in the learning stage

can see that the maximum, average and the minimum fitness are tending towards stability with increase of iteration 55. As can be seen from Fig. 3, with the increase in the number of training, fitness is quickly stabilized. The result show HGASA-ANN can avoid problems of premature convergence and escape from local optima.

3.5   Analysis of the Results

Figures 4, 5, and 6 show the observed and the forecasting by the proposed HGASA-NN model with other two models at 365 testing data points. From the graphs, we can generally see that the results of HGASA-ANN model are closer to the corresponding

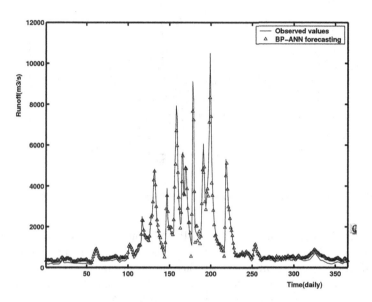

Fig. 4. Comparison between observed and predicted of BP-ANN model.

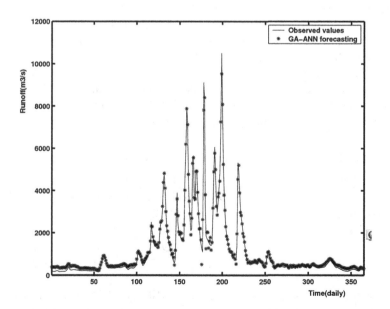

Fig. 5. Comparison between observed and predicted of GA-ANN model.

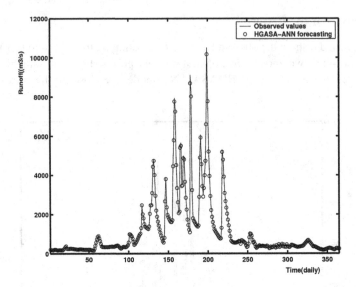

Fig. 6. Comparison between observed and predicted of HGASA-ANN models.

observed runoff values than those of the BP-ANN and GA-ANN models. Especially, the HGASA-ANN model produced better forecasting for high runoff values, while other two models were slightly overestimated or underestimated.

Table 2 illustrates the fitting and forecasting accuracy and efficiency of the model in terms of various evaluation indices for BP-ANN, GA-ANN and HGASA-ANN, respectively. As shown in Table 2, for the training data, the RMSE for BP-ANN model is 0.281, and for GA-ANN model it is 0.347; while for HGASA-ANN model, RMSE reaches 0.166. Similarly, for the testing data, the RMSE for the BP-ANN model is 0.451, and for the GA-ANN model it is 0.357; while for the HGASA-ANN model, RMSE reaches 0.121. Focusing on the MAPE indicator of the training and testing, the HGASA-ANN model is also less than the other two models. These results indicate that the errors of the HGASA-ANN are the smallest. In addition, the CE of HGASA-ANN is the maximum in all models. This also means that HGASA-ANN is capable to capture the average change tendency of the daily water level.

Table 2. Results of the proposed HGASA-ANN model with other models.

	Fitting (training)			Forecasting (testing)		
	HGASA-ANN	GA-ANN	BP-ANN	HGASA-ANN	GA-ANN	BP-ANN
RMSE	**0.166**	0.347	0.281	**0.121**	0.357	0.451
MAPE	**0.105 %**	0.238 %	0.121 %	**0.102 %**	0.256 %	0.163 %
CE	**0.978**	0.855	0.914	**0.985**	0.812	0.742
CE_{peak}	**0.985**	0.825	0.803	**0.997**	0.805	0.716

The empirical results show that the HGASA-ANN model are very promising in the runoff forecasting under the research where either the measurement of fitting performance is goodness or where the forecasting performance is effectiveness. It also can be seen that there was consistency in the results obtained between the training and testing of the HGASA-ANN model. To summarize, the HGASA-ANN model is superior to the other two models presented here in terms of RMSE, MAPE and CE for water level prediction under the same network input.

From the experiments presented in this paper, the HGASA algorithm also helps to improve the fine-tuning ability, the hill-climbing ability, the speed of convergence to the global optimum solutions or near global optimum solutions, and avoid trapping into local minimum than GA did, thus, outperform the GA-ANN model. For example, in Table 2, the CE_{peak} of the GA-ANN model is 0.805; however the CE_{peak} of the HGASA-NN model reaches 0.997 which has obvious advantages over GA-NN models. Thus, it once again reveals that HGASA algorithm is much appropriate than GA in parameter adjustments to achieve forecasting accuracy improvement by integrated into the ANN model.

The more important point in the runoff modeling which makes sure that our model is most promising when comparing with different models is the capability of the proposed model in estimating peak values. Where as the estimation of peak values is usually the most important part of the flood mitigation program for actual runoff application, as final step of the modeling process. From Table 2, the efficiency of HGASA-NN model is 0.985 in training data, and it is 0.997 in testing data. There is an identical high capability of our models for prediction of peak flows in testing data. Therefore, not only the proposed model is appropriate in monitoring peak values, but also it can be considered as most powerful tool for runoff forecasting which is necessary in the water resources systems management, where it is directly in fluencies by stream flow forecasting.

4 Conclusion

In this paper, we use a novel technique for the training of ANN beginning connection weights and thresholds by HGSA algorithm. When the fitness function value has not changed for some generations, or value changed is smaller than a predefined number, the searching process is switched to gradient descending searching according to this heuristic knowledge. The HGASA-ANN is an optimization algorithm combining the GASA with the ANN for the rainfall-runoff model, which the hybrid algorithm maintains the merit of global optimization by using HGASA process, and also the merit of local optimization by combining ANN. And also adaptive mechanisms are included to improve the searching ability for optimum solutions. This paper proposed HGASA-ANN can improve the fine–tuning ability, the hill climbing ability, and the speed of convergence to the global optimum solutions or near global optimum solutions. The new hybrid HGASA-ANN approach is compared with the other two methods with a set of benchmark mathematical functions. According to the results obtained in this paper, we can draw the following conclusions that the HGASA-ANN model can be used as an alternative tool for actual runoff forecasting application to obtain better forecasting accuracy and improve the prediction quality further in view of empirical results.

Acknowledgment. The authors would like to express their sincere thanks to the editor and anonymous reviewer's comments and suggestions for the improvement of this paper. This work was supported the Natural Science Foundation of Guangxi Province under Grant No. 2014GXNSFAA118027, and by the Guangxi Education Department under Grant No. 2013YB281 and YB2014467, and Key Laboratory for Mixed and Missing Data Statistics of the Education Department of Guangxi Province under Grant No. GXMMSL201405.

References

1. Makungo, R., Odiyo, J.O., Ndiritu, J.G., Mwaka, B.: Rainfall-runoff modelling approach for ungauged catchments: a case study of nzhelele river sub-quaternary catchment. Phys. Chem. Earth **35**, 596–607 (2010)
2. Talei, A., Chye Chua, L.H., Quek, C.: A novel application of a neuro-fuzzy computational technique in event-based rainfall-runoff modeling. Expert Syst. Appl. **37**, 7456–7468 (2010)
3. Vahid, N., Özgür, K., Mehdi, K.: Two hybrid artificial intelligence approaches for modeling rainfall–runoff process. J. Hydrol. **402**, 41–59 (2011)
4. Riad, S., Mania, J., Bouchaou, L.: Rainfall–runoff model usingan artificial neural network approach. Math. Comput. Model. **40**(7–8), 839–846 (2004)
5. Mutlu, E., Chaubey, I., Hexmoor, H., Bajwa, S.G.: Comparison of artificial neural network models for hydrologic predictions at multiple gauging stations in an agricultural watershed. Hydrol. Process. **22**(26), 5097–5106 (2008)
6. Asadi, S., Shahrabi, J., Abbaszadeh, P.: A new hybrid artificial neural networks for rainfall-runoff process modeling. Neurocomputing. **121**, 470–480 (2013)
7. Wu, J.: An effective hybrid semi-parametric regression strategy for rainfall forecasting combining linear and nonlinear regression. Int. J. Appl. Evol. Comput. **2**(4), 50–65 (2011)
8. Wu, J., Jin, L.: Study on the meteorological prediction model using the learning algorithm of neural networks ensemble based on PSO algorithm. J. Trop. Meteorol. **15**(1), 83–88 (2009)
9. Ahmadia, M.A., Ebadib, M., Shokrollahic, A.: Evolving artificial neural network and imperialist competitive algorithm for prediction oil flow rate of the reservoir. Appl. Soft Comput. **13**, 1085–1098 (2013)
10. Heckerling, P.S., Gerber, B.S., Tape, T.G., Wigton, R.S.: Use of genetic algorithm for neural network to predict community-acquired pneumonia. Artif. Intell. Med. **30**, 71–84 (2004)
11. Sedki, A., Ouazar, D., Mazoudi, E.E.: Evolving neural network using real coded genetic algorithm for daily rainfall-runoff forecasting. Expert Syst. Appl. **36**, 4523–4527 (2009)
12. Nambiar, V.P., Khalil-Hani, M., Marsono, N., Sia, C.W.: Optimization of structure and systeml atency in evolvable block-based neural networks using genetic algorithm. Neurocomputing. **145**, 285–302 (2014)
13. Yaghini, M., Khoshraftar, M.M., Fallahi, M.: A hybrid algorithm for artificial neural network training. Eng. Appl. Artif. Intell. **26**(1), 293–301 (2013)
14. Wu, C.L., Chau, K.W.: Rainfall–runoff modelling using artificial neural network coupled with singular spectrum analysis. J. Hydrol. **399**, 394–409 (2011)

A Novel Swarm Intelligence Algorithm Based on Cuckoo Search Algorithm (NSICS)

Nazanin Fouladgar[1(✉)] and Shahriar Lotfi[2]

[1] Computer Engineering Department, College of Nabi Akram, Tabriz, Iran
Nazanin_Fouladgar@yahoo.com
[2] Computer Science Department, University of Tabriz, Tabriz, Iran
shahriar_lotfi@tabrizu.ac.ir

Abstract. Cuckoo Search algorithm (CS) is swarm intelligence based algorithm motivated by nature. This algorithm is based on brood parasitism of some cuckoo species and has high capability of global search. Therefore, the global optimum can be figured out with higher probability. This paper proposes a novel meta-heuristic approach, called NSICS, based on CS. NSICS is able to explore not only the search space on global scale but also around the optimum on local scale more efficiently. Consequently, more accurate results can be obtained. To approach these purposes, three operators of Eggs laying, lévy fights and Move are applied. Experiments are studied on thirteen common benchmark functions among unimodal, multimodal, shifted and shifted rotated classes and then compared with CS, GPSO, SFLA and GSA algorithms. These algorithms are chosen from swarm intelligence based, bio-inspired based and chemistry and physics based algorithms' category. The simulations indicate the proposed algorithm has satisfactory performance.

Keywords: Optimization problems · Swarm intelligence · Meta-Heuristic · Cuckoo search

1 Introduction

Optimization is an applied science which explores the best value of the parameters of a problem that may take under specified conditions. In most simple way, it aims to obtain the relevant parameter values which enable an objective function to generate the minimum and maximum value [1]. Thus the optimization problems need to be addressed intensively.

Based on computational complexity theory, optimization problems are categorized to different classes. NP-hard belongs to one of the hardest one for which there hasn`t been found a quick and applicable solution in reasonable time yet. It is worth pointing out that a quick solution is the one that algorithms can obtain on a polynominal time. According to time-consuming troubles of large-scale problems, researchers have sufficed for nearest solutions to the best solution in a reasonable time.

In this domain, a lot of algorithms have been presented. Dividing them to categories, we can point out bio-inspired based algorithms, chemistry and physics based algorithms and swarm intelligence based algorithms [2]. The first classification belongs

© Springer International Publishing Switzerland 2015
D.-S. Huang et al. (Eds.): ICIC 2015, Part I, LNCS 9225, pp. 587–596, 2015.
DOI: 10.1007/978-3-319-22180-9_58

to bio-inspired based algorithms. These algorithms are based on some successful characteristics of biological system and are regarded as a subset of nature-inspired algorithms. The second one imitates from the basic rules in chemistry and physics. Although these two sciences are different from each other but they have many common basic rules. Although these two sciences are different from each other but they have many common basic rules. The last classification, widely applied in optimization problems, are inspired by collective behavior of insects and animals, say agents, in nature co-operating (knowingly or not) to achieve a definite goal. In these algorithms the individual agents are largely homogeneous and act asynchronously in parallel. Moreover, the communication between agents is affected by some form of stigmergy which is the name for indirect communications. These agents follow very simple rules and their behavior is under little or no centralized control [3]. Accordingly, there are four principles which swarm intelligence based algorithms assert: (1) Proximity principle: The basic units of a swarm are capable of simple computation related to its surrounding environment which means a direct behavioral response to environmental variances. This can be in form of interactions among agents for living-resource searching and nest building. (2) Quality principle: A swarm is able to response to quality elements. (3) Diverse response principle: the distribution is regarded in order to face with environmental fluctuations. (4) Stability and adaptability principle: a swarm is able to adapt with environmental fluctuations gently [4]. Regarding the characteristics of swarm intelligence based algorithms, they are among widely used algorithms in different domains like image processing, data mining, structural optimization, scheduling, clustering and etc. Moreover, they are applied in path planning in dynamic environments [5]. Basically they have been exploited in large scale optimization problems because of three major factors: performing collective behavior, parallelizing multiple agents and sharing information among agents.

According to the fact that Cuckoo Search belongs to swarm intelligence based algorithms, in this paper, a novel swarm intelligence algorithm based on cuckoo Search (NSICS), is presented. This algorithm has three operators: (1) Eggs laying: laying more than one egg for each cuckoo (2) lévy fight: exploring the search space in a global scale (3) Move: leading cuckoos through one of the better cuckoos of population than itself.

The rest of paper is organized as follows: Sect. 2 presents related works. In Sect. 3, the proposed algorithm is suggested. Section 4 shows experiments and results of comparing NSICS and four mentioned algorithms on three classes of benchmark functions. Finally in Sect. 5, the paper is concludes and the future works are dedicated.

2 Related Works

Since several decades ago, many algorithms have been suggested to solve optimization problems which belong to one of the mentioned category of Sect. 1.

In 1983, Simulated Annealing (SA) algorithm was introduced by Kirkpatrick et al. [6]. The name and inspiration comes from a technique involving heating and controlled cooling of a material. In 2001, Geem et al. presented an algorithm called Harmony Search (HS) by improvisation process of musicians [7]. Also, Rashedi et al. exploited the law of gravity and mass interactions and proposed a new optimization algorithm called Gravitational Search Algorithm (GSA) in 2009 [8]. All these algorithms are considered as chemistry and physics based algorithms.

Many of optimization algorithms are inspired from nature and biological systems. Genetic Algorithm (GA) is one of the most popular algorithms in this domain which was proposed by John Holland in 1975 [9]. It generates chromosomes as solutions using techniques inspired by the natural evolution such as mutation, crossover, selection and replacement. Also, in 1997, Stom and Price introduced Differential Evolution (DE) algorithm which is exploited from real-valued vectors as candidate solutions [10]. On the contrary to GA, Differential Evolution generates solutions using mutation and crossover techniques. Another algorithm of this area is Shuffled Frog-Leaping Algorithm (SFLA) proposed by Eusuff in 2000 [11]. This algorithm consists of a set of interacting virtual population of frogs partitioned into different memeplexes.

In 1995, Kennedy and Eberhart [12] introduced a population-based stochastic optimization technique, called Particle Swarm Optimization (PSO) which was modeled on social behaviors in animal or insects. In PSO, individual particles of a swarm represent potential solutions. The particles apply their current position and the best solution position of the population to explore the search space. In 2002, Artificial Fish-Swarm Algorithm (AFSA) was proposed by Xiao et al. inspired from collective behavior of fish swarms [13]. Moreover, in 2005 Karaboga presented Artificial Bee Colony (ABC) algorithm [14], simulating the foraging behavior of bees colony. Another prominent algorithm proposed by Yang and Deb in 2009, was Cuckoo Search (CS) algorithm [15]. It is based on brood parasitism behavior of some cuckoo species. Brood parasitism behavior is a kind of behavior in which the cuckoos will never build their own nests. In nature, they leave their young in other birds` nests such that the host bird care for it. Then the cuckoo chick attempt to dismiss the host eggs to increase the amount of food provided by the host bird. Furthermore, many animals and insects apply a special flight behavior known as lévy fight with which they explore the search space instead of simple isotropic random walks. As it has shown promising capability on optimization, cuckoos in CS algorithm apply this flight pattern. In summary, all of these recently mentioned algorithms are among many others of swarm intelligence based algorithms.

3   Proposed Algorithm

Basically, population-based algorithms start with an initial population. As the proposed algorithm of this paper is also a population-based algorithm based on Cuckoo Search, it starts with a population of cuckoos.

In nature, all cuckoos follow to lay eggs to become mother. Thus, in the proposed algorithm, each cuckoo of population lay some eggs in the search space. The number of eggs for each cuckoo is determined randomly but in a range close to what cuckoos lay in nature. This process helps the algorithm to increase the probability of exploring promising areas in search space.

Each cuckoo attemps to find a secure and well profit place in other birds` nest according to its brood parasitism. In NSICS algorithm, this place is based on the distance between its position and the best position among cuckoo population (Eq. 1). To lay eggs of each cuckoo, a space with a radius of mentioned distance is regarded in which these eggs are put randomly. Different distances of Eggs laying makes cuckoos to explore the search space in accordance with the space should be searched in persuit of better positions.

$$ROE_i = BestNest - nest_i \qquad i \in [1, \ldots, CN] \tag{1}$$

In which ROE_i is Eggs laying radius of cuckoo ith, $BestNest$ is the best position of cuckoos in population, $nest_i$ is cuckoo ith's position and CN is the size of population.

One of the behaviors observable in many birds and insects is lévy fight with which birds use a series of straight flight path determined by a sudden turn, making a special pattern of searching. To explain intensively, lévy fight is a random walk in which the step-length is drawn from a lévy distribution. According to much faster increasing of variance in lévy distribution, it can be considered to explore unknown and large-scale spaces more efficiently than other methods [16].

After Eggs laying, lévy fight for eggs starts according to Eq. 2. Now that eggs explored the search space, the best position between oviparous cuckoo and eggs new position will be survived and others will be killed. This method prevents the increase of population size generated by Eggs laying operation and makes it constant from the beginning to the end of algorithm execution.

$$Egg_j^{t+1} = Egg_j^t + \left(Rand^D \otimes l\acute{e}vy\,(\lambda)\right) \; j \in [1, \ldots, EggNumber_i], i \in [1, \ldots, CN] \tag{2}$$

In which Egg_j^{t+1} is egg jth position after lévy flights and Egg_j^t is the egg jth position before lévy flight. $Rand$ function produces a D-dimension vector of random numbers with a uniform distribution in a range [-1, 1]. $l\acute{e}vy\,(\lambda)$ is lévy flights operator [11] and $\otimes$ means entry-wise multiplications.

Some cuckoo species undertake seasonal or partial migration in nature. In both cases, they move through the best environment for breeding and reproduction. The proposed algorithm implements such behavior but with a different method. Here, each cuckoo can move through one of better cuckoo positions than itself in current population, say Target. It should be mentioned that Target is chosen randomly by each cuckoo. This operation increases the capability of local search around better positions and leads to obtain more accurate results.

$$NewNest_{i+1} = NewNest_i + \left(Rand^D \otimes (Target_i - NewNest_i)\right), i \in [1, \ldots, CN] \tag{3}$$

In which $NewNest_{i+1}$ is the cuckoo i th position after displacement and $NewNest_i$ is the cuckoo i th position before displacement. Also $Rand$ function produces a D-dimension vector of random numbers with a uniform distribution in a range [-1, 1]. $\otimes$ is entrywise multiplications and $Target_i$ shows one of better positions than cuckoo i th in current population. The Pseudo-code of the proposed algorithm is presented in follow.

4 Experiments and Results

In this Section, thirteen benchmark functions are introduced for experiments and comparing results. In Table 1, name, equations, dimensions and the search space of functions are listed.

Table 1. Benchmark functions (F_1- F_{13})

Name	Test Function	D	Search range
Unimodal Functions			
Easom	$F_1(x) = -\cos(x_1)\cos(x_2)\exp\left(-(x_1 - \pi)^2 - (x_2 - \pi)^2\right)$	2	$[-100,100]^D$
Schwefel 2.22	$F_2(x) = \sum_{i=1}^{D} \|x_i\| + \prod_{i=1}^{D} \|x_i\|$	30	$[-10,10]^D$
Step	$F_3(x) = \sum_{i=1}^{D} (\lfloor x_i + 0.5 \rfloor)^2$	30	$[-100,100]^D$
Hyper-ellipsoid	$F_4(x) = \sum_{i=1}^{D} i x_i^2$	30	$[-5.12,5.12]^D$
Sphere	$F_5(x) = \sum_{i=1}^{D} x_i^2$	30	$[-100,100]^D$
Multimodal Functions			
Six Hump Camel Back	$F_6(x) = 4x_1^2 - 2.1x_1^4 + \frac{1}{3}x_1^6 + x_1 x_2 - 4x_2^2 + 4x_2^4$	10	$[-5,5]^D$
Ackley	$F_7(x) = -20\exp\left(-0.2\sqrt{\frac{1}{b}\sum_{i=1}^{D} x_i^2}\right) - \exp(\frac{1}{b}\sum_{i=1}^{D}\cos 2\pi x_i) + 20 + e$	30	$[-32,32]^D$
Griewank	$F_8(x) = \frac{1}{4000}\sum_{i=1}^{D} x_i^2 - \prod_{i=1}^{D}\cos\left(\frac{x_i}{\sqrt{i}}\right) + 1$	30	$[-600,600]^D$
General Penalized	$F_9(x) = 0.1\left\{ \sin^2(\pi x_1) + \sum_{i=1}^{D-1}\frac{(x_i - 1)^2(1 + \sin^2(3\pi x_{i+1}))}{+(x_D - 1)^2(1 + \sin^2(2\pi x_D))} \right\}$ $+ \sum_{i=1}^{D} u(x_i, 5, 100, 4)$ $u(x_i, a, k, m) = \begin{cases} k(x_i - a)^m, & x_i > a \\ 0, & -a \leq x_i \leq a \\ k(-x_i - a)^m, & x_i < -a \end{cases}$	30	$[-50,50]^D$
Shifted & Shifted-Rotated Functions			
Shifted Ackley	$F_{10}(x) = -20\exp\left(-0.2\sqrt{\frac{1}{D}\sum_{i=1}^{D} z_i^2}\right) - \exp\left(\frac{1}{D}\sum_{i=1}^{D}\cos 2\pi z_i\right) + 20 + e$ $z = x - o^{**} \quad bias_{24} = -140$	30	$[-32,32]^D$
Shifted Griewank	$F_{11}(x) = \frac{1}{4000}\sum_{i=1}^{D} z_i^2 - \prod_{i=1}^{D}\cos\left(\frac{z_i}{\sqrt{i}}\right) + 1 + bias_{25},$ $z = x - o^{**}, \ bias_{25} = -180$	30	$[-600,600]^D$
Shifted Rotated Ackley	$F_{12}(x) = -20\exp\left(-0.2\sqrt{\frac{1}{D}\sum_{i=1}^{D} z_i^2}\right) - \exp\left(\frac{1}{D}\sum_{i=1}^{D}\cos 2\pi z_i\right) + 20 + e + bias_{29}$ $z = (x - o^{**}) \times M^* \quad bias_{29} = -140$	30	$[-32,32]^D$
Shifted Rotated Griewank	$F_{13}(x) = \frac{1}{4000}\sum_{i=1}^{D} z_i^2 - \prod_{i=1}^{D}\cos\left(\frac{z_i}{\sqrt{i}}\right) + 1 + bias_{30},$ $z = (x - o^{**}) \times M^*, \ bias_{30} = -180$	30	$[-600,600]^D$

Table 2 presents the achieved results of the proposed algorithm and four other algorithms (CS, GPSO, GSA and SFLA) on the mentioned benchmark functions in Table 1. It is worth mentioned that these four algorithms are those of swarm intelligence based, chemistry and physics based and bio-inspired based algorithms category in Sect. 1. In fact, in Table 2, the results are achieved by the mean experiments in 35 times algorithms run on maximum 200,000 function evaluations. In NSICS, the

Table 2. The Mean (Std.Dev) value with 35 times run on NSICS and four algorithms on F_1- F_{13}

Functions	NSICS	CS	G-PSO	GSA	SFLA
Unimodal Functions					
F_1	-1	-1	-1	-0.8169	-1
	(0)	(0)	(0)	(0.3238)	(0)
F_2	3.29e-21	0.11e-10	449.68	1.81e-08	5.20e-11
	(2.35e-21)	(0.06e-10)	(708.61)	(2.09e-09)	(2.59e-09)
F_3	0	0	0	0	0
	(0)	(0)	(0)	(0)	(0)
F_4	1.33E-28	0.45e-24	99.61	1.18 e-16	4.47 e-17
	(2.43E-28)	(0.63e-24)	(97.15)	(3.08 e-17)	(9.16e-17)
F_5	1.54 e-25	0.43e-23	5.16e-6	1.29e-17	1.00e-18
	(8.02e-25)	(0.34e-23)	(9.04e-6)	(2.61e-18)	(2.37e-18)
Multimodal Functions					
F_6	-1.0316	-1.0316	-1.0316	-1.0316	-1.0316
	(1.68e-4)	(6.7752e-16)	(2.24e-16)	(2.24e-16)	(2.24e-16)
F_7	2.042E-14	0.11	1.53	2.93e-09	1.14e-06
	(1.64E-14)	(0.36)	(4.84)	(3.06e-10)	(1.80e-05)
F_8	0.008	1. 60 e-9	0.02	0.008	0.01
	(0.009)	(5. 06 e-9)	(0.02)	(0.02)	(0.01)
F_9	1.006E-23	0.14e-16	0.009	3.70e-13	1.41 e-10
	(1.14E-23)	(0.38e-16)	(0.02)	(1.69e-13)	(4.20 e-10)
Shifted & Shifted-Rotated Functions					
F_{10}	-139.99	-131.08	-119.12	-139.99	-121.39
	(0.03)	(9.29)	(0.21)	(3.34e-10)	(0.22)
F_{11}	-179.97	-179.90	-81.59	-127.3782	-154.38
	(0.023)	(0.002)	(64.13)	(13.5240)	(3.38)
F_{12}	-139.30	-131.49	-119.71	-139.99	-121.96
	(0.75)	(9.79)	(1.59)	(3.32e-10)	(0.37)
F_{13}	-179.99	-179.99	-108.90	-129.39	-153.87
	(0.008)	(0.001)	(58.55)	(12.77)	(2.62)

population size is regarded 30 and the eggs number for each cuckoo in a range of [5–10]. The other algorithms parameters are based on their relevant paper.

What is inferred from Table 2 is that NSICS has better results in most cases. High capability of exploring the search space for promising areas and high convergence speed through global optimum, all are their reasons. A part of these results are observable in Fig. 1. It illustrates the results of NSICS and four comparable algorithms in Table 2 on F4 (Hyper-ellipsoid), F7 (Ackley) and F11 (Shifted Griewank) functions. It should be pointed out that these functions are selected among three different class of unimodal, multimodal and shifted and shifted rotated.

```
NSICS Algorithm
1:      for each cuckoo i ∈ [1...CN]
2:          Initialize X₁ Randomly
3:      Repeat
4:          for each cuckoo i ∈ [1 ... CN]
5:              Determine EggNumber₁
6:          for each Cuckoo i ∈ [1 ... CN]
7:              for j=1 to EggNumber₁
8:                      Lay Egg ⱼ in ROE₁ by Eq. (1)
say EggPositionⱼ
9:                      Lévy flight Eggⱼ by Eq. (2)
10:                 Choose the best among Cuckoo₁ and Eggⱼ,
                    j ∈ [1 ... EggNumber₁] say NewNest₁
11:         for each NewNest i ∈ [1 ... CN]
12:             Move NewNest₁ by Eq. (3) and replace with
Cuckoo₁
13:     until stopping criterion is met
```

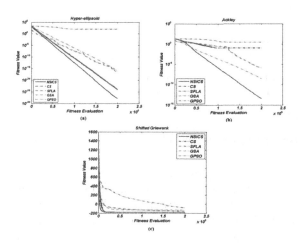

Fig. 1. The results of 35 times run with 200,000 function evaluations on (a) Hyper-ellipsoid (b) Ackley (c) Shifted Griewank benchmark functions

In two cases, this algorithm doesn't perform well. First, the results on F_8 are weaker than CS. Griewank is a multimodal function with partial difference in its previous and current peaks' height especially near the global optimum. Thus, as NSICS converges to one of these peaks, it follows to search locally on this peak according to Move operator. On the other hand, CS follows to search locally more diverse in this case by applying P_a percent randomization in population. However, the achieved results of NSICS are in an acceptable limit of accuracy.

In second case on F_{12}, GSA demonstrates better results than NSICS. The initial distribution method in GSA and consequently its effect on agents' velocity causes this result. In fact, this characteristic of GSA generates better considerable results from beginning of execution to its convergency to the global optimum in this case. Figure 2 illustrates the results of these two cases.

Testing on F_6, F_{10} and F_{13}, NSICS shows the same results as the best related comparable algorithm but with more data deviations as it is shown in Fig. 3. The reason emerges from NSICS ability to explore search space even in the time of getting close to optimum.

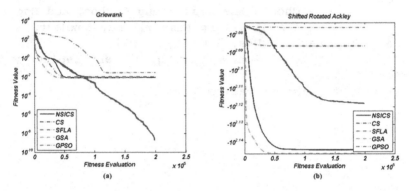

Fig. 2. The results of 35 times run with 200,000 function evaluations on (a) Griewank (b) Shifted Rotated Ackley benchmark functions

5   Conclusion and Future Works

This paper proposes a novel swarm intelligence algorithm based on Cuckoo Search, called NSICS. Increasing capability of global and local search and converging quickly to global best solution, were the motivations of this algorithm's development. To approach these purposes, three operators of Eggs laying, lévy flights and Move were applied. The proposed algorithm was tested on thirteen benchmark functions from unimodal, multimodal, shifted and shifted rotated classes and then compared with four algorithms from different classes. Finally, NSICS showed the acceptable efficiency. According to the characteristics of this algorithm, it can be exploited in global and large-scale continuous optimization problems. However, it is more applicable to add some learning and self-adapting mechanisms and also combining with other optimization algorithms.

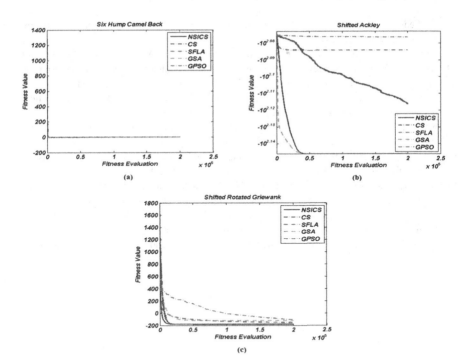

Fig. 3. The results of 35 times run with 200,000 function evaluations on a) Six Hump Camel Back b) Shifted Ackley c) Shifted Rotated Griewank benchmark functions

References

1. Civicioglu, P., Besdok, E.: A conceptual comparison of the Cuckoo-search, particle swarm optimization, differential evolution and artificial bee colony algorithms. Artif. Intell. Rev. **39**, 315–346 (2013)
2. Fister Jr., I., Yang, X.S., Fister, I., Brest, J., Fister, D.: A Brief Review of Nature-Inspired Algorithms for Optimization. http://arxiv.org.sci-hub.org/abs/1307.4186
3. Corne, D., Reynolds, A., Bonabeau, E.: Swarm Intelligence. In: Rozenberg, G., Bäck, T., Kok, J.N. (eds.) Handbook of Natural Computing, pp. 1599–1622. Springer, Heidelberg (2012)
4. Millonas, M.: Swarms, phase transitions, and collective intelligence. In: Santa Fe Institute Studies in the Sciences of Complexity-Proceedings, vol. 17, pp. 417–417. Addison-Wesley Publishing Company (1994)
5. Wang, Y., Chen, P., Jin, Y.: Trajectory planning for an unmanned ground vehicle group using augmented particle swarm optimization in a dynamic environment. In: IEEE International Conference on Systems, Man and Cybernetics, pp. 4341–4346. IEEE, San Antonio (2009)
6. Kirkpatrick, S., Gelatt, C.D., Vecchi, M.P.: Optimization by Simulated Annealing. Science **220**(4598), 671–680 (1983)
7. Geem, Z.W., Kim, J.H., Loganathan, G.V.: A new heuristic optimization algorithm: harmony search. Simul. Trans. Soc. Model. Simul. Int. **78**, 60–68 (2001)

8. Rashedi, E., Nezamabadi-pour, H., Saryazdi, S.: GSA: a Gravitational Search Algorithm. Inf. Sci. **179**(13), 2232–2248 (2009)
9. Holland, J.H.: Adaptation in Natural and Artificial Systems. MIT Press, Cambridge (1992)
10. Storn, R., Price, K.: Differential evolution - a simple and efficient adaptive scheme for global optimization over continuous spaces. J. Glob. Optim. **11**(4), 341–359 (1997)
11. Eusuff, M.M., Lansey, K.E.: Shuffled frog leaping algorithm: a memetic meta-heuristic for combinatorial optimization. Journal of heuristics (2000). (In press)
12. Kennedy, J., Eberhart, R.: Particle swarm optimization. In: Proceedings of IEEE International Conference on Neural Networks, vol. 4, pp. 1942–1948. IEEE, Perth (1995)
13. Xiao, L., Zhi, L.S., Ji, J.Q.: An optimizing method based on autonomous animals: fish swarm algorithm. Syst. Eng. Theor. Pract. **22**(11), 32–38 (2002)
14. Karaboga, D.: An idea based on Honey Bee Swarm for Numerical Optimization. Erciyes University, Engineering Faculty, Computer Engineering Department (2005)
15. Yang, X.S., Deb, S.: Cuckoo Search via Lévy flights. In: World Congress on Nature & Biologically Inspired Computing, pp. 210–214. IEEE (2009)
16. Yang, X.S.: Nature-Inspired Metaheuristic Algorithms, 2nd edn. Luniver Press, Frome (2010)

Robust PCA-Based Genetic Algorithm for Solving CNOP

Shicheng Wen, Shijin Yuan$^{(\boxtimes)}$, Bin Mu, and Hongyu Li

School of Software Engineering, Tongji University,
Shanghai, People's Republic of China
wenshicheng@yeah.net, yuanshijin2003@163.com,
{binmu,hyli}@tongji.edu.cn

Abstract. Conditional nonlinear optimal perturbation (CNOP) has been widely used in the predictability and sensitivity studies of the weather or climate models. The popular solution to the CNOP is the adjoint-based method. However, many numerical models have no adjoint models, thus bringing about a limitation to the CNOP applications. To avoid the adjoint models, we propose the robust PCA-based genetic algorithm for solving the CNOP (RGA_CNOP). To demonstrate the validity of the proposed method, it is applied to the CNOP of the Zebiak-Cane (ZC) model, and compared with the adjoint-based method. Experimental results show the RGA_CNOP can obtain approximate results to the adjoint-based method.

Keywords: Robust PCA · Genetic algorithm · CNOP · ZC model

1 Introduction

Conditional nonlinear optimal perturbation (CNOP) is a nonlinear methodology for the predictability and sensitivity of the weather or climate. It denotes the initial perturbation evolving into the largest nonlinear evolution at the prediction time [1]. So far, the CNOP has been applied to studying the optimal precursor of El Nino-Southern Oscillation (ENSO) and the effect of nonlinearity on error growth for ENSO [2, 3], determining the sensitive areas in targeting for the typhoon prediction [4, 5], conducting the ensemble forecast, and investigating the nonlinear sensitivity and stability of the ocean's THC to finite amplitude perturbations [6].

Despite of the CNOP with wide applications, how to solve the CNOP efficiently is an open issue. At present, the adjoint-based method (for short, ADJ-CNOP) is the most popular one for solving CNOP [7]. It utilizes the optimal method based on gradients which are obtained with the corresponding adjoint model of a nonlinear numerical model. The advantage of the ADJ-CNOP is to search the optimal CNOP accurately. However, many numerical models have no associated adjoint models, and implementing a new adjoint model is a huge engineering, especially for complex models. Also, the CNOP of some numerical models has the on-off cases which mean the gradient is inexistent.

To avoid adjoint models, the ensemble projecting methods were proposed in the works [8, 9]. This kind of methods learns features from a training dataset, and

© Springer International Publishing Switzerland 2015
D.-S. Huang et al. (Eds.): ICIC 2015, Part I, LNCS 9225, pp. 597–606, 2015.
DOI: 10.1007/978-3-319-22180-9_59

calculates the Jacobian matrix to further obtain the approximate gradient. However, the methods still cannot tackle with the on-off cases, due to the gradients. In addition, intelligent algorithms (IA-CNOP) were applied in the works [10] to avert the constraint brought by the gradients. As a promising kind of methods, the IA-CNOP has been applied to the Lorenz model successfully, and really addresses the on-off issues well. Yet, intelligent algorithms always encounter the convergence in the search of high dimension so that IA-CNOP is just applied to some low-dimensional models. In this paper, we follow the precious works [11, 12] to propose the robust PCA (RPCA) based genetic algorithm for solving the CNOP (RGA-CNOP). The original case in the high-dimensional space is converted to a low-dimensional case with a feature extraction method, RPCA. Then the genetic algorithm is utilized to search the CNOP in the low-dimensional space. The experimental results demonstrate that the RGA-CNOP can obtain an approximate result to the CNOP.

The rest of the paper is organized as follows: Sect. 2 introduces the background knowledge. In Sect. 3, we come up the RGA-CNOP. Section 4 reports the experimental results. This paper ends with the conclusion and future work in Sect. 5.

2   Background

2.1   CNOP

CNOP is the perturbation $u_{0\delta}^*$ which makes the target function $J(x_0)$ achieve the maximum with a condition of $\|u_0\|_\delta \leq \delta$, i.e.

$$J\left(u_{0\delta}^*\right) = \max_{\|u_0\|_\delta \leq \delta} J(u_0),$$
$$J(u_0) = \|M_\tau(U_0 + u_0) - M_\tau(U_0)\|_\delta \tag{1}$$

where $M_{t_0 \to t}$ is the propagator of a nonlinear model from the initial time to the prediction time t, $J(u_0)$ is the nonlinear evolution of the initial perturbation, u_0 is the initial perturbation, and δ is the size of uncertainty. Here, both $\|\cdot\|_\delta$ and $\|\cdot\|$ have the same meaning, the L^2 norm. For the convenience of optimization, Eq. (1) can also be converted into a minimum problem as follows:

$$J\left(u_{0\delta}^*\right) = \min_{\|u_0\|_\delta \leq \delta} -J(u_0) \tag{2}$$

2.2   Zebiak-Cane Model

The Zeabik-Cane (ZC) model [13] is treated as an air-sea coupled model with middle complexity for the Tropical Pacific, which is famous for the precise prediction of ENSO in 1986. It mainly consists of two sub-models: the atmospheric model and ocean model.

Atmospheric Model. The atmospheric model: The horizontal structure of the ZC atmospheric model can be described with the linear shallow-water equations of steady state in the equatorial β plane. Considering the moisture convergence feedback, each step of the iterative procedure in the ZC atmospheric model is dependent on the divergence of the previous step. The ZC atmospheric model is computed on the atmospheric grid in a region $101.25°E - 73.125°W$, $29°S - 29°N$.

Ocean Model. The ZC ocean model regards the mixing layer above the thermocline as being fluid, and the layer below the thermocline as being stationary. Due to the time evolution of the sea surface temperature (SST) determined by horizontal advection, upwelling and heat loss to the atmosphere, the mixing layer is thus divided into two layers, the surface and subsurface layers. The ZC ocean model is computed on the oceanic grid in a region $124°E - 80°W$, $28.75°S - 28.75°N$.

The whole region of the ZC coupled model is shown is Fig. 1. The inner rectangle in the blue dash means the integration region of SSTA for coupling the ocean and atmospheric models. The middle rectangle in the red long is the region of the ocean model. The outer rectangle of the solid line denotes the region of the atmospheric model.

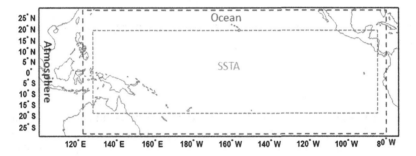

Fig. 1. The region of the ZC coupled model (Color figure online)

3 RGA-CNOP

In previous studies [14, 15], it was demonstrated that a driven dissipative system could run into a steady state composed of low-dimensional attractors after a long evolution. Based on this conclusion, we suppose that the CNOP can be projected on the space of the attractors, and principal information is preserved. Meanwhile, it is shown in the works [9], the attractors can be obtained through the feature extraction of a historical dataset. Thus we just utilize the genetic algorithm to search the optimal CNOP in the feature space.

The framework of the RGA-CNOP is described in Fig. 2. It mainly incudes two steps: the feature extraction and the genetic algorithm. Specifically, the feature extraction requires preprocessing the training dataset and then utilizing RPCA to obtain principal features. The genetic algorithm needs selecting, crossing and mutating operators. The output of the RGA-CNOP is an approximate CNOP.

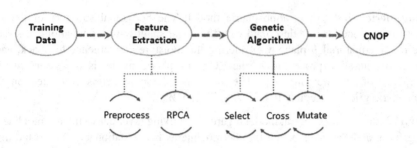

Fig. 2. The framework of RGA-CNOP

3.1   Feature Extraction

Suppose $x_1, x_2 \ldots x_n, x_i \in R^l$ $(i = 1, 2 \ldots n)$ are a training dataset taken from a long-run evolution of the models, composing a matrix $X \in R^{l \times n}$. Since each sample consists of two or more variables with different dimensions, we should preprocess these samples firstly. The dimensionless process aims to get rid of the influence of dimension to describe real physical laws better.

$$x_i = x_i/a \ (i = 1, \ldots n), \tag{3}$$

where $a = [a_1, a_2 \cdots a_l]^{\mathrm{T}}$ is a positive vector. And the centering makes the data average equal to zero:

$$x_i = x_i - \bar{x} \ (i = 1, \ldots n), \ \bar{x} = \frac{1}{n} \sum x_i \tag{4}$$

In the works [11, 13, 16], the PCA, RPCA and ONPP are all suited to the data of the ZC model. However, RPCA can show the robust character, especially in a dataset with grossly corrupted noises. The RPCA find a k-rank matrix B_0 that makes the following equation take the minimum:

$$\min_{A,B} \gamma * \|A\|_{l_1} + \|B\|_* \tag{5}$$
$$s.t. \ A + B = X$$

where A is a perturbation matrix, $\|A\|_{l_1} = \Sigma |A_{ij}|$ and $\|B\|_*$ means the sum of the singular values σ_i of B. It should be noted that the perturbation matrix, A, is sparse. As for the optimal problem like Eq. (5), we adopt the alternating direction method (ADM) to obtain the principal features, L. After getting the features, we can project the perturbations u_0 onto the feature space, so that the target function, Eq. (2) is converted into:

$$J(\omega) = \min -\|M_\tau(U_0 + L \cdot \omega) - M_\tau(U_0)\| \tag{6}$$
$$s.t. \ \|L \cdot \omega\|_\delta \leq \delta$$

where ω represents the coordinate of projection. Furthermore, the constraint, $\|L \cdot \omega\|_\delta \leq \delta$, can be further reduced as Eq. (7).

$$\|L \cdot \omega\| = (L \cdot \omega)^T (L \cdot \omega)$$
$$= \omega^T L^T L \omega \text{ (where } L^T L = I\text{)}$$
$$= \omega^T \omega$$
$$= \|w\|. \tag{7}$$

We put Eq. (7) into Eq. (6), and obtain the eventual target function:

$$J(\omega) = \min - \|M_\tau(U_0 + L \cdot \omega) - M_\tau(U_0)\|$$
$$s.t. \ \|\omega\|_\delta \leq \delta \tag{8}$$

3.2 Genetic Algorithm

With regard to the optimal problem with constraints like Eq. (8), the common approach is Lagrangian method. However, the CNOP is proved in mathematics that the optimal result always locates onto the boundary of domains [17]. So we devise a projection trick shown in Eq. (9), instead of Lagrangian method. If the current point goes out of the domain, it will be pulled back by the rule, which guarantees that all candidate points locate in the domain.

$$\omega = \begin{cases} \omega & \|\omega\| < = \delta \\ \frac{\delta}{\|\omega\|} \times \omega & \|\omega\| > \delta \end{cases} \tag{9}$$

After solving the constraint problem, we can utilize the genetic algorithm to solve the optimal problem, Eq. (8). As shown in Fig. 3, the algorithm includes six steps: initializing, calculating the fitness values, selecting operator, crossing operator, mutating operators and judging the termination conditions. The details are shown as follows:

- Initializing. Generate a number of particles randomly, each particle with k genes. If the particle w_i goes out of the domain, it should be pulled back by the Eq. (9).
- Calculating the fitness values. In the case of CNOP, the fitness values are computed by the Eq. (8).
- Selecting operator. In the process of biological evolution, the good genes are passed down by generations. Meanwhile, we do not eliminate all bad genes to keep the diversity of the swarm. Thus the strategy of the roulette is adopted here. As shown in Fig. 4, roll the roulette. The particle at the location where the pointer stops is selected. In particular, the probability of each particle selected is:

$$p_i = \frac{J_i}{\sum J_i} \tag{10}$$

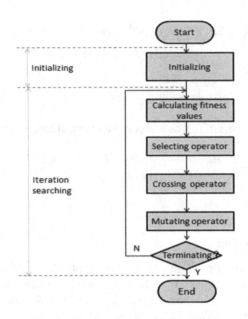

Fig. 3. The framework of RGA-CNOP

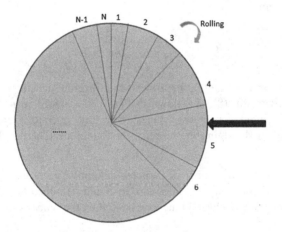

Fig. 4. The roulette

where J_i is the fitness value of the ith particle. The corresponding accumulative probability can be calculated as follows:

$$Acc_i = \sum_1^i p_i \tag{11}$$

Now, generate a random number r and_p. If $Acc_{i-1} \leq rand\_p < Acc_i(Acc_0 = 0)$, then the ith particle is selected. Furthermore, the particle with the best fitness is reserved into the next generation, called the elitist strategy.

- Crossing operator. Randomly select two particles, and then put them into the crossing equation, Eq. (12) where β is a positive parameter.

$$
\begin{aligned}
w_A^k &= \beta w_B^{k-1} + (1 - \beta) w_A^{k-1} \\
w_B^k &= \beta w_A^{k-1} + (1 - \beta) w_B^{k-1}
\end{aligned}
\tag{12}
$$

- Mutating operator. With the swarm evolving, new genes need to be generated with a probability. As for a gene of a particle, if the random probability η is less than the mutation parameter, the gene will mutate; otherwise, it will keep constant.
- Judging the termination. In view of the CNOP, the termination condition is set as the maximum iteration. If the iteration achieves the maximum, the program will end up.

4 Experiments

To demonstrate the feasibility of the RGA-CNOP, we apply it to the CNOP of the ZC model. The integration time span is 9 months. The training data are obtained through a 200-years evolution of the ZC model. To demonstrate the validity, the proposed method is compared with the ADJ-CNOP. Our experiments run on a server with 12 processors and a 64G memory.

4.1 Comparison Analysis

As the discussions in previous works [11, 12], the number of principal features has influence on the CNOP. So we take the 40 first features, the same with that in the PPSO method [11]. The measurement of the CNOP depends on the magnitude and the pattern of the CNOP. We check the magnitude of the CNOP firstly. As shown in Fig. 5, the red line denotes the RGA-CNOP, and the blue line means the ADJ-CNOP. The x-axis and y-axis represent the initial months and magnitudes, respectively. It is easy to find that two lines nearly keep the same trends. From January to February, the magnitude goes up. Then it goes down in the several months. After August, the magnitude rise again. Also, the magnitudes of the RGA-CNOP are smaller than the magnitudes of ADJ-CNOP, but their difference is not large. Therefore, the RGA-CNOP is approximate to ADJ-CNOP in the terms of the magnitude.

When discussing the magnitudes of the CNOP, we list all cases of 12 initial months. However, since a big magnitude always corresponds to a good pattern, we just need to check the pattern of the month with the smallest magnitude, August. As shown in Fig. 6, the left and right columns represent the RGA-CNOP and ADJ-CNOP, respectively. The three rows are the sea surface temperature anomaly (SSTA), thermocline height anomaly (THA) and SSTA evolution during the time span of 9 months. It can be discovered that the SSTAs have the same characters: in the equatorial areas, the left is negative while the right is positive, which is just the symbol of the El-Nino event. Also, the SSTA evolution of the RGA-CNOP is similar with the ADJ_CNOP, but the color is weaker than the ADJ_CNOP. However, the THA of the RGA_CNOP looks much smoother than the ADJ_CNOP. It is the reason that the RGA_CNOP just

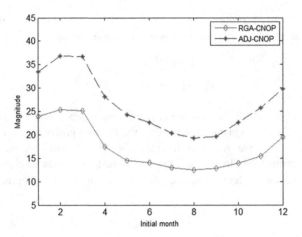

Fig. 5. The magnitudes of the CNOP. The x-axis and y-axis are the initial months and magnitudes, respectively. The red and blue lines denote the RGA-CNOP and the ADJ-CNOP (Color figure online).

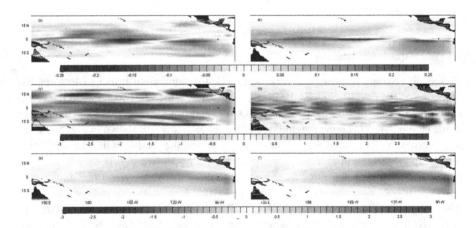

Fig. 6. The patterns of the CNOP. The three rows are the sea surface temperature anomaly (SSTA), thermocline height anomaly (THA) and SSTA evolution, respectively (Color figure online).

takes the 40 features, and ignores the remainders which maybe include some noise information.

Considering the magnitudes and patterns of the RGA_CNOP comprehensively, we can conclude that the RGA_CNOP can be treated as an approximate approach to the ADJ_CNOP.

4.2   Stability Analysis

Since the genetic algorithm is a stochastic method, the stability analysis is of significance. We repeat running 30 experiments of the RGA_CNOP, and check the distribution of the magnitudes. The initial month is still January. The results are demonstrated in Fig. 7. The x-axis and y-axis denote the magnitudes of the CNOP and the repeating times, respectively. It is easily found that all the results locate in the range of [12.0, 13.2]. There are 23 times appearing in the range of [12.6,13.2], and 15 times in a small range of [12.8,13.0]. That is to say, the RGA_CNOP performs well in a stable way. Therefore, the RGA_CNOP can always obtain an approximate CNOP to the ADJ_CNOP.

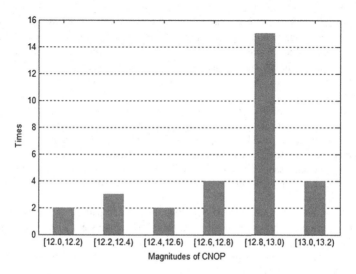

Fig. 7. Magnitude distribution. The x-axis represents the magnitude range of the CNOP, and the y-axis denotes the repeating times of CNOP.

5   Conclusion and Future Work

In this paper, we propose the RGA_CNOP for solving the CNOP. First of all, a training dataset is produce through a long evolution of the nonlinear models. Then RPCA is adopted to extract principal features. At last, search the optimal solution in the feature space. In the experiment, the comparison and stability analyses demonstrate that RGA_CNOP can obtain approximate results to the ADJ_CNOP. In addition, since the proposed method is free of adjoint models, it can expand the CNOP to the numerical models without adjoint models as well.

Acknowledgements. This work is supported by the National Nature Science Foundation of China (grant number 41405097).

References

1. Mu, M., Duan, W.S.: A new approach to studying ENSO predictability: conditional nonlinear optimal perturbation. Chin. Sci. Bull. **48**, 1045–1047 (2003)
2. Duan, W.S., Mu, M., Wang, B.: Conditional nonlinear optimal perturbation as the optimal precursors for El nino southern oscillation events. J. Geophys. Res. **109**, 1–12 (2004)
3. Duan, W.S., et al.: Behaviors of nonlinearities modulating El niño events induced by optimal precursory disturbance. Clim. Dyn. **40**, 1399–1413 (2012)
4. Mu, M., Zhou, F.F., Wang, H.L.: A method to identify the sensitive areas in targeting for tropical cyclone prediction: conditional nonlinear optimal perturbation. Mon. Weather Rev. **137**, 1623–1639 (2009)
5. Qin, X.H., Duan, W.S., Mu, M.: Conditions under which CNOP sensitivity is valid for tropical cyclone adaptive observations. Q. J. R. Meteorol. Soc. **139**, 1544–1554 (2013)
6. Mu, M., Duan, W.S.: Conditional nonlinear optimal perturbation and its applications to the studies of weather and climate predictability. Chin. Sci. Bull. **50**, 2401–2407 (2005)
7. Xu, H., Duan, W.S., Wang, J.C.: The tangent linear model and adjoint of a coupled ocean-atmosphere model and its application to the predictability of ENSO. In: IEEE International Conference on Geoscience and Remote Sensing Symposium, pp. 640–643 (2006)
8. Wang, B., Tan, X.W.: Conditional nonlinear optimal perturbations: adjoint-free calculation method and preliminary test. Am. Meteorol. Soc. **138**, 1043–1049 (2010)
9. Chen, L., Duan, W.S., Xu, H.: A SVD-based ensemble projection algorithm for calculating the conditional nonlinear optimal perturbation. Sci. China(D) (2014)
10. Zheng, Q., Dai, Y., Zhang, L., Sha, J.X., Lu, X.Q.: On the application of a genetic algorithm to the predictability problems involving "on-off" switches. Adv. Atmos. Sci. **29**(2), 422–434 (2012)
11. Mu, B., Wen, S.C., Yuan, S.J., Li, H.Y.: PPSO: PCA based particle swarm optimization for solving conditional nonlinear optimal perturbation. Computers & Geosciences. (in press)
12. Wen, S.C., Yuan, S.J., Li, H.Y., et al.: Robust ensemble feature extraction for solving conditional nonlinear optimal perturbation. International Journal of Computational Science and Engineering. (in press)
13. Zebiak, S.E., Cane, M.A.: A model El nino-southern osillation. Mon. Weather Rev. **115**, 2262–2278 (1987)
14. Osborne, A.R., Pastorello, A.: Simultaneous occurence of low-dimensional chaos and colored random noise in nonlinear physical systems. Phys. Lett. A **181**(2), 159–171 (1993)
15. Foias, C., Teman, R.: Structure of the set of stationary solution of the novier_stokes equations. Commun. Pure Appl. Math. **30**, 149–164 (1997)
16. Mu, B., Wen, S., Yuan, S.J., et al.: Orthogonal neighborhood preserving projection based method for solving CNOP. In: Proceedings of the 10th international conference on intelligent computing theory, pp. 120–126 (2014)
17. Liu, Y.M.: The optimal principle of condition nonlinear optimal perturbation. J. of East China Normal University, **2**, 131–134 (2008) (In Chinese)

Exploring Three Emotion Induction Procedures and Their Effect on E-learners' Language Learning

Zigang Ge[✉]

School of Network Education,
Beijing University of Posts and Telecommunications, Beijing, China
Shouzhou11@126.com

Abstract. This study explores the impact of positive, negative and neutral emotion induction procedures on Chinese adult e-learners' language learning. Thirty students from each of the three groups were selected as the subjects, with each group receiving one of the three treatments. The subjects attended some online lecturing sessions on English tenses and were assigned a pretest and a posttest. Data were also collected through simultaneous recording of the lecturing. The results of the posttest show that the positive treatment can generate the most facilitating impact on learning, the neutral one can also facilitate learning but not as much as the positive one, and the negative one may hinder learning.

Keywords: Emotion induction · Adult e-learners · Language learning · Online lecturing

1 Introduction

We human beings are capable of perceiving happiness, anger, sadness, bitterness, and many other emotions. These various emotions will inevitably affect our mental and behavioral actions. Among numerous human activities, learning is a very complex one, which involves too many mental and behavioral actions, so learning will also be affected by emotions.

It is generally acknowledged that positive emotions can facilitate learning while negative emotions will hinder learning [1, 2]. However, a few studies show that positive emotions can also bring negative effects [3, 4]. Some scholars think these negative effects can be explained by cognitive load theory that emotions gained in the learning experience may be an unnecessary load in working memory [5]. On the other hand, some scholars hold a balanced point of view, that is, both positive and negative moods may hinder or promote information processing [6].

This paper aims to explore the impact of three emotion induction procedures (positive, negative and neutral) on Chinese adult e-learners' language learning.

© Springer International Publishing Switzerland 2015
D.-S. Huang et al. (Eds.): ICIC 2015, Part I, LNCS 9225, pp. 607–614, 2015.
DOI: 10.1007/978-3-319-22180-9_60

2 Literature Review

Literature shows that learners' emotional state will influence their learning outcomes. For example, Isen's study (2000) indicates that positive moods can impose a happy feeling on learners, and make them show greater creativity and flexibility in solving problems and be more efficient in making decisions [7]. The research by Erez and Isen (2002) also shows that positive emotions will facilitate learners' intrinsic motivation towards learning [1]. While these studies show a positive correlation between positive emotions and learning performance, a few studies actually show that some negative effects can be brought in by positive emotions. For example, Seibert and Ellis' study (1991) shows that mood (happy or sad) can make students produce more task-irrelevant thoughts and thus exert a negative effect on their reasoning and performance [3]. On the other hand, some studies also show that positive results may be brought about by some not so positive emotions like confusion. A case in point is Craig et al.'s study (2004) about the AutoTutor group. This study indicates that learning performance is positively correlated with confusion [8].

Many emotion induction techniques have been identified; for example, a learner may listen to musical clips that can induce a particular emotional state, or watch some emotionally-charged films, or be presented with pictures, or be told some stories [9]. Even plain texts can be used to induce different moods on readers [10]. Various affective regulation methods have been explored in language teaching and learning. For example, an educational model called Community Language Learning, which was created by Charles A Curran, aims to remove learners' anxiety from learning by encouraging teachers to regard their learners as whole persons. This technique is usually used over a long period of time, until learners can move from a situation of dependence on teachers to a state of independence.

3 Research Questions

The present study tries to address the following questions by analyzing subjects' pretest and posttest scores, their responses to a questionnaire survey and the simultaneous recording of the lecturing sessions.

1: Can positive or negative emotions be induced on learners in an e-learning environment?

2: What effects do emotions have on e-learners' learning outcomes and perceptions of their learning experience?

4 Methods

4.1 Participants

90 adult e-learners were chosen as subjects of the study from a cyber education college in Beijing. They would attend a 12-h synchronous lecturing course. The course included four sessions, each of which would last three hours with two 10-min intervals.

They were randomly assigned to three groups marked A, B and C, with each group having 30 subjects. The three groups would attend the course in different time, but would be taught by the same teacher. The study had obtained consent from the subjects.

4.2 Procedure

The experiment was carried out in the second session of the course. In this session, after each knowledge point was explained, the instructor would ask several subjects to answer multiple choice questions. The instructor would try to induce neutral, positive and negative emotions on the subjects from Group A, Group B and Group C respectively. Every subject in each group would be asked at least one question. In the last hour of the session, an online posttest composed of 20 multiple choice questions about English tenses was organized. All the students should finish the test and return their answers in 30 min. A pretest was also organized in the previous session of the course to determine whether the subjects in the three groups were suitable for the experiment. The pretest also included 20 multiple choice questions about English tenses. All the teaching materials and tests were the same for the three groups. All the processes were recorded by the online teaching tool called Webex (you can find a demo in www.webex.com).

The study would try to induce different emotions by using different music and conversational patterns. Positive emotions were to be induced on the subjects of Group B and negative emotions on those of Group C. Positive emotions include excitement, satisfaction, hopefulness, curiosity, etc., and negative emotions include disappointment, confusion, frustration, etc. [11]. Two music clips were employed in the study, deemed as the positive music and the negative music respectively. The positive music is a recording of cheering and applauding sound, and the negative music is a recording of sighing. The positive music was only used for Group B, and the negative music was only for Group C.

As for Group C, the instructor would carefully choose words to bring about negative emotions like disappointment, confusion, and frustration but avoid making them feel humiliated or even quarrel with the instructor. Students could download and listen to the recorded lecturing of Group B as a remedy.

Neutral emotions would be induced on Group A subjects. The instructor would try to keep the subjects' mood in a moderate state, and no music was used.

5 Results and Data Analysis

5.1 Results of the Pretest

Subjects' scores of the pretest were put into SPSS for a descriptive analysis and an ANOVA test, which were shown by Tables 1, 2 and 3.

Table 1 shows that the mean differences among the three groups are very small, which indicates that on the whole the three groups were at the same proficiency level of English tenses. Compared with the total mark of 20, the means of the three groups are

Table 1. Descriptive analysis of the pretest

	N	Mean	Std. deviation	Std. error	Minimum	Maximum
Group A	30	6.2667	3.15062	.57522	2.00	14.00
Group B	30	6.6667	2.95172	.53891	1.00	12.00
Group C	30	6.3667	3.18924	.58227	2.00	15.00
Total	90	6.4333	3.06869	.32347	1.00	15.00

Table 2. ANOVA analysis of the pretest

			Sum of squares	df	Mean square	F	Sig.
Between groups	(Combined)		2.600	2	1.300	.135	.874
	Linear term	Contrast	.150	1	.150	.016	.901
		Deviation	. 2.450	1	2.450	.255	.615
Within groups			835.500	87	9.603		
Total			838.100	89			

just around 6, which shows that the overall proficiency level of the three groups was quite low.

The ANOVA test shown by Table 2 reveals that there are no significant differences among the three groups in their pretest scores (p value = 0.874 > 0.05). In other words, the three groups were at the same scratch line at the beginning of the experiment, and if there appeared significant differences in the results of the posttest, the differences might be caused by the three different emotion induction procedures.

5.2   Results of the Posttest

The scores of the posttest underwent the same processing as the pretest. Tables 3, 4 and 5 show the results.

The ANOVA test shown by Table 3 indicates that the p value is 0.010 < 0.05, which implies that among the three groups at least one group has shown significant differences with the other two. This implication is testified by the post hoc test shown by Table 4.

From Table 4, we can see that the p value between Group B and Group C is 0.002 < 0.05, and this means that there are significant differences between the scores of

Table 3. ANOVA analysis of the posttest

			Sum of squares	df	Mean square	F	Sig.
Between groups	(Combined)		126.289	2	63.144	4.881	.010
	Linear term	Contrast	35.267	1	35.267	2.726	.102
		Deviation	91.022	1	91.022	7.036	.009
Within groups			1125.500	87	12.937		
Total			1251.789	89			

Table 4. Multiple comparisons between the groups on the posttest

(I) Group	(J) Group	Mean difference (I-J)	Std. error	Sig.
Group A	Group B	−1.36667	.92868	.145
	Group C	1.53333	.92868	.102
Group B	Group A	1.36667	.92868	.145
	Group C	2.90000*	.92868	.002
Group C	Group A	−1.53333	.92868	.102
	Group B	-2.90000*	.92868	.002

Table 5. Descriptive analysis of the posttest

	N	Mean	Std. deviation	Std. error	Minimum	Maximum
Group A	30	8.4667	3.59821	.65694	4.00	16.00
Group B	30	9.8333	3.23860	.59128	3.00	15.00
Group C	30	6.9333	3.92106	.71588	1.00	18.00
Total	90	8.4111	3.75034	.39532	1.00	18.00

the two groups. Accordingly, we may infer that positive emotions and negative emotions will produce significant differences on e-learners' learning outcomes. Since the mean score of Group B is larger than that of Group C, we can safely conclude that positive emotions may bring much better learning outcomes than negative emotions. Table 4 also reveals that there are no significant differences between the scores of Group A and those of other two groups.

Compared with Table 1, Table 5 shows that all the three groups have increased their means after the emotional treatment, with Group B apparently increasing the most, Group C the least and Group A in between. The minimum and maximum scores of the three groups all increased except the minimum score of Group C. The above data indicate that positive emotions have produced the best results, while negative emotions would benefit little to the subjects. Neutral emotions can also improve subjects' learning a lot but not as much as positive ones.

5.3 Data Collected Through Simultaneous Recording

The lecturing session was simultaneously recorded by the online teaching tool called Webex. Hence, the recording also provides something useful for the study.

In Group B, most subjects responded to the emotional treatment by showing their gratitude or insisting on their choices. When the subjects were praised, all of them replied by saying "thank you" or "thanks". Most of them showed happiness in voice when hearing the positive music. This clearly shows their positive attitude toward the emotion induction procedure.

Things were quite different in Group C and Group A. Under the negative emotional treatment, Group C subjects tended to change their answers or keep silent, and most of their corrected answers were wrong. Compared with Group C subjects, Group A subjects showed more gratitude to the instructor and got more right answers when they changed their choices. Some students in the two groups would make apologies when their answers were wrong, but none in Group B. Most Group B students would show gratitude instead of making apologies.

In addition, 16 participants in Group B volunteered to answer the questions, compared with 2 volunteers in Group C and 8 in Group A. Group B participants posted 35 text messages, while Group C only posted 10 pieces and Group A posted 19. This shows that positive emotions can best activate learners' enthusiasm for learning.

6 Discussion

The result of the study shows that positive and negative emotional treatments could produce significant differences in e-learners' learning outcomes, while no significant differences were found in the outcome of the neutral treatment and that of the other two. This finding is consistent with the arguments of the aforementioned studies that learners' emotional state will influence their learning outcomes [1, 7], but contrary to the findings of some studies that positive and neutral emotions can also produce significant differences in learning [5]. The result of the posttest shows that the positive treatment has led to the best outcome in learning, and the neutral treatment can produce a better outcome than the negative treatment. On the other hand, although the positive treatment can produce a better outcome than the neutral one, the difference is not statistically significant.

The data collected through the simultaneous recording show that most subjects responded to the positive emotional treatment by showing their gratitude. This finding echoes Isen' study (2000) that positive moods can impose a happy feeling on learners [7]. The data also show that the learners tended to stick to their choices under positive emotional treatment while they would probably change their mind under negative treatment. This finding implies that positive emotions can make learners have more confidence in themselves and negative emotions tend to make learners doubt themselves. But on the other hand, most of those who stuck to their answers under the positive treatment actually got the wrong answers, which shows that positive emotions sometimes can also bring negative effects [3]; for example, they may be overconfident in their abilities and thus make unconsidered judgments.

In the case of Group C subjects, who received negative treatment, most of them would respond by keeping silent or abandoning their original choice. Among those who changed their mind most of them made the wrong choice. This reveals that negative emotional treatment can exert negative impact on one's reasoning. This finding is also testified by the result of the posttest, which shows that Group C subjects made the least progress among the three groups.

Things were quite different in Group A, which received neutral treatment. The subjects in Group A seemed to show a balanced reaction to the treatment. For example, 18 Group A subjects had been hesitating about their choice, but when they were inquired by the instructor about their next step, about half of them chose to stick to their original answer and half of them changed their answer. Something amazing happened here. No matter whether they changed their answer or not, most of them got the right answer. This possibly implies that the moderate emotional state is more suitable for learners to make a thoroughly considered choice. The reason may lies in the fact that the moderate emotional state can construct a more placid psychological environment for learning.

7 Conclusion

The findings of the study indicate that different emotion induction procedures can lead to different learning outcomes for e-learners. Positive emotional treatment normally will bring a better result than negative and neutral treatments while negative treatment will greatly hinder students' learning. Positive emotional treatment can make e-learners interpret their learning experience more positively and thus will facilitate their motivation in learning. Negative emotional treatment, however, often confuses or frustrates learners and learners tend to lose confidence in their abilities. Neutral emotional treatment can also produce a good result, as it may be good for the reasoning process. On the other hand, the study also shows that positive emotions can bring negative effects as well. Too much encouraging or praising may make learners overconfident in their abilities, which sometimes will lead to failure in reasoning. This implies that positive and neutral emotional treatments should be combined in teaching to produce the most desirable results.

There are some limitations of this study that need to be pointed out. First, the study bases its conclusion on the data from a three-hour online lecturing session, so it cannot be considered as a longitudinal study. The data may not fully reflect the real situations of students' learning. More definitive conclusions might have been drawn if the study had been conducted over more lecturing sessions. Second, the instructor tried to induce the same kind of emotions on 30 subjects in each lecturing session. This might have caused some unwanted effects on the subjects' reactions. What results would have come up if the three emotion induction procedures had been applied in each lecturing session? All these questions may need to be fully considered and addressed by further research.

Acknowledgements. This work was supported by the Beijing Higher Education Young Elite Teacher Project under Grant number: YETP0471.

References

1. Erez, A., Isen, A.M.: The influence of positive affect on the components of expectancy motivation. J. Appl. Psychol. **87**, 1055–1067 (2002)
2. O'Regan, K.: Emotion and e-learning. J. Asynchronous Learn. Netw. **7**, 78–92 (2003)
3. Seibert, P.S., Ellis, H.C.: Irrelevant thoughts, emotional mood states and cognitive task performance. Mem. Cogn. **19**, 507–513 (1991)
4. Oaksford, M., Morris, F., Grainger, B., Williams, J.M.G.: Mood, reasoning, and central executive process. J. Exp. Psychol. Lean. Mem. Cogn. **22**, 476–492 (1996)
5. Um, E.R., Song, H., Plass, J.: The effect of positive emotions on multimedia learning. In: Montgomerie, C., Seale, J. (eds.) Proceedings of World Conference on Educational

Multimedia, Hypermedia and Telecommunications, pp. 4176–4185. Chesapeake, AACE, VA (2007)

6. Brand, S., Reimer, T., Opwis, K.: How do we learn in a negative mood? Effects of a negative mood on transfer and learning. Learn. Instr. **17**, 1–16 (2007)

7. Isen, A.M.: Positive affect and decision making. In: Lewis, M., Haviland-Jones, J.M. (eds.) Handbook of Emotions, 2nd edn., pp. 417–435. Guilford Press, New York (2000)

8. Craig, S.D., Graesser, A.C., Sullins, J., Gholson, B.: Affect and learning: an exploratory look into the role of affect in learning with AutoTutor. J. Educ. Media **29**, 241–250 (2004)

9. Ghali, R., Frasson, C.: Emotional strategies for vocabulary learning. http://www.iro. umontreal.ca/~frasson/FrassonPub/ICALT-2010-Ghali-Frasson.pdf

10. Verheyen, C., Goritz, A.S.: Plain texts as an online mood-induction procedure. Soc. Psychol. **40**, 6–15 (2009)

11. Russell, J.A.: A circumplex model of affect. J. Pers. Soc. Psychol. **39**, 1161–1178 (1980)

Approximate Bit-Vector Algorithms for Hashing-Based Similarity Searches

Ling Wang[1(✉)], Tie Hua Zhou[2], Zhen Hong Liu[1],
Zhao Yang Qu[1], and Keun Ho Ryu[2(✉)]

[1] Department of Computer Science, School of Information Engineering,
Northeast Dianli University, Jilin, China
smile2867ling@gmail.com, ljnothingfree@hotmail.com,
qzywww@mail.nedu.edu.cn
[2] Database/Bioinformatics Laboratory,
School of Electrical and Computer Engineering,
Chungbuk National University, Chungbuk, Korea
{thzhou, khryu}@dblab.chungbuk.ac.kr

Abstract. Similarity search, or finding approximate nearest neighbors, is becoming an increasingly important tool to find the closest matches for a given query object in large scale database. Recently, learning hashing-based methods have attracted considerable attention due to their computational and memory efficiency. The basic idea of these approaches is to generate binary codes for data points which can preserve the similarity between any two of them. In this paper, we propose a novel algorithm named Approximate Bit-Vector (ABV) for hashing-based similarity search. ABV algorithm map data points into Hamming space and integrate with hash functions for fast similarity or k-NN search. Extensive experimental results over real large-scale datasets demonstrate the superiority of the proposed approach.

Keywords: Similarity search · Approximate nearest neighbors search · ABV algorithm · Hashing-based approach · Hamming space

1 Introduction

Similarity search, or finding approximate nearest neighbors, is a fundamental problem that makes efficient and fast information searching for a large scale database, has been applied to various tasks [1–3]. Given a collection of data objects D in a d-dimensional space R^d, construct a data structure with similarity measure defined between them and a query q, retrieve any object p from D that is similar to q according to the distance measure.

A lot of efficient tree-based index structures (e.g., R-tree, KD-tree) have been proposed to perform well for nearest neighbors (NN) search. Unfortunately, these approaches perform worse than a linear scan approach when the dimensionality of the space is high. Given the intrinsic difficulty of exact NN search, many hashing approaches are proposed for approximate NN search [4–11]. Particularly, efficient hashing-based methods have been widely applied to approximating NN search problem

© Springer International Publishing Switzerland 2015
D.-S. Huang et al. (Eds.): ICIC 2015, Part I, LNCS 9225, pp. 615–622, 2015.
DOI: 10.1007/978-3-319-22180-9_61

in many real applications, including image or audio retrieval, pattern recognition, information retrieval, local descriptor compression, machine learning, and many more.

The basic idea of hashing is to design compact binary codes for data points which can preserve the similarity between any two of them. More specially, each data point will be hashed into a compact binary code, and similar data in the original feature space should be hashed into close points in the binary hashing code space. These binary codes can be easily loaded into the memory in order to allow rapid retrieval of data samples. Moreover, the pairwise Hamming distance between these binary codes can be efficiently computed by using bit operations, which are well supported by modern processors, thus enabling efficient similarity calculation on large-scale datasets.

Based on this idea, we propose a novel algorithm named Approximate Bit-Vector (ABV), which map all data points into Hamming space by the weight of data feature, then integrate with the hash functions to construct buckets that all the data fall within a hamming distance of the binary code. Our contributions are summarized as follows: (1) We use feature weight of data point to support the component of vector generation; (2) We develop a novel ABV algorithm to map data vectors, which cooperating analysis for hashing techniques; (3) Extensive experiments on real datasets to demonstrate ABV is easily and nicely approach for approximate similarity search.

The organization of the paper is as follows. In Sect. 2, we provide a brief description of related work. In Sect. 3, we present the proposed ABV algorithm and analyze its properties for hashing-based similarity search. In Sect. 4, we evaluate the proposed approach with experimental results. Finally, we conclude this paper in Sect. 5.

2 Related Works

Similarity search is a fundamental problem in many search systems and also widely exists in many related application areas. The most basic but essential problem is the NN search. The majority of existing methods focus on approximate NN search that are based on the popular Locality Sensitive Hashing (LSH) algorithm. LSH is a prominent algorithm in the NN search literature [12]. Since its proposal, it has been extended and its performance has been improved for measure spaces. LSH and its variants [4, 7, 13] are the most commonly related to the problem of approximate NN search udder some measure distance.

There are a number of effective hashing methods have been developed which construct a variety of hash functions, mainly on the assumption that semantically similar data samples should have similar binary codes, such as random projection-based LSH [4], boosting learning-based Similarity Sensitive Coding (SSC) [14], and spectral hashing [9]. In this context, LSH functions are capable of creating hash keys maintaining the similarity existent between their input data, i.e., similar contents are mapped, with high probability, to similar hash identifiers [15]. Among the existent LSH functions, the Random Hyperplane Hashing (RHH) [16] is a family of LSH functions whose similarity corresponds to the cosine of the angle between vectors. According to [17], the use of the RHH function leads to a strong correlation between the cosine similarity of content vectors and the Hamming distance of their content identifiers, providing an efficient basis for the development of similarity search systems.

Recently, machine learning-based approaches have been proposed for converting the high-dimensional feature vectors used in document retrieval and machine vision (usually in Euclidean space) to low-dimensional binary vectors. For example, [18] propose a graphical model to learn a binary embedding called semantic hashing for efficient document retrieval. Semantic hashing defines the binary embedding of a document in such a way that the binary hash vector of similar documents (in terms of context) has a small Hamming distance. Spectral hashing [9] is another embedding technique that generates concise binary hashes. The binary hashes generated by the spectral hashing method are balanced, i.e., they have equal number of 0 s and 1 s. Weiss et al. [9] perform spectral hashing by thresholding a set of the eigenvectors corresponding to minimal eigenvalues of the Laplacian of the similarity graph. Chaudhry and Ivanov [19] extend spectral hashing to non-Euclidean manifolds. In [20], a binary embedding scheme is proposed that generates the hash vectors by solving an optimization problem that minimizes the error between the objects distance and the Hamming distance between their reconstructed hash vectors.

3 Proposed ABV for Similarity Search

3.1 Function Definition

Formally: Let $D = [p_1, \ldots, p_n]$ be a dataset of size n, and $D \in R^{n \times d}$. The k-NN of a data point $p \in R^d$ in the D can be denoted by $N_k(p)$. Given a query $q \in R^d$, we desire to find its k-NN $N_k(q)$ in the D. We adopt the Hamming distance $diff(\cdot, \cdot)$ as the distance measure between two points p and q: $d(p, q) = diff |p, q|$.

Definition 1 (Locality sensitive hashing [12]). Let h_H denote a random choice of a hash function from the family H. A family H of hash functions is called $(r, r(1 + \varepsilon), p_1, p_2)$–sensitive, for $d(\cdot, \cdot)$ when, for any two points $p, q \in D$, it satisfies the following:

$$\text{If } p \in R(q, r), Pr\left[h_H(q) \approx h_H(p)\right] \geq P_1 \tag{1}$$

$$\text{If } p \notin R(q, r(1 + \varepsilon)), Pr\left[h_H(q) \approx h_H(p)\right] \leq P_2 \tag{2}$$

For an LSH family to be useful, it has to satisfy $p_1 > p_2$. Given an LSH family H, the classic LSH algorithm works as follows: For each data objects $p \in D$, the bucket $g_j(p)$ is computed, for $j = 1, \ldots, l$, and then p is inserted into the corresponding bucket. To process a query q, one has to compute $g_j(q), j = 1, \ldots, l$ first, and then retrieve all points that lie in at least one of these buckets. For all candidate points retrieved, a filtering procedure is performed to calculate the distance of each point to the query.

3.2 ABV for Similarity Search

The basic idea of our work is: Assume two approximate bit-vectors that are close in the Hamming space, i.e., they differ only in few bits. If a bit-vector is extracted from each of them in the same order, then the probability of these bit-vectors being identical will be high. By the same order, we mean that, for example, if the first bit of one bit-vector corresponds to the ninth of one original space, then the first bit of the second bit-vector should correspond to the ninth bit of the second data point.

The proposed ABV algorithm for similarity search can be described as follows (as shown in Fig. 1). (1) Using ABV to map all data points into vectors space; (2) Building hash functions $h(.)$ to construct buckets that fall within a setting hamming radius. (3) Searching top k nearest candidates and expand them with their k-NN.

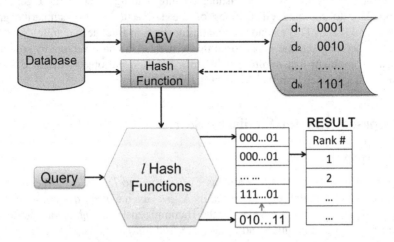

Fig. 1. Scenario for ABV similarity search

3.3   ABV Algorithm

ABV based on the sparse and dense component of vector in the high-dimensional space. And each data vector is independently rearranged followed by feature weight accordingly. The mapped data vectors $v' = (v'_1, \ldots, v'_m)$ in D satisfying the following conditions:

$$p_{v^*} = \{v_i^* \in p \mid v_1^*, v_2^* \ldots v_m^* = (v_i + v_{n+1-i})/2\} \tag{3}$$

where v is original data component of vector, and v^* is a mediator from v to v'. If the size of database is odd, the last component of vector v'_m in formulation (3) is $v_{(n+1)/2}$. And then assign the binary value to each data point according to difference value between v^* and mean value of v. If the difference is not less than 0, we assign 1 to component of vector; otherwise, 0. Then, ABV mapped all data points into Hamming space and represent the approximate bit-vector $\{0, 1\}^d$ for each data point. We summarize the ABV algorithm as shown in the following.

```
Require:
  |v|:weighting of v
  v*:mediator
  v :mean of v
Input:p=(v₁,v₂,......vₙ)
Output:p=(v₁',v₂',......vₘ')has mapped into {0.1}ᵈ
  for all data points {pᵢ|pᵢ∈D},do
  weighting |vᵢ|++
  sort v according to the weighting |v|
  vᵢ*←(vᵢ+vₙ₊₁₋ᵢ)/2
    for each refactoring v*
      if vᵢ*-vᵢ≥0
        vᵢ'←1
      else vᵢ←0
      end if
    end for
  end for
return p
```

For search problem, approximate bit-vectors accelerate the search process by using hash functions $h(.)$. By assigning each data point into l-bits hash buckets corresponding to $h(.)$. Given a query point q, we use the following three stages to perform the search: (1) coding stage: the query point is converted to a l-bits binary code using the l hash functions; (2) locating stage: all the data points in the buckets that fall within a hamming radius r of the binary code of the query are returned; and (3) linear scan stage: a linear scan over these points is conducted to return the required neighbors.

4 Experiments

In this section, we evaluate our ABV algorithm on large-scale real world datasets Reuters-21578 [21], which is a set of 21,578 news stories. Table 1 provides some important statistics of the datasets.

In the first set of experiments, we compare with Jaccard and Cosine similarity to report the similarity preserving as shown in Fig. 2. The result clearly shows the ABV keep the high quality similarity between data points in original space.

In the second set of experiments, the precision-recall curves of retrieved examples are studied in Fig. 3. We compare ABV algorithm with LSH [4] and KLSH [7] in terms of their precision-recall curves. There are six points in each curve, and each point means a precision-recall pair on a fixed bit length. Ideally, a similarity search system should be able to achieve high-quality search, which is measured by recall. Since the entire candidate objects will be ranked based on their Hamming distances to the query object and only the top k candidates will be returned.

Table 1. Statistics of Reuters-21578

Category set	# of Categories	# of w/1 + Occurrences	# of w/20 + Occurrences
EXCHANGES	39	32	7
ORGS	56	32	9
PEOPLE	267	114	15
PLACES	175	147	60
TOPICS	135	120	57

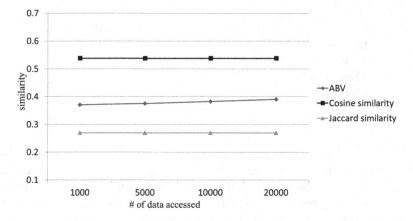

Fig. 2. Similarity preserve of ABV

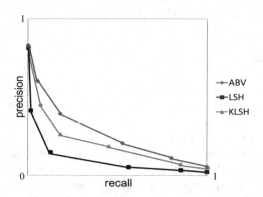

Fig. 3. Precision-recall curves

5   Conclusion

We proposed a new approximate NN search approach based on ABV in large scale database. Our approach is a classic approximation vector method based on the hashing techniques. We first map data points into $\{0, 1\}^d$, and then integrate with hash function for fast similarity search. The Experimental results on a real large-scale dataset show that the proposed ABV algorithm similarity preserving and achieves good performances on approximate similarity search.

For future work, we plan to compare with popular hashing-based approximate NN search methods and further explore more accurate hash-functions for our scheme.

Acknowledgments. This work was supported by the Science and Technology Plan Projects of Jilin city (No. 201464059), by the Ph.D. Scientific Research Start-up Capital Project of Northeast Dianli University (No. BSJXM-201319), and by Basic Science Research Program through the National Research Foundation of Korea (NRF) funded by the Ministry of Science, ICT & Future Planning (No. 2013R1A2A2A01068923).

References

1. Gao, Y., Zheng, B., Chen, G., Li, Q., Guo, X.: Continuous visible nearest neighbor query processing in spatial databases. VLDB **20**(3), 371–396 (2011)
2. Lin, Y., Jin, R., Cai, D., He, X.: Random projection with filtering for nearly duplicate search. In: Proceedings of the 26th AAAI Conference on Artificial Intelligence, pp. 641–647. AAAI Press, Toronto (2012)
3. Zhang, D., Yang, G., Hu, Y., Jin, Z., Cai, D., He, X.: A unified approximate nearest neighbor search scheme by combining data structure and hashing. In: Proceedings of the 23rd International Joint Conference on Artificial Intelligence, pp. 681–687. IJCAI Press, Beijing (2013)
4. Andoni, A., Indyk, P.: Near-optimal hashing algorithms for approximate nearest neighbor in high dimensions. Commun. ACM **51**(1), 117–122 (2008)
5. Jin, Z., Hu, Y., Lin, Y., Zhang, D., Lin, S., Cai, D., Li, X.: Complementary projection hashing. In: IEEE International Conference on Computer Vision, pp. 257–264. IEEE Press, Sydney (2013)
6. Jin, Z., Li, C., Lin, Y., Cai, D.: Density sensitive hashing. IEEE Trans. Cybern. **44**(8), 1362–1371 (2014)
7. Kulis, B., Grauman, K.: Kernelized locality-sensitive hashing. IEEE Trans. Pattern Anal. Mach. Intell. **34**(6), 1092–1104 (2012)
8. Lin, Y., Jin, R., Cai, D., Yan, S., Li, X.: Compressed hashing. In: IEEE Conference on Computer Vision and Pattern Recognition, pp. 446–451. IEEE Press, Portland (2013)
9. Weiss, Y., Torralba, A., Fergus, R.: Spectral hashing. In: Proceedings of the 22nd Annual Conference on Neural Information Processing Systems, pp. 1753–1760. Curran Associates Press, Vancouver (2008)
10. Wu, C., Zhu, J., Cai, D., Chen, C., Bu, J.: Semi-supervised nonlinear hashing using bootstrap sequential projection learning. IEEE Trans. Knowl. Data Eng. **25**(6), 1380–1393 (2013)
11. Xu, B., Bu, J., Lin, Y., Chen, C., He, X., Cai, D.: Harmonious hashing. In: Proceedings of the 23rd International Joint Conference on Artificial Intelligence, pp. 1820–1826. AAAI Press, Beijing (2013)
12. Gionis, A., Indyk, P., Motwani, R.: Similarity search in high dimensions via hashing. In: Proceedings of the 25th International Conference on Very Large Data Bases, pp. 518–529. Morgan Kaufmann Press (1999)

13. Yuan, P., Sha, C., Sun, Y.: Hashed-join: approximate string similarity join with hashing. In: Han, W.-S., Lee, M.L., Muliantara, A., Sanjaya, N.A., Thalheim, B., Zhou, S. (eds.) DASFAA 2014. LNCS, vol. 8505, pp. 217–229. Springer, Heidelberg (2014)

14. Shakhnarovich, G., Viola, P., Darrell, T.: Fast pose estimation with parameter-sensitive hashing. In: Proceedings of the 9th IEEE International Conference on Computer Vision, pp. 750–757. IEEE Press, Nice (2003)

15. Indyk, P., Motwani, R.: Approximate nearest neighbors: towards removing the curse of dimensionality. In: Proceedings of the 30th Annual ACM Symposium on Theory of Computing, pp. 604–613. ACM Press, New York (1998)

16. Charikar, M.: Similarity estimation techniques from rounding algorithms. In: Proceedings of the 34th Annual ACM Symposium on Theory of Computing, pp. 380–388. ACM Press, Montreal (2002)

17. Paula, L.B.D., Villaca, R.D.S., Magalhaes, M.F.: Analysis of concept similarity methods applied to an LSH function. In: Proceedings of the 35th Annual IEEE International Computer Software and Applications Conference, pp. 547–555. IEEE Press, Munich (2011)

18. Salakhutdinov, R., Hinton, G.: Semantic hashing. Approximate Reasoning 50(7), 969–978 (2009)

19. Chaudhry, R., Ivanov, Y.: Fast approximate nearest neighbor methods for non-euclidean manifolds with applications to human activity analysis in videos. In: Daniilidis, K., Maragos, P., Paragios, N. (eds.) ECCV 2010, Part II. LNCS, vol. 6312, pp. 735–748. Springer, Heidelberg (2010)

20. Kulis, B., Darrell, T.: Learning to hash with binary reconstructive embeddings. In: Proceedings of 23rd Annual Conference on Neural Information Processing Systems, pp. 1042–1050. Curran Associates Press, Vancouver (2009)

21. Lewis, D.D.: Reuters-21578 text categorization test collection. http://www.daviddlewis.com/resources/testcollections/reuters21578/

Dynamic Hand Gesture Recognition Using Centroid Tracking

Prashan Premaratne$^{(\boxtimes)}$, Shuai Yang, Peter Vial, and Zubair Ifthikar

School of Electrical Computer and Telecommunications Engineering,
Faculty of Engineering and Information Sciences, University of Wollongong,
North Wollongong, NSW, Australia
{prashan,peter_vial}@uow.edu.au,
{sy907,zi770}@uowmail.edu.au

Abstract. In many dynamic hand gesture recognition contexts, time information is not adequately used. The extracted features of dynamic gestures usually do not carry explicit information about time in gesture classification. This results in under-utilized data for more important accurate classification. Another disadvantage is that the gesture classification is then confined to only simple gestures. We have overcome these limitations by introducing centroid tracking of hand gestures that captures and retains the time sequence information for feature extraction. This simplifies the classification of dynamic gestures as movement in time helps efficient classification without burdensome processing.

Keywords: Dynamic hand gestures · Classification · Skin segmentation · Centroid tracking · Hidden Markov model

1 Introduction

Hand gesture recognition is considered to be in the forefront of Human Computer Interaction (HCI) which has opened a natural way to communicate with machines [1–5]. It would not only allow machines to interpret human intentions but also would allow variety of sign languages in the world to be understood and translated to multiple languages for the benefit of the mute populations across the world [1, 6]. Even though hand posture recognition (static hand gestures) has been reported to be very successful in HCI, dynamic gestures continue to have many obstacles for accurate recognition [1]. These issues coupled with poor lighting conditions, camera's inability to capture dynamic gesture in focus, occlusion due to finger movement, color variations due to lighting conditions have introduced myriad of obstacles in realizing a true HCI.

Dynamic gestures constitute hand postures (static hand gestures) coupled with hand and limb movements in order to communicate further information to others. However, due to hand and limb movements, many hand postures are occluded from the camera focus thereby loosing vital information. Since a single camera system cannot prevent occlusion, we have focused on the hand tracking without recognizing the hand posture.

Tracking using general tracking algorithms such as optical flow [7, 8] or Kalman filtering fails due to color changes of the object pixels as the hand moves and occlusion of parts of the hand as it moves. These limitations can somewhat be overcome by

© Springer International Publishing Switzerland 2015
D.-S. Huang et al. (Eds.): ICIC 2015, Part I, LNCS 9225, pp. 623–629, 2015.
DOI: 10.1007/978-3-319-22180-9_62

simplifying the process by using the centroid of the hand blob. This will result in easy tracking and is very robust to lighting or occlusion as the shape of the hand does not matter. Trajectory classification has exploited centroid in the past in order to understand human movement for surveillance [8]. The centroid of human 'blob' has been very effective as it is robust to changes in color and pixel variations. Cutler et al. has used the hand blob and its optical flow analysis to determine up to 7 gestures [9].

Another important feature of this measure is that the start of the dynamic gesture (for this research we used English alphabetical letters) or ending of one is determined by multiple overlapping centroids indicating that the hand is stationary. Another feature being that the centroid coordinate vector associated with each gesture (alphabetical letter) would indicate a timing sequence starting with 1. As a rule of thumb, we have used 5-s gestures where a complete gesture is issued in 5 s including stationary hand positions to indicate the start and the end. The vector will have values typically from 1 to 350 associated with a 5-s gesture as we have used 30 frames/second. The timing sequence will make the classification much easier using a Hidden Markov Model approach as certain gestures will start only from vertically upward motion and some will only use vertical downward, horizontal left to right or right to left motion as shown in the Fig. 1 below.

Fig. 1. Initial stroke (starting) direction of some alphabetical letters.

2 Centroid Tracking of the Hand Region

The following points highlights how the centroid tracking is implemented to separated gestures and its advantages in dynamic hand gesture recognition context:

- When there is no significant motion of the centroid, the system initiates the gesture and ends it when the centroid comes to a halt.
- The action in between is considered as a single dynamic gesture.
- The first frame of the gesture is (when movement starts from rest) is given a value one and increments (using integers) every movement of the centroid.
- When gestures are made by different individuals and the gesture path is slightly different and the timing is different, each dynamic gesture is normalized so it would be easy to apply template matching or hierarchical type classification.

As in any hand gesture recognition system, the detection of the hand region is achieved by skin detection which is very well known in computer vision [1–6]. For achieving this end, it is adequate to have any resolution that would detect a 'blob' for the hand region [10]. This is advantageous as having quite low resolution will still result in a good outcome with less computational burden on the processor. Skin segmentation is achieved when a RGB image is converted to YCbCr domain or Hue, Saturation and Value domain. However, the threshold values slightly differ under different lighting conditions (for **Incandescent**, $100 \leq Cb \leq 122$, $132 \leq Cr \leq 150$, for **Fluorescent**, $108 \leq Cb \leq 125$, $132 \leq Cr \leq 151$).

In situation where skin detected region contains more than one 'blob' due to the hand, morphological filtering is performed to remove unnecessary artifacts. These artifacts are usually due to background having color variations due to reflections but usually are smaller in size compared to the hand region. Using a structural element of appropriate size (usually smaller than the hand region and larger than the artifact) results in successful removal of such noise. Figure 2 shows color image with hand and the result of skin detection with undesirable 'holes' or artifacts in the hand region. Using morphological filtering this is removed as shown in the right image of Fig. 2.

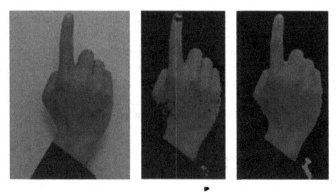

Fig. 2. Skin segmentation results and morphological processing (Color figure online)

Centroid of a binary (black and white) image region is simply the centre of mass of the hand region which is comprised of n pixels. The coordinates of the centroid is calculated as follows:

$$\bar{x} = \frac{\sum_{allhandpixels} x_i}{n}, \; \bar{y} = \frac{\sum_{allhandpixels} y_i}{n}$$

The system recognizes the initiation of a gesture when the centroid is stationary for few seconds followed by the movement and ending with centroid remaining stationary for few more seconds. The system can thus recognize individual gestures. When a gesture is started, the system adds weights to the centroid vector in its duration. For instance, once the gesture is recognized as starting from a stationary centroid, the next centroid values will have a weight such as 1,2,3,.. signifying the time sequence (as shown in Fig. 3). Since it is possible for the same user or different user to take different durations to make a sign (in our case, letters), the gesture length (duration) is normalized so that the weights are also normalized which will facilitate in later classification.

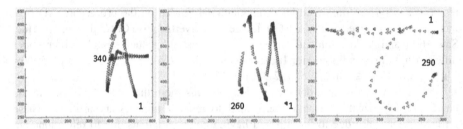

Fig. 3. The centroid of the hand region when making gestures of 'A', 'M' and 'J' (mirror effect). Here the multiple triangles indicate the centroid not moving due to hand coming to rest. Weights indicate start and the end position of the gesture as a frame number.

3   Hierarchical Classification Using Hidden Markov Models

Hidden Markov models effectively capture correlations in timing sequences. Since the dynamic gesture is a timing sequence, it can easily be classified using a HMM approach that we discuss next. HMMs are actively used in unsupervised learning environments along with expectation maximization algorithms such as Baum-Welch however, labeled observations provide by the dataset can be used to explicitly train our model [11]. But, we will use a supervised learning algorithm as we have acquired our own dataset with multiple gestures for each alphabetical letter.

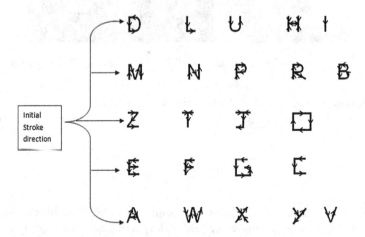

Fig. 4. Hierarchical classification strategy where the initial stroke direction dictates the possible classification candidates.

As shown in Fig. 4, hierarchical classification uses the initial stroke direction in classifying the gestures as one of five possible sequences. For instance, if the initial stroke direction is top to bottom, the gesture should be found in the top sequence (first row of Fig. 4). Then within this sequence, we use change of angle of initial stroke as a feature in a HMM classification.

As indicated in Fig. 4, initial stroke direction would allow the system to first determine the possible candidates for the stroke which is extremely useful in removing unnecessary candidates for the final HMM based classification. For instance, if the initial stroke direction is vertically upwards, the potential classification candidates will only be the second row from the top (M, N, P, R, B). Then the timing sequence will dictate what letter has been issued. For instance, M ends with final timing sequence reaching the bottom where as N results in the top. P will have its final sequence value in the middle with a horizontal right to left stroke. The R results in a final diagonal stroke (North West to South East).

Once the trajectory is determined using the tracking algorithm, features are extracted and used to compute the probability of each alphabetical letter (our gestures in this experiment) with HMM. A vector is created to describe those features which forms the input of the HMM.

HMM is a 3-step process where evaluation, decoding, and training are performed using Forward algorithm, Viterbi algorithm, and Baum-Welch algorithm, respectively [12–16]. The gesture models are trained using BW re-estimation algorithm and the numbers of states are set depending on how many bends or directional change is made in performing each letter. We used modified version of continuously variable duration HMM which is used by speech recognition for our model. This was selected as for instance as shown in Fig. 4, slight change of stroke direction would mean D, H, L, U or I and there is no probability that a D would result in I or any other letter. We believe that this model is appropriate for modeling-order-constrained time-series whose properties sequentially change over time similar to the attempt by Wang *et al.* [12]. Since the model has no backward path, the state index either increases or stays unchanged as time increases. After the training process by computing the HMM parameters for each type of gesture or alphabetical letter in this case, a given letter is recognized corresponding to the maximal likelihood of sixteen (we only recognize 16 letters) HMM models by using viterbi algorithm.

4 Experimental Results

We used a subset of the English alphabet containing 16 letters as it was determined that they would be easier to consistently sign for a video camera to record. For the entire experiment, we used 20 instances of hand signing of each letter. This process was repeated many times as we faced with many difficulties in obtaining skin segmented images due to background containing 'skin looking' regions. As described in Sect. 3, we used the initial stroke direction to determine the row of the appropriate level as shown in Fig. 4. This simplifies the entire problem. Change of direction of the stroke was calculated in order to determine the trajectory probabilities by directly calculating the angle of change using the centroid coordinates.

Then the extracted features were used in the HMM for final step of the classification. The training was accomplished by using Baum-Welch algorithm. The system testing was accomplished by signing each letter (out of the 16 letters) 20 times to determine the confusion matrix as shown in Fig. 5.

True Letter	Classification															
	D	M	N	Z	E	F	J	A	C	B	P	G	U	O	V	W
D	73													27		
M		58	42													
N		26	74													
Z				68			20							12		
E					74	26										
F					31	69										
J			10				82							8		
A								100								
C				5	3			76			16					
B		5	5							69	21					
P										21	79					
G									12			88				
U													57		30	13
O			5			6								89		
V													22		67	11
W													22		15	73

Fig. 5. Confusion matrix for 16 letters using the proposed approach

5   Conclusion

The tracking of hand signs was attempted as an experiment to determine how random hand signs can be recognized by a computer. Even though we had a mixture of positive and negative results as evident by the confusion matrix (Fig. 5), this is the first step in recognizing complex hand signs in a dynamic gesture environment. Hand postures or static gestures are much easier to interpret; yet, dynamic gestures provide the best hope of true man-machine interaction. We still believe that we have a long way to go in realizing the best strategy in tracking, feature extraction and final classification.

References

1. Premaratne, P.; Human Computer Interaction using Hand Gestures. Springer International Publisher, Singapore (2014)
2. Zou, Z., Premaratne, P., Monaragala, R., Bandara, N., Premaratne, M.: Dynamic hand gesture recognition system using moment invariants. In: Proceedings of the 2010 5th International Conference on Information and Automation for Sustainability, ICIAfS 2010, pp. 108–113 (2010)
3. Premaratne, P., Ajaz, S., Premaratne, M.: Hand gesture tracking and recognition system for control of consumer electronics. In: Huang, D.-S., Gan, Y., Gupta, P., Gromiha, M. (eds.) ICIC 2011. LNCS, vol. 6839, pp. 588–593. Springer, Heidelberg (2012)
4. Premaratne, P., Safaei, F., Nguyen, Q.: Moment invariant based control system using hand gestures. In: Huang, D.-S., Li, K., Irwin, G.W. (eds.) Intelligent Computing in Signal Processing and Pattern Recognition. LNCIS. Springer, Heidelberg (2006)
5. Premaratne, P., Nguyen, Q., Premaratne, M.: Human computer interaction using hand gestures. In: Huang, D.-S., McGinnity, M., Heutte, L., Zhang, X.-P. (eds.) ICIC 2010. CCIS, vol. 93, pp. 381–386. Springer, Heidelberg (2010)

6. Premaratne, P., Yang, S., Zou, Z., Vial, P.: Australian sign language recognition using moment invariants. In: Huang, D.-S., Jo, K.-H., Zhou, Y.-Q., Han, K. (eds.) ICIC 2013. LNCS, vol. 7996, pp. 509–514. Springer, Heidelberg (2013)
7. Saikia, P., Das, K.: Head gesture recognition using optical flow based classification with reinforcement of GMM based background subtraction. Int. J. Comput. Appl. **65–25**, 5–11 (2013)
8. Premaratne, P., Ajaz, S., Premaratne, M.: Hand gesture tracking and recognition system using Lucas-Kanade algorithms for control of consumer electronics. Neurocomputing **116**, 242–249 (2013)
9. Cutler, R., Turk, M.: View-based interpretation of real-time optical flow for gesture recognition. In: Proceedings of Third IEEE International Conference on Automatic Face and Gesture Recognition, pp. 416–421 (1998)
10. Nascimento, J.C., Figueiredo, A.T., Marques, J.S.: Trajectory classification using switched dynamical hiddn markov models. IEEE Trans. Image Process. **19**(5), 1338–1348 (2010)
11. Krevat, E., Cuzzillo, E.: Improving Off-line Handwritten Character Recognition with Hidden Markov Models. IEEE Transaction on Pattern Analysis and Machine Learning 33 (2006)
12. Wang, X., Xia, M., Cai, H., Gao, Y., Cattani, C.: Hidden Markov Models based Dynamic Hand Gesture Recognition. Math. Probl. Eng. **2012**, 1–11 (2012). doi:10.1155/2012/986134
13. Mouret, M., Solnon, C., Wolf, C.: Classification of images based on hidden Markov models. In: Seventh International Workshop on Content-Based Multimedia Indexing, 2009. CBMI 2009, pp-169–174 (2009)
14. Rabiner, L.R.: A tutorial on hidden Markov models and selected applications in speech recognition. Proc. IEEE **77**(2), 257–286 (1989)
15. Bayazit, M., Couture-Beli, A., Mori, G.: Real-time motion-based gesture recognition using the GPU. In: IAPR Conference on Machine Vision Applications (MVA2009), pp. 9–12 (2009)
16. Wang, S., Kim, H., Chen, H.: Contour tracking using centroid distance signature and dynamic programming method. In: 2009 International Joint Conference on Computational Sciences and Optimization, pp. 274–278 (2009)

Conditional Matching Preclusion Sets for an Mixed-Graph of the Star Graph and the Bubble-Sort Graph

Yunxia Ren and Shiying Wang[✉]

School of Mathematics and Information Science,
Henan Normal University, Xinxiang 453007, Henan, People's Republic of China
wangshiying@htu.cn

Abstract. The conditional matching preclusion number of a graph is the minimum number of edges, whose deletion results in a graph with no isolated vertices that has neither perfect matchings nor almost-perfect matchings. Any such optimal set is called an optimally conditional matching preclusion set. The conditional matching preclusion number is one of the parameters to measure the robustness of interconnection networks in the event of edge failure. The star graph and the bubble-sort graph are one of the attractive underlying topologies in a multiprocessor system. In this paper, we investigate a class of Cayley graphs which are combined with the star graph and the bubble-sort graph, and give all the optimally conditional matching preclusion sets for this class of graphs.

Keywords: Interconnection network · Graph · Cayley graph · Star graph · Bubble-sort graph · Conditional matching preclusion

1 Introduction

The basic feature of a multiprocessor system is to connect a number of processors and other devices by physical connections according to a pattern, so that they are able not only to process independently and simultaneously, but also to cooperate with each other to speed up the processing and improve the efficiency of calculation. With the deepening of the research on multiprocessor systems, scientists begin to realize that, science and engineering computing need not only computing speed, but also transmission speed. As the network bandwidth and the numbers of processors in multiprocessor systems increase, the cost of connecting processors increase dramatically. Therefore, achieving the functions of a multiprocessor system depends largely on the pattern by which the processors are connected. In the multiprocessor system, the pattern connecting the processors is called the interconnection networks (networks for short). The network of a multiprocessor system can be modeled by a graph, where the vertex set and the edge set correspond to the processors and the physical connections, respectively. On the contrary, a graph can be considered as a topological structure of a network. Therefore, a graph and a network are equivalent topologically.

A perfect matching in a graph is a set of edges such that every vertex is incident with exactly one edge in this set. An almost-perfect matching in a graph is a set of

D.-S. Huang et al. (Eds.): ICIC 2015, Part I, LNCS 9225, pp. 630–638, 2015.
DOI: 10.1007/978-3-319-22180-9_63

edges such that every vertex except one is incident with exactly one edge in this set, and the exceptional vertex is incident to none. So, if a graph has a perfect matching, then it has an even number of vertices; and if a graph has an almost-perfect matching, then it has an odd number of vertices. The matching preclusion number of a graph G, denoted by $mp(G)$, is the minimum number of edges whose deletion leaves the resulting graph without a perfect matching or almost-perfect matching. Any such optimal set of edges of size $mp(G)$ is called an optimal matching preclusion set. We define $mp(G) = 0$ if G has neither a perfect matching nor an almost-perfect matching. The research has begun since 2005 in work on the preclusion of a network. Brigham et al. in [3] investigated this problem. We refer the readers to [3] for details and additional references. This problem is further studied by Cheng et al. [4, 5, 14]. The following proposition is obvious.

Proposition 1.1. Let G be a graph with an even number of vertices. Then $mp(G) \leq \delta(G)$ where $\delta(G)$ is the minimum degree of G.

A matching preclusion set of a graph is trivial if all its edges are incident to a single vertex. In the event of a random link failure, it is very unlikely that all of the links incident to a single vertex fail simultaneously. Motivated by this, Cheng et al. in [6] defined the conditional matching preclusion number of a graph G, denoted by $mp_1(G)$, as the minimum number of edges whose deletion leaves a resulting graph with no isolated vertices and without a perfect matching or an almost perfect matching. Any such optimal set of edges of size $mp_1(G)$ is called an optimally conditional matching preclusion set. Define $mp_1(G) = 0$ if G has neither a perfect matching nor an almost perfect matching, or if G has no conditional matching preclusion set. For further works, see [6, 8, 9, 15–19]. It was observed in [6] that for a graph of even order, a basic obstruction to a perfect matching is the existence of a path uvw where the degrees of u and w are both one. So to produce such an obstruction set, one can pick any path uvw in the original graph and delete all the edges incident to either u or w but not v. We define $v_e(G) = \{d_G(u) + d_G(w) - 2y_G(u, w)$: there exists $v \in V(G)$ such that uvw is a path$\}$, where $d_G(.)$ is the degree function and $y_G(u, w) = 1$ if u and w are adjacent and 0 otherwise. For two disjoint vertex subsets X, Y of G, define $[X, Y] = \{xy \in E(G) : x \in X, y \in Y\}$. Let uvw be is a path in G. Then $[\{u\}, V(G) \setminus \{u, v\}] \cup [\{w\}, V(G) \setminus \{v, w\}]$ is called a trivial optimal conditional matching preclusion set. The following result is direct.

Proposition 1.2 [6]. Let G be a graph with an even number of vertices. Suppose every vertex in G has degree at least three. Then $mp_1 \leq v_e(G)$.

It was observed in [17] that for a graph of even order, the other basic obstruction to a perfect matching is the existence of a trail $vv_2v_3v_4v_5v_2$ where the degree of v are one, the degrees of v_3 and v_5 are both two. So to produce such an obstruction set, one can pick any trail $vv_2v_3v_4v_5v_2$ in the original graph, delete all the edges incident to either v, v_3 or v_5 but not v_2, and v_5, remain edges vv_2, v_2v_3, v_3v_4, v_4v_5, v_5v_2. We define $w_e(G) = \min \{d_G(v) + d_G(v_3) + d_G(v_5) - 5 - y_G(v, v_3) - y_G(v, v_5) - y_G(v_3, v_5)\}$. Let $vv_2v_3 v_4v_5v_2$ be is a trail in G. Then $[\{v\}, V(G) \setminus \{v_2, v_3, v_4, v_5\}] \cup [\{v_3\}, V(G) \setminus \{v_2, v_4, v_5\}] \cup [\{v_5\}, V(G) \setminus \{v_2, v_4\}]$ is called a trivial optimal conditional matching preclusion set by $w_e(G)$.

Let Q be a finite group, and let S be a spanning set of Q such that S has no identity element. Directed Cayley graph $Cay(S, Q)$ is defined as follow: its vertex set is Q, its arc set is $\{(g, gs) : g \in Q, s \in S\}$. Given $t \in S$, we call every arc in $\{(g, gt) : g \in Q\}$ a t-arc. If for each $s \in S$ we also have $s^{-1} \in S$, then we say that this Cayley graph is an undirected Cayley graph. Every Cayley graph in this paper is an undirected Cayley graph. Suppose that every element of S is a transposition. Then the permutation group generated by S is a subgroup of the symmetric group S_n, whose identity element is denoted by (1). It is easy to see that every undirected Cayley graph is vertex-transitive. The product $\sigma\tau$ of two permutations is the composition function τ followed by σ, that is, (1,2) (1,3) = (1,3,2). For terminology and notation not defined here we follow [11]. The transposition set could be illustrated by a simple graph, and the following concepts are introduced.

Let H be a simple connected graph whose vertex set is $\{1, 2, \ldots, n\}$ $(n \geq 3)$. Every edge of H can be considered as a transposition in S_n, and so the edge set of H corresponds to a transposition set S in S_n. In this sense, H is called a transposition simple graph. Cayley graph $Cay(S, <S>)$ is called the corresponding Cayley graph of H. By [1], $<S> = S_n$. When the transposition simple graph is a tree, it is called a transposition tree [1]. When the transposition simple graph is a path, the corresponding Cayley graph is called a bubble-sort graph [1]. When the transposition simple graph is a star, the corresponding Cayley graph is called a star graph [1].

Theorem 1.3 [18]. Let H be a simple connected graph with $n = |V(H)| \geq 3$. If H^1 and H^2 are two different labelled graph obtained by labelling H with $\{1, 2, \ldots, n\}$, then $Cay(H^1, S_n)$ is isomorphic to $Cay(H^2, S_n)$.

By Theorem 1.3, a simple connected graph H can be labelled properly. When $n \geq 4$, $Cay(S, S_n)$ can be decomposed into smaller $Cay(S*, S_{n-1})$'s as follows, where $S*$ is a spanning set of S_{n-1}. Given an integer p with $1 \leq p \leq n$, let H_i be the subgraph of $Cay(S, S_n)$ induced by vertices with i in the pth position for $1 \leq i \leq n$. We say $Cay(S, S_n)$ is decomposed along the pth position. When H is a transposition tree T, we assume that one vertex of degree one is labelled by n in T. If we decompose $Cay(S, S_n)$ along the last position, then H_i and $Cay(T - n, S_{n-1})$ are isomorphic. The edges whose end vertices in different H_i's are the cross-edges with respect to the given decomposition. We note that each vertex is incident to exactly one cross-edge and there are $(n - 2)!$ independent cross-edges between two different H_i's. For graph-theoretical terminology and notation not defined here we follow [2, 10]. The star graph and the bubble-sort graph B_n are one of the attractive underlying topologies for multiprocessor systems. In 2011, Wang et al. [18] gave the following theorem.

Theorem 1.4 [18]. Let $n \geq 3$. Then $mp_1(B_n) = 2n - 4$. When $n = 4$, optimal conditional matching preclusion sets of B_n are trivial by $w_e(G)$ or trivial. When $n \neq 4$, all optimal conditional matching preclusion sets of B_n are trivial.

In 2011, Cheng et al. [7] gave the following theorem.

Theorem 1.5. Let G be a Cayley graph obtained from a transposition generating tree on $\{1, 2, \ldots, n\}$ with $n \geq 7$. Then $mp_1(G) = 2n - 4$, and optimal conditional matching preclusion sets of G is trivial.

In this paper, we investigate a class of graphs $Cay(X_{n,m}, S_{n+m})$ which are constructed by combining the star graph with the bubble-sort graph, and give the following theorem.

Theorem 1.6. Let $n \geq 4$ and $m \geq 1$ be two integers. Then all optimally conditional matching preclusion sets of $Cay(X_{n,m}, S_{n+m})$ are trivial.

Theorem 1.6 gives that all optimally conditional matching preclusion sets of a class of Cayley graphs are trivial. Since the proof of above theorem is constructive, one can gives an algorithm.

2   Main Results

We start from some preparations.

Theorem 2.1 [13]. Let $k \geq 1$ be an integer. Then every k-regular bipartite graph has k edge-disjoint perfect matchings.

Theorem 2.2 [4]. Let T be a transposition tree of order $n \geq 4$. Suppose u and v are adjacent in $Cay(T, S_n)$. Then $Cay(T, S_n) - \{u, v\}$ has a Hamiltonian cycle.

Theorem 2.3 [12]. Let T be a transposition tree of order $n \geq 4$. If $F \subseteq E(Cay(T, S_n))$ with $|F| \leq n - 3$, then there exists a Hamiltonian path in $Cay(T, S_n) - F$ joining every pair of vertices that are in different parts of the graph.

Theorem 2.4 [18]. Let T be a transposition tree of order $n \geq 3$. Then the conditional matching preclusion number $mp_1(Cay(T, S_n)) = 2n - 4$.

Definition 2.1. A star with n vertices is denoted by X_n and the center vertex of X_n is labelled by 1. The corresponding Cayley graph of X_n is denoted by $Cay(X_n, S_n)$. Adding a new vertex $n + 1$ and an edge $(n, n + 1)$ to the star X_n, we obtain a transposition tree, denoted by $X_{n,1}$. The corresponding Cayley graph is denoted by $Cay(X_{n,1}, S_{n+1})$. Accordingly, we add m vertices, where the mth vertex is labelled by $n + m$, which is adjacent to the vertex $n + m - 1$. We obtain a transposition tree, denoted by $X_{n,m}$. The corresponding Cayley graph is denoted by $Cay(X_{n,m}, S_{n+m})$.

Theorem 2.5 [19]. Let $n \geq 4$ be an integer. Then $mp_1 Cay(X_{n,1}, S_{n+1}) = 2n - 2$ and all the optimally conditional matching preclusion sets for this class of graphs are trivial.

Let $Cay(X_{n,m}, S_{n+m})$ be defined as above, and let S_{n+m} be decomposed by the last position. For $v \in S_{n+m}$, suppose that the edges $e_1, e_2, \ldots, e_{m+n-2}, e_{m+n-1}$ of $Cay(X_{n,m}, S_{n+m})$ are incident with v, where e_{m+n-1} is an unique cross-edge. We have the following lemma.

Lemma 2.1. Let $n \geq 4$ and $m \geq 1$ be two integers. Then there exist $(m + n - 3)$ cycles of length 4 and one cycle of length 6 in $Cay(X_{n,m}, S_{n+m})$ such that every cycle contains a pair of edges e_{m+n-1} and e_i $(1 \leq i \leq m + n - 2)$.

When $n = 3$, by [17] all the optimally conditional matching preclusion sets of the bubble-sort graph B_3 are trivial. When $n = 4$, by [9] all the optimally conditional matching preclusion sets of the star graph S_4 are trivial. More generally, we have the following theorem.

Theorem 2.6. Let $n \geq 4$ and $m \geq 1$ be two integers. Then all optimally conditional matching preclusion sets of $Cay(X_{n,m}, S_{n+m})$ are trivial.

Proof The proof is by induction on m. When $m = 1$, by [19] all the optimally conditional matching preclusion sets of $Cay(X_{n,m}, S_{n+m})$ are trivial.

When $m = k$, assume that all the optimally conditional matching preclusion sets of $Cay(X_{n,m}, S_{n+m})$ are trivial. We will prove that all the optimally conditional matching preclusion sets of $Cay(X_{n,m}, S_{n+m})$ are trivial when $m = k + 1$. Let F be any optimally conditional matching preclusion set of $Cay(X_{n,k+1}, S_{n+k+1})$. Then by Theorem 2.4 $|F| = 2(n + k) - 2$. For convenience, each of F is called a faulty edge. Each of $E(Cay(X_{n,k+1}, S_{n+k+1})) \setminus F$ is called a healthy edge. Let S_{n+m} be decomposed by the last position and let H_i be defined as above. It is easy to see that all cross-edges are a perfect matching of $Cay(X_{n,k+1}, S_{n+k+1})$. It is denoted by M. Because there is at least a cross-edge in F, $|F \cap E(H_i)| \leq 2(n + k) - 3$. For every i, H_i is isomorphic to $Cay(X_{n,k}, S_{n+k})$, which is an $(n + k - 1)$-regular bipartite graph. If for every i, $|F \cap E(H_i)| \leq 2(n + k) - 2$, then by Theorem 2.1 $H_i - F$ has a perfect matching. So $Cay(X_{n,k+1}, S_{n+k+1}) - F$ has a perfect matching, a contradiction. Thus, we will discuss that faulty cross-edges $|F \cap M| \leq n + k - 1$ and there is a H_{i0} such that $|F \cap E(H_{i0})| \leq 2(n + k) - 1$ as follows. For the following cases, we prove only the case 2. Similarly, we can also prove the case 1.

Case 1. Faulty cross-edges are one ($|F \cap M| = 1$).

Case 2. Faulty cross-edges are at least two ($2 \leq |F \cap M| \leq n + k - 1$).

In this case, this is also possible for $|E(H_i) \cap F| = 2(n + k) - 4$. Thus, we further assume

(1). $|E(H_i) \cap F| = 2(n + k) - 5$ for every i.

In this case, by Theorem 2.4 $E(H_i) \cap F$ is not a conditional matching preclusion set of H_i for every i. Thus $H_i - F$ either has a perfect matching or $H_i - F$ has an isolated vertex, for each i. Because F is a conditional matching preclusion set of $Cay(X_{n,k+1}, S_{n+k+1})$, there exists at least an $(H_i - F)$ which has no perfect matching. It follows that $H_i - F$ has an isolated vertex. In order for $H_i - F$ to have an isolated vertex, we must have $|E(H_i) \cap F| \geq n + k - 1$. Because $Cay(X_{n,k+1}, S_{n+k+1})$ has at least two faulty cross-edges, at most one of these $(H_i - F)$'s has an isolated vertex. Hence we may assume that exactly one of them has an isolated vertex. For notational convenience, we may assume that it is $H_1 - F$. Since H_1 is an $(n + k - 1)$-regular bipartite graph and $|E(H_1) \cap F| \leq 2(n + k) - 5$, $H_1 - F$ has exactly an isolated vertex. Suppose that v is the isolated vertex. Let $e_1, e_2, \ldots, e_{n+k-1}$ be the edges incident with v in H_1 and e_{n+k} be the cross-edge incident with v. Since $Cay(X_{n,k+1}, S_{n+k+1}) - F$ has no isolated vertex, $e_{n+k} = vv_2$ is healthy. Let $F_1 = (F \cap E(H_1)) - e_j$ ($1 \leq j \leq n + k - 1$). Then $|F_1| \leq 2(n + k) - 6$. By Theorem 2.4, F_1 is not a conditional matching preclusion set of H_1. Combining this with the fact that $H_1 - F_1$ has no isolated vertex, it follows that $H_1 - F_1$ has a perfect matching M_1 and $e_j \in M_1$. v_2 is a vertex of another H_i, say H_2 (for notational simplicity). By Lemma 2.1, there exist $(n + k - 1)$ cycles C_j ($j = 1, 2, \ldots, n + k - 1$) such that C_j contains e_j and e_{n+k} for all j, $(n + k - 2)$ cycles of which are 4-cycles. Note that cross-edges have at most $(n + k - 1)$ faulty edges. We consider the following two cases.

Case 2.a. There exists one of the $(n + k - 1)$ cycles, the cross-edges of which are healthy.

In this case, let the cycle be a 6-cycle $vv_2v_3v_4v_5v_6v$, where $v, v_6 \in V(H_1)$, $v_2, v_3 \in V(H_2)$ and v_4v_5 is an edge of another H_i, say H_3 (for notational simplicity). Note that $2(n+k) - 2 - (n+k-1) - 2 = n+k-3$. Because H_2 has at most $(n+k-3)$ faulty edges, by Theorem 2.3 there exists a Hamiltonian path in $H_2 - F$ joining v_2 and v_3. Thus $(H_2 - F) - \{v_2, v_3\}$ has a perfect matching M_2. Similarly, $(H_3 - F) - \{v_4, v_5\}$ has a perfect matching M_3. Since $|F \cap E(H_i)| \leq 2(n+k) - 2 - (n+k-1) - 2 = n+k-3$ for $i \geq 4$, by Theorem 2.1 $H_i - F$ has a perfect matching M_i. M_1 is given as above. Therefore, $((M_1 \cup M_2 \cup \cdots \cup M_{n+k+1}) \setminus \{v_6, v\}) \cup \{vv_2, v_3v_4, v_5v_6\}$ is a perfect matching of $Cay(X_{n,k+1}, S_{n+k+1}) - F$, a contradiction. Now, let the cycle be a 4-cycle $vv_2v_3v_4v$, where $v, v_4 \in V(H_1)$, $v_2, v_3 \in V(H_2)$. Note that $2(n+k) - 2 - (n+k-1) - 2 = n+k-3$. Because H_2 has at most $(n+k-3)$ faulty edges, by Theorem 2.3 there exists a Hamiltonian path in $H_2 - F$ joining v_2 and v_3. Thus $(H_2 - F) - \{v_2, v_3\}$ has a perfect matching M_2. Since $|F \cap E(H_i)| \leq 2(n+k) - 2 - (n+k-1) - 2 = n+k-3$ for $i \geq 3$, by Theorem 2.1 $H_i - F$ has a perfect matching M_i. M_1 is given as above. Therefore, $((M_1 \cup M_2 \cup \cdots \cup M_{n+k+1}) \setminus \{v_4, v\}) \cup \{vv_2, v_3v_4\}$ is a perfect matching of $Cay(X_{n,k+1}, S_{n+k+1}) - F$, a contradiction.

Case 2.b. There exist faulty cross-edges in each of the $(n+k-1)$ cycles.

In this case, each cycle must have exactly one faulty cross-edge. Note that $Cay(X_{n,k+1}, S_{n+k+1})$ has $(n+k-1)$ faulty cross-edges and H_1 has $(n+k-1)$ faulty edges. This implies that H_i has no faulty edges for $i \geq 2$. Without loss of generality, we may suppose that the $(n+k-1)$ cycles contain some edges of H_1, H_2 and H_3. Since $n \geq 4$ and $k \geq 1$, H_4 has no edges or vertices of the $(n+k-1)$ cycles. Recalling that there are $(n+k-1)!$ cross-edges between two different H_i's, we have that there are $(n+k-1)!/2$ cross-edges between H_1 and H_4, the ends of which and v belong to different partite sets. Similarly, there are $(n+k-1)!/2$ cross-edges between H_2 and H_4, the ends of which and v_2 belong to different partite sets. Since $n \geq 4$ and $k \geq 1$, there are a healthy cross-edge such that $x \in V(H_1)$, $y \in V(H_4)$, and v, x belong to different partite sets. Similarly, there are a healthy cross-edge such that $z \in V(H_4)$, $w \in V(H_2)$, and v_2, w belong to different partite sets. It is easy to see that y and z belong to different partite sets. By Theorem 2.3, there exists a Hamiltonian path P_1 in H_1 joining v and x. Note that the edge incident with v only is faulty in P_1. Thus $(H_1 - F) - \{v, x\}$ has a perfect matching M_1. Since H_2 has no faulty edges, $(H_2 - F) - \{v_2, w\}$ has a perfect matching M_2. Similarly, $(H_4 - F) - \{y, z\}$ has a perfect matching M_4. Since H_i has no faulty edges for $i = 3$ and $i \geq 5$, by Theorem 2.1 $H_i - F$ has a perfect matching M_i. Therefore, $(M_1 \cup M_2 \cup \cdots \cup M_{n+k+1}) \cup \{vv_2, xy, wz\}$ is a perfect matching of $Cay(X_{n,k+1}, S_{n+k+1}) - F$, a contradiction.

Now, if the condition (1) is satisfied and the faulty cross-edges $|F \cap M| \geq 2$, then it leads to a contradiction that F is a conditional matching preclusion set. Thus, suppose the faulty cross-edges $|F \cap M| \geq 3$. Then $|F \cap E(H_i)| \leq 2(n+k) - 2 - 3 = 2(n+k) - 5$ for every i, satisfying Condition (1), which will lead to a contradiction. So $|F \cap M| = 2$. One of the H_i's has $2(n+k) - 4$ faulty edges, i.e., $|E(H_i) \cap F| = 2(n+K) - 4$. This can occur only once. For notational simplicity, assume it is H_1. Note that all faulty edges are given. Since H_i $(i = 2, 3, \ldots, n+k+1)$ has no faulty edges, H_i has a perfect matching M_i. If $H_1 - F$ has a perfect matching, $Cay(X_{n,k+1}, S_{n+k+1}) - F$ has a perfect matching, a contradiction. Thus, suppose that $H_1 - F$ has no perfect matching. We consider two cases.

Case 2.A. $H_1 - F$ has an isolated vertex.

Suppose that v is the isolated vertex in H_1. Since $Cay(X_{n,k+1}, S_{n+k+1}) - F$ has no isolated vertex, the cross-edge vv_2 incident with v is healthy. v_2 is a vertex of another H_i, say H_2 (for notational simplicity). Let L be the faulty edges incident with v in H_1 and let $F' = (F \cap E(H_1)) - F$. Then $|F'| = 2(n+k) - 4 - (n+k-1) = n+k-3$ in H_1. By Lemma 2.1, there exist $(n+k-2)$ cycles of length 4, which contain vv_2. Since $n \geq 4$, $k \geq 1$, and $Cay(X_{n,k+1}, S_{n+k+1})$ has two faulty cross-edges, there exist a 4-cycle $vv_2v_3v_4v$, the cross-edges of which are healthy. By Theorem 2.3, there exists a Hamiltonian path P_1 in $H_1 - F'$ joining v and v_4. Note that the edge incident with v only is faulty in P_1. Thus, $(H_1 - F) - \{v, v_4\}$ has a perfect matching M_1. Since H_2 has no faulty edges, $(H_2 - F) - \{v_2, v_3\}$ has a perfect matching M_2. Because H_i has no faulty edges for $i \geq 3$, by Theorem 2.1 $H_i - F$ has a perfect matching M_i. Therefore, $(M_1 \cup M_2 \cup \cdots \cup M_{n+k+1}) \cup \{vv_2, v_3v_4\}$ is a perfect matching of $Cay(X_{n,k+1}, S_{n+k+1}) - F$, a contradiction.

Case 2.B. $H_1 - F$ has no isolated vertex.

Since $H_1 - F$ has no perfect matchings and isolated vertices, $F \cap E(H_1)$ is a conditional matching preclusion set of H_1. Because $X_{n,k}$ is a transposition tree and H_1 is isomorphic to $Cay(X_{n,k}, S_{n+k})$, by Theorem 2.4 $mp_1(H_1) = 2(n+k) - 4$. Since $|F \cap E(H_1)| = 2(n+k+1) - 6 = 2(n+k) - 4$, $F \cap E(H_1)$ is an optimally conditional matching preclusion set of H_1. By the induction hypothesis, the optimally conditional matching preclusion set $F \cap E(H_1)$ is trivial. Thus there is a path $z_1z_2z_3$ in H_1 such that the edges incident with z_1 have $(n+k-2)$ faulty edges and the edges incident with z_3 have also $(n+k-2)$ faulty edges. Let the vertices adjacent to z_1 be $y_1, y_2, \ldots, y_{n+k-2}$ except z_2 and the vertices adjacent to z_3 be $t_1, t_2, \ldots, t_{n+k-2}$ except z_2. If the two faulty cross-edges are respectively incident with z_1 and z_3, then the all faulty edges is a trivial conditional matching preclusion set of $Cay(X_{n,k+1}, S_{n+k+1})$. Otherwise, without loss of generality, let the cross-edge z_1z_1' incident with z_1 be healthy. Since H_1 is a regular bipartite graph, by Theorem 2.1 H_1 has a perfect matching M_1 which contains z_2z_3. One of $z_1y_1, z_1y_2, \ldots, z_1y_{n+k-2}$ belongs to M_1. Without loss of generality, we may suppose that $z_1y_1 \in M_1$. Note that z_1y_1 is faulty in M_1. Assume that the cross-edge y_1y_1' is healthy. y_1' is a vertex of another H_i, say H_2 (for notational simplicity). Let $z_1', y_1' \in H_2$. Then z_1' and y_1' are adjacent in H_2. Now, $z_1y_1y_1'z_1'z_1$ is a 4-cycle. By Theorem 2.2, $H_2 - \{y_1', z_1'\}$ has a Hamiltonian cycle $\bar{C}_2$. Since H_2 has no faulty edges, C_2 contains a perfect matching M_2 of $H_2 - \{y_1', z_1'\}$. Since H_i has no faulty edges for $i \geq 3$, by Theorem 2.1 $H_i - F$ has a perfect matching M_i. Therefore, $(M_1 \setminus z_1y_1) \cup (M_2 \cup \ldots \cup M_{n+k+1}) \cup \{z_1z_1', y_1y_1'\}$ is a perfect matching of $Cay(X_{n,k+1}, S_{n+k+1}) - F$, a contradiction. Let $z_1' \in V(H_i)$ and $y_1' \in V(H_j)$ $(i \neq j)$. z_1' is a vertex of another H_i, say H_2. y_1' is a vertex of another H_i, say H_3 (for notational simplicity). Recalling that there are $(n+k-1)!$ cross-edges between two different H_i's, we have that there are $(n+k-1)!/2$ cross-edges between H_2 and H_3, the ends of which and z_1' belong to different partite sets. Since $Cay(X_{n,k+1}, S_{n+k+1})$ has two faulty cross-edges, and $n \geq 4$, $k \geq 1$, there is a healthy cross-edge r_2r_1 between H_2 and H_3, the ends of which and z_1' belong to different partite sets. Let $r_1 \in V(H_2)$, $r_2 \in V(H_3)$. It is easy to see that r_2 and y_1' belong to different partite sets. By Theorem 2.3, there exists a Hamiltonian path P_1 in $H_2 - F$ joining r_1 and z_1'. Thus $H_2 - \{r_1, z_1'\}$ has a perfect matching M_2. Similarly,

$H_3 - \{r_2, y_1'\}$ has a perfect matching M_3. Since H_i has no faulty edges for $i \geq 4$, by Theorem 2.1 $H_i - F$ has a perfect matching M_i. Therefore, $(M_1 \backslash z_1 y_1) \cup M_2 \cup \ldots \cup M_{n+k+1} \cup \{z_1 z_1', y_1 y_1', r_1 r_2\}$ is a perfect matching of $Cay(X_{n,k+1}, S_{n+k+1}) - F$, a contradiction. Next, assume that $y_1 y_1'$ is faulty and the cross-edge incident with z_3 is healthy. By Theorem 2.1 H_1 has a perfect matching M_1 which contains $z_1 z_2$. One of $z_3 t_1, z_3 t_2, \ldots, z_3 t_{n+k-2}$ belongs to M_1. Without loss of generality, we may suppose that $z_3 t_1 \in M_1$. Let the cross-edge incident with t_1 is healthy. Similar to the proof of the above paragraph, we can obtain a perfect matching of $Cay(X_{n,k+1}, S_{n+k+1}) - F$, a contradiction. Thus, let one of the cross-edges incident with z_3 and t_1 be faulty. Note that $y_1 y_1'$ is faulty. It follows that the cross-edge $y_2 y_2'$ is healthy. Recalling that $z_3 t_1, z_3 t_2, \ldots, z_3 t_{n+k-2}$ are faulty in H_1, we let $F* = \{z_3 t_2, \ldots, z_3 t_{n+k-2}\}$. Then $|F*| = n + k - 3$. Combining this with the fact that $n \geq 4$, $k \geq 1$, it follows that by Theorem 2.3 there exists a Hamiltonian path P_1 in $H_1 - F*$ joining z_1 and y_2. Since $d(z_3) = 2$ in $H_1 - F*$, P_1 contains $z_2 z_3 t_1$. $P_1 + z_1 y_2$ is a Hamiltonian cycle in $H_1 - F*$. Note that there are two faulty edges in $P_1 + z_1 y_2$. It is to see that $P_1 + z_1 y_2$ contains two perfect matching M_1, M_2 of H_1. If $z_1 y_2, z_3 t_1 \in M_1$, then $H_1 - F$ has a perfect matching, a contradiction. Without loss of generality, we may suppose that $z_1 y_2 \in M_1$ and $z_3 t_1 \in M_1$. By Lemma 2.1, there is a 4-cycle or 6-cycle which contains $z_1 z_1', y_2 y_2', z_1 y_2$. If the cycle is a 4-cycle $z_1 y_2 y_2' z_1' z_1$, then y_2' and z_1' are vertices of another H_i, say H_2 (for notational simplicity). Since H_2 has no faulty edges, by Theorem 2.2 there exists a Hamiltonian path P_2 in $H_2 - F$ joining y_2' and z_1'. Thus, $(H_2 - F) - \{y_2', z_1'\}$ has a perfect matching M_2. Because H_i has no faulty edges, by Theorem 2.1 $H_i - F$ has a perfect matching M_i. Therefore, $(M_1 \backslash z_1 y_2) \cup M_2 \cup \ldots \cup M_{n+k+1} \{z_1 z_1', y_2 y_2'\}$ is a perfect matching of $Cay(X_{n,k+1}, S_{n+k+1}) - F$, a contradiction. Let the cycle be a 6-cycle. Similar to the proof of the above paragraph, we can obtain a perfect matching of $Cay(X_{n,k+1}, S_{n+k+1}) - F$, a contradiction. The proof is complete.

By Theorems 2.4 and 2.6, we have the following theorem.

Theorem 2.7. Let $n \geq 4$ and $m \geq 1$ be two integers. Then $mp_1(Cay(X_{n,m}, S_{n+m})) = 2(n + m) - 4$ and all optimally conditional matching preclusion sets of $Cay(X_{n,m}, S_{n+m})$ are trivial.

Acknowledgements. This work is supported by the National Science Foundation of China (61370001, U1304601)

References

1. Akers, S.B., Krishnamurthy, B.: A group-theoretic model for symmetric interconnection networks. IEEE Trans. Comput. **38**(4), 555–566 (1989)
2. Bondy, J.A., Murty, U.S.R.: Graph Theory. Springer, New York (2007)
3. Brigham, R.C., Harary, F., Violin, E.C., Yellen, J.: Perfect-matching preclusion. Congr. Numerantium **174**, 185–192 (2005)
4. Cheng, E., Lipták, L.: Matching preclusion for some interconnection networks. Networks **50** (2), 173–180 (2007)

5. Cheng, E., Lesniak, L., Lipman, M.J., Lipták, L.: Matching preclusion for alternating group graphs and their generalizations. Int. J. Found. Comput. Sci. **19**(6), 1413–1437 (2008)
6. Cheng, E., Lesniak, L., Lipman, M.J., Lipták, L.: Conditional matching preclusion sets. Inf. Sci. **179**(8), 1092–1101 (2009)
7. Cheng, E., Philip, H., Jia, R., Lipták, L.: Matching preclusion and conditional matching preclusion for bipartite interconnection networks II: cayley graphs generated by transposition trees and hyperstars. Networks **59**, 357–364 (2012)
8. Cheng, E., Lipman, M.J., Lipták, L., Sherman, D.: Conditional matching preclusion for the arrangement graphs. Theor. Comput. Sci. **412**, 6279–6289 (2011)
9. Cheng, E., Lipták, L., Hsu, L., Tan, J.J.M., Lin, C.: Conditional Matching Preclusion for the Star Graph, Ars Combinatoria (to appear)
10. Curran, S.J., Gallian, J.A.: Hamiltonian cycles and paths in cayley graphs and digraphs-a survey. Discrete Math. **156**(1–3), 1–18 (1996)
11. Thomas, W.: Hungerford: Algebra. Springer-Verlag, New York (1974)
12. Li, H., Yang, W., Meng, J.: Fault-tolerant hamiltonian laceability of cayley graphs generated by transposition trees. Discrete Math. **312**(21), 3087–3095 (2012)
13. Lovász, L., Plummer, M.D.: Matching Theory. Elsevier Science Publishing Company, New York (1986)
14. Park, J.-H.: Matching preclusion problem in restricted HL-graphs and recursive circulant G (2 ^{m}, 4). J. Kiss **35**(2), 60–65 (2008)
15. Park, J.-H., Son, S.H.: Conditional matching preclusion for hypercube-like interconnection networks. Theor. Comput. Sci. **410**, 2632–2640 (2009)
16. Wang, S., Wang, R., Lin, S., Li, J.: Matching Preclusion for k-ary n-cubes. Discrete Appl. Math. **158**(18), 2066–2070 (2010)
17. Mujiangshan, W., Zhen, W., Shiying, W.: Conditional matching preclusion for bubble-sort graphs. J. Xinjiang Univ. **28**(1), 23–35 (2011). (Natural Science Edition) (Chinese Series)
18. Wang, M., Yang, W., Wang, S.: Conditional matching preclusion number for the cayley graph on the symmetric group. Acta Mathematicae Applicatae Sinica **36**(5), 813–820 (2013). (Chinese Series)
19. Wang, M., Yang, W., Wang, S.: Optimally conditional matching preclusion sets for a class of cayley graphs on the symmetric group. Chin. J. Eng. Math. **30**(6), 901–910 (2013). (Chinese Series)

A Method to Select Next Hop Node
for Improving Energy Efficiency
in LEAP-Based WSNs

Su Man Nam and Tae Ho Cho[(✉)]

College of Information and Communication Engineering,
Sungkyunkwan University, Suwon 440-746, Republic of Korea
{sm38good, thcho}@skku.edu

Abstract. In wireless sensor networks, sensors have stringent energy and computation requirements as they must function unattended. The sensor nodes can be compromised by adversaries who attack network layers such as in sinkhole attacks. Sinkhole attacks have the goal of changing routing paths and snatching data surrounding the compromised node. A localized encryption and authentication protocol (LEAP) observes different types of messages exchanged between sensors that have different security requirements to cope with the attack. Even though this original method excels in security communication using multiple keys, the data is transmitted without optimal selection of the next nodes. In this paper, our proposed method selects the optimal next node based on a fuzzy logic system. We evaluated the energy and security performances of our method against sinkhole attack. Our focus is to improve energy efficiency and maintain the same security level as compared to LEAP. Experimental results indicated that the proposed method saves up to 5 % of the energy while maintaining the security level against the attack as compared to LEAP.

Keywords: Wireless sensor networks · Security · Localized encryption and authentication protocol · Fuzzy logic

1 Introduction

In recent years, wireless sensor networks (WSNs) have been applied in ubiquitous computing systems such as remote environmental monitoring and target tracking [1, 8]. A WSN consists of a large number of sensor nodes to monitor applications and a base station to collect their sensor readings [4]. These sensors support data collection module, data process module, controls module, communication module, and energy resource module. The BS informs the sensor readings to users after collecting them. The sensors are operated in open environments and have limited hardware resources; therefore the nodes can easily be compromised by adversaries attempting sinkhole attacks [11, 12].

Figure 1 shows the sinkhole attacks which collect every report from neighboring sensor nodes on a compromised node. The compromised node forwards false broadcast messages to change routing paths to the neighboring nodes [6, 7]. The neighboring nodes receive the false message, and change their routing paths to the compromised

© Springer International Publishing Switzerland 2015
D.-S. Huang et al. (Eds.): ICIC 2015, Part I, LNCS 9225, pp. 639–648, 2015.
DOI: 10.1007/978-3-319-22180-9_64

node. As an attacked node receives a report, the node transmits the report to the compromised node. Thus, the compromised node for the sinkhole attack collects event data instead of the BS, and uses it for a malicious object.

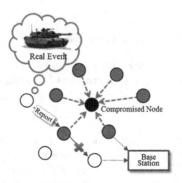

Fig. 1. Sinkhole attacks.

In order to protect against sinkhole attack, S. Zhu et al. proposed a localized encryption and authentication protocol (LEAP) using four types of keys. In LEAP, the four types of keys are individual, pairwise, cluster, and group keys. The individual key is shared with the BS, and is used to generate a message authentication code (MAC) as an event occurs. The pairwise key is shared with a node and its neighboring nodes, and is provided for secure communication. The cluster key is shared with a cluster header (CH) and nodes within a cluster to verify every data in them. The group key is shared in the sensor network to be used for encrypting a message. Thus, when the compromised node forwards the false broadcast messages, its neighboring nodes detect and drop the false messages through their keys. Although the original method detects the false broadcast messages through the keys, the effective transmission of a report should be realized from a source node to the BS.

In this paper, we propose a method to effectively select a next hop node based on a fuzzy logic system. Before transmitting a report, a source node checks a condition of the next hop node using the fuzzy logic system, and forwards the report to the node. If the next hop node's status is not good, the source node searches for a new node within its range, and transmits the report to the new node. Experimental results show that our proposed method improves energy efficiency for effectively selecting next hop nodes. Thus, our proposed method effectively selects the next hop node for improving energy efficiency against the sinkhole attack as compared to the original method.

The remainder of this paper is organized as follows: Sect. 2 introduces the background of the original method and the motivation. A detailed description of our proposed method follows in Sect. 3. In Sect. 4, a performance evaluation of the proposed method is analyzed and discussed. The conclusions are given at the end of the paper.

2 Background

2.1 LEAP

LEAP was proposed for key management in sensor networks to decrease the damage of attacks from compromised nodes. LEAP uses four types of keys in a sensor node: an individual key, a pairwise key, a cluster key, and a group key. Every node has an individual key for secure communication between the node and the BS, and shares the pairwise key for an endorsement of its neighbors within its one-hop. All nodes in a cluster share the cluster key for verifying a CH between a normal node and the cluster. Every node and the BS own a group key for encrypting and decrypting all messages in a sensor network. Five phases for generation of keys are described as follows:

- Phase 1: The BS assigns a group key, an initial key, and a master key of the BS to each node.
- Phase 2: Each node generates an individual key through the master key of the BS.
- Phase 3: The node generates its master key by using the initial key assigned from the BS.
- Phase 4: A node and its neighbors share a pairwise key using their master key.
- Phase 5: A cluster key is generated through the pairwise key between a node and a CH.

Figure 2 shows the authentication of two cases: between a normal node (ND_0) and a CH (CH_0), and between two CHs (CH_0 and CH_1). In the first case, when a real event occurs in a cluster, ND_0 detects the event and generates a report to notice it via intermediate CHs to a BS. Before ND_0 forwards the report to CH_0, they verify each other through their keys. ND_0 transmits a MAC including ND_0's CK and a nonce to CH_0. CH_0 verifies the MAC by using its CK, and makes a response by a new MAC and the nonce. After verifying the MAC through ND_0's CK, ND_0 produces a report attached to a new MAC including ND_0's IK and event data and transmits the report to CH_0. CH_0 replies with an acknowledge (ACK) message to ND_0. In the second case, before forwarding the report received from ND_0, CH_0 transmits a nonce to CH_1. After receiving the nonce, CH_1 responds with a MAC including CH_1's PK and the nonce. CH_0 then verifies CH_1's PK by using its PK. After finishing the verification, CH_0 transmits the report to CH_1. When CH_1 receives the report, an ACK is forwarded.

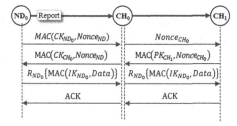

Fig. 2. Authentication between two nodes.

2.2   Motivation

In an open sensor network environment, a compromised node can forward false broadcast messages to its neighboring nodes for a sinkhole attack. LEAP detects the false broadcast messages using four types of keys between the compromised node and the neighbors against the attack. Although the original method provides an operation of the secure network from an injection of the false message, the report can be transmitted to a next hop node without consideration of its node's status. In this paper, we propose a method to select the next hop node with consideration to its condition, and transmit the report to it. Our proposed method effectively selects the next hop node with three factors based on the fuzzy logic system. Thus, the proposed method selects the effective next hop node for the enhancement of energy efficiency against the sinkhole attack as compared to LEAP.

3   Proposed Method

3.1   Assumptions

In this paper, a WSN is assumed to consist of a BS and a large number of sensor nodes (e.g., the Berkeley MICAz motes [2]). After the sensor nodes are deployed in the field, a CH node is elected in a cluster because this node is more powerful than normal nodes [4]. The topology then establishes initial routing paths by using directed diffusion [3] and minimum cost forwarding algorithms [9]. The CHs discover a routing path toward the BS based on a cluster model [5]. When a fuzzy rule-based system is computed, the CH's memory size is not considered since its hardware resources are greater than the normal node. We further assume that a compromised node could inject a false broadcast message to its neighbors within a cluster.

3.2   Overview

We propose a method to effectively select a next hop node based on a fuzzy logic system before forwarding a report in a source node. Figure 3 shows the effective selection of the next hop node between CH_0 and CH_2. As shown in Fig. 3, an initial routing path is constructed between CH_0 and CH_1. Before transmitting a report, CH_0 transmits a nonce message to check a condition of CH_1. CH_1 uses a fuzzy rule-based system with its three factors (energy level, distance to the BS, and threat of the attack) so that an output of the system is obtained. If CH_1's status is not good, CH_0 searches for a new node within its range. After getting an output from CH_2, CH_0 selects CH_1 for its next hop node instead of CH_1, and transmits the report to it. Therefore, our proposed method effectively selects the next hop node based on the fuzzy if-then rule using its three input factors.

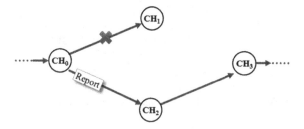

Fig. 3. Selection of next hop node.

3.3 Proposed Method Based on Fuzzy Logic

Our proposed method selects an effective next hop node based on a fuzzy rule-based system considering three input factors as shown in Fig. 4. CH_0 transmits a nonce message to CH_1 before forwarding a report. CH_1 extracts a result through its three factors as the message is received. When CH_1 transmits a message including a MAC, the result is attached in the message. If the MAC is legitimate and the result is change, CH_0 transmits the report to CH_1. In contrast, if CH_1's condition is disuse, CH_0 searches for another node within its range. CH_0 then transmits the report to the new next hop node. Thus, our proposed method effectively selects the next hop node based on the fuzzy rule-based system.

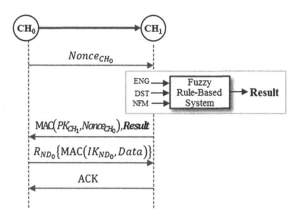

Fig. 4. Selection of next hop node based on the fuzzy rule-based system.

3.3.1 Input and Output Factors

Figure 5 show the input and output factors of the fuzzy logic system in the proposed method. The input factors are as follows:

- ENG (remaining energy level): This value is an important factor to reduce unnecessary energy consumption. If a CH's energy level is very small, the CH

Fig. 5. Input and output factors of the fuzzy system.

effectively seeks one of the neighbors to forward a report. The improvement in energy efficiency is influenced by the diminishment of the report transmission.

- DST (distance): When a report is produced, it is transmitted via multiple hops toward a BS. If a source CH is far away from the BS, energy will be consumed in each intermediate CH. The energy resource of the sensor node is influenced by high hop counts in the intermediate CH.
- FMC (filtered false message count): This factor represents the condition of the network security in each node. If a CH receives a large number of false messages for the sinkhole attack, an attacked cluster will be avoided to maintain the security transmission. The maintenance of the security level is influenced by the number of filtered false broadcast messages.

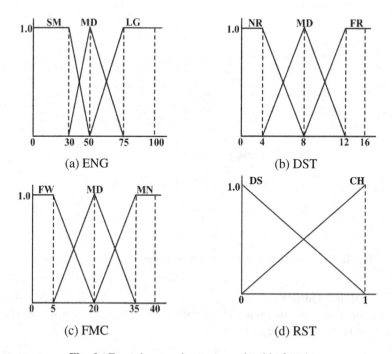

Fig. 6. Fuzzy input and output membership functions

An output factor is as follows:

- RST (result): This value determines a result (*change* or *disuse*). If the value of the next node is *change*, the report is transmitted to the node. The effective selection of the next node is influenced by three input factors based on the fuzzy rule-based system.

3.3.2 Fuzzy Membership Functions and Rules

Figure 6 shows the membership function of the fuzzy logic input and output parameters. The labels of input fuzzy variables are as follows:

- ENG = {IN (insufficient), MD (medium), SF (sufficient)}
- DST = {NR (near), MD (middle), FR (far)}
- FMC = {FW (few), MD (middle), LR (large)}

The logic output parameters are represented by a label, DS or CH. The label values represent a next node selection (CDT) as follows:

- RST = {DS (disuse), CH (change)}

Table 1 shows the rules base of the fuzzy system which is composed of 27 (=3) rules. If ENG is sufficient, DST is short, and FMC is FW; a next node is then selected because its condition is change (Rule 18). This case indicates that there is a normal state to transmit and forward a report. In contrast, if ENG is insufficient, DST is long, and FMC is few; a next node then seeks one of its neighboring nodes because its condition is disuse (Rule 6). This case indicates that there is an abnormal condition for wireless communication. A source node needs to effectively select the next node to reduce energy consumption. Thus, it is important to check the condition of the next node for improving energy efficiency.

Table 1. Fuzzy if-then rules

Rule No.	Input			Output
	ENG	DST	FMC	RST
0	IN	SH	FW	CH
1	IN	SH	MD	DS
⋮	⋮	⋮	⋮	⋮
6	IN	LN	FW	DS
⋮	⋮	⋮	⋮	⋮
18	SF	SH	FW	CH
⋮	⋮	⋮	⋮	⋮
26	SF	LN	LR	DS

4   Experimental Results

Experiments were performed for the effectiveness of the proposed method as compared to LEAP. A sensor network in the environment is 500×500 m^2. There are a total of 500 sensor nodes (50 clusters and 450 normal nodes) in the sensor network. That is, a cluster has one cluster and nine normal nodes. These sensor nodes are uniformly distributed on a 50×50 m^2 sized cluster. We randomly generated 500 events in the field. Each node consumes 16.25 µJ to transmit per byte, 12.5 µJ to receive per byte, 15 µJ to generate per byte, and 75 µJ to verify a MAC [10]. The message size of a nonce and an ACK is 1 byte, a reply message size is 12 bytes, and a report size is 24 bytes. In order to generate the sinkhole attack, we compromised a node, and the compromised node periodically forwarded false broadcast messages to its neighboring nodes.

Figure 7 shows the detection ratio of LEAP and the proposed method against the sinkhole attack as compared to LEAP. Herein, we generated false broadcast messages from a false traffic ratio of 20 % and 40 % in the compromised node. In two methods, when the false broadcast messages are forwarded, they are directly dropped. Thus, the proposed method maintains the same security level as LEAP.

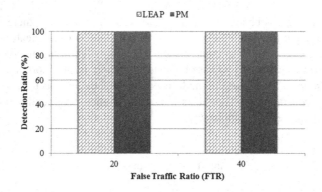

Fig. 7. Detection Ratio against false broadcast messages

Figure 8 shows the average energy consumption of the entire sensor network per event. The energy consumption of LEAP and the proposed method is almost same as 100 and 200 events are generated. A gap is created between the two methods when 300 events occurred. When 500 events are generated, LEAP and the proposed method consumed about 5,300 mJ and 5,500 mJ, respectively. The proposed method improves the energy savings up to about 5 % compared to LEAP. Therefore, our proposed method saves energy resources through the effective selection of the next hop node, and prolongs the sensor network lifetime.

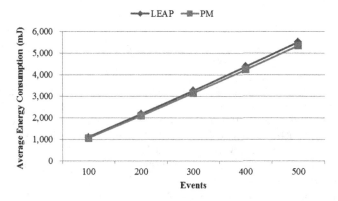

Fig. 8. Average Energy consumptions

5 Conclusions

In WSNs, a sensor node is easily compromised due to limited hardware constraints. The compromised node tries to have the sinkhole attack forward false broadcast messages to its neighboring nodes. After receiving the false message, routing paths of the neighbors are changed to go toward the compromised node. This results in reports being forwarded in the neighbor's flow in the compromised node instead of the BS. LEAP detects the false message generated from the compromised node by using multiple keys. Although LEAP maintains the secure communication between nodes, the sensor network needs effective energy savings with the selection of the next hop node. Our proposed method effectively selects the next hop node for improving energy efficiency based on the fuzzy logic system. Before forwarding a report, a source node asks for the condition of the next hop node. The next hop node draws its condition using its input factors through the fuzzy rule-based system. The drawn result is then forwarded to the source node. If the result is change, the source node transmits the report to the next hop node. Experimental results demonstrated that the proposed method saves up to 5 % of the energy resources as compared to the original method. Thus, our proposed method saves the energy resources of the sensor node, since the next hop node is effectively selected based on the fuzzy rule-based system. In addition, the sensor network's lifetime will be prolonged. In the future, we propose to apply optimization to the fuzzy logic system in the proposed method to increase the lifetime of the sensor network.

Acknowledgments. This research was supported by Basic Science Research Program through the National Research Foundation of Korea (NRF) funded by the Ministry of Education, Science and Technology (No. 2013R1A2A2A01013971).

References

1. Akyildiz, I.F., Su, W., Sankarasubramaniam, Y., et al.: A survey on sensor networks. IEEE Commun. Mag. **40**, 102–114 (2002)
2. Xbox sensor networks. http://www.xbox.com
3. Intanagonwiwat, C., Govindan, R., Estrin, D.: Directed diffusion: a scalable and robust communication paradigm for sensor networks. In: Proceedings of the 6th Annual International Conference on Mobile Computing and Networking, pp. 56–67 (2000)
4. Lee, H.Y., Cho, T.H.: Optimized fuzzy adaptive filtering for ubiquitous sensor networks. IEICE Trans. Commun. **E94.B**, 1648–1656 (2011)
5. Li, F., Srinivasan, A., Wu, J.: PVFS: a probabilistic voting-based filtering scheme in wireless sensor networks. Int. J. Secur. Netw. **3**, 173–182 (2008)
6. Ngai, E.C.H., Liu, J., Lyu, M.R.: On the intruder detection for sinkhole attack in wireless sensor networks. In: IEEE International Conference on Communications, ICC 2006, vol. 8, pp. 3383–3389 (2006)
7. Ngai, E.C.H., Liu, J., Lyu, M.R.: An efficient intruder detection algorithm against sinkhole attacks in wireless sensor networks. Comput. Commun. **30**, 2353–2364 (2007)
8. Nghiem, T.P., Cho, T.H.: A fuzzy-based interleaved multi-hop authentication scheme in wireless sensor networks. J. Parallel Distrib. Comput. **69**, 441–450 (2009)
9. Ye, F., Chen, A., Lu, S., et al.: A scalable solution to minimum cost forwarding in large sensor networks. In: Proceedings of the Tenth International Conference on Computer Communications and Networks 2001, pp. 304–309 (2001)
10. Ye, F., Luo, H., Lu, S., et al.: Statistical en-route filtering of injected false data in sensor networks. IEEE J. Sel. Areas Commun. **23**, 839–850 (2005)
11. Zhu, S., Setia, S., Jajodia, S.: LEAP: efficient security mechanisms for large-scale distributed sensor networks. In: Proceedings of the 10th ACM Conference on Computer and Communications Security, pp. 62–72. ACM (2003)
12. Zhu, S., Setia, S., Jajodia, S.: LEAP+: efficient security mechanisms for large-scale distributed sensor networks. ACM Trans. Sen. Netw. **2**, 500–528 (2006)

An Efficient Topology-Based Algorithm for Transient Analysis of Power Grid

Lan Yang[1], Jingbin Wang[2,3,4(✉)], Lorenzo Azevedo[5],
and Jim Jing-Yan Wang[6]

[1] School of Computer Science and Information Engineering,
Chongqing Technology and Business University, Chongqing 400067, China
lancy9232001@163.com
[2] Tianjin Key Laboratory of Cognitive Computing and Application,
Tianjin University, Tianjin 300072, China
jingbinwang1@outlook.com
[3] National Time Service Center, Chinese Academy of Sciences,
Xi'an 710600, China
[4] Provincial Key Laboratory for Computer Information Processing Technology,
Soochow University, Suzhou 215006, China
[5] Computer Science Department, University of São Paulo,
São Paulo, SP 05508-070, Brazil
lorenzo.azevedo@yahoo.com
[6] Computer, Electrical and Mathematical Sciences and Engineering Division,
King Abdullah University of Science and Technology (KAUST),
Thuwal 23955-6900, Saudi Arabia
jimjywang@gmail.com

Abstract. In the design flow of integrated circuits, chip-level verification is an important step that sanity checks the performance is as expected. Power grid verification is one of the most expensive and time-consuming steps of chip-level verification, due to its extremely large size. Efficient power grid analysis technology is highly demanded as it saves computing resources and enables faster iteration. In this paper, a topology-base power grid transient analysis algorithm is proposed. Nodal analysis is adopted to analyze the topology which is mathematically equivalent to iteratively solving a positive semi-definite linear equation. The convergence of the method is proved.

Keywords: Power grid · Topology · Nodal analysis · Positive semi-definite

1 Introduction

Power grid analysis is an indispensable step in modern chip simulation and verification. Due to the IR and Ldi/dt voltage drops, the actual voltages applied to the logic gates and other circuit elements are smaller than wished, which not only increases the delay of the logic gates but may also result in logic errors. As the ever decreasing supply voltages and threshold voltages make the problem worse, power grid analysis becomes an indispensable step in the simulation flow.

To capture the worst-case voltage drop and place decoupling capacitors (if necessary), DC analysis and/or transient analysis of the power grid should be performed.

© Springer International Publishing Switzerland 2015
D.-S. Huang et al. (Eds.): ICIC 2015, Part I, LNCS 9225, pp. 649–660, 2015.
DOI: 10.1007/978-3-319-22180-9_65

Frequency-domain analysis is not appropriate as the vital voltage drop in the time domain may be lost in the frequency-domain analysis. Both DC analysis and transient analysis (using forward/backward can be finally reduced to a linear solving problem.

Traditional power grid analysis only considers the IR voltage drop. As the operation frequency of chips is increasing rapidly, the Ldi/dt voltage drop should not be neglected. Thus, the power grid model should consist inductors besides resistors and capacitors. The existence of inductors changes the state-space model from order-1 to order-2, hence introduces new difficulties to the transient analysis and its topology-based algorithms. It was proved in [3] that the system matrix of the RLC model could be still positive definite, which had been thought to be impossible before the paper. However, some of its details are inaccurate. The right formulation will be proposed in Sect. 4.4, which is the basis to prove the convergence of the topology-based iterative algorithm in Sect. 4.1.

In this paper, a new topology-based algorithm is proposed. The Nodal analysis is adopted to analyze the netlist and generate positive semi-definite system model. Then a purely topology-base iterative method is introduced to solve the power grid analysis without constructing the system matrices. The convergence is proved and the complexity is analyzed.

2 Background

2.1 Problem Formulation

Power grids are usually modeled as RLC circuits, or simpler RC circuits. The current sources attached to the nodes of the power grid represent the currents drawn from the logic gates or other devices. Hence prior to the simulation of the power grid, the current pattern of the gates and devices should be obtained and modeled as piecewise linear ideal current sources. The capacitors in parallel with the currents sources can be either parasitic capacitors or decoupling capacitors. The inductors most possibly appear at the vias connecting different layers and the metal layer connecting the power grid to external power supply. In many models the inductors can be neglected.

First consider the simpler RC power grid model, which can be depicted as a state space model [4, 7–10, 28]

$$C\frac{d}{dt}v(t) + Gv(t) = -i(t) + G_0 V_{dd}, \tag{1}$$

where $v(t)$ is the vector of nodal voltages, $i(t)$ is the vector of current sources, V_{dd} is vector of dimension n with all its values equal to V_{dd}. C is a diagonal matrix whose (i,i) element is the capacitance between node i and ground. G is the conductance matrix with its diagonal (i,i) element being the total conductance connected to node i and off-diagonal (i,j) element being the opposite number of the conductance between node i and node j (it is zero if node i and node j are unconnected). G_0 is a diagonal matrix with its (i,i) element being the conductance between node i and source (if node i is connected to a source) or 0 (if node i is not connected to any source). To simplify the notation, $b(t)$ is used to represent $-i(t) + G_0 V_{dd}$. Using backward Euler method, (1) can be approximated as

$$\left(\frac{C}{h} + G\right) x(t + h) = \frac{C}{h} x(t) + b(t + h), \tag{2}$$

where h represents the time step. Thus given the initial condition $x(0)$, we can calculate all the $x(t)$ after 0. Note that $\frac{C}{h} + G$ is always invertible if the system is stable. The reason is that stability implies that all the eigenvalues of the matrix pencil (G,C) distribute in the left half complex plane and the positive real value $\frac{1}{h}$ cannot be an eigenvalue of (G,C).

If we use system matrix A to denote $\frac{C}{h} + G$ and b to denote $\frac{C}{h} x(t) + b(t + h)$, (2) can be reduced to

$$Ax = b. \tag{3}$$

In DC analysis, $A = G$ and b.

2.2   SOR-like Iterative Method

Although straightforward in theory, Eq. (2) are difficult to solve due to the extremely large size. In a standard power grid analysis problem, the number of nodes n can be hundreds of thousands or even millions, thus the dimension of the matrices can be up to millions. This causes problems owing to the limited speed and memory of computers. Direct methods based on LU decomposition requires $O(n^3)$ time for the decomposition of the system matrix (G or A) and $O(n^2)$ time for forward and backward substitution. The memory required is $O(n^2)$ even if G or A is sparse as the resulting matrices L, U may become dense. Therefore direct methods are not applicable for extremely large problems.

Alternative methods include traditional iterative methods, the widely used PCG (preconditioned conjugate gradient) method and the Monte-Carlo-like random walk method, etc. PCG method, although converges fast, may be not applicable to extremely large problems as preconditioned system matrix may become dense and thus requires prohibiting $O(n^2)$ memory. Random walk algorithm is most suitable in the case that we only want to calculate the voltages at some specific nodes, but it may be inefficient for the full-circuit analysis, which is indispensable to find the worst-case voltage drop.

An iterative method was introduced in [34], which can be regarded as an efficient implementation of the traditional Gauss-Seidel iteration method and SOR (Successive Over-Relaxation) method. The convergence of this method in the DC analysis has been proven in [34]. The advantage of this method is that it is topology-based and no matrix construction is needed. Hence it requires much less memory than PCG method. It is also shown that the method is faster than random walk in [34].

In the next sections, we will extend this SOR-like method to the transient analysis of RC and RLC power grid models and prove their convergence. In the RLC model analysis part, we will first show that the proof of positive definiteness in [3] is inaccurate and then give the right version.

3   Nodal Analysis of RLC Circuits

In the analysis of RLC circuits the most commonly used method is MNA (modified nodal analysis). It can be guaranteed that these system models are passive [14, 27, 29, 30]. By introducing extra variables for currents flowing through voltage sources and inductors, it can generate compact state-space models efficiently. However, when we perform transient analysis on such MNA models, the system matrix is not positive definite. As a result, many iterative algorithms (such as preconditioned conjugate gradient) cannot be applied since their convergence requires positive definiteness of the system matrix.

In [3], a new NA (nodal analysis) method was proposed to perform transient analysis on RLC circuits. It was proved in [3] that the resulting system matrix is guaranteed to be positive definite. Although the idea in the paper is novel and inspiring, its formulations is inaccurate. The key equation (Eq. (8) in [3]) is incorrect due to a wrong sign in Eq. (6). From another perspective, in Eq. (8), the current state of the nodal voltages only depends on the state of the previous step, which is impossible for a second-order system. In a second-order system like RLC circuits, the required initial conditions involve both $x(0)$ and $x'(0)$. In the discretized form, the state $x(t + \Delta t)$ should depend on both $x(t)$ and $x(t - \Delta t)$, which is not the case in [3]. Besides, the analysis in [3, 13, 25] did not consider voltage sources. Although Norton equivalent was used to convert voltages sources to current sources, we still wish to involve voltages sources from the beginning of the deduction, as Norton equivalent does not work for ideal voltage sources which we may want to include in the power grid model. We will give the correct deduction considering ideal voltage sources in the next subsection.

3.1   System Equation

To simplify the deduction, assume that all the negative terminals of the voltage sources are connected to ground, which is just the case in the power grid analysis (it can be extended to ground network analysis straightforwardly). Assume that there are n nodes that are neither ground nor positive terminals of voltage sources in the power grid model (called trivial nodes), numbered from 1 to n, together with p nodes being the positive terminals of voltage sources (called source nodes), numbered from 1 to p. Besides, assume there are m branches not through voltage sources, numbered from 1 to m. Define the m by n modified incidence matrix A as

$$A(i,j) = \begin{cases} +1, & \text{if trival node } j \text{ is the soruce of branch } i; \\ -1, & \text{if trival node } j \text{ is the sink of branch } i; \\ 0, & \text{otherwise.} \end{cases} \tag{4}$$

Similarly, define the m by p source incidence matrix A_s as

$$A_s(i,j) = \begin{cases} +1, & \text{if trival node } j \text{ is the soruce of branch } i; \\ -1, & \text{if trival node } j \text{ is the sink of branch } i; \\ 0, & \text{otherwise.} \end{cases} \tag{5}$$

The advantage of splitting the traditional incidence matrix to the modified incidence matrix and the source incidence matrix is that no extra current variables through the voltage sources needs to be introduced, which facilities the nodal analysis and is the basis of the positive definiteness to be proved next. By grouping the branches with resistors, inductors, capacitors and current sources together we obtain

$$
A = \begin{bmatrix} A_g \\ A_c \\ A_l \\ A_i \end{bmatrix}, A_s = \begin{bmatrix} A_{gs} \\ A_{cs} \\ A_{ls} \\ A_{is} \end{bmatrix}, v_b = \begin{bmatrix} v_g \\ v_c \\ v_l \\ v_i \end{bmatrix}, i_b = \begin{bmatrix} i_g \\ i_c \\ i_l \\ i_i \end{bmatrix}, \tag{6}
$$

where i_b is the vector of branch currents, v_b is the vector of branch voltages, g, c, l, s represent resistors, capacitors, inductors and current sources respectively.

Applying Kirchoff's current and voltage laws, we obtain

$$
\begin{aligned}
A^T i_b &= 0, \\
A v_n &= v_b - A_s v_d,
\end{aligned} \tag{7}
$$

Where v_n is the vector of nodal voltages (other than the nodes being the terminals of voltage sources) and v_d is the (constant) vector of voltage sources.

Besides, the branch currents and branch voltages follow the following branch equations (grouped by branch elements)

$$
\begin{aligned}
i_g &= \mathcal{G} v_g, \\
i_c &= C \frac{dv_c}{dt}, \\
v_l &= \mathcal{L} \frac{di_l}{dt}, \\
i_i &= I_s,
\end{aligned} \tag{8}
$$

where G, C, L are diagonal matrices with its diagonal elements being the value of corresponding branch resistance, capacitance and inductance, I_s is the vector of branch current sources.

Combine (7) and (8), we obtain $2m + n$ equations of $2m + n$ variables. Substitute the second equation of (7) to (8) and then substitute the new (8) to the first equation of (7), we have

$$
\begin{aligned}
C \frac{dv_n}{dt} + G v_n + A_l^T i_l + A_i^T I_s - G_s v_d &= 0, \\
L \frac{di_l}{dt} - A_l v_n + A_{ls} v_d &= 0.
\end{aligned} \tag{9}
$$

Here $G = A_g^T G A_g, C = A_c^T C A_c, G_s = A_g^T G A_{gs}$. Use backward Euler law to discretize (9),

$$C\frac{v_n(t+h) - v_n(t)}{h} + Gv_n(t+h) + A_i^T i_l(t+h)$$
$$+ A_i^T I_s(T+h) - G_s v_d = 0, \tag{10}$$
$$\mathcal{L}\frac{i_l(t+h) - i_l(t)}{h} - A_l v_n(t+h) + A_{ls} v_d = 0.$$

Here h is the selected time step length. Rewrite the first equation of (10) at time t and subtract it from the original equation, then substitute the second equation of (10) to it we obtain

$$\left(\frac{C}{h} + G + hL\right) v_n(t+h) = \left(\frac{2C}{h} + G\right) v_n(t) - \frac{C}{h} v_n(t-h)$$
$$+ hL_s v_d - A_i^T (I_s(t+h) - I_s(t)) \tag{11}$$

Here $L = A_i^T L^{-1} A_l$, $L_s = A_l L^{-1} A_{ls}$. L is a diagonal matrix with all its diagonal elements being nonzero, hence it is invertible. $\left(\frac{C}{h} + G + hL\right)$ is called system matrix.

3.2 Positive Definiteness

In the RLC power grid model analysis, some assumptions on the topology are made as summarized below.

1. All the voltage sources have their negative terminals as ground;
2. Any trivial node is connected to other non-ground nodes by one or more branches and such branches can not be all current sources.
3. For any pair of trivial nodes i and j, there exists at least one path from i to j and the path passes through trivial nodes only (it does not pass through source nodes).

The first assumption is straightforward and just the case in power grid models. The second assumption is also true since every node in the power grid model is connected to other non-ground nodes by at least one resistor (or capacitor, or inductor). The third assumption may be not satisfied for some special cases. However, in these cases the power grid can be separated to several sub-circuits which are independent of each other. Consider the case in Fig. 1. The circuit can be divided to two independent parts, sub-circuit 1 and sub-circuit 2. In the analysis of each sub-circuit, the third assumption is satisfied.

Lemma 1. Under the topological assumptions for the circuits, the resulting system matrix $M = \left(\frac{C}{h} + G + hL\right)$ is positive definite.

Proof. It is obvious that M is symmetric. And according to the definition of Ag, $G(i,i)$ is the sum of conductances connected to trivial node i, $\sum_{j\neq i} |G(i,j)|$ is the sum of conductances connected to trivial node i and not connected to source nodes. Hence $|G(i,i)| \geq \sum_{j\neq i} |G(i,j)|$. The same is true for C and L, which indicates the system matrix

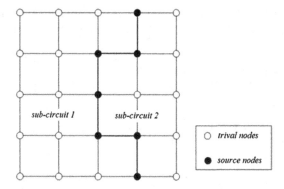

Fig. 1. A special circuit model which can be divided into two sub-circuits

is weakly diagonally dominant with nonnegative diagonal elements. Besides, according to the topological assumption, any two trivial nodes are connected by a path crossing only trivial nodes. This results in the system matrix to be irreducible. Use the well-known theorem in [2, 6, 33], we conclude that the system matrix M is positive definite.   □

4 Topology-Based Transient Analysis of Power Grid

4.1 Voltage Update

Consider a representative node in the RLC power grid model, as shown in Fig. 2. Applying Kirchoff's current law at node i we have

$$\sum_{j\in N_i^R} g_{ij}(V_i - V_j) + \sum_{j\in N_i^L} I_{ij} + I_i + C_i\frac{dV_i}{dt} = 0. \tag{12}$$

Here I_{ij} is the current from node i to node j, g_{ij} is the conductance between node i and node j, N_i^R $\left(N_i^L\right)$ is the set of indices of nodes connected to node i by a resistor

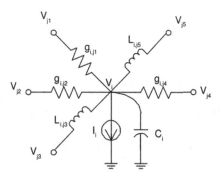

Fig. 2. A representative node of the RLC power grid

(inductor). Note that $N_i^R N_i^L$ may be nonempty. Take derivative on both sides of (12) and substitute the branch equations through inductors into it, we obtain

$$\sum_{j \in N_i^R} g_{ij} \left(\frac{dV_i}{dt} - \frac{dV_j}{dt} \right) + \sum_{j \in N_i^L} \frac{1}{L_{ij}} (V_i - V_j) + \frac{dI_i}{dt} + C_i \frac{d^2 V_i}{dt^2} = 0. \tag{13}$$

$$\left(\sum_{j \in N_i^R} g_{ij} + h \sum_{j \in N_i^L} \frac{1}{L_{ij}} + \frac{C_i}{h} \right) V_i(t+h)$$

$$= \sum_{j \in N_i^R} g_{ij} V_j(t+h) + \sum_{j \in N_i^L} \frac{h}{L_{ij}} V_j(t+h) + \left(\frac{2C_i}{h} + \sum_{j \in N_i^R} g_{ij} \right) V_i(t) \tag{14}$$

$$- \sum_{j \in N_i^R} g_{ij} V_i(t) - I_i(t+h) + I_i(t) - \frac{C_i}{h} V_i(t-h).$$

Here L_{ij} denotes the inductance between node i and node j. Using backward Euler method, (13) can be discretized to (14), as shown below. Write (14) more compactly as

$$V_i(t+h) = \sum_{j \in N_i} w_{ij} V_j(t+h) + K_i(t+h, t, t-h), \tag{15}$$

where w_{ij} represents the coefficient of $V_j(t+h)$ and where w_{ij} represents the coefficient of $V_j(t+h)$ and $K_i(t+h, t, t-h)$ is a known value corresponding to currents at $t+h$, t and nodal voltages t, $t-h$. Based on (15), we can use an iterative method to solve the power grid model. In each iteration, update the nodal voltage at node i by

$$V_i^{(k)}(t+h) = \sum_{j \in L_i} w_{ij} V_j^{(k)}(t+h) + \sum_{j \in U_i} w_{ij} V_j^{(k-1)}(t+h)$$
$$+ K_i(t+h, t, t-h). \tag{16}$$

Here $L_i(U_i)$ is defined as the set of indices that satisfy (i) the corresponding nodes are connected to node i and (ii) they are smaller (larger) than i. k denotes the iteration number.

4.2   Initial Conditions

As it can be seen from (11) and (16), the nodal voltages at $t+h$ depend on the circuit states at both t and $t-h$, which is the requirement of the "second-order" characteristic of the circuit. If the initial conditions $v_c(0)$ and $i_l(0)$ are given, we can calculate $v_n(t)$ at any time after 0 by iteratively updating nodal voltages using (16). Given the initial conditions $v_n(0)$ and $i_l(0)$, $v_n(0)$, $v_n(0)$ and $v_n(h)$ can be calculated in three steps.

1. Treat capacitors as voltage sources whose voltages are $v_c(0)$, inductors as current sources whose currents are $i_l(0)$. Perform iterative DC analysis as [34] at time 0, obtain $v_n(0)$, $i_c(0)$ and $v_l(0)$;
2. Calculate $v_c(h)$ and $i_l(h)$ according to $C\frac{v_c(h)-v_c(0)}{h} = i_c(0)$, $L\frac{i_l(h)-i_l(0)}{h} = v_l(0)$;
3. Perform iterative DC analysis at time point h and obtain $v_n(h)$.

After the three steps we obtain the voltages of trivial nodes ($v_n(0)$ and $v_n(h)$), which are the basis of the topology-based transient analysis algorithm in the next subsection.

4.3   Algorithm Description

The topology-based algorithm is proposed as Algorithm 1. No matrix construction or manipulation is required in the algorithm, which dramatically reduces the storage memory and computation time. In the algorithm, DC analysis is performed first to obtain the initial conditions. Then, the nodal voltages at each time step is solved iteratively, based on the information at the previous two time steps. The vectors $V(s)$ and $I(s)$ represent the nodal voltages and current sources at time $s \times h$. The convergence and computational complexity of the proposed algorithm are discussed in Sect. 4.4.

Algorithm 1: Topology-based transient analysis of power grid

Input: Initial capacitor voltages $v_c(0)$, inductor currents $i_l(0)$, step length h, maximum step s_{total}, currents sources $I(s)$ ($s = 0, 1, ..., s_{total}$), error tolerance tol;

Output: Nodal voltages $V(s)$ ($s = 0, 1, ..., s_{total}$);

1: Calculate $V(0)$ and $V(1)$ using iterative DC analysis;
2: **for** $s = 2,...,s_{total}$ **do**
3:     $V^{(0)}(s) = V(s-1)$;
4:     $k = 0$;
5:     **repeat**
6:         $k = k + 1$;
7:         Update $V_i^{(k)}(s)$ using equation (16);
8:     **until** $\|V^{(k)}(s) - V^{(k-1)}(s)\| < tol$
9:     $V(s) = V^{(k)}(s)$;
10: **end for**

4.4   Convergence

Theorem 1. The solution $V(s)$ ($s = 0, 1, ..., s_{total}$) of Algorithm 1 converges to the accurate nodal voltages $v_n(s \times h)$.

Proof. Compare (15) and (16) with (11), we can see that Algorithm 1 is equivalent to the Gauss-Seidel iterative solution of the matrix Eq. (11). Because the system matrix is symmetric positive definite, the Gauss-Seidel iteration converges (refer to Theorem 10.2.1 of [5]). Therefore Algorithm 1 is guaranteed to converge.   □

We can employ the SOR (successive over-relaxation) method to accelerate the con-vergence of Algorithm 1. Using the SOR-like method, the nodal voltage updating formula (16) is adapted to

$$V_i^{(k)}(t+h) = \omega \widetilde{V}_i^{(k)}(t+h) + (1-\omega)V_i^{(k-1)}(t+h). \tag{17}$$

Here $\widetilde{V}_i^{(k)}(t+h)$ is the updated voltage calculated through (16). ω is called relaxation parameter. Using (17) to update the nodal voltage, we obtain a new algorithm (named as Algorithm 2). If $0 < \omega < 2$, Algorithm 2 is guaranteed to converge. With appropriately chosen ω, the convergence procedure can be accelerated. We refer the readers to [31, 32] for the optimal choice of ω.

5 Conclusion

A topology-base power grid transient analysis algorithm has been proposed. Nodal analysis has been adopted to analyze the topology which is mathematically equivalent to iteratively solving a positive semi-definite linear equation. The convergence of the method has been proved. In the future, we will study how to solve this problem with data factorization and analysis with sparse constrains [1, 11–13, 15–24, 26], and large scale data [4].

Acknowledgements. The study was supported by the Tianjin Key Laboratory of Cognitive Computing and Application, Tianjin University, China, and the Provincial Key Laboratory for Computer Information Processing Technology, Soochow University, China (Grant No. KJS1324).

References

1. Al-Shedivat, M., Wang, J.J.Y., Alzahrani, M., Huang, J.Z., Gao, X.: Supervised transfer sparse coding. In: Proceedings of the National Conference on Artificial Intelligence. vol. 3, pp. 1665–1672 (2014)
2. Alipanahi, B., Gao, X., Karakoc, E., Li, S., Balbach, F., Feng, G., Donaldson, L., Li, M.: Error tolerant nmr backbone resonance assignment and automated structure generation. J. Bioinform. Comput. Biol. 9(1), 15–41 (2011)
3. Chen, T., Chen, C.: Efficient large-scale power grid analysis based on preconditioned Krylov subspace iterative methods. In: Proceedings of the Design Automation Conference, pp. 559–562 (2001)
4. Dai, L., Gao, X., Guo, Y., Xiao, J., Zhang, Z.: Bioinformatics clouds for big data manipulation. Biol. Direct 7, 43 (2012)
5. Golub, G., Van Loan, C.: Matrix Computations. Johns Hopkins University Press, Baltimore (1996)
6. Jang, R., Gao, X., Li, M.: Towards fully automated structure-based nmr resonance assignment of 15n-labeled proteins from automatically picked peaks. J. Comput. Biol. 18(3), 347–363 (2011)

7. Kouroussis, D., Najm, F.: A static pattern-independent technique for power grid voltage integrity verification. In: Proceedings of Design Automation Conference. pp. 99–104 (2003)
8. Lei, C.U., Wang, Y., Chen, Q., Wong, N.: On vector fitting methods in signal/power integrity applications. In: Proceedings of the International MultiConference of Engineers and Computer Scientists 2010, IMECS 2010, vol. 2, pp. 1407–1412. Newswood Limited. (2010)
9. Lim, E.G., Wang, Z., Lei, C.U., Wang, Y., Man, K.L.: Ultra wideband antennas–past and present. IAENG Int. J. Comput. sci. **37**(3), 304–314 (2010)
10. Liu, Z., Abbas, A., Jing, B.-Y., Gao, X.: Wavpeak: Picking nmr peaks through wavelet-based smoothing and volume-based filtering. Bioinformatics **28**(7), 914–920 (2012)
11. Wang, J., Gao, X., Wang, Q., Li, Y.: Prodis-contshc: Learning protein dissimilarity measures and hierarchical context coherently for protein-protein comparison in protein database retrieval. BMC Bioinform. **13**(Suppl 7), S2 (2012)
12. Wang, J.J.Y., Bensmail, H., Gao, X.: Joint learning and weighting of visual vocabulary for bag-of-feature based tissue classification. Pattern Recogn. **46**(12), 3249–3255 (2013)
13. Wang, J.J.Y., Bensmail, H., Gao, X.: Multiple graph regularized nonnegative matrix factorization. Pattern Recogn. **46**(10), 2840–2847 (2013)
14. Wang, J.Y., Almasri, I., Gao, X.: Adaptive graph regularized nonnegative matrix factorization via feature selection. In: Proceedings - International Conference on Pattern Recognition, pp. 963–966 (2012)
15. Wang, J.J.Y., Almasri, I., Shi, Y., Gao, X.: Semi-supervised transductive hot spot predictor working on multiple assumptions. Curr. Bioinform. **9**(3), 258–267 (2014)
16. Wang, J.J.Y., Alzahrani, M., Gao, X.: Large margin image set representation and classification. In: Proceedings of the International Joint Conference on Neural Networks, pp. 1797–1803 (2014)
17. Wang, J.J.Y., Bensmail, H., Gao, X.: Multiple graph regularized protein domain ranking. BMC Bioinform. **13**(1), 307 (2012)
18. Wang, J.J.Y., Bensmail, H., Gao, X.: Feature selection and multi-kernel learning for sparse representation on a manifold. Neural Netw. **51**, 9–16 (2014)
19. Wang, J.J.Y., Bensmail, H., Yao, N., Gao, X.: Discriminative sparse coding on multimanifolds. Knowl.-Based Syst. **54**, 199–206 (2013)
20. Wang, J.J.Y., Gao, X.: Beyond cross-domain learning: multiple-domain nonnegative matrix factorization. Eng. Appl. Artif. Intell. **28**, 181–189 (2014)
21. Wang, J.J.Y., Gao, X.: Semi-supervised sparse coding. In: Proceedings of the International Joint Conference on Neural Networks, pp. 1630–1637 (2014)
22. Wang, J.J.Y., Gao, X.: Max-min distance nonnegative matrix factorization. Neural Netw. **61**, 75–84 (2015)
23. Wang, J.J.Y., Huang, J.Z., Sun, Y., Gao, X.: Feature selection and multi-kernel learning for adaptive graph regularized nonnegative matrix factorization. Expert Syst. Appl. **42**(3), 1278–1286 (2015)
24. Wang, J.J.Y., Sun, Y., Gao, X.: Sparse structure regularized ranking. Multimedia Tools Appl. **74**(2), 635–654 (2014)
25. Wang, J.J.Y., Wang, X., Gao, X.: Non-negative matrix factorization by maximizing correntropy for cancer clustering. BMC Bioinform. **14**, 107 (2013)
26. Wang, J.J.Y., Wang, Y., Zhao, S., Gao, X.: Maximum mutual information regularized classification. Eng. Appl. Artif. Intell. **37**, 1–8 (2014)
27. Wang, J., Li, Y., Wang, Q., You, X., Man, J., Wang, C., Gao, X.: Proclusensem: Predicting membrane protein types by fusing different modes of pseudo amino acid composition. Comput. Biol. Med. **42**(5), 564–574 (2012)

28. Wang, Y., Lei, C.U., Pang, G.K., Wong, N.: Mfti: matrix-format tangential interpolation for modeling multi-port systems. In: Proceedings of the 47th Design Automation Conference, pp. 683–686. ACM (2010)
29. Wang, Y., Zhang, Z., Koh, C.K., Pang, G.K., Wong, N.: Peds: Passivity enforcement for descriptor systems via hamiltonian-symplectic matrix pencil perturbation. In: Proceedings of the International Conference on Computer-Aided Design, pp. 800–807. IEEE Press (2010)
30. Wang, Y., Zhang, Z., Koh, C.K., Shi, G., Pang, G.K., Wong, N.: Passivity enforcement for descriptor systems via matrix pencil perturbation. IEEE Trans. Comput. Aided Des. Integr. Circuits Syst. 31(4), 532–545 (2012)
31. Young, D.: Convergence properties of the symmetric and unsymmetric successive overrelaxation methods and related methods. Math. Comput. 24(112), 793–807 (1970)
32. Young, D.: Generalizations of property A and consistent ordering. SIAM J. Numer. Anal. 9 (3), 454–463 (1972)
33. Young, D.: Iterative Solution of Large Linear Systems. Dover Pubns, New York (2003)
34. Zhong, Y., Wong, M.: Fast algorithms for IR drop analysis in large power grid. In: Proceedings of the International Conference on Computer-Aided Design, pp. 351–357 (2005)

Implementation of Leaf Image Recognition System Based on LBP and B/S Framework

Sen Zhao[1($\boxtimes$)], Xiao-Ping Zhang[1], Li Shang[2], Zhi-Kai Huang[3], Hao-Dong Zhu[4], and Yong Gan[4]

[1] College of Electronics and Information Engineering, Tongji University, 4800 Cao'an Highway, Jiading, Shanghai, China
zhaosen_mlsb@163.com
[2] Department of Communication Technology, College of Electronic Information Engineering, Suzhou Vocational University, Suzhou 215104, Jiangsu, China
[3] College of Mechanical and Electrical Engineering, Nanchang Institute of Technology, Nanchang 330099, Jiangxi, China
[4] College of Computer and Communication Engineering, Zhengzhou University of Light Industry, Zhengzhou, China
ganyong@zzuli.edu.cn

Abstract. Plant identification system is on the basis of the previous, through continuous optimizing all aspects of the algorithm to improve efficiency and accuracy of the algorithm. For feature extraction, since the local binary pattern was proposed in the past decades, it has been widely used in computer vision to describe the feature for image classification such as image recognition, motion detection and medical image analysis. According to accuracy of the descriptor always fluctuates with different samples, some improved pattern of LBP has been presented in papers. Complete Local Binary Pattern (CLBP) is an optimized version which set an additional magnitude value to local differences. This paper shows extensive experiments of implement the LBP derivatives for plants texture identification. Finally realize an online system to identify what kind of the plant image user uploaded based on LBP descriptor.

Keywords: LBP · B/S · NNS · Texture classification · Online system

1 Introduction

Researches on plant image recognition focus on how to extract effective feature for classification. To describe texture feature has various methods, whereas selecting a lightweight approach is essential for the online system considering the performance of servers. LBP (Local Binary Pattern) is a method robust to decrease the impact of illumination variations, and the algorithm requires moderate operations to extract feature from images for texture classification in practical applications. It was first described by Ojala et al. [1]. With its ability of discrimination and simplicity, LBP operator has soon become a popular approach in many fields, especially in computer vision, so it is easy to implement in real-time systems. Although LBP can be seen as a unifying approach in texture analysis, it shows poor performance in the situation

© Springer International Publishing Switzerland 2015
D.-S. Huang et al. (Eds.): ICIC 2015, Part I, LNCS 9225, pp. 661–670, 2015.
DOI: 10.1007/978-3-319-22180-9_66

abound in random noise [2]. To tackle this issue, local ternary pattern (LTP) [3] has been presented with one additional discrimination level than LBP in order to increase the robustness against noise in uniform and near-uniform regions. Zhao et al. [5] proposed a new pattern called Local Binary Count (LBC), it has property of rotation invariant with the discard of the local binary structural information. Another derivative of LBP is that, for example, Liao et al. proposed dominant local binary patterns (DLBP) which adds an extra information derived from circularly symmetric Gabor filter.

This paper describes an online system based on Browser/Server (B/S) model, modern digital image processing, and pattern recognition theory. To implement the interaction and recognition throughout the browser, we use web development language: PHP and MySQL running on servers to offer backend processes. This system realizes generating grayscale image, binarization, noise reduction, image segmentation and contour detection functions, and can also extract the plant leaf texture feature in order to classify in the method of NN classifier for recognition, and finally through retrieving database, the information about the kind of the plant user uploaded will be presented in web browser.

Most applications of computer technology in the field of botany related to feature extraction, classification, identification. At present, image processing, computer vision and neural network can be considered useful for feature extraction, recognition, growth monitoring, and play a positive role in agriculture.

2   Brief Review of LBP

Local Binary Pattern was proposed by Ojala et al. [1] based on Gray-level Co-occurrence Matrix (GLCM) [9] texture analysis techniques. The first edition of LBP operator was that two-dimensional surface textures. Now LBP is using a circular neighborhood allow any value of radius and neighborhood pixels (Fig. 1).

First. Define the joint distribution of gray values of P (P > 1) pixels in local area as the local texture model T:

$$T = t(g_c, g_0, g_1, \cdots, g_{P-1})$$

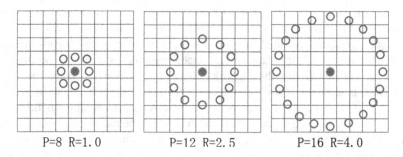

P=8 R=1.0   P=12 R=2.5   P=16 R=4.0

Fig. 1. The circular template of LBP

g_c is the gray level of the center pixel in local area, and $(g_c, g_0, g_1, \cdots, g_{P-1})$ is refer to gray level of pixels around the center pixel, which is on a circle of radius R as follows:

If the center pixel takes position of (x, y), the position of neighbor pixels g_i will be calculated by:

$$x_i = x - R \sin\left(\frac{2\pi i}{P}\right), \quad y_i = y + R \cos\left(\frac{2\pi i}{P}\right)$$

If g_i is not an integer the gray value can be obtained by interpolation.

In order to confirm the gray scale invariant, subtract the gray value of neighbor pixels g_i with the gray value of the center pixel g_c:

$$T = t(g_c, g_0 - g_c, g_1 - g_c, \cdots, g_{P-1} - g_c)$$

Next, it is assumed each distribution is independent of g_c, so the equation above can be decomposed into:

$$T = t(g_c)t(g_0 - g_c, g_1 - g_c, \cdots, g_{P-1} - g_c)$$

In fact, the complete distribution independence cannot be guaranteed, this assumption is just an approximation of joint distribution. If accept this assumption, it may lose some information, but can guarantee the gray scale invariance. $t(g_c)$ represents the global brightness, and has no help in texture analysis, therefore, joint distribution of local pixel gray value can be converted to signed gray scale differential:

$$T = t(g_0 - g_c, g_1 - g_c, \cdots, g_{P-1} - g_c)$$

To reduce the computational complexity, consider using only differential symbols, ignoring the actual value, thus implement a local texture binary encoding model:

$$T \approx t(s(g_0 - g_c), s(g_2 - g_c), \cdots, s(g_{P-1} - g_c))$$

$$s(g_c, g_i) = \begin{cases} 1 & g_i - g_c \geq 0 \\ 0 & g_i - g_c < 0 \end{cases}$$

For the purpose of reducing calculation, convert the binary encoding of local texture pattern into a real-number encoding:

$$LBP_{P,R} = \sum_{i=0}^{P-1} 2^i s(g_c, g_i)$$

The figure shows an example, the left 3×3 area corresponds to the one of the right binary texture pattern $Pattern = 11111000$, in fact, as the number of coding is $LBP = (11111000)_2 = (248)_{10}$.

6	3	1
8	6	3
7	9	8

1	0	0
1		0
1	1	1

Fig. 2. The binary code of local pixels

Such a texture of local binary pattern (LBP) depicts the changes in the neighbor-hood of gray pixels compare to the center point, it has gray scale invariance. Studies have shown that the human visual system and the average brightness of the texture perception is independent, and LBP approach attaches importance to gray scale changes, without being affected by the *average* brightness, it becomes an ideal method in texture analysis. From the definition of local binary pattern, there are 2^P different types of binary patterns (Fig. 2).

3   Design of Leaf Image Recognition Website

3.1   Website Structure

This online system is based on MVC (Model, View and Controller) architecture. The plants descriptions are saved in MySQL database, server will retrieve the detail of the species (Fig. 3).

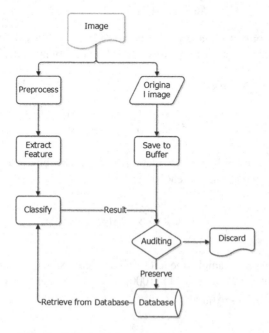

Fig. 3. Schematic diagram of the core function

Fig. 4. The homepage of online leaf recognition system

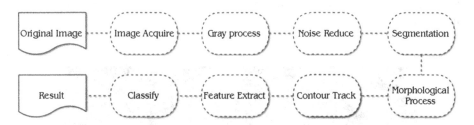

Fig. 5. Procedure of classification on server

That user uploaded after a complex backend process. The View is realized by Html and CSS to show the user interface in web browser as shown in Fig. 4. And PHP is used to send instructions to the model to refresh the model's state. It can also get parameters from view and change the view's presentation.

The procedure of identifying a leaf image contains those critical step: image acquisition, preprocessing, feature extraction and classification (Fig. 5).

3.2 Leaf Dataset

We build the system using Intelligent Computing Laboratory (ICL) Leaf Dataset from Chinese Academy of Sciences [10] as training data: (Fig. 6).

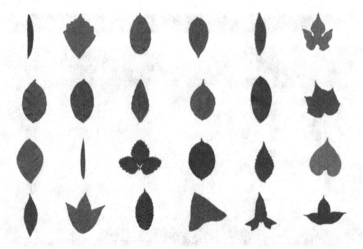

Fig. 6. Part of Intelligent Computing Laboratory (ICL) leaf images

3.3 Extract Feature of Uploaded Leaf Image with LBP and Its Derivatives

In classification, LBP collects number of each pattern into a histogram. Then a classification can simply compute histogram similarities to identify the plant. What server does is running a monitor LBP program of OpenCV to generate histogram of a leaf image (Fig. 7).

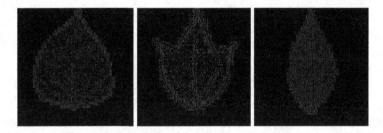

Fig. 7. LBP features images of leaves

3.4 Nearest Neighbor Classifier

The server is using Nearest Neighbor classifier to handle the feature vectors from LBP algorithm [11–13]. As illustrated in Fig. 8, in the right view, the upper left corner is uploaded image, after identifying the species' name and the corresponding picture appears in the web browser (Table 1).

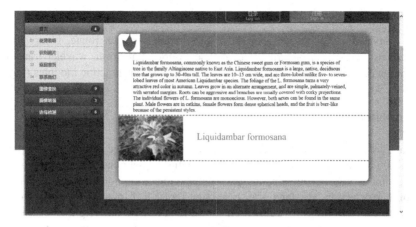

Fig. 8. Recognition result of leaf image

Table 1. Part of Intelligent Computing Laboratory (ICL) leaf details

NO.	SPECIES
·001	Common Elaeocarp
·002	Dogbane Oleander
·003	Persimmon
·004	Chinese Fevervine Herb
·005	Red leaf Cherry Plum
·006	India Mock strawberry
·007	Spica Prunellae
·008	Winter Jasmine
·009	Water elm
·010	Weeping Willow

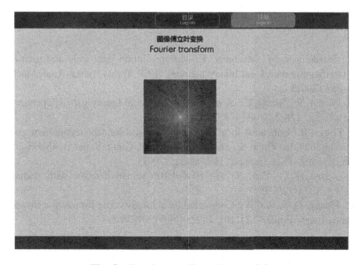

Fig. 9. Fourier transformation module

3.5   Other Functional Models

Image conversion module includes those sub-modules: resize, rotate, binary image, gray process, Fourier transformation, inverse Fourier transform, wavelet transform, Hough transform, histogram equalization and other nine sub-function modules (Fig. 9).

4   Conclusions

Plant leaf image recognition system by means of modern image processing and pattern recognition techniques, implements various functions, including leaf image gray pro-ʻcessing, image noise reduction, image segmentation, morphological processing and pre-contour detection, etc., and after extracting target leaf images' feature with LBP, the server call the embedded classifier, based on the Nearest Neighbor classifier to identify, and finally retrieves matches in the database. The software system perfor- mance indicators are:

- The number of species can be identified is: 30;
- Image acquisition costs: 1−10 ms;
- Image preprocessing costs: 20−50 ms;
- Leaf image classification costs: 1 s − 2 s;
- Leaf image recognition rate: 76 %.

Acknowledgments. This work was supported by the grants of the National Science Foundation of China, Nos. 61133010, 61373105, 61303111, 61411140249, 61402334, 61472282, 61472280, 61472173, 61373098 and 61272333, China Postdoctoral Science Foundation Grant, Nos. 2014M561513, and partly supported by the National High-Tech R&D Program (863) (2014AA021502 & 2015AA020101), and the grant from the Ph.D. Programs Foundation of Ministry of Education of China (No. 20120072110040), and the grant from the Outstanding Innovative Talent Program Foundation of Henan Province, No. 134200510025.

References

1. Ojala, T., Pietikainen, M., Maenpaa, T.: Multiresolution gray-scale and rotation invariant texture classification with local binary patterns. IEEE Trans. Pattern Anal. Mach. Intell. **24** (7), 971–987 (2002)
2. Zhou, H., Wang, R., Wang, C.: A novel extended local binary pattern operator for texture analysis. Inf. Sci. **178**(22), 4314–4325 (2008)
3. Tan, X., Triggs, B.: Enhanced local texture feature sets for face recognition under difficult lighting conditions. In: Zhou, S., Zhao, W., Tang, X., Gong, S. (eds.) AMFG 2007. LNCS, vol. 4778, pp. 168–182. Springer, Heidelberg (2007)
4. Wang, X., Han, T.X., Yan, S.: An HOG-LBP human detector with partial occlusion handling. In: ICCV (2009)
5. Zhao, Y., Huang, D.-S., Jia, W.: Completed local binary count for rotation invariant texture. IEEE Trans. Image Process. **21**(10), 4492–4497 (2012)

6. Heikkilä, M., Pietikäinen, M., Schmid, C.: Description of interest regions with local binary patterns. Pattern Recogn. **42**(3), 425–436 (2009)
7. Tan, X., Triggs, B.: Fusing gabor and LBP feature sets for kernel-based face recognition. In: Proceedings of the IEEE International Workshop on Analysis and Modeling of Face and Gesture, pp. 235–249 (2007)
8. Wang, J.-G., Yau, W.-Y., Wang, H. L.: Age categorization via ECOC with fused gabor and LBP features. In: Proceedings of the IEEE Workshop on Applications of Computer Vision (WACV), pp. 313–318 (2009)
9. Haralick, R.M., Shanmugam, K., Dinstein, I.: Textural features for image classification. IEEE Transactions on Systems, Man, and Cybernetics **SMC-3(6)**, 610–621 (1973)
10. http://www.intelengine.cn/English/dataset
11. Bremner, D., Demainem, E., Erickson, J., Iacono, J., Langerman, S., Morin, P., Toussaint, G.: Output-sensitive algorithms for computing nearest-neighbor decision boundaries. Discrete Comput. Geometry **33**(4), 593–604 (2005). doi:10.1007/s00454-004-1152-0
12. Coomans, D., Massart, D.L.: Alternative k-nearest neighbour rules in supervised pattern recognition : part 1. k-nearest neighbour classification by using alternative voting rules. Anal. Chim. Acta **136**, 15–27 (1982). doi:10.1016/S0003-2670(01)95359-0
13. Huang, D.S.: Radial basis probabilistic neural networks: model and application. Int. J. Pattern Recognit Artif Intell. **13**(7), 1083–1101 (1999)
14. Guyer, D., Miles, G., Schreiber, M., Mitchell, O., Vanderbilt, V.: Machine vision and image processing for plant identification. Trans. ASAE **29**, 1500–1507 (1986)
15. Wang, X.-F., Huang, D.-S., Xu, H.: An efficient local Chan-Vese model for image segmentation. Pattern Recogn. **43**, 603–618 (2010)
16. Huang, D.-S.: Systematic theory of neural networks for pattern recognition, vol. 28, pp. 323–332. Publishing House of Electronic Industry of China, Beijing (1996)
17. Yu, H.-J., Huang, D.-S.: Normalized feature vectors: a novel alignment-free sequence comparison method based on the numbers of adjacent amino acids. IEEE/ACM Transact. Comput. Biol. Bioinform. (TCBB) **10**, 457–467 (2013)
18. Huang, D.-S., Jiang, W.: A general CPL-AdS methodology for fixing dynamic parameters in dual environments. IEEE Trans. Syst. Man Cybern. Part B Cybern. **42**, 1489–1500 (2012)
19. Huang, D.-S., Du, J.-X.: A constructive hybrid structure optimization methodology for radial basis probabilistic neural networks. IEEE Trans. Neural Networks **19**, 2099–2115 (2008)
20. Wang, X.-F., Huang, D.-S.: A novel density-based clustering framework by using level set method. IEEE Trans. Knowl. Data Eng. **21**, 1515–1531 (2009)
21. Shang, L., Huang, D.-S., Du, J.-X., Zheng, C.-H.: Palmprint recognition using FastICA algorithm and radial basis probabilistic neural network. Neurocomputing **69**, 1782–1786 (2006)
22. Zhao, Z.-Q., Huang, D.-S., Sun, B.-Y.: Human face recognition based on multi-features using neural networks committee. Pattern Recogn. Lett. **25**, 1351–1358 (2004)
23. Huang, D., Ip, H., Chi, Z.: A neural root finder of polynomials based on root moments. Neural Comput. **16**, 1721–1762 (2004)
24. Huang, D.-S.: A constructive approach for finding arbitrary roots of polynomials by neural networks. IEEE Trans. Neural Networks **15**, 477–491 (2004)
25. Krogh, A., Vedelsby, J.: Neural network ensembles, cross validation, and active learning. Advances in neural information processing systems, pp. 231–238 (1995)
26. Freund, Y., Schapire, R.E.: Experiments with a new boosting algorithm. In: ICML, pp. 148–156
27. Li, B., Zheng, C.-H., Huang, D.-S.: Locally linear discriminant embedding: an efficient method for face recognition. Pattern Recogn. **41**, 3813–3821 (2008)

28. Do, M.N., Vetterli, M.: The contourlet transform: an efficient directional multiresolution image representation. IEEE Trans. Image Process. **14**, 2091–2106 (2005)
29. Sajedi, H., Jamzad, M.: A contourlet-based face detection method in color images. In: Third International IEEE Conference on Signal-Image Technologies and Internet-Based System, SITIS 2007, pp. 727–732. IEEE (2007)
30. Boukabou, W.R., Bouridane, A.: Contourlet-based feature extraction with PCA for face recognition. In: NASA/ESA Conference on Adaptive Hardware and Systems, AHS 2008, pp. 482–486. IEEE (2008)
31. Rahati, S., Moravejian, R., Mohamad, E., Mohamad, F.: Vehicle recognition using contourlet transform and SVM. In: Fifth International Conference on Information Technology: New Generations, ITNG 2008, pp. 894–898. IEEE (2008)
32. Xiangbin, Z.: Texture classification based on contourlet and support vector machines. In: 2009 ISECS International Colloquium on Computing, Communication, Control, and Management, pp. 521–524
33. Liu, Z., Fan, X., Lv, F.: SAR image segmentation using contourlet and support vector machine. In: Fifth International Conference on Natural Computation, ICNC 2009, pp. 250–254. IEEE (2009)
34. Wang, J., Ge, Y.: Texture feature recognition based on Contourlet transform and support vector machine. Jisuanji Yingyong/ J. Comput. Appl. 33 (2013)
35. Haralick, R.M., Shanmugam, K., Dinstein, I.H.: Textural features for image classification. IEEE Trans. Syst. Man Cybern. **3**, 610–621 (1973)
36. Vapnik, V.: The Nature of Statistical Learning Theory. Springer Science & Business Media, New York (2000)
37. Burges, C.J.: A tutorial on support vector machines for pattern recognition. Data Min. Knowl. Disc. **2**, 121–167 (1998)
38. Schöllkopf, B., Burges, C.J., Smola, A.J.: Advances in Kernel Methods: Support Vector Learning. MIT press, Cambridge (1999)
39. Soderkvist, O.J.O.: Computer Vision Classication of Leaves from Swedish Trees (2001)

Using Additive Expression Programming for System Identification

Bin Yang[(⊠)]

School of Information Science and Engineering, Zaozhuang University,
Zaozhuang, People's Republic of China
batsi@126.com

Abstract. The system identification is crucially important process, which could develop the mathematical representation of physical system from observed data. In this paper, a new model, called additive expression tree (AET) model is proposed to encode the linear and nonlinear systems. A new structure-based evolutionary algorithm and artificial bee colony (ABC) are used to optimize the architecture and parameters of additive expression tree model, respectively. Experimental results demonstrate that our proposed model and hybrid approach could identify the linear/nonlinear systems effectively.

Keywords: System identification · Additive expression tree · Artificial bee colony · Hybrid approach

1 Introduction

System identification uses statistical methods to build mathematical models of dynamical systems from input-output measured data. Mathematical models are used to represent and analyze the complex problems in various application domains, such as natural science, engineering disciplines and social sciences [1, 2]. System identification helps to study the effects of different parts, and to predict the future behavior of system [3–5]. Thus the system identification is crucially important process.

In general, there are two kinds of models: non-parametric models and parametric models. Non-parametric methods model a system directly with its responses, whereas parametric methods refer to determine the structure and estimate parameters of system [6]. During the last few years, least mean-squares (LMS) method and evolutionary algorithms have been proposed to evolve the parameters of linear/nonlinear systems, and genetic programming (GP) approach has been used to search the structures [7]. Iba used ordinary differential equation model (ODE) to reconstruct gene regulatory network. Least mean square and ordinary GP were used to identify the parameters and structure of ODE, respectively [8]. To identify dural-rate sampled-data linear systems, a hierarchical least square was presented to estimate the parameters ARMAX models [9]. Manoj Kandpal proposed a GP-based causality detection methodology, in which evolutionary computation-based procedures along with parameter estimation methods were used to derive a mathematical model of biochemical system [10].

We also proposed a new representation scheme of the additive tree models for the system identification especially identification of linear/nonlinear ODE system [11].

© Springer International Publishing Switzerland 2015
D.-S. Huang et al. (Eds.): ICIC 2015, Part I, LNCS 9225, pp. 671–681, 2015.
DOI: 10.1007/978-3-319-22180-9_67

In 2010, this model was used for identifying a system of ODEs to predict the small-time scale traffic measurements data [12]. Compared with genetic programming and gene expression programming (GEP), this model was powerful than the GEP and GP models, both in the aspects of accuracy and runtime. But like GP individuals, additive tree models are also represented and manipulated as nonlinear tree entities, which leads to some problems such as the inefficiencies of handling tree structure and the difficulties of program implementation.

In this paper, as linear variant of additive tree model, additive expression tree (AET) model is first proposed to encode the linear and nonlinear systems. We propose a hybrid evolutionary method, in which a new structure-based evolutionary algorithm is used to optimize the architecture of systems and selection of input variables and corresponding parameters are evolved by artificial bee colony.

2   Representation of Additive Expression Tree Model

We use a structure-based evolutionary algorithm to evolve the architecture of the additive expression models for the linear/nonlinear system. For this purpose, we encode the linear/nonlinear expression into an additive expression tree model as illustrated in Fig. 1. Figure 1(a) is linear/nonlinear expression of symbolic tree structure, which need be created randomly, and Fig. 1(b) is the corresponding expression tree structure.

Two instruction/operator sets I_0 and I_1 are used for generate the additive expression tree model.

$$\begin{cases} I_0 = \{+_2, +_3, \ldots, +_N\} \\ I_1 = F \cup T = \{*, /, \sin, \cos, \exp, r\log, x, R\} \end{cases} \tag{1}$$

N is an integer number (the maximum number of linear/nonlinear terms). Each term is encoded as GEP gene. I_0 is the instruction set and the root node, which returns the weighted sum of a number of linear/nonlinear terms according to the GEP gene expressions. I_1 is the operator set, where $F = \{*, /, \sin, \cos, \exp, r\log\}$ and $T = \{x, R\}$ are function and terminal sets. $*$, $/$, sin, cos, exp, r log, x, and R denote the multiplication, protected division $(\forall x, y \in R : \text{when } y = 0, x/0 = 1)$, sine, cosine, exponent, protected logarithm $(\forall x \in R, x \neq 0 : r\log(x) = \log(\text{abs}(x))$ and $r\log(0) = 0)$, system inputs, and random constant number, taking 2, 2, 1, 1, 1, 1, 0 and 0 arguments respectively [11].

A GEP gene is a string of function and terminal symbols, which is composed of a head and a tail [13–15]. The head part contains both function and terminal symbols, whereas the tail part contains terminal symbols only. The head could be created through selecting symbols randomly from the set I_1. The symbols of tail are selected from set F only. For each problem, user must determine the head length (h). The tail length (t) is computed as:

$$t = (n - 1) \times h + 1. \tag{2}$$

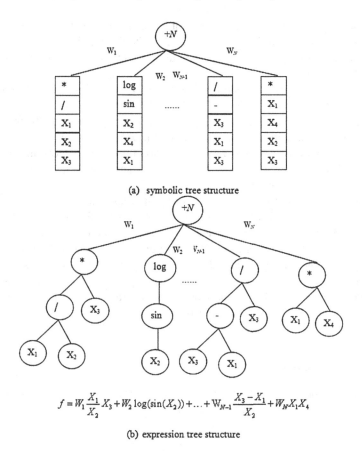

(a) symbolic tree structure

(b) expression tree structure

$$f = W_1 \frac{X_1}{X_2} X_3 + W_2 \log(\sin(X_2)) + \ldots + W_{N-1} \frac{X_3 - X_1}{X_2} + W_N X_1 X_4$$

Fig. 1. Example of a system in the form of additive expression tree model.

Where n is the maximum number of arguments of functions. According to set F, n is set as 2.

3 The Proposed Hybrid Method

In this section, the method description can be made in several sub-chapters.

3.1 Structure Optimization Methods

To search an optimal or near-optimal additive expression tree model is formulated as an evolutionary finding process. We used the structure operators as following:

(1) Mutation. We use three mutation operators to generate offsprings from the parents, which are described as following:

(a) One-point mutation. Select one point in the tree randomly, and replace it with another symbol, which selects from set I_1. Notice that in the head any symbol could be changed, but in the tail terminal the symbols are allowed to be changed only.

(b) One-gene mutation. Randomly select one GEP gene in the tree, and replace it with another newly generated gene.

(c) Change all terminal symbols. Select every terminal symbol in the additive expression tree model, and replace it with another terminal symbol.

(2) Crossover. First two parents are selected according to the predefined crossover probability P_c and select one GEP gene for each additive expression tree randomly, and then swap the selected gene.

(3) Selection. EP-style tournament selection [16] is applied to select the parents for the next generation. Pairwise comparison is conducted for the union of μ parents and μ offsprings. For each individual, q opponents are chosen uniformly at random from all the parents and offspring. For each comparison, if the fitness of individual fitness is no smaller than the one of opponent, it receives a selection. Select μ individuals out of parents and offsprings, which have most wins to form the next generation. This is repeated in each generation until a predefined number of generations or the best structure is found.

3.2   Fitness Definition

Mean square error (MSE) or root mean square error (RMSE) are used as fitness function to evaluate the performance of candidate model.

$$MSE = \frac{1}{N} \sum_{i=1}^{N} (x_1^i - x_2^i)^2$$

$$RMSE = \sqrt{\frac{1}{N} \sum_{i=1}^{N} (x_1^i - x_2^i)^2} \tag{3}$$

Where N is the number of samples, x_1^i and x_2^i are the actual and model output of i-th sample.

3.3   Parameter Optimization of Models

In the parameter learning stage, there are many learning methods, such as PSO, GA, EP and SA. In this paper, artificial bee colony algorithm is selected because its fast convergence, more accurate solution and stability [17, 18].

According to Fig. 1, we check all the parameters contained in each model, and count their number n_i ($i = 1, 2,..., M$, M is the population size of additive expression tree model). According to n_i, the parameter vector could be created randomly. ABC algorithm can be summarized as follows.

(1) Choose the initial value of algorithm parameter including maximum number of generation (*max_gen*), number of employed bees and onlooker bees (*num_employ* and *num_onlooker*), and the maximum times that food sources do not change its location (*limit*).

(2) Generate *num_employ* and *num_onlooker* solutions, calculate the fitness value, and select the best half of solutions as foods.

(3) In order to produce a candidate food position from the old one in memory, the ABC uses the following equation:

$$v_{ij} = x_{ij} + R_{ij}(x_{ij} - x_{kj}),$$ (4)

where v_{ij} is a new position of food, R_{ij} is a random number in the range $[-1,1]$, $k \in \{1, 2, 3, \ldots, N\}$ and $k \neq i$. Calculate the fitness of v_i, and when the fitness of is better than x_i, replace x_i with v_i. If x_i is not changed, *failure_i = failure_i* + 1.

(4) The probabilities for onlookers are calculated as follows.

$$p_i = \frac{fit_i}{\sum_{n=1}^{N} fit_n}$$ (5)

Where fit_i is the fitness value of i-th bee.

(5) According to p_i, onlooker bees search the area as employed bees.

(6) If *failure_i > limit*, then scout bee randomly generates a new solution according to the following equation.

$$x_{ij} = x_{min\,j} + rand(0, 1)(x_{max\,j} - x_{min\,j})$$ (6)

(7) Record the best solution at each generation.

(8) If *max_gen* is reached or a satisfactory solution is found, then stop; otherwise go to step (3).

3.4 Summary of Our Proposed Algorithm

(1) Create the initial population randomly, containing structures and their corresponding parameters.

(2) Structure optimization is achieved by the structure operators as described in Sect. 3.1. Fitness function is calculated by mean square error or root mean square error.

(3) According to fitness value, sort the population. At some interval of generations, select certain percentage of population to optimize parameters. Parameter optimization is achieved by ABC as described in Sect. 3.3. During this process, the structure of model is fixed.

(4) If the maximum number of generations is reached or a satisfactory solution is found, then stop; otherwise go to step (2).

4 Experimental Results and Analysis

To test the effectiveness of the proposed method, our method is applied to prediction of chaotic time series and identification of linear/nonlinear systems. The parameters setting in this experiment is shown in Table 1. To compare trustingly, the method using additive tree model also adopts the parameters in Table 1.

Table 1. Parameters for experiment.

Population size	10
Generation	200
Crossover rate	0.3
ABC employed bees size	15
ABC onlooker bees size	15
ABC generation	200
ABC limit	20

4.1 Prediction of Chaotic Time Series

Henon map is described as following [11]:

$$x(k+1) = 1.4 - x(k)^2 + 0.3x(k-1). \tag{7}$$

The initial conditions of $x(0)$ and $x(1)$ are created randomly from the interval $[-2.0, 2.0]$. Using Eq. (7) and initial conditions, 100 training data and 100 testing data are randomly generated. The used instruction set $I_0 = \{+2, +3, +4, +5, +6, +7, +8\}$ and $I_1 = \{*, x, R\}$.

The evolved Henon map as the best solution is obtained as following:

$$x(k+1) = 1.400006 - 1.000001x(k)^2 + 0.299998x(k-1). \tag{8}$$

Figures 2 and 3 show the actual and model outputs and prediction error for training data set and testing data set. Note that our identification model and actual model are almost identical except for slightly different parameters.

To test the validity of the additive expression tree model, we also compare prediction of chaotic time series using additive tree model. After searching the best solution, the results are listed in Table 2. From the empirical results, it is evident that the proposed method is powerful than the additive tree mode, both in the aspects of accuracy and runtime. And our method could search the best solution at 20-th generation.

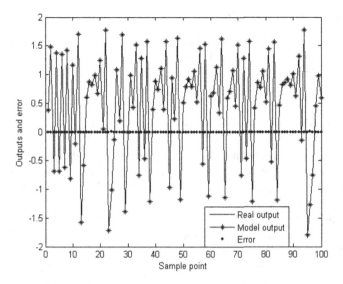

Fig. 2. The actual and model outputs for training data and prediction error.

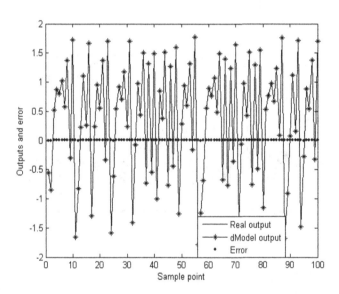

Fig. 3. The actual and model outputs for testing data and prediction error.

Table 2. Comparison results using the proposed method and additive tree model.

	Our method	Additive tree model
Runtime	4.781 s	8.25 s
RMSE	0.000004	0.0000632
Generation	20	112

4.2   Nonlinear System Identification

A second-order non-minimum phase system with gain 1, time constants 4 s and 10 s, a zero at 1/4 s, and output feedback with a parabolic nonlinearity is chosen to be identified [11]. The nonlinear system is described as following:

$$
\begin{aligned}
y(k) = &-0.07289[u(k-1) - 0.2y^2(k-1)] + 0.09394[u(k-2) - 0.2y^2(k-2)] \\
&+ 1.68364y(k-1) - 0.070469y(k-2).
\end{aligned} \tag{9}
$$

Where input signal is shown in Fig. 2 from Experiment 1, the initial conditions of x (0) and $x(1)$ are created randomly from the interval $[-1.0, 1.0]$. The used instruction set $I_0 = \{+2, +3, +4, +5, +6, +7, +8\}$ and $I_1 = \{*, x(k-1), u(k-2), y(k-1), y(k-2), R\}$. The actual and model outputs and prediction error for training data set is shown in Fig. 4. Our optimized method is applied for the testing data. The result illustrated in Fig. 5. From the simulation results, we could see that our identification model is as almost same as the targeted model.

We also make experiment using the additive tree model. The comparison results are listed in Table 3. From the simulation results, it could be seen that the proposed method performs better than the additive tree mode.

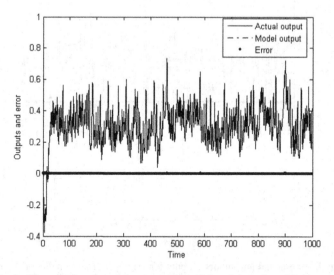

Fig. 4.   The actual and model outputs for training data and prediction error.

4.3   Reconstruction of Polynomials

To test the performance of evolving the complicated polynomial by using our model, the polynomial example is given by

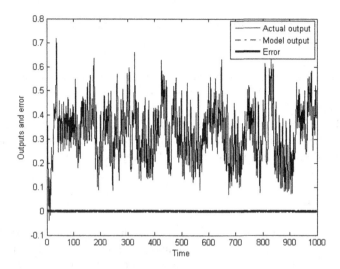

Fig. 5. The actual and model outputs for testing and prediction error.

Table 3. Comparison results using the proposed method and additive tree model.

	Our method	Additive tree model
Runtime	483.062 s	601.891 s
RMSE	0.000445	0.000985

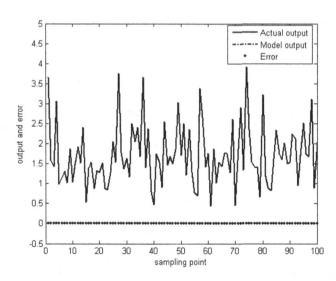

Fig. 6. The actual and model outputs for test data set and approximation error.

$$y = 0.2 + 0.3x_1 + 0.4x_2 + 0.5x_3 + 0.6x_1^2 + 0.7x_2^2 + 0.8x_3^2 + 0.9x_1x_2$$
$$+ 0.1x_1x_3 + 0.2x_2x_3. \tag{10}$$

200 training data and 100 testing data are randomly generated at interval [0,1]. The used instruction set $I_0 = \{+2, +3, +4, +5, +6, +7, +8, +9, +10, +11, +12\}$ and $I_1 = \{*, x_1, x_2, x_3, R\}$.

The RMSE results are 0.001254 and 0.001717 for training data set and test data set, respectively. The identification result is shown in Fig. 6 for test data set. It can be seen that the evolved polynomial is very close to the targeted model.

5   Conclusion

In this paper, an additive expression tree model and its design and evolved method are proposed for system identification. The additive expression tree model could be seen as linear variant of additive tree model, which is used to encode the linear and nonlinear systems. A new structure-based evolutionary algorithm is used to optimize the architecture of systems and corresponding parameters are evolved by artificial bee colony. Experiment results reveal that the proposed method perform better than the additive tree mode, both in the aspects of accuracy and runtime.

Acknowledgment. This work was supported the PhD research startup foundation of Zaozhuang University (No.1020702).

References

1. Gershenfeld, N.: The Nature of Mathematical Modeling. Cambridge University Press, England (1998)
2. Frigg, R., Hartmann, S.: Models in science. In: The Stanford Encyclopedia of Philosophy (2006)
3. Neumaier, A.: Mathematical model building, chapter 3. In: Kallrath, J. (ed.) Modeling Languages in Mathematical Optimization. Kluwer Academic Publisher, Boston (2004)
4. Sundar, S., Naresh, R., Misra, A.K., Shukla, J.B.: A nonlinear mathematical model to study the interactions of hot gases with cloud droplets and raindrops. Appl. Math. Model. **33**(7), 3015–3024 (2009)
5. Robert, M., Thomas, S.R.: Hormonal regulation of salt and water excretion: a mathematical model of whole kidney function and pressure natriuresis. Am. J. Physiol.-Renal Physiol. **306** (2), 224–248 (2014)
6. Wellstead, P.E.: Non-parametric methods of system identification. Automatica **17**(1), 55–69 (1981)
7. Cao, H., Kang, L., Chen, Y., Yu, J.: Evolutionary modeling of systems of ordinary differential equations with genetic programming. Genet. Program Evolvable Mach. **1**(40), 309–337 (2000)
8. Iba, H.: Inference of differential equation models by genetic programming. Inf. Sci. **178**(23), 4453–4468 (2008)

9. Ding, J., Ding, F., Liu, X.P., Liu, G.J.: Hierarchical least squares identification for linear SISO systems with dual-rate sampled-data. IEEE Transa. Autom. Control **56**(11), 2677–2683 (2011)
10. Kandpal, M., Kalyan, C.M., Samavedham, L.: Genetic programming-based approach to elucidate biochemical interaction networks from data. IET Syst. Biol. **7**(1), 18–25 (2013)
11. Chen, Y.H., Yang, J., Zhang, Y., Dong, J.W.: Evolving additive tree models for system identification. Int. J. Comput. Cogn. **3**(2), 19–26 (2005)
12. Chen, Y.H., Yang, B., Meng, Q.F., Zhao, Y.O., Abraham, A.: Time-series forecasting using a system of ordinary differential equations. Inf. Sci. **181**(1), 106–114 (2010)
13. Mihai, O.: A comparison of several linear genetic programming techniques. Complex Syst. **14**(4), 285–313 (2003)
14. Ferreira, C.: Gene expression programming: a new adaptive algorithm for solving problems. Complex Syst. **13**(2), 87–129 (2001)
15. Ferreira, C.: Gene expression programming in problem solving. In: Roy, R., Diplom-Phys, M.K., Ovaska, S., Furuhashi, T., Hoffmann, F. (eds.) Soft Computing and Industry: Recent Applications, pp. 635–653. Springer, Heidelberg (2002)
16. Chellapilla, K.: Evolving computer programs without subtree crossover. IEEE Trans. Evol. Comput. **1**, 209–216 (1997)
17. Cheng, X., Jiang, M.: An improved artificial bee colony algorithm based on gaussian mutation and chaos disturbance. In: Tan, Y., Shi, Y., Ji, Z. (eds.) Advances in Swarm Intelligence. LNCS, vol. 7331, pp. 326–333. Springer, Heidelberg (2012)
18. Karaboga, D., Gorkemli, B., Ozturk, C., Karaboga, N.: A comprehensive survey: artificial bee colony (ABC) algorithm and applications. Artif. Intell. Rev. **42**(1), 21–57 (2014)

Supervised Feature Extraction of Hyperspectral Image by Preserving Spatial-Spectral and Local Topology

Peng Zhang[1(✉)], Haixia He[2], Zhou Sun[1], and Chunbo Fan[1]

[1] Data Center, National Disaster Reduction Center of China, Beijing,
People's Republic of China
{zhangpeng, sunzhou, fanchunbo}@ndrcc.gov.cn
[2] Department of Satellite Remote Sensing, National Disaster Reduction
Center of China, Beijing, People's Republic of China
hehaixia@ndrcc.gov.cn

Abstract. Manifold learning, as a promising tool for nonlinear dimensionality reduction of hyperspectral image (HSI) data, has drawn great research interests in the remote sensing community. It can extract meaningful and low-dimensional features underlying complex HSI data, which is useful in classification of ground targets. However, there are two limitations with current approaches, few considerations of spatial information and lack of explicit mapping relationship. In this paper, we propose a supervised spatial-spectral local topology preserving embedding (sssLTPE) method for efficient feature extraction of HSI, which owns two merits. First, spatial and spectral information at each pixel is integrated by an intuitive strategy. Second, an explicit and nonlinear mapping relationship is provided to effectively map unlabeled data to learned feature space. Experiments conducted on benchmark data set demonstrate that high classification accuracy can be obtained by using the features extracted by sssLTPE.

Keywords: Supervised learning · Manifold learning · Feature extraction · Hyperspectral image classification

1 Introduction

In recent years, hyperspectral image (HSI) analysis has drawn great research interests in the remote sensing community. Due to advances in spectral sensors, large number of contiguous and narrow bands can be captured within a wide range of spectrum. Such rich spectral information makes HSI more capable to classify ground covers [1, 2], compared with single-spectral and multi-spectral images. Successful applications include agriculture [3], mineralogy [4], and military [5], to name just a few.

The massive spectral bands of HSI also pose great challenges for classification of ground targets. In HSI, each pixel is composed of tens or hundreds of bands recording spectral reflections of ground-covers, which form a high-dimensional vector in the sample space. Such high dimensionality issues the so-called "curse of dimensionality" or Hughes phenomenon [6], where computational complexity grows exponentially with number of dimensions. Besides, for specific target, its characteristic reflective spectrum

© Springer International Publishing Switzerland 2015
D.-S. Huang et al. (Eds.): ICIC 2015, Part I, LNCS 9225, pp. 682–692, 2015.
DOI: 10.1007/978-3-319-22180-9_68

concentrates on a very small range of bands, hence most bands in HSI are redundant, and such redundancy results in complex data structure, which adds up to the difficulties of designing an efficient classifier.

A straightforward strategy to address the above issues is to reduce the dimensionality of HSI data and map high-dimensional bands to a low-dimensional feature space, which is more compact and suitable for classification. Principal component analysis (PCA), as a classical dimensionality reduction (DR) method, has been widely applied to pre-processing and feature extraction for HSI data [7, 8]. However, due to its linear model assumption, the performance of PCA is not satisfactory for complex scenes, especially when there exist numerous complicated ground targets.

Recent research discovers that high-dimensional and contiguous spectral bands in fact concentrates in a low-dimensional space, which is nonlinear embedded in the ambient space [9–11]. Manifold learning (ML) [18, 19], as a promising tool for nonlinear dimensionality reduction, has witnessed its successful applications to processing HSI data, including feature extraction [12], segmentation [13], anomaly detection [14], and classification [15, 16]. Nevertheless, there are two issues with manifold learning methods. One is that current approaches focus on the spectral properties of data, while the natural spatial correlations among pixels are not well considered. The other is that ML methods implicitly learn low-dimensional features, which means that explicit mapping relationship is unavailable. This greatly limits the extension to testing samples in feature extraction and classification. Although linear projection constraint is introduced as an alternative [16, 17], the nonlinearity can no longer be fully preserved in such circumstance.

To address the above issues, in this paper we propose a Spatial-Spectral Supervised Local Topology Preserving Embedding method, named as sssLTPE, for nonlinear feature extraction of HSI data. We propose an intuitive approach to simultaneously consider spatial and spectral information, which is simple to implement. Based on our previous work on nonlinear and explicit mapping for manifold learning [20], we extend that to a supervised version to better fit the demand of classification. With these features, sssLTPE can efficiently learn the local topology in both spatial and spectral domain. Besides, such hidden and nonlinear geometric structure, together with inner and intra class relationships, can be optimally preserved to a low-dimensional feature space, being company with an explicit mapping. As a consequence, fast feature extraction of unlabeled pixels in the scene can be achieved and the extracted features are useful for classification task. To validate the performance of sssLTPE, we conduct experiments on benchmark data and compare it with k-nn, PCA, and its linear counterpart NPP [21, 22]. Classification results demonstrate that sssLTPE has superior performance and achieves high accuracy even with a simple one-nearest-neighbor classifier.

The rest parts of this paper are organized as follows. Section 2 describes the proposed sssLTPE method in details. Section 3 demonstrates experimental results conducted on benchmark data set and validates the effectiveness of sssLTPE. Some concluding remarks are given in Sect. 4.

2 Supervised Spatial-Spectral Local Topology Preserving Embedding

2.1 Fusion of Spatial and Spectral Information

Let $(p(r,c))$ be the pixel at the r-th row and c-th column of a hyperspectral image, which consists of B discrete values of spectral bands. We use pixel coordination $c = (r,c)$ and array of bands $b = (b_1, b_1, \cdots, b_B)$ to code spatial and spectral information for $p(r,c)$, respectively. Then a straightforward approach to fuse these two kinds of information is to integrate c and b into a coherent representation. Formally, we use vector

$$x(r,c) = (c,b)^T = (r,c,b_1,b_1,\cdots,b_B)^T = (x^1,x^2,\cdots,x^{(B+2)})^T, \qquad (1)$$

as an augmented representation for pixel $p(r,c)$. By doing so, natural spatial correlations among pixels' coordinates and contiguous spectrum can be simultaneously considered in feature extraction.

2.2 Modelling of Local Topology

For all labeled pixels, we reformulate all $x(r,c)$'s as a linear sequence by ranking along columns and denote the i-th labeled pixel by x_i. Here we use subscripts for indices of pixels (or samples) and superscripts for components of a vector. Before learning, the components of x_i are normalized by the following principle

$$x_i^j \leftarrow (x_i^j - \min_i x_i^j)/(\max_i x_i^j - \min_i x_i^j).$$

For x_i, we first identify its local neighborhood $\mathcal{N}_i$ in a supervised mode. We define $x_j \in \mathcal{N}_i$ if and only if x_j is among x_i's k nearest neighbors and they are in the same class, in other words, each sample's neighborhood is restricted within the class it belongs to.

Next, we model the local topology of x_i in the ambient space by locally linear reconstruction weights, which is introduced by the widely used locally linear embedding (LLE) method [19]. It is assumed that the data manifold is locally flat and each sample can be linearly represented by its neighbors. This is equivalent to solve the following optimization problem.

$$\begin{aligned} \min \quad & \|x_i - \sum_{x_j \in \mathcal{N}_i} W_{ij} x_j\|_2 \\ \text{s.t.} \quad & \sum_j W_{ij} = 1 \end{aligned} \qquad (2)$$

The weights $\{W_{ij}\}$ represent the linear coefficients for reconstructing x_i from its neighbors. The weights have a closed-form solution given by

$$W_i = G^{-1}e / e^T G^{-1}e, \tag{3}$$

where $W_i = (W_{i1}, W_{i2}, \cdots, W_{ik})^T$ and e is a column vector of all ones. The (j, l)-th entry of the $k \times k$ matrix G is $(x_j - x_i)^T (x_l - x_i)$. W_{ij} is non-zero if $x_j \in \mathcal{N}_i$ and vanishes otherwise, hence inter and intra class information can be coded with W_{ij}.

2.3   Manifold Learning by Preserving Local Topology

The local topology, represented by linear weights $\{W_{ij}\}$, codes intra-class similarity and inter-class separability. Then the aim of dimensionality reduction is to find a low-dimensional feature space which best preserves such weights. Let y_i be the learned low-dimensional embedding of x_i, we request that each y_i can also be linearly reconstructed by its neighbors with the same weights. This is achieved by solving the following optimization problem

$$\begin{aligned} \min \quad & \sum_i \|y_i - \sum_{x_j \in \mathcal{N}_i} W_{ij} y_j\|_2^2, \\ \text{s.t.} \quad & \sum_i y_i y_i^T = I_m \end{aligned} \tag{4}$$

where m is the dimension of the feature space and I_m is an identity matrix of order m. Let $Y = [y_1 y_2 \cdots y_N]$ and Y_i be the i-th row vector of Y, then Y_i's $(i = 1, 2, \cdots, m)$ are the eigenvectors of $M = (I_N - W)^T (I_N - W)$ corresponding to the second to the $(m + 1)$-st smallest eigenvalues. Here W is the matrix formed by W_{ij}. More details in deduction are referred to [19–22].

2.4   Nonlinear and Explicit Mapping

Standard LLE method learns low-dimensional Y implicitly. No explicit mapping relationship from x_i to y_i can be obtained after training. To extract features of unlabeled samples, the whole learning procedure needs to be redone, which greatly increases computational complexity. Although linear constraints are introduced by neighborhood preserving projection (NPP) [21, 22], where it is assumed that $y_i = U^T x_i$ with U being a linear projection, the nonlinear geometric structure is no longer preserved as a compromise.

In our previous work [20], an explicit and nonlinear mapping for LLE is proposed, which can provide accurate learning results. In this section, we apply such mapping to the aforementioned supervised model of spatial-spectral feature extraction.

Formally, we use a polynomial model to approximate the nonlinear mapping from x_i to y_i. We assume that the k-th component of y_i is a p-th polynomial of x_i that is,

$$y_i^k = \sum_{j=1}^{B+2} \sum_{l=1}^{p} c_{kjl} (x_i^j)^i, \tag{5}$$

where c_{kjl} is the polynomial coefficient. Let c_k be the $p(B+2)$-dimensional column vector formed by c_{kjl}'s and

$$
X_p^{(i)} = \begin{pmatrix} \overbrace{x_i \odot x_i \odot \cdots \odot x_i}^{p} \\ \vdots \\ x_i \end{pmatrix}, \tag{6}
$$

where $\odot$ refers to entrywise matrix multiplication. Then (5) can be rewritten as

$$
y_i = (c_1^T X_p^{(i)}, c_2^T X_p^{(i)}, \cdots, c_m^T X_p^{(i)})^T. \tag{7}
$$

By writing $X_p = [X_p^{(1)} X_p^{(2)} \cdots X_p^{(N)}]$ and substituting (7) into (4), we have the following optimization problem

$$
\begin{aligned}
\min \quad & \sum_k c_k^T X_p M X_p^T c_k \\
\text{s.t.} \quad & c_i^T X_p X_p^T c_j = \delta_{ij}
\end{aligned}, \tag{8}
$$

where δ_{ij} equals to one if $i = j$ and zero otherwise. By the Rayleigh–Ritz theorem, optimal solutions c_k ($k = 1, 2, \cdots, m$) are the eigenvectors of the following generalized eigenvalue problem corresponding to the m smallest eigenvalues:

$$
X_p M X_p^T c_k = \lambda X_p X_p^T c_k.
$$

For an unlabeled sample, its low-dimensional feature can be easily computed with Eqs. (6) and (7).

3 Experimental Results

In this section, we test the performance of the proposed sssLTPE method on Indian Pine data, which is commonly used as a benchmark data set in the literature. This data is from the AVIRIS (Airborne Visible/Infrared Imaging Spectrometer) built by JPL and flown by NASA/Ames, which can be downloaded from the website of MutiSpec project at Purdue University [23]. The scene is over an area 6 miles west of West Lafayette and covers a 2×2 mile portion of Northwest Tippecanoe County, Indiana. The data contains 220 bands with wavelength ranging from 400 nm to 2500 nm. The imageries of bands 15, 50, 120, and 200 are illustrated in grey scale in Fig. 1.

The scene is of 145 by 145 pixels, and total 10366 pixels are labeled. These labeled pixels belong to 16 classes, including various ground vegetation and infrastructures. Class numbers and their corresponding ground objects are listed in Table 1, and the image of ground truth is shown in Fig. 2. We apply k-nearest-neighbor (k-nn) classifier, PCA, NPP, and sssLTPE to classification task on this data. For PCA, NPP, and sssLTPE, trivial one-nearest-neighbor classifier is implemented in the dimension reduced feature space.

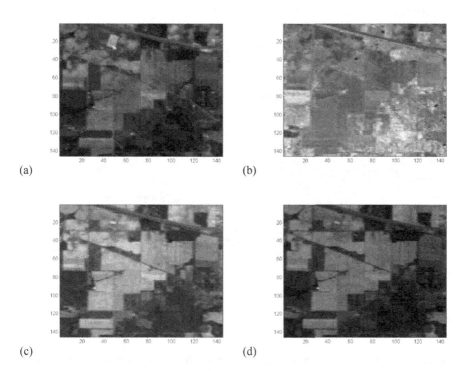

Fig. 1. Gray scale illustration of the imagery of the Indian Pine data. (a) Band 15. (b) Band 50. (c) Band 120. (d) Band 200.

Table 1. Class numbers and corresponding ground objects of the Indian Pine data.

Class number	Class name	Class number	Class name
C1	Alfalfa	C9	Oats
C2	Corn-notill	C10	Soybeans-notill
C3	Corn-min	C11	Soybeans-min
C4	Corn	C12	Soybeans-clean
C5	Grass/Pasture	C13	Wheat
C6	Grass/Trees	C14	Woods
C7	Grass/pasture-mowed	C15	Bldg-Grass-Tree-Drives
C8	Hay-windrowed	C16	Stone-steel towers

There are three parameters, which need to be tuned, for these four methods, namely, number of nearest neighbors k for k-nn, ratio r of preserved principal components for PCA, and dimension d of embedding for NPP and sssLTPE. Each parameter is tuned within a range. k varies from 1 to 9 with step length 1, r varies from 0.90 to 0.98 with step length 0.01, and d varies from 3 to 27 with step length 3. For each parameter

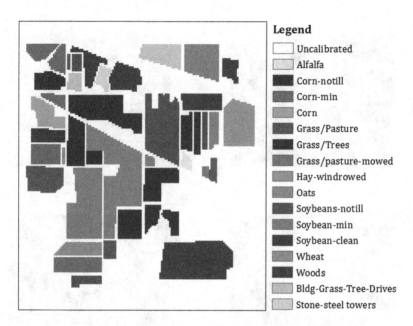

Fig. 2. Color map of labeled ground truth of the India Pine data (Colour figure online).

value, we use a 50 − 50 percent random split of labeled data to generate training and testing data, and we get 5185 training samples and 5181 testing samples. Then they are used in the classification task for the aforementioned four methods, and such random test is repeated 20 times. In each test, error rate, that is, the portion of mis-classified samples in all testing samples, is recorded.

The plots of error rate (mean ± standard deviation) versus parameter for all four methods are demonstrated in Figs. 3 and 4, and the lowest error rates together with corresponding parameter values for each method are given in Table 2.

From Figs. 3 and 4, we can see that error rate of the k-nn method goes low as k increases and the lowest value is achieved at $k = 5$. The error rate of PCA achieves the lowest value at $r = 0.91$ and goes up quickly as r increases. The same tendency occurs to NPP, where the best error rate is achieved at $d = 9$. As to the proposed sssLTPE method, the error rate quickly decreases while d increases and gets to optimal classification performance at $d = 15$, and its performance is stable along with the change of d. From Table 2, we can also observe that sssLTPE has the best performance. Its optimal error rate is much lower than those of k-nn and PCA and outperforms that of NPP. This demonstrates that by fusing spatial-spectral information and using nonlinear mapping, the feature space learned by sssLTPE owns better separating ability for hyperspectral image classification.

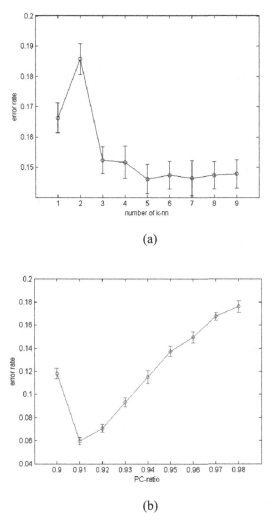

(a)

(b)

Fig. 3. Error rate versus parameter for k-nn and PCA. (a) Error rate versus number of k nearest neighbors. (b) Error rate versus PCA ratio.

Table 2. Best classification results of all four methods.

Method	Parameter	Lowest mean error (standard deviation)
knn	$k = 5$	0.1461(0.0049)
PCA	$r = 0.91$	0.0595(0.0031)
NPP	$d = 9$	0.0282(0.0027)
sssLTPE	$d = 15$	**0.0155**(0.0018)

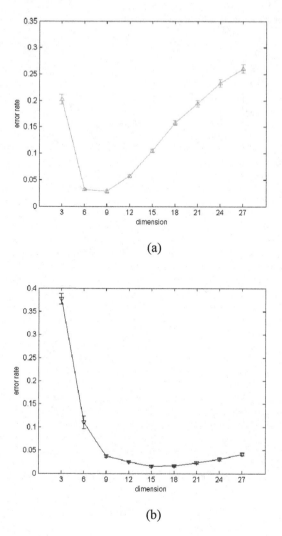

Fig. 4. Error rate versus dimension of extracted features for NPP and sssLTPE. (a) NPP. (b) sssLTPE.

4   Conclusion

In this paper, we propose a supervised feature extraction method for hyperspectral image classification based on manifold learning framework. We propose an intuitive strategy to fuse spatial and spectral information of ground objects, which is simple to implement. By introducing a nonlinear and explicit mapping from spatial-spectral space to feature space, not only hidden nonlinear structure underlying imagery pixels can be efficiently learned by preserving local topology, but also fast feature extraction for testing samples is achieved. Experiments conducted on benchmark data set show that the proposed method has high classification accuracy and superior performance over principle component analysis as well as manifold learning approach with linear projection. In our future works, we will validate the proposed method on larger and more complicated scenes and investigate strategies for fast computation.

Acknowledgement. This work was supported by the National Natural Science Foundation (NNSF) of China under Grant nos. 41201552, 41174013, and 41401605.

References

1. Zhang, L., Du, B., Zhong, Y.: Hybrid detectors based on selective endmembers. IEEE Trans. Geosci. Remote Sens. **48**(6), 2633–2646 (2010)
2. Du, B., Zhang, L.: Random selection based anomaly detector for hyperspectral imagery. IEEE Trans. Geosci. Remote Sens. **49**(5), 1578–1589 (2011)
3. Datt, B., McVicar, T., Van Niel, T., Jupp, D., Pearlman, J.: Preprocessing EO-1 hyperion hyperspectral data to support the application of agricultural indexes. IEEE Trans. Geosci. Remote Sens. **41**(6), 1246–1259 (2003)
4. Hörig, B., Kühn, F., Oschütz, F., Lehmann, F.: HyMap hyperspectral remote sensing to detect hydrocarbons. Int. J. Remote Sens. **22**(8), 1413–1422 (2001)
5. Eismann, M., Stocker, A., Nasrabadi, N.: Automated hyperspectral cueing for civilian search and rescue. Proc. IEEE **97**(6), 1031–1055 (2009)
6. Hughes, G.: On the mean accuracy of statistical pattern recognizers. IEEE Trans. Inf. Theory **14**(1), 55–63 (1968)
7. Penna, B., Tillo, T., Magli, E., Olmo, G.: Transform coding techniques for lossy hyperspectral data compression. IEEE Trans. Geosci. Remote Sens. **45**(5), 1408–1421 (2007)
8. Plaza, A., Benediktsson, J., Boardman, J., et al.: Recent advances in techniques for hyperspectral image processing. Remote Sens. Environ. **113**(1), 110–122 (2009)
9. Bachmann, C., Ainsworth, T., Fusina, R.: Exploiting manifold geometry in hyperspectral imagery. IEEE Trans. Geosci. Remote Sens. **43**(3), 441–454 (2005)
10. Bachmann, C., Ainsworth, T., Fusina, R.: Improved manifold coordinate representations of large-scale hyperspectral scenes. IEEE Trans. Geosci. Remote Sens. **44**(10), 2786–2803 (2006)
11. Lunga, D., Prasad, S., Crawford, M., Ersoy, O.: Manifold-learning-based feature extraction for classification of hyperspectral data. IEEE Signal Process. Mag. **1**, 55–66 (2014)
12. He, J., Zhang, L., Wang, Q., Li, Z.: Using diffusion geometric coordinates for hyperspectral imagery representation. IEEE Geosci. Remote Sens. Lett. **6**(4), 767–771 (2009)
13. Mohan, A., Sapiro, G., Bosch, E.: Spatially coherent nonlinear dimensionality reduction and segmentation of hyperspectral images. IEEE Geosci. Remote Sens. Lett. **4**(2), 206–210 (2007)
14. Ma, L., Crawford, M., Tian, J.: Anomaly detection for hyperspectral images based on robust locally linear embedding. J. Infrared Millimeter Terahertz Waves **31**(6), 753–762 (2010)
15. Crawford, M., Ma, L., Kim, W.: Exploring nonlinear manifold learning for classification of hyperspectral data. In: Prasad, S., Bruce, L.M., Chanussot, J. (eds.) Advances in Signal Processing and Exploitation Techniques, pp. 207–234. Springer-Verlag, London (2011)
16. Li, W., Prasad, S., Fowler, J., Bruce, L.: Locality preserving dimensionality reduction and classification for hyperspectral image analysis. IEEE Trans. Geosci. Remote Sens. **50**(4), 1185–1198 (2012)

17. Jia, X., Kuo, B., Crawford, M.: Feature mining for hyperspectral image classification. Proc. IEEE **101**(3), 676–697 (2013)
18. Tenenbaum, J.B., Silva, V., Langford, J.C.: A global geometric framework for nonlinear dimensionality reduction. Science **290**(5500), 2319–2323 (2000)
19. Roweis, S.T., Saul, L.K.: Nonlinear dimensionality reduction by locally linear embedding. Science **290**(5500), 2323–2326 (2000)
20. Qiao, H., Zhang, P., Wang, D., Zhang, B.: An explicit and nonlinear mapping for manifold learning. IEEE Trans. Cybern. **43**(1), 51–63 (2013)
21. He, X., Cai, D., Yan, S., Zhang, H.: Neighborhood preserving embedding. In: 2005 Proceedings of the IEEE International Conference on Computer Vision, Vol. 2, pp. 1208–1213 (2005)
22. Pang, Y., Zhang, L., Liu, Z., Yu, N., Li, H.: Neighborhood preserving projections (NPP): a novel linear dimension reduction method. In: Huang, D.-S., Zhang, X.-P., Huang, G.-B. (eds.) ICIC 2005. LNCS, vol. 3644, pp. 117–125. Springer, Heidelberg (2005)
23. https://engineering.purdue.edu/~biehl/MultiSpec/hyperspectral.html

A New Learning Automata Algorithm for Selection of Optimal Subset

Xinyi Guo[(✉)], Wen Jiang, Hao Ge, and Shenghong Li

Department of Electronic Engineering,
Shanghai Jiao Tong University, Shanghai, China
{sepulture, wenjiang, sjtu_gehao, shli}@sjtu.edu.cn

Abstract. A new class of learning automata for the purpose of learning the optimal subset of actions has been proposed to fulfill the demand of application such as allocation and global optimization. Learning automata are capable of dealing with multiple choice problems if some modifications are made on current algorithms. This paper discusses on how to adapt current LA algorithms to the new purpose and introduces a new kind of learning automata. The proposed automata take advantage of LELA, whose original updating schemes favor the purpose of selecting multiple actions and thus acquire faster rate of convergence than the existing automata for selecting optimal subset of actions to the best of our knowledge. Additionally, extensive simulation results are presented to compare the performance between the proposed algorithm and the existing ones. The results show that the proposed automata outperform the other automata.

Keywords: Learning automata · Optimal subset · Last-position Elimination-based learning automata · Pursuit algorithm

1 Introduction

Learning automaton (LA) is an automaton interacting with a stationary or non-stationary environment and selecting the optimal action out of a set of actions based on certain learning paradigm. LA was brought forward by Tsetlin in 1961 [1] and has been studied as a model of reinforcement learning for many years.

The learning process of a learning automaton is illustrated in Fig. 1. The automaton chooses one of the given actions according to a certain probability vector containing the probability of choosing each action. Then the environment responds the automaton with a feedback $\beta(t)$ corresponding to the chosen action. The value of the feedback indicates either reward or penalty. The automaton receives the feedback and modifies the probability vector accordingly by means of certain learning paradigm. The LAs concerned recent years fall within the variable structure stochastic automaton (VSSA). VSSA was introduced by Varshavskii and Vorontsova [2].

A typical VSSA can be defined as a quintuple <P,A,B,T> where:

- P is the set of time-variant action probability vectors.
- $A = \{a_1, a_2, \ldots, a_r\} (2 \leq r < \infty)$ is the finite set of actions. $a(t)$ denotes the action chosen at time instant t.

© Springer International Publishing Switzerland 2015
D.-S. Huang et al. (Eds.): ICIC 2015, Part I, LNCS 9225, pp. 693–702, 2015.
DOI: 10.1007/978-3-319-22180-9_69

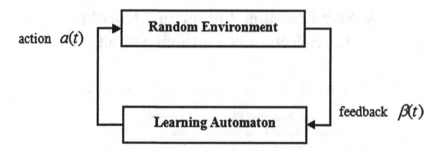

Fig. 1. Learning process of an learning automaton

- B is the set of responses from the random environment. In a P-model random environment, $B = \{0, 1\}$. $\beta(t)$ denotes the response at time instant t. $\beta(t) = 1$ indicates the reward.
- $T: P \times A \times B \rightarrow P$ is the updating rules operating on the action probability vector such that $P(t+1) = T(P(t), a(t), \beta(t))$.

Utilizing estimators and discretizing the probability space have been proved to be two effective methods of improving the convergence rate of various LAs. The discretized pursuit reward-inaction (DP_{RI}) scheme [3] is a great combination of these two methods. In the DP_{RI} algorithm, when an action is rewarded, all the actions except the optimal action that corresponds to the highest estimate decrease by a step size Δ while the optimal action increases an integral multiple of Δ.

However, the step size of the updating algorithm DP_{RI} applied decreases when the last-position actions probability vector turn to zero which does not fit the intuition when pursuing actions based on corresponding estimators. Learning should be careful in the beginning to avoid pursuing wrong actions and should gradually proceed in larger step sizes as the actions are sampled more sufficiently and the estimates become more accurate. Zhang proposed last-position elimination-based learning automata (LELAs) in [4] which adopted accelerated step sizes conforming to the intuitive learning process. In the LELA algorithm, when an action is rewarded, the active action corresponding to the lowest estimate decreases by a step size Δ while all the other active actions increase by an equal division of Δ. When the probability of one action turns to zero, the action becomes inactive. Obviously as the active actions reduce, the step sizes of the rewarded actions increase in the LELA updating scheme. LELA was proved to converge faster than DP_{RI} in all benchmark environments [4].

A great deal of algorithms have been presented and studied recent years to learn the optimal action. However, in many application fields such as allocation and global optimization, learning the best action is not the only concern. As far back as 1950, Bahadur considered multiple selection problems in statistics [5]. Learning Automata are capable of being extended to learning the optimal subset. Har-El and Rubinstein were the first to apply LA to the choice of optimal subset of numbers to the best of our knowledge [6]. However, due to the lack of development of LA back then, the algorithm is not indeed competitive. Lately, Zhang came up with several new schemes of adapting pursuit learning automata to the selection of optimal subset and brought this area back to attention [7].

Zhang adapted both continuous and discretized pursuit learning automata to the selection of optimal subset by modifying the updating rules. In continuous equal pursuit (CEP) and discretized equal pursuit (DEP) learning schemes, the estimated optimal subset is pursued based on the estimators. The probabilities of the actions in the estimated optimal subset remain equal during the learning process. In continuous unequal pursuit (CUP) and discretized unequal pursuit (DUP) learning schemes, the probabilities of the actions are updated proportionally to their previous value.

Though Zhang introduced four learning automata schemes for the selection of optimal subset, this area has not been fully investigated. The main contribution of this paper is listed as follows:

1. A more comprehensive discussion on how to adapt current learning automata to optimal subset selection is conducted. To extend the current learning automata to the selection of optimal subset does not necessarily require modifications on updating rules. Repetitive use of current automata and inheritance of estimators are effective ways of adapt current automata to the selection of optimal subset.
2. A new algorithm which extends LELA to optimal subset selection is introduced. Whether pursuit automata are the best automata fitting multiple selections remains investigation. We utilized LELA in consideration that the selection of optimal subset is implicitly embedded in the original LELA learning process of selecting the optimal action.
3. The proposed algorithm acquires faster rate of convergence than the existing automata for selecting optimal subset of actions.

The paper is organized as follows. In Sect. 2 we discuss different ways of adapting LA to the selection of optimal subset. The proposed LELAK (LELA for top k optimal actions selection) algorithm is presented in Sect. 3. Extensive simulation results that demonstrate the superiority of the proposed algorithm over the former ones are presented in Sect. 4. Finally, the conclusion is given in Sect. 5.

2 Optimal Subset Selection Scheme

To adapt current LA schemes to the selection of top k optimal actions, one simple and direct way is to repeat the LA algorithm k times.

First, learn $a_{m1} = \max_{1 \leq i \leq r} (a_i)$.

Then, learn $a_{m2} = \max_{1 \leq i \leq r, a_i \neq a_{m1}} (a_i)$ and repeat in this manner until the kth time.

$S_m = \{a_{m1}, a_{m2}, \ldots, a_{mk}\}$ denotes the optimal subset. We call this strategy naive strategy. Though the naive strategy is just repetition with respect to non-estimator LAs, estimators can be inherited for estimator LAs. When the automaton finds the first optimal action, instead of setting the estimators to zero, the automaton utilizes the estimators to further learning process. For example, naive $DP_{RI}K$ (DP_{RI} for top k optimal actions selection) is as below.

Algorithm naive DP$_{RI}$K

Parameters

m index of the maximal component of the reward estimate vector

$$\hat{d}_m(t) = \max_{1 \leq i \leq r}\{\hat{d}_i(t)\}$$

k number of actions in the optimal subset

n resolution parameter

$W_i(t)$ number of times the ith action has been rewarded up to time t, for $1 \leq i \leq r$

$Z_i(t)$ number of times the ith action has been selected up to time t, for $1 \leq i \leq r$

$\Delta = 1/r/n$ smallest step size

Method

Step 0:   Initialize $W_i(t)$ and $Z_i(t)$ by selecting each action a small number of times.

           Set $S_m = \{\}$.

Step 1:   Initialize $p_i(t) = 1/r$, for $1 \leq i \leq r$.

Step 2:   At time instant t, select an action $a(t) = a_i$ according to the probability distribution $P(t)$.

Step 3:   If $\beta(t) = 1$, update $P(t)$ according to the following equations:

$$p_j(t+1) = \max_{j \neq m}\{0, p_j(t) - \Delta\}$$

$$p_m(t+1) = 1 - \sum_{j \neq m} p_j(t+1)$$

Step 4:   Update $W_i(t), Z_i(t)$ and $\hat{d}_i(t)$ based on the feedback $\beta(t)$ from the environment.

$$W_i(t) = W_i(t-1) + \beta(t);$$
$$Z_i(t) = Z_i(t-1) + 1;$$
$$\hat{d}_i(t) = \frac{W_i(t)}{Z_i(t)};$$

Step 5:   If $|S_m| = k$ **then End**

    Else if $p_m(t+1) = 1$ **then**

$S_m = \{a_m\} \cup S_m, A = A - \{a_m\}, r = r - 1$

Goto **Step 1**

Else Goto **Step 2.**

End If

Zhang proposed several algorithms with a different strategy [7]. Among these algorithms, discretized equal pursuit reward-inaction (DEP$_{RI}$) and discretized unequal pursuit reward-inaction (DUP$_{RI}$) are in discretized manner. In both algorithms, when an action is rewarded, all the actions except the top k actions that correspond to the highest k estimates are decreased while the top k actions are increased. DEP$_{RI}$ makes sure that each probability of the action in estimated optimal subset is equal while DUP$_{RI}$ does not. DEP$_{RI}$ outperforms DUP$_{RI}$. In fact, these two algorithms apply a strategy that considers the optimal subset as a whole, as one single action and the sum of probabilities of the optimal subset represents the probability of the single action.

3   Proposed Learning Algorithm

LELA differentiates the traditional pursuit algorithm in a way that it exploits the estimators. Instead of pursuing the best estimated action, it eliminates the last-position action. With regard to the optimal subset learning scheme, the advantage of LELA over the pursuit algorithm is that selecting the top k out of r actions is embedded implicitly in the original LELA learning process of selecting the optimal actions. Before eliminating $r - 1$ actions and converging to the optimal action, LELA eliminate actions and thus get a remained subset of k actions.

In the proposed algorithm, we exploit the favorable updating schemes of LELA and apply the strategy mentioned above that considers the optimal subset as a whole to adapt LELA to learning the optimal subset. The proposed LELAK adopts the following updating rules:

If $\beta(t) = 1$ Then

$$\hat{d}_l(t) = \min_{1 \leq i \leq r} \{\hat{d}_i(t) | p_i(t) \neq 0\} \tag{1}$$

$$p_l(t + 1) = \max\{p_l(t) - \Delta, 0\} \tag{2}$$

$$S(t) = \{i | p_i(t) \neq 0, 1 \leq i \leq r\} \tag{3}$$

$$p_j(t + 1) = p_j(t) + \frac{p_l(t) - p_l(t + 1)}{|S(t)|}, i \in S(t) \tag{4}$$

The proposed algorithm of LELAK is described as follows. In each iteration, the last-position action is punished and all the other actions are rewarded according to their estimators. When the possibility of the last-position action becomes 0, then the action is eliminated and the number of active actions $|S(t)|$ decreases by 1. The iteration is terminated when the number of remaining active actions equals k and the sum of their probabilities equals 1.

Algorithm LELAK

Parameters

l	index of the minimal component of the reward estimate vector
k	number of actions in the optimal subset
n	resolution parameter
$W_i(t)$	number of times the ith action has been rewarded up to time t, for $1 \leq i \leq r$
$Z_i(t)$	number of times the ith action has been selected up to time t, for $1 \leq i \leq r$
$\Delta = 1/r/n$	smallest step size

Initialization

$p_i(t) = 1/r$, for $1 \leq i \leq r$.

Initialize $W_i(t)$ and $Z_i(t)$ by selecting each action a small number of times.

Method

Repeat

Step 1:   At time instant t, select an action $a(t) = a_i$ according to the probability distribution $P(t)$.

Step 2:   Update $W_i(t), Z_i(t)$ and $\hat{d}_i(t)$.

Step 3:   Update $P(t)$ according to the equations (1) - (4).

End *Repeat* Until $sum(P_k) = 1$. P_k is the subset of the largest k probabilities.

---

4   Simulation Results

In this section, the proposed LELAK is compared to the DEP_{RI}, DUP_{RI} automaton and naive $DP_{RI}K$. These four algorithms were simulated to be applied to five stationary random environments (E1 to E5), of which E1 to E3 are in congruity to those in [4, 8] and E4 was widely used in [3, 4, 8]. The actions' reward probabilities for each environment were taken to be as follows:

- E1: D={0.65,0.50,0.45,0.40,0.35,0.30,0.20,0.20,0.15,0.10}.
- E2: D={0.60,0.50,0.45,0.40,0.35,0.30,0.20,0.20,0.15,0.10}.
- E3: D={0.55,0.50,0.45,0.40,0.35,0.30,0.20,0.20,0.15,0.10}.
- E4: D={0.10,0.45,0.84,0.76,0.20,0.40,0.60,0.70,0.50,0.30}.
- E5: D={0.65,0.60,0.45,0.40,0.35,0.30,0.20,0.20,0.15,0.10}.

In all tests performed, an algorithm was considered to have converged if the probability of choosing an action was greater or equal to a threshold T. Before comparing the performance of the automata, numerous tests were executed to determine the

Table 1. Accuracy (number of correct convergence/number of experiments) of naive $DP_{RI}K$, DEP_{RI}, DUP_{RI} and LELAK in environments E1 to E5 corresponding to their "best" parameters when $k=2$.

	naive $DP_{RI}K$	DEP_{RI}	DUP_{RI}	LELAK
E1	0.993	0.994	0.993	0.996
E2	0.994	0.994	0.993	0.996
E3	0.993	0.994	0.993	0.997
E4	0.993	0.994	0.995	0.997
E5	0.996	0.995	0.994	0.998

Table 2. Comparison of the average number of iterations required for convergence of naive $DP_{RI}K$, DEP_{RI}, DUP_{RI} and LELAK in environments E1 to E5 when $k=2$.

		E1	E2	E3	E4	E5
naive $DP_{RI}K$	n	1814	1751	1676	1374	222
	Iterations	10042.5	10500.7	12304.0	6142.8	1864.7
DEP_{RI}	n	2524	2433	2306	1937	331
	Iterations	8776.5	8779.8	8800.2	4763.9	1148.7
DUP_{RI}	n	3199	3095	2938	2457	399
	Iterations	9782.3	9832.6	9915.3	5346.9	1250.6
LELAK	n	151	148	150	126	21
	Iterations	6092.9	6123.8	6358.3	3549.4	925.6

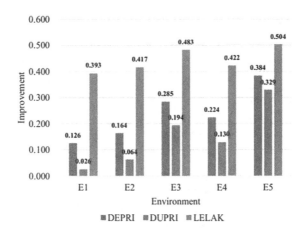

Fig. 2. The improvement of DEP_{RI}, DUP_{RI} and LELAK over naive $DP_{RI}K$

"best" parameter n for each algorithm. The parameters were considered as the "best" if they yielded to the fastest convergence and the automata converged to the correct optimal subset in a sequence of N_E experiments. We used $T = 0.999$ and $N_E = 750$ and repeated the parameter search procedure 20 times to reduce the variance coefficient of the "best" parameter n. Same methods are used in [3, 4, 8].

250000 experiments for each algorithm using the "best" parameters were then executed in order to check the accuracy of each algorithm when using its "best" parameters. In the case when $k = 2$ the accuracy (*number of correct convergence/number of experiments*) of the four algorithms when using their "best" parameters are presented in Table 1. In each environment, LELAK achieved equal or better accuracy than its counterpart naive $DP_{RI}K$, DEP_{RI} and DUP_{RI}. Therefore, the performance between LELAK and the other three algorithms is comparable under given accuracy.

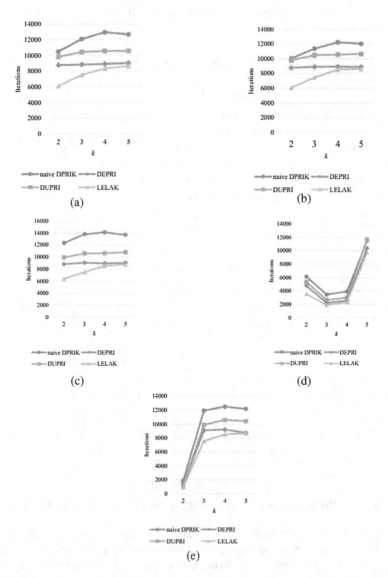

Fig. 3. The average number of iterations required for convergence of naive $DP_{RI}K$, DEP_{RI}, DUP_{RI} and LELAK in environments: (a) E1, (b) E2, (c) E3, (d) E4 and (e) E5

The estimators were initialized before starting each algorithm. All actions were sampled ten times each. The used "best" parameters n and the average number of iterations (including extra iterations for initialization) that naive $DP_{RI}K$, DEP_{RI}, DUP_{RI} and LELAK required for convergence to the top two actions when operating in environments E1 to E5 are presented in Table 2.

The results presented in Table 2 demonstrate that LELAK performs best among the four algorithms in all five environments. LELAK is approximately 27 % faster than the former best algorithm DEP_{RI}. Figure 2 graphically compares the relative performance of DEP_{RI}, DUP_{RI} and LELAK to naive $DP_{RI}K$. On average, DUP_{RI} achieves an improvement of 15 % over naive $DP_{RI}K$, DEP_{RI} achieves an improvement of 24 % and the proposed LELAK achieves an improvement of 44 %.

In addition to the superior performance, the proposed algorithm also achieves faster parameter search procedure and less variance in the "best" parameter because the parameter n is relatively small.

Experiments on larger k were executed as well. The comparison of the average number of iterations required for convergence of naive $DP_{RI}K$, DEP_{RI}, DUP_{RI} and LELAK in environments E1 to E5 was illustrated in Fig. 3. The performance advantage of LELAK over other algorithms is clear under all circumstances and we can conclude that LELA is more favorable to the optimal subset selection.

5   Conclusion

In this paper a new algorithm to the selection of optimal subset is introduced. LELAK extends the original LELA to multiple selections. The naive strategy to adapt LA to multiple selections is discussed as well as the mature structure of the optimal subset selection. Extensive simulation results have shown the superior performance of the proposed algorithm LELAK.

We are working on finding better strategies to adapt existing LA to multiple selections or introducing a novel LA that directly fits the purpose of learning the optimal subset.

References

1. Tsetlin, M.L.: On the behavior of finite automata in random media. Autom. Telemek **22**, 1345–1354 (1961)
2. Varshavskii, V.I., Vorontsova, I.P.: On the behavior of stochastic automata with variable structure. Autom. Telemek **24**(3), 327–333 (1963)
3. Oommen, B.J., Agache, M.: Continuous and discretized pursuit learning schemes: various algorithms and their comparison. IEEE Trans. Syst. Man Cybern. **31**(3), 277–287 (2001)
4. Zhang, J., Wang, C., Zhou, M.: Last-position elimination-based learning automata. IEEE Trans. Cybern. **44**(12), 2484–2493 (2014)
5. Bahadur, R.R.: On a problem in the theory of k populations. Annu. Math. Stat. **21**(3), 362–375 (1950)

6. Har-el, J., Rubinstein, Y.: Choice of optimal subset of numbers using a learning automaton. J. Optim. Theory Appl. **22**(1), 35–40 (1977)
7. Zhang, J., Li, Z., Kang, Q., Zhou, M.: A new class of learning automata for selecting an optimal subset. In: IEEE International Conference on Systems, Man and Cybernetics (SMC), pp. 3429–3434 (2014)
8. Georgios, M.S., Papadimitriou, I., Pomportsis, A.S.: A new class of ε-optimal learning automata. IEEE Trans. Syst. Man Cybern. **34**(1), 246–254 (2004)

The Recent Developments and Comparative Analysis of Neural Network and Evolutionary Algorithms for Solving Symbolic Regression

Xueshi Dong[1], Wenyong Dong[1(✉)], Yunfei Yi[1,2,3], Yajie Wang[1], and Xiaosong Xu[1]

[1] Computer School, Wuhan University, Wuhan 430072, Hubei, China
dxs_cs@163.com, dwy@whu.edu.cn
[2] College of Computer and Information Science, Hechi University,
Yizhou 546300, Guangxi, China
[3] Guangxi Key Laboratory of Hybrid Computation and IC Design Analysis,
Nanning 530006, Guangxi, China

Abstract. Symbolic regression (SR) is one of the research fields in data mining, how to use scientific and appropriate methods to study SR is a difficult problem. The traditional methods used in SR mainly focus on the models such as genetic programming (GP), the article applies the gene expression programming (GEP) and neural network (NN) to this field, in order to correctly compare the advantages and disadvantages of the three methods, some relevant works have been done. This paper first briefly introduces the NN and evolutionary algorithms including GP and GEP, their design steps and recent developments, and applies these algorithms to SR, then uses the algorithms to solve SR and makes comparison analysis, and draws some conclusions in the experiment condition: the performance of NN and evolutionary algorithms change dramatically for solving this problem; GP and GEP fluctuate greatly compared with NN, the used time is also less, and NN shows better stability and result.

Keywords: Neural network · Evolutionary algorithms · Symbolic regression · Genetic programming · Gene expression programming

1 Introduction

Symbolic regression [2, 9] can also be called function modeling, which is based on a set of given variables and function values, it finds the function relationship between the fitting function and values. In addition to these given data values, there is no other information related to the function model, so for symbolic regression, it is not only to determine the function of the structure, but also needs to determine appropriate parameters.

The traditional linear regression, polynomial regression and nonlinear regression should be specified fitting formula in advance, the parameters in the formula are determined by regression, and the SR is fitting formula without specifying, formulas form and the parameters are determined in the process of regression. Therefore, SR has

© Springer International Publishing Switzerland 2015
D.-S. Huang et al. (Eds.): ICIC 2015, Part I, LNCS 9225, pp. 703–714, 2015.
DOI: 10.1007/978-3-319-22180-9_70

much wider range of applications, such as the experience finding of scientific law and solving symbolic equations [9].

In engineering practice, it often needs to deal with a large number of experimental data, the traditional methods such as curve fitting, numerical approximation. The main characteristics of these methods are: establish one or a group of function model according to the given data, and then fit, calculate parameter and determine the model results, which is the process of parameter regression. In this process, the least square method is often used.

The principle of symbolic regression is that the related variables and groups of sample data are given, then the SR can be expressed as a symbolic expression, it makes the formula, and fits sample data in a certain precision.

However, it is difficult to carry out SR, the traditional methods such as stepwise regression algorithm with randomly generating candidate factors and GP [2] are used in this field in many papers, the algorithms applied in the field are few, this paper applies NN and GEP to SR, and does experiment simulation and comparison analysis with GP and draws some conclusions.

This paper consists of five parts: the first section displays the introduction; the second part is about the three key algorithms for symbolic regression; the third section describes the recent developments; the fourth presents the experiment simulation and analysis; the fifth part is the conclusion and future works.

2 Three Key Algorithms for Symbolic Regression

Currently, there are three key algorithms for symbolic regression including genetic programming, gene expression programming and neural network, and the more detailed introduction is in the following.

2.1 Genetic Programming

(1) The basic principles

Genetic programming is proposed by USA scholar John R. Koza in early 90's of twentieth Century, an independent domain method with machinery search procedures space. We can find detailed information about GP from the papers [2, 28]. The basic principle of algorithm is: randomly generate initial population which is applicable to the environmental issues, according to the given fitness test method, each individual has a fitness value; in accordance with the principle of survival of the fittest, after the processing of genetic operators such as reproduction, crossover and mutation, it gets the high fitness individuals, produces the next generation groups, then does the same operation to new genetic operation and repeats the same operation until the problem or approximate solutions coming up in a generation.

(2) The main steps of genetic programming

Genetic programming is mainly in the following five steps to complete the operation:

(a) Determine the expression of the individual, including function set F and terminator set T.
(b) Randomly generate initial population. There are three general methods of generating initial individuals: complete method, growth method, hybrid method.
(c) Calculate the fitness of each individual.
(d) According to the genetic parameters, make copy, crossover, and mutation operators to generate new individuals. The copy, crossover and mutation are main operations of evolutional algorithms.
(e) Repeat execution of (c) and (d) the two steps until achieving satisfactory results.

2.2   Gene Expression Programming

(1) The basic principle of GEP algorithm

More detailed information about GEP model could be found from the articles [1, 3–5, 8, 16]. Like all evolutionary algorithms, the basic steps of GEP algorithm starts from generating a number of chromosomes (initial population), and then the chromosome is expressed, based on a fitness sample set calculate the fitness of each individual; finally individuals are selected according to the fitness value, make genetic operation, produce the offspring with new characteristics. The process is repeated for several generations until finding a good solution. The core technology of GEP is to completely separate the variation process.

(2) The algorithm of GEP

The GEP algorithm structure:

> Step1 initialization, it randomly generates 50 groups, if the number is too large, it would increase the running time of the algorithm, is difficult to converge with too small number, initiate chromosomes, each chromosome consists of 5 genes;
> Step2 according to the function of fitness value, solve fitness value of each chromosome of the initial chromosome and sort, save the individual with maximum fitness value;
> Step3 implement variation, based on the number of genes which the chromosome contains, determine the number of variation gene, the paper chooses the method that each gene mutates in a gene;
> Step4 perform plug;
> Step5 perform gene insertion;
> Step6 implement single point reorganization;
> Step7 perform two points recombination;
> Step8 implement gene recombinant;

Step9 if run with maximum algebraic preset set or obtain the maximum fitness value, stop operation, output the result with graphical and save the results to a log file, or turn to Step2 to continue.

(3) Genetic Operation

GEP uses the coding of linear equation length, and genetic operation is similar to genetic algorithm. Genetic operation of GEP includes variation, insertion, single point insertion, two-point recombination and gene recombination, they are described as follows:

(a) Mutation operation: happen in any position of the chromosome. If variation is in gene head, it should reselect all the symbols, or can only choose the terminator symbol.
(b) Constant mutation operation: constant usually appear in the symbolic regression. First of all, generate specific random number in a place relevant to the problem, and then replace the original value.
(c) Insertion operation, the operation is specific genetic operator to GEP.

2.3 Neural Networks

Neural networks starts in 1950's, when the simplest neural network's architecture was presented. After the initial work in the area, the idea of neural networks became rather popular. But then the area had a crash, when it was discovered that neural networks of those times are very limited in terms of the amount of tasks they can be applied to. In 1970's, the area had another boom, the idea of multi-layer neural networks with the back propagation learning was presented. Some more information about NN can be found in the papers [10–13, 24, 26, 27].

Neural network uses standard three layers feed forward neural network, including input layer, hidden layer and output layer. Firstly, it is necessary to determine the neural network of input layer and output layer, the input neurons represent more than one variable, these attribute variables are main factors influencing the change, generally, standardize the input neurons, make the values fall into [0, 1] range, the output neurons are the target state.

3   Recent Developments

There are many improvement and optimization methods of gene programming and gene expression programming, we can utilize the improved algorithms to solve the problem, which could lead to better results, the relevant works of the field include multi-dimensional complex association rule based on artificial immune system and gene expression programming [1], research and application of genetic expression programming algorithm based on uniform-design [3], application of a novel GEP algorithm in evolutionary modeling and forecasting [4], complex function modeling based on improved gene expression programming [5], a multi-gene evolutionary algorithm based on overlapped expression [6], grid resource allocation algorithm based on parallel gene expression programming [14], M2GEP: a new evolution algorithm based on multi2layer chromosomes gene expression programming [15].

We can optimize neural network, there are many research papers in these fields including research on extreme learning of neural networks [10], research on change information recognition method of vector data based on neural network decision tree [11], the forecasting approach for short-term traffic flow based on principal component analysis and combined NN [12], decision tree based neural network design [13], fuzzy lattice constructive morphological neural network [17]. Neural network has many parameters, how to choose the proper parameters setting to make the algorithm perform well is one of most important research fields, which also has great influence on the model, we can use uniform design to solve this problem, it is one of the most valid and practical experiment methods, the relevant work in these fields containing, Guoqiu Zhang and other authors, application introduction of uniform design method [22], Yunliang Chen and other authors, research and application of genetic expression programming algorithm based on uniform-design [3].

There are other important methods to improve neural network in the following: a new neural network model based on the LVQ algorithm [25]; particle swarm optimization of ensemble neural networks with fuzzy aggregation [26]; interval type-2 fuzzy weight adjustment for back propagation neural networks [27] and so on. Deep learning is the hot research field, the research and papers of the field include hierarchical face parsing via deep learning [7], deep convolutional network cascade for facial point detection [18], learning hierarchical features for scene labeling [19], a discriminative deep model for pedestrian detection with occlusion handling [20], data-driven soft sensor development based on deep learning technique [21], some researchers use artificial bee colony programming to solve symbolic regression, and compared with genetic programming [23]; It has proved that deep learning is performing better than neural network, the future work that we can do is that, on the one hand, we can apply deep learning to symbolic regression for solving the problem, which may lead to better results; on the other hand, SR can be applied to image classification, scene labeling and visual reorganization and so on, so that it could obtain the appropriate functional models, and there are some relevant works published in the paper [28].

4 Experiment Simulation and Analysis

Experiment simulation is developed based on C#, the codes of the three algorithms mentioned above for solving SR are provided by Andrew Kirillov, a software developer (senior) of Cisco Systems from United Kingdom, training data is shown in the following tables, the experiment interfaces and results are given below. Experiment parameters settings for training data 1 are in the following.

Genetic programming parameter settings are as follows: the population: 50; selection method: rank; function set: simple; iterative function sets: 100. Gene expression parameter settings are as follows: population: 50; selection method: rank; function set: simple; iterative number: 100. The parameters of the neural network: the learning rate: 0.2; momentum: 0; sigmoid's alpha value: 3; the first layer neuron number: 25; iteration number: 1000.

Fitting error of GP, GEP and NN for solving symbolic regression with training data are shown in the following tables. Training data 1 is shown in Table 1, in the table X represents the order number of the training data, and Y stands for the each data of training data, the following tables for training data use the same rule.

Training data 2 is shown in Table 2.

Training data 3 is shown in Table 3.

Training data 4 is shown in Table 4.

Training data 5 is shown in Table 5.

Training data 6 is shown in Table 6.

Table 1. Training data 1

X	1	2	3	4	5	6	7	8
Y	1	1.5	2.5	4	6	8.5	11.5	15

Table 2. Training data 2

X	1	2	3	4	5	6	7	8	9
Y	0	4	7	9	10	9	7	4	0

Table 3. Training data 3

X	1	2	3	4	5	6	7	8	9	10	11	12
Y	22	20	18	16	14	12	10	8	6	4	2	1

Table 4. Training data 4

X	1	2	3	4	5	6	7	8	9	10	11	12
Y	12	10	8	6	4	2	4	6	8	10	12	15

Table 5. Training data 5

X	1	2	3	4	5	6	7	8	9	10	11
Y	1	3	5	7	9	11	13	15	13	11	9
X	12	13	14	15	16	17	18	19	20	21	22
Y	7	5	3	1	3	5	7	9	11	13	15

Table 6. Training data 6

X	1	2	3	4	5	6	7	8	9	10	11
Y	10	9	8	7	6	5	4	3	2	1	2
X	12	13	14	15	16	17	18	19	20	21	22
Y	3	4	5	6	7	8	9	10	9	8	7

Figures 1, 2, 3 and 4 are just some interfaces of the simulation, they just stand for some examples of many ones. In Fig. 1, from left to right, they are GP, GEP and NN symbolic regression fitting interface for training data 1, in each graph of the figure, horizontal axis is X, vertical axis is Y. From left to right in Fig. 2, GP, GEP and NN for symbolic regression are fitting interfaces of training data 2, in each graph horizontal axis is X, vertical axis is Y, other interfaces are the same.

Because of the limited place and there are too many simulation interfaces in the experiment, the experiment of every algorithm runs thirty times for training each data,

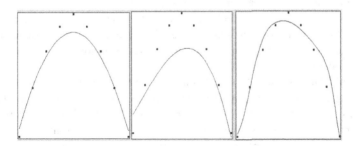

Fig. 1. The simulation interfaces of the three algorithms with training data 1

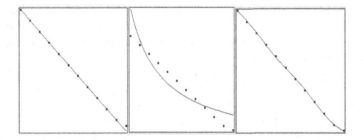

Fig. 2. The simulation interfaces of the three algorithms with training data 2

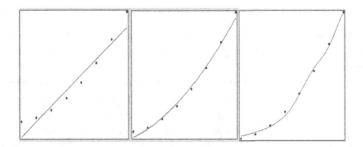

Fig. 3. The simulation interfaces of the three algorithms with training data 3

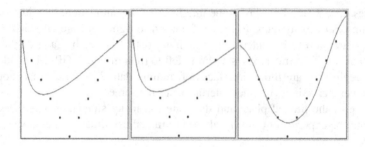

Fig. 4. The simulation interfaces of the three algorithms with training data 4

so some simulation interfaces are not listed in the paper, some representative simulation interfaces of each model for training data 1, training data 2, training data 3 and training data 4 are in the following figures:

The experiment parameters settings for training data 2: GP parameter settings are as follows: the population: 50; selection method: rank; function set: simple; iterative sets: 100. GEP parameters as follows: population: 50; selection method: rank; function set: simple; iterative number: 100. The parameters of the NN: learning rate: 0.2; momentum: 0; sigmoid's alpha value: 3; the first layer neuron number: 25; iteration number: 1000.

Fitting error of GP, GEP and NN for solving symbolic regression are as shown in tables. In the experiment, we run thirty times for each algorithm, the simulation results are in the following tables, the statistics information of experiment based on the tables is mentioned below. In the tables below, "min value" stands for the minimum value for thirty times, "max value" is maximum value without considering the values for local solutions, they are usually the maximal ones, "local solutions" are the numbers of falling into local solutions.

For GP and GEP, we use the same parameters setting mentioned above for training the following data, but NN uses different parameters to train the below data. The parameters setting of the neural network for training the following data: learning rate: 0.1; momentum: 0; sigmoid's alpha value: 3; the first layer neuron number: 25; iteration number: 1000. The experiment results are shown below.

From Tables 7, 8, 9, 10, 11, 12 and 13, we can see the fitting error fluctuation of NN for each solving symbolic regression is smaller compared to GP and GEP, GP and GEP fluctuate greatly in solving the problem, which shows that the stability of NN for solving the problem is better; from the point of the used time of the three methods for solving symbolic regression not listed in the tables, the used time of GP, GEP and NN is not big, relatively speaking, NN uses more time and shows better result, the optimal and the worst time change little.

Neural network has the ability of learning, thinking, and good at learning from data directly; evolutionary algorithms have intelligent characteristics such as self-organizing, adaptive and self-learning, and are good at solving the global optimal problem; but the general ability of neural network is not as well as evolutionary computation, and learning accuracy of evolutionary algorithms is worse than neural network in this experiment with the training data.

Table 7. The fitting error of GP, GEP and NN for SR with training data 1

GP	6.13	8.00	1.44	5.62	2.38	16.00	5.55	3.83	8.00	6.12
	1.22	1.27	16.00	7.32	2.63	6.70	4.00	5.16	10.00	3.68
	4.00	7.24	4.01	8.00	16.00	2.53	5.55	3.13	4.00	8.00
GEP	2.67	8.00	2.80	16.00	8.20	8.00	1.75	7.20	8.00	10.00
	1.67	16.00	16.00	7.72	8.00	16.00	4.00	8.00	16.00	8.00
	12.50	6.14	8.67	4.24	12.00	6.76	8.00	16.00	3.07	16.00
NN	1.15	0.98	1.01	2.07	1.14	0.52	0.63	0.95	0.72	16.46
	1.68	1.91	0.55	0.85	1.24	29.43	2.07	3.12	1.53	1.86
	0.82	2.28	1.63	1.21	1.59	0.98	2.29	1.69	1.59	1.68

Table 8. The fitting error of GP, GEP and NN for SR with training data 2

GP	19.58	27.00	27.00	10.00	9.80	27.00	27.00	27.00	27.00	27.00
	20.17	27.00	15.66	19.58	20.64	27.00	25.91	27.00	19.58	27.00
	20.53	19.67	27.00	27.00	27.00	27.00	27.00	27.00	19.58	27.00
GEP	24.81	27.00	27.00	27.00	27.00	27.00	27.00	27.00	24.74	27.00
	27.00	27.00	20.60	22.90	27.00	20.90	25.71	24.85	27.00	27.00
	27.00	19.89	21.66	21.27	27.00	23.60	19.89	22.47	27.00	19.85
NN	57.94	9.88	3.01	9.48	1.56	3.57	11.51	3.06	6.14	10.52
	4.45	0.63	5.02	6.49	9.59	4.04	13.68	1.46	1.40	2.12
	5.79	2.18	2.76	8.53	10.32	5.68	12.62	3.70	58.28	4.84

Table 9. The statistics of the data 1 and data 2

	Training data 1				Training data 2			
	Min value	Max value	Average value	Local solutions	Min value	Max value	Average value	Local solutions
GP	1.22	10.00	5.02	3	9.80	25.91	18.39	18
GEP	1.67	12.50	6.76	7	19.85	25.71	22.37	16
NN	0.52	16.46	1.94	1	1.40	13.68	5.86	2

Table 10. The fitting error of GP, GEP and NN for SR with training data 3

GP	1.22	1.00	7.00	28.13	44.53	71.00	3.60	1.00	22.42	1.00
	5.24	2.90	7.40	1.00	1.00	71.00	31.92	44.53	38.90	7.94
	2.33	1.00	1.00	1.00	1.00	1.00	1.00	1.00	4.33	37.02
GEP	45.73	71.00	22.18	29.00	35.00	14.42	45.03	25.98	41.96	71.00
	35.00	45.03	22.42	37.43	50.27	44.53	44.61	7.22	44.84	27.90
	1.00	45.30	69.24	34.42	39.70	41.00	37.00	14.47	35.00	26.34
NN	24.00	3.13	3.44	153	2.23	156	46.60	2.59	156	178
	1.99	156	1.49	153	1.61	1.53	1.89	151	2.12	151
	173	1.78	2.49	151	1.97	2.27	3.14	151	177	1.65

Table 11. The fitting error of GP, GEP and NN for SR with training data 4

GP	26.69	27.03	27.82	37.00	37.00	32.84	26.16	24.06	24.06	37.00
	37.00	23.64	29.29	26.95	26.95	37.00	33.66	27.35	25.98	26.00
	26.12	34.33	26.02	26.38	25.25	24.06	26.26	37.00	31.72	37.00
GEP	25.66	39.00	37.00	37.00	34.07	33.66	37.00	37.80	37.00	37.00
	27.07	33.66	32.84	37.00	37.00	37.00	34.14	37.40	28.20	27.59
	32.85	37.00	26.35	33.53	33.66	37.00	23.21	37.00	34.64	37.00
NN	6.79	6.75	6.82	6.64	6.70	96.76	6.16	6.49	6.68	6.55
	6.59	6.65	6.69	6.42	96.25	6.97	6.70	6.66	6.26	13.32
	6.83	6.41	7.54	6.75	10.63	6.50	6.52	96.73	6.54	6.41

Table 12. The statistics of the data 3 and data 4

	Training data 3				Training data 4			
	Min value	Max value	Average value	Local solutions	Min value	Max value	Average value	Local solutions
GP	1.00	44.53	10.76	2	23.64	34.33	27.33	7
GEP	1.00	69.24	34.36	2	23.21	34.64	30.74	15
NN	1.49	46.60	5.88	12	6.16	13.32	7.04	3

Table 13. The statistics of the data 5 and data 6

	Training data 5				Training data 6			
	Min value	Max value	Average value	Local solutions	Min value	Max value	Average value	Local solutions
GP	63.88	80.33	75.15	15	38.12	50.93	44.01	14
GEP	71.19	79.70	76.43	15	45.14	50.39	47.48	18
NN	11.18	13.98	12.51	0	6.33	8.35	7.58	0

Neural network and evolutionary algorithms can perform parallel processing of large-scale problems, which greatly improves the whole operation speed. This is the most important feature of NN and evolutionary algorithms performing better than traditional algorithms. As for evolutionary algorithms, each individual calculates the fitness value according to the function, each generation has many individuals, so compared with some methods, the computation time is relatively shorter. Evolutionary algorithms and NN have high adaptability and fault tolerance, they can use the finite solution operations to obtain the solution, and show good robustness and stability.

5 Conclusion and Future Works

From the theory analysis, GP and GEP can make the solving processes adaptive, and obtain more accurate results. In the process of testing data, using GP and GEP to fit data does not need to predetermine the structure of the equations, which can greatly simplify the data preprocessing, and NN has good stability with small error volatility of each solution in solving SR. This paper applies NN and GEP to symbolic regression compared with GP for solving the problem, then it gives some analysis and remarks for the experiment simulation and models.

There are some future works that we could do for the next time: on the one hand, training more and diverse data is necessary for the models mentioned above; on the other hand, improve and optimize the algorithms, and apply new models such as deep learning model and the diverse improved ones to the field.

Acknowledgement. We are gratefully acknowledged the financial support from the National Natural Science Foundation of China under Grant No. 61170305, No. 60873114 and the Program of Scientifc Research and Technology in Liuzhou No. 2014J020401. The authors would like to acknowledge Mr. Andrew Kirillov for providing the codes mentioned above in the paper.

References

1. Zeng, T., Tang, C., et al.: Mining multi-dimensional complex association rule based on artificial immune system and gene expression programming. J. Sichuan Univ. (Eng. Sci. Ed.) **38**(5), 136–142 (2006)
2. Wang, X., Cao, L., Gu, S.: Genetic programming and its applications in the symbolic regression. J. Tongji Univ. **29**(10), 1200–1203 (2001)
3. Chen, Y., Yang, J., Yang, J., et al.: Research and application of genetic expression programming algorithm based on uniform-design. Comput. Appl. **27**(4), 948–951 (2007)
4. Lu, X., Cai, Z.: Application of a novel GEP algorithm in evolutionary modeling and forecasting. Comput. Appl. **25**(12), 2784–2787 (2005)
5. Fang, W., Zhang, K., Shao, L.: Complex function modeling based on improved gene expression programming. Comput. Eng. **32**(21), 188–190 (2006)
6. Peng, J., Tang, C., Yuan, C., et al.: A multi-gene evolutionary algorithm based on overlapped expression. Chin. J. Comput. **30**(5), 775–785 (2007)
7. Luo, P., Wang, X., Tang, X.: Hierarchical face parsing via deep learning. In: Proceedings of IEEE Conference on Computer Vision and Pattern Recognition (CVPR), pp. 2480–2487 (2012)
8. Baylar, A., Unsal, M., Ozkan, F.: GEP modeling of oxygen transfer efficiency prediction in aeration cascades. KSCE J. Civ. Eng. **15**(5), 799–804 (2011)
9. Xia, Y., Tian, S., Wei, H., Wang, Z.: Research on symbolic regression based on genetic programming. J. China Jiliang Univ. **17**(2), 128–131 (2006)
10. Deng, W., Zheng, Q., et al.: Research on extreme learning of neural networks. Chin. J. Comput. **33**(2), 279–286 (2010)
11. Guo, T., Zhang, X., Liang, Z.: Research on change information recognition method of vector data based on neural network decision tree. Acta Geodaet. Cartographica Sin. **42**(6), 937–943 (2013)
12. Zhang, X., He, G.: The forecasting approach for short-term traffic flow based on principal component analysis and combined NN. Syst. Eng. Theor. Pract. **8**, 168–170 (2007)
13. Li, A., Luo, S., Huang, H., et al.: Decision tree based neural network design. J. Comput. Res. Devel. **42**(8), 1312–1317 (2005)
14. Deng, S., Wang, R., Zhang, Y., Zhang, J.: Grid resource allocation algorithm based on parallel gene expression programming. Acta Electronica Sin. **37**(2), 272–277 (2009)

15. Peng, J., Tang, C., Li, C., Hu, J.: M2GEP: a new evolution algorithm based on multi2layer chromosomes gene expression programming. Chin. J. Comput. **28**(9), 1459–1466 (2005)
16. Hu, J., Tang, C., Duan, L., et al.: The strategy for diversifying initial population of gene expression programming. Chin. J. Comput. **30**(2), 305–310 (2007)
17. Li, B., Dong, J., Liu, Y., Mi, S.: Fuzzy lattice constructive morphological neural network. Acta Electronica Sin. **42**(2), 319–326 (2014)
18. Sun, Y., Wang, X., Tang, X.: Deep convolutional network cascade for facial point detection. In: Proceedings of IEEE Conference on Computer Vision and Pattern Recognition (CVPR), pp. 3476–3483 (2013)
19. Farabet, C., Couprie, C., Najman, L., LeCun, Y.: Learning hierarchical features for scene labeling. IEEE Trans. Pattern Anal. Mach. Intell. **35**(8), 1915–1928 (2013)
20. Ouyang, W., Wang, X.: A discriminative deep model for pedestrian detection with occlusion handling. In: Proceedings of IEEE Conference on Computer Vision and Pattern Recognition (CVPR), pp. 3258–3265 (2012)
21. Shang, C., Yang, F., Huang, D., Lya, W.: Data-driven soft sensor development based on deep learning technique. J. Process Control **24**, 223–233 (2014)
22. Zhang, G., Wang, W.: Application introduction of uniform experiment design method. Appl. Stat. Manage. **32**(1), 89–98 (2013)
23. Karaboga, D., Ozturk, C., Karaboga, N., Gorkemli, B.: Artificial bee colony programming for symbolic regression. Inf. Sci. **209**, 1–15 (2012)
24. Chen, J., Zeng, Z., Jiang, P.: On the periodic dynamics of memristor-based neural networks with time-varying delays. Inf. Sci. **279**, 358–373 (2014)
25. Melin, P., Amezcua, J., Valdez, F., Castillo, O.: A new neural network model based on the LVQ algorithm for multi-class classification of arrhythmias. Inf. Sci. **279**, 483–497 (2014)
26. Pulido, M., Melin, P., Castillo, O.: Particle swarm optimization of ensemble neural networks with fuzzy aggregation for time series prediction of the Mexican Stock Exchange. Inf. Sci. **280**, 188–204 (2014)
27. Gaxiola, F., Melin, P., Valdez, F., Castillo, O.: Interval type-2 fuzzy weight adjustment for backpropagation neural networks with application in time series prediction. Inf. Sci. **260**, 1–14 (2014)
28. Peng, Yu.: New application of symbolic regression method based on genetic programming in power quality analysis. Electron. Des. Eng. **21**(7), 20–23 (2013)

Inferring Large Gene Networks with a Hybrid Fuzzy Clustering Method

Chung-Hsun Lin, Yu-Ting Hsiao, and Wei-Po Lee[(✉)]

Department of Information Management,
National Sun Yat-Sen University, Kaohsiung, Taiwan
wplee@mail.nsysu.edu.tw

Abstract. To tackle the scalability problem in reverse engineering gene networks, this study presents an approach with two phases: gene clustering and network reconstruction. For gene clustering, a hybrid data and knowledge-driven method is developed to calculate similarity between genes. In the network reconstruction procedure, a Boolean network model is inferred from gene clusters. A series of experiments are conducted to investigate the effect of the hybrid similarity measure in gene clustering and network reconstruction. The results prove the feasibility and effectiveness of the proposed approach.

Keywords: Gene network inference · Gene clustering · Principal component analysis · Knowledge ontology · Boolean network

1 Introduction

Gene network construction is one of the most important issues in systems biology research. Many computational methods have been proposed to automatically infer gene networks [1, 2]. However, when the number of genes involved in the interactions increases, network construction becomes more and more difficult. To solve this scalability problem, the decomposition procedure has been proposed. Clustering is a practical and useful technique for grouping genes. The goal is to identify genes that have the same functions or regulatory mechanisms. Clustering methods group those points locating closely together in the dimension-reduced space. Each gene cluster can be considered a sub-network and inferred separately then gradually combined.

The first step in gene clustering is to determine how to measure the similarity (or dissimilarity) between any two genes; that is, to define the similarity function. All genes in a dataset are then assigned by the clustering algorithms into different clusters of similar expression patterns according to a similarity/dissimilarity measure. One popular technique is to extract data features from the original gene regulatory signals, and the gene distance can then be calculated accordingly. The principal component analysis (PCA) method is a widely used technique to extract important features for dimension reduction [3]. It is a coordinate transformation in which each data is written as a linear sum over basis vectors called principal components (PCs). Usually, only more important PCs are retained to reduce the dimensionality of the data.

The other method to measure data distance is to tackle the problem with a domain knowledge mindset and to consider the gene semantic similarity [4]. Biological

© Springer International Publishing Switzerland 2015
D.-S. Huang et al. (Eds.): ICIC 2015, Part I, LNCS 9225, pp. 715–722, 2015.
DOI: 10.1007/978-3-319-22180-9_71

knowledge can be obtained from scientific literature or public databases, among which gene ontology (GO) is most popular biological knowledge source. Though GO is a useful method to derive biological similarity between genes, it is notable that the appropriate use of functional similarity measures depends on the applications [5].

In the construction of gene networks, many models have been proposed to address different levels of biological details [1, 2]. Because our work is to construct large-sized networks, we choose to use the most popular abstract model, the Boolean model, in which genes are Boolean variables that only exist in discrete states [6]. To investigate the effect of using gene clustering to assist the construction of large Boolean networks, we present an approach that includes two major phases. The first phase is gene clustering, in which a hybrid method of data and knowledge-based clustering is developed. The PCA method is adopted to derive data features and calculate data similarity, and the GO method is employed to measure the semantic similarity between genes. The two types of similarities are then assembled together in the gene clustering process. The second is a network reconstruction procedure that builds networks from the gene clusters. To evaluate our approach, experiments have been conducted. The results show that our approach can produce clusters with high interactions between genes, which can lead to better network construction performance.

2   The Proposed Method

To reconstruct large networks from gene profiles, we devise a divide-and-conquer approach: the entire set of expression data is divided into strongly correlated subsets by a clustering method, and small networks are inferred from the subsets of expression data and assembled. It is notable that gene networks derived from expression data and knowledge resources often disagree with each other. Therefore, we develop a hybrid measurement to account for both data level and knowledge level information. Once the gene groups are obtained, a reverse engineering procedure is performed for network reconstruction. The details are described in the following subsections.

2.1   Measuring Gene Distance

In a gene network, genes interact and highly co-regulate one another; they may share the same molecular function and be involved in the same biological pathway. To tackle these difficulties, we adopt the PCA technique to process time series data. In this procedure, we use the common rules for choosing how many PCs to retain: keeping enough PCs so that the cumulative variance explained by the PCs is larger than a pre-specified threshold (70 % in this work, based on a preliminary test). With these PCs retained, the original expression data can be reduced to a lower dimensional subspace. The Euclidean distance measure is then applied for gene-gene similarity calculations.

The other important part of measuring gene-gene distance is to take domain knowledge into consideration so that genes can be analyzed from the perspective of molecular function. Here, we use gene ontology for gene analysis. Two popular GO similarity measure methods are used to determine the data distance. The first method is an information content-based method adopted from Resnik's algorithm ([7]). This algorithm follows the information theory: the information content of a concept c can be quantified as $-\log p(c)$, where $p(c)$ is the probability of encountering an instance of concept c in taxonomy. The similarity of two GO terms g_1 and g_2 can be defined as

$$Sim_{GO}(g_1, g_2) = \max_{g \in S(g_1, g_2)} [-\log p(g)] \tag{1}$$

In this equation, $S(g_1, g_2)$ is the set of all common ancestor nodes that subsume both g_1 and g_2. A node with the maximum value is termed the most informative subsume. More calculation details are referred to in [7].

Differing from the above content-based method, the topology-based measures use the intrinsic topology of the GO direct acyclic graph. Wang $et\ al.$ proposed a graph-based strategy to compute semantic similarity using the topology of the GO graph structure [8]. In this method, the semantics of GO terms are encoded into a numeric format, and the different semantic contributions of the distinct relations are considered. The semantic value of a GO term A is given by

$$SV(A) = \sum_{t \in T_A} S_A(t) \tag{2}$$

In the above equation, T_A denotes the set of ancestors of term A; $S_A(t)$ is 1 if $t = A$, otherwise $S_A(t)$ is defined as $\max\{w_e \times S_A(t') | t' \in C(A)\}$, in which $C(A)$ is the set of children of term A; and w_e represents the semantic contribution factor for gene relations in the hierarchy. For two genes g_1 and g_2 with sets of GO terms T_1, T_2 that annotate g_1 and g_2, the similarity of two GO terms $t_1 \in T_1$ and $t_2 \in T_2$ is defined as

$$Sim_{GO}(t_1, t_2) = \frac{\sum\limits_{t \in T_1 \cap T_2} (S_A(t) + S_B(t))}{SV(t_1) + SV(t_2)} \tag{3}$$

Because none of the existing measures account for all aspects of GO, it is hypothesized that integrating multiple measures can improve the performance. In this work, the above two measures are averaged to obtain the semantic similarity between genes.

2.2   Gene Clustering

After defining the two types of similarity measures between genes, this section describes how they are used in the clustering algorithm to group genes. As each single gene in the network may be involved in different biological functions and interact with many other genes, a flexible strategy for gene discrimination in the clustering process provides a better choice. Therefore, we adopt the fuzzy c-means algorithm to cluster genes. The calculation details on fuzzy clustering are referred to [9].

As mentioned, using gene expression or gene ontology alone is not enough to capture the gene-gene interactions. Therefore, in our approach the similarity between any two gene elements g_1 and g_2 is calculated by a hybrid of data-based (Sim_{exp}) and knowledge-based (Sim_{GO}) measures with two weighting factors (w_1 and w_2), as below:

$$Sim(g_1, g_2) = w_1 \times Sim_{exp}(g_1, g_2) + w_2 \times Sim_{GO}(g_1, g_2) \tag{4}$$

The algorithm repeatedly generates the membership levels for each gene element, which are then used to calculate the new centers of the clusters. It terminates when no improvement is observed.

The performance of gene clustering is evaluated by the protein-protein interaction (PPI) rate. It is defined as the fraction of interacting pairs found among all gene pairs that end up in the same cluster. As a validation information resource, we use a genomic database (i.e., BioGRID, that attempts to catalogue all known PPI as a straightforward extension of the idea [10, 11]) to validate expression derived gene clusters using GO. More precisely, the PPI of a cluster c_i is defined as

$$PPI_i = \frac{\text{num of (interacting gene pairs } \in c_i)}{\text{num of (all gene pairs } \in c_i)} \times 100\% \tag{5}$$

In the above equation, i is the index of cluster c_i and where k is the total number of clusters. Taking the average across all k clusters, one may also define the global measure assessing the quality (biological significance) of the whole partition.

2.3 Network Reconstruction

As indicated, in this work we adopt the Boolean network model and use the popular software BoolNet ([12]) to infer networks from clustered expression data. BoolNet is an R package that provides tools for assembling, analyzing, and visualizing Boolean networks. We choose the synchronized mode (i.e., all genes are updated at the same time) for network reconstruction. In this tool, a novel binarization method is developed to address the gene profiles of continuous values. This method is based on a threshold determination using scan statistics, and can provide a measure of threshold validity. The main idea is to search for at least one cluster in the measurements whose probability p is lower than a specified significance level. Based on the detected clusters with high significance (i.e., a low value of p), the binarization is performed in such a way that the points within a cluster are assigned to the same binary value.

To evaluate the results of network reconstruction, different criteria often used in binary classification are taken, including *precision*, *recall*, and the F metric [10, 11]. It is notable that in a real gene network, the rate of links (i.e., regulatory relationships) between nodes (genes) is quite low, which leads to a high true negative in network inference. Therefore, as in other relevant studies, we do not measure accuracy due to the large bias caused by the true negative.

3   Results and Discussion

To evaluate our hybrid approach for gene clustering and network reconstruction, this section describes the experiments and results. Two datasets (called datasets A and B hereafter) were chosen, because they are real datasets suitable for performing both PCA and GO semantic mapping procedures, and the constructed networks can be verified (most of the network links are known). Dataset A was collected from [11] and it included 800 genes. Dataset B was collected from [10]; it described the yeast *S. cerevisiae* regulatory network and contained more than 6000 genes. Considering the constraints for dealing with large datasets by the available packages, we sampled randomly several subsets of 400 genes for experiments. The results are similar and we only report only one set here due to the space limitation.

3.1   Results of Gene Clustering

The first set of experiments investigated the effect of using different distance measurement strategies on gene clustering. Both of the datasets described above were used for evaluation. The PCA procedure was first performed on the time series expression data for feature extraction, and a set of features was selected that contained at least 70 % of the original data's information content. As mentioned in Sect. 2.1, two types of similarity measures were applied to the data using the data-based method (i.e., PCA) and the knowledge-based method. Next, the two types of similarities were combined and used for distance calculation in the clustering process.

In this set of experiments, the fuzzy c-means clustering method was used. To investigate the effects of expression data and semantic similarity on gene clustering, different combinations of w_1 and w_2 were arranged. As described, the PPI measurement was taken to evaluate the results. For each weight combination, 100 experimental trials were conducted for gene clustering, and the PPI values of the trials were averaged. The results are presented in Fig. 1.

In these figures, the x-axis indicates the weights of the two components. The indices on the left side of the x-axis (from the leftmost to the middle) represent the weights of the GO-based distance (i.e., values for w_2), and they range from 0 to 1. In this interval,

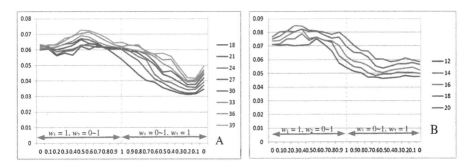

Fig. 1. PPI values of different distance combinations in gene clustering

the weight of the expression data-based distance (i.e., w_1) remains 1. Meanwhile, the indices on the right side of the x-axis are the weights of the expression data-based distance (i.e., values for w_1); the weights vary from 1 (the middle) to 0 (the rightmost), and the weight of GO-based distance (i.e., w_2) is 1 in this interval. In Fig. 1, the y-axis indicates the PPI values for different combinations of the above two components. Furthermore, the curves with different colors represent the results produced by different numbers of clusters (indicated on the right-hand side of the figure).

The results presented here demonstrate the effectiveness of using domain knowledge (i.e., semantic similarity) in gene clustering. They also suggest that certain combinations with large weights on data-based distance and relatively small weights on knowledge-based distance can have the best performance. In general, the PPI values obtained by our approach have been relatively high, in contrast to those reported in relevant studies [10].

3.2   Evaluation of Network Reconstruction

To investigate the effect of gene clustering on network reconstruction, in the second series of experiments, we reconstructed the gene networks based on the clusters obtained from previous experiments. As mentioned in Sect. 2.3, we adopted the online software BoolNet to model the genes into Boolean networks. This software includes a binarization procedure with scan statistics to transfer the real-value time series data into the binary form before network construction.

We first conducted a set of experimental trials for each dataset to verify whether the clustering results with high PPI values are beneficial to network reconstruction. In the experiments, the cluster sets with the five highest PPI values and those with the five lowest PPI values were used for network reconstruction. The results for the two datasets are presented in Fig. 2, in which the precision (p), recall (r), and the F-measure (f) are shown for each set of clusters. In the figures, the values were averaged from five trials. As can be seen, for each dataset, in network reconstruction the cluster sets with high PPIs have consistently better results (i.e., relatively higher precision, recall, and F-measure) than those with low PPIs. From these figures, we can also observe that the precision of the trials with higher PPIs is significantly better than that of the trials with lower PPIs. It is worth noting that the false positive rate in Boolean network

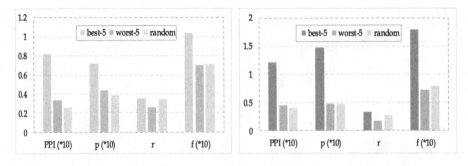

Fig. 2. Results of network reconstruction for two datasets (left: dataset A; right: dataset B)

reconstruction is often high, particularly in the reconstruction of large gene networks. This is mainly due to the disagreement between the experimental front and the computational front for link detection.

After showing that high PPI clusters can lead to better performance in network reconstruction, we conducted a second set of experimental runs to further examine whether random clustering could produce the same results. To confirm the effect of high PPI clusters, we randomly partitioned the genes in each dataset into the same numbers of clusters with the highest PPI values, and then performed the same procedure to construct networks. The results of network reconstruction are also shown in Fig. 2. As can be observed, in contrast to random clustering, the proposed gene clustering method could obtain better PPI values for all datasets and consequently resulted in better network reconstruction performance.

4 Conclusion

To tackle the problem of scalability in gene network inference, we have developed an efficient gene clustering method to perform network decomposition that is beneficial to network reconstruction. Because simply using gene expression or gene ontology alone is not enough to capture the interaction relationships among genes, we thus proposed a hybrid approach to calculate gene distance. The two types of similarities are used to calculate the gene distance in a fuzzy clustering procedure. A series of experiments have been conducted to demonstrate how the presented approach can be used to infer large networks. The experiments show that the proposed approach with a hybrid similarity measure can bring about better results in gene clustering and which consequently improves performance of network reconstruction.

References

1. Lee, W.-P., Tzou, W.-S.: Computational methods for discovering gene networks from expression data. Briefings Bioinform. **10**, 408–423 (2009)
2. Chai, L.E., Loh, S.K., Low, S.T., et al.: A review on the computational approaches for gene regulatory network construction. Comput. Biol. Med. **48**, 55–65 (2014)
3. Ma, S., Dai, Y.: Principal component analysis based methods in bioinformatics studies. Briefings Bioinform. **12**, 714–722 (2011)
4. Tan, M., Alshalalfa, M., Alhajj, R., Polat, F.: Influence of prior knowledge in constraint-based learning of gene regulatory networks. IEEE Trans. Comput. Biol. Bioinform. **8**, 130–142 (2011)
5. Mazandu, G.K., Mulder, N.J.: Information content-based gene ontology semantic similarity approaches: toward a unified framework theory. Biomed Res. Int. 2013, 1–5 (2013). 292063
6. Saadatpoura, A., Albert, R.: Boolean modeling of biological regulatory networks: a methodology tutorial. Methods **62**, 3–12 (2013)
7. Resnik, P.: Semantic similarity in a taxonomy: an information based measure and its application to problems of ambiguity in natural language. J. Artif. Intell. Res. **11**, 95–130 (1999)

8. Wang, J.Z., Du, Z., Payattakool, R., Yu, P.S., Chen, C.-F.: A new method to measure the semantic similarity of go terms. Bioinformatics **23**, 1274–1281 (2007)

9. Bezdek, J.: FCM: the fuzzy c-means clustering algorithm. Comput. Geosci. **10**, 191–203 (1981)

10. Kustra, R., Zagdanski, A.: Data-fusion in clustering microarray data balancing discovery and interpretability. IEEE/ACM Trans. Comput. Biol. Bioinform. **7**, 50–63 (2010)

11. Zainudin, S., Mohamed, N.S.: Evaluating the performance of partitioning techniques for gene network inference. In: Proceedings of International Conference on Intelligent Systems Design and Applications, pp. 1119–1124 (2010)

12. Mussel, C., Hopfensitz, M., Kestler, H.A.: Boolnet—an R package for generation, reconstruction and analysis of boolean networks. Bioinformatics **26**, 1378–1380 (2012)

A Novel Algorithm for Classifying Protein Structure Familiar by Using the Graph Mining Approach

Sun-Yuan Hsieh[1,2]($\boxtimes$), Chia-Wei Lee[1], Zong-Ying Yang[2],
Heng-Wei Wang[1], and Jun-Han Yu[1]

[1] Department of Computer Science and Information Engineering, National
Cheng Kung University, No. 1, University Road, Tainan 701, Taiwan
hsiehsy@mail.ncku.edu.tw, cwlee@csie.ncku.edu.tw,
{mathbookpeace,wirlly8888}@hotmail.com
[2] Institute of Medical Informatics, National Cheng Kung University,
No. 1, University Road, Tainan 701, Taiwan
q56001066@mail.ncku.edu.tw

Abstract. Protein structural classification is critical in bioinformatics. In this
study, a simple and connected graph was used to represent a 3D protein structure
in which each node represented an amino acid and each edge represented a
contact distance between two amino acids. The B-factor (atomic displacement
parameters) was then used to substantially reduce the number of nodes and
edges in each graph representation. A graph mining approach was applied to
determine the critical subgraphs among these graphs, which can be applied to
classify protein structural families. An experimental study was conducted in
which characteristic substructural patterns were identified in several protein
families in the SCOP database.

Keywords: Subgraph mining · Amino acid · 3D protein structure · B-factor ·
Bioinformatics · Protein classifications

1 Introduction

Recurring substructures in proteins crucial important information regarding protein
structure and function. For instance, common structural fragments can represent fixed
3D arrangements of residues that correspond to active sites or other functionally relevant features [14, 18, 19]. Numerous computational methods have been proposed to
determine motifs in proteins.

In biology, graphs are commonly used for representing chemical compounds
[8, 15], protein sequences, residue packing patterns in protein structures [9, 22], protein
networks [20], functional sites in proteins [2, 4, 5], and multiple sequence alignments
of proteins with similar structural domains [6], and can be used to provide information
about the possible common substructures and determine whether the conserved
sequence patterns in a group of homologous proteins have similar 3D arrangements.

In general, the algorithm approach for mining frequent subgraphs can be defined as
the process of finding all connected subgraphs that appears frequently in a large graph

© Springer International Publishing Switzerland 2015
D.-S. Huang et al. (Eds.): ICIC 2015, Part I, LNCS 9225, pp. 723–729, 2015.
DOI: 10.1007/978-3-319-22180-9_72

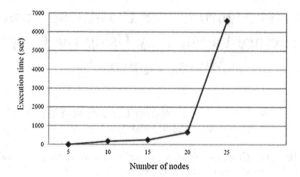

Fig. 1. Distinct numbers of nodes

database. Several methods based on the algorithm approach have been proposed to solve this problem, including the SUBDUE [7], Fast Frequent Subgraph Mining (FFSM) [8–10, 22], Graph-based Substructure Pattern Mining (gSpan) [23], Frequent Subgraph Discovery (FSG) [13], the GrAph/Sequence/Tree extractiON (GASTON) algorithms [16, 17]. However, numerous scientific and commercial applications might require obtaining frequent subgraphs in complex graph datasets with numerous vertices and edges, all of which require extensive computation time. The complexity of the frequent subgraph mining problem is evident from the diversity of approaches employed in developing various algorithms for its solution [12].

The remainder of this paper is organized as follows: The data structures and the proposed algorithm for subgraph mining are presented in Sect. 2. Section 3 presents the performance of the proposed method, using both synthetic and real-world graph data sets. Finally, the conclusion and discussion are presented in Sect. 4.

2 The Proposed Algorithm

Proteins are the most critical biological compounds necessary for life and are used for numerous purposes in the human body. They are comprised of and linked by staggered amino acids, and approximately 20 types of amino acids naturally form in nature. The simplest amino acid structure comprises four groups: (1) a central carbon atom (C atom), (2) an amino group (-NH$_2$), (3) an acid group (-COOH), and (4) an R group (side chain).

In this study, a graph was used to represent the protein structure. The nodes of the graph represented the amino acids, of which there were 20 labels. Because the protein backbone defines the overall protein conformation, the C atom was used to represent the amino acids. Regarding the computation cost, the coordinate frames were only generated from the amino acid C atom instead of all atoms, and thus the computational cost was low.

In the primary sequence, two amino acids were connected by a "bond edge" when they were consecutive. To construct 3D protein structure, "proximity edge" was used to represent the graph edge. The formal definition of "proximity edge" is provided as follows.

Definition 1 (Proximity edge). Given a protein graph with n C atoms where the distance between two atom coordinates is $D(i, j)$, where $1 \leq i, j \leq n$, if the distance between the two associated C atoms is less than or equal to the threshold, then a proximity edge exists between i and j, and $D(i, j) \equiv D(j, i)$. The experimentally determined threshold was inserted only if the distance did not exceed 12Å to reduce the graph complexity [21].

In the proposed algorithm, each protein graph each protein graph is represented by an adjacency matrix. The formal definition of an adjacency matrix is as follows.

Definition 2 (adjacency matrix). Given a graph G with n nodes, an $n \times n$ adjacency matrix M of G is defined as a triangle matrix in which $m_{i,j}$ represents the entry at the ith row and jth column in M. The diagonal entry $m_{i,i}$ for all $1 \leq i \leq n$ are filled with the labels of nodes in G, and every non-diagonal entry $m_{i,j}$, where $0 < j \leq i \leq n$, equals one if there is an edge between the two nodes $m_{i,i}$ and $m_{j,j}$, and equals zero if there is no edge between $m_{i,i}$ and $m_{j,j}$. In addition, define $code(M)$ as a string by representing M in a row major manner.

The test protein graph G was built using the canonical adjacency matrix, and then all possible subgraphs were enumerated. When an enumerated subgraph matches the tag family pattern, then the test graph G belongs to the marked pattern's family. All the canonical adjacency matrices (CAMs) can represent connected subgraphs of a test protein graph G, which can be organized using a rooted tree as follows: (1) The root of the tree is an empty matrix. (2) Level 1 of the tree contains numerous labled nodes; they are the children of the root.[1] (3) The nodes at level 2 are obtained from those at the Level 1 by adding one adjacency node to each of them. (4) After constructing Level 2 of the tree, algorithms CAM Tree-Join Case and CAM Tree-Extension Case are used to enumerate all the subgraphs until the test protein graph G matrix is generated.

Cases for CAM-Tree-Join. By building a CAM tree, the CAM-tree-join operation "superimposes" two graphs to generate a new candidate graph. Depending on the various characteristics of the graphs, the CAM-tree-join can produce one or two candidate graphs.

Given an adjacency matrix A of a graph G, A is defined as an *inner matrix* if A has at least two edge entries (two entries containing 1) in the last row. Otherwise, A is an *outer matrix*. Given two an m-by-m adjacency matrix $A = (a_{i,j})$ representing graph G_A and an n-by-n adjacency matrix $B = (b_{i,j})$ representing graph G_B, let the last non-zero entry of A be $a_{m,f} = 1$ (representing edge $(a_{m,m}, a_{f,f})$ of G_A) and the last non-zero entry of B be $b_{n,k} = 1$ (representing edge $(b_{n,n}, b_{k,k})$ of G_B). The *join operation* is used to generate another adjacency matrix using the following algorithms.

In Algorithm 1, when two outer matrix A ($m \times m$) and B ($n \times n$) are joined, a matrix with the same size $n \times n$ (Case 1a) can be obtained. It is also possible to obtain a matrix with the size $(m + 1) \times (m + 1)$ (case 1b). When joining an "inner" matrix and an "outer" matrix, the resulting matrix is generated using Algorithm 2, which similar as Algorithm 1.

---

[1] In the test protein graph G, the same labeled node is enumerated once.

Algorithm 1. CAM_TREE_JOIN_CASE_1

Input: an m-by-m outer matrix $A = (a_{i,j})$ and an n-by-n outer matrix $B = (b_{i,j})$, where
the last non-zero entry of A is $a_{m,f} = 1$ and the last non-zero entry of B is $b_{n,k} = 1$.
Output: A CAM C.

1   **if** $(m = n) \land (f \neq k) \land (a_{m,m} = b_{n,n})$ **then**
2   /* Case 1a   */
3   Output an m-by-m matrix $C = (c_{i,j})$, where
4      $c_{i,j} = a_{i,j} \lor b_{i,j}$

5   **if** $(m = n) \land (a_{m,m} \neq b_{n,n})$ **then**
6   /* Case 1b   */
7   Output an $(n + 1)$-by-$(n + 1)$ matrix $D = (d_{i,j})$, where
8

$$d_{i,j} = \begin{cases} a_{i,j} & 1 \leq i,j \leq m, \\ b_{n,j} & i = n+1, 0 < j < n, \\ 0 & i = n+1, j = n, \\ b_{n,n} & i = n+1, j = n+1. \end{cases}$$

CAM-Tree-Extension. Another enumeration technique is the extension operation performed using Algorithm 3 that generates a candidate graph G_{k+1} with $(k + 1)$ edges from a graph G_k with k edges.

Algorithm 3. CAM_TREE_EXTENSION

Input: An n-by-n outer matrix A and a protein test graph $G = (V, E)$
Output: A set of adjacency $(n + 1)$-by-$(n + 1)$ matrices

1   $S \leftarrow \emptyset$
2   **for** *each edge* $(u, v) \in E$ *such that* $(u \in (V \setminus A)) \land (v \in A)$ **do**
3

$$b_{i,j} = \begin{cases} a_{i,j}, & 1 \leq i,j \leq n, \\ 0, & i = n+1, a_{j,j} \neq v, \\ 1, & i = n+1, a_{j,j} = v, \\ u, & i = n+1, j = n+1. \end{cases}$$

4   $S \leftarrow S \bigcup \{B = (b_{i,j})\}.$

2.1   Classification of a Protein Family

First, a family of conserved sequences called *characteristic substructural family* is marked to determine whether the input test protein belongs to the characteristic substructural family. Second, each conserved sequence is transferred to a candidate subgraph. Then, we aim at determining whether a candidate subgraph belongs to subgraphs of the test protein graph. If a candidate subgraph is a subgraph of the test protein graph, then the test protein graph belongs to the characteristic substructural family.

By applying the join operations (described in Algorithm 1) and extension operation (described in Algorithm 3), the proposed algorithm utilizes the CAM tree to efficiently enumerate candidate subgraphs to determine whether a candidate subgraph belongs to subgraphs of the test protein graph. More importantly, the proposed method need not to enumerate all the subgraphs, which can significantly reduce the execution time for classification.

3   Experimental Results

To demonstrate the performance of the proposed method, the proposed method we compared with two algorithms, the Basic Local Alignment Search Tool (BLAST) [1] and the BLAST-like alignment tool (BLAT) [11].

The data sets considered for this experiment needed to be transformed to a format compatible with these executable programs. We tested compressive BLAST and BLAT on the real-world same protein dataset from the protein data bank. All experiments were performed on a 2.67 GHz PC with 3 GB memory. The proposed algorithm was implemented using the C++ programming language to ensure comparable results.

The Protein Data Bank (PDB) archive is a repository of atomic coordinates and other information describing proteins and other critical biological macromolecules. Structural biologists use methods such as X-ray crystallography, or NMR spectroscopy to determine the location of atoms relative to each other in a molecule. In this study, eight protein families were selected datasets: Bcl-xl family, E2F family, globin family, HSP family, serpin family, serine proteases family, Histone family, and Argonaute family, the protein from bacteria, eukaryota and archaea, using the PDB-file content and obtained the coordinates for all proteins.

A protein is represented by an undirected graph in which each node corresponds to an amino acid residue in the protein such that the residue type is used as the label of the node. A "peptide" edge was introduced between two residues X and Y and a "proximity" edge if the distance between the two associated C_α atoms of X and Y was below a certain threshold (10Å and 15Å in this study) and no peptide bond was observed between X and Y.

To calculate the optimal proximity edge length, the dataset Bcl-xl proteins were tested to obtain a proximity edge length at 10Å. Because the accuracy and execution time can be affected by proximity edge lengths, five distance thresholds were adopted (5Å, 10Å, 15Å, 20Å, and 25Å) to construct protein structures. For the distances of 10Å and 15Å, every nodes and edges could be successful light on the graph. As the distance threshold increased, the computation time increased because of the additional nodes and edges of each graph. Similarly, because the distinct B-factor also affected the computation time and classification accuracy, B-factors 40 and 50 were selected.

Strictly conserved residues were identified as the protein structure motifs. Based on the classification of proteins, the protein structure motifs generated structural fingerprints [3] for the various structural classes. Assume a database of protein structures the class labels of which are known, and assume diverse protein structure classes in the database; for each structure class, fingerprint were generated by mining the recurring protein structure motifs from all of the member protein structures belonging to this class. The proposed method was a pure classification scheme in that no type of structural comparison, alignment, or searching is necessary. The proposed classification was compared with BLAST and BLAT and the alignment results generated e-value, based on e-value $= 1 - e^{-10}$ ($e = 2.718$), and the classification result were examined accordingly. Artificial data sets were also created to test distinct numbers of nodes. Figure 1 shows the execution times of the five different numbers of nodes, which demonstrates that the proposed algorithm can be applied to protein graphs with large number of nodes.[2] The experimental results shows that the proposed method can be used to classify the protein structures accurately and efficiently.

4 Conclusion

This paper presented a graph mining algorithm that can be successfully applied to classify protein families. Although protein graphs are large and complex, the proposed method still can achieve high accuracy with a reasonable time. The proposed method combines two concepts, the B-factor threshold and proximity edge length, which can used to reduce the search space significantly. The experimental result demonstrated that the proposed method can classify the protein structures accurately and efficiently.

References

1. Altschul, S.F., Madden, T.L., Schaffer, A.A., Zhang, J., Zhang, Z., Miller, W., Limpman, D.J.: Gapped BLAST and PSI-BLAST: a new generation of protein database search programs. Nucleic Acids Res. **25**(17), 3389–3402 (1997)
2. Aloy, P., Querol, E., Aviles, F.X., Sternberg, M.J.E.: Automates structure-based prediction of functional sites in proteins: applications to assessing the validity of inheriting protein function from homology in genome annotation and to protein docking. J. Mol. Biol. **311**(2), 395–408 (2001)
3. Aung, Z., Tan, K.L.: Automatic protein structure classification through structural fingerprinting. In: 4th IEEE Symposium on Bioinformatics and Bioengineering, pp. 508–515 (2004)
4. Bandyopadhyay, D., Huan, J., Liu, J., Prins, J., Snoeyink, J., Tropsha, A., Wang, W.: Using Fast Subgraph Isomorphism Checking for Protein Functional Annotation Using SCOP and Gene Ontology. Technical report, The University of North Carolina at Chapel Hill Department of Computer Science (2005)
5. Bandyopadhyay, D., Huan, J., Liu, J., Prins, J., Snoeyink, J., Wang, W., Tropsha, A.: Functional neighbors: inferring relationships between nonhomologous protein families using family-specific packing motifs. IEEE Trans. Inf Technol. Biomed. **14**(5), 1137–1143 (2010)
6. Henikoff, S., Henikoff, J.G., Pietrokovski, S.: Blocks + : a non-redundant database of protein alignment blocks derived from multiple compilations. Bioinformatics **15**(6), 471–479 (1999)
7. Holder, L.B., Cook, D.J., Djoko, S.: Substructure discovery in the SUBDUE system. In: Association for the Advancement of Artificial Intelligence Workshop on Knowledge Discovery in Database (AAAI), pp. 169–180 (1994)
8. Huan, J., Bandyopadhyay, D., Wang, W., Snoeyink, J., Prins, J., Tropsha, A.: Comparing graph representations of protein structure for mining family-specific residue-based packing motifs. J. Comput. Biol. **12**(6), 657–671 (2005)

---

[2] In general, the nodes of candidate subgraph are between 10 and 15.

9. Huan, J., Wang, W., Bandyopadhyay, D., Snoeyink, J., Prins, J., Tropsha, A.: Mining protein family specific residue packing patterns from protein structure graphs. In: 8th Annual International Conference on Research in Computational Molecular Biology (RECOMB), pp. 308–315 (2004)

10. Huan, J., Wang, W., Prins, J.: Efficient mining of frequent subgraph in the presence of isomorphism. In: 3th IEEE International Conference on Data Mining (ICDM), pp. 549–552 (2003)

11. Kent, W.J.: BLAT-the BLAST-like alignment tool. Genome Res. **12**(4), 656–664 (2000)

12. Krishna, V., Suri, N.N.R.R., Athithan, G.: A comparative survey of algorithms for frequent subgraph discovery. Curr. Sci. **100**(25), 190–198 (2011)

13. Kuramochi, M., Karypis, G.: Frequent subgraph discovery. In: 1st IEEE Conference on Data Mining (ICDM), pp. 313–320 (2001)

14. Laberge, M., Yonetani, T.: Common dynamics of globin family proteins. Int. Union Biochem. Mol. Biol. **59**(8), 528–534 (2007)

15. Lam, W.W.M., Chan, K.C.C.: A graph mining algorithm for classifying chemical compounds. In: IEEE International Conference on Bioinformatics and Biomedicine (BIBM), pp. 321–324 (2008)

16. Nijssen, S., Kok, J.N.: A quickstart in frequent structure mining can make a difference. In: 10th ACM SIGKDD International Conference on Knowledge Discovery and Data Mining (KDD), pp. 647–652 (2004)

17. Nijssen, S., Kok, J.N.: Frequent graph mining and its application to molecular databases. In: IEEE International Conference on Systems, Man and Cybernetics (SMC), 5, pp. 4571–4577 (2004)

18. Petros, A.M., Olejniczak, E.T., Fesik, S.W.: Structural biology of the Bcl-2 family of proteins. Biochim. et Biophys Acta (BBA)-Mol. Cell Res. **1644**(2), 83–94 (2004)

19. Remold-O'Donnell, E.: The ovalbumin family of serpin proteins. Fed. Eur. Biochem. Societeies Lett. **315**(2), 105–108 (1993)

20. Wackersreuther, B., Wackersreuther, P., Oswald, A.: Frequent subgraph discovery in dynamic networks. In: 8th Workshop on Mining and Learning with Graphs (MLG), pp. 155–162 (2010)

21. Weskamp, N., Kuhn, D., Hllermeier, E., Klebe, G.: Efficient similarity search in protein structure databases by k-clique hashing. Bioinformatics **20**(10), 1522–1526 (2005)

22. Williams, D.W., Huan, J., Wang, W.: Graph database indexing using structured graph decomposition. In: 23th International Conference on Data Engineering (ICDE), pp. 976–975 (2007)

23. Yan, X., Han, J.: gSpan: Graph-based substructure pattern mining. In: 3th IEEE International Conference on Data Mining (ICDM), pp. 721–724 (2002)

Predicting Helix Boundaries of α-Helix Transmembrane Protein with Feedback Conditional Random Fields

Kun Wang[1], Hongjie Wu[1,3(✉)], Weizhong Lu[1], Baochuan Fu[1],
Qiang Lü[2,3], and Xu Huang[4]

[1] School of Electronic and Information Engineering,
Suzhou University of Science and Technology, Suzhou 215009, China
hongjiewu@mail.usts.edu.cn
[2] School of Computer Science and Technology,
Soochow University, Suzhou 215006, China
[3] Jiangsu Provincial Key Lab for Information Processing Technologies,
Suzhou 215006, China
[4] School of Information Engineering, Huzhou University,
Huzhou 313000, China

Abstract. Transmembrane proteins play an important role in cellular energy production, signal transmission, metabolism. Existing machine learning methods are difficult to model the global correlation of the membrane protein sequence, and they also can not improve the quality of the model from sophisticated sequence features. To address these problems, in this paper we proposed a novel method by a feedback conditional random fields (FCRF) to predict helix boundaries of α-helix transmembrane protein. A feedback mechanism was introduced into multi-level conditional random fields. The results of lower level model were used to calculate new feedback features to enhance the ability of basic conditional random fields. One wide-used dataset DB1 was used to validate the performance of the method. The method achieved 95 % on helix location accuracy. Compared with the other predictors, FCRF ranks first on the accuracy of helix location.

Keywords: Feedback · Conditional random field · α-helix transmembrane protein

1 Introduction

Transmembrane proteins play an important role in cellular energy production, signal transmission, metabolism. Currently, half of molecular drugs are related to transmembrane proteins [1], which also underlines their biological importance. In fact, 20 %–30 % of the genes in the human genome might be coded to transmembrane protein [2], which means that the count of transmembrane proteins might be 5000–6000 in total. However, due to the obstacles in expression, purification, information extraction, etc. [3], transmembrane proteins determined by experiments only account for about 1.5 % among

© Springer International Publishing Switzerland 2015
D.-S. Huang et al. (Eds.): ICIC 2015, Part I, LNCS 9225, pp. 730–736, 2015.
DOI: 10.1007/978-3-319-22180-9_73

all known-structure proteins in PDB (Protein Data Bank) [4]. Nevertheless, the topology of transmembrane proteins can provide us significant help in the analysis of membrane protein structure and function, because the function of transmembrane protein can be easily inferred from the topology (helix count, helix location, et al.).

There are two basic types of transmembrane proteins: alpha-helical and beta-barrels. Alpha helices across the lipid bilayer repeatedly, becoming a helix bundle. This paper mainly discussed the prediction of 3D structures of alpha-helical transmembrane proteins.

Many different machine learning methods have been employed to predict the topology of the membrane protein. Hidden Markov models (HMMs) are the earliest approach. TMHMM [5] and HMMTOP [6] developed latterly are both based on HMMs. Then, neural networks (NNs), support vector machines (SVMs) [7] and maximum entropy models (MEMs) are also used in topology prediction. There are also some hybrid methods, such as: OCTOPUS [8].

However, there are two problems within these existing methods: firstly, these methods are difficult to model the global correlation of protein sequences. Sequences information is important to protein related prediction. HMMs, as a kind of generative model, are based on the joint probability about the observation sequence and label sequence. Hidden Markov model independence assumptions make it cannot consider the information of context, and this limits the feature selection and modeling the long-range evolution, while the length of most of the transmember is longer than 300 and long-range mutation and co-evolution are common for transmembrane proteins. Secondly, the existing models are difficult to improve itself, because of the lack of an effective feedback mechanism. Feedback mechanism can extract features from the results, increasing the knowledge available, and it can mine deeper relationship of a long-range sequence. By timely accessing of the results of the lower level model, we can build the model according to the new features extracted from the results, and rectify the problem of too high or too low estimates, thus improve the prediction accuracy.

This paper proposed a new method based on conditioanal random field model (CRF) [9] to predict helix boundaries of alpha-helix transmembrane proteins, named feedback conditional random field model (FCRF). Firstly, we introduced the datasets. And then, the mechanism of FCRF was introduced. At last the results were compared and discussed.

2 Materials and Methods

2.1 Datasets

The dataset named as DB1 was selected from TMPDB [10], which was created in 2003 by Ikeda and is widely used, for example, it was used in Phobius [11] and MEMSAT-SVM [7]. TMPDB is a transmembrane protein database, containing information about the topology of transmembrane proteins that extracted by techniques such as X-ray, nuclear magnetic resonance (NMR), gene dissolved, etc. DB1 includes 106 different sequences.

The basic information on the dataset is shown in Table 1, the column "TMs" represents the total number of transmembrane helix that the dataset contains, the column "Residues" represents the total number of residues and the column "AVG Length" shows the average number of residues in each chain.

Table 1. Summary of the data set

Data set	Chains	TMs	Residues	AVG length
DB1	106	389	23475	221.5

2.2   Feedback Conditional Random Field

(1) Basic conditional random field

In CRFs, given an observation sequence $\mathbf{X} = (\mathbf{x1}, \mathbf{x2}, \ldots, \mathbf{xn})$, we want to get the most probable label sequence $\mathbf{Y} = (\mathbf{y1}, \mathbf{y2}, \ldots, \mathbf{yn})$, i.e. $Y^* = \mathrm{argmax}_Y P(Y|X)$. As a kind of undirected graphical model, conditional random fields are not based on the joint probability distribution P(Y,X) but the conditional probability distribution P(Y|X). HMMs obtain state sequence Y by maximizing the joint probability of X and Y, but the HMMs cannot use the remote features, which greatly limits the HMM applications. CRFs are an exponential or log-linear model, so, it can choose any one of the features to build the model. According to the theory of CRFs, conditional probability distribution between label sequence Y and observation sequence X can be given in the formula (1):

$$P(Y|X) = \frac{1}{Z(X)} \exp\left(\sum_i \sum_j \lambda_j t_j(y_{i-1}, y_i, x, i) + \sum_i \sum_j \mu_j s_j(y_i, x, i) \right) \qquad (1)$$

Where, $t_j(y_{i-1}, y_i, x, i)$ is a transition feature function of the entire observation sequence and the labels at position i and $i-1$ in the label sequence; $s_j(y_i, x, i)$ is a state feature function of the label at position i and the entire observation sequence. The index j in μ_j and s_j is feature serial number to represent different features. Parameters λ_j and μ_j correspond with feature t_j and s_j. Z(X) is a normalization factor which ensures that the distribution sums to 1. More details about CRFs can be referred from Lafferty [9].

(2) Framework of FCRF

The so-called feedback, that is, the information extracted from the lower level model will be fed back to the higher level model. The high level model uses the information as a part of features to build a new model. After getting feedback features, we appended it to the normalized files. And then by the last step of preprocessing, the training set and the test set were generated to build higher level model. The new feature type (feedback feature) was extracted, increasing the knowledge available for CRF. Through analyzing the results from lower level model continuously, higher level models got feedback timely to build a new model. Here we used the version of 0.57 of CRF++ to build CRF model. In the whole framework, one of the most crucial point is how to extract feedback features from the predicted results as features of the new model.

(3) Data preprocessing

Step 1: Label map construction
We regarded the prediction of helix boundaries of the membrane proteins as a sequence labeling problem. When the size of the label set is 3, we define L = {I, O, M}. Where, I indicates membrane-inside, O indicates membrane-outside, M indicates membrane helix. When the size of the label set is 2, we define L = {N, M}, where N indicates that the residue is not transmembrane, M indicates membrane helix. Here we adopted the latter as the label set. Because of the difference between the label set of the original dataset and that we defined here, a mapping function was used to convert.

Step 2: Feature extraction
The original dataset we used here contains two types of features. The first type is the physical properties of residues, including 5 attributes of hydrophilcity, hydrophobicity, elasticity, polarity and the transfer free energy; the second type is the evolution profile properties, which was calculated by the PSI-BlAST align tools corresponding to NR (non-redundant) database, containing 20 attributes. As the third kind of feature, feedback feature was extracted from the predicted results, and we will talk about this in detail in the following paragraph.

Although CRF can model the long-range correlation of a sequence, but the surrounding residues still have an important affect on the evolution of the current residue. Therefore, the concept of sliding window was introduced here. We extracted the attributes of surrounding residues as the features of a residue. So, when the window size is W, the count of features of a residue have except feedback features is W*25 (5 physical properties +20 profile properties).

Step 3: Feature normalization
Because the range of physical and profile feature values is inconsistent, such as the range of hydrophilcity is [−1.76, 0.73] and the range of hydrophobicity is [−1.8, 3.0], the feature values were normalized by being scaled to the interval [0, 1]. So, the feature values are in the same interval, and that will lay the foundation for the calculation and extraction of feedback features.

Step 4: Generating training and test set
Respectively, our dataset is not large. To get enough data to build FCRF model and optimize the performance of FCRF with limited data, 10-fold cross-validation was used here. We used the nine tenths of the normalized data as training set and the left as test set. Every chain was regarded as a whole and would never be divided into different data files. After ten times segmenting and merging, we got the training and test sets of the entire dataset finally.

(4) Feedback algorithm

Feedback feature extraction is the key issue in FCRFs, and the quality of feedback features affects the accuracy of predictions directly. Predicted results data contains a variety of information. In the lower model that built without feedback features, the predicted results contain five parts of the information, including predicted labels, the real labels, physical features, the profile features and residue sequences, and in the higher model that built with feedback features, the predicted results include a new type of information, that is the feedback features introduced when we built the model. In the phase of data preprocessing, the range of physical and profile feature values was mapped to the interval [0, 1] by the step of normalization. Here, we divided the interval [0, 1] into m (here, m = 10) parts equally, labeled by $[a_0, a_1), [a_1, a_2), \ldots, [a_{m-1}, a_m]$, sequentially noted as U_k. One physical feature or profile feature in lower model can be labeled by $f_{i,j}^n$, where i and j denote the row and column index of the feature matrix and n denotes the time of feedback, corresponding to the required feedback feature of higher level model noted as $t_{i,j}^{n+1}$. The feedback feature can be calculated by the formula (2):

$$t_{i,j}^{n+1} = P(M|U_{k,j}^n) \tag{2}$$

i.e. given an interval $U_{k,j}^n$, feedback feature $t_{i,j}^{n+1}$ equals the probability of predicted label M and the corresponding $f_{i,j}^n \in U_{k,j}^n$. The index j in $U_{k,j}^n$ is feature serial number to represent the interval at different column. The column is a basic unit of interval segmentation. Due to the label set we used here is L = {M,N}, $t_{i,j}^{n+1}$ can be represented by the ratio of the number of no-feedback features with predicted label M and the total number of no-feedback features in interval $U_{k,j}^n$ at column j.

3   Experiments and Discussion

3.1   Compared with Other Methods

In order to further verify the performance of FCRF, we implemented comparison of FCRF and other nine famous transmembrane protein topology predictor. They are MEMSATS-SVM [7], OCTOPUS [8], MEMSAT3 [12], ENSEMBLE [13], PHOBIUS [11], HMMTOP [6], PRODIV [14], SVMTOP [15], TMHMM [5]. The dataset used here is DB1 and the result is shown in Fig. 1. "Correct protein helix count" indicates the percent of the correct protein helix count that can be represented by the ratio of the number of chains that the number of helix is predicted correctly and the number of chains in whole dataset. "Correct helix location" indicates the percent of correct helix location that can be represented by the ratio of the number of helix with correct prediction location and the number of helix in whole dataset. The results of the other methods have been published in the paper [7]. So we used the results without any change here.

Among the nine predictor mentioned above, MEMSAT-SVM, MEMSAT3 and FCRF are fully cross-validated, with all proteins homologous to the target being removed from training sets, while results for the remaining methods were obtained from their respective web servers and consequently are not cross-validated. And MEMSAT-SVM was applied to the test set of 131 TM proteins, while the test set we used here contains 106 TM proteins. So this must have some effect on the result.

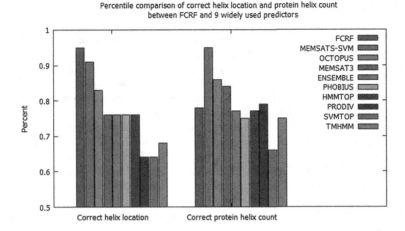

Fig. 1. Comparison of predicted results with other methods

In Fig. (1), the percent of correct helix location of FCRF ranks first, reaching 95 %.

4 Conclusions

This paper explored the ability of the feedback conditional random field model in predicting helix boundaries of alpha-helix transmembrane proteins. Helix location and protein helix count associated with protein topology were predicted. The results show that, compared with the several wide-used prediction methods, this method achieved a good result on the prediction accuracy of helix location, the percent is 95 %.

But, at the same time, we also found that FCRF did not achieve the results that we expected in predicting protein helix count. For further study, there is still some improvement can be taken based on FCRF. For example, more valuable features need to be explored. The feedback features here only contain information about the percent of residues with correct label M at different intervals, and there is no information directly related to the percent of correct protein helix count. So the basic unit of being predicted is still a label corresponding to the residue.

Acknowledgments. This paper is supported by grants no. 61170125, 61202290 under the National Natural Science Foundation of China (http://www.nsfc.gov.cn) and grants no. BK20131154 under Natural Science Foundation of Jiangsu Province. The funders had no role in study design, data collection and analysis, decision to publish, or preparation of the paper. Kun Wang wrote the software, paper and implemented the experiments, Hongjie Wu designed the algorithm and experiments, Weizhong Lu prepared the datasets, Baochuan Fu and Qiang Lü cooperated with Hongjie to improve the workflow, Xu Huang organized the paper. The authors thank Jin Wang and Shimin Chen for helping with the analysis of the experiment.

References

1. Michel, M.C., Seifert, R., Bon, R.A.: Dynamic bias and its implications for GPCR drug discovery. Nature Reviews Drug Discovery **13**, 869 (2014)
2. Mathivanan, S.: Integrated bioinformatics analysis of the publicly available protein data shows evidence for 96 % of the human proteome. J. Proteomics Bioinf. **7**, 041–049 (2014)
3. Bill, R.M., Henderson, P.J.F., Iwata, S., et al.: Overcoming barriers to membrane protein structure determination. Nat. Biotechnol. **29**(4), 335–340 (2011)
4. Laganowsky, A., Reading, E., Allison, T.M., et al.: Membrane proteins bind lipids selectively to modulate their structure and function. Nature **510**, 172–175 (2014)
5. Sonnhammer, E.L.L., Von Heijne, G., Krogh, A.: A hidden markov model for predicting transmembrane helices in protein sequences. ISMB-98 **6**, 175–182 (1998)
6. Tusnady, G.E., Simon, I.: Principles governing amino acid composition of integral membrane proteins: application to topology prediction. J. Mol. Biol. **283**(2), 489–506 (1998)
7. Nugent, T., Jones, D.T.: Transmembrane protein topology prediction using support vector machines. BMC Bioinf. **10**(1), 159 (2009)
8. Viklund, H., Elofsson, A.: OCTOPUS: improving topology prediction by twotrack ANN-based preference scores and an extended topological grammar. Bioinformatics **24** (15), 1662–1668 (2008)
9. Lafferty, J., Mccallum, A., Pereira, F.: Conditional random fields: probabilistic models for segmenting and labeling sequence data. Proc int Conf on Machine Learning, 282–289 (2001)
10. Ikeda, M., Arai, M., Okuno, M., Shimizu, T., Toshio, : TMPDB: a database of experimentally-characterized transmembrane topologies. Nucleic Acids Res. **31**(1), 406–409 (2003)
11. Lukas, K., Anders, K., Erik, L.L.S.: A combined transmembrane topology and signal peptide prediction method. J. Mol. Biol. **338**, 1027–1036 (2004)
12. Viklund, H., Bernsel, A., Skwark, M., Elofsson, A.: SPOCTOPUS: a combined predictor of signal peptides and membrane protein topology. Bioinformatics **24**, 2928–2929 (2008)
13. Martelli, P.L., Fariselli, P., Casadio, R.: An ENSEMBLE machine learning approach for the prediction of all-alpha membrane proteins. Bioinformatics **19**(1), I205–I211 (2003)
14. Viklund, H., Elofsson, A.: Best alpha-helical transmembrane protein topology predictions are achieved using hidden markov models and evolutionary information. Protein Sci. **13**, 1908–1917 (2004)
15. Lo, A., Chiu, H.S., Sung, T.Y., Lyu, P.C., Hsu, W.L.: Enhanced membrane protein topology prediction using a hierarchical classification method and a new scoring function. J. Proteome Res. **7**, 487–496 (2008)

Prediction of Protein Structure Classes

Wenzheng Bao, Dong Wang, Fanliang Kong, Ruizhi Han,
and Yuehui Chen[✉]

School of Information Science and Engineering, University of Jinan, Jinan,
People's Republic of China
baowz55555@126.com, yhchen@ujn.edu.cn

Abstract. Prediction of protein structural are crucial in Bioinformatics. More and more evidences demonstrate that an great number of prediction methods has been employed to predict these structures based on the sequences of protein and biostatistics. The accuracy of such methods, nevertheless, is strongly affected by the efficiency and the robustness of classification model and other several factors. In our present research, the features based on the correlation coefficient of dipeptide or polypeptide were put forward. For one thing, flexible neutral tree (FNT), a novel classification model which is a variable structure neural network, is employed as the base classifiers. For another, the alterable tree structure based on FNT, such model may take advantage of the selection of available information, which aimed at the improvement of efficiency. It is important to find out the tree structural of protein structure classification model. To examine the performance of such method, ASTRAL, 1189 and 640 are selected as benchmark datasets of protein tertiary structure. Fortunately, the results show that a higher prediction accuracy compared with other methods. With the selected features running in the flexible neutral tree, several redundant information of features may be cut off and the accuracy of such model may be improved in some degree and the time of running such model could be hold down.

Keywords: Protein structure classification · Flexible neural tree · Tree structural classification model

1 Introduction

The fold recognition problem is one of the fundamental issues in molecular biology and bioinformatics [1]. Moreover, clearly defined figure identification of protein spacial structure is of great significance since the main function of protein is determined by spacial structure [2]. According to an accounting report issued by Protein Data Bank (PDB), the Yearly Growth of Total Protein Structures have an increasingly growth. As the matter stands, there are two popular classes of computational methods for predicting the protein tertiary structure: initially, Template-Based Methods(TBM) and Ab initio methods. For identifying the tertiary structure of a given sequence, TBM may suggest taking advantage of the known three-dimensional structures from PDB as a template, which is world-wide protein structural database [3].

In this paper, a novel classification model of protein structure are proposed for protein fold recognition. And then flexible neutral tree(FNT), a structural alternative neutral network, is employed as the classifiers in our tree structural classification model.

© Springer International Publishing Switzerland 2015
D.-S. Huang et al. (Eds.): ICIC 2015, Part I, LNCS 9225, pp. 737–743, 2015.
DOI: 10.1007/978-3-319-22180-9_74

According to the size of feature, we take advantage of the FNT model, which may reduce redundant information of the selected features and process data mining in some degree. The tree structural classification model contains two layers: one the first layer, the protein tertiary structure may be distributed into three classes, which includes all-α classes, all-β classes and α*β class(including α + β class and α/β class). On the next layer, the model's primary role is to distributed the α*β class into α + β class and α/β class. Sharp tools make good work, the correlation coefficient of dipeptide or poly-peptide, which contain the sequence information and conformational space, were selected to the main feature in this research.

2 Corresponding Work

Ensemble method is one of the most important branches of supervised learning algo-rithm that uses multiple classifiers to obtain proper prediction accuracy and make up different classifiers in function. Guo and Gao (2008) presented two-layer ensemble classifier. In the first layer, a potential class index for every query protein in the 27-folds is identified [4]. According to this result, a 27-dimension vector is generated in the 2nd layer. And then, genetic algorithm is adopted to obtain weights for the outputs of the 2nd layer to get the final result. Comparing with those mentioned methods to classification protein structure, the solution in this research aim to reducing some less and unrelated features and discovering the inner connection between the protein structure and corresponding statically features.

3 The Dataset and Feature Vectors

3.1 Training and Test Datasets

The structural classification of proteins (SCOP) currently include eleven classes: (1) *all-α* proteins; (2) *all-β* proteins; (3) *α/β* proteins; (4) *α + β* proteins; (5) multi-domain proteins; (6) membrane and cell surface proteins; (7) small proteins; (8) coiled coils proteins; (9) low-resolution proteins; (10) peptides and (11) designed proteins. Our research only focuses on the first four categories, because they include the great majority of protein sequences and are the basis for most comparable approaches [5]. According to Levitt and Chothia's concept of protein structural classes, the tertiary structure of protein are divided proteins into four structural classes: *all-α, all-β, α + β* and *α/β* in 1976 [6]. In this research, ASTRAL dataset, only four major classes are used. The information of these datasets showed in Table 1.

3.2 Feature Vectors

So many features of protein structure has been put forward, such as amino acid compositions, dipeptide compositions, pseudo amino acid compositions, normalized Moreau-Broto auto-correlation, Moran auto-correlation, Geary auto-correlation [7].

Table 1. SCOP class distribution in the datasets used in our research

Dataset	all-α	all-β	α/β	$\alpha + \beta$	Total
ASTRAL [5]	639	661	749	764	2813
640 [6]	138	154	171	177	640
1189 [6]	223	294	334	241	1092

Nevertheless, such features could not represent both the sequence of protein structure and the conformational space of such things.

According to our previous research, several features of amino acids have been attempted [8]. Comparing with other features of protein, the correlation coefficient of general dipeptide or polypeptide may contain more information about both protein sequence and the interaction between distance animo acids residues. In the viewpoint of Computational biology and theoretical biology, the protein structure construct in some stable structure. So in such structure, the energy of animo acids and other composition may be in low-energy statement. In this research, several characteristics of animo acids residues are selected to express protein sequence, such as hydrophobicity, hydrophily, heteropolarity, polarizability, transfer free energy and information entropy of amino acids, which showed in [9].

Considering the 6 kinds of eigenvalues differ in some degree, a normalized process should be used. In this article, the process of maximum & minimum standardization could be employed.

The normalized eigenvalues are used for calculating correlation coefficient of sequence. According to the factors of dipeptide and polypeptide, novel correlation information among n amino acid residues is introduced. Equations (1) and (2) are the expressions of the correlation coefficient of dipeptides and tripeptides, respectively.

$$cc_2(\lambda, k) = \frac{1}{L - \lambda} \sum_{i=1}^{L-\lambda} \frac{P1_{i,k} \times P2_{i+\lambda,k}}{\sqrt{P1_{i,k} \times P1_{i,k}^T} \sqrt{P2_{i,k} \times P2_{i,k}^T}} \tag{1}$$

$$cc_3(\lambda_1, \lambda_2, k) = \frac{1}{L - \lambda_1 - \lambda_2} \sum_{i=1}^{L-\lambda_1-\lambda_2} \frac{P1_{i,k} \times P2_{i+\lambda_1,k} \times P3_{i+\lambda_1+\lambda_2}}{\sqrt{P1_{i,k} \times P1_{i,k}^T} \sqrt{P2_{i+\lambda_1,k} \times P2_{i+\lambda_1,k}^T} \sqrt{P3_{i+\lambda_1+\lambda_2} \times P3_{i+\lambda_1+\lambda_2,k}^T}} \tag{2}$$

With limited data processing ability for each classifier, the order of this feature is no more than 4. Firstly, λ, λ_1 and λ_2 are the distance between different position amino acids. Secondly, k is the index of eigenvalues. Thirdly, L is the amino acids length number of a protein sequence. Finally, the $A_{i,k}$, $B_{i+\lambda,k}$, $B_{i+\lambda1,k}$, $C_{i+\lambda1+\lambda2,k}$ is $No.i$, $No.\ i + \lambda$, $No.i + \lambda_1$ and $No.i + \lambda_1 + \lambda_2$ position amino acids in a sequence, respectively.

4 Supervised Learning Methods

Supervised learning is the machine learning method in which the model is learned from labeled data [10]. A supervised learner analyzes the training data and produces a model which can be used for predicting the class of test samples (or unknown examples). In the following some supervised algorithms are discussed that we have used in our proposed method.

The flexible neutral tree, which is a alternative tree structural neutral network, put forward by Chen [11]. The model is not only with the function of classification but also used to doing some data mining of corresponding feature. So the each steps of FNT model shows in Fig. 1.

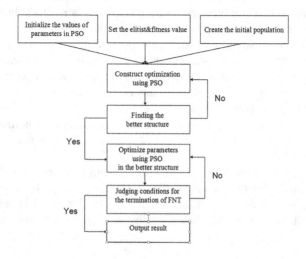

Fig. 1. The general steps of FNT

5 Classification Structure

Above mentioned of Levitt and Chothia's concept, it is no difficult to found that the classification of protein tertiary structure is a classical problem of 4-types. In this research, we put forward a tree structural model, which is a two-layer of classification model: in the 1st layer, all the protein sequences divided into 3 types(all-α, all-β and $\alpha*\beta$ which included $\alpha + \beta$ and α/β) and then in the 2nd layer, the distinguish between all-α and all-β has finished and the mainly focus on the distinguish between $\alpha + \beta$ and α/β. So the label of each types of protein structure showed in Table 2 and the structure of the model showed in Fig. 2.

To evaluate the statistics under the proposed method, our research examine the prediction success rate on low homology dataset. There are three cross-validation methods are often used to examine predictor for its effectiveness in the statistical prediction: single test-set analysis, sub-sampling test, and jack-knife test [12]. During our research, the 10-fold cross validation tests is introduced. It means that 10 % of each

Table 2. The label of each types

protein structure	all-α	all-β	α + β	α/β
Layer	1st	1st	2nd	2nd
Label	IA	IB	IC&&IIA	IC&&IIB

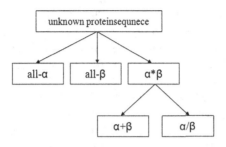

Fig. 2. The structure of novelty classification model

datasets protein sequence are treated as test data. Meanwhile, others will be treated as train data. The overall accuracy (OA) is computed for each dataset.

6   Discussion and Results

Due to the size of the feature of cc_2 and cc_3, the size of cc_2 is about to the n^2 and the size of cc_3 is about to the n^3. Among these features, several useless and irrelevant information shielded in there. So another function of the FNT model is dimensionality reduction and finding out key information to distinguish different protein structure. The FNT algorithm created tree structure based on the strategy of Breadth-First-Search (BFS). There is no hierarchical relationship among features. So the BFS strategy could be able to find out the subordination among these features. For another hand, putting forward to the novelty method will be able to decrease the size of input number, which may improve the real-time of the classification model in some degree.

Through the above process, the selected features size has greatly decreased. The selected features showed in Table 3.

There are a large variety of methods for protein structure classification. Thus the prediction efficiency and accuracy of our method are compared with other recently reported prediction methods. Other methods shown in Table 4 are used to test the four experimental datasets, and the comparison results show that our method displays the highest prediction accuracies as revealed in Table 4. However, the accuracy of some structure are lower than the other methods.

Table 3. The comparison of features selection

	cc_2	cc_3
size of before selected	36	216
size of after selected	29	157

Table 4. The accuracy of various methods

Dataset		all-α(%)	all-β(%)	α + β(%)	α/β(%)	OA(%)
ASTRAL	SCPRED	93.13	78.33	64.27	83.38	79.14
	RKS-PPSC	94.06	83.38	71.47	85.01	83.01
	PSIPRED	94.53	77.49	71.47	87.28	82.33
	Previous work	74.31	77.29	93.82	84.81	83.06
	This method1	75.41	79.29	92.77	84.37	83.37
	This method2	76.19	79.21	92.75	85.19	83.75
640	RKS-PPSC	89.10	85.10	71.40	88.10	83.10
	PSIPRED	93.72	84.01	66.39	83.53	83.44
	Previous work	76.92	81.25	83.87	94.73	84.51
	This method1	77.25	81.76	83.41	95.02	84.89
	This method2	77.52	81.74	93.46	95.27	85.03
1189	RKS-PPSC	89.20	86.70	53.80	89.60	80.60
	PSIPRED	93.72	84.01	66.39	83.53	81.96
	Previous work	72.49	82.65	77.24	93.04	82.56
	This method1	86.27	79.92	81.54	92.95	84.59
	This method2	86.51	79.85	81.97	93.01	84.76

This method1 means all the features of this research and this method2 means selected features of this research.

7   Conclusion

From the above research, several key features can be found by FNT model and the selected features have the ability to describe the main characteristics of protein sequence in some degree. And yet, the selection of features among so many kinds of information of protein sequence become an increasingly important. For another, the model of FNT should be improved, which will meet the need of the research of Bioinformatics.

Acknowledgments. Wenzheng Bao and Dong Wang contributed equally to this work and should be considered co-first authors. This research was partially supported by the Youth Project of National Natural Science Fund (61302128), the Key Project of Natural Science Foundation of Shandong Province (ZR2011FZ001), the Natural Science Foundation of Shandong Province (ZR2011FL022), the Key Subject Research Foundation of Shandong Province and the Shandong Provincial Key Laboratory of Network Based Intelligent Computing. This work was also supported by the National Natural Science Foundation of China (Grant No. 61201428, 61203105)

References

1. Ding, C.H.Q., Dubchak, I.: Multi-class protein fold recognition using support vector machines and neural networks. Bioinformatics **17**(4), 349–358 (2001)
2. Shenoy, S.R., Jayaram, B.: Proteins: sequence to structure and function-current status. Curr. Protein Pept. Sci. **11**(7), 498–514 (2010)

3. Aram, R.Z., Charkari, N.M.: A two-layer classification framework for protein fold recognition. J. Theor. Biol. **365**, 32–39 (2015)
4. Guo, X., Gao, X.: A novel hierarchical ensemble classifier for protein fold recognition. Protein Eng. Des. Sel. **21**(11), 659–664 (2008)
5. Andreeva, A., Howorth, D., Chandonia, J.M., Brenner, S.E., Hubbard, T.J.P., Chothia, C., Murzin, A.G.: Data growth and its impact on the SCOP database: new development (2007)
6. Andreeva, A., Howorth, D., Brenner, S.E., Hubbard, T.J.P., Chothia, C., Murzin, A.G.: SCOP database in 2004: refinements integrate structure and sequence family data (2004)
7. Li, Z.R., Lin, H.H., Han, L.Y., Jiang, L., Chen, X., Chen, Y.Z.: PROFEAT: a web server for computing structural and physicochemical features of proteins and peptides from amino acid sequence. Nucleic Acids Res. **34**, W32–W37 (2006) ·
8. Bao, W.Z., Chen, Y.H., Wang, D.: Prediction of protein structure classes with flexible neural tree. Bio-Med. Mater. Eng. **24**, 3797–3806 (2014)
9. Chatterjee, P, Basu, S, Nasipuri, M.: Improving prediction of protein secondary structure using physicochemical properties of amino acids [C]. In: Proceedings of the 2010 International Symposium on Biocomputing (ISB 10). ACM, New York, (2010)
10. Mohri, M., Rostamizadeh, A., Talwalkar, A.: Foundations of Machine Learning. MIT Press, Cambridge (2012)
11. Yang, B., Chen, Y.H., Jiang, M.Y.: Reverse engineering of gene regulatory networks using flexible neural tree models. Neurocomputing **99**, 458–466 (2013)
12. Chou, K.C., Shen, H.B.: Recent progress in protein subcellular location prediction. Anal Biochem **370**, 1–16 (2007)

An Integrated Computational Schema for Analysis, Prediction and Visualization of piRNA Sequences

Anusha Abdul Rahiman[1](✉), Jithin Ajitha[2], and Vinod Chandra[2]

[1] Department of Computational Biology and Bioinformatics,
University of Kerala, Thiruvananthapuram 695581, Kerala, India
anushapraveenkhan@gmail.com
[2] Computer Centre,
University of Kerala, Thiruvananthapuram 695034, Kerala, India
{emailjithin,vinodchandrass}@gmail.com

Abstract. PIWI-interacting RNAs (piRNAs) are endogenously originated predominantly germline oriented newly described class of small non-coding RNAs or transcripts. piRNAs are found to be crucial ones in rendering translational arrest of proteins thereby protecting the genome germline integrity from invasive transposable elements. piRNAs are demanding more attention because of its potential role in the process of spermatogenesis and male infertility. Though there exist several computational approaches to predict piRNA sequences, a parameter based piRNA prediction strategy was not attempted yet. Understanding this scenario, a comprehensive computational schema has been developed based on Bayes and Tree classifiers. The proposed method provides an integrated platform to analyze, predict and visualize piRNA dataset from other noncoding RNAs in a multi-threaded environment. Moreover, a comparative study of different classification algorithms applicable to piRNA predictions is presented here.

Keywords: piRNAs · Bayes classifier · Tree classifier

1 Introduction

Noncoding RNAs (ncRNAs) are small non-translational guiding RNAs in transcriptional or post transcriptional RISC mediated developmental regulations and gene silencing. Small RNA category include micro RNA (miRNA), piwi interacting RNA (piRNA), small nuclear RNA (snRNA), small nucleolar RNA (snoRNA), repeat-associated siRNA (rasiRNA) and small scan RNA (scnRNA). Comparative studies were able to differentiate one from the other based on its genomic location, biogenesis, structural and thermodynamic aspects, sequence homology, argonaute protein interaction and regulatory mechanisms.

piRNAs are 24–32 nucleotides long regulatory RNAs that associate with PIWI sub-family members of the argonaute family. piRNAs impart gene silencing, protein translational arrest, gene expression regulation, germline cell formation and maintenance, spermiogenesis and oogenesis. The genome organisation as clusters, sequence

© Springer International Publishing Switzerland 2015
D.-S. Huang et al. (Eds.): ICIC 2015, Part I, LNCS 9225, pp. 744–750, 2015.
DOI: 10.1007/978-3-319-22180-9_75

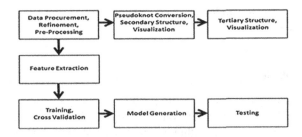

Fig. 1. Architecture of piRNA analysis, prediction and visualization

length, affinity of A at position 10 for plus (+) strand derived sequence, affinity of U at position 1 for minus (−) strand derived sequence are found to be the unique features that distinguish piRNAs from other short ncRNAs. Though the primary processing pathway and the Ping-Pong mechanism could explain piRNA biogenesis; clarifications on the type of piRNA precursor, its structural aspects and the regulatory mechanism still remain dormant.

Algorithmic analysis and approaches eased and intensified the experimental characterisation of piRNAs to a large extend. Most computational works employed in piRNA research concentrated on cluster predictions and sequence classifications based on support vector machine (SVM) and k-mer scheme [1–3]. piRPred and ProTRAC are two web based tools available for piRNA prediction and cluster detection respectively [4, 5]. A web resource, piRNA bank accumulates all reported empirical piRNA sequences of rat, mouse and human under a single data-base [6].

We focus on the development of a computational piRNA prediction method based on relevant features that characterize piRNAs from other noncoding RNAs. To facilitate parameter extraction, an analysis scheme is built with the help of an internally developed algorithm. We also attempt to incorporate an approach to predict piRNA secondary and tertiary structure along with its visualization. For the convenience, human datasets were only considered in this work. The architecture of our prediction method is depicted in Fig. 1.

2 Materials and Methods

2.1 Data Procurement and Pre-Processing

The required noncoding human RNA sequences were downloaded from public repositories such as NONCODE v3.0 and Ensembl in a single FASTA file containing 4, 11,553 entries of human alone. Since this FASTA file doesn't always follow a uniform header format, the dataset is redesigned to be consistent using sequence splitting operation. Thus after initial refinement and preprocessing, 4,11,553 entries were parsed to produce four different files containing 32,152 human piRNA dataset, 702 miRNAs, 124 snRNAs and 306 snoRNAs. As the piRNA sequences are relatively more, datasets were taken carefully in an equally distributed manner so that each class have equal no. of sequences (say 25/50/100 or higher in multiples of 5/10). Sequences which lack experimental validations were totally omitted from the dataset due to reliability constrains.

2.2 Feature Selection

The significant features that could line up piRNAs as a single entity was investigated and identified based on experimentally validated reports. Rather, to find new and better parameters, a piRNA analysis algorithm was internally designed that can bring out any physical properties, nucleotide count, hidden patterns and repeated adjacent nucleotide count within a sequence.

The relevant piRNA features under consideration and their respective calculations are listed in Table 1. Most (> 90 %) human piRNA sequence ranges between 28-33 nts. Also piRNAs generated from sense strand (+) showed strong preference for Adenine (A) at position 10 and those that arise from antisense strand (-) showed strong preference for Uracil base (U) at position 1. On this regard piRNA sequence length and its position-wise nucleotide preference were categorized as positional parameters. Based on the individual sequence pattern that could possibly contribute its secondary structures, the proportion of possible piRNA A: U pairs and G: C pairs were categorized in the structural parameters group along with the molecular weight. All the parameters considered are independent over each other and are assigned with equal weights.

2.3 Training, Cross Validation and Serialized Model Generation

Training, cross validation and prediction was implemented using two machine learning logics – a probabilistic model, Bayes and a decision model, Tree classifier. Parameters identified from different sources enabled the model building to classify piRNAs from other three small RNA classes. The datasets comprising four small RNA classes were divided into training set (70 %) and testing set (30 %). For training, the RNA sequences were read from source FASTA file containing miRNA,

Table 1. List of positional and structural parameters

Parameter Category	Parameter Names	Parameter Description	Parameter Calculations
Positional Parameters	Length	Total length of the sequence	Len(sequence) returns length as integer
	A at position 10	Affinity for nucleotide A at position 10	Pattern analysis predicts A at 10th position in the sequence
	U at position 1	Affinity for nucleotide U at position 1	Pattern analysis predicts U at 1st position in the sequence
Structural Parameters	A:U Pairs	Proportion of total A:U pairs in sequence	Applying Nussinov algorithm
	G:C Pairs	Proportion of total G:C pairs in sequence	Applying Nussinov algorithm
	Molecular Weight	Sum of molecular weights of each of the nucleotides	(329.2* no. of G)+(313.2 * no. of A)+ (304.2*no. of U)+(289.2 * no. of C)

piRNA, snRNA and snoRNA sequences and a conversion algorithm developed internally was applied to extract parameter values from each sequence. After determining the parameter values datasets were trained and cross validated to evaluate the accuracy of method. An n-fold cross validation of training set was performed where the fold condition is defined as a range of n > 2 and n < number of training set. The training dataset was partitioned randomly into complementary subsets and each subset was trained and tested (cross validation) against each other. Based on this cross validation accuracy, a serialized model was generated and rest 30 % test data was used against this model.

2.4 Pseudoknot Conversion, Secondary Structure Prediction, Visualization

RNA molecule folds to form secondary structure owing to the hydrogen bonding between complementary bases on the same strand following the wobble base pairing rule. Sequences in FASTA format is basically the primary structure. Nussinov algorithm was implemented to convert this primary sequence to its equivalent intermediate dot bracket notation called pseudoknot format which might give maximum pairs in its secondary structure [7]. Figure 2a shows the illustration of such a pseudoknot conversion of a sample sequence from the NONCODE and Fig. 2b shows the correspondingly generated secondary structure from the pseudoknot.

The VARNA application programming interface (API) facilitates visualization of predicted piRNA secondary structure generated by the pseudoknot [8]. 3-Dimensional piRNA visualization is assembled in the system using Jmol toolkit. A script using the BARNACLE python package samples RNA structures that are compatible with a given nucleotide sequence and models local conformational spaces of these structures with no energy terms [9].

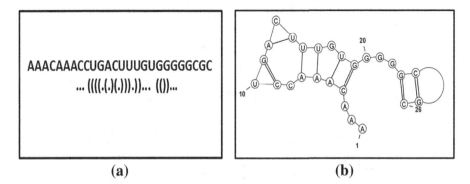

(a)   **(b)**

Fig. 2. (a) Pseudoknot conversion (b) Generated secondary structure from the pseudoknot

3 Results and Discussions

The prediction results obtained by performing cross validation of two different human datasets of size 100 and size 200 respectively using the proposed method are shown in Tables 2 and 3. The sensitivity, specificity and accuracy rate obtained shows that parameter based piRNA predictions gives a precision greater than 95 %. It also shows that batch classifiers are more suitable for piRNA prediction while considering Bayes classifiers. Since the amount of data and features are sufficiently high, Bayes Net can easily exploit the relationships lying in between the classes to produce efficient result than simple Naïve Bayes algorithm.

Further a comparative study of Bayes with seven different algorithms for tree classifiers in the piRNA prediction schema was conducted. Results show that Bayes give better precision than tree classifiers in piRNA prediction problems. Among various tree classifiers, it is found that BF, NB and Random Tree produces good accuracy than J48 Tree, J48 Graft, FT Tree and Random Forest.

3.1 Discussions

We availed human dataset in FASTA format from public databases to conduct this primary study. Experimentally verified and diversified large dataset irrespective of

Table 2. Prediction and cross validation result of datasize = (4 × 25) = 100

Classifier	Fold Size	TP	TN	FP	FN	Sensitivity	Specificity	Accuracy (x 100%)
Naïve Bayes	5	24	73	2	1	0.96	0.961426	0.97
	10	25	75	0	0	1	0.962912	1
Naïve Bayes Simple	5	25	74	1	0	1	0.96291	0.99
	10	25	75	0	0	1	0.962967	1
Naïve Bayes Updatable	5	24	73	2	1	0.96	0.961426	0.97
	10	25	75	0	0	1	0.962912	1
Dayes Net	5	25	75	0	0	1	0.96291	1
	10	25	75	0	0	1	0.962967	1
J48 Tree	5	25	74	1	0	1	0.962912	0.99
	10	25	74	1	0	1	0.962912	0.99
J48 Graft	5	25	74	1	0	1	0.962912	0.99
	10	25	74	1	0	1	0.962912	1
BF Tree	5	25	75	0	0	1	0.962912	1
	10	25	75	0	0	1	0.962912	1
NB Tree	5	25	75	0	0	1	0.962912	1
	10	25	75	0	0	1	0.962912	1
FT Tree	5	25	75	0	0	1	0.962912	1
	10	25	75	0	0	1	0.962912	1
Random Forest	5	25	75	0	0	1	0.962912	1
	10	25	75	0	0	1	0.962911	1
Random Tree	5	25	74	1	0	1	0.962937	0.99
	10	25	75	0	0	1	0.96225	1

Table 3. Prediction and cross validation result of datasize = (4 × 50) = 200

Classifier	Fold Size	TP	TN	FP	FN	Sensitivity	Specificity	Accuracy (x 100%)
Naïve Bayes	5	44	144	6	6	0.88	0.978204	0.94
	10	49	147	3	1	0.98	0.980377	0.98
Naïve Bayes Simple	5	50	148	2	0	1	0.980761	0.99
	10	50	148	2	0	1	0.980811	0.99
Naïve Bayes Updatable	5	44	144	6	6	0.88	0.978204	0.94
	10	49	147	3	1	0.98	0.980385	0.98
Bayes Net	5	50	149	1	0	1	0.980761	0.995
	10	50	148	2	0	1	0.980811	0.99
J48 Tree	5	49	150	0	1	0.98	0.980385	0.995
	10	49	150	0	1	0.98	0.980385	0.995
J48 Graft	5	49	150	0	1	0.98	0.980385	0.995
	10	49	150	0	1	0.98	0.980377	0.995
BF Tree	5	50	150	0	0	1	0.980762	1
	10	50	150	0	0	1	0.980762	1
NB Tree	5	50	150	0	0	1	0.980762	1
	10	50	150	0	0	1	0.980769	1
FT Tree	5	49	150	0	1	0.98	0.980385	0.995
	10	49	150	0	1	0.98	0.980377	0.995
Random Forest	5	50	148	2	0	1	0.980762	0.99
	10	50	149	0	1	0.980392	0.980769	0.995
Random Tree	5	50	150	0	0	1	0.980392	1
	10	50	150	0	0	1	1	1

organism or species specific can be used instead of human files for training and testing the consistency of this method based on machine capacity. But lack of clinically verified data especially snRNAs and snoRNAs in the repositories prevent us from testing with larger data set. To obtain optimal results, equally weighted feature vectors that are independent to each other are considered in the prediction strategy. To prevent increased or decreased accuracy shift towards any one class, we avoided assignment of different weights to the parameters. Non-probabilistic algorithms like support vector machines are suitable only when parameters are heavily dependent on one another. Due to the involvement of independent features in our method, we opted to follow the probabilistic approach of Bayes for training. Approaches like k-mer scheme is not good to use with probabilistic algorithms and tree classifiers since the features are more or less dependent.

4   Conclusions

A novel generic computational approach accustomed on parameters to perform classification and prediction of piRNAs from other noncoding RNAs has been designed and modeled using the variants of Bayes and Tree classifiers. Results show that this newly proposed system is compatible for other organisms and has appreciable sensitivity, specificity and accuracy (> 95 %) than similar existing method with accuracy 90 %. Also the existing k-mer scheme does not follow a systematic path for the training of different classes and adopts a sort of trial and error method till an acceptable level of accuracy is attained. On the other hand our method is probabilistic and considers independent feature with high throughput and speed. The comparative study of different algorithms in Bayes and Tree classifiers in the piRNA prediction strategy reveals that Bayes Net, BF Trees and NB Trees gives better result for these dataset than other algorithms. It is found that the set of methods and algorithms adapted for pseudo knot conversion, secondary structure prediction, and visualization does not specialize in the working for piRNA alone. Moreover, the whole architecture works in an integrated fashion within a common platform facilitating the prediction of piRNA secondary and tertiary structure along with its visualization. With the identification of specific and relevant parameters, the same method could be adopted to predict and visualize other small RNA classes also. Addition of new conversion algorithms could enhance the model performance to give optimal results. As a future extension work support for GENBANK, EMBL, and FASTAA file formats may be added and could be connected to web services provided by NCBI and NONCODE.

Acknowledgements. This study received funding from Kerala State Council for Science, Technology and Environment (KSCSTE). The authors thank Dr. Achuthsankar S. Nair, Head, Department of Computational Biology and Bioinformatics, University of Kerala for all kind of advice and support during the time of this work.

References

1. Betal, D., Sheridan, R., Mark, S.D., Sander, C.: Computational analysis of mouse piRNA sequence and biogenesis. PLoS Comput. Biol. **3**, e222 (2007)
2. Zhang, Y., Wang, X., Kang, L.: A k-mer scheme to predict piRNAs and characterize locust piRNAs. Bioinformatics **27**, 771–776 (2011)
3. Wang, K., Liang, C., Liu, J., Xiao, H., Huang, S., Xu, J., Li, F.: Prediction of piRNAs using transposon interaction and a support vector machine. BMC Bioinform. **15**, 419 (2014)
4. Brayet, J., Zehraoui, F., Jeanson-Leh, L., Israeli, D., Tahi, F.: Towards a piRNA prediction using multiple kernel fusion and support vector machine. Bioinformatics **30**, i364–i370 (2014)
5. Rosenkranz, D., Zischler, H.: proTRAC—a software for probabilistic piRNA cluster detection, visualization and analysis. BMC Bioinform. **13**, 5 (2012)
6. Lakshmi, S.S., Agrawal, S.: piRNABank: a web resource on classified and clustered Piwi-interacting RNAs. Nucleic Acids Res. **36**, D173–D177 (2008)
7. Nussinov, R., Jacobson, A.B.: Fast algorithm for predicting the secondary structure of single-stranded RNA. Proc. Natl. Acad. Sci. U.S.A. **77**, 6309–6313 (1980)
8. Darty, K., Denise, A., Ponty, Y.: VARNA: interactive drawing and editing of the RNA secondary structure. Bioinformatics **25**, 1974–1975 (2009)
9. Frellsen, J., Moltke, I., Thiim, M., Mardia, K.V., Ferkinghoff-Borg, J., Hamelryck, T.: A probabilistic model of RNA conformational space. PLoS Comput. Biol. **5**, e1000406 (2009)

Author Index

Printed in the United States
By Bookmasters